Corrosion of magnesium alloys

Related titles:

Surface engineering of light alloys: aluminium, magnesium and titanium alloys
(ISBN 978-1-84569-537-8)
This authoritative book provides a comprehensive review of the various surface engineering techniques employed to improve the properties of light alloys, focusing on titanium, magnesium and aluminium alloys. It examines surface-related degradation of light alloys and covers surface engineering technologies in detail. The book includes chapters on corrosion behaviour of Mg alloys, anodising treatments of Mg alloys, micro-arc oxidation of light alloys, physical vapour deposition of light alloys, PIII/PSII of light alloys, laser surface modification of Ti alloys, plasma nitriding of Ti and Al alloys, duplex surface treatments of light alloys and biomedical devices using Ti alloys.

Welding and joining of magnesium alloys
(ISBN 978-1-84569-692-4)
This book covers all aspects of the welding and joining of magnesium alloys. Magnesium and its alloys have been used for many years and their use is increasing due to their superior properties and light weight. Part I includes welding metallurgy, preparation for welding and welding materials. Part II covers the various welding technologies that can be used for joining magnesium alloys and Part III includes other joining technologies, weld defects and corrosion protection.

Fatigue life prediction of composites and composite structures
(ISBN 978-1-84569-525-5)
This important book addresses the highly topical subject of fatigue life prediction of composites and composite structures. Fatigue is the progressive and localised structural damage that occurs when a material is subjected to cyclic loading. The use of composites is growing in structural applications and they are replacing traditional materials, primarily metals. Many of the composites being used have only recently been developed and there are uncertainties about the long-term performance of these composites and how they will perform under cyclic fatigue loadings. The book will provide a comprehensive review of fatigue damage and fatigue life modelling.

Details of these and other Woodhead Publishing materials books can be obtained by:

- visiting our web site at www.woodheadpublishing.com
- contacting Customer Services (e-mail: sales@woodheadpublishing.com; fax: +44 (0) 1223 832819; tel.: +44 (0) 1223 499140; address: Woodhead Publishing Limited, 80 High Street, Sawston, Cambridge CB22 3HJ, UK)

If you would like to receive information on forthcoming titles, please send your address details to: Francis Dodds (address, tel. and fax as above; e-mail: francis.dodds@woodheadpublishing.com). Please confirm which subject areas you are interested in.

Corrosion of magnesium alloys

Edited by

Guang-Ling Song

Oxford Cambridge Philadelphia New Delhi

Published by Woodhead Publishing Limited,
80 High Street, Sawston, Cambridge CB22 3HJ, UK
www.woodheadpublishing.com

Woodhead Publishing, 1518 Walnut Street, Suite 1100, Philadelphia, PA 19102-3406, USA

Woodhead Publishing India Private Limited, G-2, Vardaan House, 7/28 Ansari Road,
Daryaganj, New Delhi – 110002, India
www.woodheadpublishingindia.com

First published 2011, Woodhead Publishing Limited

British Library Cataloguing in Publication Data
A catalogue record for this book is available from the British Library.

ISBN 978-1-84569-708-2 (print)
ISBN 978-0-85709-141-3 (online)

The publisher's policy is to use permanent paper from mills that operate a sustainable forestry policy, and which has been manufactured from pulp which is processed using acid-free and elemental chlorine-free practices. Furthermore, the publisher ensures that the text paper and cover board used have met acceptable environmental accreditation standards.

Typeset by Replika Press Pvt Ltd, India

Contents

Contributor contact details

(* = main contact)

Chapters 1 and 17

G.-L. Song
Chemical Science & Materials Systems Laboratory
GM Global Research and Development
General Motors Corporation
Mail Code: 480-106-224
30500 Mound Road
Warren, MI 48090
USA

E-mail: guangling.song@gm.com

Chapter 2

E. Ghali
Département de Génie des Mines, de la Métallurgie et des Matériaux
Université Laval
Pavillon Pouliot 1718A
Québec
Canada G1V 0A6

E-mail: edward.ghali@gmn.ulaval.ca

Chapter 3

A. Atrens* and M. Liu
The University of Queensland
Division of Materials
Brisbane, QLD 4072
Australia

E-mail: andrejs.atrens@uq.edu.au

N. I. Zainal Abidin
University of Malaya
Department of Mechanical Engineering
Faculty of Engineering
Kuala Lumpur 50603
Malaysia

G.-L. Song
Chemical Sciences & Materials Systems Laboratory
GM Global Research and Development
General Motors Corporation
Mail Code: 480-106-224
30500 Mound Road
Warren, MI 48090
USA

E-mail: guangling.song@gm.com

Chapter 4

T. Zhang*
Harbin Engineering University
China

E-mail: zhangtao@hrbeu.edu.cn

Y. Li
Institute of Metal Research
Chinese Academy of Sciences
China

Chapter 5

A. Gebert
Department of Electrochemical Properties of Functional Materials
Institute for Metallic Materials
Leibniz Institute for Solid State and Materials Research IFW Dresden
Helmholtzstr. 20
D-01069 Dresden
Germany

E-mail: a.gebert@ifw-dresden.de

Chapter 6

P. B. Srinivasan, C. Blawert* and D. Höche
Magnesium Innovations Centre (MagIC)
Institute of Materials Research
Helmholtz-Zentrum Geesthacht
Max-Plank-Str. 1
21502 Geesthacht
Germany

E-mail: bala.srinivasan@hzg.de
carsten.blawert@hzg.de

Chapter 7

M. Jönsson* and D. Persson
SWEREA KIMAB,
Drottning Kristinas väg 48
SE-114 28 Stockholm
Sweden

E-mail: martin.jonsson@swerea.se
dan.persson@swerea.se

Chapter 8

A. Atrens*
The University of Queensland
Division of Materials
Brisbane, QLD 4072
Australia

E-mail: andrejs.atrens@uq.edu.au

N. Winzer
Fraunhofer Institute for Mechanics of Materials, IWM
Freiburg
Germany

E-mail: nicholas.winzer@iwm.fraunhofer.de

W. Dietzel and P. B. Srinivasan
Helmholtz-Zentrum Geesthacht
Max-Plank-Str. 1
21502 Geesthacht
Germany

E-mail: wolfgang.dietzel@hzg.de

G.-L. Song
Chemical Sciences & Materials Systems Laboratory
GM Global Research and Development
General Motors Corporation
Mail Code: 480-106-224
30500 Mound Road
Warren, MI 48090
USA

E-mail: guangling.song@gm.com

Chapter 9

Y. B. Unigovski and E. Gutman*
Ben-Gurion University of the Negev
Faculty of Engineering Sciences
Department of Materials Engineering, Building #59, Room 036
P.O.B. 653
Beer-Sheva 84105
Israel

E-mail: gutman@bgu.ac.il

Chapter 10

F. Witte*
Laboratory for Biomechanics and Biomaterials
Hannover Medical School
Anna-von Borries-Str. 1-7
30625 Hannover
Germany

E-mail: witte.frank@mh-hannover.de

N. Hort and F. Feyerabend
Helmholtz-Zentrum Geesthacht
Institute of Materials Research
Max-Planck-Str. 1
21502 Geesthacht
Germany

E-mail: norbert.hort@hzg.de
frank.feyerabend@hzg.de

C. Vogt
Institute of Inorganic Chemistry
Callinstrasse 9
30167 Hannover
Germany

E-mail: vogt@acc.uni-hannover.de

Chapter 11

G.-L. Song*
Chemical Sciences & Materials Systems Laboratory
GM Global Research and Development
General Motors Corporation
Mail Code: 480-106-224
30500 Mound Road
Warren, MI 48090
USA

E-mail: guangling.song@gm.com

David H. StJohn
CAST Cooperative Research Centre
Division of Materials
School of Mechanical and Mining Engineering
The University of Queensland
Brisbane, QLD 4072
Australia

Chapter 12

A. Atrens* and Z. Shi
The University of Queensland
Division of Materials
Brisbane, QLD 4072
Australia

E-mail: andrejs.atrens@uq.edu.au
zmshi@uq.edu.au

G.-L. Song
Chemical Sciences & Materials Systems Laboratory
GM Global Research and Development
General Motors Corporation
Mail Code: 480-106-224
30500 Mound Road
Warren, MI 48090
USA

E-mail: guangling.song@gm.com

Chapter 13

D. Aurbach* and N. Pour
Department of Chemistry
Bar Ilan University
Ramat-Gan Israel 52900

E-mail: aurbach@mail.biu.ac.il

Chapter 14

W.-T. Tsai*
Department of Materials Science and Engineering
National Cheng Kung University
Tainan
Taiwan

E-mail: wttsai@mail.ncku.edu.tw

I.-W. Sun
Department of Chemistry
National Cheng Kung University
Tainan
Taiwan

E-mail: iwsun@mail.ncku.edu.tw

Chapter 15

B. L. Luan, D. Yang and X. Y. Liu
Industrial Materials Institute
National Research Council of Canada
800 Collip Circle
London, Ontario
Canada N6G 4X8

G.-L. Song*
Chemical Sciences & Materials Systems Laboratory
GM Global Research and Development
General Motors Corporation
Mail Code: 480-106-224
30500 Mound Road
Warren, MI 48090
USA

E-mail: Guangling.song@gm.com

Chapter 16

G.-L. Song
Chemical Sciences & Materials Systems Laboratory
GM Global Research and Development
General Motors Corporation
Mail Code: 480-106-224
30500 Mound Road
Warren, MI 48090
USA

E-mail: guangling.song@gm.com

Z. Shi
The University of Queensland
Division of Materials
Brisbane, QLD 4072
Australia

E-mail: zmshi@uq.edu.au

Preface

Magnesium (Mg) is the eighth most abundant element on Earth, especially in the ocean, and is a huge worldwide resource. Pure Mg has relatively low yield stress, elongation and modulus. However, for its amazingly low density, Mg has a strength–weight or stiffness–weight ratio higher than steel or aluminum alloys. Pure Mg is chemically active: it can be ignited in air and reacts violently with many media. Alloying is an effective way of improving Mg's physical and chemical performance. Generally speaking, an Mg alloy has significantly improved mechanical properties compared with pure Mg, and its mechanical performance, in most cases, can be further improved through proper heat treatment.

In addition to their low density and high strength to weight ratio, Mg alloys have many unique physical and chemical properties such as impressive damping, superb castability, good thermal conductivity, excellent electrical shielding effect, low heat capacity, negative electrode potential, zero magnetization, outstanding recyclability, excellent biocompatibility and non-toxicity to human body and environment. These properties have been attracting considerable research interest, and some of them have made Mg alloys irreplaceable in many applications. The first historically practical application of Mg could have been in fireworks. However, the largest use of Mg so far is for alloying aluminum. In the metallurgical industry, Mg can be used to remove sulfur during cast iron production and as a reducing reagent in the production of titanium. Mg is even an important additive for production of some organics in the chemical industry. It used to be an anode material in non-rechargeable batteries because of its negative potential and high energy density in aqueous electrolytes.

In practice, Mg alloys are much more popular than pure Mg in industrial applications. One of the well-known uses of Mg alloys is for cathodic protection, and some Mg alloys are used as sacrificial anodes due to their negative potential. Although their rapid rate of corrosion is a disadvantage, Mg alloys are superior to aluminum and zinc anodes in some environments such as in soil and water. The most important application of Mg alloys is for aerospace and military purposes, and some high-strength and creep-resistant rare earth-containing Mg alloys are critical materials in these industries. These applications normally require high corrosion resistance, which is still a big challenge for Mg alloys today. The electronics industry is a new emerging market for Mg alloys. Their low density, excellent castability, desirable damping performance, satisfactory thermal conductivity, great electrical shielding effect and good recyclability, as well as non-magnetism and non-toxicity, allow Mg alloys to replace plastic cases for many electronic devices.

However, the main driving force for the development of Mg alloys since the 1990s is the automotive industry. Although Mg alloys have been used in this industry for about 80 years, the drive to increase use of Mg alloy parts in vehicles in order to improve fuel economy has never been so strong. Mg alloys stand out as a promising alternative to Al alloys and steels. However, since the service environment of automotive components is very complex, the corrosion protection of Mg alloys is commonly recognized as a critical issue in this industry.

Mg alloys may also have great potential in the energy industry, particularly in the manufacture of batteries. As an example, the Mg/non-aqueous electrolyte battery is believed to be a future complement to the lithium (Li)-ion battery due to its favorable cost, energy density, safety and recyclability. It may fill the gap between the high-tech Li-ion battery and some low-tech rechargeable battery systems. A challenge in this area is how to avoid the passivity of the Mg anode, which is also an interesting topic in Mg corrosion science. Another example is the use of Mg alloys as hydrogen storage materials. In theory, Mg can store about 7.6 wt% of hydrogen. Alloying is a possible approach to lower the dehydrogenation temperature to improve the feasibility of hydrogen storage. Although many Mg alloy systems, such as the Mg_2X series alloys, have been investigated intensively for this purpose, no practical solution has been found to the slow dehydrogenation problem.

Another potential application of Mg alloys is in the medical field where Mg alloys may be used as a biodegradable implant material. The biocompatibility of Mg was first known in the early twentieth century. However, the intrinsic rapid corrosion and hydrogen evolution phenomena associated with Mg forced researchers to give up the first clinical trials. Recently, a greater understanding of corrosion mechanisms and development of innovative corrosion protection techniques are reviving the research interest in biodegradable Mg alloys.

Despite its great potential, Mg alloys have some limitations that cannot be overlooked. Some of the applications listed above have clearly revealed that corrosion is a critical issue due to the high chemical reactivity of base Mg, which can frequently be a 'show stopper' for their applications when exposed to the environment. The importance of the corrosion performance of any Mg alloy cannot, therefore, be overemphasized. A critical understanding of the corrosion mechanisms involved is needed and robust cost-effective solutions for its prevention are required.

Sustained efforts to combat the corrosion problems associated with Mg alloys have become prominent only in the past two decades. Increasing resources have been directed into the investigation of corrosion mechanisms and to the development of protection techniques for Mg alloys. These efforts are clearly shown in the large number of recent publications, proposals and theories in the public domain discussing Mg corrosion science and protection engineering. New Mg alloys and applications are emerging as a result of these sustained efforts. As much of this knowledge is disseminated in various journals, symposia and forums, it is important that a systematic understanding of the corrosion behavior of Mg alloys and protection techniques be synthesized and consolidated, which will provide the necessary insight to guide technologists for the further applications of Mg alloys.

This book is a compilation of comprehensive overviews of recent research by distinguished experts in the field of Mg corrosion and protection research. The book consists of four linked parts covering the main themes in the study of corrosion and protection of Mg alloys: Part I Fundamentals, Part II Metallurgical effects, Part III Environmental influences and Part IV Corrosion protection. Each part has several related chapters focusing on the important topics in its chosen area.

Part I provides a foundation for understanding the corrosion performance of Mg alloys in service environments. Since corrosion is essentially an electrochemical reaction, the electrochemistry of Mg is key to understanding Mg alloy corrosion mechanisms. Correspondingly, by analyzing electrochemical activity and passivity, various corrosion behaviors of Mg and its alloys in different environments can be clearly revealed.

Since the corrosion performance of a metal is determined primarily by its chemical composition and microstructure, an understanding of the metallurgical effects on corrosion is critically essential. Part II considers the role of typical microstructures and alloying elements of Mg alloys in corrosion. Metallic glass Mg alloys and some other innovative magnesium alloys are also discussed for their interesting atypical compositions or microstructures. This part provides an extension and development of the fundamentals presented in Part I.

As corrosion is a surface degradation process resulting from the interaction between a metal and its environment, it can be significantly influenced by environmental conditions. Part III, therefore, describes the corrosion behavior of Mg alloys in various natural and service environments. In addition to more familiar environmental conditions, there are special media such as moist air and coolants which are likely to be the service environments in industrial applications. Environmental media also include body fluids (or simulated body fluids) and ionic liquids. Mg biocompatibility has now become one of the most interesting research areas. The non-aqueous Mg battery is also being actively studied and developed. A summary of pioneering work in these areas will make researchers aware of relevant corrosion results and may be helpful to transplanting new ideas and advances across other areas in the field of Mg corrosion and protection. Moreover, applied stress and other metals in contact with Mg alloys can also be included in the environmental factors that have a significant influence on the corrosion of Mg alloys. Therefore, Part III covers atmospheric corrosion, coolant-induced corrosion, biodegradation, non-aqueous electrochemical dissolution, stress corrosion cracking, fatigue corrosion and galvanic corrosion. The understanding gained from this part is closely associated with industrial applications and also supplements the information presented in Part I.

Finally, Part IV focuses on the techniques for protecting Mg alloys from corrosion attack. This is an area rapidly gaining attention from corrosion scientists and engineers. Three typical surface treatment and coating techniques are selectively presented in this part:

1. Al electro-deposition,
2. conversion and electrophoretic coatings, and
3. anodization.

They are representative of numerous research publications in this area. The key message delivered in this part is that preventing Mg alloys from corrosion in a cost-effective manner is a challenging task.

In summary, the systematic presentation of recent research by distinguished experts in this book is not simply a comprehensive review or overview on progress in the field of corrosion and protection of Mg alloys. It also provides a further development in knowledge about Mg and its alloys in general. It is the editor's intention that the contents of this book will be useful to all researchers who are interested in corrosion science and engineering of Mg alloys. It will be a great reward for the editor and all the chapter contributors if this book can serve as an informative resource for corrosion scientists, metallurgists, engineers and designers working in industry, or for professional researchers and students studying in universities and research organizations.

Acknowledgements

The editor would like to take this opportunity to sincerely thank all the chapter contributors for sharing their work and knowledge with the readers of this book. With their great efforts and time devoted to this book, we now have a deep insight into so many interesting research areas and a comprehensive overview of the recent progress in the field of corrosion and protection of Mg alloys. The editor also gratefully appreciates the assistance from Adam

Hooper, Benjamin Hilliam, Bonnie Drury, Francis Dodds and Rob Sitton at Woodhead Publishing. Their hard work significantly sped up the publication of this book. Moreover, the editor's gratitude also goes to his colleagues for their great support in editing this book.

Guang-Ling Song

Part I

Fundamentals

1
Corrosion electrochemistry of magnesium (Mg) and its alloys

G.-L. SONG, General Motors Corporation, USA

Abstract: Magnesium (Mg) alloys are light, structural and functional engineering materials with a high strength to weight ratio which are increasingly being used in the automotive, aerospace, and electronic and energy industries. However, magnesium is chemically active with an electrochemistry differing from most conventional engineering metals. This chapter presents electrochemical reactions and corrosion processes of Mg and its alloys. First, an analysis of the thermodynamics of magnesium and possible electrochemical reactions associated with Mg are presented. After that an illustration of the nature of surface films formed on Mg and its alloys follows. To comprehensively understand the corrosion of Mg and its alloys, the anodic and cathodic processes are analyzed separately. Having understood the electrochemistry of Mg and its alloys, the corrosion characteristics and behavior of Mg and its alloys are discussed, including: self-corrosion reaction, hydrogen evolution, the alkalization effect, corrosion potential, macro-galvanic corrosion, the micro-galvanic effect, impurity tolerance, influence of the chemical composition of the matrix phase, role of the secondary and other phases, localized corrosion and overall corrosivity of alloys.

Key words: Magnesium (Mg), corrosion, electrochemistry.

1.1 Introduction

Magnesium (Mg) and its alloys have many outstanding properties relative to other engineering materials such as: low density, high strength, great damping capability, excellent fluidity for casting, good electric shielding effect, non-magnetic, satisfactory heat conductivity, low heat capacity, negative electrochemical potential, acceptable recyclability and non-toxicity. These properties make Mg and its alloys attractive to many industries. Particularly in the automotive and aerospace industries, where the strength/weight ratio is a critical issue, Mg alloys have been regarded as a promising alternative to aluminum alloys (Aghion and Bronfin, 2000; Makar and Kruger, 1993; Song, 2005b, 2006; Song and Atrens, 2000; Song *et al.*, 2005c) and have already found many applications (Aghion and Bronfin, 2000; Polmear, 1996). It is anticipated that a much wider application in the automotive, aerospace and electronic industries will be seen in the twenty-first century.

Currently, however, more ambitious Mg alloy applications in the automotive, aerospace and electronic industries are still unrealistic because of the poor corrosion resistance of the existing Mg alloys (Aghion and Bronfin, 2000; Bettles *et al.*, 2003a,b). Before effective solutions to the corrosion problems of Mg alloys become available, a further expansion of Mg alloy applications appears to be unlikely. To date, a large number of studies have been carried out to address corrosion issues and to improve the corrosion performance of Mg alloys (Blawert *et al.*, 2006; Hawkin, 1993; Hills, 1995; Gray and Luan, 2002; Jia *et al.*, 2003a; Liu *et al.*, 2008, 2009a; Nakatsugawa, 1996; Polmear, 1981; Shi *et al.*, 2003b, 2006a,b; Skar and Albright, 2002; Shreir, 1965; Song, 2004a,b, 2005a,b, 2006, 2008a, 2009d; Song and Atrens, 1999, 2003, 2005, 2007; Song and Shi, 2006; Song and Song, 2006a,b,c; Song and StJohn, 2002, 2004, 2005a,b; Song *et al.*, 1999, 2000, 2003, 2004a,c, 2005b,c, 2006b,c, 2007, 2010; Tawil, 1987; Wan *et al.*, 2006; Wang *et al.*, 2007; Zhao *et al.*, 2008b; Zhu and Song, 2006; Zhu *et al.*, 2005). The published results have clearly suggested that the corrosion of Mg is quite special in terms of its electrochemical behavior. In nature, the corrosion of Mg and its alloys is an electrochemical process and their corrosion performance or characteristics can be ultimately attributed to their electrochemical behavior. Therefore, revealing the electrochemical reactions involved in the corrosion process can provide a theoretical basis for understanding the characteristic corrosion phenomena for Mg and its alloys.

This chapter systematically summarizes the electrochemical characteristics and relevant corrosion behaviors of Mg and its alloys, in order to better understand their corrosion performance.

1.2 Thermodynamics

The thermodynamics of pure Mg is a foundation for understanding the electrochemical corrosion of Mg and its alloys. The stability of Mg in various environments can also provide clues for estimating the corrosion performance of Mg alloys in typical environments which is helpful towards understanding the thermodynamic behavior of Mg alloys. This section briefly touches upon the thermodynamics of pure Mg.

1.2.1 Thermodynamic tendency

Thermodynamically, Mg is very active. The standard Gibbs free energy changes (ΔG^0) for the following Mg oxidation reactions are quite negative (Ott and Boerio-Goates, 2000; Perrault, 1978; Wall, 1965; Weast, 1976–1977).

$$Mg + O_2 + H_2 = Mg(OH)_2 \qquad \Delta G^0 = -833 \text{ kJ/mol} \tag{1.1}$$

$$Mg + {}^1/_2\, O_2 = MgO \qquad \Delta G^0 = -569 \text{ kJ/mol} \tag{1.2}$$

$$Mg + 2H_2O = Mg(OH)_2 + H_2 \qquad \Delta G^0 = -359 \text{ kJ/mol} \tag{1.3}$$

This means that Mg in natural environments has a great tendency to spontaneously transform into its oxidized states. Therefore, when Mg is exposed to environments containing oxygen or water, its surface always tends to be rapidly oxidized, thereby forming an oxide or hydroxide surface film (Alves *et al.*, 2000; Nordlien *et al.*, 1997).

The fact that Mg in an oxidized state is more stable than in its metallic state is also supported by the thermodynamic data of Mg compounds and species listed in Table 1.1. Corrosion is an oxidation process where various oxidized Mg species and compounds can be generated depending on the exposure media. The corrosion of Mg according to the tabulated data is a spontaneous process and thus Mg in most practical environments is thermodynamically unstable, as Mg in its oxidized states Mg^{2+}, MgO or $Mg(OH)_2$ has a much more negative chemical potential. The more negative chemical potential of $Mg(OH)_2$ than that of Mg^{2+} or MgO also suggests that in a solution $Mg(OH)_2$ is a more stable corrosion product than Mg^{2+} or MgO. Table 1.1 also indicates that the MgH_2 free energy of formation is negative, signifying that MgH_2 in Mg is stable at ambient conditions if water is not present. This implies that the release of hydrogen is a hurdle in the application of Mg as a hydrogen storage material for the power industry.

1.2.2 Stability in aqueous environments

Mg is thermodynamically unstable and can corrode in pure water (Avedesian and Baker, 1999; Perrault, 1978; Song 2005b). The thermodynamic data (referring to Table 1.1) suggest that Mg will ultimately oxidize into $Mg(OH)_2$ and while doing so give off hydrogen gas according to Equation (1.3). In fact, water is critical to the corrosion of Mg as it is an essential environmental element for the electrochemical reactions (Song 2005b, 2006).

The stability of Mg in water can be theoretically predicted by an *E*–pH diagram. There are many possible reactions occurring to Mg in water and as such, various formats of *E*–pH diagrams have been presented based on

Table 1.1 Chemical potential of Mg and its compounds in various states at 25 °C (Perrault, 1974, 1978)

Species	Oxidation state	State	μ_o (kcal/mol)
Mg	0	Solid	0
Mg^+	+1	Ion	–61
Mg^{2+}	+2	Ion	–109
$Mg(OH)_2$	+2	Solid	–199
MgH	–1	Gas	+34
MgH_2	–2	Solid	–8
MgO	+2	Solid	–136

Note: the more negative a species is, the more stable it is.

different considerations of reaction possibilities (Avedesian and Baker, 1999; Pourbaix, 1974). However, taking only the most probable substances into account, the stability of Mg in an aqueous solution can be summarized in one *E*–pH diagram (see Fig. 1.1) which is a combination of those published Pourbaix and *E*–pH diagrams (Avedesian and Baker, 1999; Perrault, 1978); Pourbaix, 1974). In this diagram, some relatively unstable and intermediate substances are excluded, such as H^-, MgOH, MgH. This simplification leads to a very large corrosion domain, a narrow negative potential region of immunity (much more negative than its equilibrium potential) and a possible high alkaline (pH > 10.5) passive range. The diagram (Fig. 1.1) clearly shows that Mg in most *E*–pH regions tends to be oxidized into ions, oxides or hydroxides. Only in region 'Mg, MgH_2, H_2' is Mg relatively stable, although it has a tendency to be reduced into MgH_2. This forms a theoretical

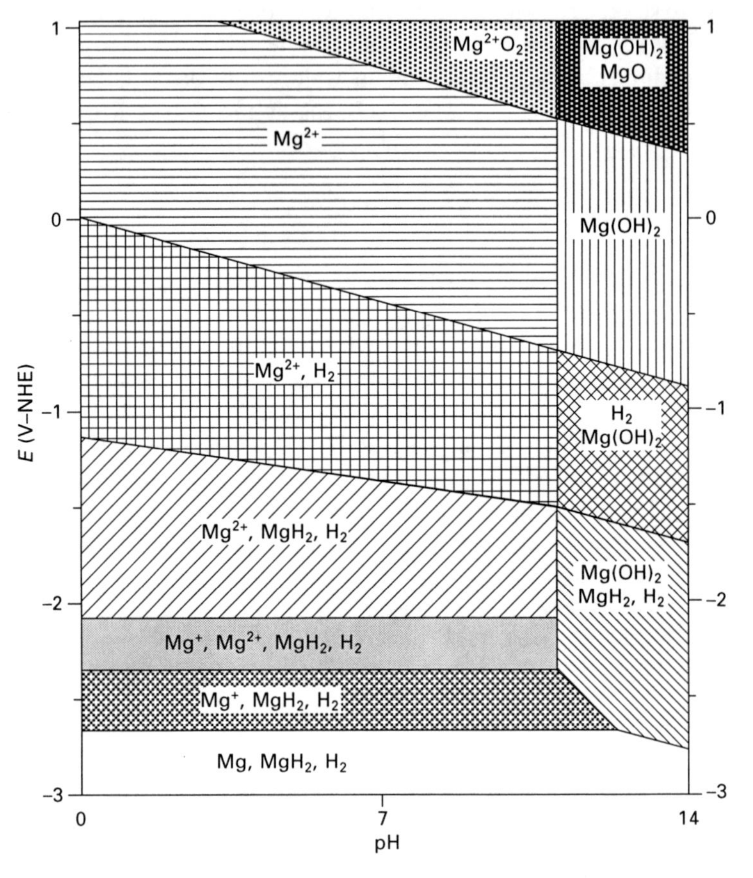

1.1 *E*–pH diagram with possible stable substances in a Mg–H_2O electrochemical system (data source: Avedesian and Baker, 1999; Perrault, 1978; Song, 2005b, 2006).

basis for cathodic protection of Mg alloys (Song, 2005a,b, 2006; Song *et al.*, 1997b). Under a natural condition metallic Mg in theory is not stable in an aqueous environment. Figure 1.1 also shows that $Mg(OH)_2$ is stable at a high pH value, thus there is a possibility that Mg becomes passive in a very basic solution. Unfortunately, most common environments are not sufficiently alkaline for Mg self-passivate.

It should be noted that the *E*–pH diagram can only predict the thermodynamic stability or tendency for corrosion of Mg in water. It has nothing to do with its kinetic processes, detailed reaction steps or intermediate substances. Unfortunately, the corrosion behavior of Mg is determined to a greater extent by its kinetics. Moreover, the *E*–pH diagram provides an overall chemical stability of substances, assuming all the substances (or phases) are uniform. This is not true in a practical Mg corrosion system. For example, the local pH value of the solution adjacent to Mg surface can be very different from that in the bulk solution (Nazarrov and Mikhailovskii, 1990; Song, 2006). Furthermore, this *E*–pH diagram does not contain any information on the effect of chemical composition of solution on corrosion process. For example, it cannot tell us the different corrosion rates of Mg in two aqueous solutions having the same pH value but with different NaCl contents.

In practice it is quite rare for Mg to be exposed to pure water at a potential more negative than its equilibrium potential. It is more often that the environment is not sufficiently alkaline and the natural open-circuit potential of Mg is much more positive than its equilibrium potential. Under these conditions Mg usually falls within the corrosion region in the *E*–pH diagram. Furthermore, in the passive region, if the aggressive species is present in the solution, Mg will not be passivated even though it is indicated so according to the *E*–pH diagram. For example, in an NaCl containing solution Mg does not exhibit passivity at a pH value as high as 13 (Song *et al.*, 1997a).

1.2.3 Critical electrochemical and chemical reactions

In an aqueous solution, many reactions may occur to Mg. The following are a few which may critically impact the corrosion of Mg (all the potentials if not specified in this chapter are relative to the standard hydrogen electrode (NHE)) (Perrault, 1974):

$$Mg = Mg^{2+} + 2e^- \qquad E^0 = -2.363\,V \qquad 1.4$$

$$Mg + 2OH^- = Mg(OH)_2 + 2e^- \qquad E^0 = -2.689\,V \qquad 1.5$$

$$Mg = Mg^+ + e^- \qquad E^0 = -2.659\,V \qquad 1.6$$

$$Mg + OH^- = MgOH + e^- \qquad E^0 = -3.140\,V \qquad 1.7$$

$$Mg^+ = Mg^{2+} + e^- \qquad E^0 = -2.067\,V \qquad 1.8$$

$$Mg^+ + 2OH^- = Mg(OH)_2 + e^- \quad E^0 = -2.702\,V \quad 1.9$$

$$Mg^+ + 2H_2O = Mg(OH)_2 + 2H^+ + e^- \quad E^0 = -1.065\,V \quad 1.10$$

$$MgOH + OH^- = Mg(OH)_2 + e^- \quad E^0 = -2.240\,V \quad 1.11$$

$$MgH_2 = Mg^{2+} + 2H^+ + 4e^- \quad E^0 = -1.114\,V \quad 1.12$$

$$MgH_2 = Mg^+ + H_2 + e^- \quad E^0 = -2.304\,V \quad 1.13$$

$$MgH_2 = Mg^+ + 2H^+ + 3e^- \quad E^0 = -0.768\,V \quad 1.14$$

$$MgH_2 = Mg^{2+} + H_2 + 2e^- \quad E^0 = -2.186\,V \quad 1.15$$

$$MgH_2 + 2OH^- = Mg(OH)_2 + 2H^+ + 4e^- \quad E^0 = -1.256\,V \quad 1.16$$

$$MgH_2 + 2OH^- = Mg(OH)_2 + H_2 + 2e^- \quad E^0 = -2.512\,V \quad 1.17$$

$$MgH_2 + OH^- = MgOH + 2H^+ + 3e^- \quad E^0 = -0.928\,V \quad 1.18$$

$$Mg + 2H^+ + 2e^- = MgH_2 \quad E^0 = +\,0.177V \quad 1.19$$

$$Mg + 2H_2O + 2e^- = MgH_2 + 2OH^- \quad E^0 = +\,0.177\,V \quad 1.20$$

$$Mg^{2+} + 2OH^- = Mg(OH)_2 \quad lg[Mg^{2+}] = 16.95\text{–}2pH \quad 1.21$$

$$Mg + H_2 = MgH_2 \quad \Delta G = -8.17\,kcal/mol \quad 1.22$$

$$MgH_2 + 2H^+ = Mg^{2+} + 2H_2 \quad \Delta G = -92.42\,kcal/mol \quad 1.23$$

$$MgH_2 + 2H_2O = Mg(OH)_2 + 2H_2 \quad \Delta G = -296.074\,kcal/mol \quad 1.24$$

Most of the electrochemical reactions have a negative standard equilibrium potential, suggesting that they can be an anodic process relative to an electrochemical hydrogen process in the corrosion of Mg. The relatively positive standard equilibrium potentials of reactions (1.19) and (1.20) imply that they can be a cathodic process in the corrosion of Mg. Moreover, reaction (1.22) indicates that MgH_2 may be formed if Mg is exposed to a hydrogen environment. However, if the environment contains proton (H^+) or moisture (H_2O), the hydrogen-stored Mg will become unstable and Mg may be dissolved according to reactions (1.23) and (1.24).

In an aqueous environment, water itself has the following reactions that can contribute to the corrosion of Mg:

$$H_2 = 2H^+ + 2e^- \quad E^0 = 0\,V \quad 1.25$$

$$H_2 + 2OH^- = 2H_2O + 2e^- \quad E^0 = -0.826\,V \quad 1.26$$

These two reactions can be cathodic processes coupling with the anodic dissolution of Mg.

In summary, the above 26 reactions may play an important role in the corrosion of Mg, some can be directly responsible for the corrosion damage, and some may even critically affect the corrosion process.

In practice there are many factors that can affect the above reactions and thus also influence the corrosion of Mg. For example, the presence of oxygen in water can lead to greater or less oxygen reduction on Mg which may have some influence on the cathodic and anodic polarization behaviors of Mg, particularly in atmospheric environment or under good aeration conditions where the supply of oxygen is sufficient (Ferrando, 1989; Makar and Kruger, 1990). However, the influence of oxygen is not critical (Froats *et al.*, 1987; Song, 2006).

1.3 Surface film

A surface film can, thermodynamically, and kinetically, be formed on an electrode surface. If a metal can spontaneously transform into its oxides or hydroxides, then in a natural environment the metal surface is usually covered by an oxide or hydroxide film. The surface film can also be an intermediate product existing on the metal surface, as long as the growth rate of the film is kinetically equal to or greater than its dissolution rate (Song *et al.*, 1994, 2005a). Mg and its alloys tend to be dissolved and oxidized in most practical environments. Subsequent corrosion products can deposit and form a surface film on its surface. Even under a condition (for example, pH~7) where a stable film does not seem to exist according to the *E*–pH diagram, there are some mechanisms leading to a kinetically stable surface film. For example, due to the dissolution of Mg and hydrogen evolution, the Mg surface becomes more alkaline than the bulk solution and this can result in deposition of $Mg(OH)_2$ on the surface (Nazarrov and Mikhailovskii, 1990; Song, 2006). A surface film on Mg or its alloy may not be really protective, but it can significantly influence all the reactions on the surface. To better illustrate the cathodic and anodic electrochemical reactions involved in the corrosion of Mg and its alloys, it is important to understand the surface film.

1.3.1 Composition and microstructure

A surface film on Mg or its alloys can vary in composition and microstructure depending upon the metallic substrate, environment and formation conditions. To date, the naturally formed surface film on pure Mg is not yet well understood.

From a thermodynamic point of view, the surface film should be composed of $Mg(OH)_2$ and MgO. $Mg(OH)_2$ is the main constituent of the surface film as it is more stable than MgO in an aqueous solution, whereas in dry atmospheric conditions MgO is the main composition. If water vapor is present in air, a more stable hydrated oxide (containing hydroxyl or hydroxide species) will be formed on Mg (Alves *et al.*, 2001; Fuggle *et al.*, 1975; Peng and Barteau, 1990). As most atmospheric environments contain some moisture, the surface films of Mg formed in natural atmospheric environments often contain both MgO and $Mg(OH)_2$. That is why $Mg(OH)_2$ can often be detected in a surface film formed in air (Splinter, 1993, 1994). After Mg is immersed in an aqueous solution, even if its original surface film is formed in very dry air, it will rapidly turn into $Mg(OH)_2$ to a great degree (Bradford *et al.*, 1976).

The composition and structure of a surface film can be significantly influenced by its formation process and the environment. In dry air, the MgO surface film is relatively thin but compact with some amorphous characteristics, allowing water to penetrate into the film and for the formation of an additional amorphous hydrated layer. The carbon dioxide in air can then combine with moisture to form a carbonate that further reacts with $Mg(OH)_2$ to produce Mg carbonate. The adsorption of atmospheric CO_2 can also occur directly during atmospheric corrosion of Mg (Baliga and Tsakiospolous, 1992; Froats *et al.*, 1987). Therefore, it is quite common that $MgCO_3$ is detected within the surface film (Fournier *et al.*, 2002). A practical surface may be a mixture of $MgCO_3{\cdot}3H_2O$, $MgCO_3{\cdot}5H_2O$ and $MgCO_3{\cdot}Mg(OH)_2{\cdot}9H_2O$, apart from the main composition of MgO and $Mg(OH)_2$. Industrial pollutants, such as SO_2, can also react with $Mg(OH)_2$ and generate a S-containing film on the Mg surface, such as $MgCO_3{\cdot}6H_2O$ and $MgSO_4{\cdot}6H_2O$.

After a Mg specimen is immersed into a solution, the original surface film will react with water and the outer layer of the surface film will become mainly $Mg(OH)_2$. Chlorides present in the solution can be combined into the film, resulting in the formation of $5Mg(OH)_2{\cdot}MgCl_2$, $MgCl_2{\cdot}6H_2O$, or $Mg_3(OH)_5Cl{\cdot}4H_2O$. If carbon oxide is dissolved in the solution, the surface film could change into $Mg(OH)HCO_3$. For a fluoride-containing solution, the surface film will be mainly MgF_2 which is insoluble in HF acids and can provide some corrosion protection for Mg.

When the Mg substrate contains trace or alloying elements, their oxides or hydroxides will more or less become the constituents of the surface film (Song *et al.*, 1998). For example, Mg and Al hydroxides have been reported to be two of the main components of the surface film on AZ91D formed in an aqueous solution (Chen *et al.*, 2008). $Mg_6Al_2(OH)_{16}CO_3{\cdot}4H_2O$ can be found on AZ31B. Various ions have unique affinities for oxygen and hydroxyls and different mobility within the surface film. This can lead to a variation in the ratio for a specific ion in the surface film as compared to the substrate. Mg typically has much stronger affinity for oxygen and hydroxide than do its alloying elements, such as Al, Zn, Mn and Zr. Thus, the main constituents of the surface film formed on a Mg alloy in air are Mg oxides and hydroxides. Recent research results (Liu *et al.*, 2009b) indicate that the surface films on some Mg–Al intermetallics have a lower Al/Mg ratio in the films than in the intermetallic substrate. A possible reason could be the stronger affinity of Mg to O and hydroxide as compared to aluminum's affinity.

In some cases, the solubility of the oxides and hydroxides derived from Mg and/or its alloying elements in an aqueous solution can also affect the composition of the surface film. The lower solubility of Al oxides and hydroxides compared with Mg oxides and hydroxides should contribute to the Al/Mg ratios within the film, too.

Generally speaking, the surface film formed in the air is relatively thin. However, the surface film can grow with time. It has been reported that the thickness of a film on Mg in air in the first 10 seconds is only about 2 nm (McIntyre and Chen, 1998). However, extended exposure in humid air or water can result in a surface film thickness of 100–150 nm (Nordlien *et al.*, 1995).

It is also believed that the microstructure of a surface film is more complicated than a simple layer. One concept is that the surface film on Mg consists of multi-layers. A schematic diagram illustrating a multi-layer structure of surface film on Mg has been proposed based on transmission electron microscopy (TEM) observations (Nordlien *et al.*, 1995). For example, a platelet-like morphology has been suggested for the surface film on Mg in water (Vermilyea and Kirk, 1969). It is postulated that a very thin and compact MgO may be present next to the Mg substrate and that the relatively thick and non-compact (porous) outer layer is mainly $Mg(OH)_2$.

There are relatively few studies on the effect of alloying elements on the microstructure of a surface film. One such study (Nordlien *et al.*, 1995, 1997) reports that when the Al content in the Mg substrate is greater than 4%, the Al concentration can reach up to 35 wt% in the inner layer of an air-formed surface film. In fact, after the Al content exceeds a threshold concentration in the Mg alloy substrate, a continuous Al_2O_3 skeletal amorphous structure may be formed in the surface film as a film matrix. It has also been proposed that the surface film on an Al-containing Mg alloy has an inner layer rich in Al oxides and outer layer rich in Mg oxides and hydroxides (Nordlien *et al.*, 1995, 1997).

1.3.2 Stability and protectiveness

Since metallic Mg has a propensity to form an oxide or hydroxide surface film as does metallic Al, an interesting question to investigate is then, 'Why isn't Mg as corrosion resistant as Al?'

A straightforward answer is that the Mg surface film is not as protective as the Al film, although the films on Mg and Al have a similar microstructural model with a compact inner layer and a non-compact (porous) outer layer. However, in further explaining the differences between the protection provided by the surface films on Al and Mg, a misleading mechanism is sometimes postulated by many researchers purely based on the volume ratio (the Pilling–Bedworth ratio) of metal oxide to metal. The Pilling–Bedworth ratio of MgO/Mg is smaller than 1 (0.81; Pilling and Bedworth, 1923), whereas the ratio of $Al_2O_3/(2Al)$ is greater than 1 (1.28, refer to Table 1.2). In that explanation the MgO film is believed to be relatively porous compared with Al_2O_3 and it does not fully cover the substrate surface to offer sufficient protection for Mg alloys.

The above explanation based on the Pilling–Bedworth ratio reasonably interprets the oxidation performance of Mg. In dry air at ambient or low temperatures the MgO film is thin and sufficiently ductile to fully cover the Mg surface and thereby offers limited protection for Mg (Alves *et al.*, 2001). However, when the film grows too thick at a high temperature, it will crack due to the small Pilling–Bedworth ratio and become non-protective (Alves *et al.*, 2001; Gulbransen, 1954; Song, 2005b).

In an aqueous solution the Pilling–Bedworth ratio obviously does not explain the corrosion behavior of Mg. Previously, it has been suggested that the surface film on Mg in an aqueous solution is mainly $Mg(OH)_2$. The Pilling–Bedworth ratio of $Mg(OH)_2$ to Mg is actually greater than 1 (which can be estimated from Table 1.2). Thus, there must be another reason for the poor protection of the surface film on Mg.

In an aqueous solution, $Mg(OH)_2$ is more difficult to dissolve than MgO, but it is still theoretically unstable in an acidic, neutral or weak alkaline aqueous solution according to the *E*–pH diagram. Even if there is such a film on Mg, it cannot be thick or compact and, thus, offers only a limited level of protection for Mg. For the case of an Al substrate, its surface film consists of Al_2O_3, $Al(OH)_3$ or a combination thereof which is electrochemically stable from weak acidic to weak alkaline. As a result, it can exist as a stable form on an Al surface and offer good corrosion protection for Al in many aqueous solutions. Therefore, the different thermodynamic stabilities of the surface films on Mg and Al in a natural aqueous solution should be the main reason for their different corrosion resistances.

Furthermore, it has been reported that the presence of chlorides in solution can dramatically accelerate the corrosion of Mg (Song and Song, 2006b). Table 1.3 lists the solubilities of $Mg(OH)_2$ in select solutions and from this it can be seen that the solubilities of $Mg(OH)_2$ in KOH and KCl are not significantly different. This means that the thermodynamic stability or solubility of the surface film on Mg alone cannot sufficiently account for its poor corrosion performance in an alkaline solution. The kinetic process of film formation

Table 1.2 Mole volumes of Mg, Al and their oxides and hydroxides

	Mg	MgO	$Mg(OH)_2$	Al	Al_2O_3	$Al(OH)_3$
Mole weight (g/mole)	24.4	40.4	58.4	27	102	78
Density (g/cm^3)	1.74	3.58	2.4	2.7	3.97	2.4
Mole volume (cm^3/mole)	14.02	11.28	24.33	10	25.69	32.50

Table 1.3 Solubility of $Mg(OH)_2$ in various media (Perrault, 1978; Weast, 1976–1977)

Media	Solubility (mol/L)
Cold water	0.000225
Hot water	0.001
10^{-3} mol/L KOH	0.00022
1mol/L KCl	0.00033
1mol/L KBr	0.00029
1mol/L KI	0.00032
0.05mol/L K_2SO_4	0.0009

and dissolution should be taken into account. For example, steel in some acidic solutions is not thermodynamically stable, but a chemically unstable surface film in the solution can actually be formed (Song *et al.*, 2005a), because the growth rate of the passive film is faster than its dissolution rate in the solution. The absolution rates of the film growth and dissolution are very low. Thereby, the presence of a film can effectively passivate steel. The passivating mechanism is also applicable to Al. On Mg, $Mg(OH)_2$ film can to some extent stay on the Mg surface due to the surface alkalization effect (this will be discussed later). As discussed earlier, the formation of a surface film on Mg to a great degree depends on the surface alkalization. In nature a kinetic process of OH^- diffusion in solution adjacent to the surface is relatively fast. Therefore, the presence of the film which accompanies a fast anodic dissolution process in solution can significantly accelerate the dissolution kinetics of the film, leading to a much higher corrosion rate of Mg. In other words, the protectiveness of the surface film on Mg is a function of the detailed kinetics of the film in a solution.

In reality, there are always some oxides already formed on a Mg surface in air. After immersion into an aqueous solution, the MgO formed in air will transform into a film consisting mainly of $Mg(OH)_2$. There may be two possible approaches for the air formed MgO film being replaced by $Mg(OH)_2$: (1) hydration of Mg directly and (2) dissolution of MgO and deposition of $Mg(OH)_2$. In the first approach, MgO is hydrated immediately after immersion in a process whereby the cubic MgO lattice is converted into hexagonal $Mg(OH)_2$ which has a volume twice that of MgO. It is believed that the volume expansion from MgO to $Mg(OH)_2$ can lead to a disruption of the surface film (Shaw, 2003; Xia *et al.*, 2003; Yao *et al.*, 2000) and may explain the porous microstructure of the $Mg(OH)_2$ layer in the surface film. In the second approach, not only the MgO, but also the metallic Mg substrate, can be dissolved resulting in the deposition of $Mg(OH)_2$ on the Mg surface because of the lower solubility of $Mg(OH)_2$ in solution (see Table 1.3). The deposited $Mg(OH)_2$ is normally quite loose and cannot offer significant protection for the Mg substrate. Therefore, even if an inner MgO layer exists within the surface film in an aqueous solution, it will become discontinuous and in some areas converted or dissolved into porous $Mg(OH)_2$ (see Fig. 1.2). These areas are defect sites ideally suited to initiate the corrosion of Mg.

After a surface film breaks down and corrosion is initiated in a specific area, it is relatively difficult to have the surface film repair itself. Although the dissolved Mg^{2+} can react with OH^- and form $Mg(OH)_2$ deposited on the Mg surface, the loosely deposited $Mg(OH)_2$ film does not necessarily deposit onto the broken areas. Simultaneously, the hydrogen bubbles being generated from the corroding areas (hydrogen evolution from corroding areas will be discussed in a later section) can stir the deposited $Mg(OH)_2$ and prevent the corroding areas from being fully covered by the deposited $Mg(OH)_2$. Therefore, the deposition or formation rate of $Mg(OH)_2$ film in general cannot be greater than that of the dissolution of $Mg(OH)_2$, and thus the corrosion cannot be easily self-inhibited.

The presence of specific ions in the solution can significantly influence the dissolution kinetics of a surface film. As discussed earlier, chloride is a very detrimental ion. If some passivating reagents, such as chromates and fluorides, are contained in a solution the reagents can react with Mg to form a very thin, compact, nearly insoluble, chromate or fluoride-containing surface film. This film can effectively protect Mg from corrosion attack (Song, 2005b). The beneficial ions also include dichromate, molybdate, nitrate, phosphates, metal-phosphate, vanadates, tartrate and, oxalate (Ghali, 2006; Schmutz and Guillaumin, 2003).

Mg alloying elements can be to a greater or lesser extent incorporated into the surface film and thereby modify the composition and structure and affect the stability and dissolution kinetics of the surface film. It is postulated (Nordlien *et al.*, 1995, 1997) that the surface film on a Mg–Al alloy is rich in Al_2O_3 and when the Al_2O_3 content in the film has attained a sufficiently elevated level, the film itself becomes an Al_2O_3 matrix. This matrix is dominated by an amorphous microstructure and the oxidation of the film becomes determined predominantly by Al_2O_3 which has superior corrosion

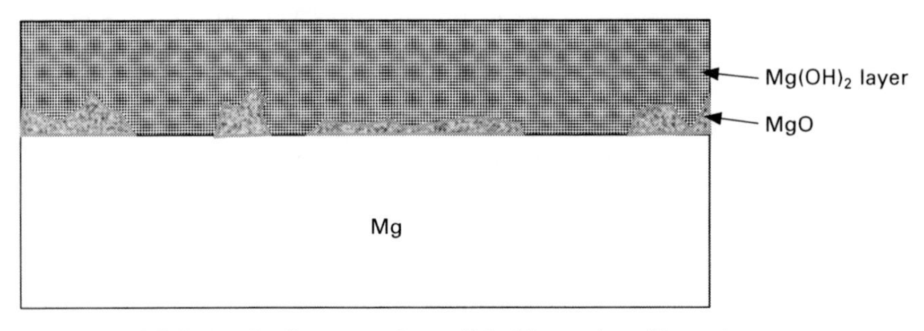

1.2 Schematic diagram of possible Mg surface film microstructure.

protection properties compared with MgO or $Mg(OH)_2$. More importantly, this species in the film can increase the activation energy for Mg transportation through the film and hence significantly improve the corrosion resistance of the film.

The presence of a film on Mg and its alloys has a direct effect on the anodic electrochemical processes. This will be discussed in the following two sections.

1.4 Anodic process

The anodic process of Mg is complicated and exhibits some 'special' behaviors. This is due to not only the presence of surface film, but also the unique electrochemistry of Mg.

1.4.1 Negative difference effect (NDE)

The difference effect is defined as the difference, Δ, between the hydrogen evolution rate at the open-circuit potential, I_o (polarization current density is zero), and the hydrogen evolution rate, I_H, at a given anodic polarization potential (or current density), such that:

$$\Delta = I_O - I_H \quad 1.27$$

When $\Delta < 0$, the phenomenon is termed the negative difference effect (NDE).

For most metals, such as iron, copper and nickel, an anodic increase in applied potential can cause an increased anodic dissolution rate and simultaneously a decreased cathodic hydrogen evolution rate. This is a result of the anodic reaction being accelerated and the cathodic reaction decelerated when the polarization potential becomes more positive.

However, for Mg the hydrogen evolution rate actually increases when the polarization potential or current density becomes more positive in the anodic region (James *et al.*, 1963, 1964; Straumanis 1958; Straumanis and Bhatia, 1963). Figure 1.3 shows the hydrogen evolution of Mg and Mg alloys in corrosive solutions. The hydrogen evolution rates of Mg and its alloys under anodic polarization conditions all increase dramatically as the anodic polarization current density increases. In fact, the NDE is a common phenomenon for Mg alloys (Song, 2005b, 2006; Song *et al.*, 1997a,b, 1998, 2004a). The basic feature of the NDE phenomenon, i.e. the hydrogen evolution rate (HER) increasing as polarization potential or current density becoming more positive, is an unusual anodic phenomenon that Mg and its alloys typically exhibit.

If the onset of an increasing hydrogen evolution with increasing potential on the polarization curves as shown in Fig. 1.3 is examined carefully, it can

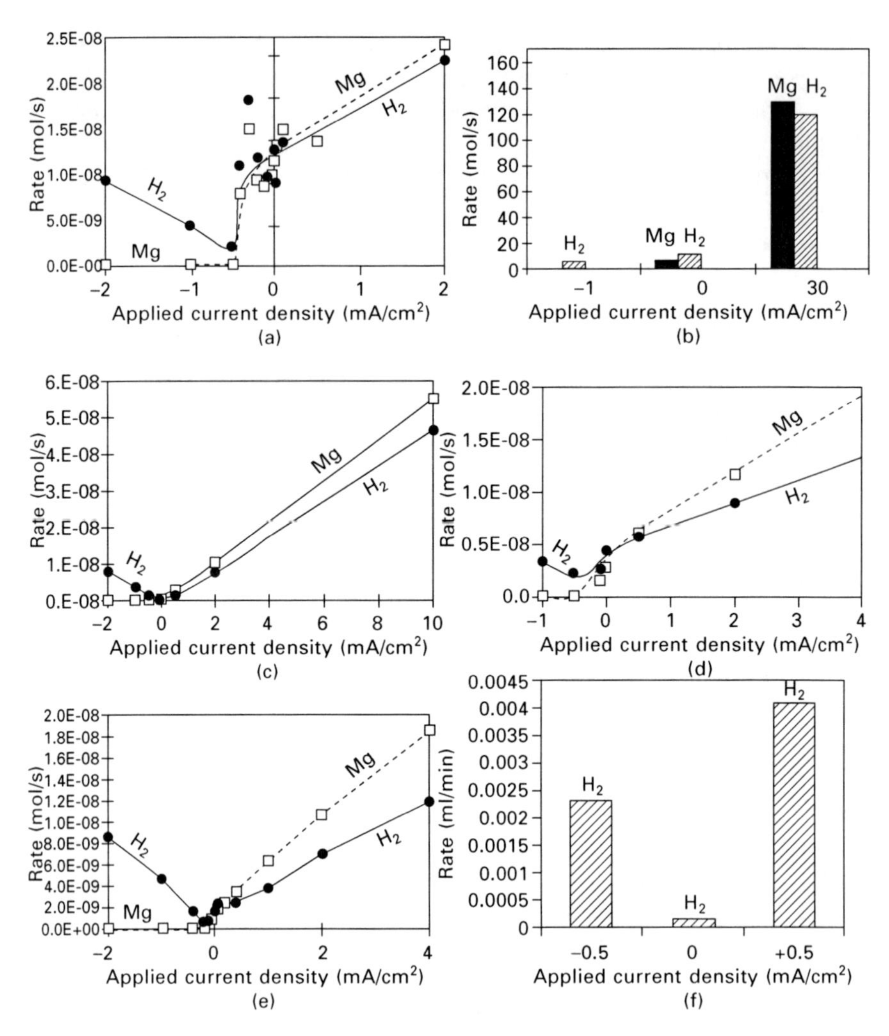

1.3 Hydrogen evolution and Mg dissolution rates: (a) Mg in 1 N NaCl (pH 11); (b) Mg in 1 N Na_2SO_4 (pH 11); (c) AZ21 (matrix phase) in 1 N NaCl (pH 11); (d) AZ91 ingot in 1 N NaCl (pH 11); (e) diecast AZ91 in 1 N NaCl (pH 11); (f) sand-cast MEZ in 5 wt% NaCl (based on Song *et al.*, 1997a,b, and Song, 2006).

be found that this unusual hydrogen evolution behavior actually starts at a cathodic potential or current density for a Mg or its alloy in 1 N or 5 wt% NaCl. In Fig. 1.4, the onset point is shown to be more negative than the corrosion potential, which means that the NDE can occur not only under an anodic polarization as specified in the definition but also under a cathodic polarization condition.

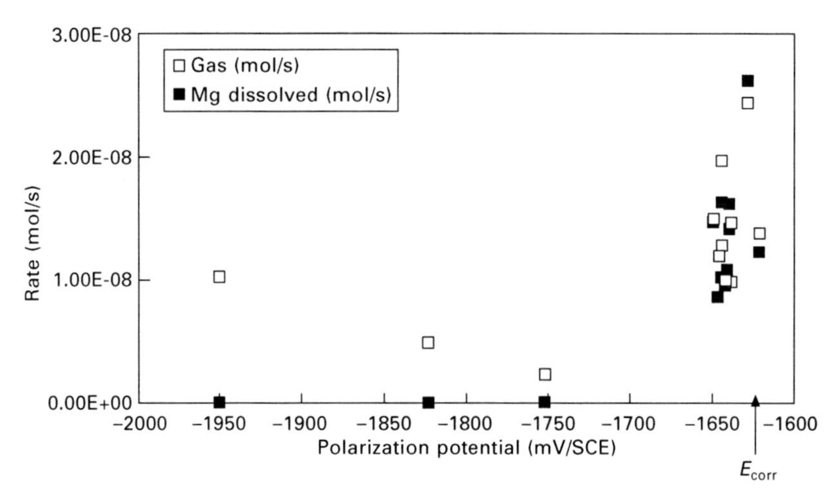

1.4 Dependence of rates of hydrogen evolution and Mg dissolution on polarization potential for Mg in 1 N NaCl (pH 11) (data source: Song *et al.*, 1997b).

1.4.2 'Anodic hydrogen evolution' (AHE)

The NDE phenomenon can be indicated by hydrogen evolution at anodic polarization. For example, the NDE behavior as demonstrated in Fig. 1.3 is caused by a 'strange' hydrogen evolution behavior when the polarization potential is more positive than a certain potential. In fact, a hydrogen evolution process is always involved in the anodic dissolution of Mg (Song, 2007a; Song *et al.*, 2003). This 'strange' hydrogen evolution is defined (Song, 2005b; 2006a) as 'anodic hydrogen evolution' (AHE) to distinguish it from the normal cathodic hydrogen evolution (CHE) resulting from cathodic reaction under cathodic polarization.

The AHE behavior is schematically illustrated in Fig. 1.5. The HER (see curve H_2^a in Fig. 1.5) first decreases and then increases as the polarization potential or current density changes from a negative to a positive value. It is different from a normal CHE (see curve H_2^c) for a conventional metal Me. The different trends of the dependence of HER on polarization potential or current density are caused by two different hydrogen evolution processes on Mg; normal CHE and AHE.

It should be noted that the equilibrium potential of the normal CHE is much more positive than the corrosion potential of Mg or its alloy. Even in the anodic region, the normal hydrogen reaction is strongly cathodically polarized and CHE is likely to occur. Certainly, the CHE rate decreases with increasing potential. Thus, in the anodic potential region, CHE does contribute to the hydrogen evolution behavior, but it cannot be the reason for the NDE phenomenon.

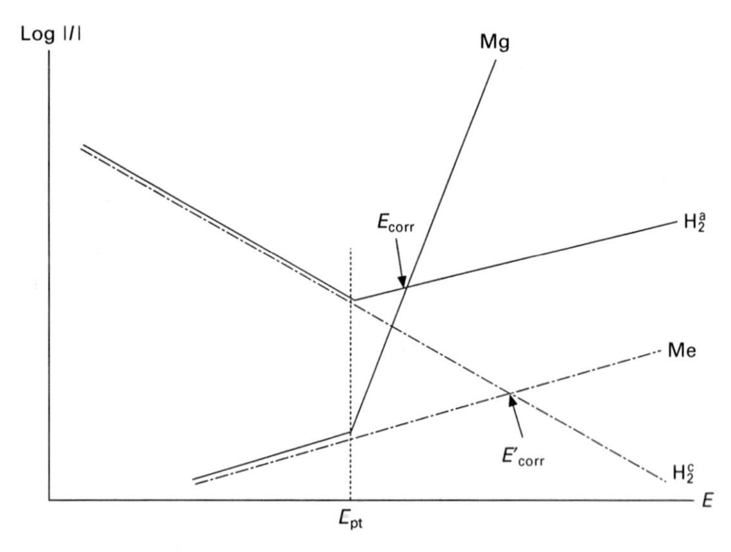

1.5 Schematic diagram for dissolution and hydrogen evolution from Mg or its alloy (Mg) compared with those from a normal metal (Me).

The above description of the 'strange' AHE behavior can be further supported by an *in situ* observation (Song *et al.*, 2004a) of a polarized MEZ Mg alloy (1.25 wt% Ce, 0.34 wt% Nd, 0.9 wt% Pr, 0.65 wt% La, 0.54 wt% Zn) in a 5 wt% NaCl solution. It was found (Song, 2005b, 2006; Song *et al.*, 2004a) that when the MEZ was cathodically polarized to –1.8 V/SCE, hydrogen evolution was observed from sites (e.g. particles) in grains or grain boundaries. This is normal CHE (see H_2^c curve in Fig. 1.5) resulting from a normal electrochemical hydrogen reduction reaction under a cathodic polarization condition. If the polarization potential was increased to a value more positive than the corrosion potential, e.g. –1.4 V/SCE, then the CHE from those sites became very slow. Meanwhile, in another area where corrosion was occurring, intense hydrogen evolution was observed. The hydrogen evolution became more intense as the applied potential became more positive (see curve H_2^a in Fig. 1.5). The hydrogen evolution occurring in a corroding area is an AHE process, and is closely associated with the anodic dissolution of Mg in that location. The differences in experiment between AHE from a corroding area and CHE from a non-corroding area are clearly demonstrated in Fig. 1.6.

These phenomena strongly suggest that in general the hydrogen evolution from a corroding Mg or Mg alloy electrode consists of both 'anodic' and 'cathodic' processes (Song, 2005a,b, 2006). The CHE emanates mainly from the non-corroded area of a Mg alloy and is responsible for the cathodic polarization behavior as represented by the cathodic branch of the polarization curve for the alloy. The anodic dissolution of a Mg alloy is closely associated

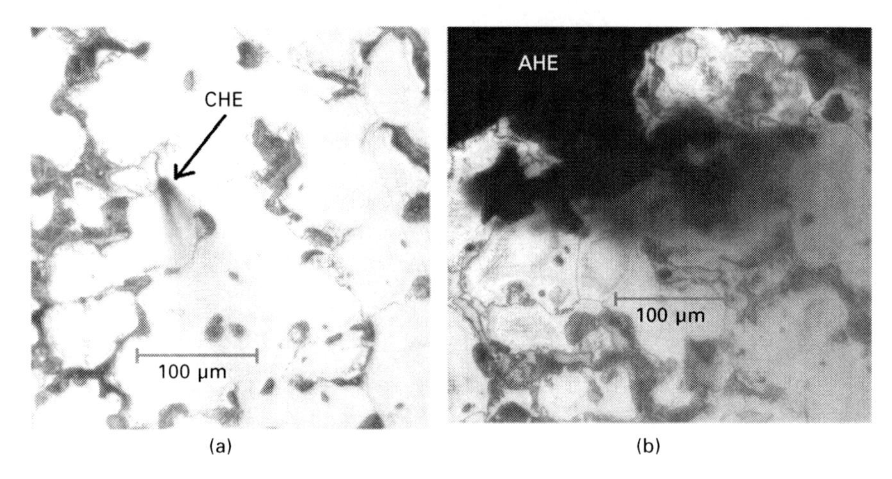

1.6 (a) 'Cathodic hydrogen evolution' (CHE) and (b) 'anodic hydrogen evolution' (AHE) from AZ91E in 5 wt% NaCl solution (Song, 2006).

with the AHE from the corroding areas of the alloy. The anodic dissolution and AHE are responsible for the anodic polarization behavior of the Mg alloy.

1.4.3 Apparent valence and anodic dissolution efficiency

When metallic Mg is dissolved and becomes ionic, electrons are generated and result in a Faradic current, i_F. The dissolution of Mg can be related to a Faradic current i_F through the so-called Faradic law:

$$Q_F = i_F\, t = nFW/W_q \qquad 1.28$$

where Q_F is the Faraday charge, F is the Faraday constant, t is the period of dissolution time, W is the mass of Mg dissolved, W_q is the mole weight of Mg, and n is the number of electrons generated in the electrochemical dissolution per Mg atom.

If the dissolution of Mg is a two electron transfer electrochemical reaction (1.4), then $n = 2$. If the dissolution concerns only a one electron reaction (1.6), then $n = 1$. However, in the anodic dissolution process involving AHE, if the AHE is also assumed to be an electrochemical process, then it will consume some electrons generated from the anodic dissolution of Mg, which will result in an apparent n lower than 2. It has been summarized in Table 1.4 that the apparent valence of Mg in many electrolytes is actually between 1.26 and 1.90 (Perrault, 1978; Song, 2005b).

The apparent valence of Mg in anodic dissolution at values less than two can also be reflected by the low anodic dissolution current efficiency. Theoretically, each mole of Mg atoms after dissolving into Mg^{2+} ions will

Table 1.4 Apparent valence of Mg anodically dissolved in various electrolytes (Perrault, 1978; Song, 2005b)

Solution	Applied current density (nA/cm^2)	Potential (V/NCE)	*n*
0.05 mol/L $MgCl_2$	2.5–30	–1.65	1.26
2.3 mol/L $MgCl_2$	7.7–160	–1.7	1.26
0.05 mol/L $MgBr_2$	4.2–50.5	–1.60	1.30
0.5 mol/L $MgBr_2$	12.5–112	–1.65	1.30
0.005 mol/L $MgSO_4$	6.2–41	–1.5	1.33
0.05 mol/L $MgSO_4$	10–85	–1.55	1.33
0.5 mol/L $MgSO_4$	2.5–100	–1.6	1.33–1.37
0.5 mol/L $MgSO_4$+ 0.05 mol/L K_2CrO_4	2.5–100	–1.5	1.27–1.32
0.5 mol/L $KClO_4$	6.4	–1.3	1.64
	10.9	–1.2	1.77
	100	–0.85	1.9
2 mol/L $MgSO_4$ + 0.5 mol/L K_2CrO_4	>–30	>0.3	1.86

Table 1.5 Anodic dissolution efficiency of Mg in various electrolytes (Glicksman, 1959)

Solution	Efficiency (%)
1 mol/L $MgCl_2$	65
0.1 mol/L LiCl	56
1 mol/L NaBr	59
0.1 mol/L LiCl	65
1 mol/L LiCl	66

give out tow moles of electrons, thereby generating $2F$ coulombs of electricity. However, since the hydrogen evolution consumes some of the electrons, the measured electric charge of the current generated from the dissolution of Mg will be smaller than the theoretically expected value. The relationship between the apparent valence n and anodic dissolution efficiency η can be established as follows (Song, 2006):

$$\eta = (n/2) \times 100\% \qquad 1.29$$

When $n = 2$, η is equivalent to the theoretical value, meaning that Mg dissolution is utilized 100% for current generation. If n is smaller than 2, for example, $n = 1.2$, then η will be as low as 60%, implying that only some of the dissolution of Mg has contributed to the Faraday current generation and the remaining consumed in other processes involved in the dissolution. Table 1.5 lists the anodic dissolution efficiency of Mg in some electrolytes. The apparent valence appears to be a function of the electrolyte solutions.

1.4.4 Anodic polarization resistance and 'passivity'

If the AHE from corroding areas is a normal hydrogen evolution process like CHE, it will consume anodic current resulting from the dissolution of Mg, making the anodic current density lower than that expected theoretically. Consequently, the anodic current density will not increase dramatically with increasing anodic polarization potential. However, in many aqueous solutions (acidic, neutral and weak alkaline), Mg and its alloys exhibit a dramatically increasing anodic current density with increasing anodic polarization potential. It has been measured (Song, 2004b, 2005a, 2007a; Song and Atrens, 2003; Song and StJohn, 2000, 2002, 2005a; Song *et al.*, 1998, 2003, 2004a,c) that the anodic polarization curves of Mg and its alloys always exhibit a low slope (i.e. low $|\Delta E/\Delta \log I|$). Figure 1.7 clearly shows that the anodic current densities of Mg and Mg–Al alloys (except the β phase) in corrosive NaCl solutions increase dramatically with increasing anodic polarization potential to a significantly greater extent as compared with the dependence of their cathodic polarization current densities on polarization potential.

It is always observed that hydrogen evolution is accompanying the anodic process during such measurements. If these polarization curves are examined carefully, a sudden change in current density on a polarization curve can be identified (Fig. 1.7). Such a sudden change in current density occurring on an anodic polarization curve can easily be recognized as an onset of dramatic increase in anodic current density (see Fig. 1.7(b)) and denoted as E_{pt}. When the sudden change in current density occurs in the cathodic region, it may be overwhelmed by the changing current density in the weak cathode polarization region (see Fig. 1.7(d)). In this case, a sudden decrease in cathodic current density should be carefully selected. Figure 1.8 displays a typical polarization curve of AZ91 in a corrosive solution with a sudden change in current density on the cathodic polarization curve. Experimentally, E_{pt} also always corresponds to a sudden increase in hydrogen evolution from the Mg surface. Therefore, such a change in current density is also recognizable by observation of the electrode surface.

The sudden increase in hydrogen evolution actually corresponds to a sudden increase in anodic dissolution of Mg (see Figs 1.3 and 1.4) in some local areas. It is in some sense similar to 'pitting' damage in morphology, but it is not a real pitting process which will be further explained later. In many corrosive environments, E_{pt} is more negative than the corrosion potential E_{corr}. Therefore, the experimentally measured anodic polarization curves of Mg and its alloys are actually intense 'pitting' processes. Consequently, the anodic current density increases dramatically with increasing polarization potential and Mg and its alloys normally display low anodic polarization resistance.

Since the low anodic polarization resistance of a Mg alloy originates from the fact that the 'pitting' potential E_{pt} of the alloy is more negative than its open-circuit potential E_{corr}, i.e. $E_{pt} > E_{corr}$, one would expect to see a relatively high polarization resistance in the potential region between E_{corr} and E_{pt}. However, a relatively high anodic polarization resistance does not always mean that the Mg alloy has a true passivating behavior. Only when the surface film effectively retards the anodic process and the anodic polarization current density is sufficiently low, e.g. a few μA/cm², can the anodic dissolution rate be neglected and the anodic polarization phenomenon

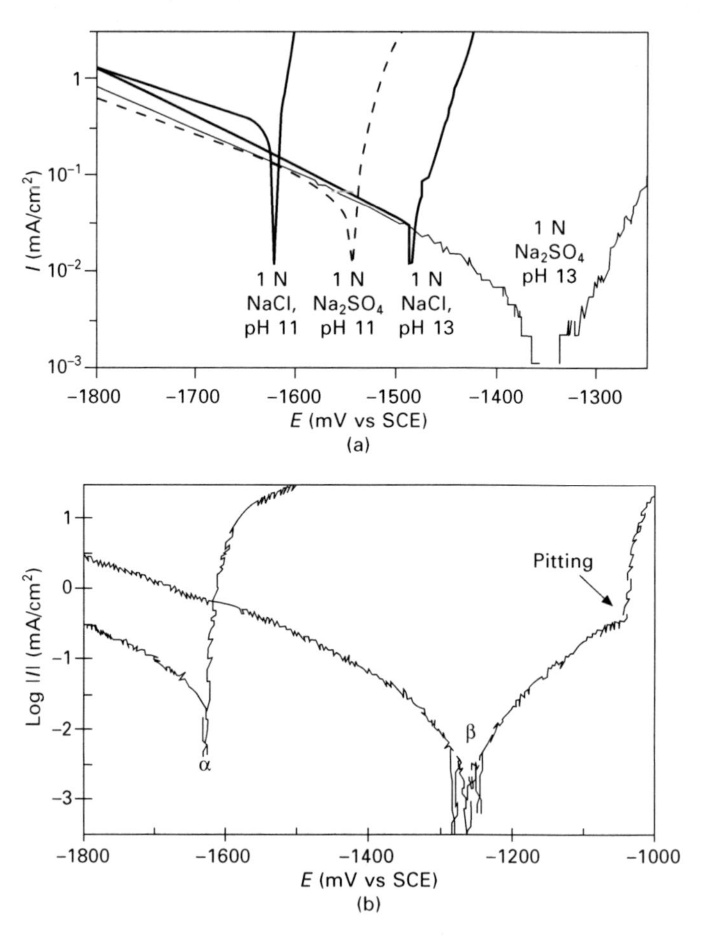

1.7 Polarization curves of Mg and Mg alloys in corrosive solutions: (a) Mg in 1 N NaCl and Na_2SO_4 at pH 11 and pH 13 (Song *et al.*, 1997a); (b) Mg–Al matrix phase and secondary phase in NaCl at pH 11 (Song *et al.*, 1998); (c) Mg–Al single phase in $Mg(OH)_2$ saturated 5 wt% NaCl (based on Song and StJohn 2000); and (d) diecast AZ91 and sand-cast MEZ alloys in 5 wt% NaCl at pH 11 (Song *et al.*, 2004a).

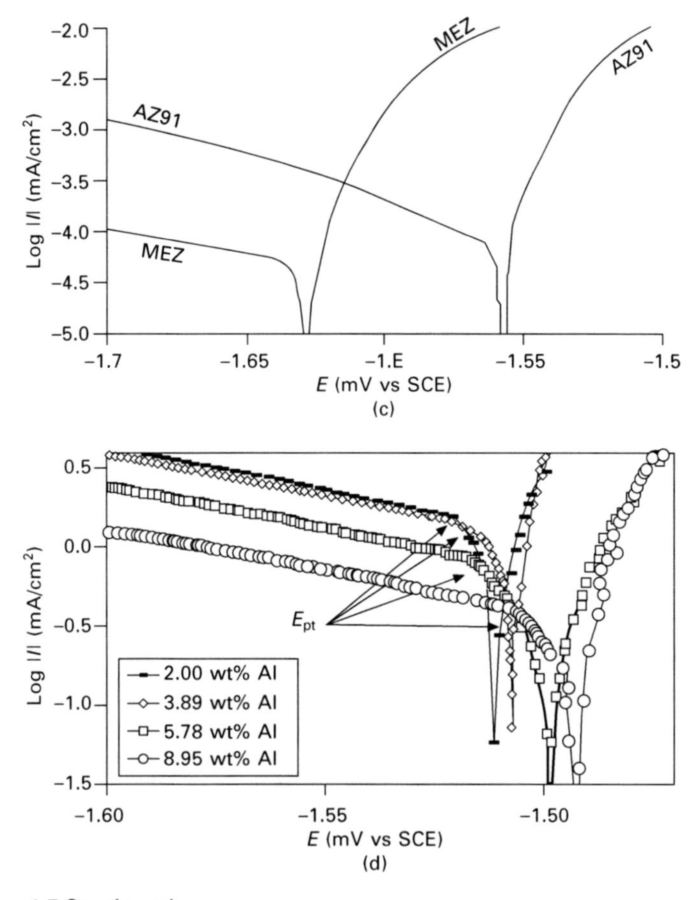

1.7 Continued

be termed as 'passivity'. Otherwise, the anodic process only shows a tendency toward 'passivity' in the potential region between E_{corr} and E_{pt}; the system actually has not got into its real passive state.

It should be noted that Mg and its alloys do not show a typical active peak on their anodic polarization curves. The 'passivity' is displayed in the region between E_{pt} and E_{corr}. The width of the 'passive' region (the difference between E_{corr} and E_{pt}) depends on the alloying elements and environmental solution conditions. Generally speaking, a Mg alloy can have a clear 'passive' region if it contains sufficiently stronger passivating alloying elements. Therefore, only some particular Mg alloys can exhibit relatively high polarization resistance or 'passivity' in the potential region from E_{corr} to E_{pt}. For example, the β phase alloy as shown in Fig. 1.7(b) has an E_{pt} more positive than its E_{corr} and shows some 'passivity'. Some Al containing intermetallic phases in Mg alloys have similar 'passivity'

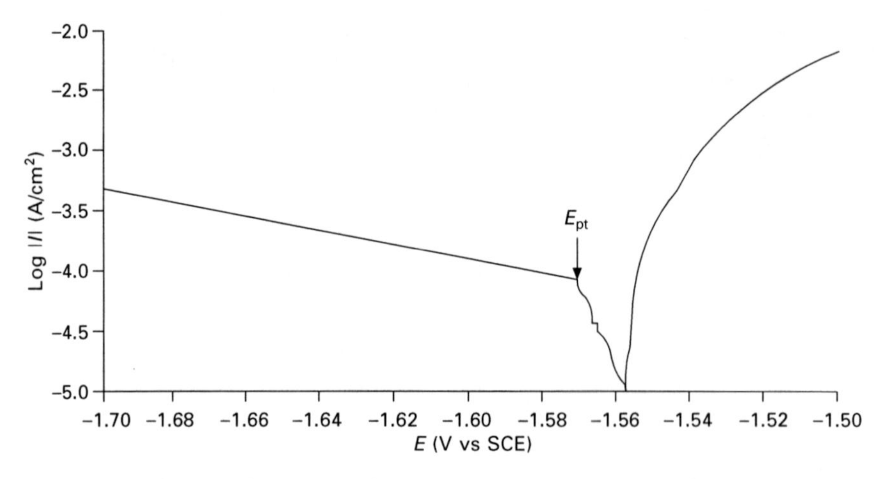

1.8 Polarization curve of sand-cast AZ91 in 5 wt% NaCl at pH 11 (Song and Atrens *et al.*, 2003).

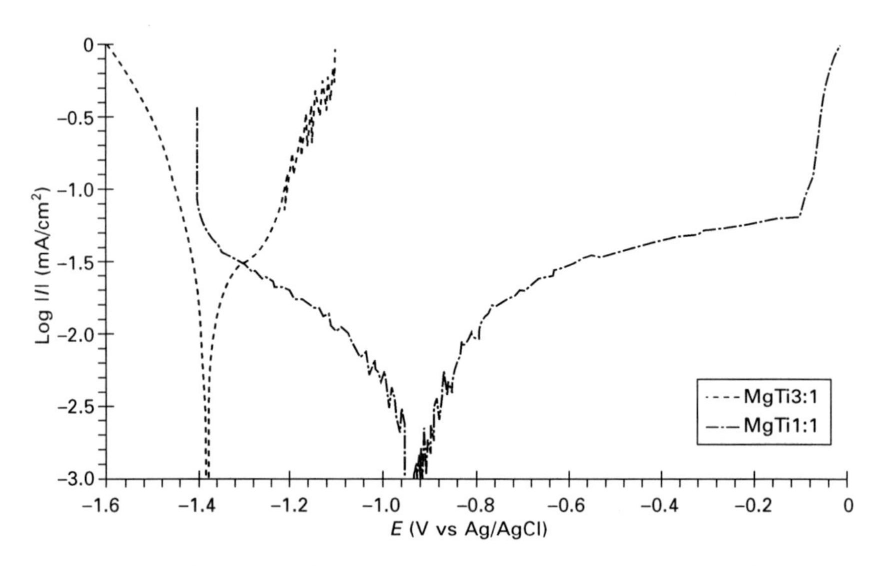

1.9 Polarization of magnetic sputtered Mg Ti3:1 (75 atm% Mg, 25 atm% Ti) and MgTi 1:1 (50 atm% Mg, 50 atm% Ti) in 0.1 N NaCl+1 N Na_2SO_4 + saturated $Mg(OH)_2$.

due to their high Al contents. Mg alloys containing rare earth elements and zirconium sometimes can also show 'passivity' (Nakatsugawa *et al.*, 1996). Ti is a very strong passivating element, but difficult to alloy with Mg (Song and Haddad, 2010). If alloyed, it can make Mg passive (referring to Fig. 1.9). The effect of solution composition on 'passivity' of a Mg alloy is also significant.

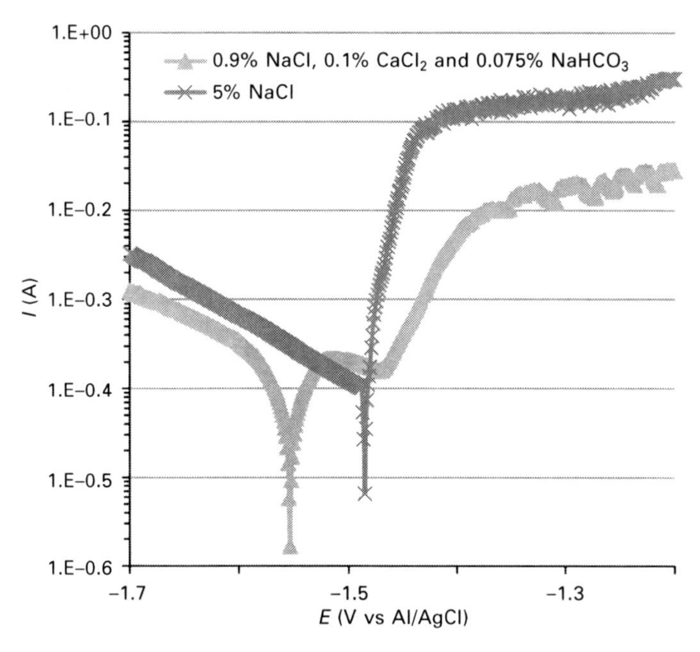

1.10 Polarization of AZ31B in 5% NaCl solution and 0.8% NaCl+0.1% $CaCl_2$+0.075% $NaHCO_3$ solution respectively.

Apart from the alloying element, environmental solution is another critical factor determining the passivity of a Mg alloy. Figure 1.10 shows that AZ31B has a very low anodic polarization resistance in 5% NaCl, but in a solution of 0.8% NaCl+0.1% $CaCl_2$+0.075% $NaHCO_3$ it displays a 'passive' region. Usually, the 'passivity' of a Mg alloy increases with increasing pH value of the environmental solution, and a 'passive' Mg alloy can lose its 'passivity' at low buffered pH values (Shaw and Wolfe, 2005). In nature, the 'passivity' is closely associated with the protectiveness of surface film. Since the quality of a surface film determines the AHE process, the 'passivity' can also be reflected by an AHE process.

1.4.5 A comprehensive anodic dissolution model

The 'strange' anodic polarization behaviors (negative difference effect, lower apparent valence, low anodic dissolution efficiency, low anodic polarization resistance and poor 'passivity') are closely associated with the AHE process which is further related to the onset of localized corrosion or 'pitting'. A comprehensive anodic dissolution model can be employed to understand these.

From a normal anodic dissolution perspective, the general anodic dissolution of Mg in an aqueous solution can be simply described as follows:

$$Mg \rightarrow Mg^{2+} + 2e^- \quad 1.30$$

The detailed anodic dissolution process of Mg involves intermediate steps. It is well known that one electron transfer is much easier than two electron transfers in an electrochemical reaction. The above reaction is more likely to realize through an intermediate step involving mono-valence Mg^+:

$$Mg \rightarrow Mg^+ + e^- \quad 1.31$$

Mg^+ is not stable and can rapidly change into the more stable Mg^{2+} through three possible approaches as follows.

Further anodic oxidization

The Mg^+ is directly anodized into Mg^{2+} via an anodic electrochemical reaction:

$$Mg^+ \rightarrow Mg^{2+} + e^- \quad 1.32$$

It can occur immediately on the surface of Mg. The overall dissolution of Mg will be

$$Mg \rightarrow Mg^{2+} + 2e^- \quad 1.33$$

It is a normal anodic dissolution process (the same as reaction (1.30)).

Disproportionation reaction

Owing to the high instability in solution, Mg^+ may change into other intermediate species through a disproportionation reaction:

$$2Mg^+ \rightarrow Mg + Mg^{2+} \rightarrow Mg \cdot Mg^{2+} \quad 1.34$$

The generated $Mg \cdot Mg^{2+}$ can be in the form of very fine metallic Mg particles deposited on the Mg surface, which is more active than the Mg electrode surface and will further react with water to form the final dissolution products:

$$Mg \cdot Mg^{2+} + 2H^+ \rightarrow 2Mg^{2+} + H_2 \quad \text{(in acidic solution)} \quad 1.35$$

$$Mg \cdot Mg^{2+} + 2H_2O \rightarrow Mg(OH)_2 + Mg^{2+} + H_2 \quad \text{(in neutral or basic media)} \quad 1.36$$

Therefore, the overall anodic dissolution will be:

$$Mg + H^+ \rightarrow Mg^{2+} + \tfrac{1}{2}H_2 + e^- \quad \text{(in acidic solution)} \quad 1.37$$

$$Mg + H_2O \rightarrow \tfrac{1}{2}Mg(OH)_2 + \tfrac{1}{2}Mg^{2+} + \tfrac{1}{2}H_2 + e^- \quad \text{(in neutral or basic media)} \quad 1.38$$

Direct hydration

Mg^+ can also react with water and generate hydrogen:

$$2Mg^+ + 2H^+ \rightarrow 2Mg^{2+} + H_2 \quad \text{(in acidic solutions)} \qquad 1.39$$

$$2Mg^+ + 2H_2O \rightarrow 2OH^- + 2Mg^{2+} + H_2 \quad \text{(in neutral or basic solutions)} \qquad 1.40$$

The overall anodic dissolution reactions will be:

$$Mg + H^+ \rightarrow Mg^{2+} + \tfrac{1}{2}H_2 + e^- \quad \text{(in acidic solutions)} \qquad 1.41$$

$$Mg + H_2O \rightarrow OH^- + Mg^{2+} + \tfrac{1}{2}H_2 + e^- \quad \text{(in neutral or basic solutions)} \qquad 1.42$$

The above-mentioned anodic dissolution processes (1.30)–(1.42) should mainly occur on bare Mg surface without a surface film. On a film-covered surface, Mg^+ is unlikely to survive, thus these processes can be neglected. Figure 1.11 shows a schematic diagram summarizing the above anodic dissolution reactions.

Overall anodic dissolution reaction

It should be noted that the further anodic reaction does not produce hydrogen while the hydration and disproportionation approaches can both lead to hydrogen evolution, and as such they can be treated as one group of reactions. In fact, these two approaches have exactly the same overall anodic dissolution reactions (see reactions (1.37), (1.38), (1.41) and (1.42)).

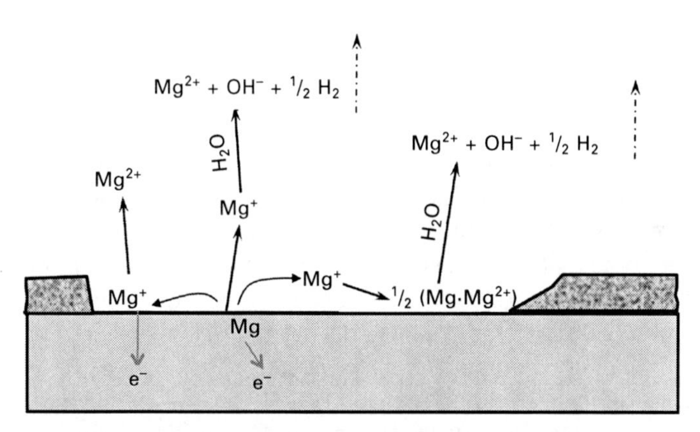

1.11 Schematic diagram for anodic dissolution of Mg or the matrix phase of a Mg alloy.

If the ratio of the further anodic reaction over the hydrogen producing reactions (disproportional reaction and hydration reaction) is assumed to be 'y', then the total anodic dissolution of Mg can be written as (Song, 2006):

$$Mg + [1/(1 + y)]H^+ \rightarrow Mg^{2+} + [1/(2 + 2y)]H_2 + [(1 + 2y)/(1 + y)]e^- \quad \text{(in acidic solution)} \qquad 1.43$$

$$Mg + [1/(1 + y)]H_2O \rightarrow Mg^{2+} + [1/(1 + y)]OH^- + [1/(2 + 2y)]H_2 + [(1 + 2y)/(1 + y)]e^- \quad \text{(in neutral or basic media)} \qquad 1.44$$

After the above overall anodic dissolution of Mg was proposed by Song in 2006 (Song, 2006), a similar Mg dissolution process was also presented (Atrens and Dietzel, 2007), in which another parameter 'k' was used instead of parameter 'y'. Nevertheless, the overall anodic dissolution reaction in nature is exactly the same as Song's equation.

1.4.6 Understanding of anodic phenomena

The overall anodic dissolution process (1.43) or (1.44) can successfully explain the 'strange' anodic behavior of Mg. The equations indicate that hydrogen evolution is involved in the anodic process. In fact, the mechanism of AHE concerns the involvement of Mg^+ in the anodic dissolution of Mg (Perrault, 1978; Song, 2005b, 2006, 2007a; Song and Atrens, 2003) thus the hydrogen evolution is not only a phenomenon at corrosion potential but occurs under various polarization conditions as well. According to these total anodic reactions, AHE would be accelerated by a positively increased polarization potential, truly illustrating the basic feature of NDE.

As to the phenomenon that NDE only occurs at potentials more positive than E_{pt} (see Figs 1.3, 1.4 and 1.5), the breakdown of surface film at E_{pt} should be taken into consideration. From a kinetic point of view, the surface film can exist on Mg even at a relatively negative potential (probably between the corrosion potential and the equilibrium potential of Mg). The film can be originally air-formed oxides, deposited hydroxides, a MgH_2 layer or even a mixture of them. It is assumed that the number or the total area of film-free sites is potential-dependent and can become larger at a more positive potential. The assumption is to some extent similar to the mechanism of a passive film breaking down at potentials more positive than a critical value. When the polarization potential is more negative than E_{pt}, the full surface of Mg is protected by a film and reaction (1.43) or (1.44) is negligible on the film-covered surface. As a result, the anodic dissolution of Mg and AHE are undetectable (CHE can still occur). After the polarization potential becomes more positive than E_{pt}, the surface film breaks down in some areas and reaction (1.43) or (1.44) becomes intense. This explains the sudden increases in anodic dissolution and ACE at E_{pt}, as well as the NDE behavior.

From reaction (1.43) or (1.44), we also have:

$$n = (1 + 2y)/(1 + y) \quad 1.45$$

which suggests that the apparent valence n should be a value between 1 and 2. When the ratio of further anodic reactions to the hydrogen-producing reactions is negligible, i.e. $y \to 0$, then $n \to 1$. If the further anodic reactions overwhelm the hydrogen producing reactions, i.e., $y \to \infty$, then $n \to 2$.

Correspondingly, according to equation (1.29), the anodic dissolution efficiency will be between 50% and 100%:

$$\eta = [(1 + 2y)/(2 + 2y)] \times 100\% \quad 1.46$$

When the further anodic dissolution reaction is dominating the whole anodic process, $\eta \to 100\%$. If Mg is dissolved mainly through the hydrogen-producing reactions, then $\eta \to 50\%$.

According to the comprehensive anodic dissolution model, the further anodic reaction (1.32) is a potential-dependent electrochemical process which can become faster at a more positive polarization potential. This means that y will increase and therefore n should also increase as the potential becomes more positive. This prediction has been experimentally verified; n starting at 1.2 approached a value of 1.8 after the anodic polarization potential was more positive than –1 V/NHE (Perrault, 1978). Because the hydrogen production reactions in the anodic process cannot fully stop at anodic potentials, n cannot reach its theoretical value of 2, which is also consistent with experimental observation (Perrault, 1978).

The low anodic polarization resistance of Mg and its alloys can be ascribed to the breakdown of the film. At potentials more negative than E_{pt}, the surface is fully covered by a surface film according to the model, and thus the anodic polarization current density does not dramatically increase with increasing potential. The presence of surface film on the surface also accounts for the fact that no active dissolution peak appears on the anodic polarization curve. After the polarization potential becomes more positive than E_{pt} and there are considerable film-free sites or areas, the Mg^+ generation and the subsequent further anodic oxidation, disproportionation and hydration of Mg^+ become significant. Therefore, the anodic current density dramatically increases.

Alloying can affect the anodic behavior according to this model. For example, Al alloying makes the surface film more stable and results in more positive E_{pt} (see Fig. 1.7(d)). Al in the matrix phase of the substrate can also stabilize Mg and decelerate reaction (1.31), thereby slowing down the entire anodic process (see Fig. 1.7(d)). This explains the anodic behavior of the β phase as shown in Fig. 1.7(b), which has an E_{pt} more positive than the corrosion potential. High polarization resistance of this phase is displayed prior to E_{pt} on its anodic polarization curve as a result of the considerably high content of Al in this phase. For the same reason, the composition of the solution can also affect the stability of the surface film and the anodic reactions. Hence, the 'passivity' of Mg and Mg alloys also varies with solution compositions.

1.5 Cathodic process

Oxygen reduction and hydrogen evolution are two typical cathodic processes for a conventional corroding metal in an aqueous solution. In a neutral or alkaline solution, oxygen reduction is usually believed to play a more important role than hydrogen evolution. However, this is not the case for Mg and its alloys.

1.5.1 Contribution of oxygen reduction

The equilibrium potential of Mg is far more negative than the hydrogen potential and as such the hydrogen reaction is strongly polarized cathodically on Mg. Therefore, hydrogen evolution always dominates the cathodic process in an aqueous environment. Certainly, in a neutral or basic solution oxygen reduction can still occur; however, the concentration of oxygen dissolved in the solution is limited by its diffusion within the solution, which is significantly slow compared with the hydrogen evolution reaction on Mg. It has been generally accepted (Froats *et al.*, 1987; Hur and Kim, 1998; Song, 2005b, 2006), that dissolved oxygen does not play an important role and hydrogen evolution is the main cathodic process involved in the corrosion of magnesium. Electrochemical experiments (Baril and Pebere, 2001) also show that the reaction rate of Mg electrode does not affect its AC impedance spectrum, suggesting that oxygen diffusion has a very limited effect on the cathodic process on Mg. Therefore, in practice, the reduction of oxygen can usually be ignored and hydrogen evolution is considered to be the only significant cathodic reaction in an aqueous environment.

1.5.2 Possible cathodic reactions

There are a number of possible detailed cathodic hydrogen processes on Mg and its alloys.

Normal cathodic hydrogen evolution

In this mechanism, hydrogen evolution on Mg generally follows the same process as exhibited by other conventional metals: a proton from the solution is first adsorbed onto the Mg surface and becomes an intermediate adsorbed hydrogen atom H_{ad}:

$$H^+ + e^- \rightarrow H_{ad} \quad \text{(in an acidic solution)} \qquad 1.47$$

or

$$H_2O + e^- \rightarrow H_{ad} + OH^- \quad \text{(in a neutral or basic solution)} \qquad 1.48$$

The intermediate adsorbed hydrogen atoms then combine to form hydrogen molecules or gaseous H_2 through two possible paths:

$$2H_{ad} \rightarrow H_2 \qquad 1.49$$

or

$$H_{ad} + H^+ + e^- \rightarrow H_2 \qquad 1.50$$

Therefore the overall cathodic reaction is:

$$2H^+ + 2e^- \rightarrow H_2 \quad \text{(in an acidic solution)} \qquad 1.51$$

or

$$2H_2O + 2e^- \rightarrow H_2 + 2OH^- \quad \text{(in a neutral or basic solution)} \qquad 1.52$$

This reaction can in theory continue as long as the potential is negative. In practice, the hydrogen over-potential can significantly affect the hydrogen evolution rate. Although there is no accurate or reliable hydrogen over-potential measured for Mg yet, this value is sometimes assumed to be very large (Makar and Kruger, 1993). Nevertheless, according to this mechanism hydrogen evolution should occur at a potential much more noble than the corrosion potential of Mg.

Intermediate Mg^+-catalyzed hydrogen evolution

Mg^+ participates in the cathodic process. Mg^{2+} is first reduced into intermediate Mg^+ on the Mg surface:

$$Mg^{2+} + e^- \rightarrow Mg^+ \qquad 1.53$$

The generated Mg^+ then reacts with water to generate hydrogen and itself is oxidized back to Mg^{2+} by water:

$$2Mg^+ + 2H_2O \rightarrow 2Mg^{2+} + 2OH^- + H_2 \qquad 1.54$$

The overall cathodic reaction is:

$$2H_2O + 2e^- \rightarrow 2OH^- + H_2 \qquad 1.55$$

No net Mg^+ is generated or consumed in the process. The equilibrium potential of reaction (1.53) is –2.067 V (see reaction (8)), which is more negative than the normal standard hydrogen evolution potential and the corrosion potentials of Mg and its alloys. Hence, the Mg^+ intermediate mechanism is likely to operate only when the Mg electrode is strongly cathodically polarized and dissolved Mg^{2+} present in the solution.

In addition to reaction (1.53), there may be other theoretical cathodic reactions on the Mg electrode involving $Mg(OH)_2$ to generate Mg^+, such as:

$$Mg(OH)_2 + 2H^+ + e^- \rightarrow Mg^+ + 2H_2O \quad \text{(in an acidic solution)} \qquad 1.56$$

or

$$Mg(OH)_2 + e^- \rightarrow Mg^+ + 2OH^- \quad \text{(in a neutral or basic solution)} \qquad 1.57$$

The equilibrium potential of reaction (1.56) is much more positive than reactions (1.53) and (1.57) (see reactions (1.8)–(1.10)). Thus rcaction (1.56) is much more significant. However, in acidic solutions where $Mg(OH)_2$ can barely be formed, its rate cannot be high due to the limited amount of $Mg(OH)_2$. Reaction (1.57) can only occur when the Mg electrode is cathodically polarized to a potential more negative than that of reaction (1.53). Nevertheless, both reactions (1.56) and (1.57) cannot last long on the Mg surface as they will terminate as soon as the $Mg(OH)_2$ is consumed. Only reaction (1.53) can reach its steady state and still be a real cathodic process on the Mg surface.

Intermediate MgH_2 catalyzed hydrogen evolution process

Similar to the intermediate Mg^+ mechanism, in an intermediate MgH_2 catalyzing process, Mg^{2+} or $Mg(OH)_2$ film is reduced together with protons on the surface of Mg to form intermediate MgH_2:

$$Mg^{2+} + 2H^+ + 4e^- \rightarrow MgH_2 \qquad 1.58$$

or

$$Mg(OH)_2 + 2H^+ + 4e^- \rightarrow MgH_2 + 2OH^- \qquad 1.59$$

The formed MgH_2 thus can be further oxidized particularly in a solution containing oxidizing reagents to generate hydrogen gas and Mg^{2+}:

$$MgH_2 + 2H_2O \rightarrow Mg^{2+} + 2OH^- + 2H_2 \leftrightarrow Mg(OH)_2 + 2H_2 \qquad 1.60$$

Therefore, an overall cathodic reaction can be expressed as:

$$H^+ + H_2O + 2e^- \rightarrow OH^- + H_2 \qquad 1.61$$

Which is equivalent to:

$$2H_2O + 2e^- \rightarrow 2OH^- + H_2 \qquad 1.62$$

This is a hydrogen evolution reaction for neutral and basic solutions. In an acidic solution, it can be rewritten as:

$$2H^+ + 2e^- \rightarrow H_2 \qquad 1.63$$

There is no net amount of MgH_2 generated or consumed in this mechanism.

Cathodic reactions (1.58) and (1.59) can occur at potentials ($E^0 = -1.114\,V$ and $E^0 = -1.256\,V$ respectively) which are more positive than reaction (1.53). When the cathodic polarization potential is not too negative, the cathodic hydrogen process may follow the MgH_2-catalyzed hydrogen evolution mechanism in addition to the normal cathodic hydrogen evolution mechanism. However, four electrons are involved in reactions (1.58) and (1.59) and as such, the probability for these reactions is very low in practice.

Possible H ingress and MgH_2 formation processes

Apart from the cathodic processes and mechanisms discussed previously, there may be other possible cathodic processes. For example, reactions (1.13)–(1.15) and (1.17)–(1.19) in the cathodic direction can lead to the generation of MgH_2 which can by reacting with water result in cathodic hydrogen evolution. However, reactions (1.13), (1.15) and (1.17) require hydrogen gas that must come from reaction (1.25). If reaction (1.25) is substituted into these reactions, then they are actually reactions (1.12), (1.14) and (1.16). Reactions (1.12) and (1.16) have been included in mechanisms II and III. Reactions (1.14) and (1.18) require Mg^+ and Mg(OH). These species can come from reactions (1.8)–(1.11). Reactions (1.10) and (1.11) under certain conditions are equivalent to reactions (1.8) and (1.9) and have been considered in the intermediate Mg^+ and MgH_2 processes and will not be repeated here. Therefore, only reactions (1.19) and (1.20) are unique and should be considered. These reactions have relatively positive equilibrium potentials:

$$Mg + 2H^+ + 2e^- \rightarrow MgH_2 \quad \text{(in an acidic solution)} \qquad 1.64$$

or

$$Mg + 2H_2O + 2e^- \rightarrow MgH_2 + 2OH^- \quad \text{(in a neutral or alkaline solution)} \qquad 1.65$$

The MgH_2 formed on the Mg surface is not stable in contact with water and as such can react with water to decompose into H_2 and Mg^{2+} or $Mg(OH)_2$ according to equations (1.23) or (1.24) respectively:

$$MgH_2 + 2H^+ = Mg^{2+} + 2H_2 \quad \text{(in an acidic solution)} \qquad 1.66$$

or

$$MgH_2 + 2H_2O = Mg(OH)_2 + 2H_2 \quad \text{(in an alkaline or neutral solution)} \qquad 1.67$$

Hence, the overall cathodic reaction should be:

$$Mg + 4H^+ + 2e^- = Mg^{2+} + 2H_2 \quad \text{(in an acidic solution)} \qquad 1.68$$

or

$$Mg + 4H_2O + 2e^- = Mg(OH)_2 + 2OH^- + 2H_2 \quad \text{(in an alkaline or neutral solution)} \qquad 1.69$$

suggesting that the Mg dissolves or transforms into $Mg(OH)_2$ in a cathodic process and that the rate of 'cathodic dissolution' of Mg will increase as the applied potential becomes more negative, which is abnormal and contradictory to observed experimental phenomena. It has been reported that when the polarization potential is more negative than a certain potential, the dissolution of Mg will stop (Song *et al.*, 1997b). Figure 1.4 shows the rate dependence of hydrogen evolution and Mg dissolution on the polarization potential for pure Mg immersed in 1 N NaCl (pH = 11) solution. In this case, when the potential becomes more negative than −1.7 V/saturated cabmel electrode (SCE), the Mg dissolution nearly stops while the hydrogen evolution rate is increasing. Therefore, the cathodic process with electrochemically formed MgH_2 on Mg can be excluded.

There is one more possibility that Mg hydride formation does not necessarily occur via an electrochemical reaction on the Mg surface as do reactions (1.64) or (1.65). It can be a hydrogen absorption or storage process for Mg starting from the adsorption of hydrogen atoms onto the Mg surface. The normal hydrogen evolution process should be involved in the above reactions, i.e., H^+ or H_2O has to be reduced into H_{ad} on the Mg surface first.

$$H^+ + e^- \rightarrow H_{ad} \qquad 1.70$$

or

$$H_2O + e^- \rightarrow H_{ad} + OH^- \qquad 1.71$$

Apart from the combination of H_{ad} and production of H_2 as illustrated in the normal hydrogen evolution process, the intermediate hydrogen atom H_{ad} on the Mg surface, before it is combined into a hydrogen molecule on the Mg surface, may also react with Mg to form MgH_2 on the Mg surface or diffuse into the Mg matrix (dissolved in Mg as H_{Mg}) and then combine with Mg to form MgH_2 in the Mg:

$$2H_{ad} + Mg \rightarrow MgH_2 \qquad 1.72$$

$$H_{ad} \rightarrow H_{Mg} \qquad 1.73$$

or

$$2H_{Mg} + Mg \rightarrow MgH_2 \qquad 1.74$$

In practice, the diffusion of H_{Mg} in Mg could be much faster than some theoretical calculations, because even in pure Mg there are a large number of defects such as dislocations and grain boundaries that could act as quick paths for H_{Mg} to diffuse into Mg (Winzer *et al.*, 2005, 2007a, b, 2008a,b,c).

In this process, the overall cathodic reaction will be:

$$2H^+ + Mg + 2e^- \rightarrow MgH_2 \qquad 1.75$$

which is actually reaction (1.19). The final cathodic product is MgH_2.

It should be noted that hydrogen molecules generated in a normal cathodic hydrogen evolution process cannot get into the Mg matrix because their decomposition into hydrogen atoms requires more energy. The contribution of generated hydrogen gas to the MgH_2 cathodic process on Mg can be neglected.

The generated MgH_2 may stay inside Mg, so its reaction with water could be insignificant. However, on the Mg surface, the generated MgH_2 from reaction (1.75) can react with water and produces hydrogen:

$$MgH_2 + 2H_2O \rightarrow Mg^{2+} + 2OH^- + 2H_2 \rightarrow Mg(OH^-)_2 + 2H_2 \qquad 1.76$$

In this case, the overall cathodic reaction of this mechanism becomes (1.68) or (1.69).

The above analysis suggests that the generated MgH_2 mainly stay inside Mg and not on the surface in direct contact with water. It can be deduced that the MgH_2 in this mechanism is formed via reaction (1.73) by ingress of H_{ad}, and not by the simple combination of $2H_{ad}$ on the Mg surface through reaction (1.74). This differs from the intermediate MgH_2-catalyzed mechanism, in which MgH_2 is only an intermediate species formed on the surface.

H ingress and MgH_2 formation in Mg is relatively slow compared with the normal, intermediate Mg^+-catalyzed, and intermediate MgH_2-catalyzed hydrogen evolution processes. Usually, the H ingress and MgH_2 formation cathodic processes are neglected. However, in the case of hydrogen-assisted cracking (SCC) occurring to Mg or its alloys, this process could be critical (Winzer *et al.*, 2005, 2007a,b, 2008a,b,c).

1.5.3 Practical cathodic processes

The possible Mg cathodic processes discussed above can be summarized in a schematic diagram (Fig. 1.12). It should be noted that different mechanisms may operate under different conditions and in practice there could be only one mechanism dominating the cathodic process at a time.

As stated earlier, H ingress and MgH_2 formation is not a significant

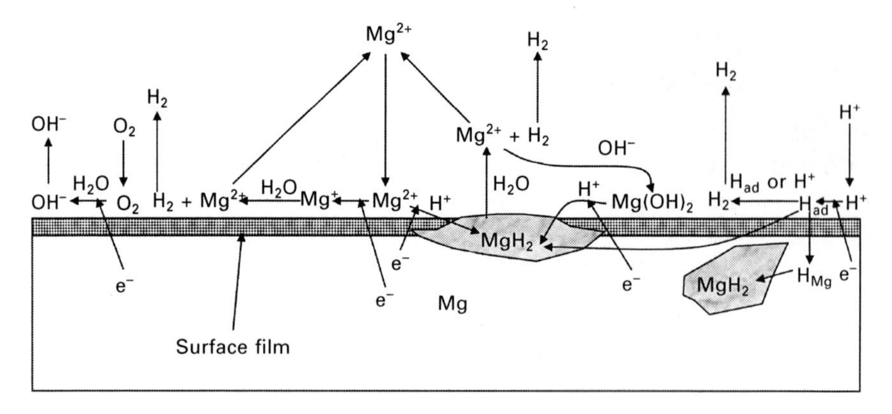

1.12 Schematic illustration of the cathodic reactions on Mg surface.

cathodic process and normally not considered in corrosion. The other three mechanisms can all lead to the same overall cathodic hydrogen evolution process:

$$2H^+ + 2e^- \rightarrow H_2 \quad \text{(in an acidic solution)} \quad 1.77$$

$$2H_2O + 2e^- \rightarrow H_2 + 2OH^- \quad \text{(in an akaline or neutral solution)} \quad 1.78$$

Although oxygen in solution can affect the cathodic process on the surface of Mg, the contribution of oxygen to the cathodic process is not significant. In fact, there is no oxygen-diffusion controlled plateau on the cathodic polarization curves of Mg and its alloys (Shi *et al.*, 2001; Song, 2007a; Song and Atrens, 2003, 2007; Song and Song, 2006b; Song and StJohn, 2000, 2002, 2005a; Song *et al.*, 1999, 2003, 2006a). The cathodic polarization curves always exhibit a typical Tafel region. In a very wide potential range on a cathodic polarization curve, the log plot of cathodic current density increases linearly as polarization potential becomes more negative, suggesting that the hydrogen evolution in the cathodic region follows the Tafel equation.

In theory, all three cathodic hydrogen evolution mechanisms, i.e., normal, Mg^+-catalyzed and MgH_2-catalyzed processes which all have the same overall hydrogen reactions, can result in the Tafel behavior. However, the latter two processes are relatively unlikely to dominate the cathodic process, since they have equilibrium potentials much more negative than the normal hydrogen equilibrium potential and since the four electron reactions (1.58) and (1.59) very rarely occur. Therefore, the most practical hydrogen evolution mechanism should be the normal hydrogen evolution process.

Mg alloying elements do not have a significant effect upon the cathodic process. This is substantiated by the insignificant cathodic Tafel slope difference between Mg and its alloys (see Fig. 1.7), which is also indirect evidence for the normal cathodic hydrogen evolution. Nevertheless, alloying elements may affect the surface film and consequently influence the cathodic current density. It has been reported that Al alloying can make the surface film thinner and more stable (Song *et al.*, 1998, 2004a) which is favorable for electron transfer as in the cathodic hydrogen evolution reaction (1.77) or (1.78). Therefore, the single phase Mg–Al alloy with a higher level of Al exhibits higher cathodic current density (see Fig. 1.7(b) and (d)). Obviously, the surface state of a Mg alloy can also influence the cathodic hydrogen process (Song and Xu, 2010).

1.6 Corrosion mechanism and characteristic processes

Based on the electrochemistry of Mg discussed in the prior sections, the corrosion mechanism and characteristics of Mg and its alloys can be illustrated as follows.

1.6.1 Self-corrosion of Mg matrix phase

Self-corrosion is basically an anodic dissolution process at a rate equal to that of the coupling cathodic process on an electrode. Under a steady state self-corrosion condition, all the electrons generated by anodic reactions are consumed by cathodic reactions on an elecrode. On Mg the overall cathodic process is reaction (1.77) or (1.78) and the overall anodic process is reaction (1.43) or (1.44), respectively. When they are coupled, ie. having the same rate, the overall corrosion of Mg can be written as:

$$Mg + 2H^+ \rightarrow Mg^{2+} + H_2 \quad \text{(in acidic solution)} \qquad 1.79$$

or

$$Mg + 2H_2O \rightarrow Mg^{2+} + 2OH^- + H_2 \quad \text{(in neutral or basic media)} \qquad 1.80$$

which is the widely reported overall corrosion reaction of Mg in aqueous solutions.

The detailed anodic and cathodic reactions under the self-corrosion condition are schematically presented in Fig. 1.13. In general, anodic reactions occur mainly in film-free areas whereas in film-covered areas anodic dissolution is negligible. Cathodic reactions with hydrogen evolution as a main process, can take place in both the film-free and film-covered areas, although in a film-free area much faster than on a film-covered surface.

The corrosion mechanism of Mg discussed above is applicable to both pure and the matrix phase (a single phase Mg alloy) of Mg alloys. The matrix phase has the same crystalline structure as Mg and similar electrochemical behavior, the presence of alloying elements in the matrix phase may alter

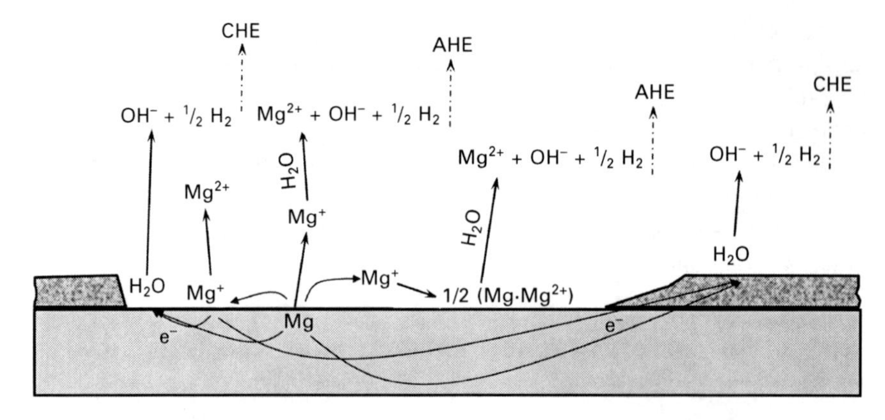

1.13 Schematic illustration of anodic and cathodic reactions involved in the self-corrosion of Mg.

the rates of reactions or steps involved in the corrosion process, but it does not change the corrosion approach or mechanism. Since the matrix is the major phase of a Mg alloy and corrosion of the alloy mainly occurs in the matrix phase, the corrosion process and behavior of the matrix phase to a great extent will represent the corrosion performance of the alloy. The other phases in the alloy basically act as a galvanic couple or a barrier to accelerate or decelerate the corrosion process of the matrix phase. Their effects on the corrosion of the matrix phase can be figured out if the corrosion mechanism of the matrix phase is well understood. Therefore, Fig. 1.13 also presents the corrosion mechanism of the Mg matrix phase.

1.6.2 Characteristic corrosion phenomena

According to the corrosion mechanism of the Mg matrix phase illustrated above, the corrosion of Mg and its alloys have the following characteristic processes.

Hydrogen evolution

According to the above analyses, both the cathodic reaction (1.77) or (1.78) and anodic reaction (1.43) or (1.44) can lead to hydrogen evolution. The overall corrosion reaction (1.79) or (1.80) suggests that hydrogen evolution always accompanies the dissolution of Mg, regardless which anodic and cathodic mechanisms (normal cathodic hydrogen evolution, intermediate Mg^+-catalyzed hydrogen evolution, intermediate MgH_2 catalyzed hydrogen evolution, further anodic oxidation, disproportionation and direct hydration of Mg^+, etc.) are involved in the corrosion process. The effect of hydrogen ingress and MgH_2 formation on reaction (1.79) or (1.80) is negligible.

Undermining of Mg particles that may take place in a very severe corrosion process has been shown not to influence reaction (1.79) or (1.80) (Song *et al.*, 1997b) either. Therefore, the hydrogen evolution phenomenon in an aqueous solution is one of the most fundamental and important corrosion features of Mg and its alloys.

According to reaction (1.79) or (1.80), if there is one mole of hydrogen evolved, there must be one mole of Mg dissolved (Song *et al.*, 2001). Measuring the volume of evolved hydrogen is equivalent to measuring the weight loss of Mg in corrosion. According to the relationship between hydrogen evolution and corrosion process, a simple hydrogen evolution measurement technique was first employed by Song (Song *et al.*, 1997b) to estimate the corrosion rate of Mg. The set-up for measuring hydrogen evolution is very simple (Song *et al.*, 2001) and consists of a burette, a funnel and a beaker which can also be combined into an electrolytic cell for electrochemical measurements (Song and StJohn, 2002) (see Fig. 1.14). The theoretical error associated with this technique has been analyzed to be less than 10% (Song, 2005a,b, 2006; Song *et al.*, 2001) and the measured corrosion rates of various Mg alloy specimens have been found to be in good agreement with those by weight loss (the correlation coefficient is 1.0) (Song *et al.*, 2001). This technique has now been widely used on Mg alloys (Bonora *et al* 2000; Eliezer *et al.*, 2000; Hallopeau *et al.*, 1997; Krishnamurthy *et al.*,

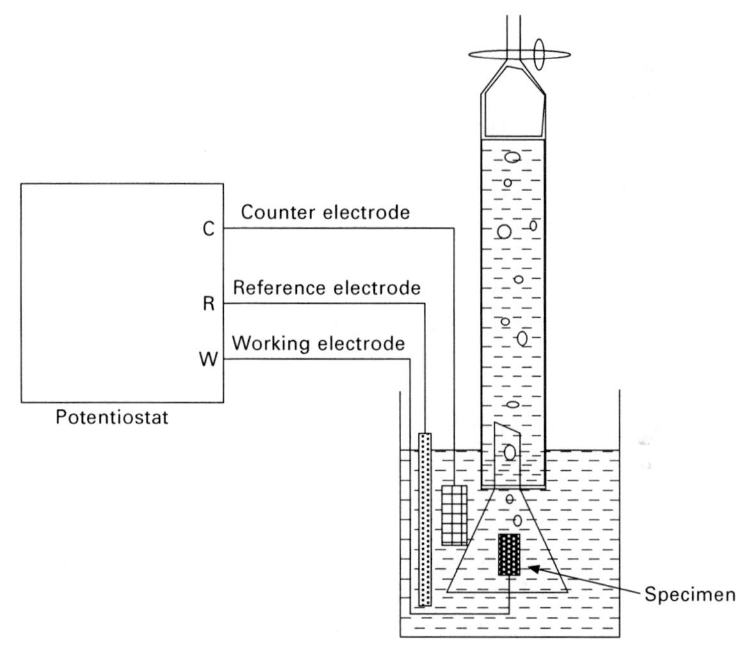

1.14 Schematic illustration of experimental equipment for measuring the volume of hydrogen evolved.

1988; Mathieu *et al.*, 2000; Song, 2005a, 2007a; Song and Atrens, 2003; Song and StJohn 2002; Song *et al.*, 1998, 1999).

The hydrogen evolution method has several advantages over the traditional weight loss measurement in estimating corrosion damage or corrosion rate of a Mg alloy:

- smaller theoretical and experimental errors are introduced into the final estimated corrosion rate;
- easy to set up and operate;
- suitable for monitoring the corrosion of Mg and its alloys; and
- no need for removal of corrosion products.

The hydrogen evolution method is also superior to the estimation of corrosion rate from polarization curve in terms of measurement accuracy. The application of the traditional Tafel extrapolation in estimating the corrosion rates of Mg and its alloys is actually questionable and in many cases can lead to a misleading result, although it has been employed to investigate or evaluate the corrosion performance of Mg and its alloys in some studies (Bonora *et al.*, 2000; Eliezer *et al.*, 2000; Hallopeau *et al.*, 1997; Krishnamurthy *et al.*, 1988; Mathieu *et al.*, 2000).

In theory, to estimate corrosion rates from the Tafel regions on a polarization curve, the anodic and cathodic reactions should be governed by only one anodic and one cathodic Tafel equation in a wide potential range including the corrosion potential. However, according to earlier analyses (referring to Fig. 1.5), there are more than one anodic and cathodic electrochemical reactions involved in the anodic and cathodic processes of a corroding Mg or its alloy around its corrosion potential E_{corr}. When E_{pt} is more negative than E_{corr} and the polarization potentials are more negative than E_{pt}, the cathodic reaction is mainly hydrogen evolution governed by one cathodic Tafel equation on the film covered surface. However, around E_{corr} which is more positive than E_{pt}, the cathodic reaction will be dominated by hydrogen evolution from film-free sites which does not follow the same cathodic Tafel equation as the cathodic hydrogen evolution from a film covered surface in the potential more negative than E_{pt}. Thus, the Tafel equation obtained from the strong cathodic polarization region cannot be applied to the potential region around E_{corr}.

When the electrode is anodically polarized, the anodic current density resulting from the dissolution of Mg in the film-free areas will be substantially influenced by the current density resulting from the cathodic hydrogen evolution reaction process even in a strong anodically polarized potential region. Consequently, the overall anodic process cannot follow a single anodic Tafel equation around the corrosion potential and in the anodic polarization region. Therefore, the corrosion rate at the corrosion potential cannot be estimated via either cathodic or anodic Tafel extrapolation. In other words, hydrogen evolution accompanies the corrosion process and complicates the theory of anodic and cathodic

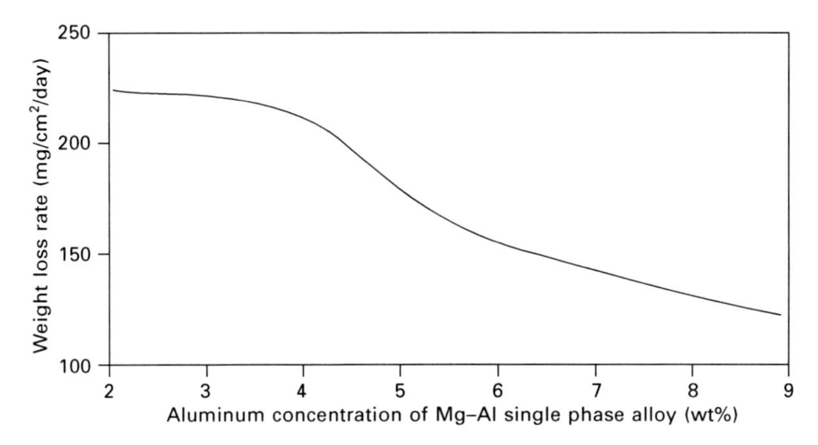

1.15 Average corrosion rates for Mg–Al single phases in 5 wt% NaCl for 3 hours (based on Song *et al.*, 2004a).

polarization around E_{corr}. It was found (Song *et al.*, 2001) that the corrosion rates of sand-cast MEZ_U and diecast AZ91D (Al9Zn0.8Mn0.3) were comparable if estimated by the Tafel extrapolation of their polarization curves. However, the weight loss measurements showed that the corrosion rate of MEZ_U was approximately 25 times higher than the AZ91D corrosion rate. Similarly, if the polarization curves of Mg–Al single phase alloys as shown in Fig. 1.7(d) are Tafel-extrapolated to their corrosion potentials, it may be concluded that their corrosion rates are in an increasing order: 2.00%Al<3.89%Al<5.78% Al<8.95%Al. However, experimentally measured corrosion rates exhibit a reversed order (Song and StJohn, 2000c) (see Fig. 1.15), demonstrating the unreliability of the Tafel extrapolation method. Similar unreliability has also been reported for Am70 and AT72 Mg alloys (Song, 2009d).

If E_{pt} is more positive than E_{corr}, the electrode surface does not change dramatically around E_{corr} and the single Tafel equation can be applied to describe the anodic and cathodic polarization behaviors around E_{corr}, respectively. In this case, the Tafel extrapolation may be applicable (Song and Radovic, 2010).

Alkalization

Reaction (1.79) or (1.80) suggests that there is a correlation between dissolved Mg ions and generated hydroxyls or consumption of protons during the corrosion of Mg and its alloys, implying that the corrosion of Mg and its alloys will result in an increased pH value of the solution. In fact, the surface of a corroding Mg alloy always experiences an alkalization process. When hydroxyls are generated or protons are consumed on the surface, they take time to migrate away from the surface into the bulk solution. It has been estimated that the local pH value of the solution adjacent to a Mg surface can be 10.5 even if the bulk solution is acidic (Nazarrov and Mikhailovskii 1990; Song, 2005b, 2006) (see Fig. 1.16).

The increase in pH value of the solution levels off at a pH of ~10.5 even though the corrosion continues to proceed at this pH value. This is because the deposition of $Mg(OH)_2$ (reaction (1.21)) dominates at this or higher pH levels (Song, 2005b) and thus the additional hydroxyls if generated in corrosion are consumed by dissolved Mg^{2+} via formation of $Mg(OH)_2$ deposits which stabilize the pH value of the solution. A more significant alkalization effect is likely to be achieved with a larger Mg specimen in a smaller volume of solution. For example, in an atmospheric environment, only a small amount of aqueous drops or a thin aqueous film can stay on the surface of the Mg specimens. Hence, atmospheric corrosion is slow compared with immersion corrosion because of the stronger surface alkalization effect (Song *et al.*, 2004c).

The surface alkalization property can even be employed to generate coating for Mg alloys. For example, Song (2009a,b,c, 2010b) deposited an E-coating layer on Mg alloy surfaces without applying any current or voltage simply by utilizing the high surface alkalinity of the alloys. The alkalization effect has also been utilized to monitor the corrosion rates of Mg alloys (Hur *et al.*, 1996; Weiss *et al.*, 1997). However, it is not as reliable as the hydrogen evolution technique:

- The alkalization is dependent upon the volume of the testing solution. It is difficult to compare corrosion rates of Mg alloys having different surface/volume ratios.

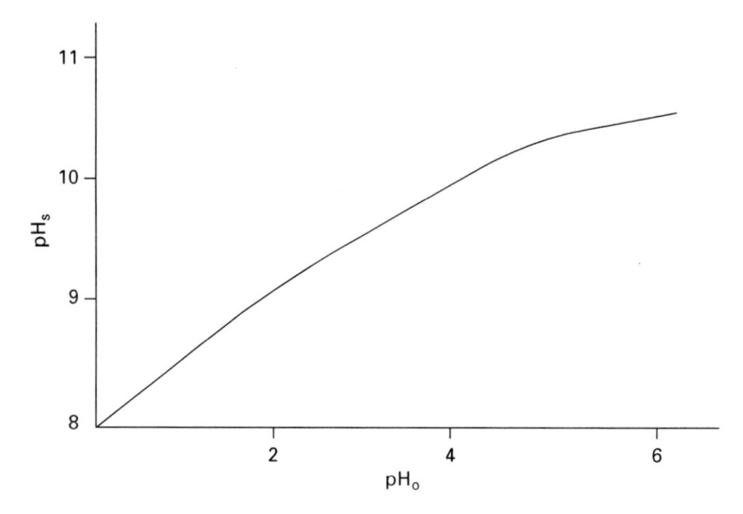

1.16 Estimated alkalinity (pH_s) of solution adjacent to Mg surface vs bulk solution pH_o (based on Nazarrov and Mikhailovskii, 1990).

- In the later stages of corrosion, the corrosion rate may be still high, but the pH value of the testing solution stabilizes at 10.5, failing to indicate the corrosion rate.
- The dissolution of carbon dioxide from air could influence the pH value of the solution and result in a faulty corrosion rate reading based on the affected pH value.
- In a strong acidic or basic solution, the variation of pH is not sufficiently sensitive to reflect the change of corrosion rate.

Negative potential

The high chemical activity of Mg corresponds to a negative standard equilibrium potential of Mg/Mg^{2+}, about –2.4 V (Perrault, 1978; Song, 2005b). This standard equilibrium potential is more negative than any other engineering metal (including coatings) (Bard and Faulkner, 1980). A negative equilibrium potential does not mean a negative corrosion potential. For example, it is well known that Al, Cr and Ni have relatively positive corrosion potentials in many aqueous solutions although their equilibrium potentials are fairly negative (Bard and Faulkner, 1980). A stable and protective passive film forming on the surfaces of these metals significantly inhibits the anodic process and thus results in a positive corrosion potential. Unfortunately, the surface films formed on Mg and its alloys are generally unprotective and the anodic polarization resistance is low. Therefore, in a corrosive environment, the corrosion potentials of Mg and its alloys are as negative as –1.7– –1.6 V/NHE. Alloying and solution composition can affect the corrosion potential of a Mg alloy. For example, the corrosion potential of Mg can be ~2 V–NHE in an acidic solution, and can also become ~1 V–NHE in a basic $Ca(OH)_2$ or $Ba(OH)_2$ saturated solution (Perrault, 1978). Al alloying usually can enhance the corrosion potential. Nevertheless, in the same aqueous solution the corrosion potential of a Mg alloy is usually much more negative than any other engineering metal.

1.6.3 Galvanic effect

The galvanic effect is one of the most important corrosion phenomena of Mg alloys in practical applications and is responsible for the majority of corrosion damage of Mg alloys in engineering.

With negative corrosion potentials, Mg and its alloys will always act as anodes if they are in contact with other engineering metals and suffer from a galvanic corrosion attack. The corrosion potential difference between the anode and cathode is one of the important factors determining the galvanic corrosion rate i_g. In an NaCl solution, the corrosion potential of Mg is over 600 mV more negative than the second active engineering metal Zn (Song,

2007a), implying that the galvanic corrosion driven by the potential difference between Mg and other metals is particularly significant. The galvanic corrosion rate is also dependent upon the anodic and cathodic polarization resistance. The low anodic polarization resistance of Mg and its alloys as discussed earlier means that in general the galvanic corrosion of Mg and its alloys in contact with the other metals will be very severe. In theory, if the electronic resistance is large enough as when the anode and cathode are separated by an insulator, the macro-galvanic corrosion will be stopped. Unfortunately, the anode and cathode in practice are often connected electrically.

For a given Mg alloy in a given solution, the corrosion potential and anodic polarization resistance of the alloy are already fixed. In this case, the corrosion potential and the cathodic polarization resistance of a coupled cathode metal will have a decisive influence on the severity of the galvanic corrosion (Song *et al.*, 2004c). A metal with a noble potential and low cathodic polarization resistance will severely accelerate the corrosion of a Mg alloy if the metal is in contact with the alloy. It is well known that Fe, Co, Ni, Cu, W, Ag and Au, etc. are much nobler than Mg alloys in an aqueous solution and that they have relatively low hydrogen evolution overpotentials (lower than 500 mV). They can form galvanic corrosion couples with Mg alloys resulting in serious galvanic corrosion damage to Mg alloys. In practice aluminum, steel, galvanized steel and sometimes copper may be used together with Mg alloys. Among these metals, steel always has the most detrimental effect upon the corrosion of a Mg alloy and Al the least (Avedesian and Baker 1999; Song *et al.*, 2004c).

So far, many applied studies are focused on the compatibility of the materials (including fasteners) with Mg components based on the overall galvanic corrosion performance (Boese *et al.*, 2001; Gao *et al.*, 2000; Hawke, 1987; Senf *et al.*, 2000; Skar, 1999; Starostin *et al.*, 2000; Teeple, 1956). However, in many practical applications, the distribution of galvanic current density I_g over a component surface will be a big concern. For a simple one-dimensional galvanic couple of Mg alloy, an analytical prediction of galvanic current density distribution is possible (Gal-Or *et al.*, 1973; Kennard and Waber, 1970; McCafferty, 1976, 1977; Melville, 1979, 1980; Waber, 1955; Waber and Rosenbluth, 1955). It has been reported that the galvanic current density has an exponential distribution (Song *et al.*, 2004c) which was confirmed experimentally by directly measuring the distributions of galvanic current densities of a Mg alloy in contact with another metal under a standard salt spray condition using a specially designed 'sandwich'-like galvanic corrosion probe (Song, 2008b; Song *et al.*, 2004b,c; Jia *et al.*, 2006). An exponential distribution of galvanic current density (I_g) with distance x can be theoretically deduced using a linear-transfer model (Song, 2010a; Song *et al.*, 2004c). The theoretically predicted decreasing galvanic current as a function of the thickness of an insulating spacer (Song, 2010a) was

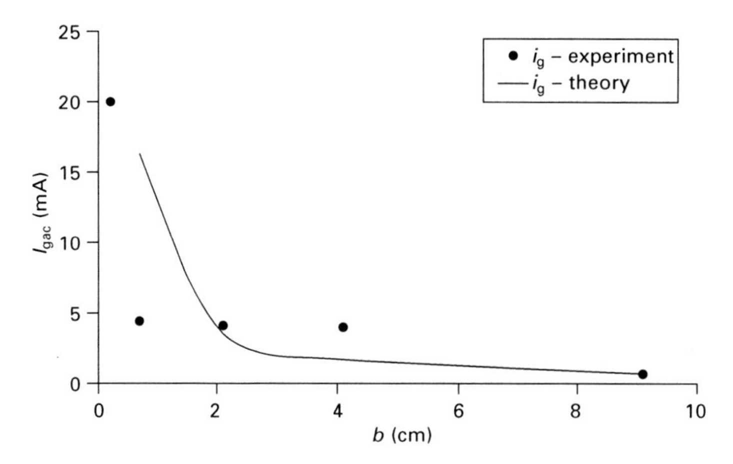

1.17 Theoretical and experimental dependence of Ig on the thickness of an insulating spacer (based on Song *et al.*, 2004c, and Song, 2009d).

measured and experimentally confirmed (Song *et al.*, 2004c) (referring to Fig. 1.17), showing that galvanic corrosion attack can still occur even though the thickness of the insulating spacer between a Mg alloy and steel is as wide as 9 cm under the standard salt spray condition (Song, 2005b, 2006; Song *et al.*, 2004c). This result can also be derived by analyzing the non-linear polarization curves of the coupled anode and cathode (Song *et al.*, 2004c).

When the geometry of the galvanic couple becomes complicated, a computer modeling technique is an option to estimate the galvanic corrosion damage (Song and Atrens, 2002; Klingert *et al.*, 1964; Doig and Flewitt, 1979; Fu, 1982; Sautebin *et al.*, 1980; Helle *et al.*, 1981; Kasper and April, 1983; Munn, 1982; Munn and Devereux 1991a,b; Miyaska *et al.*, 1990; Aoki and Kishimoto, 1991; Hack, 1997). There are already a few successful studies on computer modeling of the galvanic corrosion of Mg alloys (Jia *et al.*, (2003b, 2004, 2005a,b, 2007). It should be noted that due to the negative difference effect, alkalization effect, 'poisoning' effect, and 'short-circuit' effect (Song, 2005b, 2006; Song *et al.*, 2004c), it is not surprising that significant deviations are observed between computer-modeled data and experimentally measured galvanic corrosion results.

1.6.4 Corrosion of Mg alloy

The galvanic corrosion can also occur within an alloy if micro-anodes and cathodes are present in the alloy. Mg alloys are not uniform in terms of their composition, microstructure and even crystalline orientation. These differences can result in various electrochemical activities within a Mg alloy and thereby generate galvanic couples on a micro-scale.

Basically, the corrosion reactions over a Mg alloy are those anodic and cathodic processes discussed earlier. The activity of these reactions may vary from area to area, grain to grain and phase to phase. Some areas, grains, phases and particles may act as anodes while some as cathodes. Specifically, the Mg matrix with a lower content of alloying elements is always a micro-anode and is preferentially corroded (Song, 2007a; Song and Atrens, 2003). Many constituents, including impurity containing particles, intermetallic particles, secondary phases, and the matrix phase with a higher concentration of solid solutes, can act as micro-cathodes. Even in a single phase Mg alloy (i.e. Mg matrix phase), different sites, areas and grains can also act as micro-galvanic anodes and cathodes. Figure 1.18 schematically summarizes the micro-galvanic processes involved in the corrosion of a Mg alloy. Various micro-galvanic cells dominate the corrosion behavior of a Mg alloy.

1.6.5 Micro-galvanic cell in a Mg alloy

Generally speaking, the following factors should be considered in analyzing the micro-galvanic cell related corrosion of Mg alloys.

Grain orientation

Different grains can show varied corrosion rates. The corrosion depths of different crystal planes were found to vary on the three low index planes (Liu *et al.*, 2008). The corrosion rate of a metal can be to some extent correlated

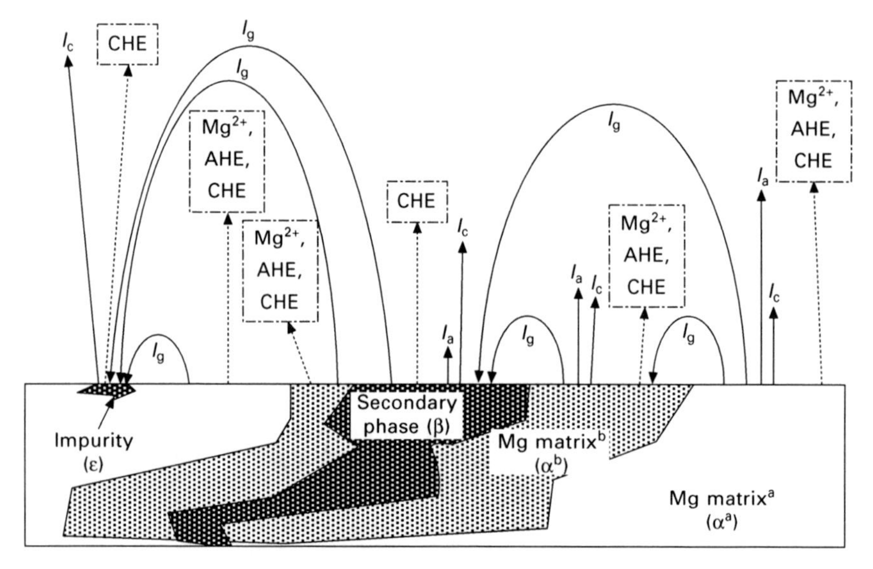

1.18 Corrosion model of Mg alloy with various micro-galvanic cells.

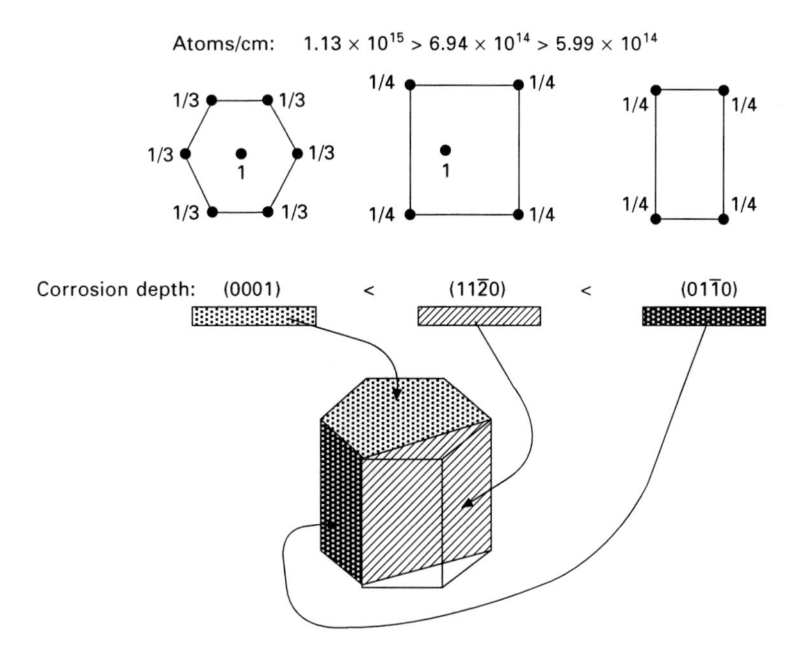

1.19 Corrosion resistance of a few typical crystal planes and their atomic densities (based on Liu *et al.*, 2008).

with its surface energy (Abayarathna *et al.*, 1991; Ashton and Hepworth, 1968; Buck and Henry, 1957; Konig and Davepon, 2001; Weininger and Breiter, 1963) which is somewhat associated with the atomic density of a given crystal plane. For example, the lowest index plane (0001) has a significantly higher atomic density than (*hki*0) planes, thus it has the lowest surface energy and as such should be dissolved slower than the other surfaces (Liu *et al.*, 2008) (refer to Fig. 1.19).

A similar effect of grain orientation on corrosion rate is also found on AZ31B sheet (Song *et al.*, 2010). Figure 1.20 shows that the rolling surface and the cross-section surface of AZ31B sheet have different hydrogen evolution rates. This can be due to the different grain orientations over the surface and cross-section. The rolling surface mainly consists of the lowest index plane (0001) while the cross-section surface has many (10$\bar{1}$0) and 11$\bar{2}$0) orientated gains.

Solid solution concentration

In a Mg alloy, the matrix phase is a major constituent and always preferentially corroded. Solutes in the matrix have a significant influence on the corrosion behavior of the phase. For example, an Al containing Mg matrix phase can become more passive as the Al content increases (Song, 2005a; Song *et*

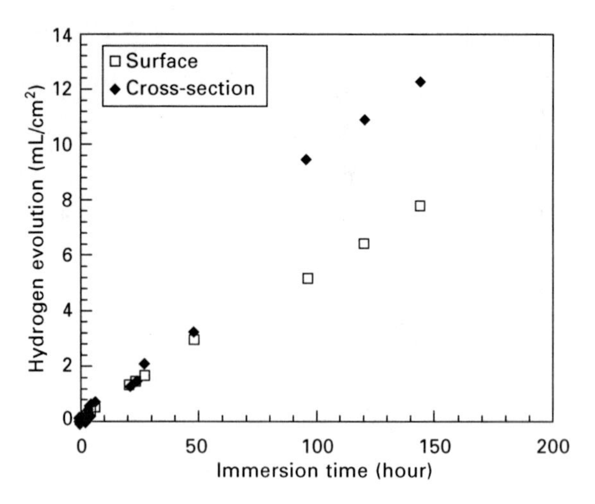

1.20 Hydrogen evolution over the surface and cross-sections of AZ31B sheet in 5 wt% NaCl solution.

al., 2004a, 1998, 1999) and consequently, the corrosion rate of the Mg–Al matrix phase decreases as the Al content increases (see Fig. 1.15). The Al content in solid solution can vary from 1.5 wt% at the grain center to about 12 wt% along the grain boundary in the vicinity of the β phase (Dargusch *et al.*, 1998). The difference in solid solution concentration can result in differences in corrosion potential and anodic/cathodic activity as shown in Fig. 1.7(d), which can lead to a micro-galvanic cell within a grain, causing preferential corrosion in the grain center (referring to Fig. 1.21) (Song and Atrens, 2002).

Different corrosion damage can be observed in the group of non-Al containing alloys, but attributed to the same galvanic effect. The matrix of the alloys in this group typically contains some Zr as grain refiner. The distribution of Zr focused in the central area of a grain. Thus, the alloy is less corrosion resistant along the grain boundaries (referring to Fig. 1.22).

Secondary phases

Almost all the Mg intermetallic phases are more noble and also more stable than the Mg matrix in commercial Mg alloys (Nisancioglu *et al.*, 1990b; Song, 2007a; Song and Atrens, 2003; Song and StJohn, 2002; Song *et al.*, 1998, 1999). These phases can act as micro-galvanic cathodes to accelerate the corrosion of the matrix as well as acting as a corrosion barrier to retard or confine the corrosion development in a Mg alloy (Ambat *et al.*, 2000; Lunder *et al.*, 1989, 1993, 1995; Nisancioglu *et al.*, 1990a; Uzan *et al.*, 2000; Yim *et al.*, 2001). Hence, it is proposed (Song, 2005a, 2007a; Song and Atrens,

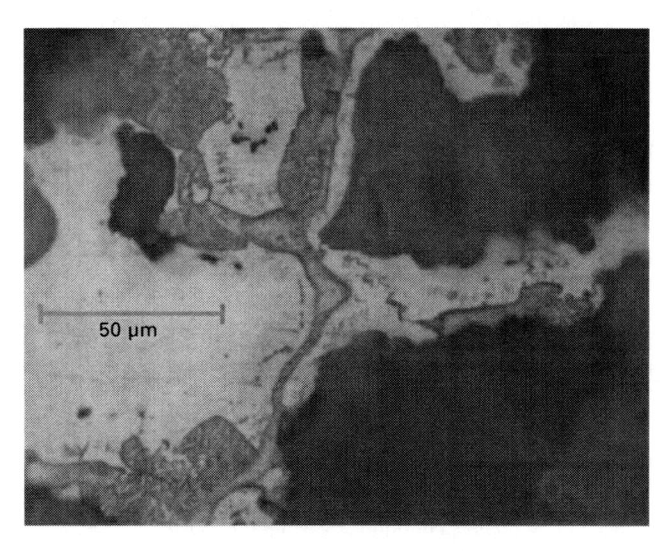

1.21 Corrosion morphologies of AZ91E after 4 hour immersion in 5% NaCl (based on Song, 2005a).

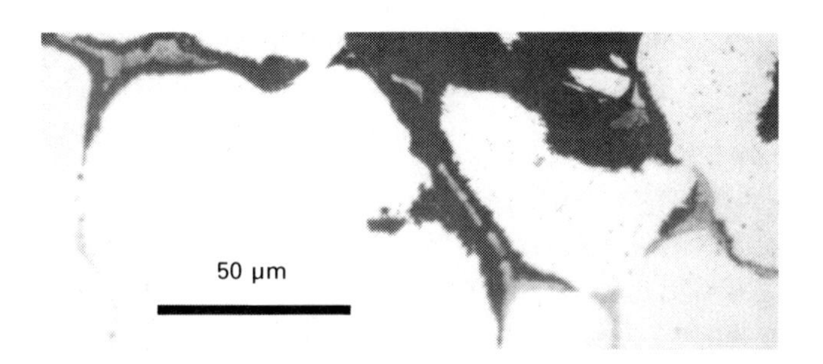

1.22 Cross-section of ZE41 after 75 hours of immersion in 5 wt% NaCl solution.

1998, 2003; Song *et al.*, 1999, 2004a) that the secondary phase plays a dual role in the corrosion of a Mg alloy (referring to Fig. 1.23).

For example, the β-phase in an AZ alloy can be either a corrosion barrier to retard corrosion or a galvanic cathode to accelerate corrosion, depending on the amount, distribution and continuity of the β-phase. Finely and continuously distributed β-phase is more effective in stopping the development of corrosion in an alloy, whereas the presence of a small amount of discontinuous β-phase accelerates the corrosion.

For non-Al-containing Mg alloys, the corrosion is more likely to occur adjacent to the secondary phases and the corrosion damage is easier to be confined within grains by a continuous secondary phase network along the grain boundaries (Song and StJohn, 2000, 2002). This dual-role model

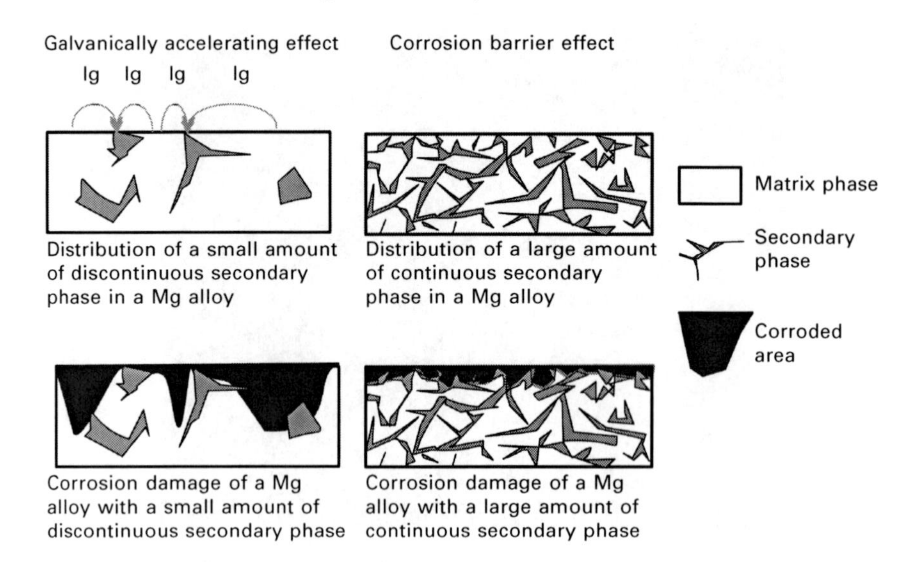

1.23 Schematic illustration of the dual-role model of the secondary phase in a Mg alloy in its corrosion.

for the secondary phase has now been widely employed to interpret the corrosion performance of many Mg alloys under various environmental conditions.

Impurity

A very small addition of impurities of Fe, Ni, Co or Cu can dramatically increase the corrosion rate of Mg or a Mg alloy through the micro-galvanic effect as discussed earlier (Hanawalt *et al.*, 1942; Hillis and Murray, 1987; Lunder *et al.*, 1995; Nisancioglu *et al.*, 1990a). It is well known that higher purity can lead to higher corrosion resistance for Mg and Mg alloys (Aune, 1983; Avedesian and Baker, 1999; Busk, 1987; Emley, 1966; Frey and Albright, 1984; Froats *et al.*, 1987; Hillis, 1983).

Impurities have a tolerance limit, below which their detrimental effect is insignificant. For an impurity, its tolerance limits can be associated with its solubility in a Mg alloy. When these impurities are below a critical concentration (e.g. their solubility in the matrix phase), they are present in the form of solutes in Mg solid solutions. No micro-galvanic cells between the impurities and Mg matrix can be formed and hence they have no significant detrimental effect. It has been claimed that there is a rough correspondence between critical concentrations and the solubility of some elements in Mg alloys (Sheldon Roberts, 1960). Liu *et al.* (2008) calculated that Fe tolerance limit in Mg corresponds well to the solubility of Fe in Mg.

Alloying elements present in Mg can shift the eutectic point and thus alter the impurity tolerance limit (Emley, 1966). For example, with a higher addition of Al in Mg, Fe and Al can combine to form Fe–Al phase (ie. FeAl3) particles which precipitate and act as galvanic cathodes to accelerate the corrosion of Mg–Al alloys (Hawke, 1975; Linder *et al.*, 1989; Loose, 1946). This is why the tolerance limit is about 5 ppm Fe for a Mg alloy with 7% Al while the limit becomes too low to be determined when the Al concentration is increased to 10 wt% (Loose, 1946). It also accounts for different Mg alloys having different impurity tolerance limits (Albright, 1988; Song, 2007a). If the addition of Mn or Zr in a Mg alloy exceeds the solubility, the corrosion can accelerate due to the galvanic effect of their precipitates (Song, 2005b, 2006). However, an adequate addition of Mn or Zr can effectively reduce the detrimental effect of impurities in Mg alloys (Emley, 1966; Froats *et al.*, 1987; Hillis, 1983; Makar and Kruger, 1993; Polmear, 1992). Fe/Mn ratio has been found to be a critical factor (Hillis and Shook, 1989; Zamin, 1981) determining the tolerance limit. A nearly direct proportionality was observed between the Fe/Mn ratio and the corrosion rate (Nisancioglu *et al.*, 1990a). Similarly, the addition of Zr can lead to a higher purity and hence a more corrosion resistant Mg alloy.

In practice, impurities do not always exist inside a Mg alloy. They can be external contaminants on the surface. It is found that such a surface contamination can lead to over 30-fold higher corrosion rate for AZ31B alloy (Song and Xu, 2010).

It should be stressed that some tiny intermetallic phase particles that may or may not contain the impurity elements may also have a similar detrimental effect if their corrosion potential is noble enough. For example, it was recently found that the presence of tiny Al–Mn(Fe) intermetallic particles in AZ31 can significantly influence the corrosion performance of the alloy (Song and Xu, 2010).

1.6.6 Corrosion performance

In applications, the biggest concern about corrosion of Mg alloys is the localized or non-uniform damage.

Owing to the micro-galvanic effect or electrochemical non-uniformity inevitable in Mg and its alloys, the corrosion damage cannot be uniform. The differences in orientation of grain and distribution of alloying element in the matrix phase, and the presence of secondary phase and impurity particles, can result in an uneven distribution of micro-anodes and micro-cathodes in a Mg alloy, leading to nonuniform or localized corrosion damage. The localized corrosion or non-uniform damage can also be caused by more intense anodic and cathodic reactions in a corroding area than in an uncorroded area, and thus corrosion progresses non-uniformly in some local areas. Severe localized corrosion can even result in particle undermining which is often observed for

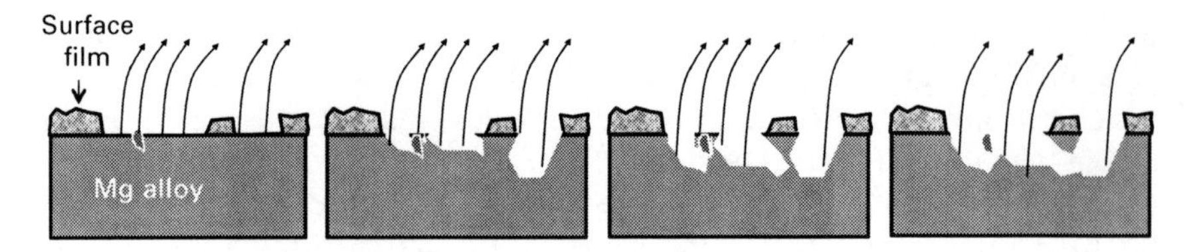

1.24 Schematic diagram of non-uniform (localized) corrosion and particle undermining.

corroding Mg and its alloys. Figure 1.24 schematically illustrates a localized corrosion-induced particle undermining process.

Although localized corrosion is an important corrosion feature of Mg and its alloys, it does not mean that this type of corrosion is a pitting process. The localized corrosion of Mg and its alloy in most environments is initiated from some film-free sites as discussed earlier and typically in the form of tiny irregular localized pits which then spread laterally over the surface. In contrast to typical pitting corrosion, the localized corrosion of Mg or a Mg alloy does not penetrate deeply. This is a result of the alkalization effect at the tips of the corroding pits which prevents the solution at the tips from acidifying or forming occlusive auto-catalytic cells. In other words, the corrosion of Mg and its alloys leads to alkalization of the solution and the alkalization in turn reduces the anodic dissolution of Mg at the corrosion (pitting) tips (see Fig. 1.7(a)). Therefore, there is no way for an auto-catalytic corrosion cell to establish at the corrosion tips to maintain a pitting process. The inherent self-limiting corrosion development tends to result in relatively widespread corrosion damage although the spreading degree of corrosion can vary markedly from alloy to alloy. That is why the 'pitting' corrosion of Mg and its alloys, as stated earlier, is not a real pitting process.

For a conventional metal, it is quite common that an inter-granular zone or the secondary phase is more active than a grain interior and hence inter-granular corrosion damage can result. That is why some Al alloys and stainless steels are highly susceptible to the inter-granular corrosion. However, this type of inter-granular corrosion does not occur to many Mg alloys, because the inter-granular phases or the phases distributed along the grain boundaries are more corrosion resistant than the Mg alloy matrix. Sometimes, the matrix phase adjacent to the grain boundaries may be more severely corroded for some Mg alloys with Zr as a grain refiner. However, this is not real inter-granular corrosion as the inter-granular phases (secondary phases) are still intact. What is corroded is the matrix phase of the grains.

For conventional metals, particularly high passivity metals, the difference in oxygen concentration in solution can sometimes lead to severe crevice corrosion damage. However, this does not happen to Mg and its alloys since they are not sensitive to crevice corrosion attack (Song, 2005b, 2006; Song and Atrens, 2000). According to corrosion reaction (1.79) or (1.80), hydrogen evolution is mainly responsible for the corrosion, not oxygen. Hence, even if there is a difference in oxygen concentration inside and outside a crevice; it cannot lead to a significant galvanic effect and to result in significant crevice corrosion damage. In practice, more severe corrosion may sometimes be observed in crevices of a Mg alloy. However, this is mainly a result of the accumulation of moisture in the crevices, thus the section of the Mg alloy within a crevice is exposed to the corrosive solution significantly longer as compared to the section outside the crevice. In this case, the corrosion in the crevices actually does not follow the traditional crevice corrosion mechanism that involves oxygen depletion and acidification inside a crevice.

Apart from the localized corrosion damage, the overall corrosion performance is another concern in Mg applications. Different Mg alloys should have different corrosion rates even in the same environment because of their different compositions and microstructures.

Table 1.6 lists corrosion rates of some typical Mg alloys which have a very wide variety of corrosion resistance. Alloys even with similar levels of alloying elements have largely scattered corrosion rates. This is not surprising considering that their impurity levels, distributions of chemical composition and phase may be considerably different while their chemical compositions are the same. It is well known that without changing the purity and chemical composition, T4, T5 or T6 heat treatment can significantly modify the corrosion resistance of some Mg alloys (Shi *et al.*, 2003a, 2005, 2006c; Song, 2005b, 2006; Song *et al.*, 2004a,b; Zhao *et al.*, 2008a,b,c). Generally speaking, high-purity Mg or a Mg alloy has relatively good corrosion performance. For example, a high-purity Mg alloy in air can sometimes be even better than mild steel in terms of the general corrosion resistance.

From an environmental perspective, the chemical composition and pH value of an electrolyte are most critical to the corrosion performance of Mg and its alloys. Any other factors which affect these two variables can certainly influence the corrosion. A general rule is that Mg and its alloys are more corrosion resistant in a high alkaline and low chloride concentration solution. Many chemicals in solution can significantly affect the corrosion resistance of Mg alloys which has been summarized previously (Song, 2006; Song and Atrens, 1999) and will not be repeated in this chapter.

During atmospheric exposure, the corrosion of Mg and its alloys is relatively uniform while under an immersed condition the damage tends to be localized. This is a result of the micro-galvanic effect within a Mg alloy having greater significance under a condition of immersion versus an atmospheric condition. Mg and its alloys normally suffer severe corrosion damage under a salt immersion test than under a salt spray test (assuming the same concentration of NaCl solution). Figure 1.25 shows the difference in corrosion rates of Mg alloys under salt spray and salt immersion test

Table 1.6 Typical corrosion rates of Mg alloys (Liu *et al.*, 2008)

Alloy	Corrosion test	Corrosion rate ($mg/cm^2/d$)
LP Mg	3.5% NaCl immersion	310
CP Mg	5% NaCl immersion	12
LP Mg	1 N NaCl (pH 11) immersion	24
Pure Mg	0.1M NaCl immersion	0.03–0.095
HP Mg	1 N NaCl (pH 11) immersion	0.52
HP Mg	1 N NaCl immersion	0.44
HP Mg	5% NaCl immersion	3.6
Cast HP Mg	3% NaCl immeriosn	0.4
Cast HP Mg	3% NaCl immersion	0.8
HP Mg, 1d 550 °C	3% NaCl immersion	3.6
HP Mg, 2d 550 °C	3% NaCl immersion	4.5
Mg–Al alloys	5% NaCl immersion	125–225
Mg1Al	5% NaCl immersion	~2.4
Mg–1Al	3% NaCl immersion	20
Mg–2Al	3% NaCl immersion	40
Mg–2%Al	1 N NaCl (pH 11) immersion	52
AZ31	3.5% NaCl immersion	5.5
Mg–4Al	3% NaCl immersion	40
Mg–6Al	3% NaCl immerson	40
Mg–9Al	3% NaCl immersion	40
Mg–12Al	3% NaCl immersion	50
Mg5Al	5% NaCl immersion	~100
AZ80	3.5% NaCl immersion	0.074
Mg–9%Al	1 N NaCl (pH 11) immersion	46
Mg–9Al	3.5% NaCl immersion	4.4
Mg9Al0.2Ho	3.5% NaCl immersion	0.3
Mg9Al0.4Ho	3.5% NaCl immersion	0.3
HP AZ91	1 N NaCl (pH 11) immersion	5.6
AZ91	Standard salt spray	7.05
AZ91	5% NaCl immersion	1–2
AZ91	3.5% NaCl immersion	2.7
AZ91	Standard salt spray	7.4
AZ89	3.5% NaCl immersion	0.2
AZ91D	3.5% NaCl immersion	2.7
AZ91D	Standard salt spray	0.5
AZ91D	5% NaCl immersion	1–3
AZ91D	3.5% NaCl immersion	1.4
AZ91D	1 N NaCl (pH 11) immersion	0.3–2.6
AZ91D	3.5% NaCl immersion	0.089
AZ91D	1 N NaCl immersion	3.1
AZ91D-DC	Standard salt spray	0.01
AZ91D-T6	Standard salt spray	3
Mg10Al	5% NaCl immersion	~130
Mg22Al	5% NaCl immersion	~5
Mg41Al	5% NaCl immersion	~0
Mg–0.5Al	3% NaCl immerison	20
ZE41	1 N NaCl immersion	5.9
MEZu	5% NaCl immersion	~10
MEZ_R	5% NaCl immersion	~1

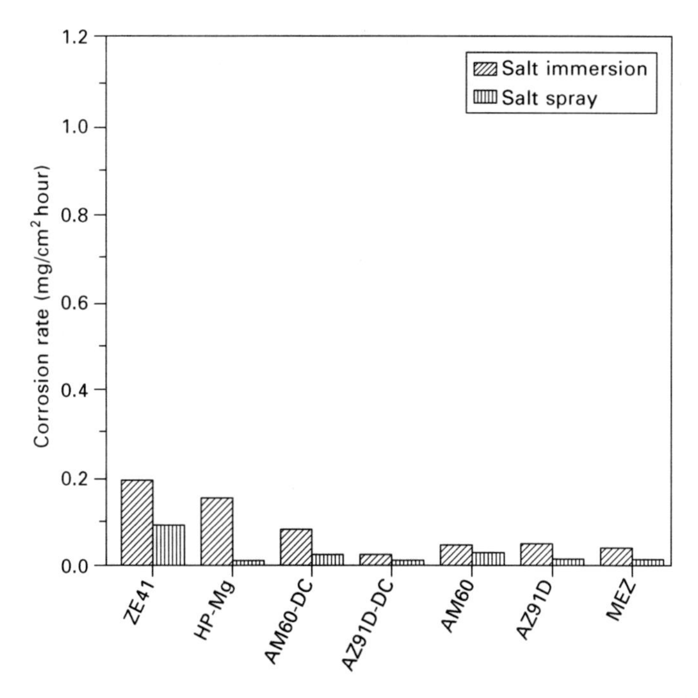

1.25 Corrosion rate of Mg alloys under salt (5 wt% NaCl) immersion test (SIT) condition for 6 hours and salt spray test (SST) condition for 21 hours (based on Shi *et al.*, 2006c).

conditions. It is different from the corrosion behavior of a conventional metal. For example, steel has a higher corrosion rate in a salt spray test compared with an immersion test. The different corrosion behaviors between conventional metals and Mg alloys by salt spray and salt immersion methods originate from their different cathodic processes. Oxygen reduction is the main cathodic reaction in the corrosion of conventional metals and this cathodic process can be significantly enhanced under a salt spray condition. For Mg and its alloys, as discussed earlier, oxygen reduction is not as important as hydrogen evolution during corrosion. An immersion condition is certainly more favorable for hydrogen evolution than salt spraying.

Another possible reason for the relatively high corrosion rate of Mg or a Mg alloy under salt immersion as compared to salt spray testing could be the difficulty for $Mg(OH)_2$ to deposit on the surface under an immersion condition. The solubility of $Mg(OH)_2$ is not very high, so $Mg(OH)_2$ is easy to deposit on a Mg surface under spray conditions because only limited amount of solution can stay on the surface, whereas under an immersion condition, dissolved Mg^{2+} can migrate relatively easily from the surface into the bulk solution, particularly when the migration is further facilitated by hydrogen evolution from a corroding surface. Hence, the formation or deposition of

$Mg(OH)_2$ on the surface is relatively difficult. Figure 1.26 shows evidence of this difference. Under the salt spray immersion the corrosion rate increased significantly with time because of an enlarged corroding surface area, but under the salt spray condition the tendency for increased corrosion became insignificant and in some cases even decreased with time due to the deposition of $Mg(OH)_2$.

Recently, biodegradable Mg alloys have become a hot topic and some progress has been made in this area (Song, 2007a; Witte *et al.*, 2006). Because of their toxicity to the human body, many traditional alloying elements have to be abandoned. However, it has been proposed (Song, 2007a) that the

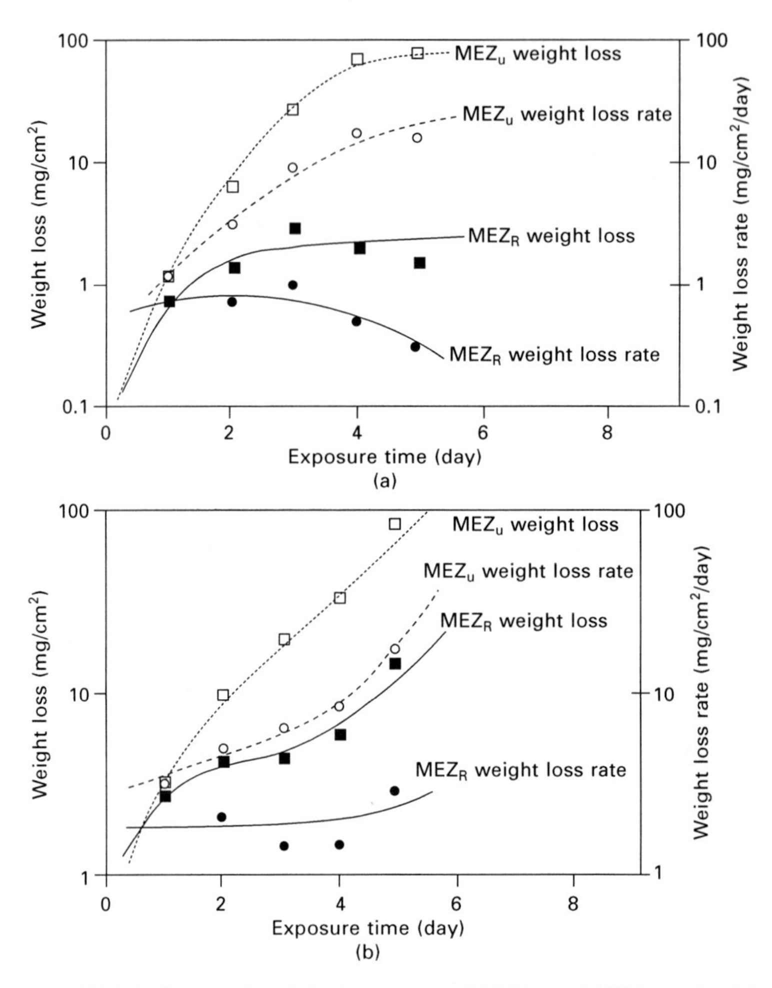

1.26 Weight loss and weight loss rates of MEZ_U and MEZ_R under (a) salt (5 wt% NaCl) spraying and (b) immersion conditions (Song and StJohn, 2002).

Mg–Zn–Mn series alloys should have potential to develop into a successful biodegradable implant material. It is also suggested that an anodized coating on a Mg alloy can be employed to control the biodegradation and hydrogen evolution rates (Song, 2007b; Song and Song, 2007). However, the corrosion mechanism of the new alloys in the body fluid has not been well understood, and requires further systematic investigation.

1.7 References

Abayarathna D., Hale E.B., O'Keefe T.J., Wang Y.M., Radovic D. (1991), *Corrosion Science* **32**: 755.

Aghion E., Bronfin B. (2000), *Materials Science Forum* **350–351**: 19–28.

Albright D.L. (1988) 'Relationship of microstructure and corrosion behavior in Mg alloy ingots and castings', in H.G. Paris and W.H. Hunt (eds), *Advances in Mg Alloys and Composites*, International Mg Association and the Non-Ferrous Metals Committee, The Minerals, Metals, and Materials Society, Phoenix, Arizona, January **26**: pp 57–75.

Alves H., Koster U., Eliezer D. (2000), 'Environmental Behaviour of Mg-Alloys', in E. Aghion and D. Eliezer (eds), *Mg 2000, Proceedings of the second israeli International Conference on Mg Science and Technology*, Dead Sea Israel, pp 347–355.

Alves H., Koster U., Aghion E., Eliezer D. (2001), *Materials Technology: Advanced Performance Materials* **16**(2): 110–164.

Ambat R., Aung N.N., Zhou W. (2000), *Corrosion Science* **42**: 1433–145.

Aoki S., Kishimoto K. (1991), *Mathematical and Computer Modelling* **15**: 11.

Ashton R.F., Hepworth M.T. (1968), *Corrosion* **24**: 50.

Atrens A., Dietzel W. (2007), *Advanced Engineering materials* **9**: 292–297.

Aune T.K. (1983), 'Minimizing base metal corrosion on Mg products. The effect of element distribution (structure) on corrosion behavior', *Proceedings of the 40th World Mg Conference*, Toronto, June.

Avedesian M.M., Baker H. (eds) (1999), *ASM Specialty Handbook, Mg and Mg Alloys*, Chapter 'Cleaning and finishing' pp 138–162, Chapter 'Corrosion behaviour' pp 194–210, Chapter 'Stress Corrosion' pp 210–215, Chapter 'Fatigure and Fracture Resistance' pp 173–176, ASM.

Baliga C.B., Tsakiospolous P. (1992), 'Design of Mg alloys with improved corrosion properties', conference proceedings Garmich Partenkirchen, April, pp 119–126.

Bard A.J., Faulkner L.R. (1980), *Electrochemical Methods, Fundamentals and Applications*, John Wiley and Sons, Inc.

Baril G., Pebere N. (2001), *Corrosion Science* **43**: 471–484.

Bettles C.J., Forwood C.T., StJohn D., *et al.* (2003a), 'AMC-SC1: an elevated temperature magnesium alloy suitable for precision sand casting of powertrain components', in H. Kaplan (ed.), *Magnesium Technology*, TMS, pp 223–226.

Bettles C.J., Forwood C.T., Griffiths J.R., *et al.* (2003b), 'AMC-SC1: new magnesium alloy suitable for powertrain applications', SAE Technical paper# 2003-01-1365, SAE World Congress, Detroit, MI.

Blawert C., Dietzel W., Ghali E., Song G. (2006), *Advanced Engineering Materials* **8**(7): 511–533

Boese E., Gollner J., Heyn A., Strunz J. (2001), *Materials and Corrosion* **52**: 247

Bonora P.L., Andrei M., Eliezer A., Gutman E. (2000), 'Corrosion Behaviour of Stressed Mg Alloys', in E. Aghion and D. Eliezer, (eds), *Mg 2000, Proceedings of the Second*

Israeli International Conference on Mg Science and Technology, Dead Sea, Israel, pp 410–416.

Bradford P.M., Case B., Dearnaley G., Turner J.F., Woolsey I.S. (1976), *Corrosion Science* **16**: 747–766.

Buck I.W.R., Henry L. (1957), *Journal of The Electrochemical Society* **104**: 474.

Busk R.S. (1987), *Magnesium Products Design*, International Magnesium Association.

Chen J., Wang J., Han E., Dong J., Ke W. (2008), *Corrosion Science* **50**: 1292–1305.

Dargusch M.S., Dunlop G.L., Pettersen K. (1998), 'Elevated temperature creep and microstructure of die cost Mg-Al-Alloys', in *Mg Alloys and Their Applications*, Werkstoff-Informations GmbH, Wolfsburg, Germany: pp 277–282.

Doig P., Flewitt P.E.J. (1979), *Journal of the Electrochemical Society*, **126**: 2057.

Eliezer A., *et al.* (2000), 'Dynamic and static corrosion fatigue of Mg alloys in electrolytic environment', in E. Aghion and D. Eliezer, (eds), *Mg 2000, Proceedings of the Second Israeli International Conference on Mg Science and Technology*, Dead Sea, Israel, 356–362.

Emley E.F. (1966), *Principles of Magnesium Technology*, Chapter XX, Pergamon Press.

Ferrando W.A. (1989), *Journal Engineering Materials* **11**: 299–313.

Fournier V., Marcus P., Olefjord I. (2002), *Surface and Interface Analysis* **34**: 494–497.

Frey D., Albright L.L. (1984), 'Development of a Mg alloy structural truck component', *Proceedings of the 41st World Mg Conference*, London, June.

Froats A., Aune T.Kr., Hawke D., Unsworth W., Hillis J. (1987), 'Corrosion of Mg and Mg Alloys', in *A.S.M. Metals Handbook* (formerly 9th Ed), Vol.13 *Corrosion*, pp 740–754.

Fu J.W. (1982), *Corrosion* **38**: 295.

Fuggle J.C., Watson L.M., Fabian D.J. (1975), *Surface Science.* **49**: 61–76.

Gal-Or L., Raz Y., Yahalon J. (1973) *Journal of the Electrochemical Society*: **120**: 598.

Gao G., Cole G., Richetts M., Balzer J., Frantzeskakis P. (2000), 'Effects of fastener surface on galvanic corrosion of automotive Mg components', in E. Aghion and D. Eliezer (eds), *Mg 2000, Proceedings of the second Israeli International Conference on Mg Science and Technology*, Dead Sea, Israel, pp 321–338.

Ghali E. (2006), 'Some aspects of corrosion resistance of magnetism alloys', in M.O. Pekgueryuz and L.W.F. Mackenzie (eds), *International Symposium on Magnesium Technology in the Global Age*, Montreal, Canada, Canadian Institute of Mining, Metallurgy and petroleum, pp 271–293.

Glicksman R. (1959), *Journal of the Electrochemical Society* **106**: 83.

Gray, J.E. Luan B. (2002), *Journal of Alloys and Compounds* **336**: 88–113.

Gulbransen E.A. (1954), *Transactions of the Electrochemical Society* **87**: 589.

Hack H.P. (1997), *Corrosion Review* **15**: 195.

Hallopeau X., Beldjoudi T., Fiaud C., Robbiola L. (1997), 'Electrochemical behaviour and surface characterisation of Mg and AZ91E alloy in aqueous electrolyte solutions containing XOyn- inhibiting ions', in G.W. Lorimer (ed.), *Proceedings of the Third International Mg Conference*, The Institute of Materials, London, pp. 713–724.

Hanawalt J.D., Nelson C.E. , Peloubet J.A. (1942), *Transactions of the Amtricun Institute Mining Metal Engineers* **147**: 273.

Hawke D. (1975), 'Corrosion and wear resistance of Mg die castings', *SYCE 8th International Die Casting Exposition and Congress*, Detroit, Paper No. G–T 75–114.

Hawke D.L. (1987), 'Galvanic corrosion of Mg', *SDCE 14th International Die Casting Congress and Exposition*, Toronto, Canada, May, Paper No. G-T 87–004.

Hawkin J.H. (1993), 'Assessment of protective finishing systems for Mg', presented at *International Mg Association's 50th Annual World Mg Conference*, Washington D.C., May 11–13.

Helle H.P.E., Beck G.H.M., Ligtelijin J.TH. (1981), *Corrosion* **37**: 522.

Hillis J.E. (1983), 'The effects of heavy metal contamination on Mg corrosion performance', SAE Technical Paper # 830523, Detroit, March.

Hillis J.E., Murray R.W. (1987), 'Finishing alternatives for high purity Mg alloys', *SDCE 14th International Die Casting Congress and Exposition*, Toronto, Paper No. G–T 87–003.

Hillis J.E., Shook S.O. (1989), 'Composition and performance of an improved Mg AS41 alloy', SAE Technical Paper Series # 890205, Detroit.

Hills J.E. (1995), Chapter 45 'Mg', in R Baboian (ed), *Corrosion Tests and Standards: Application and interpretation*, ASTM, pp 438–446.

Hur B.Y., Kim K.W. (1998), *Corrosion reviews* **16**: 85–94.

Hur B.Y., Kin K., Ahn H., Kim K. (1996), 'A new method for evaluation of pitting corrosion resistance of Mg alloys', in G.W. Lorimer (ed.), *Proceedings of the Third International Mg Conference*, Manchester, pp 557–564.

James W.J., Straumanis M.E., Bhatia B.K., Johnson J.W. (1963), *Journal of the Electrochemical Socical* **110**: 1117.

James W.J., Straumanis M.E., Johnson J.W. (1964), *Corrosion* **23**: 15.

Jia J., Song G.-L., Atrens A. (2003a), 'Modeling and experimental study of the galvanic corrosion of AZ91D magnesium alloy in contact with steel in NaCl solution', *13th Asian-Pacific corrosion control conference (APCCC)*, Osaka, Nov. 2003 paper No. K–14.

Jia J., Song G.-L., Atrens A., Chandler G., StJohn D. (2003b), 'Prediction and experimental measurement of galvanic corrosion for a magnesium–steel couple in corrosive water', in A. Dahle (ed.) *Proceedings of the Light Metals Technology Conference 2003*, Brisbane pp 397–400.

Jia J., Song G.-L., Atrens A., StJohn D., Baynham J., Chandler G (2004), *Materials and Corrosion* **55**: p 845–852.

Jia J., Song G.-L., Atrens A. (2005a), *Materials and Corrosion* **56**(4): 259–270.

Jia J., Atrens A., Song G.-L., *et al.* (2005b), *Materials and Corrosion* **56**(7): 468–474.

Jia J., Song G.-L., Atrens A.(2006), *Corrosion Science* **48**: 2133–2153.

Jia J., Song G.-L., Atrens A. (2007), *Advanced Engineering Materials* **9**(1–2): 65–74.

Kasper R.G., April M.G. (1983), *Corrosion* **39**: 181.

Kennard E., Waber J. (1970), *Journal of the Electrochemical Society* 117: 880.

Klingert J.A,, Lynne S., Tobias C.W. (1964), *Electrochimica Acta* **9**: 297.

Konig U., Davepon B. (2001), *Electrochimica Acta* **47**: 149.

Krishnamurthy S., Robertson E., Froes F.H. (1988), 'Rapidly solidified Mg alloys containing rare earth additions', in H.G. Paris W.H. Hunt (eds), *Advances in Mg Alloys and Composites* (International Mg Association and the Non-Ferrous Metals Committee, The Minerals, Metals, and Materials Society), Phoenix, Arizona, January **26**: pp 77–89.

Linder O., Lein J.E., Aune T.Kr., *et al.*, (1989), *Corrosion* **45**(9): 741–747.

Liu M., Qiu D., Zhao M., Song G.-L., Atrens A. (2008), *Scripta Materialia* **58**: 421–442.

Liu M., Uggowitzer P.J., Nagasekhar A.V., Schmutz P., Easton M., Song G.-L., Atrens A. (2009a), *Corrosion Science* **51**: 602–619.

Liu M., Zanna S., Ardelean H., *et al.* (2009b), *Corrosion Science* **51**, 1115–1127.

Loose W.S. (1946), 'Corrosion and protection of magnesium', in L.M. Pidgeon, J.C. Mathes, N.E. Woldman, J.V. Winkler, W.S. Loos (eds), *Magnesium*, A series of five educational lectures on magnesium presented to members of the A.S.M. during the twenty-seventh national metal congress and exposition, Cleveland, February 4–8, American Society for Metals, pp 173–260.

Lunder O., Lein J.E., Aune T.Kr., Nisancioglu K. (1989), *Corrosion* **45**(9): 741–748.

Lunder O., Nisancioglu K., Hansen R.S. (1993), 'Corrosion of diecast Mg–aluminium alloys, SAE Technical Paper Series #930755, Detroit.

Lunder O., Videm M., Nisancioglu K. (1995), 'Corrosion resistant Mg alloys', SAE 1995 Transactions, Journal of Materials and Manufcturing, Section 5-Volume 104: pp 352–357, paper#950428.

Makar G.L., Kruger J. (1990), *Journal of the Electrochemical Society* **13**: 414–421.

Makar G.L. Kruger J. (1993), *International Materials Reviews* **38**(3): 138–153.

Mathieu S., Hazan J., Rapin C., Steinmetz P. (2000), 'Corrosion behaviour of die cast and thixocast AZ91D Alloys – the use of sodium linear carboxylates as corrosion inhibitors', in E. Aghion and D. Eliezer (eds), *Mg 2000, Proceedings of the Second Israeli International Conference on Mg Science and Technology*, Dead Sea, Israel, pp 339–346.

McCafferty E. (1976), *Corrosion Science* **16**: 183.

McCafferty E. (1977), *Journal of the Electrochemical Society* **124**: 1869.

McIntyre N.S., Chen C. (1998), *Corrosion Science* **40**(10): 1697–1709.

Melville P.H. (1979), *Journal of the Electrochemical Society* **126**: 2081.

Melville P.H. (1980), *Journal of the Electrochemical Society* **127**: 864.

Miyaska M., Hashimoto K., Kishimoto K., Aoki S. (1990), *Corrosion Science* **30**: 299.

Munn R.S. (1982), *Materials Performance August* 29.

Munn R.S., Devereux O.F. (1991a), *Corrosion*: **47**: 612.

Munn R.S., Devereux O.F. (1991b), *Corrosion* **47**: 618.

Nakatsugawa I. (1996), 'Surface modification technology for Mg products', *Proc IMA 53*, '*Mg – A Material Advancing to the 21st Century*', Yamaguchi, June, pp 24–29.

Nakatsugawa I. Kamado S., Kojima Y., Ninomiya R., Kubota K. (1996), 'Corrosion behavior of magnesium alloys containing heavy rare earth elements', in Lorimer G.W. (ed.), *3rd International Magnesium Conference*, Manchester, pp 687–698.

Nazarrov A.P., Mikhailovskii Y.N. (1990), *Protection of Metals* **26**(1): 9–14.

Nisancioglu K., Lunder O., Aune T.Kr. (1990a), 'Corrosion mechanism of AZ91 Mg alloy', *Proceedings of 47th World Mangesium Association*, Mcleen, Virginia, pp 43–50.

Nisancioglu K., Lunder O., Aune T., (1990b), 'Corrosion mechanism of AZ 91 Mg alloy', *47th World Mg Conference*, Cannes, France.

Nordlien J. H., Sachiko O., Noburo M., Kemal N. (1995), *Journal of the Electrochemical Society* **142**: 3320–3322.

Nordlien J. H., Ono S., Masuko N., Nisancioglu K. (1997), *Corrosion Science* **39**: 1397–1414.

Ott B.J, Boerio-Goates J. (2000), *Chemical Thermodynamics: Advanced Applications*, Academic Press.

Peng X.D., Barteau M.A. (1990), *Surface Science* **233**: 283–292.

Perrault G.G. (1974), *Journal of Electroanalytical Chemistry*, **51**: 107.

Perrault G.G. (1978), 'Magnesium', in A.J. Bard (ed.) '*Encyclopedia of Electrochemistry of the Elements*', Marcel Dekker, pp 263–319.

Pilling N.B., Bedworth R.E. (1923), *Journal of the Institute of Metals* **29**: 529.

Polmear I.J. (1981), '*Light Alloys: Metallurgy of the light metals*', Chapter 5, Edward Arnold.

Polmear I.J. (1992), *Physical Metallurgy of Mg Alloys*, DGM Informationsgesellschaft, Oberursel, Germany.

Polmear I.J. (1996), *Metallugy of Light Alloys*, Halsted Press, NY.

Pourbaix M. (1974), Atlas of *Electrochemical Equilibria in Aqueous Solutions* National Association of Corrosion Engineers.

Sautebin R., Froidewaux H., Landolt D. (1980), *Journal of the Electrochemical Society* **127**: 1096.

Schmutz P., Guillaumin V. (2003), *Journal of the Electrochemical Society* **150**(4): B99–110.

Senf J., Broszeit E., Gugau M., Berger C. (2000), 'Corrosion and galvanic corrosion of die casted Mg alloys', in H.I. Kaplan, J. Hryn and B. Clow (eds), *Mg Technology 2000*, TMS Nashville, March, pp 137–142.

Shaw A. (2003), in S.D. Cramer and B.S. Covino Jr. (ed), *Corrosion of Magnesium and Magnesium-Base Alloys*, Vol.13A, ASM International, pp 692–696.

Shaw A., Wolfe C. (2005), in S.D. Cramer and B.S. Covino Jr. (eds), *Corrosion of Magnesium and Magnesium-Base Alloys*, Vol.13A ASM International, pp 205–227.

Sheldon Roberts C. (1960), 'Mg alloy systems' in *Mg and its Alloys*, John Wiley & Sons, pp 42–80.

Shi Z., Song G.-L., StJohn D. (2001), 'A method for evaluating the corrosion resistance of anodised magnesium alloys', *Corrosion and Prevention 2001*, ACA, Newcastle, paper #058.

Shi Z., Song G.-L., Atrens A. (2003a), 'Efffects of zinc on the corrosion resistance of anodised coaitng on magnesium alloys', in A. Dahle (ed.), *Proceedings of the Light Metals Technology Conference 2003*, Brisbane, pp 393–396.

Shi Z., Song G.-L., Atrens A. (2003b), 'The influence of anodising parameters on the corrosion performance of an anodised coating on magnesium alloy', *13th Asian-Pacific Corrosion Control Conference (APCCC)*, Osaka, Nov. 2003, Paper No. K–06.

Shi Z., Song G.-L., Atrens A. (2005), 'Influence of the β phase on the corrosion performance of anodised coatings on magnesium–aluminum alloys.', *Corrosion Science* **47**: 2760–2777.

Shi Z., Song G.-L., Atrens A. (2006a), *Surface and Coatings Technology*, **201**: 492–503.

Shi Z., Song G.-L., Atrens A. (2006b), *Corrosion Science* **48**: 1939–1959.

Shi Z., Song G.-L., Atrens A. (2006c), *Corrosion Science* **48**: 3531–3546.

Shreir L.L. (1965), *Corrosion* Vol.1 *Metal/Environment Reactions*, Newnes-Butterworths.

Skar J.I. (1999), *Materials and Corrosion* **50**: 2.

Skar J.I., Albright D. (2002), 'Emerging trends in corrosion protection of Mg diecastings', in H.I. Kplan (ed.) *Mg Technology 2002*, TMS, pp 255–261.

Song G.-L. (2004a), *Environment-friendly, Non-toxic and Corrosion Resistant Magnesium Anodisation* 2004904949 (Australia).

Song G.-L. (2004b), *Journal of Corrosion Science and Engineering (JCSE)*, **6**, Paper C104.

Song G.-L. (2005a), *Materials Science Forum* 488–489: 649–652.

Song G.-L. (2005b), *Advanced Engineering Materials* **7**(7): 563–586.

Song G.-L. (2006), '*Corrosion and Protection of Magnesium Alloys*', Chemical Industry Press (in Chinese).

Song G.-L. (2007a), *Corrosion Science* **49**: 1696–1701.
Song G.-L. (2007b), *Advanced Materials Research* **9**(4): 1–5.
Song G-L. (2008a), 'An electroless e-coating bath sealing technique for anodized and conversion coated magnesium alloys', MPL-641(GM confidential report).
Song G.-L. (2008b), 'One-dimensional galvanic corrosion-potential and current', MPL-666 (GM confidential report).
Song G.-L. (2009a), *Surface and Coatings Technology* **203**: 3618–3625.
Song G.-L. (2009b), *Materials Science Forum* **618–619**: 268–271.
Song G.-L. (2009c), *Electrochemical and Solid-State Letters*, **12**(10): D77–D79.
Song G.-L. (2009d), *Corrosion Science* **51**: 2063–2070.
Song G.-L. (2010a), *Corrosion Science* **52**: 455–480.
Song G.-L. (2010b), *Electrochimica Acta* **55**: 2258–2268.
Song G.-L., Atrens A. (1998), 'Corrosion behaviour of skin layer and interior of diecast AZ91D', *International Conference on Magnesium Alloys and their Applications*, Wolfsburg, pp 415–419.
Song G.-L., Atrens A. (1999), *Advanced Engineering Materials* **1**(1): 11–33.
Song G.-L., Atrens A. (2000), 'Corrosion of non-ferrous alloys. III. Magnesium alloys', Volume 19B *Corrosion and Environmental Degradation*, in Wiley-VCH series 'Materials Science and Technology', Vol. **2**: 131–171.
Song G.-L., Atrens A. (2002), *Corrosion Science and Technology* **31**(2): 103–115.
Song G.-L., Atrens A. (2003), *Advanced Engineering Materials* **5**(12): 837–857.
Song G.-L., Atrens A. (2005), 'Understanding the corrosion of Mg alloy', *16th International Corrosion Congress*, Beijing 2005, paper# 03–Mg–key–note.
Song G.-L., Atrens A. (2007), *Advanced Engineering Materials* **9**(3): 177–183.
Song G.-L., Haddad D. (2010), *Materials Chemistry and Physics*, **125**: 548–552.
Song G.-L., Radovic D. (2010), 'Lab evaluation and comparison of corrosion performance of Mg alloys', *SAE International 2010*, 2010–01–0728.
Song G.-L., Shi Z. (2006), 'Characterisation of anodized coatings on a magnesium alloy', M.O. Pekguleryuz and L.W.F. Mackenzie (eds), *Magnesium Technology in the Global Age, 45th Annual Conference of Metallurgists of CIM*, Montreal, Quebec, October, pp 385–395.
Song G.-L., Song S. (2006a), *Acta Physico Chimica Sinica* **22**(10): 1222–1226.
Song G.-L., Song S. (2006b), 'Corrosion characteristics and Bio-compatibility of magnesium', *2006 Beijing International Materials Week (2006BIMW)*, C-MIRS, June, Beijing 2006, A-56.
Song G.-L., Song S. (2006c), 'Magnesium as a possible degradable bio-compatable material', in M.O. Pekguleryuz and L.W.F.Mackenzie (eds), *Magnesium Technology in the Global Age, 45th Annual Conference of Metallurgists of CIM*, Montreal, Quebec, October pp 345–358.
Song G.-L., Song S. (2007), *Advanced Engineering Materials* **9**(4): 298–302.
Song G.-L., StJohn D. (2000), *International Journal of Cast Metals Research* **12**: 327–334.
Song G.-L., StJohn D. (2002), *Journal of Light Metals* **2**: 11–16.
Song G.-L., StJohn D. (2003), *Corrosion Science and Technology* **32**(1): 30–35.
Song G.-L., StJohn D. (2004), *Corrosion Science* **46**: 1381–1399.
Song G.-L., StJohn D. (2005a), *Materials and Corrosion* **56**(1): 15–23.
Song G.-L., StJohn D. (2005b), 'Corrosion inhibition of magnesium alloys in coolants', N.R. Neelameggham, H.I. Kaplan and B.R. Powell, (eds)TMS, *Magnesium Technology 2005*, pp 469–474.

Song G.-L., Xu Z. (2010), *Electrochimica Acta* **55**: 4148–4161.

Song G.-L., Cao C.-N., Lin H.-C. (1994), *Corrosion Science* **36**(9): 1491–1497.

Song G.-L., Atrens A., StJohn D.H, Wu X., Nairn J. (1997a), *Corrosion Science* **39**(10–11): 1981.

Song G.-L., Atrens A., StJohn S., Nairn J., Li Y. (1997b), *Corrosion Science* **39**(5): 855–857.

Song G.-L., Atrens A., Wu X., Zhang B. (1998), *Corrosion Science* **40**(10): 1769–1791.

Song G.-L., Atrens A., Dargusch M. (1999), *Corrosion Science* **41**: 249–273.

Song G.-L., Atrens A., StJohn D., Zheng L. (2000), 'Corrosion behaviour of the microstructural constituents of AZ alloys', in K.U. Kainer (ed), *Magnesium Alloys and their Applications*, Wiley-VCH, pp 426–431.

Song G.–L., Atrens A., StJohn D. (2001), 'An hydrogen evolution method for the estimation of the corrosion rate of magnesium alloys', in J. Hryn (ed.), *Magnesium Technology 2001*, TMS, pp 255–262.

Song G.-L., Hapugoda S., Ricketts N., Dias-Jayasinha S., Frost M., Polkinghorne K. (2003), 'Influences of environmental temperature and humidity on the degradation of the surfaces of magnesium alloys', in A. Dahle (ed), *Proceedings of the Light Metals Technology Conference 2003*, Brisbane, pp 389–392.

Song G.-L., Bowles A.L., StJohn D.H. (2004a), *Materials Science and Engineering* **A366**: 74–86.

Song G.-L., Bowles A., StJohn D. (2004b), 'Effect of aging on yield stress and corrosion of diecast magnesium alloy', *International Conference on Magnesium – Science, Technology and Applications*, Beijing 2004, paper E31, pp 709–712.

Song G.-L., Jonhannesson B., Hapugoda S., StJohn D.H. (2004c), *Corrosion Science* **46**: 955–977.

Song G.-L., Cao C., Chen S. (2005a), *Corrosion Science* **47**(2): 323–339.

Song G.-L., StJohn D., Abbott T. (2005b), *International Journal of Cast Metals Research* **18**(3): 174–180.

Song G.-L., StJohn D., Bettles C., Dunlop G. (2005c), *Journal of the Minerals, Metals & Materials Society (TMS)* **57**(5): 54–56.

Song G.-L. Hapugoda S., StJohn D. (2006a), *Corrosion Science* **49**: 1245–1265.

Song G.-L., Hapugoda S., StJohn D., Bettles C. (2006b), 'Simulation of atmospheric environments for storage and transport of magnesium and its alloys', in A.A. Luo, N.R.Neelameggham and R.S. Beals (eds), *Magnesium Technology*, TMS, pp 3–6.

Song G.-L., Shi Z., Hinton B., McAdam G., Talevski J., Gerrard D. (2006c), 'Electrochemical evaluation of the corrosion performance of anodized Mg alloys', *14th Asian–Pacific Corrosion Control Conference*, October, Shanghai, paper# Keynote-11.

Song G.-L., Song S., Li Z. (2007), 'Corrosion control of magnesium as an implant biomaterial in simulated body fluid', *Ultralight 2007 – 2nd International Symposium on Ultralight Materials and Structures* Beijing, September.

Song G.-L., Mishra R., Xu Z. (2010), *Electrochemistry Communications*, **12**: 1009–1012.

Song G.-L., Wang Y.-M., Guo H. (2010), 'Electroless E-coating for AZ91D magnesium alloy', *SAE International 2010*, 2010-01-0727.

Splinter S.J. (1993), *Surface Science* **292**(1–2): 130–144.

Splinter S.J. (1994), *Surface Science* **314**(2): 157–171.

Starostin M., Smorodin A., Cohen Y., Gal-Or L., Tamir S. (2000), Galvanic corrosion of Mg alloys, in E. Aghion and D. Eliezer (eds), *Mg 2000, Proceedings of the second*

Israeli International Conference on Mg science and technology, Dead Sea, Israel, pp 363–370.

Straumanis M.E. (1958), *Journal of the Electrochemical Society* **105**: 284.

Straumanis M.E., Bhatia B.K. (1963), *Journal of the Electrochemical Society*, 110: 357.

Tawil D.S. (1987), 'Corrosion and surface protection developments', *Proceedings of Conference 'Mg Technology'*, pp 66–74.

Teeple H.O. (1956), Atmospheric Galvanic Corrosion of Mg Coupled to Other Metals, ASTM, STP 175, pp 89–115.

Uzan P., Frumin N., Eliezar D., Aghion E. (2000), 'The role of composition and second phases on the corrosion behaviour of AZ Alloys', in E. Aghion and D. Eliezer (eds), *Mg 2000, Proceedings of the 2nd Israeli International Conference on Mg Science and Technology*, Dead Sea Israel, pp 185–191.

Vermilyea D.A., Kirk C.F. (1969), *Journal of the Electrochemical Society* **116**: 1487–1492.

Waber J.T. (1955), *Journal of the Electrochemical Society* 102: 220.

Waber J.T., Rosenbluth M. (1955), *Journal of the Electrochemical Society* **102**: 344.

Wall F.T. (1965), *Chemical Thermodynamics, a Course of Study*, 2nd edition, W.H. Freeman and Company.

Wan Y., Tan J., Song G., Yan C. (2006), *Metallurgical and Materials Transactions*, **37A**(7): 2313–2316.

Wang H., Estrin Y., Fu H.M., Song G., Zúberová Z. (2007), *Advanced Engineering Materials* **9**(11): 967–972.

Weast R.C. (ed), (1976–1977), *Handbook of Chemistry and Physics*, 57th edition, Section F, CRC Press.

Weininger J.L., Breiter M.W. (1963), *Journal of the Electrochemical Society* **110**: 484.

Weiss D., Bronfin B., Golub G., Aghion E. (1997), 'Corrosion resistance evaluation of Mg and Mg alloys by an ion selective electrode', in E. Aghion and D. Eliezer (eds) *Mg 97, Proceedings of the First Israeli International Conference on Mg Science and Technology*, Dead Sea, Israel, pp 208–213.

Winzer N., Atrens A., Song G.-L., Ghali E., Dietzel W., Kainer K.U., Hort N., Blawert C. (2005), *Advanced Engineering Materials* **7**(8): 659–693.

Winzer N., Atrens A., Dietzel W., Song G.-L., Kainer K.U. (2007a), *Materials Science and Engineering-A* **447**: 18–31.

Winzer N., Atrens A., Dietzel W., Song G.-L., Kainer K.U. (2007b), *Journal of the Minerals, Metals & Materials Society (TMS)* August: 49.

Winzer N., Atrens A., Dietzel W., Song G.-L., Kainer K. (2008a), *Advanced Engineering Materials* **10**(5): 453–458.

Winzer N., Atrens A., Dietzel W., Song G.-L., Kainer K.U. (2008b), *Materials Science and Engineering-A*, **472**(1–2): 97–106.

Winzer N., Atrens A., Dietzel W., Song G.-L., Kainer K.U. (2008c), *Metallurgical and Materials Transactions A – Physical Metallurgy and Materials Science* **39**(A): 1157–1173.

Witte F., Fisher J., Nellesen J., Crostack H-A, Kaese A., Pisch A. (2006), *Biomaterials* **27**,7: 1013–1018.

Xia S.J., Birss V.I., Rateick Jr. R.G. (2003), *Electrochemical Society Proceedings* **25**: 270–280.

Yao H.B., Li Y., Wee A.T.S. (2000), *Applied Surface Science* **158**: 112–119.

Yim C., Ko E., Shin K. (2001), 'Effect of heat treatment on corrosion behaviour of

an AZ91HP Mg alloy', in *Proceedings of the 12th Asia Pacific Corrosion Control Conference 2001*, Seoul, Vol.2, pp 1306–1306–7.
Zamin M. (1981), *Corrosion* **37**(11): 627.
Zhao M., Liu M., Song G.-L.. Atrens A. (2008a), *Corrosion Science, Advanced Engineering Materials* **10**(1–2): 104–111.
Zhao M., Liu M., Song G.–L., Atrens A. (2008b), *Corrosion Science* **50**: 1939–1953.
Zhao M., Liu M., Song G.-L., Atrens A. (2008c), *Advanced Engineering Materials*, **10**, (1–2): 93–103.
Zhu L., Song G.-L. (2006), *Surface and Coatings Technology* **200**(8): 2834–2840.
Zhu L., Liu H., Li W., Song G.-L. (2005), *Journal of Beijing University of Aeronautics and Astronautics* **31**(1): 8–12.

2
Activity and passivity of magnesium (Mg) and its alloys

E. GHALI, Université Laval, Canada

Abstract: Thermodynamic equilibrium cannot exist for magnesium (Mg) in aqueous solution: however, this is possible if the hydrogen overpotential is about 1 V and the pH is greater than 5. Corrosion potential is slightly more negative than –1.5 V/standard hydrogen electrode (SHE) in dilute chloride solutions. The passive film formed in water after 48 h of immersion contains a hydrated inner cellular layer (0.4–0.6 μm), dense intermediate layer (20–40 nm), and outer layer with plate-like morphology (1.8–2.2 μm). The Pilling–Bedworth ratio of MgO/Mg is 0.81, while that of brucite '$Mg(OH)_2$' is 1.77. The influence of pH, oxygen, Cl^- and the negative difference effect on corrosion resistance is discussed as well as some promising factors that can improve the quality of the passive state. An overview of the influence of the active and passive states on the seven forms of corrosion of magnesium and its alloys in aqueous media is given. Localized galvanic, filiform and pitting corrosion are three frequent types of the passive magnesium. Magnesium is promising as a biodegradable material for human implants. The performance of pure and alloyed magnesium as sacrificial anode is commented for some key uses. Evaluation of the sacrificial behavior is also discussed.

Key words: passive film, *E*–pH diagram, galvanic corrosion, pitting corrosion, localized corrosion, open circuit potential, corrosion rate, biodegradable Mg, sacrificial Mg.

2.1 Active and passive behaviors of magnesium (Mg) and its alloys

Magnesium (Mg) has interesting active and passive behaviors; however, in several conditions of service, its passive properties do not meet the required efficiency.

Magnesium is a very base metal and the whole domain of stability of magnesium in de-aerated water is far below that of water. Thus it acts as a powerful reducing agent for hydrogen ions (Pourbaix, 1974).

2.1.1 Pourbaix and Perrault *E*–pH diagrams

Mg dissolution is considered traditionally in the divalent state:

$$Mg^{2+} + 2e^- \rightarrow Mg \qquad E^{\circ} = -2.37\,V \qquad 2.1$$

The probable primary overall corrosion reaction for magnesium in aqueous solutions is:

$$Mg(s) + 2H_2O(l) \rightarrow Mg(OH)_2(s) + H_2(g) \quad 2.2$$

Makar and Kruger (1993) considered the key reactions leading to the diagram in Fig. 2.1 originated by Perrault (1974, 1978). This diagram corresponds to the Mg–H_2O system in the presence of H_2 molecules at 25 °C and considers the reactions of hydride-ion and hydride-hydroxide formation:

$$MgH_2 \rightarrow Mg^{2+} + 2H^+ + 4e^- \qquad E^0 = -1.114 \text{ V/SHE} \quad 2.3$$

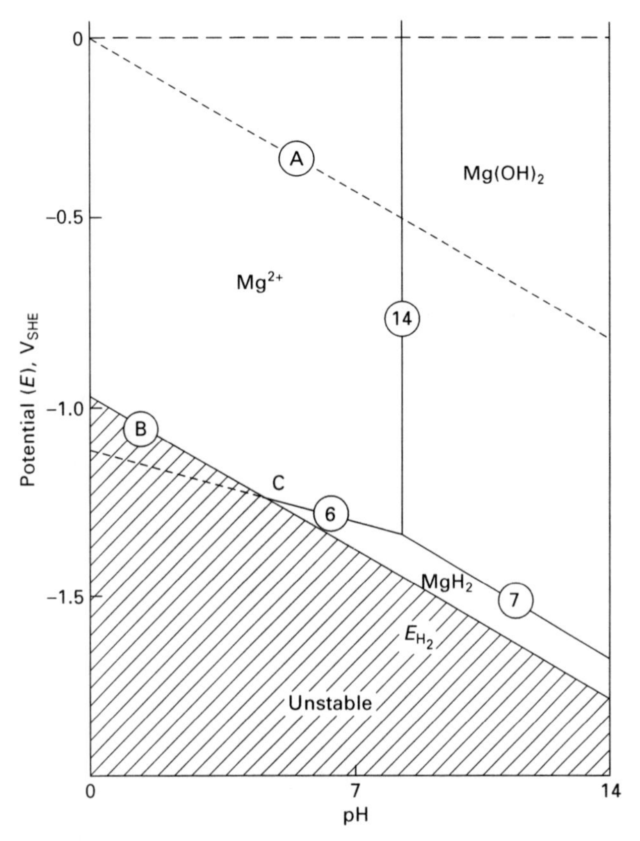

2.1 *E*–pH diagram of Mg showing the stability domains of the Mg compounds in aqueous solutions with hydrogen overvoltage of 1 V at 25 °C and atmospheric pressure (Perrault, 1974, 1978). Line A represents potentials below which hydrogen is evolved, line B is parallel to A above which hydrogen is evolved, C is the intersection of reaction 6 with the line B. 6 and 7 concern the electrochemical reactions of MgH_2 to give magnesium ions or magnesium hydroxide, respectively as a function of pH. 14 is the chemical stability reaction of Mg^{2+} and $Mg(OH)_2$ as a function of pH.

$$MgH_2 + 2OH^- \rightarrow Mg(OH)_2 + 2H^+ + 4e^- \quad E^0 = -1.256 \text{ V/SHE} \qquad 2.4$$

He assumed that thermodynamic equilibrium cannot exist for a magnesium electrode in contact with aqueous solutions. Such equilibrium is, however, possible if the hydrogen overpotential is about 1 V and the pH is greater than 5 (Fig. 2.1). This is supported by the fact that magnesium has a standard electrode potential at 25 °C of –2.37 V, and the corrosion potential of magnesium is slightly more negative than –1.5 V in dilute chloride solution or a neutral solution with respect to the standard hydrogen electrode due to the polarization of the formed $Mg(OH)_2$ film. This indicates that the metal corrodes with an accompanying fairly stable film of rather low conductivity even in acidic solutions. Dissolution could occur through the pores of the passive hydroxide film (if present) or the dissolution of the film itself. There is no equilibrium because of kinetic considerations that depend on the metal, the properties of the interface metal/solution, the level of open circuit corrosion potential, the presence of an oxidant as hydrogen peroxide, and cathodic or anodic polarization (Perrault, 1974, 1978).

The cathodic reaction – the hydrogen evolution reaction – is shown in the E–pH diagram (Fig. 2.1):

$$2H^+ + 2e^- \rightarrow H_2 \qquad E = -0.0592 \text{ pH} \qquad 2.5$$

At higher pH values, this reaction can be expressed as a function of an intermediate adsorption step as follows: $H_2O + e^- \rightarrow H_{ads} + OH^-$.

The influence of oxygen as de-polarizer of the cathodic reaction in solutions saturated with atmospheric oxygen, as well as another strong oxidant can be expressed as:
In acidic medium:

$$2H_{ads} + \tfrac{1}{4}O_2 \rightarrow \tfrac{1}{2}H_2 + \tfrac{1}{2}H_2O \qquad 2.6a$$

and/or in alkaline one:

$$H_{ads} + \tfrac{1}{2}H_2O + \frac{1}{4}O_2 + e^- \rightarrow \frac{1}{2}H_2 + OH^- \qquad 2.6b$$

There is also the evident chemical reaction of formation of hydroxide instead of oxide; however, this does not exclude the possibility to find oxides in the passive layer (Fig. 2.1):

$$Mg^{2+} + 2OH^- \rightarrow Mg(OH)_2 \qquad \log(Mg^{2+}) = 16.95 - 2\text{pH} \qquad 2.7$$

Two important considerations can be mentioned from the E–pH diagram.

The Mg^+ ion

The admittance of the possible existence of the monovalent Mg ion and its domain of existence as function of pH are considered as follows:

$$Mg^+ \rightarrow Mg^{2+} + e^- \quad E^0 = -2.067 \text{ V/NHE} \tag{2.8}$$

$$Mg^+ + 2OH^- \rightarrow Mg(OH)_2 + e^- \quad E^0 = -2.720 \text{ V/NHE} \tag{2.9a}$$

$$Mg^+ + 2H_2O \rightarrow Mg(OH)_2 + 2H^+ + e^- \quad E^0 = -1.065 \text{ V/NHE} \tag{2.9b}$$

The kinetic steps of the formation and oxidation of monovalent Mg ion can be (Ardelean *et al.*, 1999):

$$Mg \rightarrow Mg^+_{ads} + e^- \text{ or } Mg + OH^- \rightarrow MgOH_{ads} + e^- \tag{2.10a}$$

$$Mg^+_{ads} + H_2O \rightarrow Mg^{2+} + \tfrac{1}{2}H_2 + OH^- \tag{2.10b}$$

The presence of the divalent magnesium hydride

The following reactions of electrochemical dissolution of hydride to monovalent and divalent ions are not shown in the E–pH diagram because of instability or the overpotential.

$$MgH_2 \rightarrow Mg^+ + H_2 + e^- \quad E^0 = -2.304 \text{ V/SHE} \tag{2.11}$$

$$MgH_2 \rightarrow Mg^{2+} + H_2 + 2e^- \quad E^0 = -2.186 \text{ V/SHE} \tag{2.12}$$

$$MgH_2 + 2OH^- \rightarrow Mg(OH)_2 + H_2 + 2e^- \quad E^0 = -2.512 \text{ V/SHE} \tag{2.13}$$

The overall electrochemical reaction cannot explain its corrosion rate in frequent conditions. The electrochemical dissolution of magnesium can be carried out through two successive steps: the electrochemical formation of Mg^+ and its oxidation to a divalent ion through hydrogen ion reduction (chemical reaction) or the electrochemical formation of divalent ion directly or both. The mechanism of dissolution can depend also on the stability of magnesium dihydride as a function of pH at the interface.

The E–pH diagram shows that there is possible protection of magnesium at high pH values starting at 8.5 when the activity of Mg ions is equal to 1 mol at 25 °C under atmospheric pressure while the aluminum oxide film is stable in the pH range of 4.0 to 9.0 (Pourbaix, 1974). At acidic and neutral pH, the barrier layer on magnesium is difficult to detect; however, at pH 9, a thick white precipitate of magnesium hydroxide begins to form on the outside of the inner film. This surface film protects magnesium in alkaline environments and poorly buffered environments where the surface pH can increase gradually at least until the relatively high pH of magnesium hydroxide formation (10.45) allows magnesium to resist strong bases (Hawke *et al.*, 1999).

Thermodynamic and practical nobility of magnesium

The Nernst scale of 'solution potentials' permits the metals to be classified in order of 'thermodynamic nobility' according to the value of the equilibrium potential of their reaction of dissolution in the form of a simple given ion considered in the standard state (1 ion g/L). The more positive the reduction potential of metal at 25 °C and atmospheric pressure, the more the metal will be stable according to thermodynamic classification.

Considering arbitrarily that the passive films are perfectly protective, it is admitted that a metal is practically more noble the greater the surface common both to the total of the immunity and passivation domains and to the stability domain of water. This is known as 'practical nobility' (Pourbaix, 1974). It has been accepted that this 'practical nobility' is greater the more the immunity and passivation domains extend below and above the stability domain of water, and the more these domains overlap the section of the diagrams which corresponds to pH values between 4 and 10 (most frequently met with in practice). The classification of the 43 elements on one hand according to 'thermodynamic nobility', and on the other according to 'practical nobility', was then made. ΔN is the difference between practical nobility of an element and its thermodynamic nobility classification order. When ΔN is negative, this means that the metal loses its place in considering active–passive behavior (practical nobility) as compared with 'thermodynamic nobility'. Iron, for example, when it is positive, gains higher classification because of its better projected passive domains. This is especially true for passive metals such as aluminum and titanium.

It is interesting to compare the order of nobility of magnesium with some other current metals such as Al, Ti and Fe. Figure 2.2 gives ΔN, i.e. expresses the difference between the practical nobility of 9 metals and the

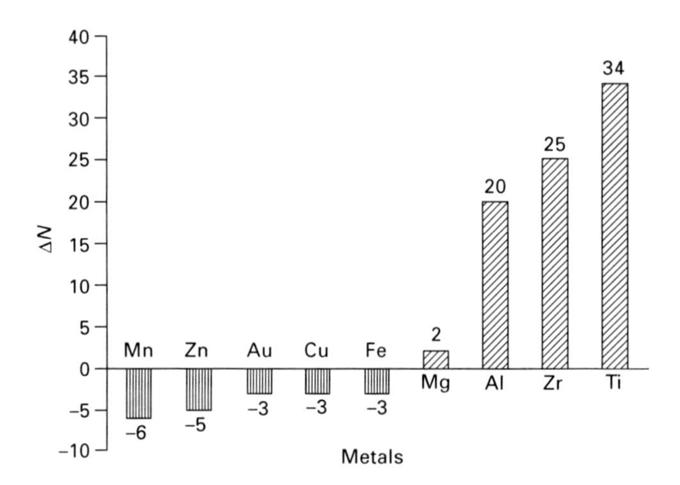

2.2 Comparison of ΔN (practical nobility – thermodynamic nobility) of Mg and that of eight other elements Au, Cu, Fe, Al, Zn, Mn, Zr and Ti as deduced from Pourbaix *E*–pH diagrams (Ghali, 2010). (Positive ΔN indicates better possible projected performance due to passivity.)

thermodynamic order of nobility inspired from Pourbaix classification of order of nobility of 43 metals.

Gold or the top metal goes down to the fourth place and loses three points as well as another relatively less noble metal as copper (ΔN). Also, active metals such as iron, zinc and manganese equally lose their order of nobility in practice. It is interesting to note that aluminum gains 20 points (from 39 in thermodynamic nobility to 19 in practical nobility) influenced largely by its active–passive behavior and this is also the trend of zirconium and titanium. Magnesium gains and is up two steps from 43 to 41 in the order of nobility but far below that of the mentioned three metals. Hopefully, a breakthrough in research could lead to better passive Mg alloys (Ghali, 2010).

2.1.2 Formation and properties of the barrier film

Magnesium exposed to air is covered by a gray oxide film, which can offer considerable protection to magnesium exposed to atmospheric corrosion in rural, most industrial and marine environments. The morphology and structure of oxide films on magnesium, formed by immersion in distilled water after 48 hours, have been shown to be composed of three layers. Transmission electron microscopy (TEM) of naturally formed oxide cross-sections on pure magnesium indicates that the oxide formed in dry air was apparently thin, dense and stable. In humid air, a hydrated layer forms between the metal and the initial layer as a result of water ingress and metal oxidation. The film formed in water after 48 h of immersion contains a hydrated inner cellular layer (0.4–0.6 μm), an apparently dense intermediate layer (20–40 nm), and an additional outer layer with plate-like morphology (1.8–2.2 μm) (Fig. 2.3) (Nordlien *et al.*, 1995, 1996, 1997).

2.2 Passive properties and stability

2.2.1 High temperature oxidation of magnesium

The Pilling–Bedworth ratio (PBR) of MgO/Mg is 0.81, meaning that the scale is formed under tension and tends not to be protective, while that of brucite '$Mg(OH)_2$' is 1.77, which indicates a resistant film in compression (Ghali, 2000). Due to the PBR value (less than 1), the oxide scale cannot form a compact layer generally. The inherent strength of the thin MgO film in which stress is operating is an essentially two-dimensional system; and the oxide can withstand the tensile stress necessary to adapt to the dimensions of the metal. Rupturing occurs only after the film exceeds a critical thickness as a function of longer times of exposure to high temperatures.

The diffusivity D of Mg within the MgO lattice at 400 °C is as low as $2.24 \times 10^{-18}\,m^2/s$, justifying negligible weight gains. The oxidation rate of

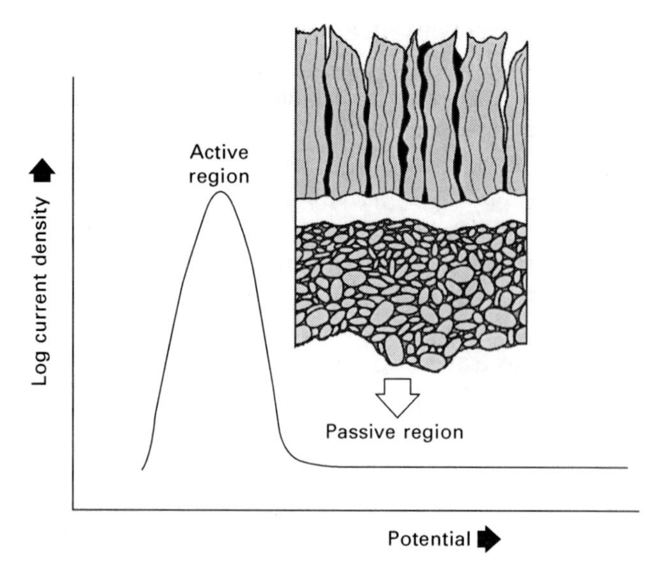

2.3 Schematic showing the three passive layers on polished strips of pure magnesium after immersion in distilled water for 48 h (Nordlien *et al.*, 1995).

magnesium in oxygen increases with temperature and at elevated temperature (approaching melting), the oxidation rate of magnesium in oxygen is a linear function of time (Shaw and Wolfe, 2005). The commercial AZ91D Mg alloy tested at 197 °C did not increase its weight over time periods as long as 10 h. The measurement conducted at 437 °C revealed an accelerated weight gain after ~ 30 min of the reaction and this indicates that the protective capacity of the oxide barrier disappears progressively at a critical temperature zone (Czerwinski, 2003).

Dry chlorine, iodine, bromine and fluorine cause little or no corrosion of magnesium at room or slightly elevated temperature. The presence of a small amount of water causes pronounced attack by chlorine, some attack by iodine and bromine, and negligible attack by fluorine. The presence of BF_3 or SF_6 in the ambient atmosphere is particularly effective in suppressing high-temperature oxidation up to and including the temperature at which the alloy normally ignites.

2.2.2 Aqueous media

The corrosion product film on magnesium starts at neutral pH and is stabilized with the increase of pH. This causes the lustrous metal to assume a dull gray appearance when exposed to air. The film is amorphous and exposure to humid air or to water leads to the formation of a thick hydrated amorphous film that has an oxidation rate less than 0.01 mm/yr. The corrosion rate of chemically pure magnesium in salt water is in the range of 0.30 mm/yr (Nordlien *et al.*, 1996). The quasi-passive hydroxide film on magnesium is much less stable than the passive films formed on aluminum or stainless steels but better than that of iron and low carbon steels (Song and Atrens, 1999).

The film formed in air immediately after scratching the metal surface is initially thin, dense, amorphous and relatively dehydrated. The oxide thickness on pure magnesium after exposure for only ~10 seconds at ambient is 2.2 ± 0.3 nm (~ seven monolayers of MgO) and increases slowly, linearly with the logarithm of exposure during a test of a period of 10 months (McIntyre and Chen, 1998). Continuing exposure to humid air or to water leads to the formation of a thicker hydrated film adjacent to the metal; e.g. for exposure to humid air with ~ 65% relative humidity during 4 days gives 100–150 nm thickness (Nordlien *et al.*, 1995). In case of aluminum, the air-formed film on the surface is amorphous aluminum oxide and appears to reach a terminal thickness of 2 to 4 nm at room temperature after 1 hour exposure (McIntyre and Chen, 1998).

2.2.3 Critical evaluation of the passive layer

The passive film has the advantage of effective stability for a broad zone of pH in the alkaline medium which is not the case of aluminum alloys that are vulnerable to cathodic corrosion and alkaline pitting. The solubility of the metal as related to the concentration of log Mg^{++} in solution decreases linearly with pH starting at pH 8.5 approximately. The advantages of the mechanism of passivation of magnesium and its alloys are that the pH goes to alkaline, stabilizing the passive film and self-limiting serious types of corrosion as pitting corrosion, etc. However, the passive region is not perfect especially at low buffered pH values (Shaw and Wolfe, 2005). There are some main processes of concern of the film quality of protection against general or localized corrosion as follows:

- As MgO readily reacts with water to form $Mg(OH)_2$, a partial blocking of the pores occurs at the beginning accompanied by an increase in the film resistance since the molar volume of $Mg(OH)_2$ is larger than that of MgO. Further formation of $Mg(OH)_2$ inside the oxide film could change the mechanical stresses within the oxide film and the film undergoes compressive rupture, causing some cracks to develop. The oxide film eventually fails by spilling or flaking (Froes *et al.*, 1989; Shaw, 2003; Xia *et al.*, 2003).
- Even at high pH, the corrosion product film of magnesium hydroxide (brucite) is only semi-protective. The outer layer of the passive film with plate-like morphology allows for the ingress of electrolyte to the

metal underneath. The corrosion resistance of the protective passive film is considered to be highly insensitive to the oxygen concentration but depends enormously on the sites of hydrogen discharge composed frequently of alloyed metals and different phases that can create local active galvanic corrosion cells (Shaw, 2003).

- There is the possibility of conversion of the protective surface film during atmospheric corrosion to soluble films such as bicarbonates, sulfites and sulfates that can be washed away by acid rain, agitation or any other flowing liquids (Ferrando, 1989). However, carbon dioxide and sulfur dioxide play an important role in the stability and composition of the film. A mixture of crystalline hydroxycarbonates of magnesium hydromagnesite, $MgCO_3 \cdot Mg(OH)_2 \cdot 9H_2O$, nesquehonite $MgCO_3 \cdot 3H_2O$ and lansfordite $MgCO_3 \cdot 5H_2O$ is reported to be an oxidation product of magnesium; hydromagnesite and hydrotalcite $Mg_6A_2(OH)_{16}CO_3 \cdot 4H_2O$ are formed on AZ31B. In an industrial atmosphere with high SO_2 content, traces of $MgSO_4 \cdot 6H_2O$ and $MgSO_3 \cdot 6H_2O$ were detected in addition to the hydroxy-carbonate products for unalloyed ingot (Hillis, 1995).
- Passivity of magnesium is destroyed by several anions, including chloride, bromide, sulfate, nitrate and perchlorate. Chlorides, even in small amounts, usually break down the protective film on magnesium. The ability of an anion to reduce the Mg potential appears to depend on the solubility of its Mg salt. In the presence of salt solutions of these anions, the open circuit potential (OCP) of magnesium or alloys becomes several tenths of a volt active to the hydrogen electrode potential. Hydrogen discharge becomes the controlling factor on effective sites as elements of low hydrogen overvoltage other than that of the hydroxide film itself because of its poor electronic conduction (Ferrando, 1989). Robinson and King (1961) suggested that anions are carried by electrochemical transport to anodic sites on the metal surface, where they form magnesium salts which are acidic to the magnesium hydroxide film. The rapid uniform corrosion rate observed in 3 M $MgCl_2$ at a less noble electrode potential supports this mechanism.

2.3 Improvements and promising avenues of the passive behavior

2.3.1 Alloying and passivity

Alloying can increase passivity by incorporation of some components which stabilize the oxide as in the case of aluminum. Improvements in corrosion resistance have been found to correlate with an increase in the concentration of the alloying element or its oxide in the passive film and upgrade the passive behavior of Mg and its alloys (Shaw and Wolfe, 2005).

Such components either have appreciable solid solution solubility with the magnesium matrix (aluminum for example) or they are only sparingly soluble such as manganese and the rare earth metals (Nordlien *et al.*, 1996). On the other hand, reduction of the heavy metal content and application of appropriate heat treatments have already, led at least partially, to increased corrosion resistance of certain Mg alloys. Amorphous oxides are in general considered to have much better passive properties than crystalline ones and stability of the inner layer on magnesium is responsible for the passivity of the surface in aqueous environments.

It has been stated that an increase in the Al content of the alloy to or above 4% causes an important increase of the Al content of the inner layer, leading to an Al/Mg weight ratio of about 35%. It also reduces the thickness of the inner layer by almost an order of magnitude but not as dramatically as the thickness of the platelet layer and consequently the inner layer becomes the thicker layer of an overall thinner film. It has been proven that the reduced thickness leads to reduced hydration and better corrosion resistance (Nordlien *et al.*, 1996). It will be interesting to examine the stability of these oxides for high pH values above 11.5, for example where MgO can be more stable than Al_2O_3. Frequently, the important presence of magnesium and its oxide in the passive film leads to less noble electrode potentials in alkaline medium.

2.3.2 Influence of inhibitors

General uniform corrosion on pure magnesium has been drastically reduced by the use of inhibitors such as chromate, dichromate, molybdate, nitrate, phosphates and vanadates that promote the formation of a protective layer, tend to retard corrosion (Ghali, 2006). Addition of substances that can form soluble complexes as tartrate, metaphosphate, etc. or insoluble salts as oxalate, carbonate, phosphate, fluoride, etc. is efficient at reducing corrosion. Adding soluble chromates, neutral fluorides or rare earth metal salts is effective in reducing magnesium–base metal corrosion (Schmutz *et al.*, 2003).

At room temperature, ethylene glycol solutions produce negligible corrosion of magnesium that is used alone or galvanically connected to steel; at elevated temperatures, such as 115 °C (240 °F), the rate increases, and corrosion occurs unless proper inhibitors are added. The corrosion of magnesium in ethylene glycol can be effectively inhibited by addition of fluorides that react with magnesium and form a protective film on the surface (Kaesche, 1974; Uhlig, 1970).

At room temperature, an insoluble passive MgF_2 film was generated on the surface of Mg alloy activated in HF solution and the mass of magnesium alloy treated in different concentrations of HF acid (10–70%) increased as function of time. When the mass ratio of Mg/F in the film was 11.3:1, the mass of the deposited MgF_2 film reached a constant value. The passive film can protect the alloy in fluoride solutions through the adsorption of HF_2^-, $H_2F_3^-$ and $H_3F_4^-$ ions. However, no such film can protect the substrate in non-fluoride solutions (Li *et al.*, 2009).

2.3.3 Anodized and oxidized films

A thick and protective oxide–hydroxide layer on magnesium is formed by different techniques. The characteristics of the oxide film formed on an Mg-based WE43 alloy using an AC/DC anodization technique in an alkaline silicate solution at 30 mA/cm^2 for 5 min, followed by 25 min decreasing current, have been investigated. The anodic oxide film formed in alkaline silicate anodizing bath is composed of MgO, $Mg(OH)_2$, SiO_2 and MgF_2, with the molar ratio of MgO to $Mg(OH)_2$ being close to 2:1. The interpretation of the AC impedance technique and OCP measurements of corrosion kinetics suggests the presence of three stages: the initial hydration of MgO, the blocking of film pores as MgO begins hydrated, and then the formation of cracks and dissolution of $Mg(OH)_2$ into pores and the solution (Shaw, 2003; Xia *et al.*, 2003). Also, fluoride anodizing of magnesium and its alloys is frequently chosen when complete removal of the contaminant is essential (Shaw, 2003).

Micro-arc oxidation (MAO) or micro-plasma oxidation is a surface treatment method used to form an oxide coating by anodic spark deposition on Al, Ti and Mg alloys in a suitable electrolyte with a high anodic applied voltage (400–600 V). During the MAO process, a plasma environment is generated by spark discharge at a high anodic applied voltage giving rise to a series of plasma thermochemical interactions between the substrate and the electrolyte. Bai and Chen (2009) considered different electrolytes such as phosphate, silicate or both in the presence of two inhibitors to form anodic films on AZ91D alloy. The best electrolyte composition during the processing of Mg alloy AZ91D was: potassium hydroxide 1.5 M, sodium citrate 0.04 M, phosphoric acid 0.1 M and sodium silicate 0.08 M. The breakdown voltages during anodization in an MAO process were changed from 400 to 200 V to anodize oxide coating on AZ91D alloy samples, using an alkaline silicate phosphate electrolyte instead of the simple alkaline phosphate electrolyte. The lower breakdown voltage gave anodized oxide coatings with better composition, lower porosity, smaller pores and fewer cracks, and thus better anti-corrosion ability.

Hexamethylenetetramine (0.1 M) was a useful additive for removing cracks and reducing the size of pores in an oxide coating and the surface roughness (R_a or center line average) was reduced from 10.12 to 1.22. Sodium borate (0.1 M), accompanying additive, changed the composition of the Na content in an oxide coating on the AZ91D alloy without influencing the morphology or roughness value of the oxide coating. The anti-corrosion abilities of the different oxide coatings were examined in a 5 wt% NaCl solution using the potentiodynamic polarization. The thickness of the oxide coatings was controlled to be approximately 20–30 μm for all the specimens formed in the different electrolytes, with the fixed 60 min MAO process preparation period and the different breakdown potentials. Tafel slopes of the oxide coating on AZ91D were performed with a scan rate of $1\,mV\,s^{-1}$ from −2 V to 0.8 V in 5 wt% NaCl solution. The corrosion current of the oxide coating changed from 3.6×10^{-5} to $4.2 \times 10^{-7}\,A/cm^2$ when the additives were added to the plating solution.

2.3.4 Cathodic charging

Nakatsugawa (2001) succeeded to create a pseudo-passive film of MgH_2 on magnesium surface and developed a method to create hydrogen-rich layer onto AZ91D by way of cathodic charging. MgH_2 decomposes gradually to form the hydroxide $Mg(OH)_2$ in aqueous environment. The treated Mg or Mg–Al alloys show a pseudo-passive behavior in the anodic region in 5% NaCl saturated with sodium hydroxide solution and an increase of the Tafel slope in the cathodic region (Fig. 2.4). The corrosion resistance of this

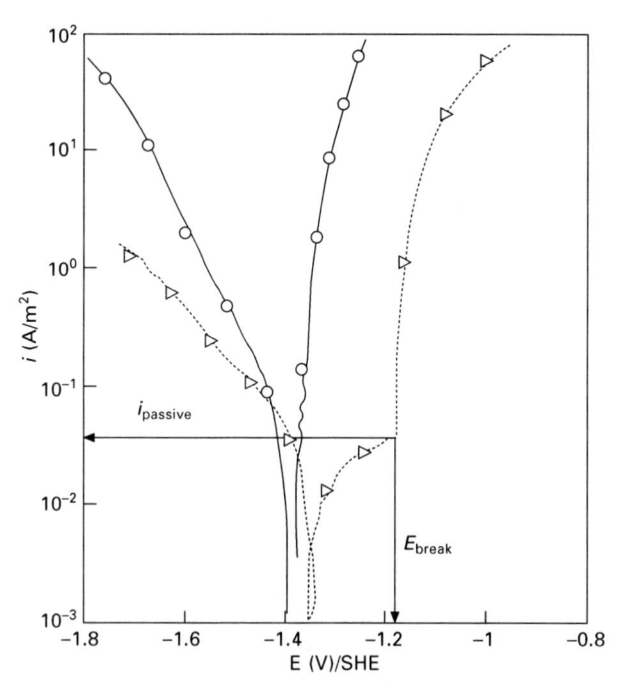

2.4 Potentiodynamic polarization curves of untreated and H-coated AZ91D alloy in 5% NaCl at pH 10.5 solution from OCP to anodic and cathodic directions at scan rate of $1\,mV\,s^{-1}$, (Δ) H-coated, (○) untreated (Nakatsugawa *et al.*, 1999).

coating is superior to Cr^{6+}-based conversion coating and has a fairly good adhesion to paint (Nakatsugawa *et al.*, 1999).

2.4 Specific factors characterizing corrosion behavior

2.4.1 Influence of pH on polarization curve in de-aerated solutions

In alkaline solutions of pH >10.45, Mg tends to passivate in alkaline solutions at pH>10.45, and at more alkaline pH such as 13 (0.1 M NaOH) a truly passive behaviour is expected (Fig. 2.5). An average current density of a few μA/cm² is observed in the passive region at a relatively high scan rate 1 mV/s.

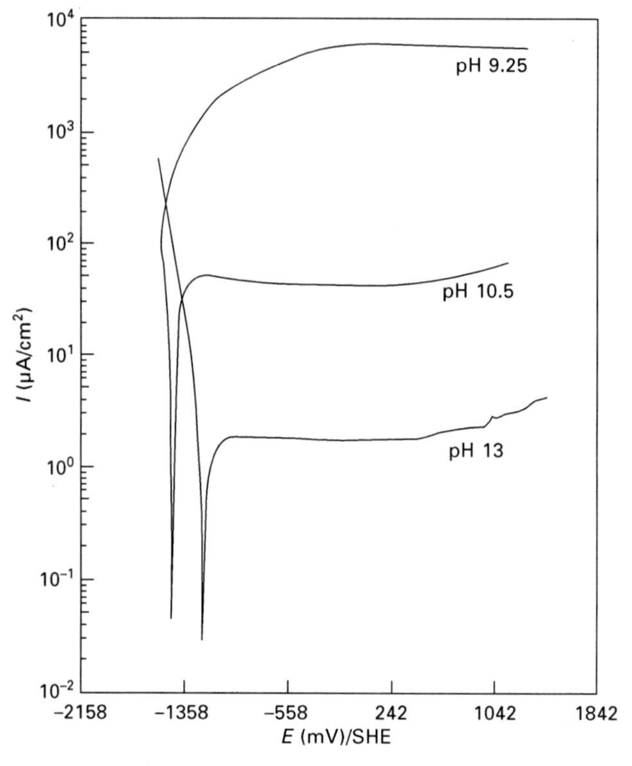

2.5 Potentiodynamic polarization characteristics of magnesium as a function of solution pH in de-aerated 0.1 M NaOH solution with scan rate 1 mV/s (Mitrovic-Scepanovic and Brigham, 1990).

2.4.2 Theoretical schematic approach to active–passive behavior

Mitrovic-Scepanovic and Brigham (1990) considered a schematic and theoretical approach for the dissolution of magnesium in de-aerated alkaline solution at pH 13. In this approach the divalent state dissolution has been considered; the dissolution of magnesium in the monovalent state accompanied by the oxidation–reduction reaction of hydrogen ions leading to the negative difference effect on the anodic curve and the formation and stability of magnesium hydride at this pH are not considered.

They suggested three open circuit potentials (OCP) E_A, E_B and E_C for pure magnesium at pH 13 (Fig. 2.6). Regarding the active behavior of magnesium, there is an E_A value considering equal mixed control for V-logi (Evans diagram deduced from anodic and cathodic polarization curves). Regarding the shift of the polarization curve to more noble values on the active–passive state of the metal, the OCP oscillates between an active relatively less active value E_B and a more positive one E_C at the beginning of the passive region. Practically, Mitrovic-Scepanovic and Brigham (1990) found that the potential oscillates between these two values for AZ31 alloy as a function of time in de-aerated 0.1N NaOH solution (pH 13). This has been found in the active

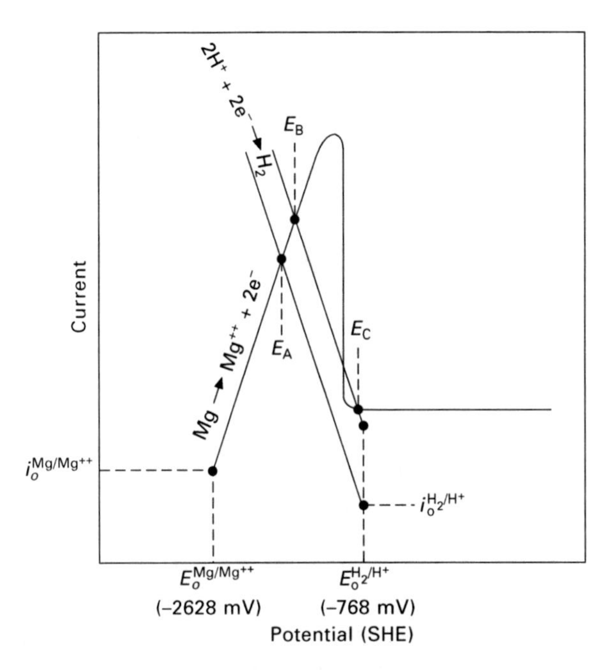

2.6 Schematics of theoretical polarization curve of magnesium in deareated 0.1N NaOH solution (pH 13) (Mitrovic-Scepanovic and Brigham, 1990).

behavior of other metals also such as Fe in nitric acid solution with specific concentration corresponding to the breakdown and repair of the passive film. E_B can be described as active unstable potential before passivation while E_C corresponds to the beginning of the passive zone of magnesium. It is worth mentioning that Mitrovic-Scepanovic and Brigham have derived the value of ~ –0.768 V/SHE practically when a very pure magnesium specimen was coupled to a small piece of Pt and the specimen was shifted to the passive state because of the extreme depolarization of hydrogen ion reduction. Also, it is likely in the presence of an efficient oxidant such as a solution saturated with atmospheric oxygen or containing perchlorate ions and for alloys with certain alloying elements that can shift the cathodic curve to more positive values and cross the anodic curve or the passive region in the fourth suggested point. This can give rise to an OCP value more positive than the other three suggested corrosion potential values.

Bonora *et al.* (2002) compared the potentiodynamic curves in 0.05 M sodium tetraborate solution (pH = 9.7) of two unstressed Mg alloys to that of pure Mg (Fig. 2.7). It can be observed that the shape of the polarization curves is almost the same for the AM and AZ alloys. In addition, a current plateau of passivation begins at –1.2 and –1.3 V for AM50 and AZ91D, respectively, the current values being quite similar and relatively high. This type of behavior can be considered as 'pseudo-passivation' at this buffered pH. Pure magnesium showed more active OCP, and much higher current in the passive region, meaning that the quality of passivation is much lower.

The influence of mechanochemical effects (MCEs) and creep on corrosion parameters were studied by Bonora *et al.* (2002). They stated that deformation increases the anodic current densities and shifting the potentials to more active values. It was also shown that the AZ91D alloy has a high corrosion rate in the deformed state than AM50 alloy under stress, while in the non-loading state the corrosion rate was found to be higher for the AM50 alloy. This behavior confirms the MCE theory and the behavior of these alloys at creep.

2.4.3 Influence of solution agitation

Agitation or any other means of destroying or preventing the formation of a protective film leads to corrosion. When magnesium is immersed in a small volume of stagnant water, its corrosion rate is negligible. When the water is completely replenished, the solubility limit of $Mg(OH)_2$ is never reached and the corrosion rate may increase. When agitation (erosion) destroys or depletes the surface film, corrosion can be significantly increased. Figure 2.8 illustrates the cathodic and anodic curves in an aerated 0.5 M Na_2SO_4 using a rotating electrode and these give a conceivable Tafel slope for cathodic reaction of reduction of the water molecules (b_c = 180 mV/log I) which

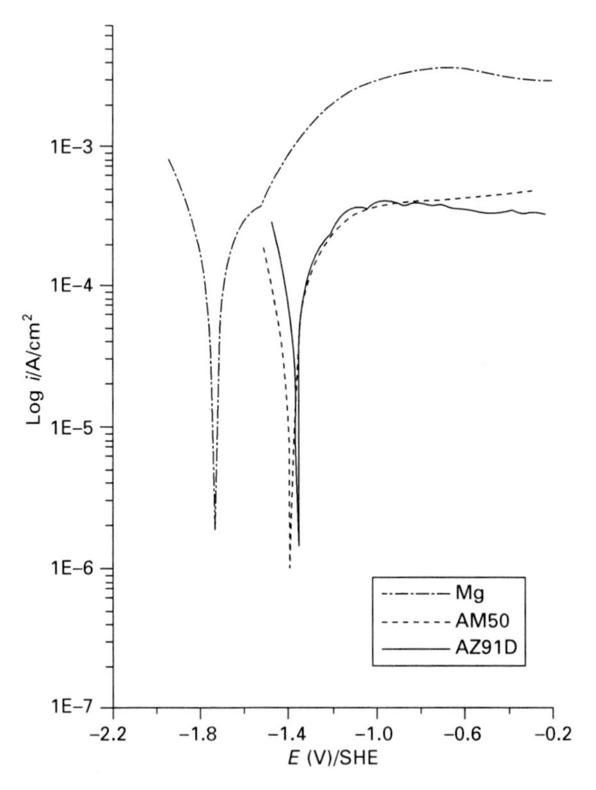

2.7 Potentiodynamic curves of pure magnesium, and AM50 and AZ91D alloys in 0.05 M sodium tetraborate solution buffered at pH = 9.7 (scan rate 0.2 mV/s) (Bonora *et al.*, 2002).

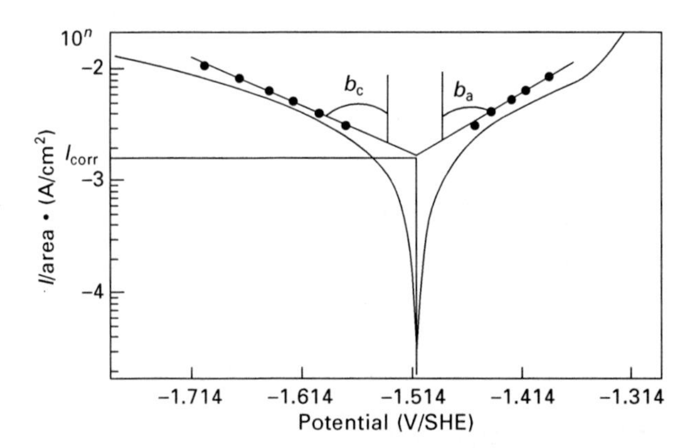

2.8 Tafel extrapolation of the rotating disc electrode (RDE) polarization curve of pure magnesium in an aerated 0.5 M Na_2SO_4 (rotation speed: 1500 rpm, potential scan rate: 0.5 mV/s) (corrected for the ohmic drop) (Ardelean *et al.*, 1999).

is different from 120 or 210 mV/log I found by others. These have been considered by some authors to derive an approximate corrosion rate for the sake of comparison only. The active–passive transition in the anodic curve is not so evident although the pH of the solution changed from 6.2 to 9.6, very possibly because of the high rotation speed of the electrode (Fig. 2.8). Generally, the anodic Tafel slope is not currently considered to confirm the corrosion rate calculations because of the active–passive behavior (Ardelean *et al.*, 1999).

In stagnant distilled water at room temperature, Mg alloys rapidly form a protective film that prevents further corrosion. Small amounts of dissolved salts, in water, particularly chlorides or heavy metal salts will break the protective film locally, which usually leads to pitting (Froats *et al.*, 1987).

Baril and Pébère (1999, 2001) carried out impedance measurements and determined the diagrams after different hold times at E_{corr} in aerated 0.01 M (Fig. 2.9). The diagram is characterized by two well-defined capacitive loops, at high and medium frequencies, followed by an inductive loop in the low-frequency range. The same trend of these diagrams was found in 0.1 M Na_2SO_4 or in de-aerated medium (0.01 M). The increase in immersion time at E_{corr} led to an increase in both high- and medium-frequency capacitive loops. The high-frequency loop appears to result from both charge transfer and a film effect. The medium-frequency capacitive loop is related to relaxation of the mass transport in the solid phase, i.e. in an aggregating layer whose thickness increases with immersion time. The inductive loop is fundamentally associated with the breakdown of the passive film on the magnesium surface (Song *et al.*, 1997a). The influence of rotation rate on the impedance diagram was negligible and so the process was not influenced by oxygen diffusion

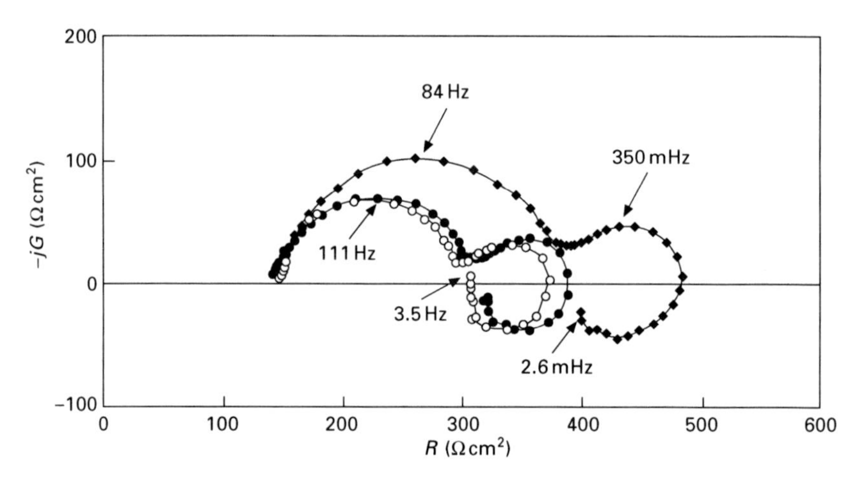

2.9 Impedance diagrams determined after different preliminary hold times at E_{corr} in 0.01 M Na_2SO_4 (○) 3 h 30 min; (●) 6 h; (◆) 21 h in contact with air at 25 °C (Baril and Pébére, 1999).

in the liquid phase. Chen *et al.* (2007) observed the inductive loop of AZ91 alloy in Na_2SO_4 solution after an initial immersion period of 1 hour to the presence of metastable Mg^+ ions, while that observed after longer period (181 h) was attributed to pitting corrosion. Jin *et al.* (2007) suggested that the inductive loop at low frequency is an indication of pitting corrosion of an Mg alloy in chloride solution.

2.4.4 Influence of oxygen and some active ions

The influence of oxygen concentration in aqueous media on Mg corrosion seems to depend on the presence of halide ions, bicarbonate as examples, more positive OCPs, cathodic phases for hydrogen evolution in the alloy, etc.

Influence of oxygen on general corrosion

Dissolved oxygen plays no major role in the corrosion of magnesium in either freshwater or saline solutions (Froats *et al.*, 1987). In acidic solutions, and at more negative OCPs, it seems that hydrogen reduction is the main cathodic reaction (Hur and Kim, 1998), especially in presence of active cathodic sites for hydrogen evolution (Hanawalt *et al.*, 1942). On the other hand, the presence of oxygen is an important factor in atmospheric corrosion (Makar and Kruger, 1993). The most positive potentials are observed in pure water and alkaline solutions containing sub-critical amounts of certain anions. These potentials are usually near the hydrogen electrode reversible potential or readily rises thereto by application of a very small anodic current. Only in environments of this type, and then under good aeration, does oxygen reduction play a significant role (Ferrando, 1989; Ghali, 2010).

Baril and Pébère (2001) studied the corrosion behavior of pure magnesium in aerated and de-aerated solutions (0.01 and 0.1 M) by steady-state current–voltage measurements using rotating electrode (1000 rpm) and electrochemical impedance measurements. It was shown that the anodic current densities were lower and the resistance values higher in de-aerated media. They have stated that the presence of oxygen does not influence the cathodic reaction and so oxygen has no effect on Mg corrosion; however, the shift of the potential in the cathodic direction in aerated solutions and higher anodic corrosion current densities can be explained by the presence of bicarbonate ion in natural conditions (40 mg HCO_3^-/L). The HCO_3^- increased the rate of dissolution by formation of soluble salts.

Pitting corrosion of passive films

There is a synergetic effect of oxygen and Cl^- on pitting of passive alloys. A partially protective surface film plays an important role in the electrochemical dissolution processes for magnesium in NaCl, Na_2SO_4 and NaOH solutions. The presence of Cl^-, for example, makes the surface films more active or increases the broken area of the film, and also accelerates the electrochemical reaction rate from magnesium to magnesium univalent ions (Vargel, 2004). The solubility of air and oxygen in saline solutions decreases with increasing concentration of the salt, but salt increases the solution conductivity. The two effects combine in oxygen reduction cathodic systems to produce increasing corrosion rates up to about 3.5 wt% sodium chloride solutions and decreasing corrosion rates above that (Baloun, 1987). It has been shown also that oxygen plays a major role in the initiation of pitting of AZ91, HK31 and some Mg–Zn alloys in 5 wt% sodium chloride solution at room temperature at relatively high corrosion potentials (Reboul and Canon, 1977).

2.4.5 Active and passive states at high-temperature aqueous media

The effect of increasing the temperature is to increase the severity of attack. Magnesium alloys do not have adequate corrosion resistance for applications above ambient temperature (Danielson, 2001; Song *et al.*, 1997b). Pure magnesium (99.5% + % purity < 10 ppm (Fe + Ni + Cu)) immersed in distilled water, from which acid atmospheric gases have been excluded, is also highly protected. However, this good resistance to corrosion in water at room temperature decreases with increasing temperature, corrosion becoming particularly severe above 100 °C. Pure magnesium and alloy ZK60A corrode excessively at 100 °C with rates up to 25 mm/yr (Song and Atrens, 1998). Water vapor in air or in oxygen sharply increases the rates of oxidation of magnesium and its alloys above 100 °C, but BF_3, SO_2 and SF_6 are effective in reducing the oxidation rates (Nisancioglu *et al.*, 1990).

The increasing rate of corrosion, with increase in temperature of ternary alloys, is higher than that of pure magnesium and may be due to the activation of some impurities in the ternary alloy at higher temperatures. It appears that the onset of pitting in a given alloy and in certain media depends on a critical pitting temperature below which only uniform corrosion is encountered. Increasing temperatures sometimes precipitates protective salts, such as calcium carbonate, which decreases corrosion rates in normal-to-hard waters (Amira *et al.*, 2007; Shaw and Wolfe, 2005).

2.5 Active and passive behaviors and corrosion forms

The first three forms of corrosion: uniform or quasi-uniform general corrosion, galvanic corrosion and localized corrosion (pitting, crevice and filiform) have no clear separation. The oxide–hydroxide passive layer can play the role of a coating and show filiform corrosion (a type of localized corrosion that occurs under coatings or paints) in the same time of uniform corrosion or pitting. Metallurgically and microbiologically influenced corrosion can lead to the appearance of intensive corroded localized zones. Mechanically assisted corrosion (especially corrosion fatigue) and stress corrosion cracking (expressed also as environmentally induced corrosion (EIC)) are commented generally as specific localized defaults of the passive film due to the existence of electrochemical galvanic cell (Ghali 2010).

2.5.1 General corrosion

Sivaraj *et al.* (2006) determined the corrosion rates in 5% NaCl solution using salt-spray testing (ASTM B117). Other techniques are employed in parallel (weight loss, electrochemical DC polarization titration and hydrogen evolution). The weight loss method can be considered as the best method to determine corrosion rates of magnesium and its alloys, in spite of the fact that loss of non-corroded material during etching is always a concern. Polarization experiments are quick and practical to carry out, but a number of factors, such as scan rate, cell geometry, influence of the Nernst diffusion layer, agitation or circulation of the electrolyte, influence the reproducibility and reliability of the corrosion rate determined by the polarization technique. In the titration method, the pH of the solution is held constant and it can be tailored to suit the requirements of the corrosion environment. In general, the titration test results are more reproducible. A lack of accuracy and experimental difficulty (e.g. proper gas sealing) are the main difficulties with the hydrogen evolution method and hence the variability in the comparative corrosion rates among experiments is high.

Influence of negative difference effect (NDE) and Mg^+ on corrosion rate

The corrosion resistance of commercial Mg alloys does not significantly exceed that of pure Mg. Appropriate quantitative agreement between weight loss and corrosion rates derived from electrochemical polarization measurements has been found for most metals. Instead of having a decrease of hydrogen evolution during anodic polarization, it has been observed that over a restricted potential range of anodic polarization of Mg the rate and amount of hydrogen evolution actually increase as the potential of the metal is shifted to more noble or positive values. This is defined as the negative difference effect (NDE) (Stampella *et al.*, 1984). It has been suggested that there is a chemical reaction corresponding to hydrogen reduction and oxidation of monovalent magnesium to divalent magnesium. This is governed by the laws of mass action, solubility product and chemical equilibrium that involve oxidation–reduction processes (Hawke *et al.*, 1999).

Song and Atrens (2003) and Song (2005) proposed an appropriate new mechanism for NDE and stated that the reason is the anodic dissolution of magnesium in the surface-film broken areas giving then the metastable Mg^+ that is oxidized chemically. Hawke *et al.* (1999) mentioned that exposing active metal by mechanical and chemical attacks of the protective film, formation of magnesium hydride and loss of metal by disintegration (chunk effect) are also possible causes. MgH_2 is also considered as an intermediate of an anodic dissolution process. Also, this does not exclude the direct dissolution of magnesium as bivalent ion (Song, 2005). Based on AC impedance spectra, Song (2007a) believes that the anodic reaction of pure magnesium takes place in two steps: the first is the electrochemical formation of monovalent magnesium that can be followed by the chemical oxidation reduction reaction through hydrogen ion reduction or by electrochemical oxidation. Atrens and Dietzel (2007) suggested that at free corrosion potential, these two steps could be expressed as follows:

$$Mg \rightarrow Mg^+ + e^- \text{ (anodic partial reaction)} \qquad 2.14$$

$$kMg^+ \rightarrow k\, Mg^{2+} + ke^- \text{ (anodic partial reaction)} \qquad 2.15$$

$$(1-k)\, Mg^+ \,(1-k)\, H^+ \rightarrow (1-k)\, Mg^{2+} + (1-k)/2H_2$$

$$\text{(chemical reaction)} \qquad 2.16$$

$$(1+k)\, H^+ + (1+k)\, e^- \rightarrow (1+k)/2H_2 \text{ (cathodic total reaction)}$$

$$2.17$$

This mechanism of dissolution is supported by the fact that the corrosion rate of magnesium and its alloys evaluated from weight loss agrees within an error of less ± 10% with the corrosion rate independently measured from the hydrogen evolution. Also, it has been reported frequently for certain experimental conditions that the corrosion rate evaluated by Tafel extrapolation from polarization curves does not agree with that evaluated from weight loss and hydrogen evolution. It has been mentioned that the relative errors in the evaluation of the corrosion rate from Tafel slope extrapolation can be 25 times as high in certain experimental conditions (Song *et al.*, 2001). It has also been stated by Shi *et al.* (2010) that the relative errors in the evaluation of the corrosion rate from Tafel extrapolation method ranged from ~50 to 90%. These large relative errors are much larger than the precision of the electrochemical method and thus indicate that there is a need for careful consideration of the use of Tafel examination. Simple measurement methods such as weight loss rate, hydrogen evolution rate or dissolved Mg^{2+} ions are recommended as complementary or checking methods.

Liu and Schlesinger (2009) recognized the presence of three situations. When the anodic overpotential ($E_{applied} - E_{corrosion}$) = ΔE is low, the anodic reaction for magnesium is mostly the formation of Mg^+ and this is followed by the slow chemical reaction of oxidation to divalent ion Mg^{++}. When ΔE is high, the anodic reaction is mostly the formation of Mg^{++} directly, while for intermediate values of ΔE, both Mg^{+} and Mg^{++} can exist. They expressed quantitatively the microgalvanic corrosion, especially the negative difference effect of magnesium based on these three assumptions.

Sacrificial magnesium in the active state

Magnesium alone and its alloys have been employed intensively as sacrificial anode for cathodic protection and are now considered for numerous present and future promoting applications (see Section 2.4).

2.5.2 Galvanic corrosion or bimetallic corrosion

Galvanic corrosion or bimetallic corrosion is important to consider since most of the structural industrial metals and even the metallic phases in the microstructure alloys create galvanic cells between them and/or the α Mg anodic phase. However, these secondary particles which are noble to the Mg matrix, can in certain circumstances enrich the corrosion product or the passive layer, leading to a decrease or a control of the corrosion rate. Severe corrosion may occur in neutral solutions of salts of heavy metals, such as copper, iron and nickel. The heavy metal, the heavy metal basic salts or both plate out to form active cathodes on the anodic magnesium surface. Small amounts of dissolved salts of alkali or alkaline-earth metal (chlorides, bromides, iodides and sulfates) in water will break the protective film locally and usually lead to pitting (Froats *et al.*, 1987; Shaw and Wolfe, 2005).

Under conditions where the corrosion product is not continuously removed or under conditions of high cathodic current density where the surroundings may become strongly alkaline, both the magnesium and an amphoteric contacting metal such as aluminum may suffer severe attack. Such attack destroys compatibility in alloys containing significant iron contamination. Cathodic corrosion of aluminum is much less severe in seawater than in NaCl solution because of the buffering effect in seawater. Aluminum alloys containing appreciable magnesium, such as 5052, 6053 and 5056, are least severely attacked in chloride media when galvanically coupled (Froats *et al.*, 1987). Hydrogen evolution and strong alkalinity generated at the cathode can damage or destroy organic coatings applied to fasteners or other accessories coupled to magnesium. Alkali-resistant resins are necessary, but under severe conditions, such as salt spray or salt immersion, which do not simulate adequately a real application, the coatings may be simply blown off by hydrogen, starting at small voids or pores. Some new processing methods such as new rheocasting processes, rapid solidification processes, ion implantation and vapor deposition enhance corrosion resistance by producing a more homogeneous microstructure and/or by increasing the solubility limits of alloying additions (Shaw and Wolfe, 2005). This can be lead to more corrosion-resistant passive states.

There is considerable interest in the use of magnesium alloys in metal–matrix composites (MMCs). However, this is a strenuous application for magnesium, considering the extreme galvanic nobility of many composite materials, such as graphite. In the case of AZ91 combined with alumina fibres, the corrosion rate of the MMC is seven times higher than that of the bulk alloy, showing the paramount importance of galvanic effects (Shaw and Wolfe, 2005). Hihara and Kondepudi (1993, 1994) investigated the galvanic corrosion behavior of Mg MMCs of the two matrices: pure Mg and ZE41A-Mg alloy in contact with SiC monofilament (MF) or pure SiC particles. The results showed evidence of higher galvanic corrosion rates of the matrices in both cases; however, ZE41A–Mg matrix showed a better corrosion resistance than pure Mg. Hall (1987) observed the evidence of galvanic corrosion for carbon fiber/magnesium MMCs in a normal laboratory atmosphere of about 60% relative humidity at 20 °C. The determined rate of penetration was about 100 μm per year. Bakkar and Neubert (2009) showed that the galvanic coupling with C-fibers leads to severe corrosion of Mg matrix composites and invalidates the virtual effect of alloying elements on corrosion resistance.

2.5.3 Localized corrosion of the passive state

The passive state of magnesium and its alloys is generally more vulnerable to localized corrosion than the active state. In natural atmospheres, the corrosion of magnesium can be localized depending on alloy composition, microgeometry of the surface, distribution of different metallurgical phases of the microstructure, ionic species and temperature of the electrolyte. Weak conductivity of the electrolyte and small anode/important cathode relative area ratios increase localized corrosion. Localized corrosion of magnesium and its alloys as pitting, filiform and crevice forms of corrosion of magnesium and its alloys have some differences in occurrence, kinetics, morphology and mechanism compared with other metals. For example, the oxygen differential cell is not always the principal factor in case of magnesium. The acid pH at the pit or crevice due to hydrolysis and the retention of humidity could control the corrosion kinetics.

Pitting and filiform corrosion can initiate simultaneously in NaCl solutions and filiform corrosion is observed frequently as the front runner. Filiform then develops into cellular or pitting corrosion. Filiform corrosion occurs on some uncoated extruded magnesium alloys but not on bare pure Mg. Its occurrence on bare Mg–Al alloys indicates that highly resistant oxide films can be naturally formed. In chloride solutions, such as seawater, attack on the metal usually results in pitting of some areas only, while for a reactive metallic surface, by sand blasting for example, attack may be so rapid that uniform dissolution is observed (Froats *et al.*, 1987; Lunder *et al.*, 1990).

Galvanic cells develop and some areas become anodic to other areas and as corrosion proceeds at the anodic areas, pitting develops. There is no evidence of initiation at particle-free areas, indicating that hydrogen evolution on the particle is the predominant controlling cathodic reaction (galvanic corrosion). Stable corrosion pits initiate at flaws adjacent to a fraction of the intermetallic particles present as a result of the breakdown of passivity. Pitting can also occur in non-passivating alloys with protective coatings or in certain heterogeneous corrosive media (Ghali, 2000).

After the initiation period of corrosion pits, filiform corrosion dominates the morphology as narrow semi-cylindrical corrosion filaments project from the pit. Lunder *et al.* (1990) observed that propagation of the filaments occurs with voluminous gas evolution at the head while the body immediately behind passivates. Electrochemical transport of chloride ions to the head of the filament appears to be an essential component as is precipitation of insoluble $Mg(OH)_2$ by the anodic reaction with Mg^{2+} ions elsewhere along the filament (Ghali *et al.*, 2004).

The corrosion behavior of AZ91D Mg alloy in alkaline chloride solution was investigated by electrochemical noise (EN). The noise resistance (R_n), power spectral density (PSD) and wavelet transform were considered to analyze the EN data. It was revealed that there exist three different stages of corrosion for AZ91D Mg alloy in alkaline chloride solution: the anodic dissolution process accompanying the growth, absorption and desorption of hydrogen bubbles, the development of pitting corrosion and the possible inhibition process by protective MgH_2 film (Zhang *et al.*, 2007).

The atmospheric corrosion occurs frequently under thin electrolyte layers (TELs) or even adsorbed layers. The thickness of electrolyte has important role on corrosion phenomena such as the mass transport of dissolved oxygen, the accumulation of corrosion products, and the hydration of dissolved metal ions. The corrosion behavior of pure magnesium was investigated under aerated and de-aerated TELs with various thicknesses by means of cathodic polarization curve, electrochemical impedance spectroscopy (EIS) and EN measurements (Zhang *et al.*, 2008).

After 20 min of immersion of pure Mg in 0.05 M NaCl + 0.5 M Na_2SO_4 solution saturated by Mg $(OH)_2$, the cathodic polarization curve of the expected passive state was scanned from the open circuit potential to –2.2 V/Ag, AgCl with a scan rate of 1 mV/s. Cathodic polarization studies showed that the cathodic current density was higher in absence of air. EIS measurements were carried out for different thicknesses by scanning frequency ranged from 100 kHz to 10 mHz and the perturbing AC amplitude was of 5 mV for different thicknesses of the electrolyte (Fig. 2.10). Effectively, with the

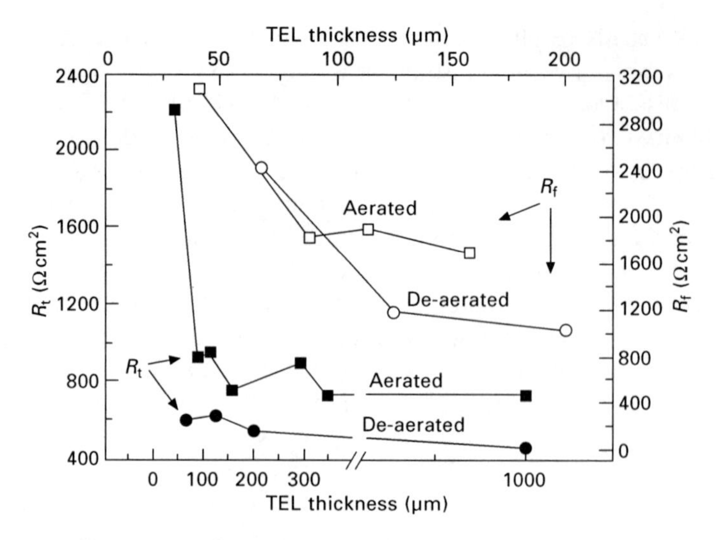

2.10 Charge transfer resistance R_t and film resistance R_f of pure magnesium under thin electrolyte layers with various thicknesses in dilute alkaline solution in the aerated and de-aerated conditions (Zhang *et al.*, 2008).

decrease of TEL thickness, both R_t and R_f of pure magnesium exhibited an increase or increment trend in the aerated and de-aerated conditions. Also, the R_t and R_f values in aerated media were higher than those in de-aerated condition, confirming then the same trend of polarization studies. Considering that the cathodic process of pure magnesium under TEL is dominated by hydrogen evolution reaction, the presence of oxygen had an inhibited effect on the kinetics of this reaction. A better passive state in presence of oxygen is achieved through the inhibition of the anodic reaction by oxygen (Zhang *et al.*, 2008). However, it is very possible that other gases in the air could assist also in the observed inhibited effect.

The electrochemical noise current was measured as the galvanic coupling current between two identical electrodes kept at the same potential. EN data were instantaneously recorded with time for 12 h. Each set of EN records, containing 8196 data points, was recorded with a data sampling interval of 0.25 s. Based on shot noise theory and stochastic theory, the EN measurements were quantitatively analyzed by using the Weibull and Gumbel distribution function, respectively. TEL had two distinctive effects on anodic process of pure magnesium corrosion. On one hand, the pit initiation rate was inhibited. The frequency of corrosion events under TEL is greatly lowered compared with that in the bulk solution. On the other hand, the pit growth probability was increased, meaning that the metastable pit on pure magnesium has a higher probability to become larger pit cavity during shorter time interval than that in bulk solution (Zhang *et al.*, 2008).

2.5.4 Metallurgically influenced corrosion

It has been observed that the corrosion resistance of diecast Mg alloys is a function of the polishing depth of the specimen. Effectively, removing certain surface layers during mechanical polishing expose the surface of the interior skin with less contaminants and somewhat different microstructure that can lead to an improvement in the active and passive behaviors of the specimen. The corrosion resistance of diecast and freely solidified or electromagnetically stirred thixocast AZ91D alloy has been studied using EN technique and EIS in dilute chloride solution saturated with atmospheric oxygen to assess the influence of the microstructure on corrosion kinetics and morphology. At depths between 10 and 50 μm (skin), all specimens showed general non-uniform corrosion with the lowest corrosion resistance. Between 100 and 200 μm (interior skin), the observed corrosion was accompanied by superficial undefined pits due to metastable pitting (Lafront *et al.*, 2008). There is then an advantage to removing the superficial skin since the interior provides the best possibility for better passive performance. Hot- or grit-blasted surfaces often exhibit poor corrosion performance not from induced cold work but from embedded contaminants. Acid pickling to a depth of 0.01 to 0.05 mm can be used to remove reactive contaminants, but re-precipitation of the contaminant should be avoided such as steel shot residues (Shaw, 2003).

The corrosion behavior of the skin of diecast AZ91 Mg alloy has been examined as a function of the thickness of the cast alloy in 3.5% NaCl solution saturated with $Mg(OH)_2$ at room temperature. It was found that the corrosion resistance of cast specimens with relatively more important thicknesses was higher than that of the less thick ones. This was explained in terms of the increasing amount of Al and β phase (Mg_{17} Al_{12}) in their skins (Aghion and Lulu, 2009). It has been also stated that the morphology, the level of porosity and the composition of the passive layer especially in this passive alkaline medium were showing a better corrosion resistance.

A hot-chamber diecast AZ91D thin plate with a die chill skin on its surface was severely corroded in 5 wt% chloride solution (I_{corr} ~ 1600 μA/cm^2), whereas a plate with a die skin layer etched in an HF/H_2SO_4 aqueous solution to remove interdendritic phases had a substantially lower corrosion rate (3 to ~16 μA/cm^2). The die-chill skin was composed of a thin layer of chill zone and a thick layer composed of interdendritic Al-rich α-Mg/$Al_{12}Mg_{17}$ β-phase particle/α-Mg grain composite. The chill zone (4 ± 1 μm in thickness) had fine columnar and equiaxed grains and contained a distribution of submicron Mg–Al–Zn intermetallic particles. The removal of the primary β-phase from the diecast sample surface did not improve the corrosion performance of the specimen (Uan *et al.*, 2008).

2.5.5 Microbiological influenced corrosion

Magnesium undergoes two different corrosion phenomena: microbiologically influenced corrosion (MIC) and rational biodegradation. MIC occurs in the biosphere and even in oxygen-free media where microorganisms are present. The second phenomenon occurs inside the human or animal body where the immune system prevents microorganism colonization. Physiological fluids containing water and high rates of chloride are principally involved in this degradation process.

Recently, magnesium and some of its alloys were investigated as suitable biodegradable materials. Metals which consist of trace elements existing in the human body such as magnesium are promising candidates for temporary implant materials. These implants are needed temporarily to provide mechanical support during the healing process of the injured or pathological tissue. $Mg(OH)_2$ accumulates on the underlying matrix as a corrosion protective layer in water as long as the chloride concentration is not above 30 mmol/L. The chloride content *in vivo* is about 150 mmol/L and so severe pitting corrosion can exist. Galvanic corrosion can be observed in alloys having favorable microstructural cathodic sites for hydrogen evolution and this causes potential local gas cavities *in vivo*. Alongside pure magnesium, Be, Ni and Al-free alloys are recently recommended for use in humans. For biomedical stent *in vivo* development, it seems that rare earth Mg alloys are targeted; however, mischmetal in the alloy should be examined for some undesired element concentrations (Witte *et al.*, 2008).

Simulated corrosion studies for biomaterial use

Two solutions are currently used for biomaterial corrosion studies: 'Hank's' and simulated blood plasma (SBP) solutions. Hank's solution is a balanced salt solution with a pH very close to the blood 7.4. Potentiodynamic curves of the magnesium sample in Hank's and SBP solutions with scan rate of 0.5 mV/s were carried out after 1 h in OCP at 37 °C (Fig. 2.11) (Yang and Zang, 2009). The OCP of extruded Mg 1wt% Mn and 1.0 wt% Zn was –1.21/SHE after 1 hour immersion in SPB and was more positive (noble) than that in Hank's solution (–1.27). Corrosion resistance measurements as predicted from polarization curves after 1 hour immersion in both testing solutions show one order of magnitude higher corrosion resistance R_p in SPB than in Hank's solution, while the corrosion current (I_{corr}) in Hank's solution was nearly double that in SBP. The weight loss studies confirmed the trend of electrochemical measurements in these two solutions with volume/surface area (SV/SA) = 6.7 (Yang and Zhang, 2009) (Fig. 2.12).

The values of pH after 300 h of immersion were sensitive to the ratio of the solution SV/SA. Keeping the surface of the specimen constant ($6 cm^2$), the

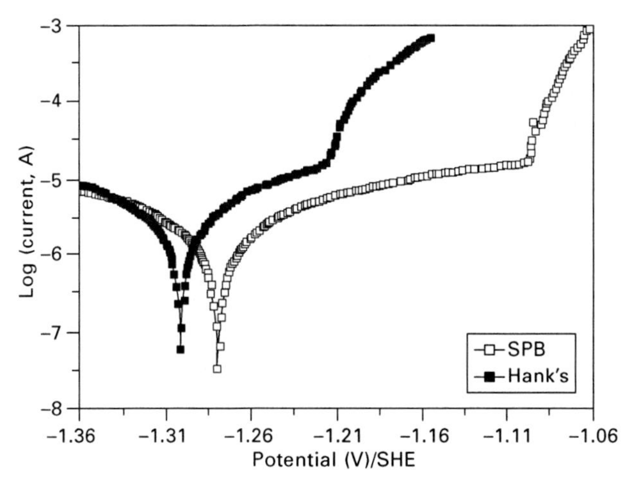

2.11 Potentiodynamic curves of the magnesium extruded specimen ($2\,cm^2$) in 350 ml of Hank's and simulated blood plasma solutions at 37 °C (Yang and Zhang, 2009).

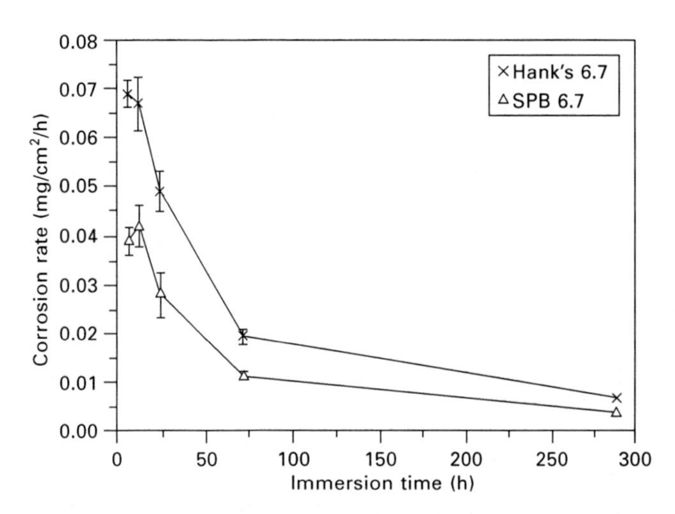

2.12 Average corrosion rates from weight loss measurements of Mg–0.1 Mn and 0.1 wt% Zn alloy in both biomaterial testing solutions during 300 h for the same SV/SA ratio of 6.7 (Yang and Zhang, 2009).

solution volume varied from 4 to 40 and to 400 ml, leading to SV/SA relative values = 0.67, 6.7 and 67, respectively. The pH values after 300 h immersion in Hank's solution were ~9.2 for 67, 10.3 for 6.7 and 10.8 for 0.67 SV/SA. These values show high alkalinization for more vigorous attack as function of progressive relative small solution volumes. Also the shift of pH in SBP was ~ 9 for 6.7 SV/SA less than that of Hank's solution (10.3), indicating less vigorous dissolution reaction for SBP because of the presence of some inhibiting or film-forming ions (Yang and Zhang, 2009). The corrosion rates derived from weight loss measurements confirmed the same trend deduced from polarization curves and pH measurements. Effectively, magnesium specimens showed lower corrosion rate in SBP solution as compared with that in Hank's one (Fig. 2.12). It is interesting to note the prompt decrease in the corrosion rate during the first 72 hours of immersion, followed by a gradual decrease up to 3000 h immersion and that for both solutions.

Mg–Zn alloys as biodegradable materials

It has been reported frequently that Mg alloys containing aluminum (Al) and/or rare earth (RE) elements are not recommended for biosafety and health considerations of the human body. It is desirable then to develop novel degradable Mg alloy without aluminum and rare earth metals. Certain attention has been given to calcium or manganese as candidate alloying elements; however, zinc is one of the effective alloying elements for improving corrosion. Song (2007a) found that the addition of 1–2 wt% Zn into pure Mg led to a significantly reduced biodegradation rate in a simulated body fluid (SBF). Zhang *et al.* (2010) examined the influence of heat treatments on the *in vitro* and the *in vivo* degradation behavior of Mg–6 wt% Zn alloy. A solid solution of this alloy was treated at 350 °C for 2 h followed by quenching in water. The heat-treated alloy was also hot extruded at ~ 250 °C with an extrusion of 8:1.

A kind of SBF was used as a testing solution. The pH value of the SBF was adjusted before measurements to 7.4 and the temperature was maintained at 37 °C. Weight loss methods and electrochemical measurements, including potentiodynamic polarization and EIS were employed. The grain size was refined and a uniform single phase of the alloy was obtained after solid solution and hot working. This led to lower corrosion rate than pure magnesium. The corrosion products on the surface of the Mg–Zn alloy were hydroxyapatite (HA) and other Mg/Ca phosphates in SBF, and they act as protective layer. The tensile strength and elongation achieved of the alloy were ~ 280 MPa and 18.8%, respectively, showing its suitable mechanical properties for implant applications. Mg–Zn rods were implanted into the femoral shaft of rabbits, and the radiographs illustrated that the magnesium alloy could be gradually absorbed *in vivo* at about 2.32 mm/yr degradation rate. The *in vitro* cytotoxicity of Mg–6% Zn alloy was found to be equivalent to the grade 0-1, indicating that it is an implantable material with good biocompatibility (Zhang *et al.*, 2010).

Alkaline treatment for better biodegradation

Mg degradation rate is extremely high at physiological pH 7.4–7.6 where the metal is in the active state especially for agitated conditions. However, the high degradation rate can be effectively inhibited by a suitable surface treatment. Song (2007b) reported that a 30 minute anodizing treatment in 1.6 wt% K_2SiO_3+1 wt% KOH stopped magnesium from biodegrading; no corrosion damage was detected after 30 days of immersion in an SBF. Kannan and Singh (2010) suggested that a pre-treatment of magnesium in 1 M NaOH for two different durations (for 24 h and 48 h) prior to *in vitro* tests can reduce the biodegradation rate. *In vitro* electrochemical tests were carried out on the diecast AZ91 magnesium alloy soaked at body temperature (36.5 °C) for 1 hour in SBF (Hank's solution). A passivation-kind of behavior is observed in the anodic polarization curve of alkali-treated alloys. The passive-like region is extended over about 90–130 mV above E_{corr} before breakdown appears and this was slightly more important for the 48 h alkali-treated surface. The Nyquist plot of impedance measurements of AZ91 Mg alloy, untreated and alkali-treated, soaked in Hank's solution show that the alkaline treatment improved the polarization resistance of the alloy by an order of magnitude in comparison with that of the untreated alloy, very possibly because of the prior formed passive hydroxide film. Also, it has been found that the body temperature significantly decreases the corrosion resistance of the alloy, and that chloride ions alone minimize the corrosion resistance of the alloy with increase in immersion period, whereas the other constituents of SBF such as phosphate, calcium and carbonate enhance the film-forming tendency and hence the corrosion resistance.

2.5.6 Mechanically assisted corrosion

Erosion-corrosion (as impingement or water drop corrosion) is a serious problem since it keeps the active state of magnesium at low pH values or prevents the formation and stability of the passive layer in alkaline solutions. Preventive measurements of erosion–corrosion and fretting fatigue corrosion include better passive surfaces that can be achieved by inhibitors, surface treatments and selected coating to improve wear resistance and wear corrosion of Mg alloys.

Galvanic corrosion of magnesium decreases resistance to corrosion fatigue crack initiation. Seawater has a greater corrosive effect than tap water because of chloride ions and localized corrosion. Corrosion fatigue strengths of magnesium alloys can be as low as 10% of those in air and no fatigue limit could be determined because of accelerated crack initiation and propagation.

2.5.7 Stress corrosion cracking

Mg alloys that contain neither aluminum nor zinc are the most resistant to stress corrosion cracking (SCC). Aluminum content above a threshold level of 0.15 to 2.5% is reported frequently for SCC or EIC observation in aluminum-containing Mg alloys. Mg alloys are very susceptible to SCC through pitting. Pitting is frequently accelerated and initiated through the local active cell in a porous passive film. Second phase particles even as small as 40 nm at the grain boundaries that are cathodic to α-Mg present a deficiency in the passive film and create active galvanic cells that can lead to SCC. For example, slow cooling from the solution-treating temperature, along with certain thermomechanical treatments, can produce $Mg_{17}Al_{12}$ grain-boundary precipitates, which increase the susceptibility of an alloy to intergranular SCC.

2.6 Performance of sacrificial magnesium (Mg) and its alloys

2.6.1 Introduction to sacrificial metal and its alloys

Magnesium is used as a high driving voltage anode; it is universally applicable and is generally little affected by its environment; however, packaging and backfilling techniques are necessary for certain applications. Considering a current use such as steel protection in seawater, the three metals Mg, Al and Zn can be used. However, for massive applications in seawater, such as in the offshore industry, the high driving voltage of magnesium is sometimes required and can be used to great advantage. On the other hand, the pure Mg anode cannot generally be used economically to give a long life because of its high current densities in certain electrolytes and is comparatively expensive if compared with other sacrificial anode metals. A commercial anode, such as 'Ultramag' is a typical high-potential Mg anode that contains: Al (0.01% max), Mn (0.5–1.3%), Cu (0.02% max), Si (0.05% max), Ni (0.001% max) and not more than 0.05% max for other each metal (ASTM B843, 2003) (Farwest Corrosion Control Company, 2009). High-purity magnesium, such as 'Galvomag', is not advantageous generally in seawater anodes.

Extruded Mg anodes can be obtained in many diameters and in lengths up to ~7.6 m long. The anodes are available in either rod form (Galvarod, trademark of the Dow Chemical Co.) alloy which contains Al (2.5–3.5%), Mn (0.20% min), Zn (0.7–1.3%), Si (0.05% max), Cu (0.01% max), Ni (0.001% max), Fe (0.002% max) and other impurities (0.05% each or 0.30 total) or in the high-potential pure metal. Extruded Mg ribbon such as Galvoline (trademark of the Dow Chemical Co.) is a flexible anode material generally used in high-resistivity soils and water and can provide higher current outputs than other cast Mg anodes due to its greater surface area to weight ratio (Farwest Corrosion Control Company, 2009). For underground aggressive soils with low resistivity or for other similar uses, it is preferable to use alloyed magnesium. Alloyed Mg anodes with one or more metals are currently considered. Aluminum is advantageous as less active metal and zinc can lead to better control of the cathodic hydrogen evolution reaction with its high overpotential (Morgan, 1993).

2.6.2 Critical parameters for efficiency

The consumption of the metal will be 7.71 kg per ampere year. Wastage of the anode metal on replacement of the anode stub will increase this consumption by 1.36 or 1.81 kg per ampere year. If magnesium is considered to corrode by a divalent reaction then its theoretical electrochemical equivalent is 2206 ampere hours per kg or ~ 4 kg per ampere year. There is some doubt on this point if magnesium corrodes by univalent reaction only giving ~ 1103 ampere hours per kg or ~ 8 kg per ampere year. Efficiencies of more than 50% are seldom reported though several reliable cases of efficiencies up to 60% tend to detract from the univalent theory. The subsequent oxidation of the univalent salts to the divalent form must also occur, but this reaction would not contribute to the useful anode current (Morgan, 1993). This statement suggests that the magnesium dissolution through the monovalent ion, followed by the oxidation–reduction chemical reaction of hydrogen in the active region, is not the only mechanism of dissolution.

Three factors influence the efficiency of magnesium as an anode: the current density, the composition and the environment.

Current density

Pure magnesium should have a driving potential of ~ 850 mV to protect steel but in practice the metal corroded very rapidly with a very low efficiency. The metal suffers low polarization in the presence of chloride or sulfate ions and produce highly soluble chloride and sulfate salts. These ions are usually artificially introduced into the electrolyte as a backfill when a deficiency is expected, the hydroxide which is preferentially formed because of its low solubility becomes enriched with the backfill anions and itself functions as a backfill. Uniform general corrosion can then be obtained and well-designed inserts help to keep most or all of the anode metal available for sacrificial consumption. In freshwater or electrolytes which contain none of these ions, the hydroxide and carbonate may form, but these do not seriously polarize the anode (Morgan, 1993).

The anode efficiency, that is the useful ampere hours per kg, increases at high current densities; composition has some effect upon this, but generally maximum efficiency is obtained at current densities above ~1.08 A/m^2; Fig. 2.13 shows this relationship. The extrapolation of these curves could suggest a certain corrosion rate even when it is not acting as sacrificial anode; however, a relatively high efficiency can be maintained at low

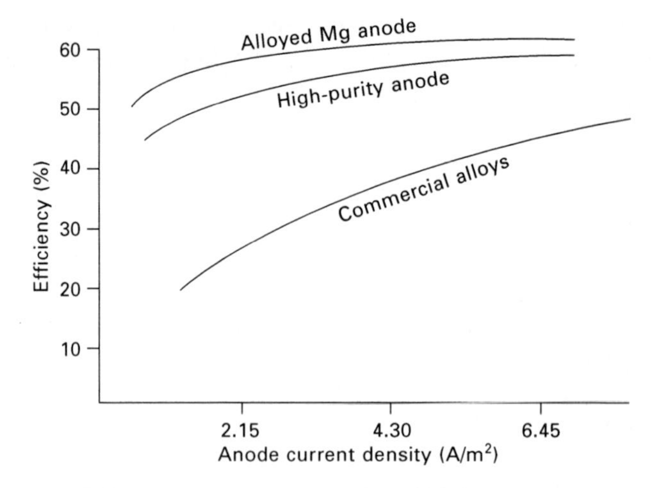

2.13 Efficiency vs. current density of high-purity magnesium as compared with that of currently employed cathodic protection anodes and commercial Mg alloys in saturated $CaSO_4$ solution (Morgan, 1993).

current densities in electrolytes rich in chloride or sulfate ions. An Mg anode corrodes uniformly on macroscopic scale at all current densities; however, the low current density attack is characterized by a considerable amount of pitting. This process has been attributed to a parasitic corrosion, though selective corrosion would give the same effect. At high current densities the anode surface becomes smooth and this is most pronounced in seawater. In solutions devoid of chlorides or sulfates, but containing large amounts of phosphate, carbonate or hydroxyl ions, the working potential of the anode is considerably reduced, and in some environments the anode driving voltage may be halved at a current density of ~ 5.4 A/m^2 (Morgan, 1993).

Anode composition and microstructure

The 6% aluminum and 3% zinc alloy is usually one of the most cost-effective magnesium alloys. The early experience with magnesium and its alloys showed that although the pure metal had a higher driving voltage, some 150 mV greater than its alloys, certain alloys, particularly the 6% Al–3% Zn, give much greater efficiency in a variety of electrolytes and the composition of the employed Mg alloy is not as critical as in other applications. It has been suggested that Mg alloys suffer from 'parasitic corrosion' (selective or better expressed as localized corrosion) at much lower corrosion rates than commercial magnesium. It seemed probable that the more noble metals, particularly those of low hydrogen overvoltage, found in commercial magnesium, would cause the localized type of attack mentioned above and copper, nickel and iron are impurities of this class found in commercial magnesium. Iron is one of the major causes of parasitic corrosion but it was known that manganese had a very powerful scavenger effect upon it.

Nickel can be tolerated up to 0.003% before it has any effect upon anode efficiency; however, this can depend upon the total amount of all the minor impurities. The noble metals with a high hydrogen overvoltage, such as lead, tin, cadmium and zinc seem to have little effect upon the anode performance (Morgan, 1993). Generally, the Mg alloy can suffer severely from rapid corrosion rates in the presence of surface contamination or high impurity content of less active metals and more noble secondary particles to Mg matrix. For example, in sodium chloride solutions (3–6%), the OCPs of mild steel, nickel and copper are much more positive (noble): ~ –0.78, –0.14 to –0.22, respectively while that of Mg corrosion potential is more negative than ~ –1.67 V/SHE (Hawke and Olsen 1993; Hawke *et al.*, 1988; Song and Atrens, 1999).

The effects of alloying elements on electrochemical properties of magnesium-based sacrificial anodes were evaluated in a saturated calcium sulfate–magnesium hydroxide solution that simulates the frequently employed backfill. Not all anodes underwent passivation but demonstrated only active behavior. Corrosion morphology was changed from localized to uniform attack by alloying. The efficiency of Mg–Al anodes was improved up to ~6% Al addition. The addition of zinc increased the efficiency of Mg–Al–Zn anodes compared with that of Mg–Al anodes, but the reversal of this behavior happened as the zinc content exceeded ~ 3%. The increase in the efficiency of Mg–Al and Mg–Al–Zn anodes was accompanied by a decrease in the driving potential that might have resulted from a somewhat resistive film on the surface, which hindered the transport of ions. The increase in corrosion resistance generally improved anode efficiency (Kim and Koo, 2000).

In commercial anodes the iron content of the alloy is restricted to 0.03% and the manganese is added to a minimum of 0.15%. In the presence of Mn, the iron content is reduced by the settling of iron from the melt, and the iron remaining in the alloy is surrounded by manganese, preventing then its accelerating role as cathodic site (Morgan, 1993). Manganese has a beneficial effect upon copper impurities somewhat similar to that found with iron and, with the manganese content required to counter the effect of iron, copper can be tolerated up to 0.1% in seawater and to a 0.03% elsewhere. Addition of manganese to magnesium anodes yielded increased driving potential and efficiency (Kim and Koo, 2000). The other impurities, tin, silicon and lead, show little detrimental effect at their normal levels of impurity; silicon should be restricted to 0.1%, lead to 0.04% and tin to less than 0.005% (Morgan, 1993).

Two series of anode alloys are generally considered. The first is the high-purity anode for use in soils with high resistivity (wt% 0.02 Cu, 0.003

Fe, 0.002 Ni, 0.1 Si as maximum values and 0.15 minimum Mn), giving, in these conditions, slightly lower efficiencies than the 6% Al–3% Zn alloy. The second is the lower-purity anode for use in seawater (wt% 0.03 Cu, 0.03 Fe as maximum values and 0.1 minimum Mn). The production of a high-purity Mg anode, with about 1% of manganese in place of the aluminum and zinc, has been examined. The anode has a higher driving voltage, about 200 mV more than the 6% Al–3% Zn alloy, and so gives about 25% higher driving voltage to polarized steel. The efficiency of the material is slightly lower than the alloyed Mg anode but as the anode is primarily intended for use in high-resistivity soils the increased current density of the anode will enable it to work at efficiencies comparable with the best of the alloys. Figure 2.13 shows the trend of efficiencies generally reported for three categories of sacrificial anodes in saturated calcium sulfate solution. The success of the high-voltage alloy lies in its microstructure and the addition of considerable manganese to the metal that influence the critical current impurities (Fe and Cu) (Morgan, 1993).

Addition of calcium to the Mg–Mn sacrificial anodes enhanced the anode efficiency in the simulated backfill electrolyte by promoting the uniformly distributed corrosion along the grain boundaries and increased the driving potential by intrinsic electronegative potential of α-Mg. The reason for the uniform intergranular corrosion is that Mg_2Ca precipitate (cathode) at grain boundaries is galvanically coupled to α-Mg matrix (anode). Larger anodic area limits the localized nature of the intergranular attack (Kim and Kim, 2001).

Environment

Mg anodes operate at high corrosion rates in seawater because of the low resistivity, high chloride content and the ease with which the corrosion product is washed off by water movement. In soils which contain sulfate ions, most agricultural land does in the form of gypsum, the anode operates well; good drainage into the anode is necessary and this is usually obtained automatically where the anode is buried below the water table. In dry soils and those denuded of sulfates and chlorides the anode tends to polarize and as these soils will display a high resistivity, anode efficiency is reduced by the low current density achieved. Similar conditions exist in freshwater though polarization is reduced by sedimentation of the corrosion product (Morgan, 1993).

The magnesium anode can be obtained as a bare or packaged anode. If packaged, the anode is delivered in a prepared backfill consisting currently of 75% gypsum, 20% bentonite and 5% sodium sulfate contained in a cloth bag (Farwest Corrosion Control Company, 2009). Where the soil conditions are poor it is customary to surround magnesium anodes with a mixture of chemicals or backfill. The chemicals used are gypsum, sodium sulfate and common salt, the last, having a detrimental effect on plants, is used sparingly, but gypsum is common and cheap, so mixtures of this with a smaller amount of sodium sulfate are popular. There are two possible methods of applying the backfill: the chemicals and bentonite may surround the anode in a porous cotton bag and the whole be lowered into ground as a unit, or the chemicals may be mixed on site with local soil or imported clay and poured as a slurry into the anode hole around the magnesium (Morgan, 1993).

2.7 Mechanism of corrosion of sacrificial anodes

2.7.1 Potential difference

In the presence of a good electrolyte, as little as 15 mV difference in corrosion potential of the two metals can have an effect, and if the difference is 30 mV or greater the anodic material will definitely corrode sacrificially to protect the contacting cathodic metal at least partially (Ghali, 2010). When exposed to a given environment, the potential of a metallic material is determined by many factors, such as temperature, liquid flow rate and the level of aeration. Although the potential difference is the origin and the cause of the existence of the galvanic cell, it gives no information about the kinetics of galvanic corrosion. Kinetics depends on the current flowing between the two metals in the couple and this is function of polarization phenomena (mainly electron transfer, concentration of appropriate ionic species and conductivity of the electrolyte) (Roberge, 2006). The severity of galvanic sacrificial action is determined by the galvanic current and can be expressed as follows:

$$I = (E_k - E_a)/(R_m + R_e) \qquad 2.18$$

where E_k and E_a are the polarized measured potentials of the cathode and anode, respectively, and R_m and R_e are the resistances of the metal-to-metal contact and the electrolyte portions of the circuit, respectively. In many practical applications, R_m is negligibly small and is a function of electrical and mechanical factors and R_e (the electrolyte resistance) then becomes the controlling factor in the circuit resistance. The conductivity and composition of the medium at the interface of both metals are controlling factors in the rate of sacrificial corrosion.

2.7.2 Anode/cathode surface area ratio of sacrificial anodes

The electric resistance between two conductors frequently becomes low because of oxide formation accompanied by diffusion control and the polarization of the two electrodes, etc. (Ghali, 2000; Mears, 1976). If the surface area of the anode is very low with respect to the cathodic surface, the rate of general corrosion will be very high at the limited anodic surfaces, and this can lead to severe localized corrosion and perforation. The larger the cathode surface as compared with that of the anode, the more oxygen reduction or other cathodic reaction can occur, and hence the greater the galvanic current. Under static or slow-flow conditions, where the galvanic-corrosion current is often dependent on the rate of diffusion of dissolved oxygen to the cathode, the total amount of galvanic corrosion is independent of the size of the anode and proportional to the area of the cathodic–metal surface. Thus for a sacrificial anode with a constant cathodic area, the total amount of corrosion of the anode is constant, but the corrosion per unit area of the anode increases as its area is decreased, due frequently to blockage by excessive corrosion products. Seawater, with a low resistivity (a few ohms/ cm^2), is particularly prone to showing pitting and crevice types of corrosion when the solution is stagnant (Roberge, 2006).

2.7.3 Sacrificial magnesium and cathodic corrosion of aluminum alloys

Magnesium and its alloys are definitely anodic to the Al alloys and, thus, contact with aluminum increases the corrosion rate of magnesium. For example, in sodium chloride solutions (3–6%), the potential of Mg alloys is ~ –1.67 V/SHE while that of Al–12%Si and pure aluminum are –0.83 to –0.85, respectively. However, such contact is also likely to be harmful to aluminum, since magnesium may send sufficient current to the aluminum to cause cathodic corrosion in alkaline medium. Aluminum oxide is amphoteric and so it is soluble in acid as well as in alkaline solutions. The standard reduction potentials of these two half-reduction reactions are (–1.66 V/SHE) and (–2.35 V/SHE), respectively. Alkaline reaction of the possible existence of aluminum phase in sacrificial Mg anodes is:

$$H_2AlO_3^- + H_2O + 3e^- \rightarrow Al + 4\,OH^- \; (-\,2.35\,V/SHE) \qquad 2.19$$

Cathodic corrosion of aluminum is much less severe in seawater than in NaCl solution since the buffering effect of Mg ions reduces the equilibrium pH from 10.5 to about 8.8 (Arsenault and Ghali, 2006). Exposure of iron aluminum intermetallic particles (e.g. $FeAl_3$) engages in separate galvanic activity with magnesium. For this reason AA 5056 rivets (aluminum base alloy) have been extensively employed in assembling Mg alloy structures since its tolerance is up to 1000 ppm of Fe content, while for other aluminum alloys a maximum of 200 ppm Fe has been already stated (Ghali, 2000; Mears, 1976).

2.8 Examples of actual and possible uses

2.8.1 Pipe line protection

The usual form of the anode is a prefabricated bracelet which is either made of segments of anode attached to a pair of steel split rings, or the anode itself is cast as a hallow semi-cylinder and the two halves are joined together. Two typical types of bracelet are shown in Fig. 2.14. The bracelet is usually made to conform to the weight coating on the line but occasionally the weight coating is insufficiently thick and then the anode is tapered. By doing this it will pass over the rollers and can be laid with the pipe. To ensure continuity it is usual to attach a jumper lead from the anode bracelet to the pipe. This is sometimes done by thermite welding and sometimes by other techniques, including welding to pads or lugs fabricated in the pipe shop (Morgan, 1993).

2.8.2 Water heaters

Pure magnesium is used in private and apartment steel reservoirs of hot water. Their service life duration for domestic water reservoirs can exceed 10 years but checking is necessary since accelerated corrosion rates are functions of water composition. Domestic tanks are of two kinds; either they

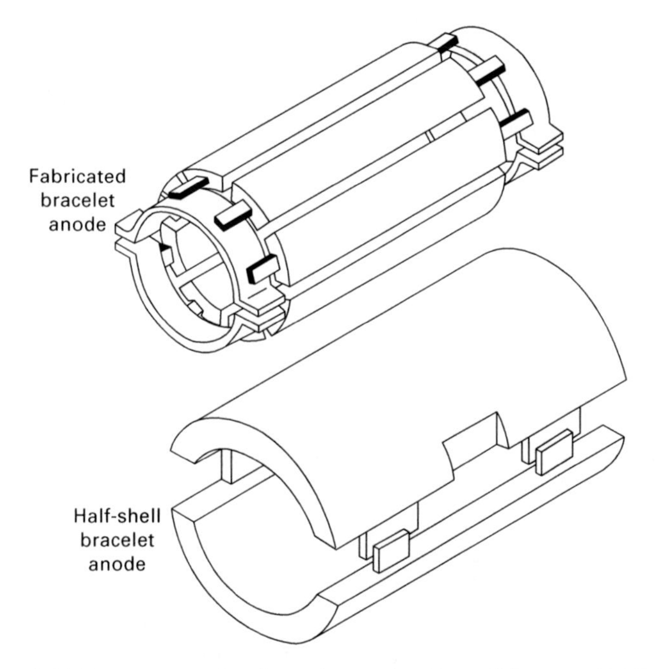

2.14 Bracelet anodes for subsea pipe lines (Morgan, 1993).

are rectangular of 94.6 to 151.4 liter capacity for the average household, or they are cylindrical and somewhat larger (Morgan, 1993). The rectangular type can be protected by an anode mounted on to the inspection plate or bolted to the tapped holes in the tank. Larger cylinder tanks are best protected by a long magnesium rod which can have a threaded insert in a steel core (Fig. 2.15) or can be attached directly to a pipe fitting, often the inlet pipe being modified to hold the anode.

Mg anodes can deliver sufficient current density 21.5 to 32.28 mA/m² to protect the inner wall of a galvanized steel hot water tank. It is reported that galvanized hot water tanks either corrode within few years or, if they last longer than this, remain sound for 15 to 20 years. This is supported by the fact that scale formation has a preservative or corrosion preventive effect of galvanized steel in hard water. Soft water can prevent the formation of this scale due to localized galvanic corrosion that can lead to perforation. It is suggested that if a tank is cathodically protected for the first year or

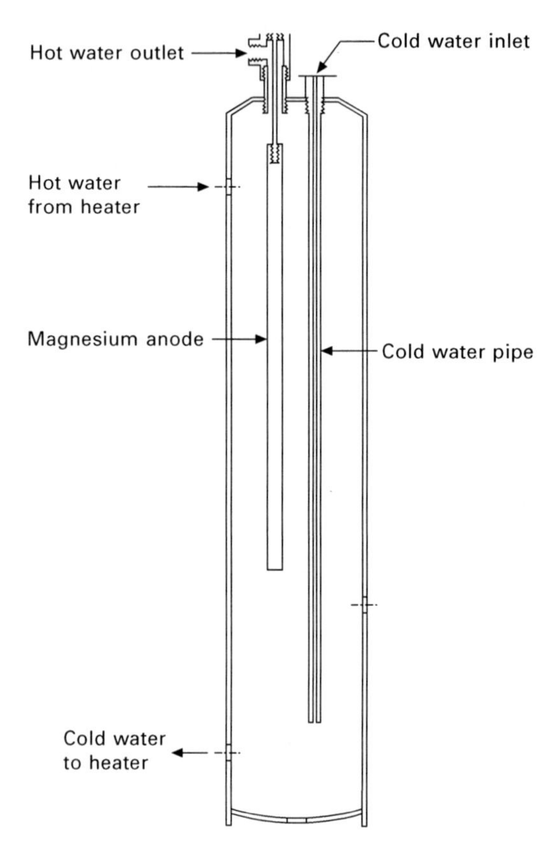

2.15 Cathodically protected hot water tank with Mg anode (Uhlig and Revie, 1985).

so to form a deposit of chalk upon its surface, then it will last for a further 15 to 20 years even though the anode is no longer giving current (Morgan, 1993).

The most important factor in whether a water heater fails is the condition of its sacrificial anode. Generally, a 6 year warranty residential tank will have one, while a 12 year warranty tank will have two or an extra-large primary anode. Commercial tanks have from one to five. Special Al/Zn sacrificial anodes or powered anodes may be used to resolve odor problems caused by bacteria in some water. Magnesium is frequently preferred to aluminum for several reasons principally because aluminum produces less driving current between anode and cathode and shows less protection especially in soft water. Also, aluminum produces about a thousand times its original volume in corrosion by-product, most of which falls into the bottom of the tank as a sort of jelly or chunk, and adds to sediment buildup there, or can occasionally float out the hot water port (http://www.waterheaterrescue, 1995–2009).

2.8.3 Cathodic protection of steel in concrete using Mg alloy anode

Corrosion of steel (0.01 m diameter and 1 m long) embedded in concrete structures and bridges can be prevented by cathodic protection, employing impressed current system. Use of sacrificial anode system for the above purpose is very limited. An Mg alloy anode containing 0.184% manganese, designed for 3 years' life, was installed at the center of reinforced concrete slab, containing 3.5% sodium chloride with respect to the weight of the cement. The shift in potential towards less negative values and the decrease in chloride content with time at any distance from anode confirmed the protection offered to the embedded steel. The potential of the embedded steel shifted from more negative values to less negative plateau, at all distances from the anode. The current flowing in the concrete decreased with increase in time (42 months) (Parthiban *et al.*, 2008).

2.9 Evaluation of the sacrificial behavior

2.9.1 General considerations

The effect of applied current, testing time and microstructure on the electrochemical properties of magnesium-based sacrificial anode in potable water was evaluated by Andrei *et al.* (2003). The Al–Zn Mg alloy AZ63 alloy was utilized as the Mg sacrificial anode for use in potable water considered as a high-resistivity electrolyte. The anodes were tested in order to evaluate the main anode properties such as efficiency, current capacity and charge loss (by hydrogen reaction). The results showed relatively low efficiency and current capacity, especially at low current densities and short testing time. The processes that contribute to the anode wastage are: (1) hydrogen evolution from micro-cathodic sites on the alloy surface, termed local cell action (LCA) and (2) mechanical loss of metal, known as the chunk effect (CE). Each phenomenon was evaluated with its equivalent delivered current.

The corrosion process of a sacrificial Mg alloy is evaluated in the laboratory in either 3% NaCl solution or a solution containing 5 g $CaSO_4$. $2H_2O$ and 0.1 g Mg $(OH)_2$/L), as recommended after the ASTM G97 standard. Beside the ASTM procedure, there is also the Mexican test method (NMX-K-109-1977, 'Magnesium anodes used in cathodic protection') that considers a test environment made of artificial seawater. Both standards consider galvanostatic tests, in which a known direct current is passed through test cells connected in series in order to determine efficiency of sacrificial anode materials (Guadarrama-Mu–oz *et al.*, 2006).

2.9.2 Polarization and impedance methods of investigation

Polarization curves and EIS were used by Guadarrama-Mu–oz *et al.* (2006) to evaluate three sacrificial anodes. The quantity of important elements of the three specimens of Mg alloys M1, M2 and M3 is given respectively in wt%: Al (<0.01, 0.01, 7.2); Cu (0.001, 0.001, 0.13); Fe (0.01, 0.002, 0.006); Mn (0.07, 0.75, 0.18); Ni (0.001, 0.001, 0.004); Zn (0.003, 0.01, 3.92). The magnesium content of these three anodes was in wt% (99.1, 94.94, 85.94) respectively. Polarization curves and EIS were used (Guadarrama-Mu–oz *et al.*, 2006). The weight loss during the test period can be compared to theoretical weight loss calculated based on coulometer measurements for thorough comparison or as a reference.

Potentiodynamic polarization tests

Polarization curves were carried out, in each testing solution, after 6 h of exposure time. This period of time showed near steady-state conditions of the E_{oc} ($\Delta E \leq 10$ mV for 60 min) are reached. All polarization curves were recorded potentiodynamically with a sweep rate of 10 mV/min. A potential range of ± 300 mV, referred to as E_{oc}, was selected.

The 3% NaCl solution is more aggressive to the Mg alloy specimens than the ASTM solution. It can be stated also that, during all the exposure time, M3 alloy shows the more noble values of E_{oc}; M1 specimen shows the more active values of E_{oc}; while M2 gives values of E_{oc} between those of M1 and M3 values. The measured data indicate that the M1 sample is the only one that fulfills the value of open circuit potential as stated in the

NMX standard. Figure 2.16 shows the obtained polarization curves in the simulated backfill solution. It is clear that the slope of each polarization curve is different for each specimen in each environment. This difference can be associated with the different corrosion rates as related to the chemical composition of the magnesium alloys. The polarization curves show that the pure Mg anode (M1) is more active by ~36 mV if compared with that of M3. The calculated corrosion rates of pure Mg M1 and M2, M3 alloys (i_{corr} in mA/cm^2) are 0.0079, 0.0072 and 0.046, showing the accelerating influence of impurities (alloy M3) and the beneficial content of manganese combined with fewer impurities (M2).

EIS measurements

The Nyquist representation of the impedance shows two clear semicircles that can be associated to two time constants (Fig. 2.17), therefore to two

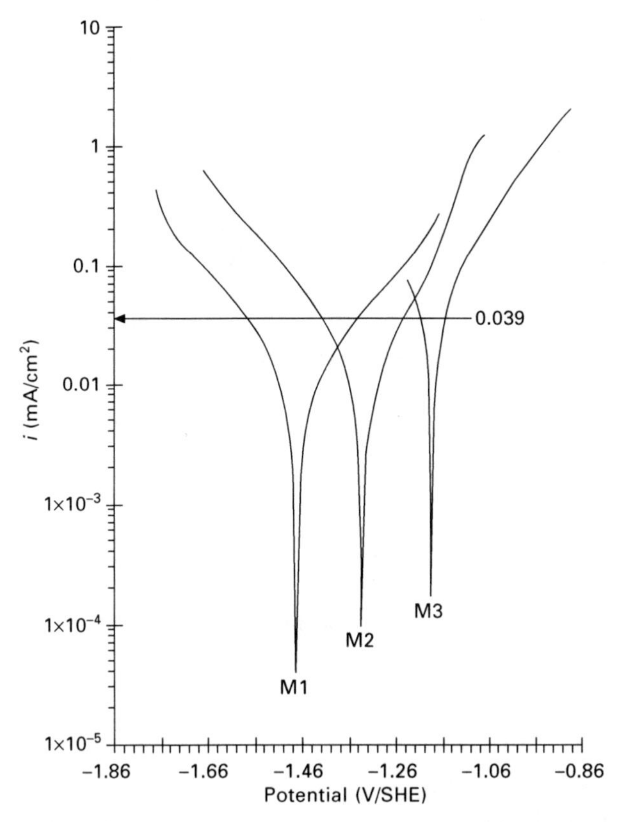

2.16 Polarization curves for M1, M2 and M3 Mg alloy specimens; $CaSO_4 \cdot 2H_2O$–$Mg(OH)_2$ solution (ASTM G97-97), after 2.5 h of immersion time (Guadarrama-Muñoz *et al.*, 2006).

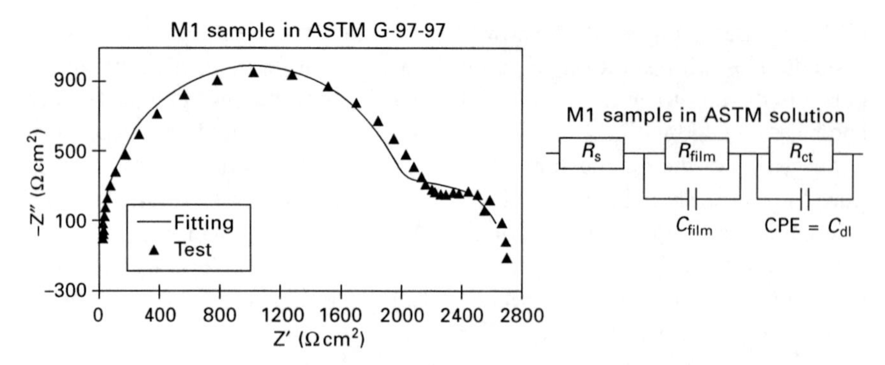

2.17 EIS spectra, Nyquist representation, at 6 h immersion of the sacrificial alloy M1; with the corresponding equivalent circuit (ASTM solution) (CPE = constant phase element) (Guadarrama-Muñoz *et al.*, 2006).

capacitors. This can be described by a simple electrical equivalent circuit made of: one resistor, associated to the solution resistance (R_s), in series with two parallel resistor–capacitor (*R–C*) circuits, both connected in series. One circuit representing the charge transfer process taking place on the surface of the alloy R_{ct}–C_{dl} and a second circuit associated to the film of corrosion products covering the surface of the electrode R_{film}–C_{film}.

It was found that, when an Mg alloy, intended to work as a sacrificial anode, is polarized at a constant anodic potential, near the OCP (E_{oc}), the dissolution process can be described by an electrical equivalent circuit similar to the one described above (Guadarrama-Mu–oz *et al.*, 2006).

When an Mg alloy does not fulfill the chemical composition specified for a sacrificial magnesium anode, features as inductive loops at lower frequencies appear in the Nyquist representation of the measured impedance. As the magnesium alloy is polarized further away from its E_{oc}, in the anodic direction, the Nyquist representation of the impedance exhibits inductive loop behavior (Fig. 2.18). This fact leads to the consideration of an inductor component in the corresponding electrical equivalent circuit. This inductive loop can be associated with the adsorption and desorption phenomena occurring on the surface of the sample and leading to the process of formation of the corrosion product layer on the surface of the electrode (Guadarrama-Mu–oz *et al.*, 2006).

2.10 Future trends

Recent progress in science and technology is leading to more corrosion-resistant alloys and is promising to discover the real passive Mg alloys. Original new creep-resistant alloys for structural purposes and new appropriate surface preparation, pre-treatments and coatings will open the door to this metal for

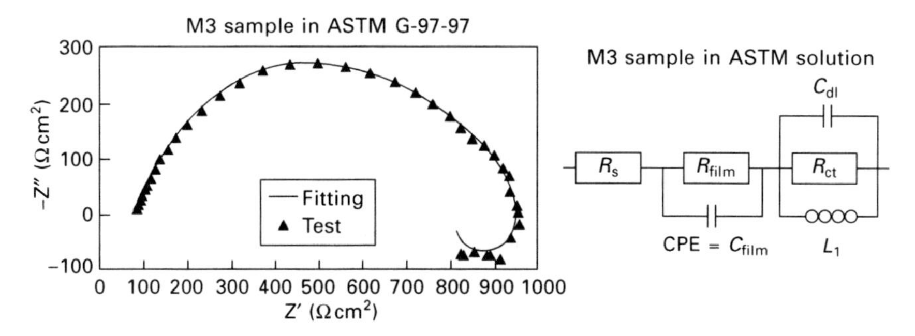

2.18 EIS spectra, Nyquist representation, at 6 h immersion; with the corresponding equivalent circuit of the sacrificial alloy M3 (ASTM solution) (Guadarrama-Muñoz *et al.*, 2006).

innovative applications. The use of magnesium and its alloys as sacrificial anode for cathodic protection is gaining importance. Magnesium or its alloys can be employed as biodegradable material for human body implants for example but they need better corrosion control for lower corrosion rates. The mechanism of dissolution needs to be better understood. Corrosion fatigue, fretting corrosion, fretting corrosion fatigue and corrosion-erosion need to be further explored for the metal and its alloys in considering the active and passive states of the material. Considering the importance of MMCs of magnesium for certain applications, their microstructural effects on galvanic corrosion can be investigated with innovative methods of surface treatment and coating, accompanied by appropriate electrochemical techniques of investigation.

2.11 Acknowledgements

The author appreciates the effort of Carl Moniz, final year student in chemical engineering program at Laval University, for his help throughout the progress of this chapter, in the bibliographic documentation study, and originality in the conception of figures.

2.12 References and further reading

Aghion E and Lulu N (2009), 'The effect of the skin characteristics of die cast AZ91 magnesium', *Journal of Materials Science*, **44**, 4279–4285.

Amira S, Lafront A-M, Dubé D, Tremblay R and Ghali E (2007), 'Corrosion behaviour of die-cast and thixocast AXJ530 magnesium alloy in chloride medium', *Advanced Engineering Materials*, **9**(11), 973–980.

Andrei M, di Gabriele F, Bonora P L and Scantlebury D (2003), 'Corrosion behavior of magnesium sacrificial anodes in tap water', *Materials and Corrosion*, **54**, 5–11.

Ardelean H, Ives M B, Fiaud C and Marcus P (1999), 'Electrochemical study of the effects of a chemical cerium nitrate treatment on the corrosion behaviour of magnesium', in *Environmental Degradation of Materials and Corrosion Control in Metals*, Elboujdaini M and Ghali E ed., Montreal, Canada, METSOC, 38th Annual Meeting of Metallurgists of CIM, 90.

Arsenault B and Ghali E (2006), 'Prevention of environmentally assisted cracking of structural aluminum alloys by Al and Al–5Mg thermal sprayed coatings using different surface preparation techniques', in *Emerging Materials, Processes, and Repair Techniques*, Proceedings of the Aerospace Symposium, 45th Conference of Metallurgists (METSOC), Montreal, No IMI 2006–112595-G, Montreal, Canada, 61–74.

ASM International Handbook (1995), *Handbook of Corrosion Data*. 2nd edition, Materials Park, OH, ASM International, 24–26.

ASTM B296 (1994), *Annual Book of ASTM Standards, Vol. 02.02, Aluminum and magnesium Alloys*, Philadelphia, PA, ASTM, 288–289.

ASTM B843 (2003), *Annual Book of ASTM Standards, Vol. 02.02, Aluminum and magnesium Alloys*, Philadelphia, PA, ASTM, 600–601.

Atrens A and Dietzel W (2007), 'The negative difference effect and unipositive Mg^{+}', *Advanced Engineering Materials*, **9**, 292–297.

Bai A and Chen Z-J (2009), 'Effect of electrolyte additives on anti-corrosion ability of micro-arc oxide coatings formed on magnesium alloy AZ91D', *Surface & Coatings Technology*, **203**, 1956–1963.

Bakkar A and Neubert V (2009), 'Corrosion behaviour of carbon fibres/magnesium metal matrix composite and electrochemical response of its constituents', *Electrochimica Acta*, **54**(5), 1597–1606.

Baloun C H (1987), *ASM Metals Handbook,* 9th edition Materials Park, OH, ASM International, Vol. 13, 207–208.

Baril G and Pébère N (1999), 'Investigation of the corrosion of pure magnesium in aerated sodium sulfate solutions', in *Environmental Degradation of Materials and Corrosion Control in Metals*, Elboujdaini M and Ghali E, ed, Montreal, Canada, METSOC, 38th Annual Meeting of metallurgists of CIM, 61.

Baril G and Pébère N (2001), 'The corrosion of pure magnesium in aerated and dearated sodium sulphate solutions', *Corrosion Science*, **43**, 471–484.

Berger C, Eppel K, Ellermeier J, Troβmann T, Dilthey U, Masny H and Woeste K (2006), *7th International Conference on Magnesium Alloys and their Applications*, Weinheim, Germany, Wiley-VCH Gmb H & Co. KGaA, 734–742.

Bonora P L, Andrei M, Eliezer A and Gutman E M (2002), 'Corrosion behaviour of stressed magnesium alloys', *Corrosion Science*, **44**, 729–749.

Chen J, Wang J, Han E, Dong J and Ke W (2007), 'AC impedance spectroscopy study of the corrosion behaviour of an AZ91 magnesium alloy in 0.1M sodium sulphate solution', *Electrochimica Acta*, **52**, 3299–3309.

Czerwinski F (2003), 'The oxidation of magnesium alloys in solid and semisolid state', in *Magnesium Technology* 2003 Proceedings Kaplan H I, ed, (The Minerals, Metals & Materials Society), Warrendale, PA, 39–42.

Danielson M J (2001), 'Magnesium Alloys', in *Environmental Effects on Engineered Materials*, Jones R H, ed, New York, NY, Marcel Dekker Inc., 253–274.

Delahay P, Pourbaix M and Russelberghe V (1951), *Diagramme d'équilibres Potentiel-pH de quelques éléments*, Berne, C. R. 3e réunion du CITCE.

Farwest corrosion control company (2009), www.farwestcorroson.com, Cardena, CA.

Ferrando W A (1989), 'Review of corrosion and corrosion control of magnesium alloys and composites', *Journal Engineering Materials*, **11**, 299–313.

Froats A F, Aune T Kr, Hawke D, Unsworth W and Hillis J (1987), 'Corrosion of magnesium and magnesium alloys' in *Corrosion*, Vol. 13, Korb L J, Olson D L and Davis J R, ed., Materials Park, OH, ASM International, 740–754.

Froes F H, Kim Y and Krishnamurthy S (1989), 'Rapid solidification of lightweight metal alloys', *Materials Science and Engineering*, **117**, 19–32.

Ghali E (2000), Chapter 40 'Aluminum and aluminum alloys' and Chapter 44 'Magnesium and magnesium alloys', *Uhlig's Corrosion Handbook*, Second edition, R Winston Revie, ed, New York, NY, John Wiley & Sons Inc, 677–715; 793–830, respectively.

Ghali E (2006), 'Some aspects of corrosion resistance of magnesium alloys', in *M.O. International Symposium on Magnesium Technology in the Global Age: Magnesium in the Global Age, Pekguleryuz and L.W.F. Mackenzie, ed*, Montreal, Canada, Canadian Institute of Mining, Metallurgy and Petroleum, 271–293.

Ghali E (2010), 'Active and passive behaviors of aluminum and magnesium and their alloys', in *Corrosion Resistance of Aluminum and Magnesium Alloys, Understanding, Performance and Testing*, New York, John Wiley & Sons Inc., chapter 3 (2010, 78–120).

Ghali E, Dietzel W and Dietzel K-U (2004), 'Testing of general and localized corrosion of magnesium alloys: a critical review', *Journal of Materials Engineering and Performance*, **13**(5), 517–529.

Guadarrama-Mu–oz F, Mendoza-Flores J, Duran-Romero R and Genesca J (2006), 'Electrochemical study on magnesium anodes in NaCl and $CaSO_4$–$Mg(OH)_2$ aqueous solution', *Electrochimica Acta*, **51**, 1820–1830.

Hall I W (1987), 'Corrosion of carbon/magnesium metal matrix composites', *Scripta Metallurgica*, **21**, 1717–1721.

Hanawalt J D, Nelson C E and Peloubet J A (1942), 'Corrosion studies of magnesium and its alloys', *Transactions of American Society of Mining and Metallurgical Engineering*, **147**, 273–299.

Hawke D and Olsen A (1993), 'Corrosion properties of new magnesium alloys', *Proceedings of the SAE*, 79–84.

Hawke D L, Hillis J E and Unsworth W (1988), 'Technical Committee Report', International Magnesium Association, p. 8.

Hawke D L, Hillis J, Pekguleryuz M and Nakatsugawa I (1999), 'Corrosion behaviour', in *Magnesium and Magnesium Alloys*, Avedesian M M and Baker H, ed, Materials Park, OH, ASM International, 194–210.

Hihara L H and Kondepudi P K (1993), 'The galvanic corrosion of SiC monofilament/ZE41 Mg metal–matrix composite in 0.5M $NaNO_3$', *Corrosion Science*, **34**, 1761–1772.

Hihara L H and Kondepudi P K (1994), 'Galvanic corrosion between SiC monofilament and magnesium in NaCl, Na_2SO_4 and $NaNO_3$ solutions for application to metal–matrix composites', *Corrosion Science*, **36**, 1585–1595.

Hillis J E (1995), *Corrosion Testing and Standards: Application and Interpretation*, Baboian R, ed, ASTM Manual Series: MNL 20, Materials Park, OH, ASM International, 438–446.

http://www.waterheaterrescue.com/pages/WHRpages/English/Longevity/water-heater-anodes.html, Copyright 1995–2009 by Randy Schuyler, Marina, CA.

Hur B Y and Kim K W (1998), 'A new method for evaluation of pitting corrosion resistance for magnesium alloys', *Corrosion Reviews*, **16**, 85–94.

Inoue H, Sugahara K, Yamamoto A and Tsubakino H (2002), 'Corrosion rate of magnesium and its alloys in buffered chloride solutions', *Corrosion Science*, **44**, 603–610.

Jin S, Amira S and Ghali E (2007), 'Electrochemical impedance spectroscopy evaluation of the corrosion behaviour of die cast and tixocast AXJ530 Mg alloy in chloride solution', *Advanced Engineering Materials*, **9**, 75–83.

Kaesche H (1974), *Pitting Corrosion of Aluminum and Intergranular Corrosion of Aluminum Alloys*, in localized corrosion, proceedings Staehle R, Brown B F, Kruger J and Agrawal A, ed, Houston, Texas, NACE International, 516–525.

Kannan M B and Singh R (2010), 'A mechanistic study of *in vitro* degradation of magnesium alloy using electrochemical techniques', Journal of Biomedical Materials Research Part A, published online: 14 Sep 2009 (http:/www3.interscience.wiley.com). **93**A, 3, 1050–1055.

Kim J-G and Kim Y M (2001), 'Advanced Mg–Mn–Ca sacrificial anode materials for cathodic protection', *Materials and Corrosion*, **52**, 137–139.

Kim J-G and Koo S-J (2000), *Effect of Alloying Elements on Electrochemical Properties of Magnesium-based Sacrificial Anodes* Corrosion Science Section, NACE International, Houston, Texas, 380–388.

Lafront A-M, Dubé D, Tremblay R, Ghali E, Blawert C and Dietzel W (2008), 'Corrosion resistance of the skin and bulk of die cast and thixocast AZ91D alloy in Cl^- solution using electrochemical techniques', *Canadian Metallurgical Quarterly*, **47**, 459–468.

Li J-Z, Huang J-G, Tian Y-W and Liu C-S (2009), 'Corrosion action and passivation mechanism of magnesium alloy in fluoride solution', *Transactions of Nonferrous Metals Society of China*, **19**, 50–54.

Lindström R, Johansson L-G, Thompson G E, Skeldon P and Svensson J-E. (2004), 'Corrosion of magnesium in humid air', *Corrosion Science*, **46**, 1141–1158.

Liu L J and Schlesinger M (2009), 'Corrosion of magnesium and its alloys', *Corrosion Science*, **51**, 1733–1737.

Lunder O *et al.* (1990), 'Filiform corrosion of a magnesium alloy', Paper presented at the 11th Annual Corrosion Congress, Florence, Italy, 5.255–5.262.

Makar G L and Kruger J (1993), 'Corrosion of magnesium', *Journal of International Materials*, **38**, 3, 138–153.

McIntyre N S and Chen C (1998), 'Role of impurities on Mg surfaces under ambient exposure conditions', *Corrosion Science*, **40**, 1697–1709.

Mears R B (1976), 'Aluminum and aluminum alloys', in *Corrosion Handbook*, Uhlig H H, ed, Pennington, NJ, The Electrochemical Society pp 39–55.

Mitrovic-Scepanovic V and Brigham R J (1990), 'A fundamental corrosion study of magnesium', progress report No. 1, CANMET, Metals Technology Laboratory, Energy, Mines and Resources , Ottawa, Canada.

Morgan J (1993), *Cathodic Protection,* second edition, Houston, Tx, NACE, 126–131; 259; 353; 367–368.

Nakatsugawa I (2001), 'Cathodic protection coating on magnesium or its alloys methods of production', Canada: 218,983, USA: 6,291,076.

Nakatsugawa I, Renaud J and Ghali E (1999), 'Protective coating for Mg alloys', in *Environmental Degradation of Materials and Corrosion Control of Metals*, Elboujdaini M and Ghali E, ed, Montreal, Canada, METSOC, 38th Annual Meeting of Metallurgists of CIM.

Nisancioglu K, Lunder O and Aune T (1990), *Proc 47th World Magnesium Conference*, Canne, IMA, 43.

Nordlien J H, Ono S, Masuko N and Nisancioglu K (1995), 'Morphology and structure of oxide films formed on magnesium by exposure to air and water', *Journal of the Electrochemical Society*, **142**, 3320–3322.

Nordlien J H, Ono S, Masuko N and Nisancioglu K (1996), 'Morphology and structure of oxide films formed on MgAl alloys by exposure to air and water', *Journal of the Electrochemical Society*, **143**, 2564–2571.

Nordlien J H, Ono S, Masuko N and Nisancioglu K (1997), 'A TEM investigation of naturally formed oxide films on pure magnesium', *Corrosion Science*, **39**, 1397–1414.

Parthiban G T, Parthiban T, Ravi R, Saraswathy V, Palaniswamy N and Sivan V (2008), 'Cathodic protection of steel in concrete using magnesium alloy anode', *Corrosion Science*, **50**, 3329–3335.

Perrault G G (1974), 'The potential–pH diagram of the magnesium–water system', *Electroanalytical Chemistry and Interfacial Electrochemistry*, **51**, 107–119.

Perrault G G (1978), *Encyclopedia of Electrochemistry of the Elements*, New York, Marcel Dekker Inc., Vol. 8, 263–319.

Pourbaix M (1974), *Atlas of Electrochemical Equilibria in Aqueous Solutions*, Centre Belge d'Étude de la Corrosion, NACE International and CEBELCOR, 100–145; 168–175.

Reboul H and Canon R (1977), 'Corrosion galvanique de l'aluminium Mesures de protection', *Revue de l'aluminium*, France, 403–426.

Roberge P (2006), 'Corrosion by water and steam', in *Corrosion Basics An Introduction*, second edition, Houston, TX, NACE, 125–136.

Robinson J L and King P F (1961), 'Electrochemical behaviour of the magnesium anode', *Journal of the Electrochemical Society*, **108**, 36–41.

Schmutz P, Guillaumin V, Franpel G S, Lillard R S and Lillard J A (2003), 'Influence of dichromate ions on corrosion processes on pure magnesium', *Journal of Electrochemical Society*, 150 (4), B99–B110.

Shaw B A (2003), 'Corrosion resistance of magnesium alloys', *Corrosion* Vol. 13A, Cramer S D and Covino Jr. B S, ed, Materials Park, OH, ASM International, 692–696.

Shaw B A and Wolfe C (2005), 'Corrosion of magnesium and magnesium-base alloys', *Corrosion*, Vol. 13B, Cramer S D and Covino Jr. B S, ed, Materials Park, OH, ASM International, 205–227.

Shi Z, Liu M and Atrens A (2010), 'Measurement of the corrosion rate of magnesium alloys using Tafel extrapolation', *Corrosion Science*, **52**, 579–588.

Sivaraj D, McCune R, Mallick P K and Mohanty (2006), 'Aqueous corrosion of experimental creep-resistant magnesium alloys', 2006 SAE World Congress, SAE Paper No. 2006–01–0257.

Song G L (2005), 'Recent progress in corrosion and protection of magnesium alloys', *Advanced Engineering Materials*, **7**, 308–317; 563–586.

Song GL (2007a), 'Control of biodegradation of biocompatablemagnesium alloys', *Corrosion Science*, **49**, 1696–1701.

Song GL (2007b), 'Control of degradation of biocompatible magnesium in a pseudo-physiological environment by a ceramic like anodized coating', *Advanced Materials Research*, **29–30**, 95–98.

Song G and Atrens A (1998), *Corrosion behavior of skin layer and interior of die cast AZ91D*, Magnesium Alloys and their Application, Germany, Journal of Werkstoff-Informations gesellschaft, 415–419.

Song G L and Atrens A (1999), 'Corrosion mechanisms of magnesium Alloys', *Advanced Engineering Materials*, **1**, 11–33.

Song G and Atrens A (2003), 'Understanding magnesium corrosion', *Advanced Engineering Materials*, **5**, 837–858.

Song G L, Atrens A, St-John D, Wu X and Nairn J (1997a), 'The anodic dissolution of magnesium in chloride and sulphate solutions', *Corrosion Science*, **39**, (10–11), 1981–2004.

Song G L, Atrens A, St-John D, Nairn J and Li Y (1997b), 'The electrochemical corrosion of pure magnesium in 1N NaCl', *Corrosion Science*, **39**, 855–875.

Song G L, Atrens A and St-John D (2001), 'A hydrogen evolution method for the estimation of the corrosion rate of magnesium alloys', *Magnesium Technology 2001*, TMS Annual Meeting, New Orleans, 255–262.

Stampella R S, Procter R P M and Ashworth V (1984), 'Environmentally-induced cracking of magnesium', *Corrosion Science*, **24**, 325–341.

Strehblow H H (1995), *Corrosion Mechanisms in Theory and Practice*, New York, Marcel Dekker Inc., 201–237.

Uan J-Y, Li C-F and Yu B-L (2008), 'Characterization and improvement in the corrosion performance of a hot-chamber diecast mg alloy thin plate by the removal of interdendritic phases at the die chill layer', *Metallurgical and Materials Transactions A*, **39A**, 703–715.

Uhlig H H (1970), *Corrosion et protection*, Paris, France, (Voeltzel J, translator) 98–108, 136–143; 148–157.

Uhlig H H and Revie R W (1985), *Corrosion and Corrosion Control*, third edition, John New York, NY, Wiley & Sons, Inc., 222.

Vargel C (2004), *Corrosion of Aluminum*, Boston, MA, Elsevier, 75–85; 181–195.

Witte F, Fisher J, Nellesen J, Crostack H-A, Kaese A, Pisch A *et al.* (2006), '*In vitro* and *in vivo* corrosion measurements of magnesium alloys', *Biomaterials*, **27**, 7, 1013–1018.

Witte F, Hort N, Vogt C, Cohen S and Kainer K U (2008), 'Degradable materials based on magnesium corrosion', *Current Opinion in Solid State and Materials Science*, **12**, 63–72.

Winzer N, Atrens A, Song G, Ghali E, Dietzel W, Kainer K U, Hort N and Blawert C (2005), 'A critical review of the stress corrosion cracking (SCC) of magnesium alloys', *Advanced Engineering Materials*, **7**(8), 659–693.

Xia S J, Birss Va I and Rateick Jr. R G (2003), 'Anodic oxide film formation at magnesium alloy WE43', *Electrochemical Society Proceedings*, **25**, 270–280.

Yang L and Zhang E (2009), 'Biocorrosion behaviour of magnesium alloy in different simulated fluids for medical application', *Materials Science and Engineering C*, **29**, 1691–1696.

Zhang S, Zhang X, Zhao C, Li J, Song Y, Xie C, Tao H, Zhang Y, He Y, Jiang Y, Bian Y (2010), 'Research on an Mg–Zn alloy as a degradable biomaterial', *Acta Biomaterialia*, **6**, 2, 626–640.

Zhang T, Shao Y, Meng G and Wang F (2007), 'Electrochemical noise analysis of the corrosion of AZ91D magnesium alloy in alkaline chloride solution', *Electrochimica Acta*, **53**, 561–568.

Zhang T, Chen C, Shao Y, Meng G, Wang F, Li X and Dong C (2008), 'Corrosion of pure magnesium under thin electrolyte layers', *Electrochimica Acta*, **53**, 7921–7931.

Part II

Metallurgical effects

3 Corrosion of magnesium (Mg) alloys and metallurgical influence

A. ATRENS and M. LIU, The University of Queensland, Australia, N. I. ZAINAL ABIDIN, University of Malaya, Malaysia and G.-L. SONG, General Motors Corporation, USA

Abstract: An overview is provided of the following aspects which determine the form and rate of corrosion of typical two-phase magnesium (Mg) alloys: (i) measurement details; (ii) concentration of the impurity elements iron, nickel, copper and Co; (iii) volume fraction, size, distribution and electrochemistry of second phases; (iv) heat treatment; (v) propensity of the environment to cause surface film formation and breakdown; and (vi) the composition of the α-Mg matrix. Our understanding of the Mg corrosion mechanism is based on research using chloride solutions, which are appropriate for applications such as auto service. These chloride solutions are quite aggressive. The chloride ions tend to break down the partly protective film on the Mg alloy surface. The corrosion rate increases with exposure time until steady state is reached, which may take several weeks. This understanding can elucidate corrosion for the other application of growing importance to Mg alloys as biodegradable medical implants. Solutions that elucidate medical applications tend to form surface films and the corrosion rate decreases with immersion time. The surface films increase the resistance between micro-galvanic elements in the microstructure, so that there is less acceleration of the corrosion rate by second phase particles. Surface condition is an additional influence on the corrosion of Mg components. This is of significantly greater importance for auto applications than for medical applications because cost effectiveness is much more important for auto applications.

Key words: Mg alloy, corrosion, metallurgy.

3.1 Introduction

This chapter provides an overview of the corrosion mechanisms of typical Mg alloys, most of which contain two or more phases. Magnesium (Mg) is a reactive metal and corrosion protection is an issue of importance [1] particularly for the automobile industry. The rapid increase in Mg use is due to its lightweight and good casting capabilities, particularly its ability to be diecast into large, thin sections. Typical examples are automobile seats, instrument panels, computer cases, etc. Reviews [1–4] and our early research [5–9] have indicated that the poor corrosion resistance of Mg alloys results from (a) the high intrinsic dissolution tendency of magnesium, which is only weakly inhibited by corrosion product films, and (b) the presence of impurities and second phases acting as local cathodes and thus causing local galvanic acceleration of corrosion. The cathodic reaction is hydrogen liberation. This chapter deliberately focuses on understanding the corrosion mechanisms.

3.2 Measurement details

Much of our understanding of Mg corrosion [1–4] has come from studies in chloride solutions, typically 3% NaCl. These solutions are used to understand corrosion associated with applications such as automobiles where corrosion conditions can be aggressive because of road salt or because of proximity to the ocean. The corrosion rate for Mg alloys is typically more than 1 mm/yr in common testing environments like 3% NaCl solution. No alloying element has been discovered [1–3] that produces a solid solution Mg alloy with a corrosion rate less than that of high-purity (HP) Mg in 3% NaCl. Consequently, it is useful to include HP Mg as a standard in any comparative study of corrosion of Mg alloys as illustrated in Fig. 3.1 for binary two-phase Mg–Y alloys [10] (see also [2,11,12]). Moreover, it is important to use HP Mg as the standard; commercial purity (CP) Mg can have a corrosion rate more than 50 times that of HP Mg [2,8] and so CP Mg should be designated as low-purity (LP) Mg. This is illustrated in Fig.

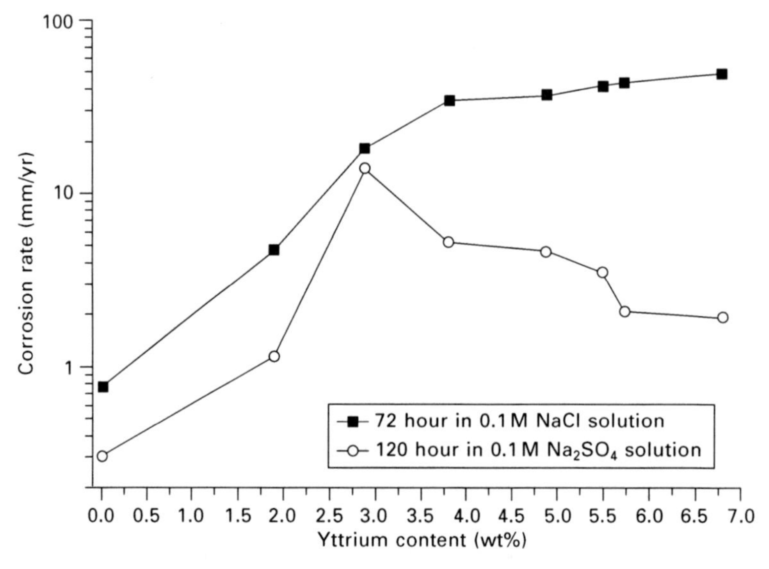

3.1 Average corrosion rate of binary two-phase Mg–Y alloys as a function of yttrium content is greater than the corrosion rate of high-purity Mg which is used as a standard for comparison [10].

3.2 [13], which presents the corrosion behaviour for immersion in 3% NaCl, as revealed by hydrogen evolution, for LP Mg (280 ppm Fe), HP Mg (45 ppm Fe) and Mg–Al–Fe alloys. The behaviour of the Mg–Al–Fe alloys was dominated by the Fe impurity. Note the significant difference in corrosion behaviour between HP Mg and LP Mg. HP Mg in the corrosion context means an alloy in which all the impurity elements have concentrations below their concentration-dependent tolerance limits [13]. Figure 3.1 also indicates that the corrosion rate of HP Mg is much lower than 1 mm/yr in solutions that are less corrosive than 3% NaCl.

The corrosion of HP Mg and Mg alloys typically starts on part of the exposed surface and slowly extends over the whole surface. The corrosion rate changes with immersion time. Figure 3.3 [10] illustrates a typical example where hydrogen evolution was used to characterise the corrosion of binary Mg–Y alloys immersed in 0.1 M NaCl. Low corrosion rates are typically measured during early exposure times in chloride-containing solutions; such data are partly dominated by the breakdown of whatever surface film is on the surface at the start of the immersion test. Thus specimen preparation is important. It is fairly standard to abrade Mg specimens to 1200 grit some time before solution immersion. The corrosion rate can be evaluated at any desired time from the slope of the hydrogen evolution data such as that illustrated in Fig. 3.3, or an average corrosion rate can be evaluated from the average hydrogen evolution over a particular time period.

The critical review of Mg corrosion by Song and Atrens [2] in 2003 indicated that, for Mg alloys, Tafel extrapolation had not estimated the corrosion rate reliably. Shi *et al.* [14] examined this issue further. The hypothesis that the corrosion of Mg alloys can be adequately estimated using Tafel extrapolation of the polarisation curve was termed the electrochemical measurement hypothesis for Mg. In principle, any hypothesis can be disproved by a single valid counter-example. Shi *et al.* [14] found that the literature shows that, for Mg alloys, corrosion rates evaluated by Tafel extrapolation from polarisation curves have not agreed with corrosion rates evaluated from weight loss and hydrogen evolution. Typical deviations have been ~50% to 90%. These are much larger than the precision of the measurement methods and indicate a need for careful examination of the use of Tafel extrapolation for Mg. For research that nevertheless does intend to use Tafel extrapolation to elucidate corrosion of Mg associated with service, it is strongly recommended that these measurements be complemented by the use of at least two of the three other simple measurement methods: (i) weight loss rate, (ii) hydrogen evolution rate and (iii) rate of Mg^{++} leaving the metal surface. There is much better insight for little additional effort.

Shi *et al.* [15] carried out a study to evaluate electrochemical measurements for HP Mg, the simplest Mg alloy. For measurements of Mg alloy corrosion and for Mg alloy electrochemical measurements, it is common [1,2,8,10–13]

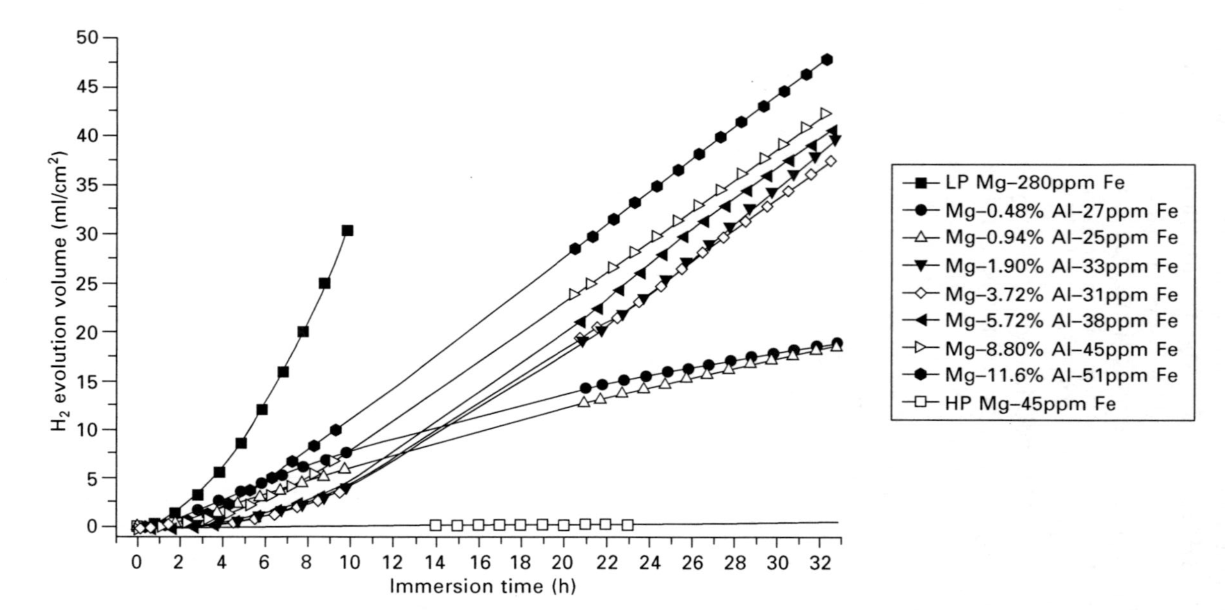

3.2 Corrosion behaviour from immersion in 3% NaCl, as revealed by hydrogen evolution, for LP Mg (280 ppm Fe), HP Mg (45 ppm Fe) and Mg–Al–Fe alloys [13]. Note that the behaviour of the Mg–Al–Fe alloys is dominated by the Fe impurity. Note also the significant difference in corrosion behaviour between HP Mg and LP Mg.

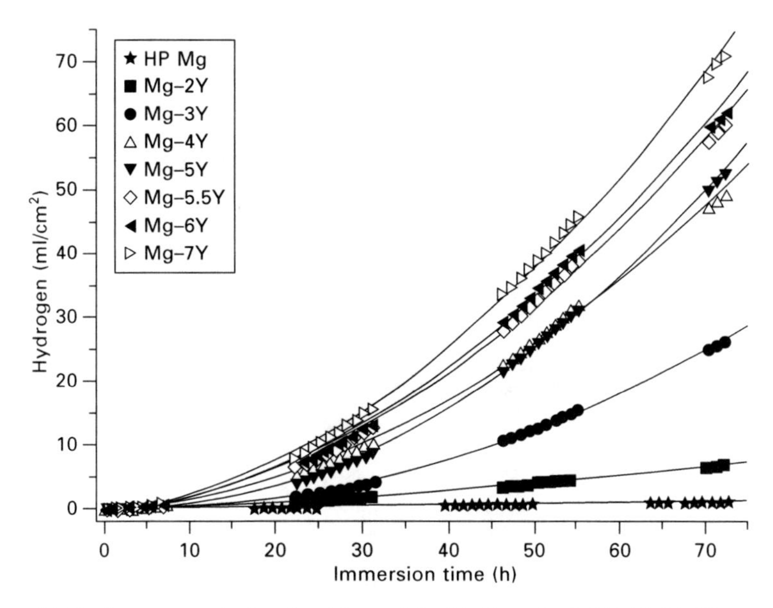

3.3 Hydrogen evolution for binary Mg–Y alloys in 0.1 M NaCl. The corrosion rate can be evaluated as an instantaneous value from the slope of the hydrogen evolution curve or can be evaluated as an average value over a selected time period [10].

to encapsulate specimens so that one defined surface is exposed to the solution such as the arrangement illustrated in Fig. 3.4. This method allows the surface to be prepared to the desired surface finish: (i) e.g. grinding to 1200 grit for corrosion immersion measurements, and (ii) e.g. diamond polishing when it is desired to study the details of the corrosion morphology and its development. Electrical connections through the back of the specimen allow electrochemical measurements. However, Tafel extrapolation from polarisation curves measured with such specimens has predicted corrosion rates [14] that were not in good agreement with the actual corrosion rate as determined by hydrogen evolution or by weight loss, which typically agree within 10%.

For HP Mg, crevice corrosion was identified as causing an issue by Shi *et al.* [15]. Polarisation curves were not reproducible for HP Mg specimens encapsulated as illustrated in Fig. 3.4 and immersed in 3.5% NaCl saturated with $Mg(OH)_2$, and such cathodic curves predict a rather high corrosion rate. With this specimen arrangement, crevice corrosion starts in the crevice between the Mg and the mounting medium. The crevice is also often the initiator of corrosion on the exposed Mg surface. Moreover, whenever there is crevice corrosion, the corrosion rate is substantially higher than in the absence of crevice corrosion. Crevice corrosion in other systems (Fe, stainless steels, Al alloys) is considered to be driven by an oxygen concentration cell, with

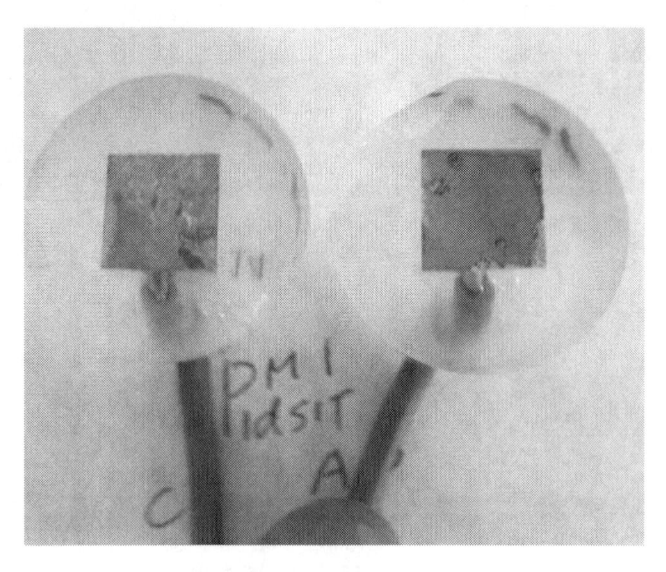

3.4 Typical encapsulation of Mg alloys for electrochemical experiments. One Mg surface is exposed to the solution. This surface can be prepared to the desired surface finish [15].

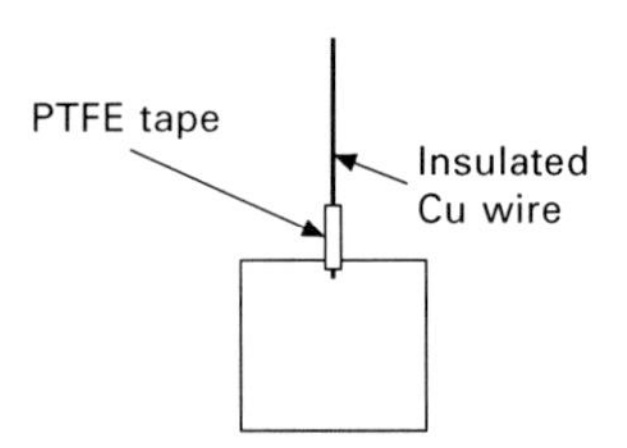

3.5 Specimen configuration which had essentially no crevice corrosion. An insulated copper wire with a bared end was forced into a slightly undersized hole machined into the edge of the Mg sample. Poly(tetrafluoroethylene) (PTFE) tape was wound around the end of the insulated copper wire to ensure there was no crevice between the sample and the insulated copper wire [15].

oxygen reduction the cathodic reaction. Oxygen reduction is not considered to be important for Mg corrosion [1–3], so crevice corrosion of Mg is totally unexpected and the mechanism is totally unexplored at this stage.

Figure 3.5 provides a specimen configuration that has allowed corrosion immersion experiments to be carried out without crevice corrosion for HP Mg immersion in 3.5% NaCl saturated with $Mg(OH)_2$. This specimen configuration allows measurement of reproducible polarisation curves without any significant crevice corrosion. Tafel extrapolation was used to estimate the corrosion rate at the corrosion potential, i_{corr}, which was used to estimate the corrosion rate, P_i. This specimen configuration also allows simultaneous measurement of the evolved hydrogen, which allows an independent evaluation of the corrosion rate, designated as P_H. Figure 3.6 shows a comparison of the corrosion rate evaluated from hydrogen evolution, P_H, compared with the corrosion rate, P_i, evaluated from i_{corr}, obtained by Tafel extrapolation of polarisation curves using the specimen configuration of Fig. 3.5. There was good agreement between P_H and P_i. Moreover, P_i was less than P_H as expected from the negative difference effect. The apparent Mg valence, evaluated as the quantity $2(P_i/P_H)$, had an average value of 1.4 in good agreement with the value of 1.5 measured by Petty *et al.* [16]. It is also stressed that the value of $P_H \sim 0.6$ mm/yr was significantly lower than prior measurements which typically evaluate the corrosion rate of pure Mg = ~ 1 mm/yr, for measurements using the specimen configuration illustrated in Fig. 3.4 where there is typically crevice corrosion, indicating that crevice corrosion causes an apparent corrosion rate about double that which is measured with no crevice corrosion.

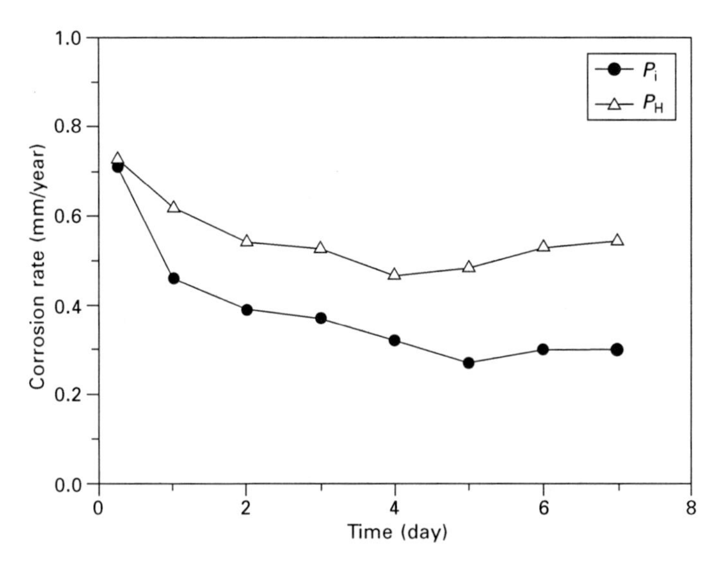

3.6 Corrosion rate measured from Tafel extrapolation, P_i, from cathodic polarisation curves measured with the specimen configuration shown in Fig. 3.5 for pure Mg immersed in 3.5% NaCl saturated with $Mg(OH)_2$ [15]. Simultaneous measurements of hydrogen evolution gave independent measurements of the corrosion rate, designated as P_H. There was excellent agreement between the two independent measurements of the corrosion rate, P_i was less than P_H as expected from the negative difference effect (NDE), and the apparent valence of Mg was equal to 1.4 from the average value of $2P_i/P_H$.

3.3 Second phase effect

3.3.1 Micro-galvanic corrosion

The active nature of Mg means that galvanic corrosion is always an issue. Typical values of the corrosion potential E_{corr} are given in Table 3.1 for common engineering alloys. Mg is more active than all other engineering alloys, and consequently Mg is the anode and corrodes preferentially in any galvanic couple. Macro-galvanic corrosion is illustrated in Fig. 3.7(a) in terms of coupling with a less active metal, while Fig. 3.7(b) illustrates micro-galvanic corrosion in which the second phase causes galvanic acceleration of the corrosion of the alpha-Mg matrix.

Second phases have a pronounced influence on the corrosion of Mg, because most elements influence only the corrosion of Mg alloys after the formation of a second phase. For example, a high Al content alloy like AZ91 has an appreciable amount of β, $Mg_{17}Al_{12}$ along the α-grain boundaries. Figure 3.8 shows the microstructure of as-cast AZ91D. $Mg_{17}Al_{12}$ is cathodic with respect to the matrix (see Table 3.2[17]) and exhibits passive behaviour over a wider pH range than either of its components Al and Mg [18] (Fig. 3.9). $Mg_{17}Al_{12}$ was inert in chloride solutions in comparison with the surrounding Mg matrix and could act as a corrosion barrier [19] so that its distribution determined the corrosion resistance of the Mg–Al alloys [8,19]. The resistance of the $Mg_{17}Al_{12}$ to corrosion is due to the presence of a thin protective film on its surface. However, there was also an opposite opinion that $Mg_{17}Al_{12}$ is detrimental to the Mg matrix. It was suggested [20] that the absence of the $Mg_{17}Al_{12}$ could enhance the corrosion resistance of the Al-rich Mg–base alloys by eliminating the micro-galvanic effects. The Mg–Al-based alloy was reported to corrode predominantly by galvanic action between the magnesium matrix and $Mg_{17}Al_{12}$ phase [21].

3.3.2 Electrochemistry of second phases

Song *et al.* [7,8] measured the polarisation curves for the α-phase and the β-phase as shown in Fig. 3.9. The α-phase has a pitting potential about 15 mV

Table 3.1 Typical corrosion potential values for common engineering alloys

Metal	E_{corr}, V_{SCE}
Mg	−1.65
Zn	−1.02
Al-7075	−0.88
Al-1xxx	−0.73
Fe	−0.50
Cu	−0.12
Ni	+0.01

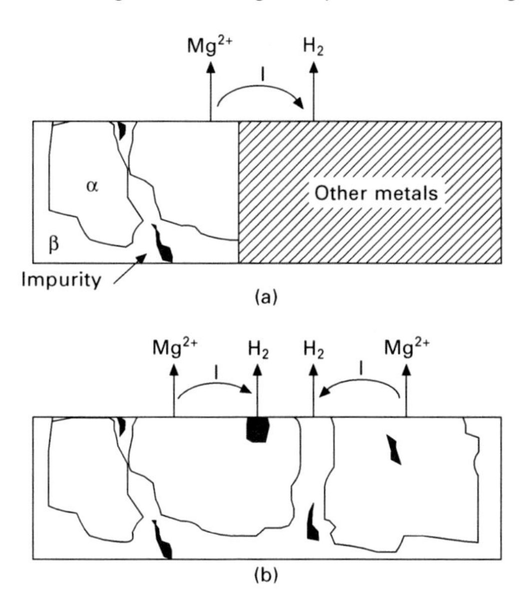

3.7 (a) Macro-galvanic corrosion. (b) Micro-galvanic corrosion [2,8].

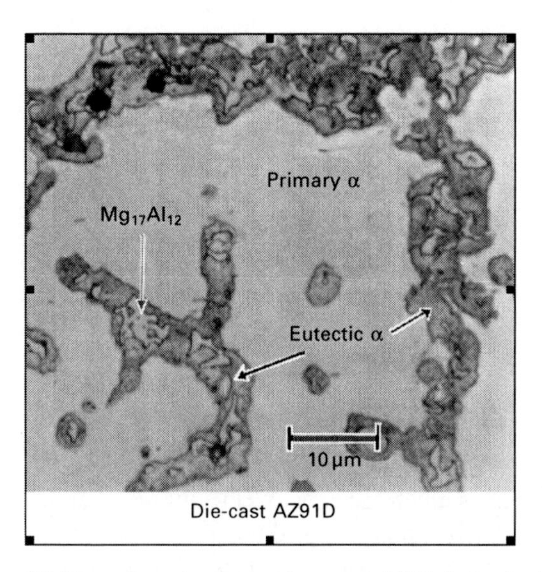

3.8 The microstructure of as-cast AZ91 (a typical two-phase alloy) consists of primary α-Mg and the eutectic micro-consituent which contains eutectic-α plus the β phase (the intermetallic $Mg_{17}Al_{12}$) [2,8].

more negative than its free corrosion potential, consistent with observations of pitting corrosion for the α-phase exposed to 1 M NaCl at the free corrosion potential. The β-phase has a higher cathodic activity than the α-phase, and

Table 3.2 Typical corrosion potential values [17] for common magnesium second phases (after 2 h in de-aerated 5% NaCl solution saturated with $Mg(OH)_2$ (pH 10.5))

Metal	E_{corr}, V_{SCE}
Mg	−1.65
Mg_2Si	−1.65
Al_6Mn	−1.52
Al_4Mn	−1.45
Al_8Mn_5	−1.25
$Mg_{17}Al_{12}$ (β)	−1.20
Al_8Mn_5(Fe)	−1.20
β-Mn	−1.17
Al_4RE	−1.15
Al_6Mn(Fe)	−1.10
Al_6(MnFe)	−1.00
Al_3Fe(Mn)	−0.95
Al_3Fe	−0.74

RE = rare earth elements.

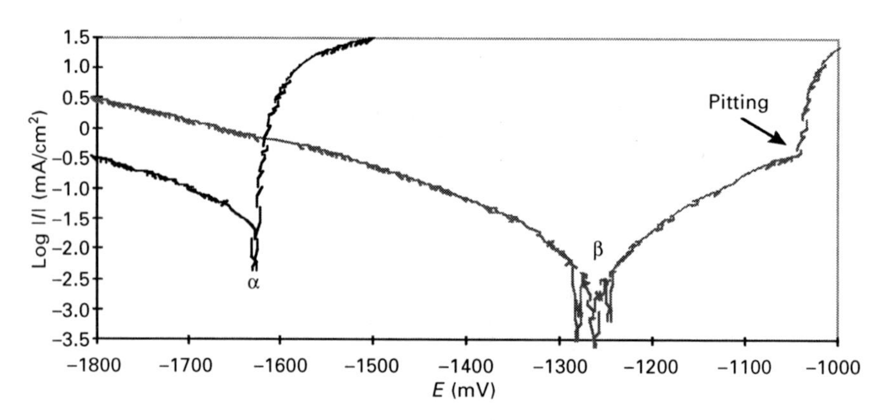

3.9 Polarisation curves for the α-phase and the β-phase in 1 M NaCl at pH 11 [2,8].

the anodic dissolution rate is much lower below the pitting potential than that of the α-phase. At the corrosion potential the corrosion current density of the β-phase is much lower than that of the α-phase. These findings indicate that the β-phase is more stable in NaCl and is more inert; the β-phase is, however, an effective cathode. Consequently, the β-phase has two influences on corrosion, as a barrier and as a galvanic cathode, depending on the volume fraction and distribution of β. The β-phase serves as a galvanic cathode and accelerates the corrosion of the α-matrix if the volume fraction of β-phase is small. However, for a high-volume fraction, the β-phase can act as a barrier to inhibit corrosion if the β-phase is essentially continuous.

In addition to the β-phase, the most potent cathodes in Mg–Al alloys were thought to be the iron-rich phases; the iron–aluminium intermetallic phase FeAl is one of the most detrimental cathodic phases in Mg–Al alloys on the basis of its potential and low hydrogen over-voltage [22]. AlMn [17,18] is also detrimental, and Mg_2Si [18] seems to have no influence, while Mg_2Pb facilitates localised corrosion and leads to a negative difference effect.

The matrix α-phase in Mg alloys is typically anodic to the second phases and is preferentially corroded. Song *et al.* [7,8] suggested that the primary α and eutectic α-phases, which have different aluminium contents, have different electrochemical behaviour. Both the primary and eutectic α can form a galvanic corrosion cell with the β-phase, as illustrated in Fig. 3.10. There are therefore two kinds of corrosion morphology:

- the primary α grain is preferentially dissolved; and
- the β-phase is undermined because of the dissolution of the eutectic α.

Actual exposure of diecast AZ91D in NaCl illustrated both corrosion morphologies (Fig. 3.11). Furthermore, Song *et al.* [8] found that the casting method influenced the corrosion performance through microstructure differences. The skin of diecast AZ91D showed a corrosion resistance significantly better (by nearly a factor of ten) than its interior, Table 3.3. This was attributed to a higher volume fraction of β, and more continuous β around finer α-grains. If the α-grains are fine and the β-fraction is not

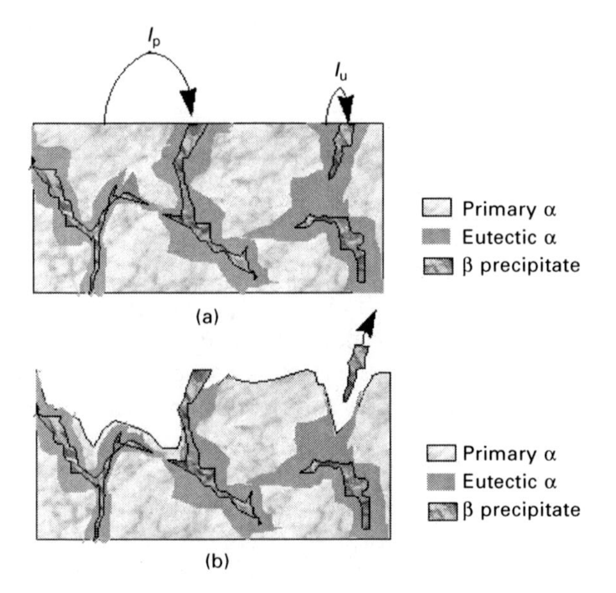

3.10 Schematic representation of the galvanic corrosion between the β-phase and the α-phases [2,8].

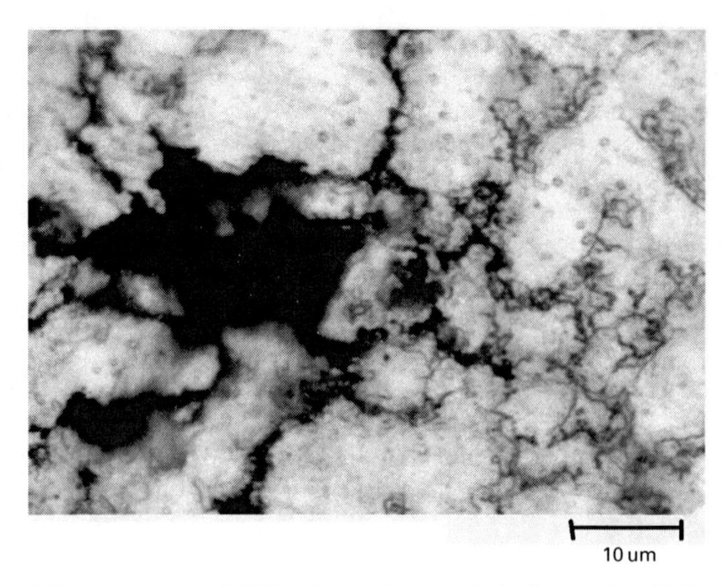

3.11 Appearance of AZ91 after a short period of corrosion [2,8].

Table 3.3 Corrosion rates under open circuit conditions in 1 M NaCl [8]

Sample	Rate (mm/yr)	Mechanism
LP Mg (240 ppm Fe)	53	Impurity accelerated corrosion
HP AZ91 sand-cast	12	β accelerates corrosion
AZ91D – diecast (interior)	5.7	β accelerates corrosion
HP Mg	1.1	Standard for comparison
AZ91D – diecast (surface)	0.66	β protects
β-phase	0.30	
Z-phase	?? 10^{-6} ??	Needs to be more passive than the β-phase and have a lower ability to liberate hydrogen on its surface

too low, then the β-phase is nearly continuous like a net over the α-matrix, and the β-phase particles do not easily fall out by undermining. Instead, the α-matrix is much more easily corroded. Also if the α-grains are fine, the gaps between β-particles are narrow and the β-phase is nearly continuous. The corrosion of the β-phase is then quite easily obstructed by corrosion products on its surface, and so the corrosion is greatly retarded. Figure 3.12 schematically illustrates this case. On the other hand, if the grain size is large, the β-phase is agglomerated and the distance between the β-phase is large, then the corrosion of the α-phase is not effectively blocked either by the β-phase or by corrosion products, and the β-phase accelerates the corrosion of the α-phase.

The active nature of Mg means that micro-galvanic effects are always an

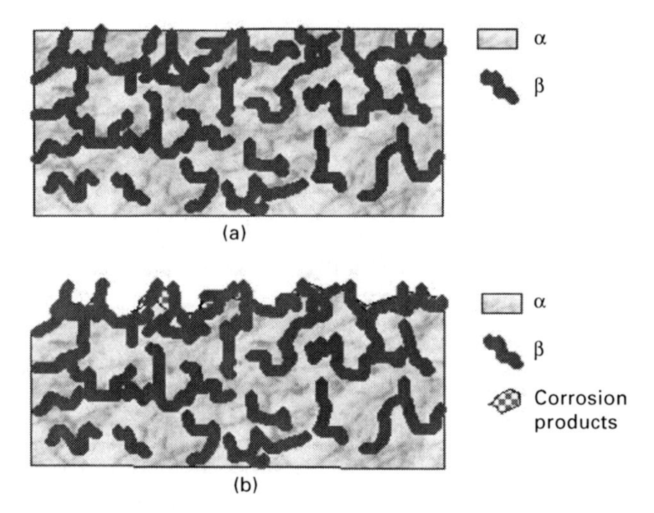

3.12 Schematic representation of protection of the corroding surface (top surface) by the β-phase when the β-phase is essentially continuous [2,8].

issue. Typical values of E_{corr} are given in Table 3.2 for Mg second phases [2,18]. Mg is more active than all known second phases, and consequently the Mg matrix is the anode and corrodes preferentially in any galvanic couple. Acceleration of corrosion of two-phase Mg alloys by second phases has been reported for AZ91 [8,12,23,24], ZE41 [11,25], NK30K [26,27], Mg–xGd–3Y–0.4Zr (x = 6, 8, 10, 12%) [28], Mg–10Gd–3Y–0.4Zr [29], Mg–Zn–RE [30], Mg–6Zn–1Y–0.6Zr [31], Mg–Zn–Mn–Si–Ca [32,33], Mg–8Li [34], Mg–RE (0.5–5% La, 0.5–5% Ce, 0.5–4% Nd) [35]. Birbilis *et al.* [35] showed that the corrosion rate increased with volume percent intermetallic for Mg–Ce, Mg–La and Mg–Nd alloys; the second phases were $Mg_{12}Ce$, $Mg_{12}La$ and Mg_3Nd.

3.3.3 Distribution of second phases

The possible types of behaviour are summarised by the corrosion rate data in Table 3.3 for HP Mg and Mg alloys corroding at their free corrosion potentials in 1 M NaCl at pH 11. HP Mg, taken as standard for comparison, showed a corrosion rate of 1.1 mm/yr. A higher corrosion rate was shown by the interior of diecast AZ91D and by the HP sand-cast AZ91. The β-phase accelerated the corrosion. In contrast, the surface of diecast AZ91D had a corrosion rate lower than that of HP Mg. The β-phase provided protection as is clear from the still lower corrosion rate shown by pure β.

However, the corrosion rate of the β-phase is nevertheless quite high. Practitioners of corrosion protection working with aluminium (Al) alloys find [36] that the corrosion properties of Al alloys are severely degraded by this same phase, the β-phase. This points to the methodology by which it would be possible to increase the corrosion resistance of Mg alloys by the alloying to produce a radically different second phase; let us designate this new phase the Z-phase. Much better corrosion for the alloy becomes possible if this new second phase, the Z-phase, has a corrosion resistance much greater than that of the β-phase. Such a Z-phase should be more passive than the β-phase and also have a much lower ability to allow hydrogen evolution on its surface.

3.3.4 Heat treatment of AZ91

Zhao *et al.* [12,24] studied the influence of the microstructure, particularly the morphology of the β-phase, on the corrosion of Mg alloys using AZ91 as a model Mg alloy. Corrosion was characterised for five different types of microstructure produced by heat treatment of as-cast AZ91 as illustrated in Fig. 3.13. The influence of microstructure can be understood from the interaction of the following three factors:

- The surface films can be more or less effective in hindering corrosion and more or less effective in controlling the form of corrosion as uniform corrosion or localised corrosion.
- The second phase (the β-phase in AZ91) can cause micro-galvanic acceleration of corrosion.
- The second phase can act as a corrosion barrier and hinder corrosion

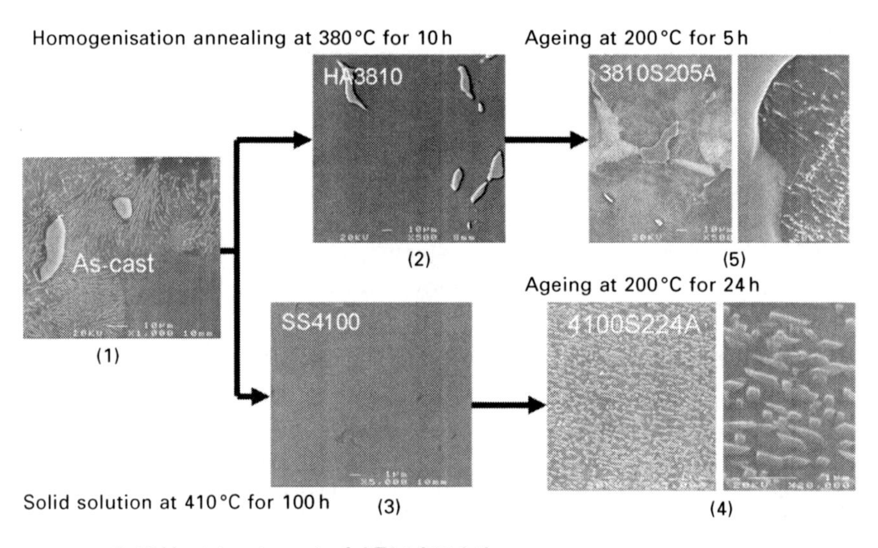

3.13 Heat treatment of AZ91 [12,24].

propagation in the matrix, if the second phase is in the form of a continuous network.

It is expected that these factors are important for all multiphase Mg alloys because all known second phases have corrosion potentials more positive than that of the α-phase. A particular example of the corrosion barrier effect is provided by the fine (α + β) lamellar micro-constituent; when a β-phase plate nucleates this micro-constituent, the β-phase plate acts as a corrosion barrier. In contrast, nano-sized β precipitates, produced by ageing, caused micro-galvanic corrosion acceleration of the adjacent α-phase.

A homogenisation annealing (HA) heat treatment was proposed [24] for property enhancement for AZ91; HA for 10 h at 410 °C caused an improvement in hardness, ultimate tensile strength and ductility without loss of corrosion properties. The improvement of the mechanical properties is due to the absence of a continuous easy crack path. The influence on the corrosion behaviour of the microstructure was studied (including (i) α-Mg and fine lamellar α + β plus coarse β particles, (ii) α-Mg and isolated coarse β particles and (iii) α-Mg and isolated fine β particles). The fine lamellar α + β do not cause preferential corrosion but act as a corrosion barrier to a certain extent. The isolated coarse β particles cause a significant micro-galvanic corrosion of the α-Mg matrix but do not act as a significant corrosion barrier, resulting in relatively poor corrosion properties for the moderate strength condition (HA for 5 h at 380 °C). In contrast, the isolated fine β-particles do not lead to an obvious loss of corrosion resistance because they act as a small cathode connected to a large anode (α-Mg matrix) so that the peak aged condition (HA for 10 h at 410 °C) has negligible micro-galvanic corrosion.

3.3.5 Heat treatment of Mg–(2–15)Gd

Hort *et al.* [37] explored the use of Gd as an alloying element for Mg alloys for biodegradable implant applications. The Mg–Gd phase diagram indicates that Gd has a large solubility of 23.5% at the eutectic temperature of 548 °C. Gd lowers the melting temperature of Mg and so improves castability. Gd can also be used to improve strength by precipitation hardening. Feyerabend *et al.* [38], based on the examination of cytotoxcity, indicated that Gd is an interesting alloying element for medical implant applications. Hort *et al.* [37] studied HP Mg–2Gd, Mg–5Gd, Mg–10Gd, Mg–15Gd alloys made as castings. The alloys were investigated in the as-cast condition (F), solutionised 24 h at 525 °C plus water quenched (T4), and solutionised and aged 6 h at 250 °C (T6). All alloys contain the eutectic phase Mg_5Gd at grain boundaries in the as-cast state. After T4 treatment, most of these particles dissolve although some remained. The T6 treatment leads to precipitation of fine nano-sized particles in all alloys except for Mg–2Gd. Selected area diffraction indicated that these particles are the metastable β′ and β″. The β′ phase was homogeneously distributed throughout the matrix whereas β″ was only in limited areas.

The corrosion rate of the alloys was measured [37,39] by immersion in 1% NaCl (Fig. 3.14). The corrosion rate decreased with increasing Gd content up to 10%. In particular the corrosion rate for Mg-10Gd in the T4 condition was 0.7 mm/yr, which appears to be somewhat smaller than that of pure Mg. The corrosion rate in the T6 condition was even smaller at 0.4 mm/yr despite the presence of nano-sized precipitates; these appear not to have had an adverse influence on the corrosion rate. The corrosion rate for Mg–15Gd was much higher, consistent with micro-galvanic acceleration by the second phase.

3.3.6 Heat treatment of Mg–10Gd–3Y–0.4Zr

Peng *et al.* [29] studied the corrosion of Mg alloy GW103K (Mg–10Gd–3Y–0.4Zr) in 5% NaCl, in the as-cast (F), solution treated (T4) and aged (T6) conditions. Figure 3.15 shows (i) the age hardening curve for GW103K aged at 250 °C and (ii) transmission electron microscopy (TEM) micrographs of the precipitates. The hardness increased rapidly after ~ 1 h, the peak hardness occurred at ~ 16 h, there was a plateau from 10 h to 60 h and further ageing led to a slow decrease in hardness. Corrosion was studied for ageing times of 0.5 h, 16 h, 193 h and 500 h, which corresponded to under-aged, peak-aged, slightly over-aged and significantly over-aged conditions. The precipitates, after ageing 0.5 h (T6-0.5 h), were mainly meta-stable β″, ~ 40 nm in size.

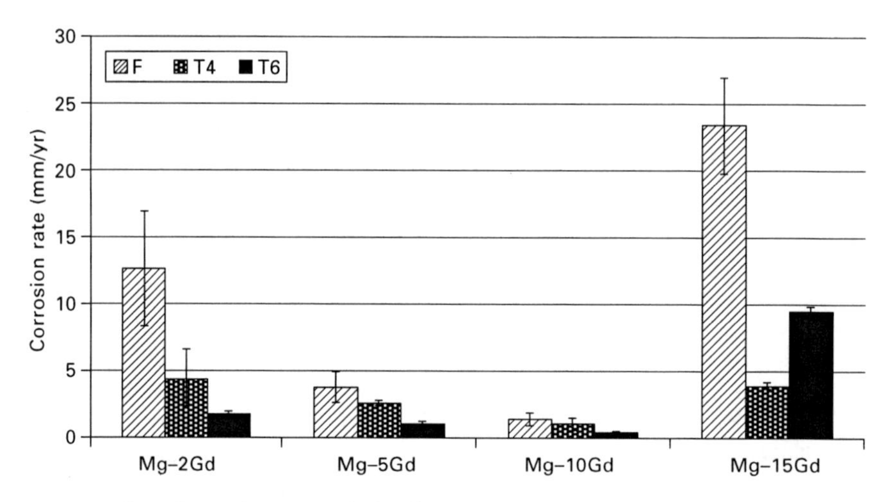

3.14 Corrosion rate of Mg–Gd alloys evaluated by immersion in 1% NaCl. The unit of mm/a is identical to mm/yr [37].

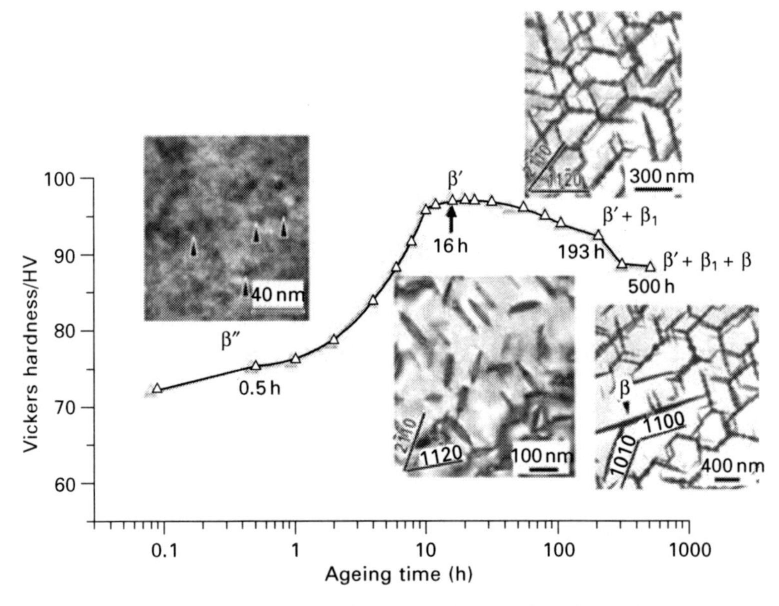

3.15 Age hardening curve for GW103K (Mg–10Gd–3Y–0.4Zr) aged at 250 °C and TEM micrographs of the precipitates [29].

The peak-aged (T6-16 h) condition contained predominantly meta-stable β′ phase. The over-aged conditions (193 h and 500 h) contained β′ and β_1, with a minor fraction of β phase for 500 h ageing.

Optical microscopy was carried out for GW103K in the as-cast (F), solution treated (T4) and aged (T6-193 h) conditions. The as-cast (F) microstructure consisted of the Mg matrix and the eutectic, distributed along the dendrite boundaries, especially at the triple points. The solution treated (T4) microstructure was single grained α-Mg; the Gd and Y in the eutectic had dissolved in the Mg matrix. Optical microscopy did not reveal any change in microstructure between the solution treated (T4) condition and the microstructure after ageing 0.5 h (T6-0.5 h). However, precipitates were visible after ageing 16 h or longer; the optical micrographs were similar after ageing at 16 h, 193 h and 500 h. The optical microstructure of the alloy aged 193 h (T6-193 h) consisted of small cuboid-shaped phases distributed unevenly in the matrix and boundaries and Zr-rich cores in the centre of most grains.

Figure 3.16 presents the corrosion rates for GW103K in the different heat treatment conditions, measured by immersion in 5% NaCl solution for 3 days. The as-cast (F) condition had the highest corrosion rate due to micro-galvanic corrosion of the α-Mg matrix by the eutectic. Solution treatment led to the lowest corrosion rate, attributed to the absence of any second phase and a relatively compact protective surface film. Ageing at 250 °C increased the

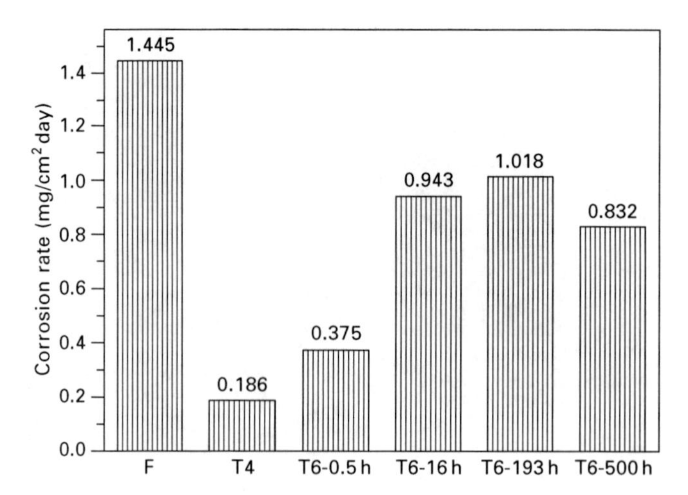

3.16 Corrosion rates for GW103K (Mg–10Gd–3Y–0.4Zr) in the different heat treatment conditions, measured by immersion in 5% NaCl solution for 3 days [29].

corrosion rate with increasing ageing time to 193 h attributed to increasing micro-galvanic corrosion acceleration of the Mg matrix by increasing amounts of the precipitates. The T6-0.5 h condition already indicated micro-galvanic acceleration, even though the precipitates were only ~40 nm in size. Ageing for longer periods caused a decrease in the corrosion rate attributed to some barrier effect by a nearly continuous second-phase network. The weight loss rate, ΔW (mg/cm^2/day) can be converted to a penetration rate, P_W (mm/yr) by

$$P_W = 2.1\ \Delta W$$

The corrosion rate in the T4 condition corresponds to 0.4 mm/yr, somewhat less than that usually measured for HP Mg, and the corrosion rate for the T6-0.5 h at 0.8 mm/yr is also somewhat less than that of HP Mg.

3.3.7 Mg–Y alloys

Figure 3.3 presents the measurement of corrosion [10] of the Mg–Y alloys in 0.1 M NaCl for 72 h. HP Mg was also tested as a comparison alloy containing 0% Y. The hydrogen evolution volume is plotted against time. For short times, the corrosion rate of each alloy was low. As the immersion time increased, the corrosion rate of Mg–2Y increased slowly, whereas the corrosion rates of the other alloys accelerated and localised corrosion was observed on the sample surface. The rate of acceleration increased with increasing Y content. HP Mg showed a constant low corrosion rate during the whole period.

After 72 h immersion in 0.1 M NaCl, optical microscopy revealed filiform corrosion on the surface of Mg–2Y with little corrosion on part of the surface. The other Mg–Y alloys had suffered significant corrosion and their surfaces were covered by a thick white corrosion product, which was assumed to be $Mg(OH)_2$. After the surface corrosion products were removed by the chromate acid cleaning solution, Mg–2Y revealed areas of corroded surface and there were also some areas of the original flat surface with traces of filiform corrosion. This filiform corrosion was typical for these Mg alloys in 0.1 M NaCl. Mg–3Y revealed several separate small flat areas with traces of filiform corrosion whereas Mg–4Y had only a few tiny areas. The other Mg–Y alloys did not have these flat surfaces because their surfaces were significantly corroded. These surface observations indicated the degree of surface damage increased with increasing Y content.

Figure 3.1 presents the average corrosion rate over the 72 h immersion test in 0.1 M NaCl and over 120 h in 1 M Na_2SO_4. In both solutions, Mg–2Y had a low corrosion rate, which was almost constant during the whole immersion period. Mg–3Y had a higher corrosion rate, which increased with immersion time. The corrosion rate of the Mg–Y alloys increased in NaCl with increasing Y content in the range of 0–7% Y, but decreased in Na_2SO_4 for Y contents 3–7% Y. In both solutions HP Mg had a corrosion rate lower than the Mg–Y alloys.

Figure 3.1 shows that, in 0.1 M NaCl, the corrosion rate of the Mg–Y alloys increased with increasing Y content when the Y content exceeded a threshold value of 2% and there was a two-phase microstructure containing the intermetallic $Mg_{24}Y_5$ with a matrix containing increasing amounts of Y. Although there was Y_2O_3 in the surface layer, this film was too weak to prevent penetration by Cl^- ions. The increase in the corrosion rate with increasing Y content is attributed to the acceleration of the micro-galvanic corrosion caused by an increasing volume fraction of Mg–Y intermetallics.

Figure 3.1 also shows that in 0.1 M Na_2SO_4, the corrosion rate decreased with increasing Y content in the range 3% to 7% despite the fact that these were two-phase alloys with increasing amounts of second phase. In this case, the decrease in the corrosion rate with increasing Y content is attributed to an increasingly protective surface film by the incorporation of more Y in the surface film, due to a greater Y content in the α-Mg matrix resulting from non-equilibrium cooling.

For Mg–Y alloys, Y in the surface layer can increase the protectiveness of the surface film. The Mg–Y intermetallic $Mg_{24}Y_5$ can cause micro-galvanic corrosion, so Y can have a dual effect on the corrosion of a Mg alloy. Which effect is more important depends on the electrolyte. In 0.1 M NaCl, the chloride ions can penetrate the surface film, and localised corrosion initiated as filiform corrosion. The important effect is the micro-galvanic corrosion. The corrosion rate increases with increasing Y content once the Y content exceeds the Y solid solubility and the microstructure contains a second phase:

the Mg–Y intermetallic $Mg_{24}Y_5$. There is filiform corrosion and localised corrosion. In 0.1 M Na_2SO_4, the surface film can be protective and prevent serious corrosion. Despite the presence of the Mg–Y intermetallic $Mg_{24}Y_5$, the corrosion resistance increased with increasing Y content.

3.3.8 Stainless Mg

The films formed to date using alloying elements such as Zr, Nd and Y have not been particularly successful, but they do show one way to seek further improvement in the corrosion performance of Mg alloys. It was proposed [2] that a step improvement is expected for an Mg–X (or Mg–A–B–X) alloy where (a) the alloying element(s) is (are) in solid solution, and (b) the alloy forms a stable passive film based on an oxide of X rather than the Mg-based oxide/hydroxide formed on conventional Mg–base alloys [40]. This is the same approach as in Fe–Cr (and Fe–Ni–Cr, Fe–Cr–Ni–Mo–N...) alloys, the basis of stainless steels. In Fe–Cr alloys, the nature of the film formed in aqueous solutions changes from one based on Fe to a film based on Cr (see e.g. [41,42]). This change in film characteristics occurs at about 12% Cr, and concomitantly the corrosion resistance increases by orders of magnitude. The corrosion behaviour is similar to that of Cr, even though there may be only 12% Cr in the alloy. Such abrupt changes in corrosion behaviour (between low corrosion resistance and high corrosion resistance) occur in other systems, such as Co–Cr [43] , Ni–Cr [44], Fe–Si [45] and Fe–Ti [46].

It was suggested [2] that the most exciting passivating elements to study for Mg–base alloys are Cr, Ti and Si. Cr is well known for its passivating properties in the production of highly corrosion resistant stainless steels (Fe–Cr alloys), with corrosion resistance increasing with increasing Cr content above the threshold value of about 12% Cr. Cr is similarly effective in corrosion resistant Ni–base alloys and Co–base alloys. Similarly Ti forms corrosion resistant Ti–base alloys and Si is important in imparting high corrosion resistance in high Si cast irons. Furthermore, it is suggested that it is useful to explore the behaviour for such alloys with the Cr, Ti or Si in complete solid solution; it is expected that the Cr, Ti or Si can act as passivators, in contrast to the action they might have when in a second phase such as Mg_2Si.

Alloys containing significant quantities of Cr, Ti and Si in complete homogeneous solid solution have not been produced. They cannot be produced by conventional ingot metallurgy because the melting point (MP) of each alloying addition is above the boiling point (BP) of Mg, and all have low solubilities in Mg [47,48] (BP of Mg = 1090 °C, MP of Cr = 1857 °C, MP of Ti = 1667 °C, MP of Si = 1412 °C). Such alloys have also not been produced by rapid solidification processing [1,2,48]. However, production of such Mg–base alloys containing significant amounts (5–40 at.%) of Cr,

Ti or Si in complete homogeneous solid solution could be possible with the bulk-amorphous alloys.

Zberg *et al.* [49] studied a range of glassy $Mg_{60+x}Zn_{35-x}Ca_x$ alloys (x = 0–15 at.%) prepared by melt-spinning to produce thin ribbons of 0.050 mm thickness and copper mould injection-casting to 0.5 mm thick plates. Crystalline samples (Mg and $Mg_{98}Zn_2$) and the Zn-poor glassy alloys had significant corrosion rates. The Zn-rich glassy alloys had low corrosion rates. Animal studies with glassy $Mg_{60}Zn_{35}Ca_5$ showed no hydrogen evolution, no inflammatory reaction, and good biocompatibility. Thus Zberg *et al.* [49] showed that glassy $Mg_{60+x}Zn_{35-x}Ca_x$ alloys ($0 \leq x \leq 7$) had potential as a new generation biodegradable implant.

3.4 Impurity concentration

3.4.1 Manifestation

The impurity effect is a special case of a second phase accelerating the corrosion of a Mg alloy. It warrants separate treatment because it is important and is wholly negative. Four impurity elements (Fe, Ni, Cu and Co) were found by Hanawalt *et al.* [50] to have a large accelerating effect on saltwater corrosion of Mg binary alloys. Corrosion rates were accelerated 10–100-fold when their concentrations were increased [51]. These impurity elements have extremely deleterious effects because of their low solid-solubility limits in α-Mg and their abilities to serve as active cathodic sites [52]. When their concentrations exceed their tolerance limits, they serve as active catalysts for electrochemical corrosion [53]. For each of these elements a tolerance limit can be defined as illustrated by Fig. 3.17. When the impurity content exceeds the tolerance limit, the corrosion rate is greatly accelerated, whereas, when the impurity content is lower than the tolerance limit the corrosion rate is low.

Studies [54,55] have shown dramatic improvements in corrosion resistance

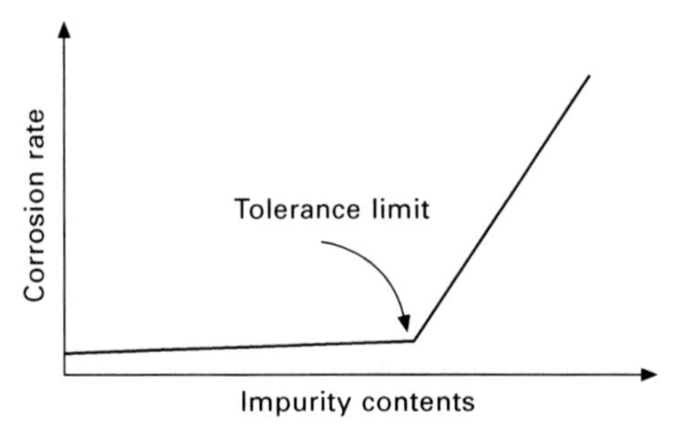

3.17 Generalised curve for the influence of the impurity elements Fe, Ni, Co and Cu [1,2].

through the control of these impurity elements. The critical tolerance level [56] for commercial alloys were defined and applied to the production specifications for AZ91 and AM60. Dow recommended the following contaminant limits to ensure optimum saltwater corrosion performance in AZ91: Fe < 50 ppm, Ni < 5 ppm, Cu < 300 ppm. This provides an effective way to produce corrosion resistant Mg alloys and is of great significance from a practical point of view. For example Hillis and Reincheck [57] showed that AZ91 containing low levels of impurities had corrosion rates lower than the cast aluminium alloy 380 and steel for salt spray testing and for atmospheric exposures at the Brazos River Site on the Texas Gulf Coast.

3.4.2 Mg–Fe phase diagram

The mechanistic possibilities for the impurity effect were summarised [2] as follows. In each case, a key part of the corrosion acceleration is easy hydrogen evolution on a new phase in contact with the Mg alpha matrix.

- Accelerated galvanic corrosion of the Mg alloy matrix due to general dissolution of the Mg alloy followed by the re-precipitation of metallic Fe, Ni and Cu on the alloy surface. The re-precipitation of metallic Fe, Ni and Cu on the alloy surface is the key part of this mechanism, which is similar to the acceleration of corrosion by alloyed copper for aluminium alloys.
- Galvanic acceleration of corrosion when any of the impurity elements Fe, Ni or Cu exceeds its solubility limit and precipitates as a separate phase in the alloy, so that the alloy has a two-phase microstructure.

In order to understand the impurity tolerance limits, Liu *et al.* [13] calculated Mg phase diagrams using the Pandat software package (database PanMg7) [58]. The phase diagrams so calculated used the thermodynamic data in the latest available Pandat Mg database. The thermodynamic data had not been optimised for these calculations. Thus, it is likely that particular numerical values may be somewhat in error, but the trends can be assumed to be valid. Furthermore, the database was not complete; the database included data for Fe and Cu, but not Ni or Co. Thus phase diagrams were calculated to understand the tolerance limits for Fe and Cu. Understanding the tolerance limits for Ni and Co needs an extension of the data in the database. Also the database did not include Pr and La, so it was not possible to fully explore the effect of rare earth addition as in AE42.

Figure 3.18 presents the calculated Mg–Fe phase diagram: a eutectic phase diagram with an eutectic temperature of ~650 °C, and with α-Mg having a maximum Fe solubility of ~10 ppm Fe. Cooling of an Mg–Fe alloy containing more than 180 ppm causes precipitation, from the melt, of a body centred cubic (BCC) phase, essentially pure Fe with little Mg in solid solution. For

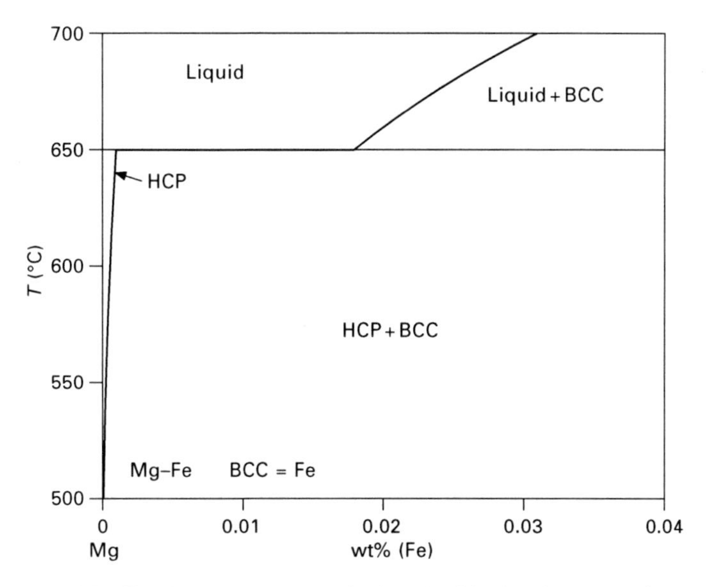

3.18 Mg–Fe phase diagram calculated with the Pandat software [13] (BCC = body centred cubic, HCP = hexagonal close packed).

an Fe content of 180 ppm, the calculated phase diagram predicts that on cooling, the liquid Mg alloy undergoes eutectic solidification at 650°C to form α-Mg containing about 10 ppm iron in solid solution plus the BCC phase. However, the two-phase region (liquid Mg + α-Mg) is extremely narrow, so that it would be expected that the pre-eutectic and eutectic reactions would be suppressed during normal (non-equilibrium) cooling of a Mg ingot or casting, so that a Mg alloy containing less than 180 ppm Fe would solidify to a single α-Mg phase with Fe in supersaturated solid solution in the Mg lattice. The implication is that castings produce single-phase Mg up to a critical Fe concentration of 180 ppm and that Mg-castings contain a BCC phase (rich in Fe) for iron contents greater than 180 ppm. This value of 180 ppm Fe was compared [13] with the tolerance level for Fe in pure as-cast Mg reported to be 170 ppm [1,5,59,60] or 150 ppm [61]. The calculated phase diagram thus offers an explanation of the Fe tolerance limit: the tolerance level corresponds to the minimum content of Fe in a cast Mg alloy for which a BCC phase precipitates from the melt before final solidification.

The phase diagram allows estimation of how much second phase forms for a given alloy in equilibrium. For a Mg alloy containing 280 ppm Fe, the fraction of the primary BCC phase, $f_{BCC/casting}$, can be calculated using the lever rule as follows:

$$f_{BCC/casting} = (0.028 - 0.018)/(100 - 0.018) = 0.01\%$$

If this BCC phase is responsible for increasing the corrosion rate by a factor of 100 above that of HP Mg, then the BCC phase needs to evolve hydrogen 10^6 times faster than on α-Mg.

Figure 3.18 provides a prediction regarding heat-treated cast alloys and regarding wrought alloys. A Mg alloy casting containing 40 ppm Fe would be expected to have a low corrosion rate because the Fe content is below the Fe tolerance limit of 170 ppm Fe measured for castings. What happens to the corrosion performance if this cast Mg alloy is heat-treated at just below 650 °C? Figure 3.18 indicates the precipitation of some BCC phase; the volume faction can be determined, by applying the lever rule, as follows.

$$f_{\text{BCC/casting heat treated}} = \frac{0.0040 - 0.0010}{100 - 0.001} = 0.0031\%$$

Is this amount of the Fe-rich BCC phase sufficient to cause the heat-treated HP Mg to have a corrosion rate higher than the as-cast condition? This issue was studied experimentally. The results are shown in Fig. 3.19, which presents the corrosion behaviour, measured as hydrogen evolution volume, for (i) as-cast HP Mg, (ii) HP Mg heat treated for 1 day at 550 °C and (iii) HP heat treated for 2 days at 550 °C. The average corrosion rate for as-cast HP Mg in Fig. 3.19 was 1.8 mm/yr consistent with other measurements [8,11]; this gives confidence in the experimental measurements. There was a significant increase in corrosion rate after heat treatment at 550 °C, consistent with the precipitation of the BCC Fe-rich phase as predicted, Fig.

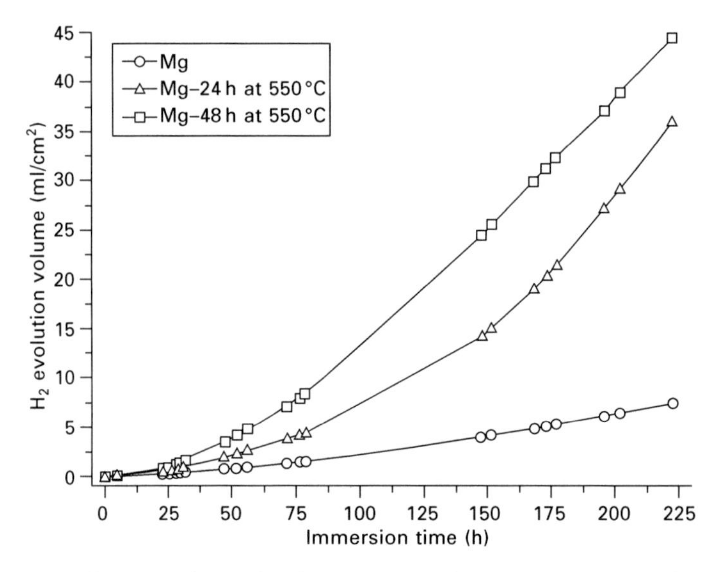

3.19 The corrosion behaviour, measured as hydrogen evolution volume, for as-cast HP Mg, HP Mg heat treated for 24 h at 550 °C and HP heat treated for 48 h at 550 °C [13].

3.18. This indicates that the Fe tolerance limit is indeed ~5–10 ppm Fe for heat-treated pure Mg.

3.4.3 Mg–Al–Fe

Figure 3.20 presents an isothermal section calculated for 651 °C, slightly above the eutectic temperature, through the Mg–Al–Fe phase diagram. This section indicates that the eutectic point is shifted to lower Fe contents and thus the Fe tolerance limit decreases rapidly with increasing Al content. The trend of decreasing Fe tolerance limit with Al alloying has been previously documented [13]. Figure 3.20 provides a numerical prediction and in particular indicates that there is a significant decrease with a few tenths of percent of Al.

It is reported [1,2,62,63] that increasing Al concentration in Mg–Al alloys has a beneficial effect on the corrosion behaviour in chloride media, but the specific mechanism and influence of Al is still not well understood. For instance, Song *et al.* [8] found that the corrosion rate of Mg–9%Al at 101 mm/yr was slightly smaller than that of Mg–2%Al at 114 mm/yr, however, the difference was small and may not be significant because of the high level of Fe content above the tolerance limit. Similarly, Song *et al.* [62] found that increasing Al concentrations in the range 2–9 wt% Al in Mg–Al single phase alloys did correlate with a decrease in the corrosion rate in 5% NaCl. These alloys contained a high Fe content and the corrosion rates were

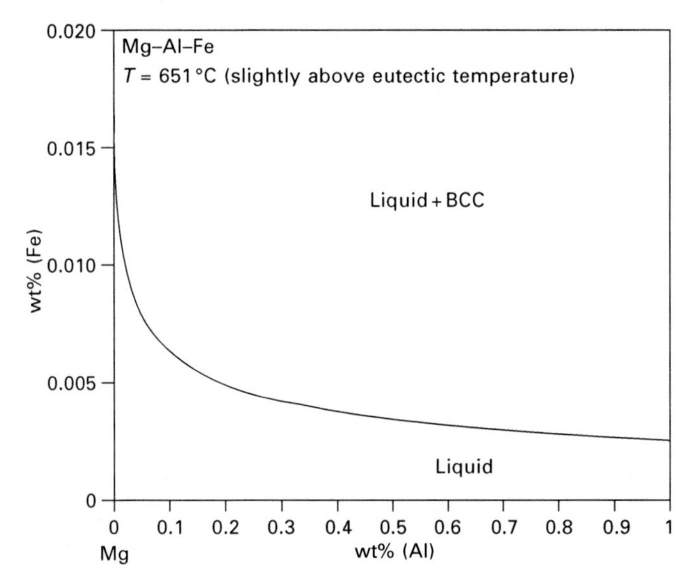

3.20 Calculated isothermal section through the Mg–Al–Fe phase diagram [13].

much higher than for high-purity alloys; thus the decrease in corrosion rate appears to be related to a decrease in the Fe impurity level with increasing Al content rather than to the Al content itself. Therefore, to investigate the influence of Al on the corrosion of Mg, the Fe content should be kept to a low concentration below the tolerance limit.

3.4.4 Wrought-cast

Previous work has typically measured a corrosion rate of ~1 mm/yr in continuous immersion in 3% NaCl solution for HP as-cast Mg. The one outstanding exception is the recent work by Prado *et al.* [64,65] who reported a much higher corrosion rate: a corrosion rate of ~ 600 mm/yr [64,65] and ~ 90 mm/yr [65] for wrought HP Mg (containing 40 ppm). The material used by Prado *et al.* was wrought rather than the as-cast material used in the prior studies. Figure 3.18 implies that wrought Mg has a Fe tolerance limit ~ 5–10 ppm, much lower than the Fe tolerance level of 130 ppm Fe measured for as-cast material.

Figure 3.18 implies that the Fe tolerance limit might be ~ 5–10 ppm for wrought pure Mg, i.e. the implication is that wrought pure Mg has a tolerance limit significantly lower than that of cast pure Mg. Does this also apply to wrought Mg alloys? The answer appears to be that wrought alloys containing Mn, as do typical commercial Mg alloys, have a Fe tolerance limit that is not too dissimilar to that of their cast counterparts. The presence of Mn seems to dominate over the effect implied from Fig. 3.18. This evaluation is based on the work of Ben-Haroush *et al.* [66] who found that extruded AM60 acted consistent with being high purity and did not show extreme sensitivity to Fe as is inferred from Fig. 3.18; they found that the Al_8Mn_5 phase did not cause any significant micro-galvanic acceleration of corrosion in the wrought alloy. It is also consistent with our recent work with extruded AZ31 and AM30 [67].

3.4.5 High-purity castings

The controlled casting experiments of Hillis and Reichek using AZ91 [60], AM60 [68] and AS41 [69] demonstrated that HP alloys could be produced from LP alloys by control of the casting temperature. For AZ91 experiments, they melted ~40 kg of HP AZ91–base alloys containing ~9% Al, 0.5% Zn, ~390 ppm Fe, <10 ppm Ni and < 100 ppm Cu. The melt was equilibrated with 0.2% Mn (trial 1), 0.4% Mn (trial 2) or 0.8% Mn (trial 3) at 750 °C; die-castings were made at 750 °C and after equilibration for 10–15 min at the three lower nominal temperatures: 725 °C, 690 °C and 650 °C (in each case they measured the actual temperature of the melt at the time of casting, and these actual temperatures were somewhat different from the nominal temperatures). Chemical analysis of the resulting die-castings revealed the chemical composition of the melt just before casting. Chemical analysis of the resulting die-castings of trial 1 indicated a decrease in only the Fe content as the melt temperature was decreased from 750 °C to 690 °C consistent with the precipitation from the melt of an iron rich phase containing little Mn. In contrast, trial 3 (0.8% Mn) indicated a decrease in both the Fe content and the Mn content between 750 °C to 690 °C consistent with the precipitation from the melt of an iron-rich phase containing substantial Mn. Each of these die castings had a Fe content above the Mn-dependent tolerance limit and had a high corrosion rate in the salt spray test. The trial 1 die-casting at 650 °C had a lower Fe composition and a lower Mn composition indicating the precipitation of both these elements from the melt between 690 °C and 650 °C, which could be by the precipitation of a single Fe_aMn_b compound (which could also contain Mg or Al) or by the precipitation of two compounds, one Fe-rich, the other Mn-rich. The trial 1 650 °C casting had a Fe content below the tolerance limit and a low corrosion rate in the salt spray test. The other trials with AZ91 [60], and the subsequent trials with AM60 [68] and AS41 [69], revealed similar trends. Table 3.4 presents the values of the measured critical melt temperature, T_M, at which castings were produced with measured low corrosion rates and above which there was precipitation of both Fe and Mn from the melt. Scanning electron microscopy/energy dispersive X-ray spectroscopy (SEM/EDX) analysis, of the sludge of the casting trials using AZ91 [60], revealed that the precipitate from the Mg melt contained Fe–Mn–Al in variable amounts; the particulates

Table 3.4 Values of the measured critical melt temperature, T_M, measured for AZ91 [60], AM60 [68] and AS41 [69]

Alloy	Al (wt%)	$[Mn]_{750°C}$(wt%)	T_M(°C)	T_C(°C)	T'_C(°C)
AZ91	9	0.2	654	640	650
AZ91	9	0.4	694	690	705
AZ91	9	0.8	> 750	760	755
AM60	6	0.2	640	620	630
AM60	6	0.4	650	670	675
AM60	6	0.8	720	720	730
AS41	4	0.2	< 660	620	625
AS41	4	0.4	658	630	660
AS41	4	0.8	–	685	710

T_M is the temperature at which low corrosion rates were measured and above which there was precipitation of both Fe and Mn from the melt. $[Mn]_{750°C}$ is the Mn concentration in the starting melt at 750 °C; the die-castings typically contained a lower Mn concentration. T_C is the temperature at which the calculated Mg–Al–Mn–Fe phase diagrams (calculated with 0.02% Fe) predict that solidification of a casting would lead to no BCC phase in the casting. T'_C is the corresponding temperature from the phase diagram calculated with 0.005% Fe.

were often Fe and/or Mn-rich in the core with the surrounding or bridging regions containing Al + Mn.

Figure 3.21 presents a pseudo-binary section calculated [13] through the Mg–Al–Mn–Fe phase diagram at 0.4% Mn and 0.02% Fe; this section was calculated to understand the controlled casting experiments carried out by Hillis and co-workers with AZ91 [60], AM60 [68] and AS41 [69]. Figure 3.21 indicates that cooling of an alloy containing 6% Al causes initially the precipitation of an iron-rich BCC phase labelled BCC_B2. Between ~675 and ~660 °C there is a three phase region of liquid + BCC_B2 + Al_8Mn_5. From the experiments of Hillis and co-workers [60,68,69] it is assumed that the BCC_B2 phase settles out as the cast is slowly cooled and that the alloy as-cast at 660 °C contains no BCC_B2 phase but only Mg liquid + Al_8Mn_5. If it is assumed that the phase Al_8Mn_5 is passive (i.e. it is no more effective as a cathode than HP Mg) then a 6% Al alloy cast at 660 °C has an Fe content below the Fe tolerance limit and would be expected to show a low corrosion rate. Thus, Fig. 3.21 predicts the critical temperature to be 660 °C at which the alloy can be cast to produce a casting with a low corrosion rate, this calculated critical temperature, T_C, is included in Table 3.4 (this is the alloy with 6% Al–0.4% Mn). Similarly, Fig. 3.21 predicts critical temperatures of 690 °C for 9% Al (corresponding to AZ91) and 630 °C for 4% Al (corresponding to AS41).

Similarly the pseudo-binary sections were calculated [13] through the

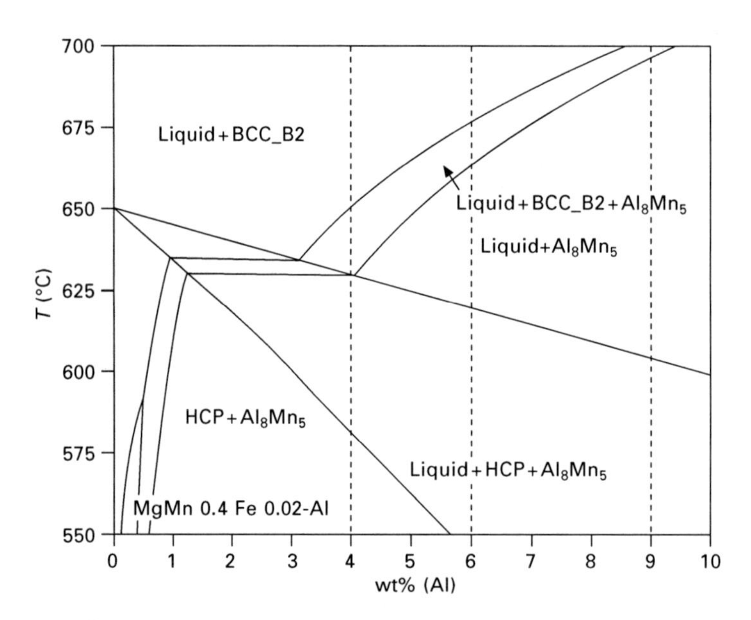

3.21 Calculated pseudo-binary section through the Mg–Al–Mn–Fe phase diagram at 0.4%Mn + 0.02%Fe [13].

Mg–Al–Mn–Fe phase diagram at 0.2% Mn + 0.02% Fe and 0.8% Mn + 0.02% Fe to allow comparison with the experiments of Hillis and co-workers starting with 0.2% Mn and 0.8% Mn, respectively. The critical temperatures have been included in Table 3.4. There is good agreement between the measured critical temperature, T_M, and the calculated critical temperature, T_C, estimated from the calculated phase diagrams, indicating the phase diagrams to be an extremely useful tool for the prediction of tolerance limits as well as processing parameters.

The phase diagram, such as Fig. 3.21, was calculated [13] using 0.02% Fe (200 ppm Fe); this value is reasonable to allow comparison with the experiments of Hillis and co-workers, they used 390 ppm in the starting melt for their experiments with AZ91 (the actual Fe content was typically 200 ppm or lower in most of their castings) and 200 ppm Fe for the starting melt for their experiments with AM60 and AS41.

However, the phase diagram might be sensitive to the Fe content. The phase diagram calculated for 0.4% Mn and 0.04% Fe was essentially identical to the one with 0.4% Mn and 0.02% Fe; thus there is little change at higher Fe contents. What about lower Fe contents? Lower Fe contents are relevant to the experiments of Hillis and co-workers and to the production of high-purity castings from low-purity feedstock, in which case the Fe is allowed to settle out, and consequently the Fe content decreases; this situation is addressed by the phase diagram calculated [13] for 0.2% Mn + 0.005% Fe. There is an influence of Fe content; the calculated temperature T'_C corresponding to T_C, is included in Table 3.4; there is even better agreement with T_M.

The calculated phase diagrams could explain the controlled casting experiments of Hillis and co-workers [60,68,69] based on the assumption that the BCC_B2 phase (an Fe–Mn–Al phase, essentially pure Fe containing some Mn and Al, and little Mg in solid solution) is allowed to settle out when the temperature of the melt is maintained at temperatures of 725 °C, 690 °C and 650 °C; there is also agreement if the Al_8Mn_5 phase settles from the melt. An estimate of 5.9 g/cm^3 (= 0.333 × 7.84 + 0.333 × 7.21 + 0.333 × 2.7) can be made for the density of the BCC_B2 phase with an equal number of Fe, Mn and Al atoms on the assumption that the density is simply related to the mass of the constituent atoms. (The density of Fe is 7.84 g/cm^3, the density of Mn is 7.84 g/cm^3 and the density of Al is 2.7 g/cm^3). Similarly the density is estimated to be 5.0 g/cm^3 for Al_8Mn_5 containing an equal number of Al and Mn atoms. While these are rough estimations of the density of BCC_B2 and Al_8Mn_5, these values are considerably in excess of the density of liquid Mg at 1.58 g/cm^3. Thus it would indeed be expected for there to be a tendency for both these phases to settle from the melt. The 0.8% trials are consistent with the settling out of the BCC_B2 phase and the calculated composition for this phase. The 0.2% Mn trials imply that equilibrium does not occur and an Fe-rich BCC_B2 phase settles out between 750 and 690 °C.

Hillis and co-workers [60,68,69] found experimentally using industrial trials that the temperature range from 750 to 650 °C is the important temperature range in purifying the melt and decreasing the Fe content. This temperature range corresponds to the solidification of BCC_B2. The calculated volume fraction of the BCC_B2 phase at ~ 650 °C is about 0.0005. If an Fe content of 40% is assumed, and if it is also assumed that all this volume fraction of BCC_B2 settles out, then the amount of Fe settling out corresponds to 0.02%, the amount of Fe in the alloy. Thus a back-of-the-envelope mass balance calculation indicates that the solidification of the BCC_B2 phase and its settling out from the melt between 750 °C and 650 °C is sufficient to purify the melt.

The SEM/EDX analysis of the sludge of the casting trials using AZ91 [60] revealed that the precipitate from the Mg melt contained Fe–Mn–Al in variable amounts; the particulates were often Fe and/or Mn-rich in the core, with the surrounding or bridging regions containing Al + Mn. A Fe or Mn-rich core is consistent with the precipitation of the BCC_B2 phase, the surrounding regions containing Al + Mn most probably corresponds to Al_8Mn_5, which contained little Fe because the Fe has already been incorporated in the BCC_B2 by the time of solidification of the Al_8Mn_5.

It is perhaps worth restating that the above analysis rests on the assumptions that the BCC_B2 phase is a good hydrogen cathode and that Al_8Mn_5 is not.

The calculated phase diagrams can explain the production of HP castings by means of control of melt conditions; this has significance for the production of quality castings from recycled Mg. As shown by the work of Hillis and co-workers [60,68,69] using commercial high pressure die cast (HPDC), HP castings can be produced by control of the melt, particularly the melt temperature. The alternative approach is to develop tailored alloys that are less sensitive to the impurity elements as, for example, by the research of Scharf *et al.* [70] and Fechner *et al.* [71].

3.4.6 Surface activation for CP Mg

Figure 3.3 indicates that the corrosion rate changes with immersion time. Initially the corrosion rate is low. During this period the corrosion behaviour is dominated by whatever film is on the sample surface and by the breakdown of the surface film. Williams and McMurray [72] have shown that breakdown can be relatively rapid, as is illustrated in Fig. 3.22 for CP Mg (280 ppm Fe, i.e. Mg containing an Fe content above the tolerance limit) in 5% NaCl, pH 6.5. Figure 3.22 presents the distribution of the normal anodic current density, j_Z, measured using a scanning vibrating electrode technique (SVET). Local breakdown occurred between 1 and 6 min after immersion in the solution. The first scan after immersion, Fig. 3.22(a), recorded initiation

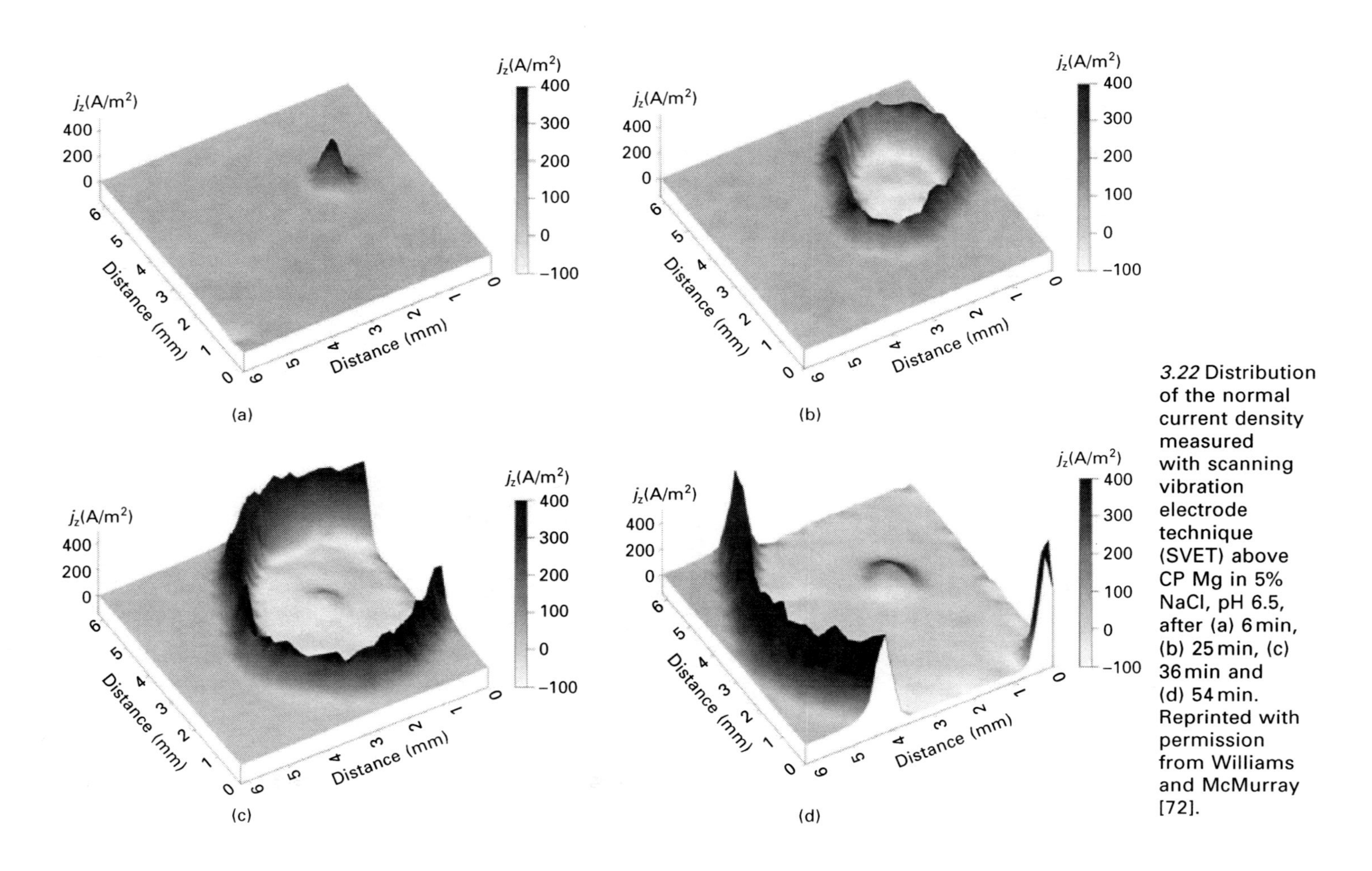

3.22 Distribution of the normal current density measured with scanning vibration electrode technique (SVET) above CP Mg in 5% NaCl, pH 6.5, after (a) 6 min, (b) 25 min, (c) 36 min and (d) 54 min. Reprinted with permission from Williams and McMurray [72].

of intense corrosion as a region of intense anodic current in the top right quadrant. Subsequent scans show that the corrosion disc expanded radially. There were intense anodic currents at the circumference while the interior was strongly cathodic. Any currents from the intact film were negligible in comparison. The local anodic current, at the boundary of the spreading corrosion, increased significantly with the size of the corroding disc. The local anodic current balanced the cathodic current in the interior of the corroding disc. The local cathodic currents within the central portion of the disc remained approximately constant in magnitude while the corrosion expanded over the whole surface. Thereafter the magnitude of the anodic and cathodic currents decreased significantly. The majority of the surface remained cathodic at protracted immersion times, Fig. 3.23, while local anodic currents were confined to the sample edges and a single central location, possibly a location of a Fe-rich particle.

Williams [73] carried out similar experiments for a 1 cm^2 HP Mg (100 ppm Fe) sample in 5% NaCl, pH 6.5 and SVET maps were obtained at 2, 8.4, 17 and 28.3 h after specimen immersion. This sample took significantly longer than the CP Mg to show the first sign of breakdown – about 2 h as opposed to a few minutes for CP Mg. The point of breakdown was visually identified as a black dot on the surface with some associated hydrogen evolution. The local current density was over an order of magnitude lower than that obtained with CP Mg at initial breakdown. Thereafter the HP Mg showed features akin to filiform corrosion. *In situ* SVET showed that the

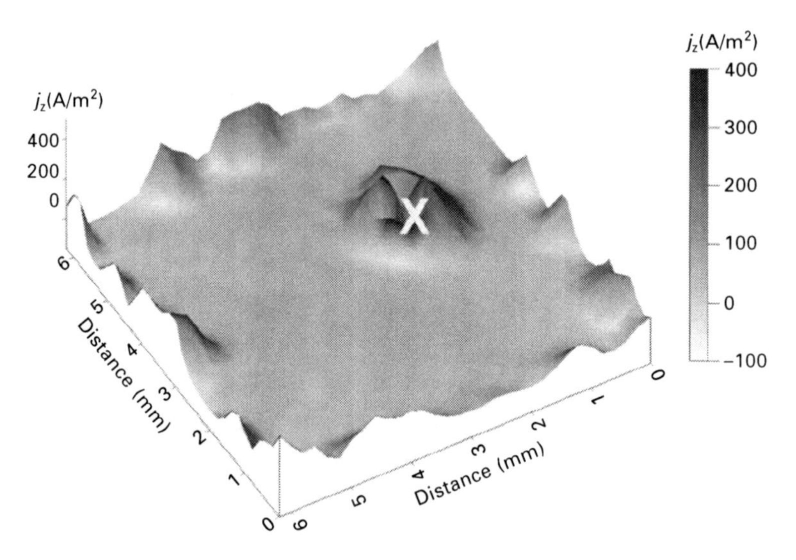

3.23 Distribution of the normal current density measured with SVET above CP Mg in 5% NaCl, pH 6.5, after 80 min. Reprinted with permission from Williams and McMurray [72].

fronts of the black corrosion tracks were net local anodes, while the corroded region left in their wake as they traversed the Mg surface were cathodically activated compared with the intact surface which was electrochemically inert. The local current densities were quite low (0 to 40 A/m^2 anodic, 0 to –2 A/m^2 cathodic). With increasing time, more of the surface was covered by the local anodes, and their number increased. Figure 3.24 shows a typical example after 17 h immersion.

3.4.7 Surface activation for two-phase alloys

Surface activation also occurs for two phase Mg alloys [11,12,24,25,27,74]. Corrosion typically initiates in the α-phase as illustrated [11] in Fig. 3.25 and spreads out over the specimen surface as illustrated in Fig. 3.26. This is one reason that the corrosion rate increases with immersion time as illustrated [11] in Fig. 3.3.

3.5 Surface condition

Song and Xu [75] studied AZ31 sheet because it is currently one of the most promising candidate materials for the closure inners in some vehicles. In the automobile industry, a Mg–base metal part must undergo various mechanical and chemical surface treatments, heat treatments and mechanical deformations. Figure 3.27 presents the influence on corrosion behaviour (as characterised by hydrogen evolution) of the following surface conditions: (i)

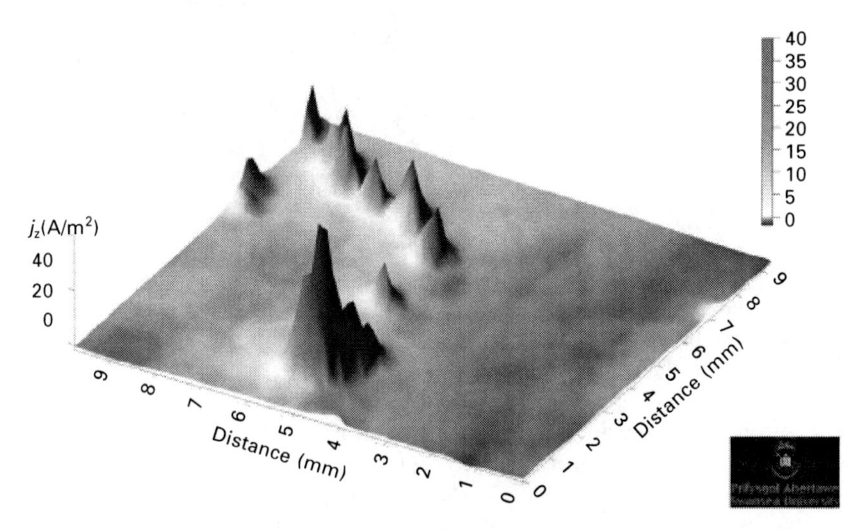

3.24 Distribution of the normal current density measured with SVET above HP Mg in 5% NaCl, pH 6.5, after 17 h. Reprinted with permission from Williams [73].

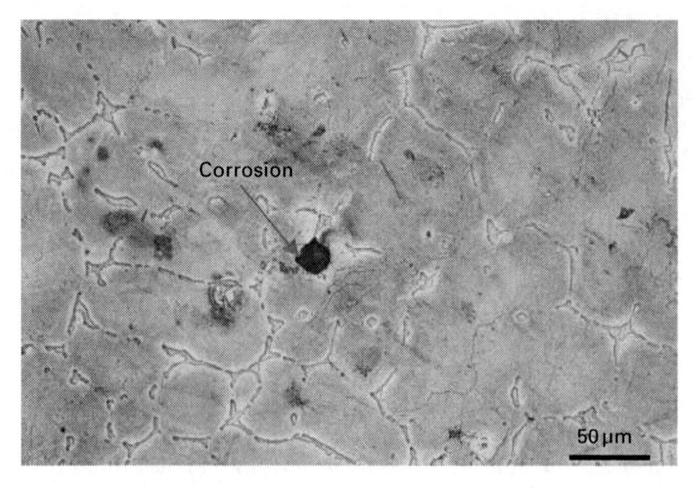

3.25 Corrosion initiation in two-phase alloys like ZE41 often occurs in the α-phase adjacent to the second phase as shown here for ZE41 in 1 M NaCl after 10 min immersion [11].

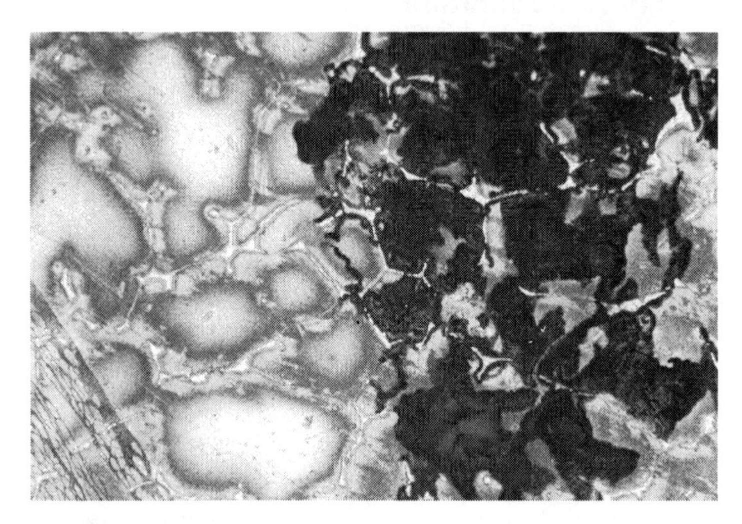

3.26 After corrosion initiation in two-phase alloys like ZE41, the corrosion spreads over the surface in the α-phase leaving the second phase uncorroded as shown here for ZE41 in 1 M NaCl after 30 min immersion [11].

AR – as-rolled mill finished sheet containing surface oxides and Fe-containing contaminants from the rolling mill; (ii) HT – heat-treated in air at 450 °C for 10 min (which did not remove the surface embedded Fe-containing particles); (iii) HT + sandblast – heat-treated and sandblasted with glass beads to remove surface oxide and produce a shiny surface (which did not remove the surface embedded Fe-containing particles but produced a more uniform distribution of finer Fe-containing particles indicating Fe pick-up during sandblasting); (iv) AR + grinding – AR surface mechanically ground with 400, 800 and 1200 grit SiC paper (removed the surface embedded Fe-containing particles as did also each of the following); (v) HT + grinding – HT surface mechanically ground with 400, 800 and 1200 grit SiC paper; (vi) HT + acid clean: HT sample immersed in 10% H_2SO_4 for 20 s at room temperature to remove surface contaminants, rinsed in deionised water and ethanol and dried; (vii) HT + grinding + ageing – the ground HT sample was heat treated in air for 8 h at 250 °C. Figure 3.28 presents mass loss data.

Figures 3.27 and 3.28 show that the as-rolled sheet had a high corrosion rate attributed to the Fe-containing particles in the surface from the rolling process. Heat treatment led to a slight increase in corrosion rate. Sandblasting led to a further increase in corrosion rate consistent with Fe-particle pick-up during sandblasting. Both acid cleaning and grinding effectively removed the surface contaminants and dramatically improved the corrosion resistance.

Song *et al.* [76] measured the mass loss rate in 3.5% NaCl for HP Mg as cast and after being subjected to a number of equal channel angular pressings (ECAP). They found that the corrosion rate increased with number of ECAP passes. Based on the work of Song and Xu [75], the increase in corrosion rate with ECAP passes observed by Song *et al.* [76] is consistent with surface pick-up of Fe-containing particles in each ECAP pass.

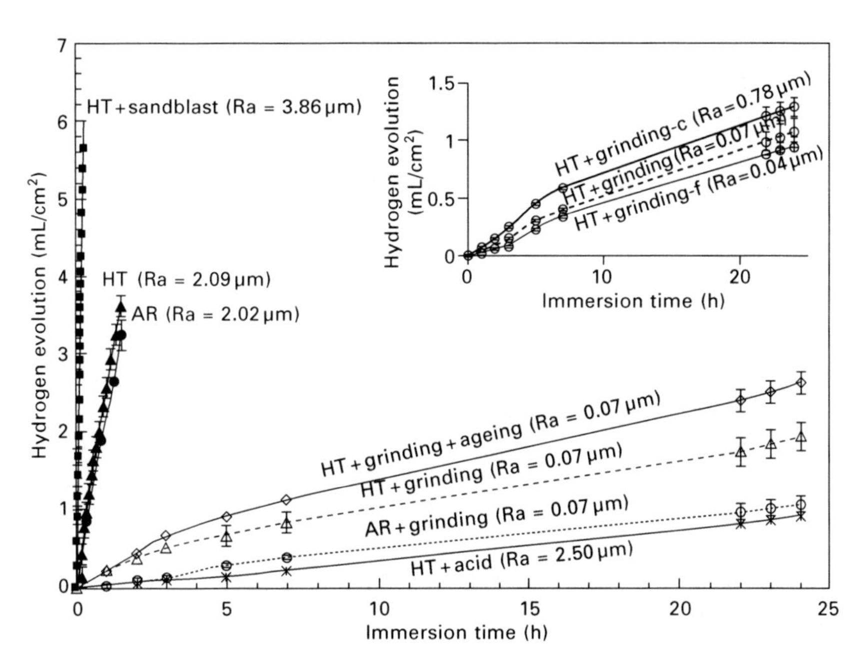

3.27 Hydrogen evolution from pre-treated AZ31 specimen immersed in 5% NaCl. The surface roughness values (Ra) are given [75].

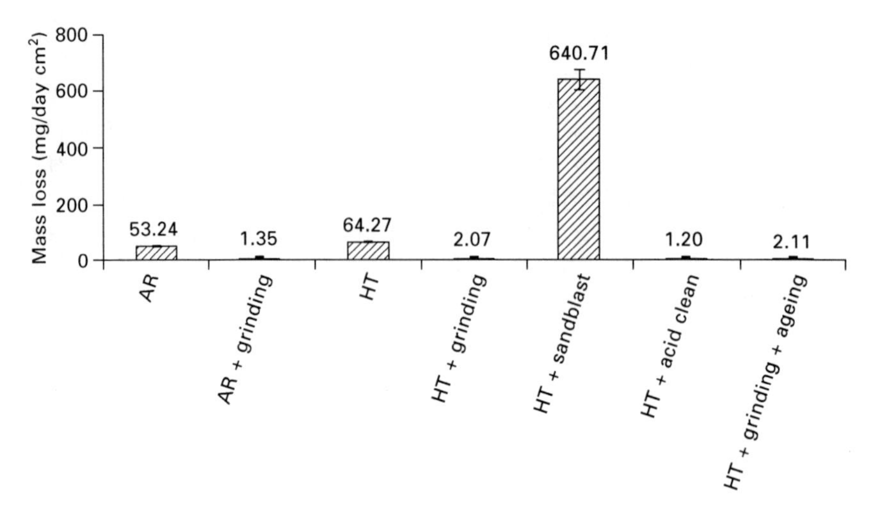

3.28 Mass loss for pre-treated AZ31 specimen immersed in 5% NaCl for 24 h [75].

Crystallographic orientation can also influence corrosion. Song *et al.* [77] showed that (0001) crystallographic planes of AZ31 were more electrochemically stable and corrosion resistant than $(10\bar{1}0)$ and $(11\bar{2}0)$ planes in 5 wt% NaCl. The different corrosion performance was attributed to their different surface energy levels or surface potentials [77]. This was similar to the prior finding by Liu *et al.* [78] that (0001) planes were the most corrosion resistant for pure Mg corroding in 0.1 M HCl.

3.6 Medical implant applications

3.6.1 *In vitro* testing

This section deals with how our understanding of the Mg corrosion mechanism can be used to help understand the behaviour of Mg alloys for the other application of growing importance to Mg alloys as biodegradable implants for medical applications. Solutions that elucidate these applications tend to form surface films and the corrosion rate decreases with immersion time.

Witte *et al.* [79] found that the corrosion rates of AZ91D and LAE442 measured *in vivo* were orders of magnitude lower than those measured in substitute ocean water prepared according to ASTM-D1141-98 [80]. Subsequently, *in vitro* testing has been carried out in a variety of solutions including: (i) Hank's solution, (ii) simulated body fluid (SBF), (iii) artificial plasma (AP), (iv) phosphate buffered saline (PBS) and (v) minimum essential medium (MEM, Invitrogen).

Figure 3.29 shows hydrogen evolution data for a range of Mg alloys in Hanks's solution at 37 °C [81]. In agreement with the trends discussed above,

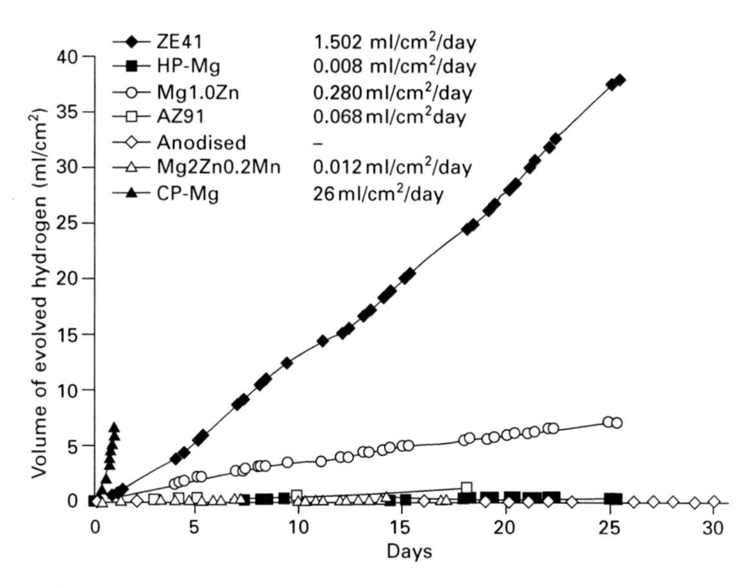

3.29 Hydrogen evolution and average corrosion rates for Mg alloys in Hank's solution at 37 °C [81].

HP Mg had the lowest corrosion rate, and the corrosion rate increased with increasing amount and effectiveness of the second phase as an effective cathode to cause micro-galvanic corrosion. Mg–2Zn–0.2Mn had a corrosion rate of 0.03 mm/yr, which was thought to be close to the corrosion rate of 0.02 mm/yr, which was thought to be acceptable for medical implants. If the corrosion rate was too high, then Song [81] suggested that the corrosion rate could be decreased with an anodised coating.

Figure 3.30 shows hydrogen evolution data for four wrought Mg alloys in SBF [82], showing the tendency for such solutions to produce surface films that are partially protective and for the corrosion to decrease with immersion time. This tendency for the corrosion to decrease with time is often reported [83–90], and examples are presented in Figs. 3.31 [84] and 3.32 [87]. Figure 3.31 presents the average corrosion rate for Mg–1.0Mn–1.0Zn in Hank's solution and SBP as a function of immersion time and indicates that Hank's solution is the more aggressive. Figure 3.32 presents the average corrosion rate for AZ31 in Hank's solution as a function of immersion time. The AZ31 was in the following three metallurgical conditions: (i) SC – squeeze cast, α-phase + some second phase particles, (ii) HR – hot rolled, fine-grained α, 15 μm, (iii) ECAP – equal channel angular pressing, finer-grain α, 2.5 μm. The higher corrosion rate is the squeeze cast condition, is attributable to the presence of the second phase particles.

Figure 3.30 shows [82] that there was a significant difference in corrosion performance of ZQ30 compared with the other alloys ZW21, WZ21 and

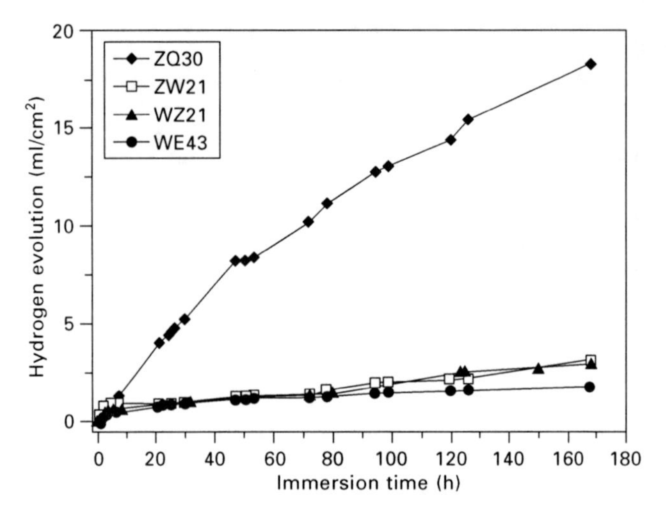

3.30 Hydrogen evolution data for four wrought Mg alloys immersed in SBF at 37 °C. Reprinted with permission from Haenzi *et al.* [82].

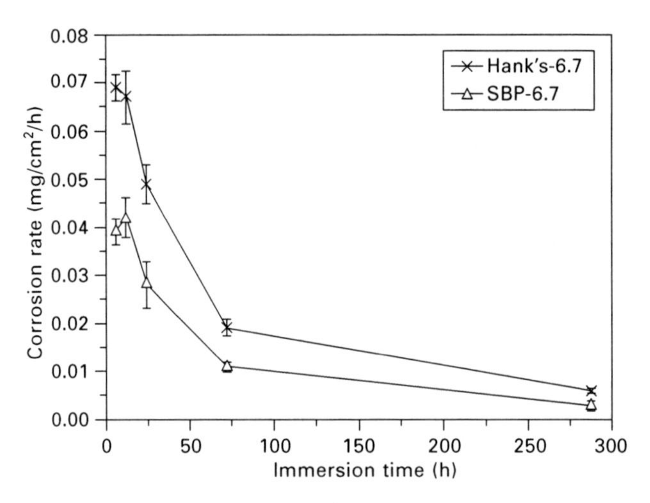

3.31 Average corrosion rate for Mg–1.0Mn–1.0Zn in Hank's solution and SBP as a function of immersion time. Reprinted with permission from Yang and Zhang [84].

WE41; this difference was caused by a greater tendency of ZQ30 to suffer significant localised corrosion attributed to a less protective surface film. Thus the filming tendencies and film properties are important considerations in the performance of Mg alloys in these solutions.

Figure 3.33(a) shows the influence of testing media on the corrosion of WZ21 and Fig. 3.33(b) the corresponding pH change. WZ21 corroded most in SBF, whereas little hydrogen was evolved in PBS and there was little

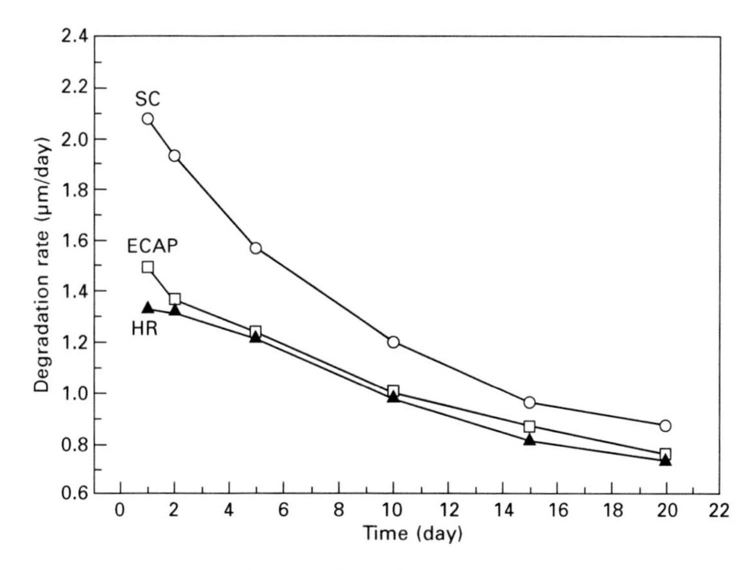

3.32 Average corrosion rate for AZ31 in Hank's solution as a function of immersion time. The AZ31 was in the following three metallurgical conditions: (i) SC – squeeze cast, α phase + some second phase particles, (ii) HR – hot rolled, fine-grained α, 15 mm, (iii) ECAP – equal channel angular pressing, finer-grain α, 2.5 mm. Reprinted with permission from Wang *et al.* [87].

degradation in MEM. This behaviour is explicable when the change in pH is considered (Fig. 3.33(b)). In the non-buffered MEM, even though it is one of the most widely used synthetic cell culture media, the pH increased rapidly and reached a value of 8 within a few hours of immersion. This significantly lowered the corrosion rate. In PBS, the pH also increased, though less significantly. To keep the conditions as constant as possible, the solution was changed after the first 24 h immersion, and then after every 48 h. This explains the drops in pH value in Fig. 3.33(b).

Haenzi *et al.* [83] found that the general hydrogen evolution versus time plot was sigmoidal in shape: there was often an incubation period with a low rate of hydrogen evolution, then a period of accelerated hydrogen evolution followed by a deceleration of hydrogen evolution. The initial incubation period was attributed to the breakdown of the original surface film and the subsequent deceleration of hydrogen evolution is attributed to the partially protective surface film formation.

The surface film has been found to contain magnesium carbonate and hydroxyapatite (magnesium apatite) and it has been suggested [84,91,92] that these form by the interaction of Mg ions with the calcium, carbonate and phosphate ions which are constituents of Hank's solution and SBF, by the following precipitation reactions:

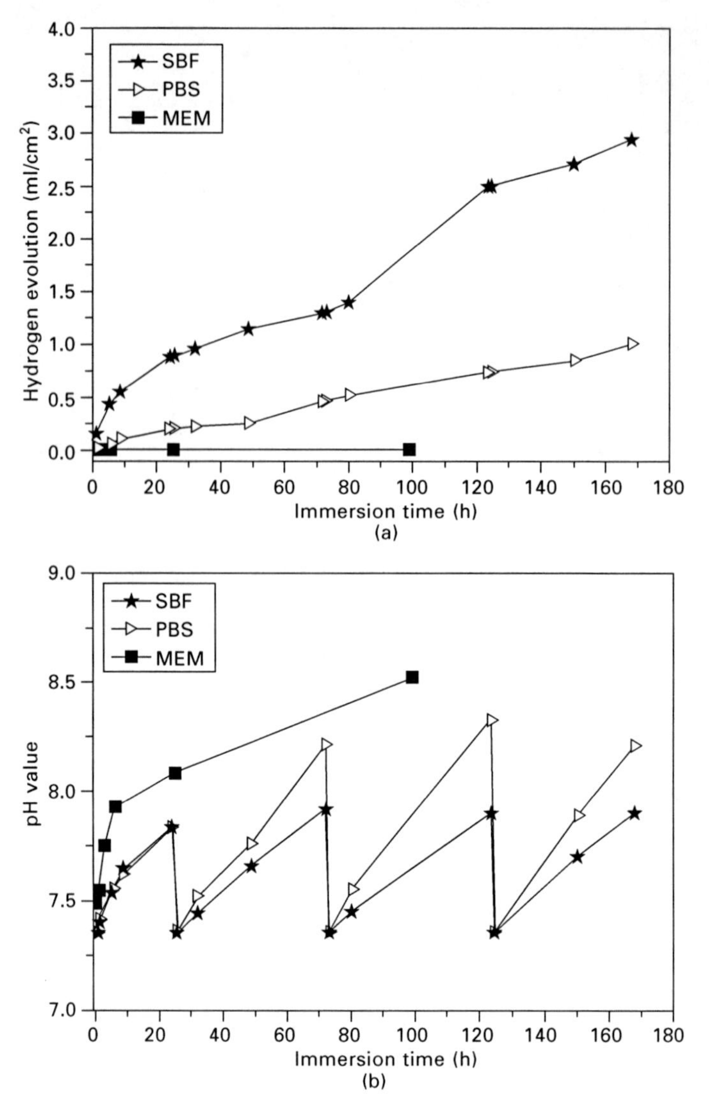

3.33 (a) Hydrogen evolution data for WZ21 in three physiological media at 37 °C; (b) corresponding pH value. Note that the PBS and SBF solutions were exchanged after 24 h, 72 h and 120 h, explaining the pH drop at these times. Reprinted with permission from [82].

$$3Mg^{2+} + 3Ca^{2+} + 4(PO_4)^{3-} = Ca_3Mg_3(PO_4)_4$$

$$Mg^{2+} + (CO_3)^{2-} = MgCO_3$$

Song *et al.* [92] found that the surface film degenerated for long exposure times and proposed a more complex series of precipitation reactions for surface film formation.

Because of the partly protective surface film, it has been found [120] that microstructure effects are less pronounced than in chloride solutions. Nevertheless, micro-galvanic acceleration of corrosion by second phases has been reported [81,86,89,93–96]. Song [97] proposed to control biodegradation with an anodised coating. The biodegradation rate can be controlled to a desired level by varying the thickness and integrity of this coating. This opens up a way to convert many high corrosion rate Mg alloys into biocompatible materials.

3.6.2 *In vivo* experience

The good performance of biodegradable magnesium implants from 1878 to 1981, was reviewed by Witte [98]. Mg has been used for the following applications: (i) cardiovascular (wires and other designs for ligature; connectors for vessel anastomosis; wires for aneurysm treatment), (ii) musculoskeletal (Mg sheet between bone surfaces to restore joint motion in knees in animals and humans; fixator pins, nails, wires, pegs, clamps, sheets and plates), and (iii) general surgery (sheets and plates for suturing organs such as liver and spleen; Mg tubes as connects of intestine anastomosis; wire as sutures). Nearly all patients benefited from the treatment with Mg implants. Although most patients experienced subcutaneous gas cavities caused by rapid implant corrosion, most patients had no pain and almost no infections were observed during the postoperative follow-up. Andrews [99] indicated that Mg would be ideal for ligatures and sutures for deep wound closure. The use of absorbable Mg alloy screws and plates in bone surgery was proposed by McBride [100].

Bone biocompatability

Good bone biocompatibility was found by (i) Witte *et al.* [101] for AZ31, AZ91, WE43 and LAE442 and (ii) Witte *et al.* [102] for AZ91D and LAE442 implanted intramedullary into the femoral of guinea pigs; (iii) Li *et al.* [103] for Mg–1Ca implanted into the left rabbit femoral shaft; (iv) Xu *et al.* [104] for Mg–1.2Mn–1.0Zn, and (v) Zhang *et al.* [105] for Mg–1Zn–0.8Mn into femora of rats.

Magnesium scaffolds

An open AZ91D scaffold was formed by (i) slowly infiltrating molten AZ91D at 600–670 °C into dry NaCl crystals in a core box placed in the chill mould of a low-pressure die-casting system, (ii) machining to shape, and (iii) dissolving the NaCl crystals in a sodium hydroxide solution with a pH above 11.5 which protected the AZ91D. The scaffolds were implanted

[106,107] into the distal femur condyle of rabbits. After 3 months, most of the original Mg alloy had disappeared and a fibrous capsule enclosed the operation site. There was no significant harm to neighbouring tissue. The observed effects indicate that the Mg scaffolds are attractive for musculoskeletal implants.

Skin sensitivity

No skin sensitisation was caused by AZ31, AZ91, WE43 and LAE442 in a study by Witte *et al.* [108] in which solutions and chips of these materials were prepared and tested in 156 guinea pigs according to the Magnusson–Kligman test. These materials are therefore suitable as musculoskeletal implant materials.

Soft tissue

Good biocompatibility was found [82] for WZ21 implanted in Göttingen minipigs in the liver, lesser omentum, rectus abdominis muscle and subcutaneous tissue of the abdominal wall. There was homogeneous degradation and only limited gas formation.

Mg stents

Promising first results were obtained [109] in animal experiments and with 20 patients with a drug-eluting bio-absorbable Mg stent. Good results were reported [110] from the implantation of a biodegradable 3 mm Mg stent into the left pulmonary artery of a preterm baby when the baby weighed 1.7 kg. Reperfusion of the left lung was established and persisted throughout the 4 month follow-up period during which the gradual degradation process of the stent was completed. The degradation process was clinically well tolerated despite the small size of the baby. Hermawan *et al.* [111] review good results from additional studies and there is more information available in [112–115].

3.6.3 Alloy development

Figure 3.34 [89] presents corrosion rates in MEM at 37 °C for a range of experimental and commercial alloys. High-purity Mg was < 40 ppm Fe. G indicates general corrosion, P indicates localised corrosion and X indicates extremely localised corrosion.

Figure 3.35 [88] presents relative corrosion rates (measured from the amount of Mg dissolved in the solution) in SBF at 37 °C for as-rolled Mg1X alloys. Note that these alloys were produced using commercially pure Mg so

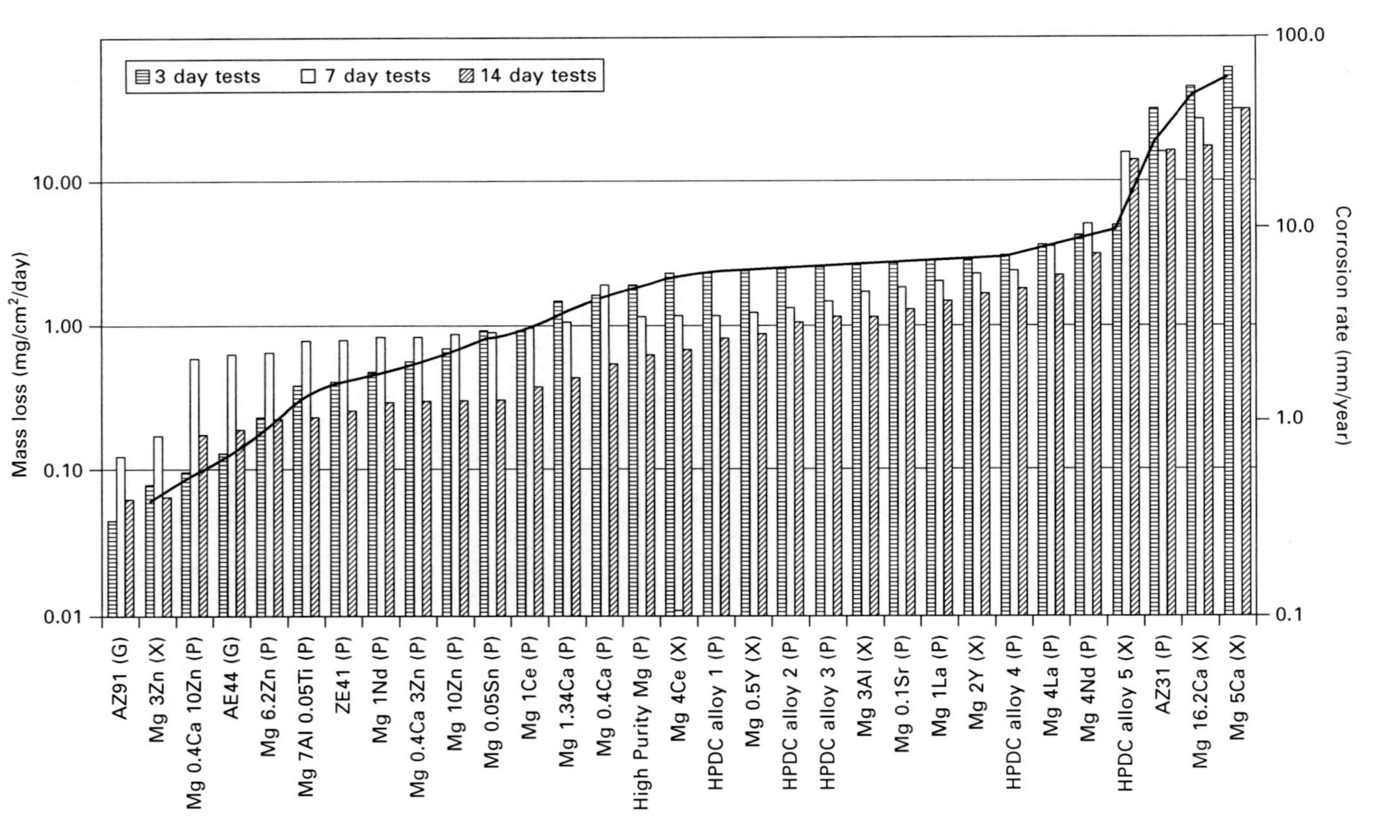

3.34 Corrosion rates in MEM at 37 °C for a range of experimental and commercial alloys. High-purity Mg was < 40 ppm Fe. G indicates general corrosion, P indicates localised corrosion and X indicates extremely localised corrosion. Reprinted with permission from Kirkland *et al.* [89].

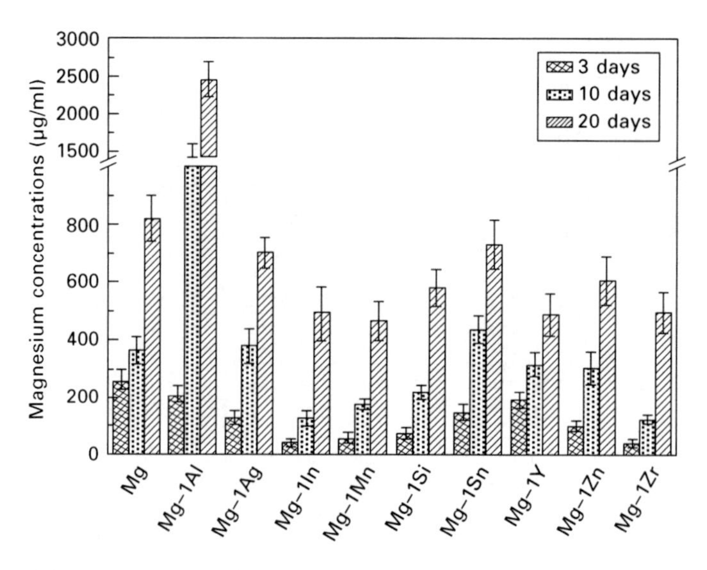

3.35 Relative corrosion rates (measured from the amount of Mg dissolved in the solution) in SBF at 37 °C for as-rolled Mg1X alloys. Note that these alloys were produced using commercially pure Mg so the interpretation of these data might take into account the corrosion rates in Fig. 3.2. Reprinted with permission from Gu *et al.* [88].

the interpretation of these data might take into account the corrosion rates in Fig. 3.2.

Alloy developments that are particularly noteworthy include: (i) Mg–2Zn–0.2Mn [81] which had a corrosion rate of 0.03 mm/yr, which was thought to be close to the corrosion rate of 0.02 mm/yr which was thought to be acceptable for medical implants, (ii) Mg–6Zn [116], (iii) Mg–10Gd [37] and (iv) WZ21 [82,83]. Mg–10Gd shows great promise; however, there is significant development research to carry out. The development of WZ21 by Haenzi *et al.* [82,83,117,118] represents elegant research; the alloy WZ21 has high promise. Witte *et al.* [119] provided some suggestions as to the design and selection of magnesium alloys for biodegradable applications.

3.7 Concluding remarks

This chapter has built on the prior reviews and has provided a succinct overview of the corrosion mechanisms. It has attempted to show that understanding of the corrosion of Mg alloys builds upon our understanding of the corrosion of HP Mg. This understanding of the mechanism of corrosion of HP Mg provides a basis for understanding the various manifestations of corrosion of Mg and its alloys. It is hoped that this deep understanding will allow the production of Mg alloys much more resistant to corrosion than the present alloys. Much has already been achieved, but there is much room for improvement. There is still vast scope both in fundamental understanding of corrosion processes and on the engineering usage of Mg and also on the corrosion protection of Mg alloys in service.

3.8 Acknowledgements

Support for this work comes from the CAST CRC and the Australian Research Council Centre of Excellence in Design of Light Alloys. CAST CRC was established under, and is supported in part by the Australian Government's Cooperative Research Centres Scheme.

3.9 References

1 G Song and A Atrens, *Advanced Engineering Materials* **1** (1999) 11–33.
2 GL Song and A Atrens (Invited Review), *Advanced Engineering Materials* **5** (2003) 837.
3 G Song and A Atrens, *Advanced Engineering Materials* **9** (2007) 177–183.
4 *ASM Handbook,* Vol 13, *Corrosion* ASM International, fourth printing (1992).
5 G Song, A Atrens, S StJohn, J Nairn and Y Lang, *Corrosion Science* **39** (1997) 855.
6 G Song, A Atrens, D StJohn, X Wu and J Nairn, *Corrosion Science* **39** (1997) 1981–2004.
7 G Song, A Atrens, X Wu and B Zhang, *Corrosion Science* **40** (1998) 1769–1791.
8 G Song, A Atrens and M Dargusch, *Corrosion Science* **41** (1999) 249–273.
9 G Song, A Atrens and DH StJohn, An hydrogen evolution method for the estimation of the corrosion rate of magnesium alloys, in J Hryn ed, *Magnesium Technology 2001*, New Orleans, TMS (2001) 255–262.
10 M Liu, P Schmutz, PJ Uggowitzer, G Song and A Atrens, Influence of yttrium (Y) on the corrosion of Mg-Y binary alloys, *Corrosion Science*, **52** (2010) 3687–3701.
11 MC Zhao, M Liu, G Song and A Atrens, *Advanced Engineering Materials* **10** (2008) 104.
12 MC Zhao, M Liu, G Song and A Atrens, *Corrosion Science* **50** (2008) 1939.
13 M Liu, PJ Uggowitzer, AV Nagasekhar, P Schmutz, M Easton, G Song and A Atrens, *Corrosion Science* **51** (2009) 602–619.
14 Z Shi, M Liu and A Atrens, *Corrosion Science*, **52** (2010) 579–588.
15 Z Shi and A Atrens, An innovative specimen configuration for the study of Mg corrosion, *Corrosion Science*, **53** (2011) 226–246.
16 RL Petty, AW Davidson, J Kleinberg, *Journal of the American Chemical Society* **76** (1954) 363–366.
17 O Lunder, K Nisancioglu and RS Hanses, SAE Technical Paper #930755 (1993).
18 K Nisancioglu, O Lunder and T Aune, 'Corrosion mechanisms of AZ91 magnesium alloy', *Proc 47th World Magnesium Conference*, Cannes IMA, (1990) 43–50.
19 N Pebere, C Riera and F Dabosi, *Electrochimica Acta* **35**(2) (1990) 555–561.
20 S.K. Das and L.A. Davis, *Materials Science Engineering* **98** (1988) 1.

21 D Tawil, 'Protection of magnesium components in military applications', Paper No 90445, *Corrosion 90*, NACE (1990).
22 DL Hawke, 'Galvanic corrosion of magnesium', *SDCE 14th International Die Casting* Congress and Exposition, Toronto (1987), paper No. G–T87–004.
23 G Song, AL Bowles and DH StJohn, *Materials Science and Engineering* A366 (2004) 74–86.
24 MC Zhao, M Liu, G Song and A Atrens, *Advanced Engineering Materials* **10** (2008) 93–103.
25 MC Zhao, M Liu, GL Song and A Atrens, *Corrosion Science* **50** (2008) 3168–3178.
26 JW Chang, PH Fu, XW guo, LM Peng and WJ Ding, *Corrosion Science* **49** (2007) 2612–2627.
27 JW Chang, XW Guo, PH Fu, A Atrens, LM Peng, WJ Ding and XS Wang, *Journal of Applied Electrochemistry* **38** (2008) 207.
28 J Chang, X Guo, S He, P Fu, L Peng and W Ding, *Corrosion Science* **50** (2008) 166–177.
29 LM Peng, JW Chang, XW Guo, A Atrens, WJ Ding and YH Peng, *Journal of Applied Electrochemistry* **39** (2009) 913–920.
30 M Yamasaki, N Hayashi, S Izumi and Y Kawamura, *Corrosion Science* **49** (2009) 255–262.
31 Y Song, D Shan, R Chen and EH Han, *Corrosion Science* **52** (2010 1830–1837.
32 Y Lisitsyn, G Ben-Hamu, D Eliezer and KS Shin, *Corrosion Science* **52** (2010), 2280–2290.
33 Y Lisitsyn, G Ben-Hamu, D Eliezer and KS Shin, *Corrosion Science* **51** (2009) 776–784.
34 Y Song, D Shan, R Chen and EH Han, *Corrosion Science* **51** (2009) 1087–1094.
35 N Birbilis, MA Easton, AD Sudholz, SM Zhu and MA Gibson, *Corrosion Science* **51** (2009) 683.
36 GS Frankel, private communication.
37 N Hort, Y Huang, D Fletcher, M Stormer, C Blawert, F Witte, H Drucker, R Willumeit, KU Kainer and F Feyerabend, *Acta Biomaterialia*, **6** (2010), 1714–1725.
38 F Feyerabend, J Fischer, J Holtz, F Witte, R Willumeit, H Drucker, C Vogt and N Hort, *Acta Biomaterialia*, **6** (2010), 1834–1842.
39 KU Kainer, N Hort, R Willumeit, F Feyerabend and F Witte, Magnesium alloys for the design of implants. In: M Niinomi, M Morinaga, M Nakai, N Bhatnagar, TS Srivatsan, (Eds.): *Processing and Fabrication of Advanced Materials*, Proceedings of 18th International Symposium, PFAM 18. Vol. 1 Sendai (J), 12.-14. 12.2009 (2009) 975–984.
40 GL Song, A Atrens, X Wu and B Zhang, *Corrosion Science* **40** (1998) 1769–1791.
41 P Bruesh, K Miller, A Atrens and H Neff, *Applied Physics A* **38** (1985) 1–18.
42 S Jin and A Atrens, *Applied Physics* A **42** (1987) 149–165.
43 AS Lim and A Atrens *Applied Physics A* 54 (1992) 270.
44 AS Lim and A Atrens *Applied Physics A* 54 (1992) 343.
45 AS Lim and A Atrens *Applied Physics A* 53 (1991) 273.
46 AS Lim and A Atrens *Applied Physics A* 54 (1992) 500.
47 EA Brandes ed, *Smithels Metal Reference Book*, Butterworths (1983).

48 IJ Polmear, *Light Alloys*, Arnold (1995).
49 B Zberg, PJ Uggowitzer and JF Loeffler, *Nature Materials* **8** (2009) 887.
50 JD Hanawalt, CE Nelson and JA Peloubet, *Transactions of the American Institute and Mining and Metal Engineers* **147** (1942) 273.
51 JE Hillis and RW Murray, 'Finishing alternatives for high purity magnesium alloys', *SDCE 14th International Die Casting Congress and Exposition*, Toronto (1987), Paper No. G–T87–003.
52 AL Olsen, 'Corrosion characteristics of new magnesium alloys', Translation of paper presented at the '*Deutscher Verband Fur Materialforshung u. Prufung e.v.' Bauteil'91*. Berlin (1991) 1–21.
53 WE Mercer II and JE Hillis, SAE Technical Paper Series # 920073, Detroit (1992).
54 D Frey and LL Albright, 'Development of a magnesium alloy structural truck component', *Proceedings of the 41st World Magnesium Conference*, London, June (1984).
55 TK Aune, 'Minimizing base metal corrosion on magnesium products. the effect of element distribution (structure) on corrosion behavior', *Proceedings of the 40th World Magnesium Conference*, Toronto, June (1983).
56 JD Hanawalt, CE Nelson and JA Peloubet, *Transactions of the American Institute of Mining and Metal Engineers*, 147:273, (1942).
57 JE Hillis and KN Reichek, Soc Auto Engineers Technical Paper Series #860288, Detroit.
58 http://www.computherm.com/pandat.html
59 GL Makar and J Kruger, *International Materials Reviews* **38** (1993) 138.
60 KN Reichek, KL Clark and JE Hillis, SAE Technical Paper 850417 (1985).
61 A Froats, TK Aune, D Hawke, W Unsworth and JE Hillis, *Metal Handbook*, 9th ed., *Corrosion* Vol 13, ASM International, Materials Park, OH (1987) pp. 740–754.
62 G Song, AL Bowles and DH StJohn, *Materials Science and Engineering A* **366** (2004) 74.
63 R Ambat, NN Aung and Z Zhou, *Corrosion Science* **42** (2000) 1433.
64 A Pardo, MC Merino, AE Coy, R Arrabal, F Viejo E and Matykina, *Corrosion Science* **50** (2008) 823.
65 A Prado, MC Merino, AE Coy, F Viejo, R Arrabal and E Matykina, *Electrochimica Acta* **53** (2008) 7890.
66 MB Haroush, CB Hamu, D Eliezer and L Wagner, *Corrosion Science* **50** (2008) 1766.
67 MC Zhao, P Schmutz, S Brunner, M Liu, G Song and A Atrens, An exploratory study of the corrosion of Mg alloys during salt spray testing, *Corrosion Science*, **51** (2009) 1277–1292.
68 WE Mercer II and JE Hillis, SAE Technical Paper 920073 (1992).
69 JE Hillis and SO Shook, SAE Technical Paper 890205 (1989).
70 C Scharf, A Ditze, A Shkurankov, E Morales, C Blawert, W Dietzel and KU Kainer, *Advanced Engineering Materials* 7 (2005) 1134.
71 D Fechner, C Blawert, N Hort and KU Kainer, *Light Metals Technology* **618–619** (2009) 459–462.
72 G Wiliams and HN McMurray, *Journal of the Electrochemical Society* **155** (2008) C340–C349.
73 G Williams, private communication, Jan 2009.
74 MC Zhao, P Schmutz, S Brunner, M Liu, G Song and A Atrens, *Corrosion Science*, **51** (2009) 1277–1292.

75 GL Song and ZQ Xu, *Electrochimica Acta* **55** (2010) 4148–4161.
76 D Song, AB Ma, J Jiang, D Yang and J Fan, *Corrosion Science* **52** (2010) 481–490.
77 G-L Song, R Mishra and Z Xu, *Electrochemistry Communications*, **12** (2010), 1009–1012.
78 M Liu, D Qiu, MC Zhao, G Song and A Atrens, *Scripta Materialia* **58** (2008) 421–424.
79 F Witte, J Fischer, J Nellesen, H-A Crostack, V Kaese, A Pisch, F Beckmann and H Windhagen, *Biomaterials* **27** (2006) 1013.
80 ASTM-D1141-98: Standard practice for the preparation of substitute ocean water. ASTM, 2004.
81 G Song, *Corrosion Science* **49** (2007) 1696–1701.
82 AC Haenzi, I Gerber, M Schinhammer, JF Loeffler and PJ Uggowitzer, *Acta Biomaterialia* **6** (2010), 1824–1833.
83 AC Haenzi, P Gunde, M Schinhammer and PJ Uggowitzer, *Acta Biomaterialia* **5** (2009) 162–171.
84 L Yang and E Zhang, *Materials Science and Engineering C* **29** (2009) 1691–1696.
85 Y Wang, M Wei, J Gao, J Hu and Y Zhang, *Materials Letters* **62** (2008) 2181–2184.
86 Z Li, X Gu, S Lou and Y Zheng, *Biomaterials* **29** (2008) 1329–1344.
87 H Wang, Y Estrin, H Fu, G Song and Z Zuberova, *Advanced Engineering Materials* **7** (2007) 967–972.
88 X Gu, Y Zheng, Y Cheng, S Zhong and T Xi, *Biomaterials* **30** (2009) 484–498.
89 NT Kirkland, J Lespagnol, N Birbilis and MP Staiger, *Corrosion Science* **52** (2010) 287–291.
90 J Luvesque, H Hermawan, D Dube and D Mantovani, *Acta Biomaterialia* **4** (2008) 284–295.
91 NC Quach, PJ Uggowitzer and P Schmutz, *Comptes Rendus Chimie* **11** (2008) 1043.
92 Y Song, D Shan, R Chen and EH Han, *Materials Science and Engineering C* **29** (2009) 1039–1045.
93 MB Kannan and RKS Raman, *Biomaterials* **29** (2008) 2306–2314.
94 X Gu, Y Zheng, S Zhong and T Xi, *Biomaterials* **30** (2009) 484–498.
95 MB Kannan, *Materials Letters* **64** (2010) 739–742.
96 E Zhang and L Yang, *Materials Science and Engineering A* **497** (2008) 111.
97 G Song, *Advanced Materials Research* **29–30** (2007) 95–98.
98 F Witte, *Acta Biomaterialia* **6** (2010), 1680–1692.
99 EW Andrews, *Journal of the American Medical Association* **LXIX** (1917) 278–281.
100 ED McBride, *Journal of the American Medical Association* **111** (1938) 2464–2567.
101 F Witte, V Kaese, H Haferkamp, E Switzer, A Meyer-Lindenberg, CJ Wirth and H Windhagen, *Biomaterials* **26** (2005) 3557–3563.
102 F Witte, J Fischer, J Nellesen, HA Crostack, V Kaese, A Pisch, F Beckmann and H Windhagen, *Biomaterials* **27** (2006) 1013–1018.
103 Z Li, X Gu, S Lou and Y Zheng, *Biomaterials* **29** (2008) 1329.
104 L Xu, G Yu, E Zhang, F Pan and K Yang, *Journal of Biomedical Research* **83** (2007) 703.

105 E Zhang, L Xu, G Yu, F Pan and K Yang, *Journal of Biomedical Research* **90** (2009) 882.
106 F Witte, H Ulrich, M Rudert and E Willbold, *Journal of Biomedical Research* **81** (2007) 748.
107 F Witte, H Ulrich, M Rudert and E Willbold, *Journal of Biomedical Research* **81** (2007) 757.
108 F Witte, I Abeln, E Switzer, V Kaese and A Meyer-Lindenberg, *Journal of Biomedical Research* **86** (2008) 1041.
109 C Di Mario, H Griffiths, O Goktekin, N Peeters, J Verbist, M Bosiers, K Deloose, B Heublein, R Rohde, V Kaese, C Ilsley and R Erbel, *Journal of Intervention Cardiology* **17** (2004) 391.
110 P Zartner, R Cesnjevar, H Singer and M Weyand, *Catheterization and Cardiovascular Interventions* **66** (2005) 590.
111 H Hermawan, D Dube and D Mantovani, *Acta Biomaterialia* **6** (2010), 1693–1697.
112 P Peeters, M Bosiers, J Verbist, K Deloose and B Heublein, *Journal of Endovascular Therapy* **12** (2005) 1–5.
113 Z Peter, C Robert, S Helmut and W Michael, *Catheterization and Cardiovascular Interventions* **66** (2005) 590–594.
114 S Dietmar, Z Peter, M-B Ina and A Hakan, *Catheterization and Cardiovascular Interventions* **67** (2006) 671–673.
115 R Erbel, C Di Mario, J Bartunek, J Bonnier, B de Bruyne, FR Eberli, P Erne, M Haude, B Heublein, M Horrigan, C Ilsley, D Böse, J Koolen, TF Lüscher, N Weissman and R Waksman, *The Lancet* **369** (2007) 1869–1875.
116 S Zhang, X Zhang, C Zhao, J Li, Y Song, C Xie, H Tao, Y Zhang, Y He, Y Jiang and Y Bian, *Acta Biomaterialia* **6** (2010) 626–640.
117 AC Haenzi, AS Sologubenko and PJ Uggowitzer, *International Journal of Materials Research* **100** (2009) 1127.
118 AC Haenzi, FH Dalla Torre, AS Sologubenko, P Gunde, R Schmid-Fetzer, M Kuehlein, JF Loeffler and PJ Uggowitzer, *Philosophical Magazine Letters* **89** (2009) 377–390.
119 F Witte, N Hort, C Vogt, S Cohen, KU Kainer, R Willumeit and F Feyerabend, *Current Opinion in Solid State and Materials Science* **12** (2008) 63.
120 NI Zainal Abidin, D Martin and A Atrens, *Corrosion Science*, 10.1016/j.corsci.2010.10.008.

4
Role of structure and rare earth (RE) elements on the corrosion of magnesium (Mg) alloys

T. ZHANG, Harbin Engineering University, China and Y. LI, Institute of Metal Research, Chinese Academy of Sciences, China

Abstract: This chapter discusses the effect of microstructure and rare earth (RE) elements on the corrosion of magnesium (Mg) alloy. Firstly, this chapter discusses the effect of β-phase and microcrystallization on the corrosion behavior of magnesium. Secondly, it describes the roles of RE elements on the corrosion behavior of Mg alloys.

Keywords: β-phase, negative difference effect, microcrystallization, rare earth (RE) element, thin electrolyte layer.

4.1 Introduction

In materials science, the relationship between the properties, microstructure, composition and manufacture is described as tetrahedroid. This chapter discusses the factors of microstructure and composition (rare earth (RE) elements) on the corrosion process of magnesium (Mg) alloy.

4.2 Role of structure on the corrosion process of magnesium (Mg) alloy

4.2.1 Roles of β-phase on the corrosion process of Mg alloy

Many researchers believe that the β-phase might serve as a galvanic cathode and accelerate the corrosion rate of the α-matrix if the volume fraction of the β-phase was small. However, for a high volume fraction, the β-phase might act as an anodic barrier to inhibit the overall corrosion of the alloy. However, owing to the negative difference effect (NDE), hydrogen atoms would diffuse into Mg alloys and enrich in the β-phase during the corrosion process of AZ91D alloy. The presence of hydrogen in metals has a significant influence on the metals' electrochemical behavior. Therefore, the synergistic effect of the β-phase and hydrogen diffusion on the corrosion of AZ91D alloy should be noted.

The NDE of AZ91D alloy was investigated by a hydrogen evolution experiment. The kinetics of hydrogen evolution of the cast and T4-treated AZ91D alloys at different anodic current are illustrated in Fig. 4.1. It can be seen that hydrogen evolution rate (HER) increased with the increasing anodic current, and the discrepancy of HER between T4 alloy and AZ91D alloy decreased. The difference ratios of the two alloys and the HER ratios of AZ91D alloy to T4 alloy calculated from Fig. 4.1 are shown in Fig. 4.2. The hydrogen evolution results indicated that the β-phase had great influence on the NDE of AZ91D alloy.

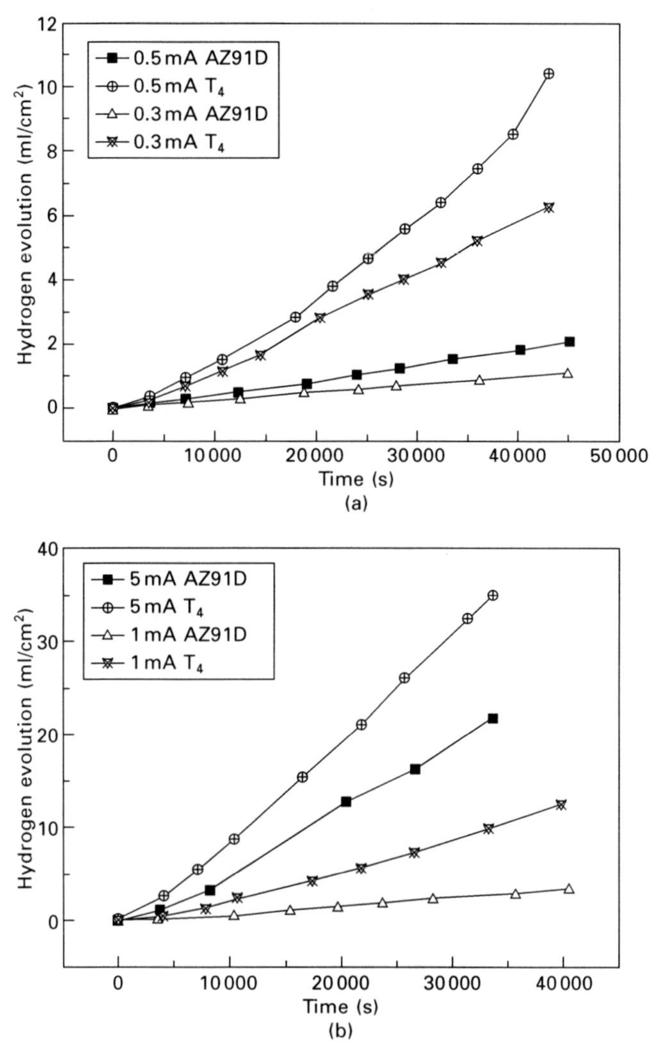

4.1 Kinetics of hydrogen evolution on AZ91D or T4 alloy at different anodic current in 1 mol/L NaCl solution.

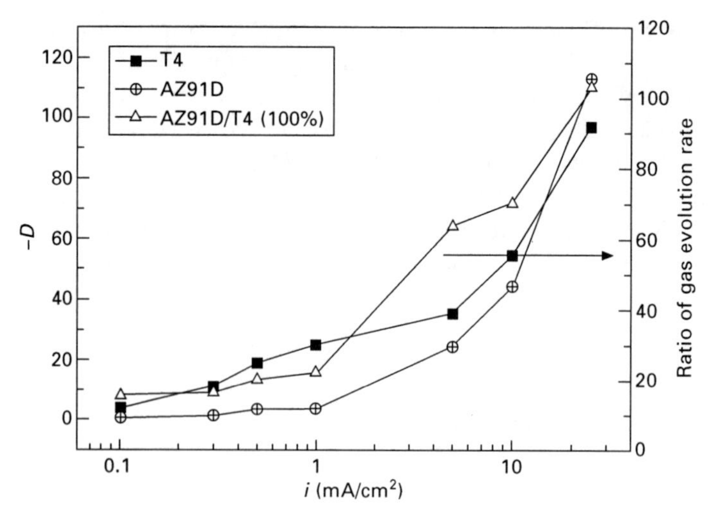

4.2 Difference ratio (right axis) and HER ratio (left axis) of AZ91D to T4 alloy at different anodic current in 1 mol/L NaCl solution.

At an anodic current, $I = I_{corr} + 5\,mA$, the electrochemical impedance spectroscopy (EIS) of AZ91D and T4 alloy is shown in Fig. 4.3. To better understand the effect of hydrogen diffusion into the β-phase on NDE, a model for effect of β-phase on NDE is presented as follows.

Cao [1] believed that if diffusion was not involved in the Faradic process for reaction sequence, the admittance Y (or impedance Z) for the electrode could be expressed as:

$$1/Z = y = j\omega C_{dl} + Y_F \qquad 4.1$$

where y is admittance, ω represents angular frequency ($\omega = 2\pi f$, f is the frequency), $j = \sqrt{-1}$; C_{dl} is capacitance of double layer and Y_F is Faraday admittance.

If the electrode reaction was controlled by one surface state variable X, then Cao [1] thought that the faradic admittance Y_F could be expressed as

$$Y_F = 1/R_t + \frac{(\partial I/\partial X)_{SS}(\partial X'/\partial E)_{SS}}{(j\omega - \partial X'/\partial X)} \qquad 4.2$$

Correspondingly

$$1/Z = j\omega C_{dl} + 1/R_t + \frac{(\partial I/\partial X)_{SS}(\partial X'/\partial E)_{SS}}{(j\omega - \partial X'/\partial X)} \qquad 4.3$$

There was a capacitive loop in the high-frequency range due to the double layer capacitance, C_{dl}. There was a capacitive loop in the low-frequency range caused by the faradic electrochemical reaction process when $(\partial I/\partial X)_{SS}$

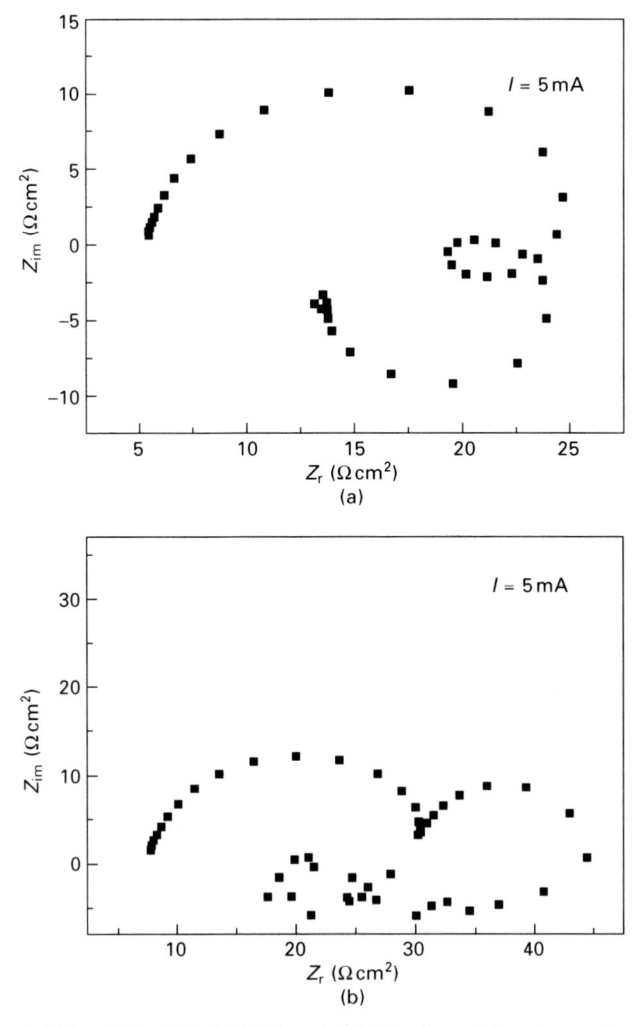

4.3 The EIS of (a) AZ91D and (b) T4 alloy at 5 mA anodic current in 1 mol/L NaCl solution.

$(\partial X'/\partial E)_{SS} < 0$. In contrast, there was an inductive loop in the low-frequency range when $(\partial I/\partial X)_{SS}\,(\partial X'/\partial E)_{SS} > 0$.

If there are two surface state variables, X_1 and X_2, which controlled the electrode processes and the two variables X_1 and X_2 were independent of each other, then it can be shown [1,2] that the admittance of the faradic and non-faradic processes is:

$$1/Z = j\omega C_{dl} + 1/R_t + \frac{(\partial I/\partial X_1)_{SS}(\partial X_1'/\partial E)_{SS}}{(j\omega - \partial X_1'/\partial X)} + \frac{(\partial I/\partial X_2)_{SS}(\partial X_2'/\partial E)_{SS}}{(j\omega - \partial X'/\partial X_2)} \tag{4.4}$$

Partially protective film plays an important role in the corrosion process of AZ91D alloy [3, 4]. The corrosion process of AZ91D alloy involves three independent surface state variables: area fraction of the partially protective film (θ) on the alloy surface, the concentration of intermediate species Mg^+ (C_m), and the continuous changing of potential on β phase (E_β).
Let

$$X_1 = \theta$$

$$X_2 = C_m$$

$$X_3 = E_\beta$$

In the corrosion process, there are capacitive loops and inductive loops in the intermediate frequency range induced by area fraction of the partially protective film (θ) on the alloy surface, the concentration of intermediate species Mg^+ (C_m), respectively [5].

Magnesium and its alloys are kinds of hydrogen storage materials, and the storage capacity of alloy is usually higher than that of pure magnesium. With the increasing potential, more hydrogen atoms might be absorbed on the alloy surface, so that hydrogen diffusion rate increased. During the diffusion process, the atomic hydrogen enriches in the β-phase [6]. The increasing hydrogen content evokes the decreasing surface potential of the β-phase [7]. Therefore

$$dE_\beta/dE < 0 \qquad 4.5$$

At an anodic potential, the current of the Faradic process could be expressed as:

$$I_F = (1 - \theta)\{(1 - f)I_\alpha \exp[(E - E_\alpha)/b_\alpha\} - fI_\beta \exp[-(E - E_\beta)/b_\beta]\} \qquad 4.6$$

where f is the area fraction of the β-phase, I_α and I_β are exchange currents of α-phase and β-phase, E_α and E_β are equilibrium potentials of α-phase and β-phase.

So

$$(\partial I/\partial E_\beta)_{SS} = -fI_\beta \exp[-(E - E_\beta)/b_\beta]/b_\beta < 0 \qquad 4.7$$

Combining Eq. (4.5) with (4.7), predicted:

$$(dE_\beta/dE)(\partial I/\partial E_\beta)_{SS} > 0 \qquad 4.8$$

which indicates that the Nyquist plot should contain an inductive loop related to the surface state variable E_β.

From the above theoretical calculation, the EIS of AZ91D alloy in NaCl aqueous solution consists of one capacitive loop in the high-frequency range (double layer), a capacitive loop (surface state variable C_m) and an inductive loop (surface state variable θ) in the intermediate frequency range, and one inductive loop in the low frequency range (E_β).

The above theoretical calculation indicated that the diffusion of hydrogen into the β-phase should be the proper reason why the EIS of AZ91D alloy was different from that of T4 alloy. If it was reasonable, the β-phase of the charged AZ91D alloy should have no effect on the EIS of the alloy, which could be explained by the fact that the surface potential of the β-phase became constant because the β-phase was saturated by hydrogen atoms. In other words, the EIS of the charged alloy should contain two surface state variables, similar to that of T4 alloy. The above discussion was confirmed by the results of Fig. 4.4. After being charged for 6 h, the inductive loop in the low-frequency range of the EIS seemed to disappear.

With the hydrogen saturation of the β-phase, no hydrogen further diffused into the β-phase. The HER of AZ91D alloy should approach that of T4 alloy. A higher applied anodic current implied a higher HER. It was expected that the β-phase would be saturated by hydrogen in a shorter time by a higher anodic current in comparison with a lower applied anodic current. So, the difference of HER between AZ91D and T4 alloy should decrease with the increasing of applied current, as was claimed by Fig. 4.2. The HER ratio of AZ91D alloy to T4 alloy increased from 16.3% (I = 0.3 mA) to 70.1% (I = 10 mA). For the same reason, the D-ratio of AZ91D alloy to T4 alloy also increased with the increasing applied current (Fig. 4.2).

Moreover, some literature has pointed out that the increasing hydrogen composition was of benefit to the improvement of corrosion resistance of magnesium [7]. It was reported that hydrogen entering the film would be

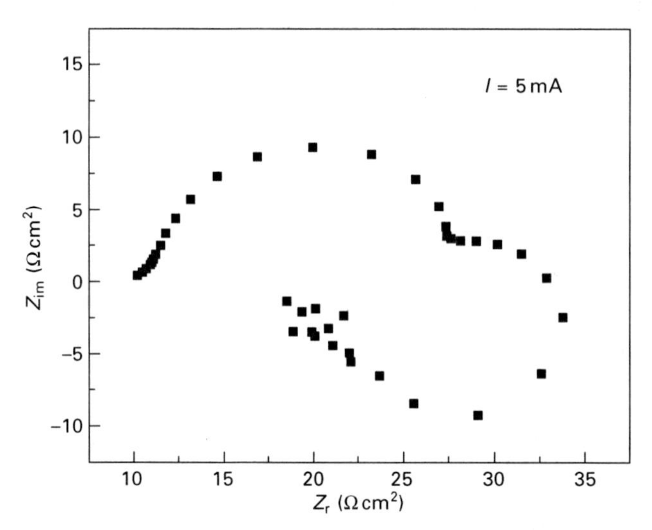

4.4 The EIS of AZ91D alloy at 5 mA anodic current after 6 h charging in 1 mol/L NaCl solution.

ionized, and an MgH_2 film formed on the alloy surface [7]. In brief, owing to the NDE, the HER increased with the increasing potential, hydrogen diffused into the product film and therefore the corrosion resistance of AZ91D alloy was enhanced.

Some researchers [3,4] believe that the β-phase might serve as a galvanic cathode and accelerate the corrosion rate of the a matrix if the volume fraction of the β-phase was small. On the other hand, for a high volume fraction, the β-phase might act as an anodic barrier to inhibit the overall corrosion of the alloy. However, their results [3,4] were not in contradiction with our results concerning the hydrogen diffusion from the β-phase to product films. The corrosion rate of AZ91D alloy was determined by the synergistic effect of the volume fraction of the β-phase and hydrogen diffusion from the β-phase to the product film. For a small volume fraction, the β-phase has two opposite effects on corrosion rate: galvanic cathode (accelerated) and hydrogen diffusion (decelerated). The corrosion rate was determined by the competition between the galvanic cathode and the hydrogen diffusion. For a high volume fraction, either the anodic barrier or the hydrogen diffusion might induce the decrease of the corrosion rate of AZ91D alloy.

In summary, in the initial corrosion process of AZ91D alloy, atomic hydrogen diffused into the β-phase, which evoked the decreasing of hydrogen evolution rate and the weakening of the NDE. Owing to the diffusion of hydrogen into the β-phase, the surface potential of the β-phase decreased. At the initial stage, the corrosion process of AZ91D alloy was controlled by three surface state variables: the area fraction of partially protective film (θ) on alloy surface, the concentration of intermediate species Mg^+ (C_m) and continuous change of potential on the β-phase (E_β). However, with time increasing, owing to the saturation of atomic hydrogen in alloy, the corrosion process was controlled by area fraction of the partially protective film (θ) on the alloy surface and the concentration of intermediate species Mg^+ (C_m).

During the diffusion process, hydrogen entering the film would be ionized, and an MgH_2 film formed on the alloy surface. As a result, the corrosion resistance of AZ91D alloy was improved.

4.2.2 Roles of microcrystallization on the corrosion behavior of Mg alloy

Poor corrosion resistance is one important reason why magnesium alloy is not widely used. During the last few years, some investigations have been carried out to study the corrosion resistance of sputtering magnesium and its alloys [8–10]. The microcrystalline (MC)/nanocrystalline (NC) structure and the change of constituent phases should be the main factors responsible for the improvement of corrosion resistance of sputtering AZ91D alloy.

A transmission electron microscopy (TEM) image of the sputtering film is shown in Fig. 4.5, which indicates that the grain size of sputtering alloy is about 300 nm and the diffraction pattern confirmed that crystal grain has a (hexagonal close packed, HCP) α-phase structure.

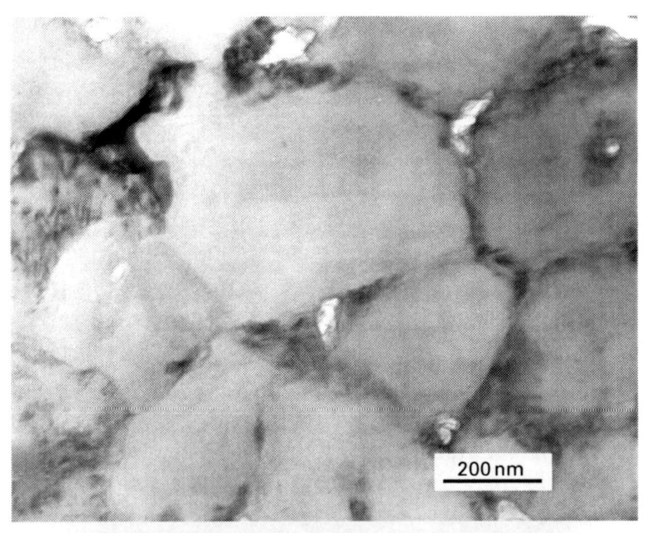

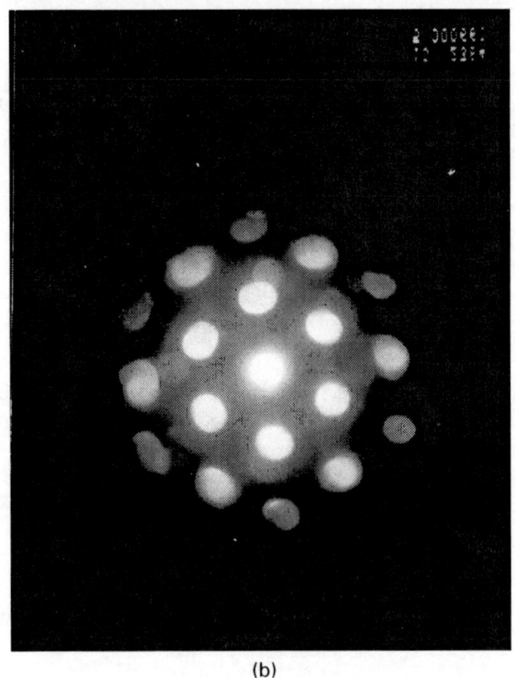

(b)

4.5 TEM images of sputtering AZ91D alloy coating.

It is well known that the overall corrosion reaction of magnesium in aqueous solution at its corrosion potential can be expressed as follows:

$$Mg + 2H_2O \rightarrow Mg(OH)_2 + H_2 \uparrow \qquad 4.9$$

This means that the dissolution of one mole of Mg atoms generates one mole of hydrogen gas. So, in theory, measuring the volume of hydrogen evolved is equivalent to measuring the weight loss of magnesium dissolved and the measured hydrogen evolution rate is equal to the weight loss rate if both of them have been converted into the same unit. Therefore, the corrosion rate of the Mg alloys can be evaluated by the volume of hydrogen evolution from them *in situ*. Figure 4.6 gives the hydrogen evolution from MC alloy, cast AZ91D alloy and T4 alloy in NaCl aqueous solution as a function of immersion time. All the measurements were duplicated three times with a good reproducibility. We can claim that at the beginning (time A), the corrosion rate of all the specimens can be ranked as an increasing series: cast alloy (α + β-phases) < MC alloy (α-phase) < T4 alloy (α-phase). However, as the immersed time increased, the corrosion rate of MC alloy gradually became lower than that of cast alloy. At time B, the corrosion rate of all the specimens should be ranked as: MC alloy (α-phase) < cast alloy (α + β-phases) < T4 alloy (α-phase).

According to previous works [5,11,12], there should be a product film formed on the alloy surface in NaCl aqueous solution. But for the cast alloy

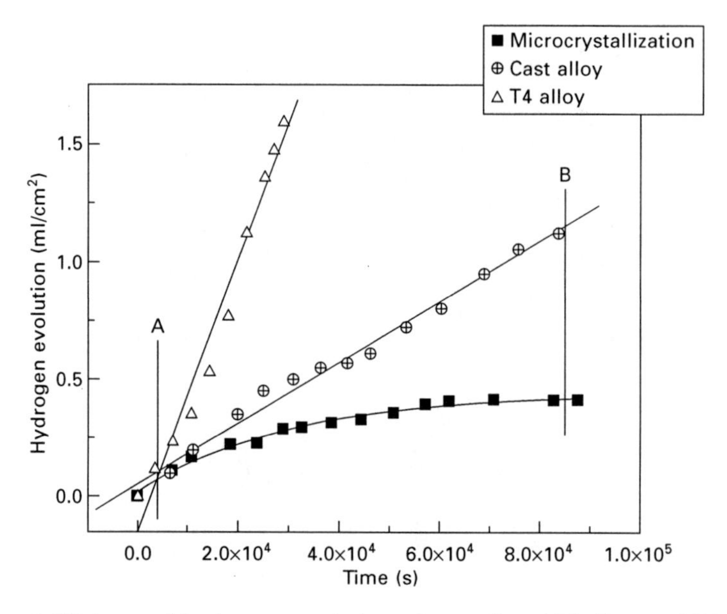

4.6 Volume of hydrogen evolution of cast alloy, MC alloy and T4 alloy in 1 mol/L NaCl aqueous solution.

and T4 alloy, the hydrogen evolution volume increased lineally with time, which implied the product film on the alloy surface was less protective. On the other hand, for the MC alloy, the HER decreased with time and the hydrogen evolution volume approached a maximum in the end. A protective product film formed on the MC alloy surface might be the reason why the hydrogen evolution rate decreased so obviously. At time B, two different types of surface morphologies for the above specimens were illustrated by scanning electron microscopy (SEM). The corrosion of cast and T4 alloys initiated in the form of pits at several locations and then spread over the surface in a cellular fashion (Fig. 4.7(a) and (b)), while there was a feather-shaped product film-covered MC alloy surface (Fig. 4.7(c)) in the same condition.

The polarization curves of MC alloy, T4 alloy and cast AZ91D alloy in NaCl aqueous solution after immersion for B time were measured, respectively, and are shown in Fig. 4.8. It can be seen that at B time, the corrosion rate of all the specimens should be ranked as: MC alloy (β-phase) < cast alloy (α + β-phases) < T4 alloy (α-phase). The result agreed with the amount of hydrogen produced. The hydrogen production and potentiodynamic polarization curves for AZ91D alloys illustrated that the phase change has a significant influence on the corrosion resistance of Mg alloy. We can confirm that the corrosion resistance would deteriorate with the absence of the β-phase because the corrosion rate of T4 alloy with only the α-phase was much higher than that of the cast alloy with both the α and β-phases. On the other hand, the grain size of alloy also plays a big role in the corrosion resistance of an Mg alloy. The corrosion rate of MC alloy is much lower than that of T4 alloy, which has the same phase but the grain size is different. It seems that changing grain size is more effective than changing phase to improve the corrosion resistance of an Mg alloy because the corrosion rate of the MC alloy was reduced to less than that of the cast alloy when the grain size is lowered to nanoscale even through no β-phase exists in it.

Polarization curves (Fig. 4.8) also meant that the anodic process of the MC alloy was inhibited. It attributed to a protective product film formed on MC alloy surface. Therefore, for thorough understanding of the microcrystallization effect on corrosion resistance ofAZ91D alloy, it is necessary to study the nature of the corrosion product film on MC alloy.

The results of TEM (see Fig. 4.9) showed the grain size of the product film scaled off from cast alloy was about 200 nm, while that of MC alloy was less than 10 nm. The diffraction pattern confirmed that there was a MgO film (face centered cubic, FCC) formed on both cast alloy and MC alloy (Fig. 4.9(c)).

Figure 4.10 described the electronic energy band model for a semiconductor film formed on AZ91D alloy in NaCl aqueous solution. The proposed model has two interfaces: (I) substrate alloy/film and (II) film/electrolyte. At interface

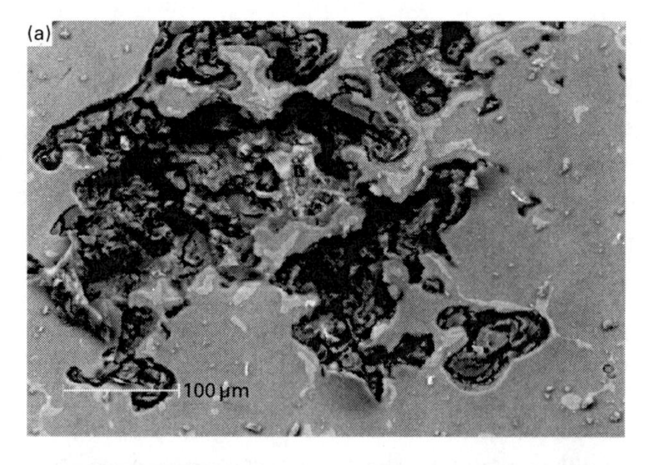

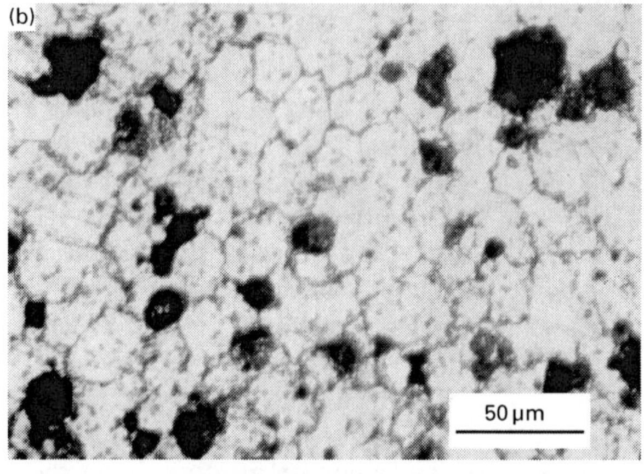

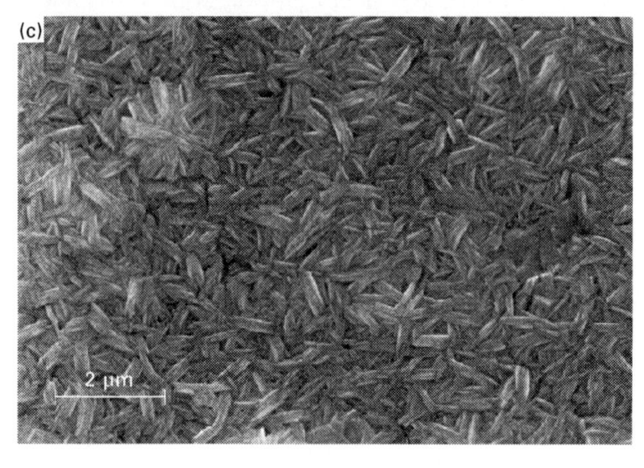

4.7 Scanning electron microscopy (SEM) images of cast, T4 and MC alloy after 24 h immersion in 1 mol/L NaCl solution: (a) cast alloy; (b) T4 alloy; (c) MC alloy.

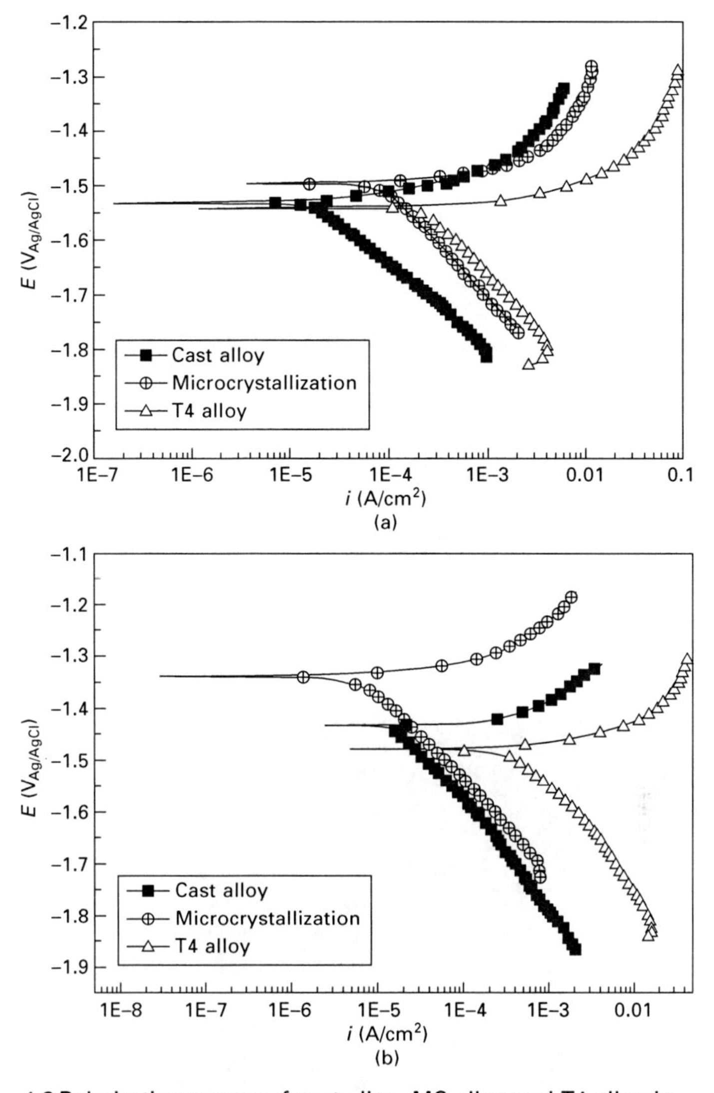

4.8 Polarization curves of cast alloy, MC alloy and T4 alloy in 1 mol/L NaCl aqueous solution after (a) 0.5 h immersion and (b) 24 h immersion.

(I), with the increasing of anodic potential, the energy band slopes in the semiconductor film descended toward the film/substrate interface. At interface (II), electronic transition was carried out from valence band to Fermi level (E_f) and vacancies were generated in valence band. Several investigations had demonstrated that the energy shift in the band gap, ΔE, as a function of particle size can be predicted by the three-dimensional confinement model based on the effective mass approximation [13–15]:

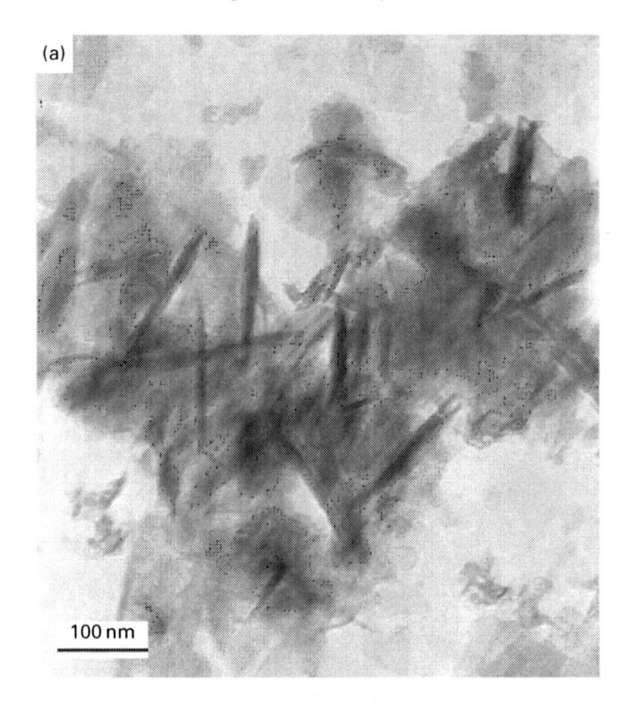

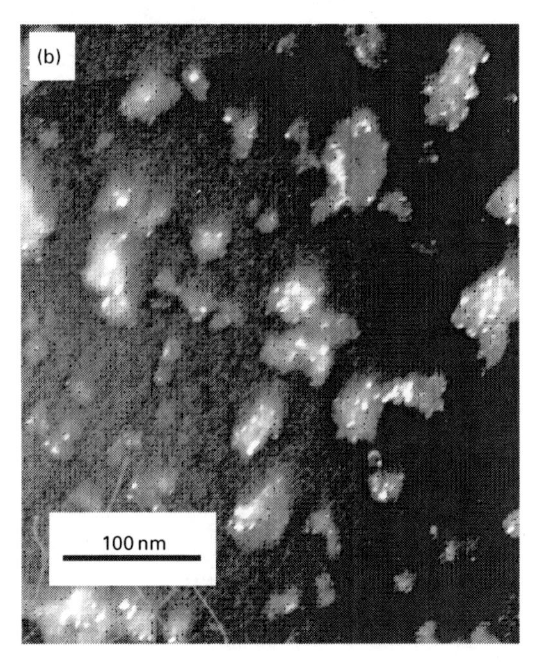

4.9 The TEM of the product film on (a) cast and (b) MC alloy. The diffraction pattern is shown in (c).

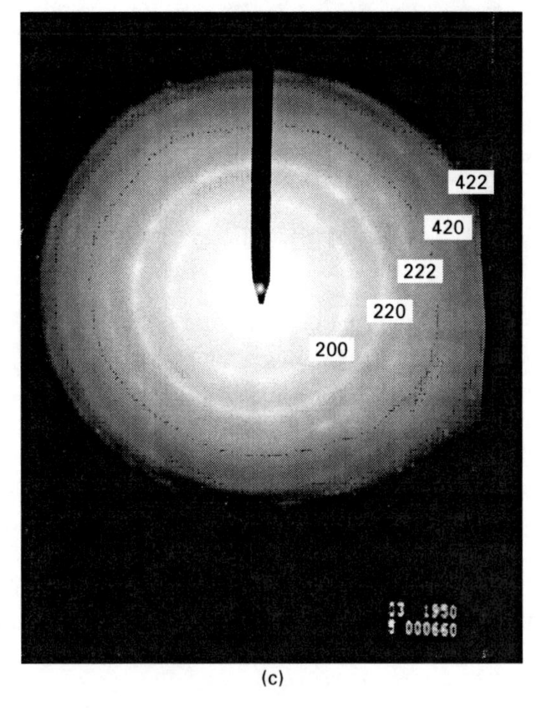

(c)

4.9 Continued

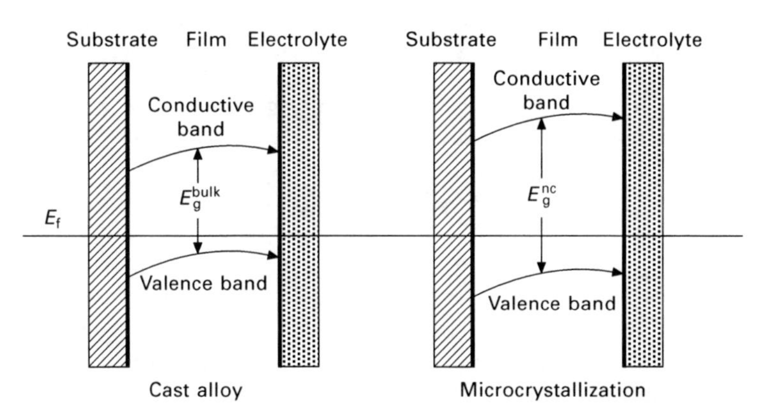

4.10 Schematic presentation of the electronic band structure of product films formed on cast alloy and MC alloy in NaCl aqueous solution.

$$\Delta E = \frac{h^2\pi^2}{2R^2}\left[\frac{1}{m_e^*} + \frac{1}{m_h^*}\right] - \frac{1.786e^2}{\varepsilon R} - 0.248E_{RY} \qquad 4.10$$

where R is the grain size, m_e^* and m_h^* the effective masses for electrons and holes, ε the dielectric constant, and E_{RY} the effective Rydberg energy given as $e^4/2\varepsilon^2h^2(1/m_e^* + 1/m_h^*)$. So, the energy shift in the band gap was:

$$\Delta E_g^{nc} > \Delta E_g^{bulk} \tag{4.11}$$

where E_g^{nc} and E_g^{bulk} are the energy gap of the microcrystallization alloy and bulk, respectively.

This indicated:

$$E_g^{nc} > E_g^{bulk} \tag{4.12}$$

Therefore, it could be expected that the energy band of the product film on MC AZ91D alloy increased, suggesting that it became difficult for electronic transition from valence band to Fermi level (see Fig. 4.10), which might be the proper explanation for the increasing of corrosion resistance of the product film on MC alloy.

From the above results and discussions, we can confirm that the phase change did not appear to have a significant influence, but the grain size of alloy plays a key role on the corrosion resistance. There was a more protective product film formed on the MC alloy. The main reason for the more protective product film might be attributed to the smaller grain size of the product film on MC alloy, which was favorable to the widening of energy band.

4.3 Role of rare earth (RE) elements on the corrosion process of magnesium (Mg) alloy

4.3.1 Role of RE elements on the corrosion of an Mg alloy under thin electrolyte layers

For various Mg alloys, RE-containing alloys are known to show excellent creep resistance [16]. Recently, Mg alloys containing gadolinium and yttrium were found to exhibit higher strength than WE54 alloy at both room and high temperatures. Many work revealed that Mg–Gd–Y alloy could offer good combination of strength, elongation and creep resistance in the peak hardness by artificial aging [17]. Despite the fact that Mg–Gd–Y alloys as structural materials are mostly used in atmospheric environments, the literature on atmospheric corrosion of Mg–Gd–Y alloys is scarce [18–20]. The role of RE elements on the atmospheric corrosion of magnesium alloy is still unclear. In the other word, the detailed corrosion mechanism of an Mg–RE alloy in case of atmospheric condition awaits clarification.

The atmospheric corrosion is the most widely spread form of corrosion. It occurs under thin electrolyte layers (TEL), or even adsorbed layers. A change in the thickness of the electrolyte layer affects a number of processes, such as the mass transport of dissolved oxygen, the accumulation of corrosion products and the hydration of dissolved metal ions. Thus, the thickness of electrolyte has an important role on the atmospheric corrosion of metals.

During the last five decades, many investigations on metals or alloys corrosion under TEL were carried out [21–26]; however, the fundamental understanding is still short. The main reasons are the errors in the electrochemical measurement of corrosion rates that arise from the ohmic drop between the reference and working electrodes and the uneven current distribution over the working electrodes. Moreover, in the conventional method, the Luggin capillary is used to measure the potential of the working electrode, which will change the thickness and composition of TEL. Thereafter, a scanning Kelvin probe (SKP) was applied to the investigations of the corrosion in TEL [27,28]. But this technique still has its disadvantages, such that the voltage potential measured by SKP is not entirely identical to the corrosion potential under all conditions and the oscillations required by SKP during measurement can cause significant convective effects in the TEL on the substrate, aiding O_2 transport across the TEL. Thus in many recent investigations, the conventional electrochemical method to study the atmospheric corrosion process has regained interest [29]. A new experimental arrangement was development to study the TEL corrosion of metals, the common difficulty in the TEL electrochemical measurement is considered and minimized by design and correct testing method.

The experimental arrangement is shown in Fig. 4.11. Concerning the design of the experimental setup, an important issue to take into account is to assure that the TEL formed on the working electrode is even and stable. The working electrode is inserted in the center of a Teflon cylinder, which can be fixed firmly in the cell. The electrochemical cell is put on a horizontal stage, which can be adjusted according to a water level. After the electrochemical cell is set up and adjusted to a horizontal level (Fig. 4.11), an even TEL with certain thickness can be obtained on the whole electrode surface. One of the advantages of this arrangement is that, even if the TEL is ultra-thin on the working electrode, the counter and reference electrodes are still immersed in the bulk electrolyte, which can minimize the ohmic drop between the reference electrode and the working electrode.

The thickness of the TEL is measured by the arrangement also shown in Fig. 4.11. A very fine platinum needle is welded on a micrometer screw and the micrometer is fixed at a position right above the electrode. When the cell is arranged, the Pt needle is adjusted slowly toward the electrode. Once the needle touches the electrode surface, there will be a sudden value shown on the ohmmeter, then the move of the needle is stopped and the position of the micrometer is recorded. Afterwards the micrometer is moved backward until some distance is kept between the electrode and the tip of the needle, and then the electrolyte is poured into the cell to form a relative thick electrolyte film on the electrode. Then the micrometer is moved towards

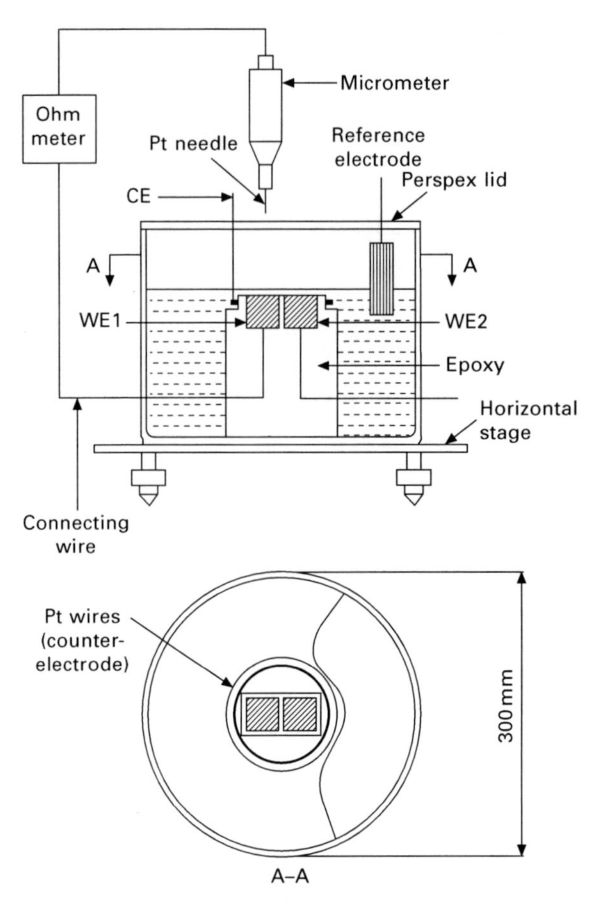

4.11 Schematic diagram of the experimental arrangement for thin electrolyte film corrosion study.

the electrode again, when the Pt needle touches the TEL surface, the ohmic meter will show a sudden ohmic value too, and the position of the micrometer is recorded again. The thickness of the TEL will be calculated micrometer from the two values. In order to form the needed TEL thickness on the electrode, a set of syringes was used in this study to remove the electrolyte. This experiment setup allows the measurement of the TEL thickness with an accuracy of 10 μm. In the design of this arrangement, the diameter of the cell is much greater than that of the electrode (the inner diameter of the cell is 30 cm while the electrode is only 1 cm), which can minimize the change of TEL thickness due to evaporation, because the cell contains large amount of electrolyte and the air–electrolyte interface is large. To keep long-term stability of the TEL, the cell is covered with a Perspex lid. The Perspex lid has three holes for the Pt needle, counter electrode and reference electrode.

After the measurement of the thickness, the holes are sealed with Vaseline to prevent the evaporation of the electrolyte.

Figure 4.12 presents the cathodic polarization curves of Mg–Gd–Y alloy under TEL with various thicknesses. The electrochemical parameter was calculated from Fig. 4.12 and illustrated in Table 4.1. Our previous work revealed that the cathodic process of magnesium was dominated by hydrogen reduction process [30]. In this study, the cathodic current density of Mg–Gd–Y alloy ranged from 10^{-1} to ~10^{-3} A/cm^2, which was two orders higher than that dominated by oxygen diffusion process (10^{-4}~10^{-5}A/cm^2) [31]. Therefore, it implied that the cathodic process of Mg–Gd–Y alloy was dominated by the hydrogen reduction process under TEL. Moreover, the cathodic current density of Mg–Gd–Y alloy decreased with decreasing TEL thickness. It was a unique phenomenon because, for most metals, such as iron, copper and aluminum, the cathodic current density increased significantly with the decrease of TEL thickness [29–31]. It also revealed that TEL thickness had

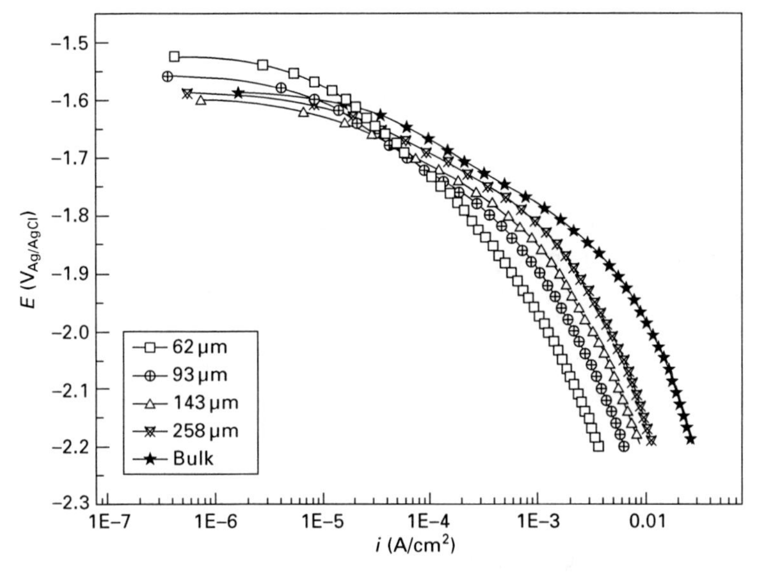

4.12 Cathodic polarization curves of GW102K under various thin layer thicknesses in 0.05 M NaCl + 0.5 M Na_2SO_4 solution saturated by $Mg(OH)_2$.

Table 4.1 The fitted electrochemical parameter for the cathodic polarization curve of Mg–Gd–Y alloy under various thin layer thicknesses

Thickness (μm)	62	93	143	258	Bulk
E_{corr} ($V_{Ag/AgCl}$)	−1.524	−1.559	−1.600	−1.589	−1.587
I_{corr} (A/cm^2)	6.383E−7	6.082E−7	1.097E−6	1.008E−6	2.283E−6

significant inhibition influence on the cathodic process of Mg–Gd–Y alloy with the decrease in TEL thickness.

Table 4.1 demonstrated that the open circuit potential of Mg–Gd–Y alloy shifted to the noble direction with the decrease in TEL thickness. Generally, the increased corrosion potential attributed to the acceleration of cathodic process or the inhibition of anodic process. The polarization curve results implied that the cathodic process was significantly inhibited under the TEL condition. So, it could be predicated that the inhibition of anodic process under TEL might be the proper explanation for the increasing of corrosion potential.

The EIS was carried out at open circuit potential (Fig. 4.13), which was aimed at confirming the results of cathodic polarization curve and at demonstrating the ohmic drop between working and reference electrode. The EIS of Mg–Gd–Y alloy under TEL consisted of two capacitive loops in the high and low frequency ranges.

The equivalent circuit in Fig. 4.14 was presented to fit the EIS results in Fig. 4.13, and the parameters of fitted results for Fig. 4.13 are listed in Table 4.2. For the equivalent circuit, R_s was the solution resistance, CPE_{dl} was the double layer capacitance, R_t was the charge transfer resistance of cathodic reduction reaction, CPE_f was the capacitance of protective film and R_f was the film resistance.

It should be noted in Fig. 4.13(b) that all of the phase shift exceed –45°, thus, the current distribution on the working electrode is even and uniform during these EIS tests, the effect of solution resistance could be considered to be minimized [30,31]. The R_s value ranged from 6 to 47 $\Omega\,cm^2$ with the decreasing TEL thickness (see Table 4.2). It meant that the experimental setup could effectively minimize the ohmic drop between working electrode and reference electrode in the case of TEL. It can be seen from Table 4.2 that the value of the double layer capacitance CPE_{dl} was varied between 10 and 16 $\mu F\,cm^2$, which implied that the TEL was continuous and covered most of the working electrode surface.

When the thickness of TEL was as low as 37 μm, the R_f value increased from 1345 to 4096 $\Omega\,cm^2$, which revealed that there was a protecting film formed on the surface of Mg–Gd–Y alloy and the corrosion resistance of the film enhanced with the decreasing of TEL thickness. It was likely that this had to do with faster accumulation of corrosion products. In other words, the corrosion of Mg–Gd–Y alloy was inhibited under TEL condition, which agreed with the polarization results.

According to the cathodic polarization curve and EIS results, it was clear that the cathodic process of the corrosion of Mg–Gd–Y alloy was dominated by the hydrogen evolution reaction in the case of TEL. With the decreasing of TEL thickness, the corrosion of Mg–Gd–Y alloy was retarded.

The corrosion morphology of Mg–Gd–Y alloy after the immersion time

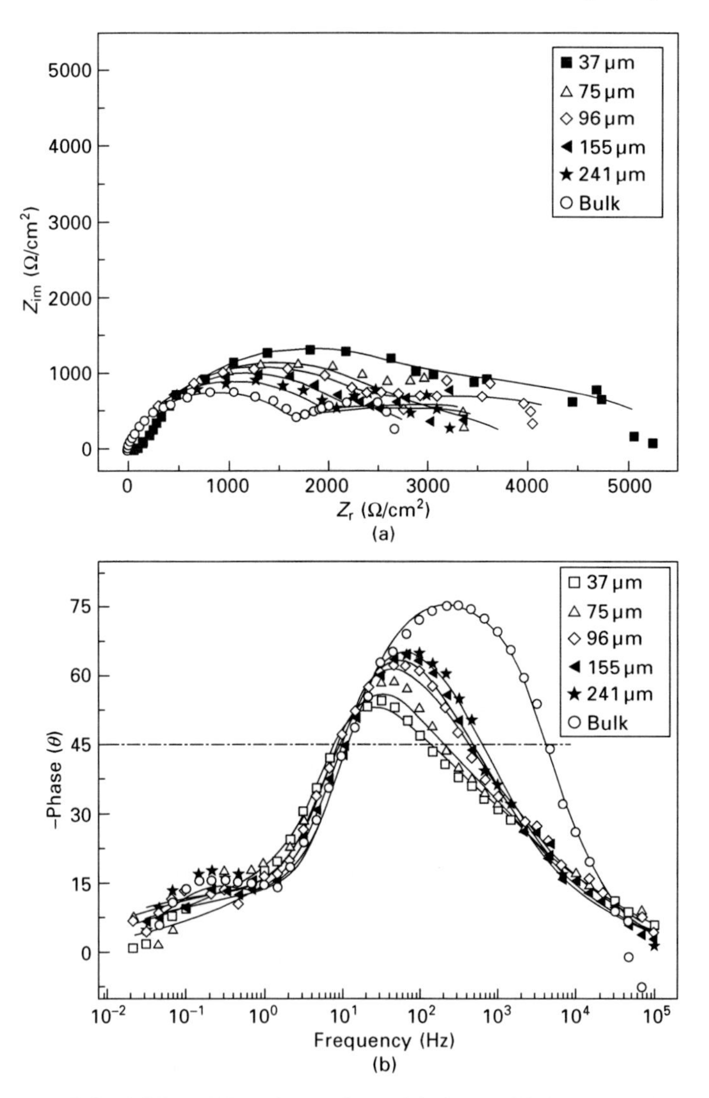

4.13 EIS of GW102K under various thin layer thickness at the open circuit potential in 0.05 M NaCl+0.5 M Na_2SO_4 solution saturated by $Mg(OH)_2$.

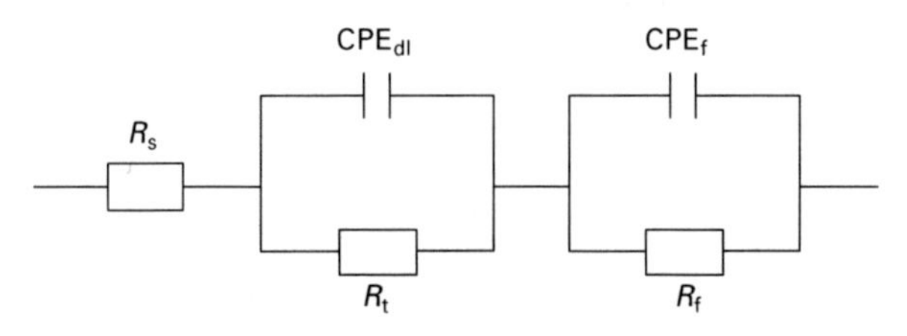

4.14 Equivalent circuit for Fig. 4.13.

Table 4.2 The fitted electrochemical parameter for EIS of Mg–Gd–Y alloy under various thin layer thicknesses

Thickness (μm)	R_s (Ωcm^2)	CPE_{d1} $(Y_0)_{dl}$ ($\mu F cm^2 Hz^{1-n}$)	n_1	R_t (Ωcm^2)	CPE_f $(Y_0)_{dl}$ ($\mu F cm^2 Hz^{1-n}$)	n_2	R_f (Ωcm^2)
37	46.83	16.5	1	1657	171	0.4861	4069
75	37.68	14.3	1	1637	220	0.4848	2360
96	26.96	12.8	0.9803	1747	366	0.4785	3380
155	28.83	11.7	0.9849	1689	453	0.4905	2535
241	23.92	10.4	0.9902	1524	525	0.4868	2881
bulk	5.75	13.5	0.9142	1637	793	0.8305	1345

of 120 h under TEL and bulk solution is shown in Fig. 4.15. It was observed that the number of pits was reduced with the decreasing of TEL thickness. Furthermore, the pit size under TEL was smaller than that in bulk solution. This meant that TEL had a significant influence on the anodic process of the corrosion of Mg–Gd–Y alloy. Further analysis of the anodic process under TEL should be carried out.

The susceptibility of pit corrosion could be attributed to two aspect factors: pit initiation and pit growth [32]. In the other word, the pit corrosion is dominated by the synergistic effect of the pit initiation and pit growth, which determine how it grows and whether it becomes stable or not. Therefore, a system undergoing serious uniform corrosion can have both higher pit initiation rate and lower pit growth probability. In contrast, the corrosion would be rare since lower both pit initiation rate and pit growth probability are dominant. Finally, in the case of lower pit initiation rate and higher pit growth probability, the corrosion would be rather localized over the surface.

Shot noise theory is based on the assumption that the signals are composed of packets of data departing from a base line. This theory can be applied to the analysis of electrochemical noise data from corrosion systems, the current noise signals being considered as packets of charge [33]. Among noise-generating processes, shot noise is caused by the fact that the current is carried by discrete charge carriers. Consequently, the number of charge carriers passing a given point will be a random variable.

Provided that the individual events are independent of other events such as the stochastic processes, it has been known that the shot noise analysis is applicable to the individual events. In the recent literature, the shot noise theory has been applied to the analysis of electrochemical noise signals [33]. If we assume that shot noise is produced during breakdown of the passive film, pit initiation and hydrogen evolution, the average corrosion current $\overline{I_{corr}}$ is defined as

$$\overline{I_{corr}} = q \times f_n \qquad 4.13$$

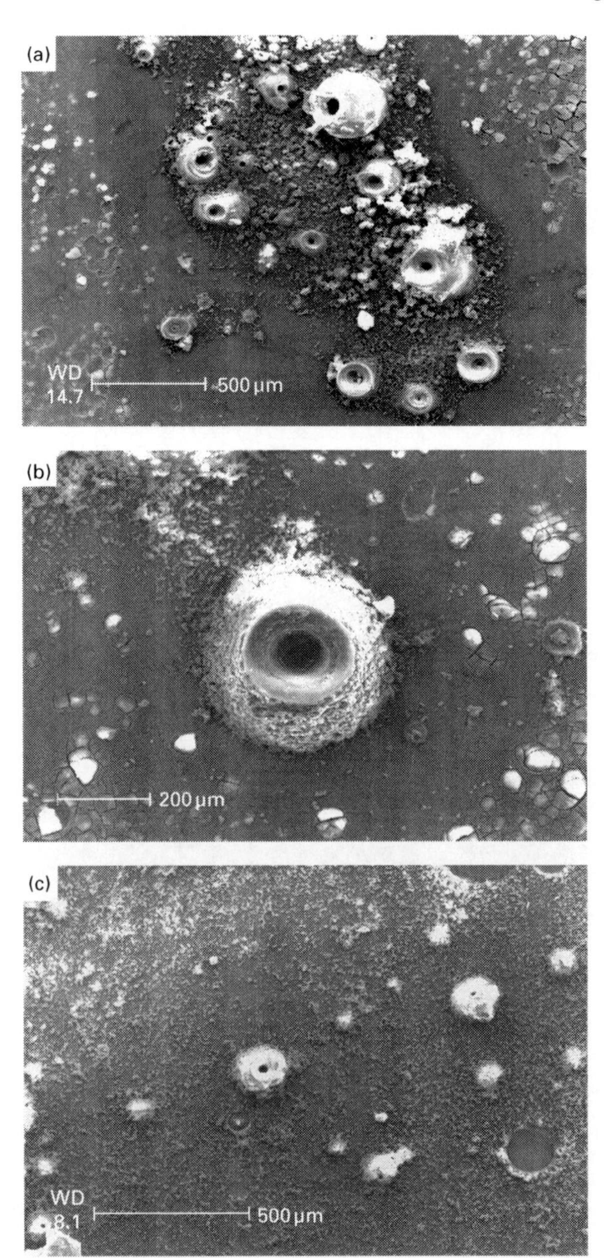

4.15 SEM images of corrosion morphology of GW102K after the immersion of 120 h under various thin layer thicknesses: (a) small magnification and (b) large magnification in bulk solution; (c) small magnification and (d) large magnification under 233 μm thin layer; (e) small magnification and (f) large magnification under 164 μm thin layer; (g) small magnification and (h) large magnification under 65 μm thin layer.

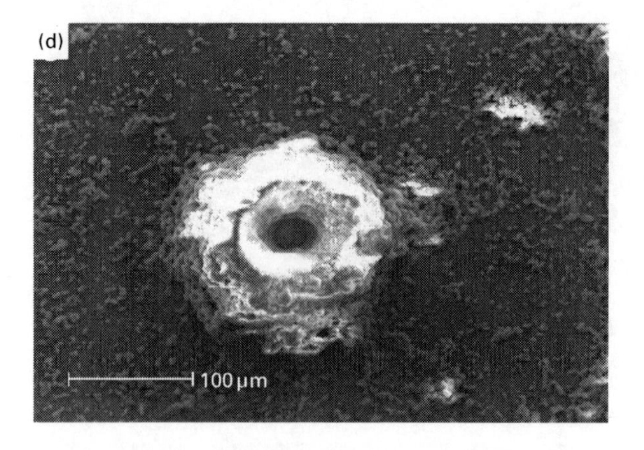

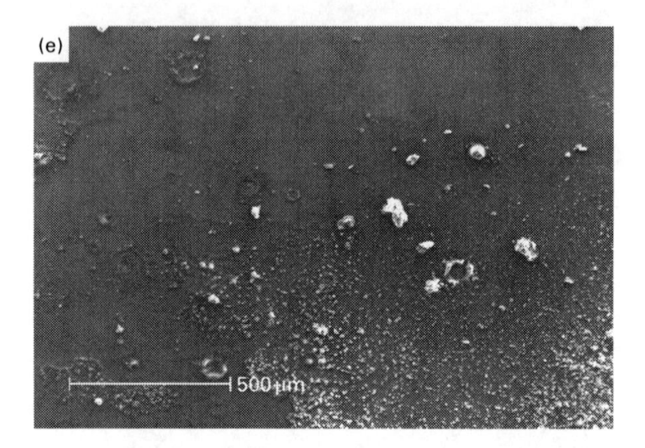

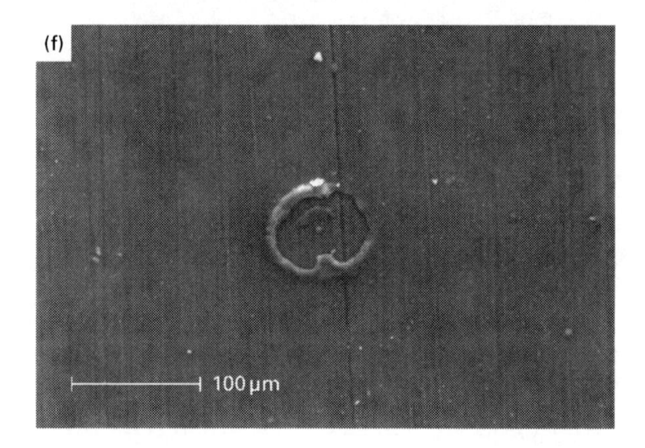

4.15 Continued

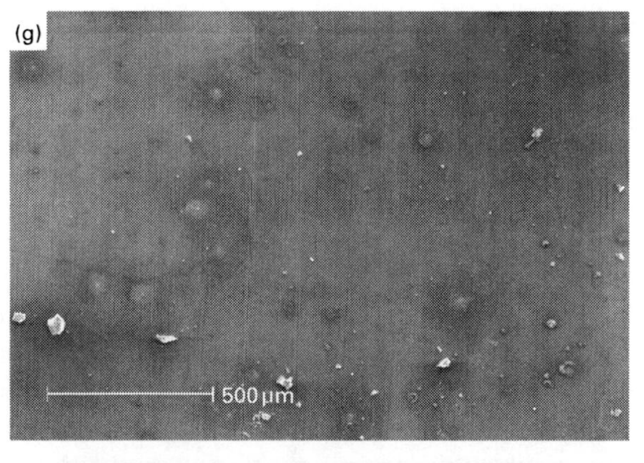

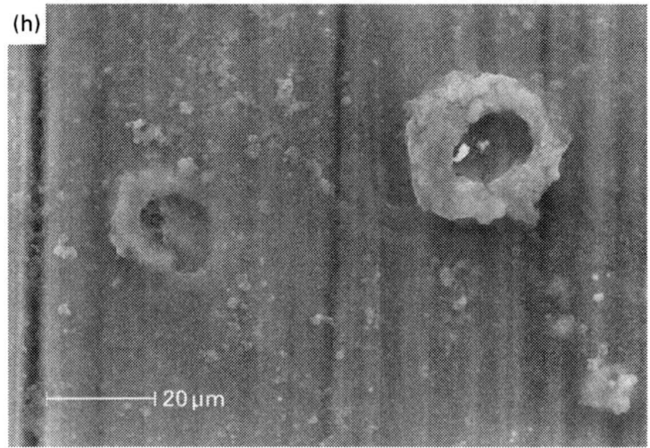

4.15 Continued

where q is the average charge in each event and f_n the frequency of events. f_n can be estimated from the following relations based upon shot noise theory. Taking into account that $I_{corr} = B/R_p$ and assuming that R_p is equal to R_n, where R_n is the noise resistance. The following expression is acceptable:

$$B = I_{corr} \times R_n$$

Since $f_n = B^2/(\psi_E \cdot s)$ [33] where s is the sample area and $s = 1\,cm^2$, ψ_E is the low frequency limit (0.01Hz) of power spectral density of potential, respectively. f_n can be estimated in function of I_{corr}, R_n and ψ_E:

$$f_n = (\overline{I_{corr}} \times R_n)^2/\psi_E \quad 4.14$$

The cumulative probability $F(f_n)$ is plotted against the frequency of events f_n in Fig. 4.16 for Mg–Gd–Y alloy under various TEL thicknesses. The procedure for determining the cumulative probability $F(f_n)$ from f_n data is described as follows: first, all calculated f_n data are arranged in order from the smallest and then the cumulative probability $F(f_n)$ is calculated as

$M/(N + 1)$, where M is the rank in the ordered f_n data and N the total number of f_n data.

It was found that the distribution of f_n shifted to a lower frequency region with the decrease in TEL thickness. Considering that high-frequency events would tend to occur all over the alloy surface, the corrosion of Mg–Gd–Y alloy would be less localized as high-frequency events became dominant. In contrast, the corrosion would be rather localized over the surface as low-frequency events were dominant.

Since the space of the distribution function should be the positive t axis, the plots of the cumulative probability in Fig. 4.16 were transformed from the f_n domain to the mean free time t_m domain in order to investigate the low-frequency events associated with localized corrosion in more detail based upon a stochastic theory.

Weibull distribution function is one of the widely used cumulative probability functions for predicting lifetime in reliability test [34]. This is because it can easily approximate the normal distribution, logarithmic normal distribution and exponential distribution functions. In addition, it is also possible to analyze data even when two or more failure modes are present at the same time. The cumulative probability $F(t)$ of a failure system can be introduced just as Weibull distribution function based upon a 'weakest-link' model [34], which is expressed as:

$$F(t) = 1 - \exp(-t^m/n) \quad 4.15$$

where m and n are the shape and scale parameters, respectively. From rearrangement of Eq. 4.15:

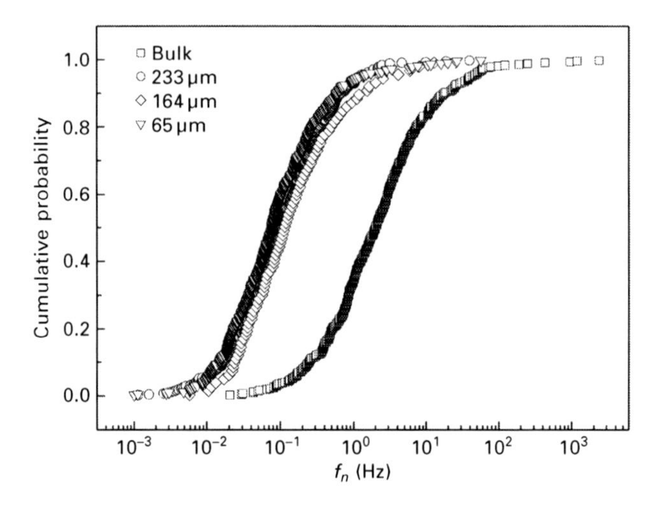

4.16 Cumulative probability plots for the frequency of events, f_n, under various thin layer thicknesses.

$$\ln\{\ln[1/(1 - F(t))]\} = m\ln t - \ln n \quad 4.16$$

The resulting plots of the cumulative probability $F(t_m)$ versus t_m were rearranged according to Eq. 4.16, which gave the Weibull probability plots.

Figure 4.17 described the Weibull probability plots for Mg–Gd–Y alloy under the TEL with various thicknesses. The plots showed satisfactorily good two linear regions in one plot, which indicated that two failure modes existed, depending upon t_m. Considering that only uniform and localized corrosion would occur during the noise measurement, it was suggested that the slopes in the relatively shorter t_m range were associated with uniform corrosion. On the other hand, the slopes in the longer t_m range were responsible for localized corrosion such as pit initiation. From Fig. 4.17, the values of the shape parameter m and scale parameter n were determined for pit corrosion of Mg–Gd–Y alloy as a function of TEL thickness, which are listed in Table 4.3.

The conditional event generation rate $r(t)$ is employed as a kind of failure rate in reliability engineering, which is defined as [34]:

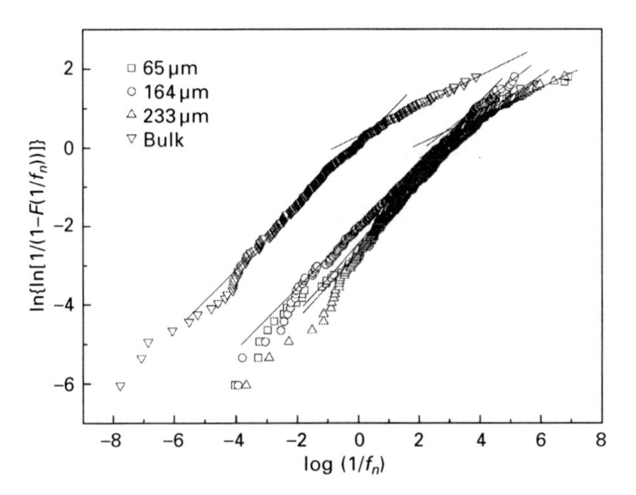

4.17 Weibull probability plots under various thin layer thicknesses.

Table 4.3 Weibull distribution parameters for the pit corrosion of Mg–Gd–Y alloy under various thin layer thicknesses

Thickness (μm)	Shape parameter m [–]	Scale parameter n [s^m]
65	0.538	4.104
164	0.567	3.347
233	0.382	2.067
bulk	0.419	0.788

$$r(t) = \frac{m}{n} t^{m-1} \tag{4.17}$$

The conditional event generation rate $r(t)$ was calculated as a function of TEL thickness by inserting the values of the shape and scale parameter m and n into Eq. 4.17. The result of $r(t)$ was illustrated as a function of time t in Fig. 4.18. The value of $r(t)$ represents the formation rate of pit initiation in the next unit time for the specimens, in which pit initiation has not yet been generated when t has elapsed. It was observed that the value of $r(t)$ under TEL was remarkably lowered compared with that in the bulk solution at a given time t. This indicated that pit initiation was inhibited under TEL.

Pit depth growth is modeled using a nonhomogeneous Markov process. The way to link the initiation and growth stages when multiple pits are considered is proposed for the first time [32]. To do this, the theoretical foundations of extreme value statistics have been employed. It is shown that the solution of the Kolmogorov forward equations, governing the growth of an individual pit, is in the domain of attraction of the Gumbel distribution [32]. The Gumbel extreme value distribution is used to model the behavior of the deepest pits [32].

The extreme value statistics analysis can be estimated according to the following procedure [35]: first, all calculated extreme value data are arranged in order from the smallest and then, the probability $F(Y)$ is calculated as $1-[M/(N + 1)]$, where M is the rank in the ordered extreme value and N the total number of extreme value data. The reduced variant (Y) can be calculated by the formula $Y = -\ln \{-\ln [F(Y)]\}$.

The probability that the largest value of pit depth $\leq x$ is described by a

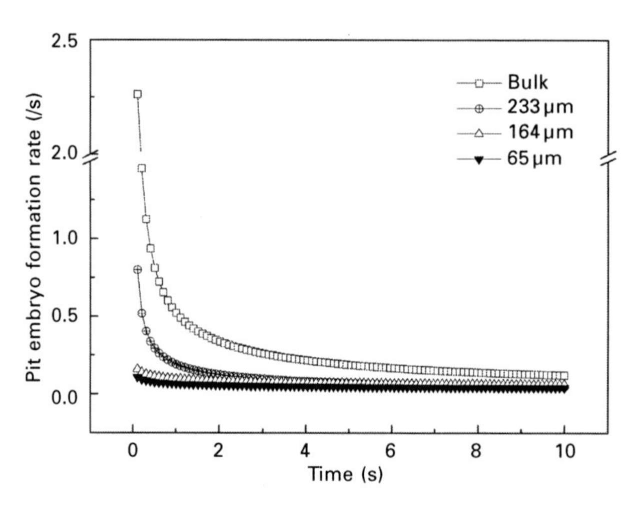

4.18 Plots of the pit initiation rate under various thin layer thicknesses.

double exponent (Gumbel type extreme value distribution) can be calculated by the following form [35]:

$$Pit_{max} = \mu + \alpha \ln S \tag{4.18}$$

Probability of pitting size

$$= 1 - \exp\left\{-\exp\left[\frac{-\text{pitting size}-[\mu + \alpha \ln S]}{\alpha}\right]\right\} \tag{4.19}$$

where μ is the central parameter (the most frequent value), α is the scale parameter, which defines the width of the distribution, and S is the area of the large system.

The integration of current transient with time could be used to determine the charge passed for each current transient spike. This charge is the result of the formation of a pit and can be related to the physical volume of the pit by using Faraday's equation, Eq. 4.20, which was based on the correlation between optical pit size and anodic current transient charge [36]. If the pits are assumed to be hemispherical the pit radius/depth can be calculated, using Eq. 4.21:

$$\text{Volume (cm3)} = \frac{\text{Charge passed} \times \text{molecular mass}}{\text{Faraday constant} \times n \times \text{density}} \tag{4.20}$$

$$\text{Pit radius } (\mu\text{m}) = \left(\sqrt[3]{\frac{3 \times \text{volume}(\text{cm}^3)}{2\pi}}\right) \times 10\,000 \tag{4.21}$$

where molecular mass of Mg–Gd–Y alloy is 27.8 g/mol, Faraday constant is 96 500 and density of magnesium is 1.74 g/cm^3. The largest pit sizes within each of the electrochemical noise (EN) segments were calculated and the values were subjected to extreme value statistics analysis.

The values of the reduced variant were plotted against the ordered pit sizes (Fig. 4.19), which indicated that both metastable pit and stable pit occurred on the alloy surface. The observation of the straight line confirmed that the data did in fact fit the Gumbel distribution. The values of α and μ are the scale and location parameters for the distribution of the largest pits, respectively. These values are analogous to the standard deviation and average, and describe the shape and center of the probability distribution of maximum metastable and stable pit sizes expected from electrodes identical to those used for the measurements and treated in the same manner for the same period of time. The scale and location parameters measured under the TEL with various thicknesses are shown in Table 4.4. The probability of a given pit size occurring under the TEL with various thicknesses was calculated using Eq. 4.19 and the results were shown in Fig. 4.20. The probabilities can be converted into an expected time for pit cavity with a

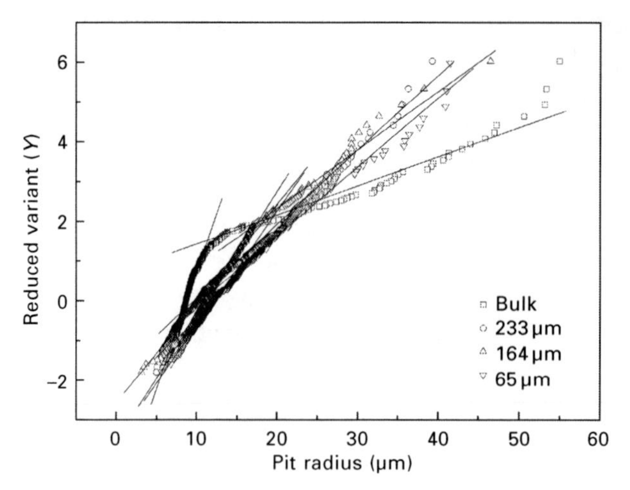

4.19 Gumbel probability plots under various thin layer thicknesses.

Table 4.4 Gumbel distribution parameters for GW102K under various thin layer thicknesses

Thickness (μm)	Metastable pit		Stable pit	
	Scale parameter α	Location parameter μ (μm)	Scale parameter α	Location parameter μ (μm)
65	3.788	12.47	5.814	10.49
164	3.311	10.97	6.369	5.268
233	3.559	10.02	5.438	9.989
bulk	1.658	8.755	12.50	–5.836

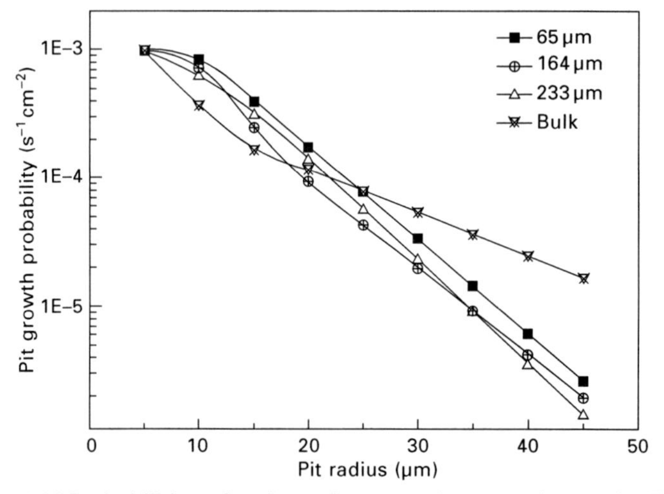

4.20 Probabilities of various diameter pits occurring under TEL with various thicknesses.

particular size to occur by taking the reciprocal of the probability. That is, calculating the time it takes for the cumulation of the probabilities to equal unity. For example, a 30 μm stable pit radius in the bulk solution will occur on average after 5.46 h, but under 233 μm TEL the average time for the same 30 μm stable pit to occur is 12.41 h, under 164 μm it is 13.96 h and under 66 μm it is 8.3 h. These results indicated that the metastable pit formed on Mg–Gd–Y alloy has a lower probability to become stable ones and, finally, developed into larger pit cavity during longer time interval under TEL than that in bulk solution.

The above discussion revealed that TEL had significant influence to the pit susceptibility of Mg–Gd–Y alloy. In case of TEL, both the pit initiation rate and the pit growth probability were decreased.

In summary, the corrosion behavior of Mg–Gd–Y under TEL was significantly different from that in bulk solution. The experimental results indicated that the corrosion rate of Mg–Gd–Y alloy decreased significantly with the decrease in TEL thickness. The cathodic process of the corrosion of Mg–Gd–Y alloy was dominated by hydrogen evolution reaction not only under TEL but also in bulk solution. With the decrease in TEL thickness, the cathodic process was retarded. Under TEL condition, TEL had significant influence on the anodic process of Mg–Gd–Y alloy corrosion. The pit initiation rate was inhibited. The frequency of corrosion events under TEL is greatly lowered compared with that in the bulk solution. Furthermore, the pit growth probability was decreased. The metastable pit on Mg–Gd–Y alloy has a lower probability to become larger pit cavity during shorter time interval than that in bulk solution.

4.3.2 Influence of RE element on passivity behavior of AZ91 alloy

RE elements have been used in Mg alloys as well as AZ91 alloy in recent years. The beneficial effect of RE elements in AZ91 alloy has been investigated. Rosalbino *et al.* [37] considered that the improved corrosion resistance of the Mg–Al–Er alloys could be attributed to the incorporation of erbium in the $Mg(OH)_2$ lattice. Zhou *et al.* [38] studied the influence of Ho addition on corrosion resistance of AZ91 alloy. They believed that the improvement of corrosion resistance of the Mg–9Al–Ho alloys could be explained by the fact that the deposited Ho-containing phases were less cathodic and the corrosion product films could restrain further corrosion. Wu *et al.* [39] found that 1 mass% Ca and 1 mass% RE element (the mixture of Ce and La) together added to AZ91 alloy could decrease the corrosion rate due to the formation of the reticular Al_2Ca phase, which acted as an effective barrier against corrosion. Huang *et al.* [40] and Liu *et al.* [41] studied the influence of Ce to AZ91 alloy corrosion resistance. They found that as the microstructure of as-cast AZ91 alloy was refined, the β-phase volume and distribution could be changed and rod-like Al_4Ce phase was formed, which led to the decrease of corrosion current density in AZ91 alloy.

Generally, wrought Mg alloys have superior mechanical properties to as-cast Mg alloys [42], due to such mechanisms as the refinement of grains, the elimination of cast defects and homogenization of microstructure during the plastic deformation process. Comparing with conventional wrought Mg alloys like AZ31, AZ61 and ZK60, AZ91 alloy is likely to have better combination of cast ability, plasticity and cost [43]. Although the improved corrosion resistance after addition heavy (Er or Ho) and light (Ce or La) RE elements were studied, RE elements effect on passivity behavior in wrought AZ91 alloy has not been clear.

In this section, the wrought AZ91 alloy was fabricated as follows: the cast ingots were extruded at 230–240 °C with a horizontal water press, the rate was 2–3 m/min and the extrusion ratio was 18:1.

The potentiodynamic anodic polarization plots of wrought AZ91 alloy without and with Ce in 0.01 M NaOH aqueous solution are shown in Fig. 4.21. It can be seen that the anodic polarization plot of wrought AZ91 alloy without Ce shows an active–passive behavior. A decrease in current density at potential above $-1.456\,V_{SCE}$ is observed. Within the potential range of passivity, a relative small passive current density (about 4×10^{-6} A/cm^2) is observed, which is potential-dependent.

The polarization plot of wrought AZ91 alloy with Ce differs from that of wrought AZ91 alloy without Ce. For wrought AZ91 alloy with Ce, the

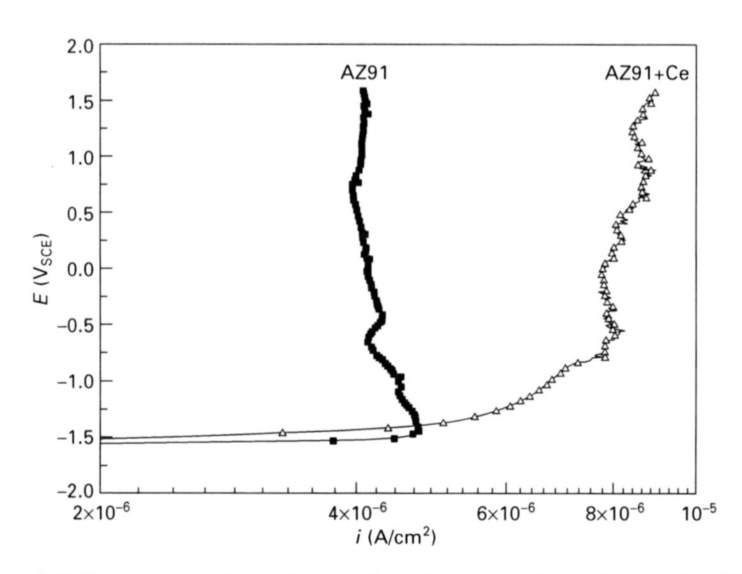

4.21 The potentiodynamic anodic polarization plots of wrought AZ91 alloy without and with Ce in 0.01 M NaOH.

anodic current density increases with the increment of applied potential. At low anodic potential, the polarization plot is characterized by a progressive increase of current density with increasing anodic potential, thereby indicating an appreciable tendency for metal dissolution, which may be caused by Ce rapid dissolution because of its high chemical activity. When the potential increases around $-0.5\,V_{SCE}$, the passivity behavior appears and the passive current density is between 8×10^{-6} and 9×10^{-6} A/cm^2 at the potential range of -0.5 to $1.6\,V_{SCE}$. The passive current density of wrought AZ91 alloy with Ce is about twice that of wrought AZ91 alloy without Ce, which indicates that the stability of the passive film becomes worse with addition of Ce.

Variation of current with time at $0\,V_{SCE}$ was measured. According to the point defect model [44], the relationship between current density and time is as follows:

$$\log(t_s) = k \log(t) + C \qquad 4.22$$

where k represents the slope of double-logarithmic plot for potentiostatic polarization. $k = -1$ indicates the formation of a compact, highly protective passive film on the electrode surface and a high field-controlled oxide layer growth. $k = -0.5$ indicates the presence of a loose film and a diffusion-controlled oxide layer growth [45,46].

Fig. 4.22 shows the double-log plots of current density with time for two alloys. For passive films on two alloys, $k = -0.5$ is obtained and there is no obvious difference in the k value. Therefore, the passive films on two alloys are loose and passive film growths are at the control of diffusion process.

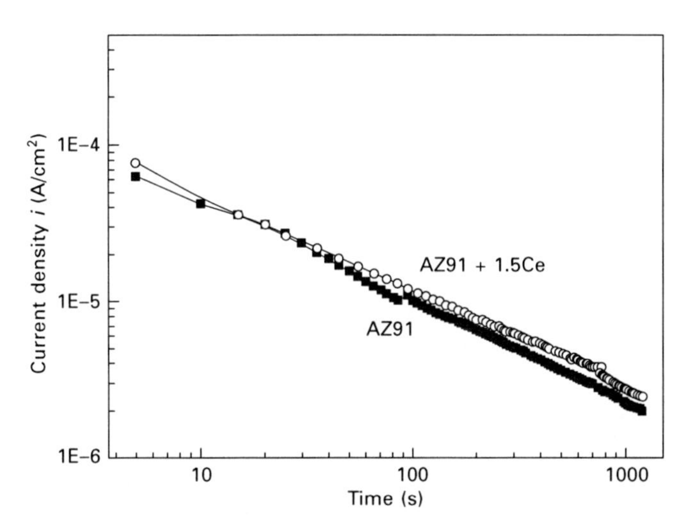

4.22 Potentiostatic polarization plots for wrought AZ91 alloy without and with Ce at $0\,V_{SCE}$ in 0.01 M NaOH solution.

X-ray photoelectron spectroscopy (XPS) is an effective way to ascertain film composition, thus to provide an additional access for understanding passive behavior. XPS depth profiling (DP-XPS) spectra of Mg_{1s}, Al_{2p} and O_{1s} of the passivity film on wrought AZ91 alloy without Ce are shown in Fig. 4.23(a) [47]. The lower bond energy (BE) peak at ~1303.8 eV is assigned to metallic Mg. The broader higher BE peak is centered between ~1305.5 and ~1307 eV, indicating the co-existence of Mg oxide and Mg hydroxide in the film [48]. The broaden peak indicates that the passivity film is loose, which is in good agreement with the potentiostatic polarization result. The lower BE peak at ~73.5 eV is assigned to metallic $Mg_{17}Al_{12}$ and the higher BE peaks at ~75.1 are attributed to Al oxide and $Al(OH)_3$ [49], respectively. Figure 4.23(a) shows that two O_{1s} peak at ~531.8 eV (assigned to oxide) and at ~534.3 eV (assigned to hydroxide) exist in the passive film of wrought AZ91 alloy without Ce. From the result of DP-XPS, it can be concluded that the passive has double layers: $Mg(OH)_2$ and $Al(OH)_3$ predominate at the outer layer of passive film, while MgO and Al_2O_3 are present in the inner layer. With sputtering, the peak intensities of $Mg(OH)_2$ and $Al(OH)_3$ decrease while those of MgO and Al_2O_3 exhibit the opposite behavior.

DP-XPS spectra in the Mg_{1s}, Al_{2p}, Ce_{3d} and O_{1s} for the passive film on wrought AZ91 alloy with Ce are shown in Fig. 4.23(b). The components of the passive film are $Mg(OH)_2$ and $Al(OH)_3$ at the outer layer, which is almost the same as those of the film on wrought AZ91 alloy without Ce. No Ce hydroxide is incorporated into the outer layer of the surface film, which is the same as the result obtained by Yao *et al.* [50]. They found that passive film composition on Mg–13.2 at.% Y alloy was mainly of $Mg(OH)_2$ mixed with small amount of MgO and no Y was incorporated into the outer layer of the surface film formed in 0.01 M NaCl solution (pH 12). When sputtering time increased to 300 s, the $3d_{3/2}$ and $3d_{5/2}$ double peaks of Ce appeared. The double BE peaks at ~883.4 and ~902.2 eV are attributed to CeO_2. The inner film component of wrought AZ91 alloy with Ce is the mixture of MgO, Al_2O_3 and CeO_2.

From DP-XPS result of wrought AZ91 alloy with Ce, there are no Ce hydroxides at the outer layer of the passivity film. Ce only exists in the inner layer. From potential–pH diagram of cerium–water system [51], it can be seen that the Ce^{3+}, $Ce(OH)_3$ and CeO_2 are the most stable species in the water. The Ce^{3+} species is stable at higher acidic region ($pH < 1$) at all potential conditions, but as the pH increases $Ce(OH)_3$ or CeO_2 becomes stable. Therefore, $Ce(OH)_3$ or CeO_2 was stable in 0.01 M NaOH aqueous solution (pH 12). It is known that the radii of Mg^{2+}, Al^{3+} and Ce^{3+} are 0.65, 0.50 and 1.69 Å, respectively [52]. The radius of Ce^{3+} is almost 1.5 and 2 times those of Mg^{2+} and Al^{3+}, respectively. In this case, Ce^{3+} outside diffusion in the passivity film will become very difficult. At the initial stage of passivity, Ce hydroxide together with Mg and Al hydroxides could easily form on the

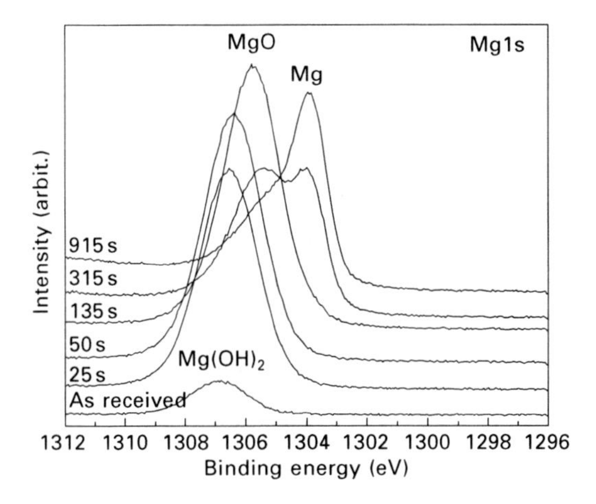

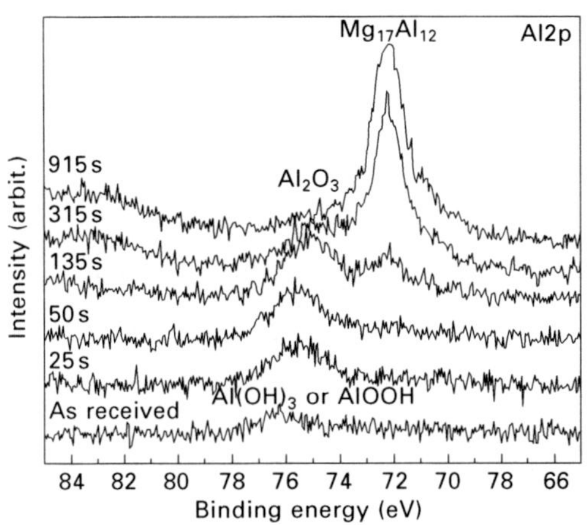

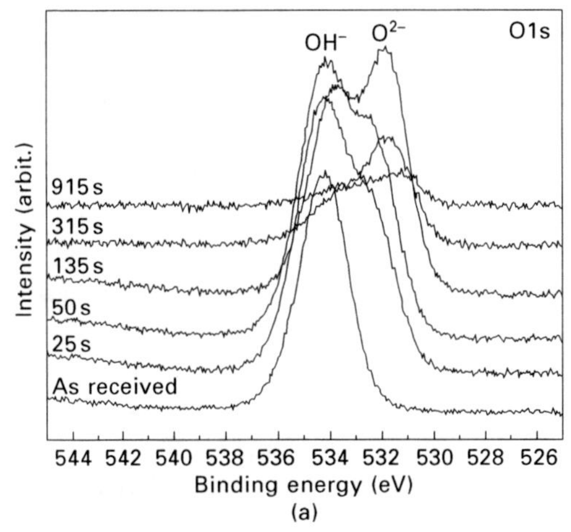

4.23 XPS spectra of Mg1s, Al2p and O1s for wrought AZ91 alloy without Ce (a), Mg1s, Al2p, Ce3d and O1s for wrought AZ91 alloy with Ce (b).

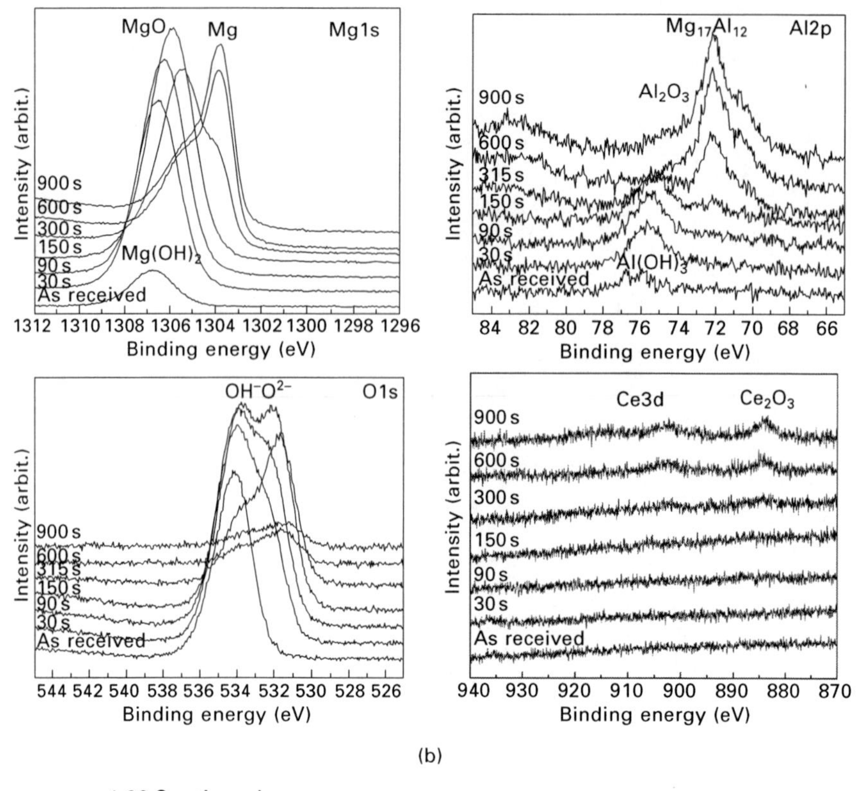

4.23 Continued

surface of the electrode. Ce hydroxide movement outside the interface of film and solution was almost impossible because the radius of Ce^{3+} is larger than those of Mg^{2+} and Al^{3+}. Therefore, Ce hydroxide existed only in the inner film. As the passivity went on, the Ce hydroxides had less opportunity to contact H_2O, which could lose H_2O from hydroxide and become the metal oxides (Eq. 4.26). Therefore, there are no Ce hydroxides at the outer layer of the passivity film for wrought AZ91 alloy with Ce.

In order to ascertain the effect of Ce oxides in inner layer on the passivity for wrought AZ91 alloy with Ce, EIS was measured at $0\,V_{SCE}$ in the passive range. Figure 4.24 shows the experimental EIS spectra of wrought AZ91 alloys without and with Ce. For wrought AZ91 alloy without Ce, a high-frequency capacitive loop and Warburg impedance (*W*) at low-frequency region are observed. For a wrought AZ91 alloy with Ce, the Nyquist plot consists of a high frequency capacitive loop and a line with the slope larger than 1 at low frequency region, which indicates a tangent hyperbolic diffusion (*T*) existing in the passive process. In other words, Ce addition to wrought AZ91 alloy made the diffusion process change from the Warburg impedance (*W*) to the

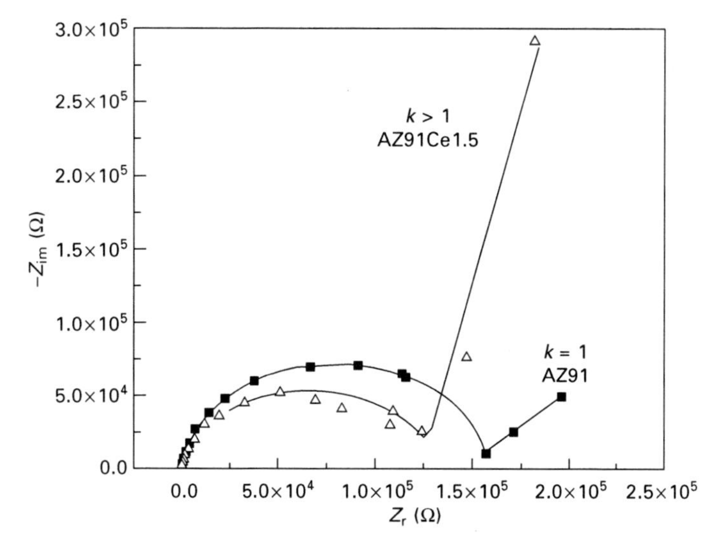

4.24 EIS spectra for wrought AZ91 alloy without and with Ce at 0 V_{SCE} in 0.01 M NaOH solution.

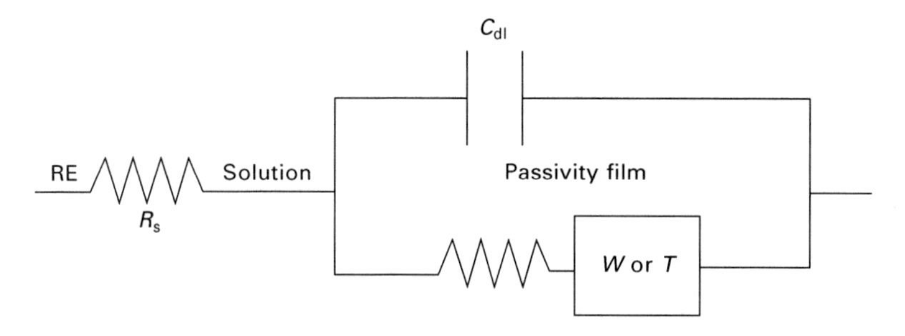

4.25 An equivalent circuit model for EIS spectra.

Table 4.5 The fitting results of EIS of Fig. 4.27

	R_s (Ω cm^2)	CPE$_{dl}$ (Y_0)$_{dl}$ (μF cm^2 Hz^{1-n})	n_1	R_t (kΩ cm^2)	*W* or *T*
AZ91	91.8	0.3426×10^{-5}	0.9347	159.6	*W*
AZ91Ce1.5	137.2	0.4344×10^{-5}	0.8436	136.8	*T*

tangent hyperbolic diffusion (*T*). Therefore, Ce oxides in the inner layer of the passivity film had the barrier layer effect.

EIS was fitted based on the equivalent circuit (Fig. 4.25) and the results are listed in Table 4.5. The transition resistance (R_t) of the passive film for wrought AZ91 alloy without Ce is larger than that of wrought AZ91 alloy with Ce. This result again proves that the passive current density of wrought AZ91 alloy without Ce is lower than that of wrought AZ91 alloy with Ce.

Mott–Schottky (M-S) plots of passive films on wrought AZ91 alloys without and with Ce are shown in Fig. 4.26(a). As it can be seen, the positive slopes are obtained on wrought AZ91 alloy without and with Ce, which indicate n-type semiconductors of passive films for two alloys. Namely, Ce in wrought AZ91 alloy does not change the type of semiconductor.

The slope of $1/C^2$ versus V plot is inversely proportional to carrier concentration (donor or accepted). The slope of wrought AZ91 alloy without

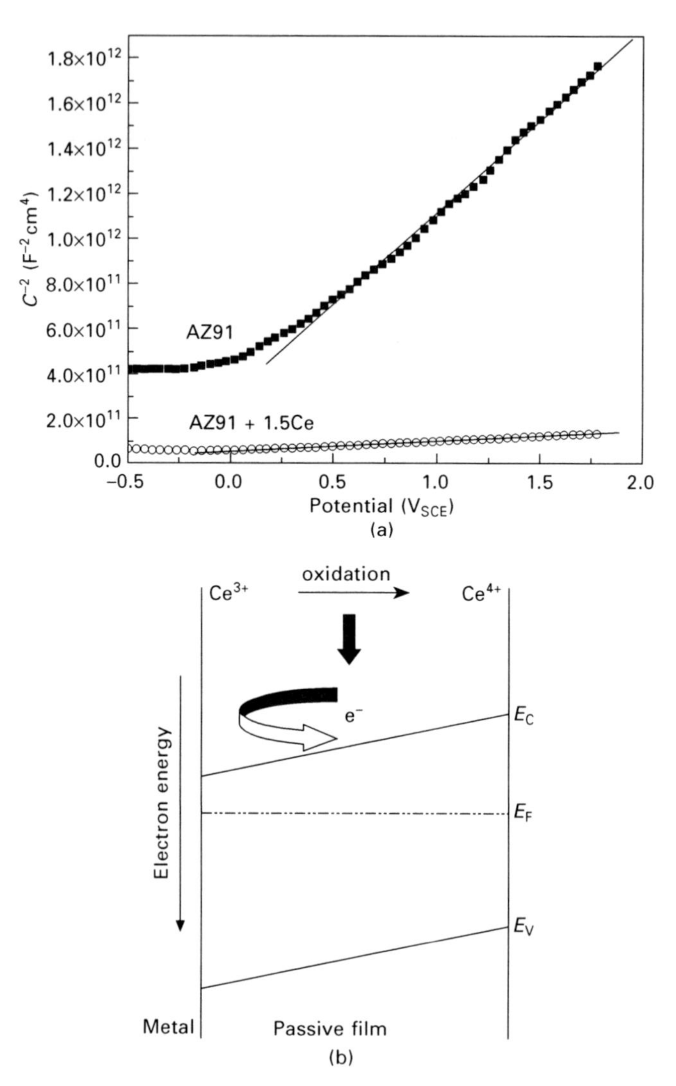

4.26 M-S plots of the passive films for wrought AZ91 alloy without and with Ce (a), sketch presentation of the electronic structure of n-type passive film at anodic potential (b).

Ce is bigger than that of wrought AZ91 alloy with Ce, suggesting donor concentration in the passive film of wrought AZ91 alloy without Ce is lower than that of wrought AZ91 alloy with Ce. N_d can be obtained from the slopes: 9.07×10^{20} cm^{-3} for wrought AZ91 alloy without Ce and 1.67×10^{22} cm^{-3} for wrought with Ce.

The results of the potentiostatic polarization and EIS have showed that the passive film formation processes for two alloys are diffusion-controlled processes, which means the diffusion stage is slower than the electrochemical stage during the passive film formation process. The metal ionization process at the interface of metal and film was the electrochemical process. The process was more rapid than the diffusion process, which led to metal ions or oxygen vacancies predominated in the passive film. This kind of passive film has an n-type semiconductor character. Therefore, the passive films of wrought AZ91 alloy without and with Ce addition present n-type semiconductors.

Figure 4.26(b) shows a schematic diagram of the energy band structure of an n-type semiconductor. It can be seen that the Fermi level (E_F) is near to the conductive band energy (E_C). If an anode polarization is applied to the passive electrode, E_F of the electrode is decreased, which causes the energy band to bend upwards to the side of solution. In the initial passivity, Ce^{3+} comes from metal ionization in wrought AZ91 alloy and Ce might react with OH^- to form the Ce hydroxides:

$$Ce^{3+} + 3OH^- \rightarrow Ce(OH)_3 \qquad 4.23$$

As the passivity process went on, $Ce(OH)_3$ is oxidized to CeO_2 [53]:

$$Ce(OH)_3 + OH^- \rightarrow CeO_2 + 2H_2O + e^- \qquad 4.24$$

The electron (e^-) comes into the conductive band (Fig. 4.26(b)), which would result in one magnitude increase of N_d in the passive film with Ce. The more donor the passive film has, the easier the anodic reaction proceeds. In other words, the passive current density of wrought AZ91 alloy with Ce is larger than that of wrought AZ91 alloy without Ce. From the above results, Ce oxide could form in the inner layer of the passivity film for wrought AZ91 alloy with Ce, which increased the passive current density and led to the formation of unstable passive film.

In summary, short rod-like Al_4Ce phases were formed in wrought AZ91 alloy with 1.5 mass% Ce, and DRX had been retarded. The passive current density increased and the passive film stability decreased after Ce addition into wrought AZ91 alloy.

The passive film formed on two alloys had double layers. The outer layer was the metal hydroxides and the inner layer was the metal oxides. No Ce hydroxides were found at outer layer and Ce in the form of CeO_2 existed only in the inner layer of the passive film for wrought AZ91 alloy with Ce.

The existence of CeO_2 in the inner layer had two effects on the passive behavior of wrought AZ91 alloy. One is that CeO_2 acted on the barrier film and made mass transport through the passive film follow tangent hyperbolic (*T*) impedance for wrought AZ91 alloy with Ce instead of Warburg impedance (*W*) for wrought AZ91 alloy without Ce. The other is that transformation from $Ce(OH)_3$ to CeO_2 led to the increase of N_d in the passive film, which was the main reason of passive current density increasing for wrought AZ91 alloy with Ce.

4.4 References

1. Cao C N (1990), 'On the impendence plane displays for irreversible electrode reactions based on the stability condition of the steady-state', *Electrochim. Acta* **35**, 837–844.
2. Cao C (1994), *Corrosion Electrochemistry* Beijing, Chemical and Industrial Press.
3. Song G L, Atrens A and Dargusch M (1999), 'Influence of microstructure on the corrosion of diecast AZ91D', *Corros. Sci.* **41**, 249–273.
4. Eliezer D, Uzan P and Aghion E (2003), 'Effect of second phases on the corrosion behavior of magnesium alloys', *Mater. Sci. Forum* **419–422**, 857–866.
5. Song G L, Atrens A, St. John D and Wu X (1997), 'The anodic dissolution of magnesium in chloride and sulphate solution', *Corros. Sci.* **39**, 1981–2004.
6. Chen J, Wang J, Han E, Dong J and Ke W (2008), 'States and transport of hydrogen in the corrosion process of an AZ91 magnesium alloy in aqueous solution', *Corros. Sci.* **50**, 1292–1305.
7. Zhang T, Shao Y, Meng G, Li Y and Wang F (2006), 'Effect of hydrogen on the corrosion of pure magnesium', *Electrochim. Acta* **52**, 1323–1328.
8. Yamamoto A, Watanabe A, Sugahara K, Tsubakino H and Fukumoto S (2001), 'Improvement of corrosion resistance of magnesium alloys by vapor deposition', *Script. Mater.* **44**, 1039–1042.
9. Lee M H, Bae I Y, Kim K J, Moon K M and Oki T (2003),'Formation mechanism of new corrosion resistance magnesium thin films by PVD method', *Surf. Coatings Technol.* **169–170**, 670–674.
10. Miller P L, Shaw B A, Wendt R G and Moshier W C (1993), 'Improving corrosion resistance of magnesium by nonequilibrium alloying with yttrium', *Corros.* **49**, 947–950.
11. Song G L, Atrens A, St. John D, Nairn J and Li Y (1997), 'The electrochemical corrosion of pure magnesium in 1N NaCl', *Corros. Sci.* **39**, 855–875.
12. Song G L, Atrens A, Wu X and Zhang B (1998), 'Corrosion behavior of AZ21, AZ501 and AZ91 in sodium chloride', *Corros. Sci.* **40**, 1769–1791.
13. Brus L E (1983), 'A simple model for the ionization potential electron affinity and aqueous redox potentials of small semiconductor crystallites', *J. Chem. Phys.* **79**, 5566–5571.
14. Brus L E (1984), 'Electron–electron and electron–hole interactions in small semiconductor crystallites – the size dependence of the lowest excited electronic state', *J. Chem. Phys.* **80**, 4403–4409.
15. Kayanuma Y (1988), 'Quantum size effects of interacting electrons and holes in semiconductor microcrystals with spherical shape', *Phys. Rev. B* **38**, 9797–9805.

16. Suzuki M, Kimura T, Koike J and Maruyama K (2003), 'Strengthening effect of Zn in heat resistant Mg–Y–Zn solid solution alloys', *Script. Mater.* **48**, 997–1002.
17. He S, Zeng X, Peng L, Gao X, Nie J and Ding W (2006), 'Precipitation in a Mg–10Gd–3Y–0.4Zr (wt.%) alloy during isothermal ageing at 250 °C', *J. Alloys Comp.* **421**, 309–313.
18. Lindström R, Johansson L, Thompson G, Skeldon P and Svensson J (2004), 'Corrosion of magnesium in humid air', *Corros. Sci.* **46**, 1141–1158.
19. Jönsson M, Persson D and Thierry D (2007), 'Corrosion product formation during NaCl induced atmospheric corrosion of magnesium alloy AZ91D', *Corros. Sci.* **49**, (2007) 1540–1558.
20. Song G, Hapugoda S and St John D (2007), 'Degradation of the surface appearance of magnesium and its alloys in simulated atmospheric environments', *Corros. Sci.* **49**, 1245–1265.
21. Nüñez L, Reguera E, Corvo F, González E and Vazquez C (2005), 'Corrosion of copper in seawater and its aerosols in a tropical island', *Corros. Sci.* **47**, 461–484.
22. Itoh J, Sasaki T and Ohtsuka T (2000), 'The influence of oxide layers on initial corrosion behavior of copper in air containing water vapor and sulfur dioxide', *Corros. Sci.* **42**, 1539–1551.
23. Corvo F, Minotas J, Delgado J and Arroyave C (2005), 'Changes in atmospheric corrosion rate caused by chloride ions depending on rain regime', *Corros. Sci.* **47**, 883–892.
24. Aastrup T, Wadsak M, Schreiner M and Leygraf C (2000), 'Simultaneous infrared reflection absorption spectroscopy and QCM measurements for *in situ* studies of the metal/atmospbere interface', *Corros. Sci.* **42**, 957–967.
25. Wadsak M, Aastrup T, Wallinder I, Leygraf C and Schreiner M (2002), 'Multianalytical *in situ* investigation of the initial atmospheric corrosion of bronze', *Corros. Sci.* **44**, 791–802.
26. Mendoza A, Corvo F, Gómez A and Gómez J (2004), 'Influence of the corrosion products of copper on its atmospheric corrosion kinetics in tropical climate', *Corros. Sci.* **46**, 1189–1200.
27. Stratmann M, Streckel H, Kim K T and Crockett S (1990), 'On the atmospheric corrosion of metals which are covered with thin electrolyte layers – iii. The measurement of polarisation curves on metal surfaces which are covered by thin electrolyte layers', *Corros. Sci.* **30**, 715–734.
28. Frankel G, Stratmann M, Rohwerder M, Michalik A, Maier B, Dora J and Wicinski M (2007), 'Potential control under thin aqueous layers using a Kelvin probe', *Corros. Sci.* **49**, 2021–2036.
29. Nishikata A, Yamashita Y, Isatayama H, Tsuru T, Usami A, Tanabe K and Mabuchi H (1995), 'An electrochemical impedance study on atmospheric corrosion of steels in a cyclic wet–dry condition', *Corros. Sci.* **37**, 2059–2069.
30. Zhang T, Chen C, Shao Y, Meng G and Wang F (2008), 'Corrosion of pure magnesium under thin electrolyte layers', *Electrochim. Acta* **53**, 7921–7931.
31. Cheng Y, Zhang Z, Cao F, Li J, Zhang J, Wang J and Cao C (2004), 'A study of the corrosion of aluminum alloy 2024-T3 under thin electrolyte layers', *Corros. Sci.* **46**, 1649–1667.
32. Valor A, Caleyo F, Alfonso L, Rivas D and Hallen J (2007), 'Stochastic modeling of pitting corrosion: a new model for initiation and growth of multiple corrosion pits', *Corros. Sci.* **49**, 559–579.
33. Sanchez-Amaya J, Cottis R and Botana F (2005). 'Shot noise and statistical parameters for the estimation of corrosion mechanisms', *Corros. Sci.* **47**, 3280–3299.

34. Na K and Pyun S (2008), 'Comparison of susceptibility to pitting corrosion of AA2024–T4, AA7075–T651 and AA7475–T761 aluminium alloys in neutral chloride solutions using electrochemical noise analysis', *Corros. Sci.* **50**, 248–258.
35. Gumbel E J (1957), *Statistics of Extremes*, Columbia University Press, NY.
36. Pride S, Scully J and Hudson J (1994), 'Metastable pitting of aluminum and criteria for the transition to stable pit growth', *J. Electrochem. Soc.* **141**, 3028–3040.
37. Rosalbino F, Angelini E and De Negri S (2005), 'Effect of erbium addition on the corrosion behaviour of Mg–A1 alloys', *Intermetallics* **13**, 55–60.
38. Zhou X H, Huang Y W and Wei Z L (2006), 'Improvement of corrosion resistance of AZ91D magnesium alloy by holmium addition', *Corros. Sci.* **48**, 4223–4233.
39. Wu G H, Fan Y and Gao H T (2005), 'Differentiation of bone marrow mesenchymal stem cells transplanted into the brain in rats with cerebral infarction and its effect on the recovery of nerve functions', *Mater. Sci. Eng. A* **408**, 255–257.
40. Huang Z H, Guo X F and Zhang Z M (2005), 'Effects of Ce on corrosion resistance of AZ91D magnesium alloy', *Acta Metall. Sin.* **18**, 129–136.
41. Liu S F, Liu L Y and Huang S Y (2006), 'Using Hajós construction to generate hard graph 3-colorability instances', *J. Chin. Earth Soc.* 24, 211–225.
42. Sajuri Z B, Miyashita Y and Hosokai Y (2006), 'Effects of Mn content and texture on fatigue properties of as–cast and extruded AZ61 magnesium alloys', *Inter. J. Mechan. Sci.* **48**, 198–209.
43. Ding H L, Liu L F and Kamado S (2008), 'Study of the microstructure, texture and tensile properties of as-extruded AZ91 magnesium alloy', *J. Alloy. Compound* **456**, 400–406.
44. Chao C Y, Lin L F and Macdonald D D (1981), 'Point defect model for anodic passive films', *J. Electrochem. Soc.* **128**, 1187–1194.
45. Gebert A, Wolff U and John A (2001), 'Stability of the bulk glass-forming Mg65Y10Cu25 alloy in aqueous electrolytes', *Mater. Sci. Eng. A* 299, 125–135.
46. Subba Rao R V, Wolff U and Baunaclk S (2003), 'Corrosion behaviour of the amorphous Mg65Y10Cu15Ag10 alloy', *Corros. Sci.* **45**, 817–832.
47. Zhang T, Li Y and Wang F H (2006), 'Roles of β phase in the corrosion process of AZ91D magnesium alloy', *Corros. Sci.* **48**, 1249–1264.
48. http://srdata.nist.gov.
49. Liu M, Zanna S, Ardelean H, Frateur I, Schmutz P, Song G, Atrens A and Marcus P (2009), 'A first quantitative XPS study of the surface films formed, by exposure to water, on Mg and on the Mg–Al intermetallics: Al_3Mg_2 and $Mg_{17}Al_{12}$', *Corros. Sci.* **51**, 1115–1127.
50. Yao H B, Li Y and Wee A T S (2003), 'Passivity behavior of melt-spun Mg–Y alloys'. *Electrochim. Acta* **48**, 4197–4024.
51. Yu P, Hayes S A and Okeefe T J (2006), 'The phase stability of cerium species in aqueous systems', *J. Electrochem. Soc.* **53**, C74–C79.
52. Lange A (2003), *Lange Chemistry Handbook*, Science Publishing Company 3–118.
53. Yao H B, Li Y, Wee A T S and Pan J S (2001), 'Correlation between the corrosion behavior and corrosion films formed on the surfaces of Mg82–*x*Ni18Nd*x* (x = 0, 5, 15) amorphous alloys', *Appl. Surf. Sci.* **173**, 54–61.

5 Corrosion behaviour of magnesium (Mg)-based bulk metallic glasses

A. GEBERT, Leibniz Institute for Solid State and Materials Research, Germany

Abstract: Magnesium (Mg)-based alloys offer a high potential for use as lightweight structural materials in the automotive and aerospace industry, but up to now their application has been limited because of insufficient corrosion resistance. In general, improvement of corrosion properties can be obtained by proper alloying as well as by homogenization of the microstructure applying advanced preparation techniques. A fully glassy or amorphous alloy state should principally yield a high corrosion resistance due to its ideally single-phase chemically homogeneous and defect-free nature. However, the glassy state is metastable, which gives rise to an enhanced reactivity. Those fundamental aspects will be critically assessed.

In recent years, much progress has been made in the development of multicomponent amorphous Mg-based alloys with large glass-forming ability (GFA) exhibiting outstanding mechanical performances. In particular for bulk metallic glasses the alloy composition strictly defines the GFA. Therefore, the type and concentration of constituents cannot be easily modified for tailoring their corrosion properties. Highlights of the bulk glassy alloy design will be briefly reviewed. Fundamental studies on the corrosion behaviour of selected amorphous alloys and metallic glasses in different aqueous environments are reported. It is demonstrated that the particular new alloy compositions and structures can yield much lower free corrosion reactivity and improved passivity compared with Mg and conventional crystalline alloys. The passive layer properties are characterized in detail by means of surface analytical methods. However, for amorphous Mg-based alloys chloride-induced localized corrosion processes, i.e. pitting and filiform corrosion, are critical issues. Mechanisms of those processes, including initiation conditions, are described. Another important aspect is the hydrogen reactivity of Mg-based glasses. Under cathodic control these alloys can absorb large amounts of hydrogen which has an effect on the stability of the amorphous phase and on phase transformation reactions and can lead to severe embrittlement.

Key words: bulk metallic glass, amorphous alloy, Mg-based alloy, general corrosion, passivity, filiform corrosion, hydrogen.

5.1 Introduction

For two decades a new generation of multi-component metallic alloy systems, which are capable of solidifying from the melt in fully amorphous structural states at relatively low cooling rates of ≤1000 K/s, has been the focus of intensive research activities. By careful compositional selections following basic empirical rules of (i) multi-component systems with ≥3 elements, (ii) significant atomic size ratios of ≥12% and (iii) negative heats of mixing between the main constituent elements, a retardation of crystallization processes during conventional copper mould casting can be achieved. This way, amorphous samples can be obtained in bulk shape with thickness values of ≥1 mm. These new alloy types exhibit a characteristic thermal behaviour with glass transition T_g and wide undercooled liquid regime ΔT_x before crystallization and are therefore more commonly known as *bulk metallic glasses* (BMGs). Their unique short- and medium-range ordered amorphous structures yield properties which are superior to those of conventional crystalline materials, e.g. very high mechanical strengths, large elastic elongation and good soft magnetic properties. This makes BMGs very promising as new class of engineering materials [1–4].

Due to their ideal single-phase homogeneous nature, bulk amorphous alloys or BMGs have gained the reputation of being highly corrosion-resistant materials. However, many experimental studies revealed a limited corrosion stability of different alloy types depending on the environment. This is mainly explained with the particular compositions of alloys for which high glass-forming ability (GFA) can be achieved and by the defective nature of real cast samples [5]. Therefore, for bulk metallic glasses the analysis and the understanding of corrosion phenomena are of decisive importance on their way to applications.

Among the first glass-forming systems studied were those on Mg–base [1]. Starting from ternary Mg–TM–RE systems (TM = transition metal, RE = rare earth), e.g. Mg–Ni–La which was firstly published in 1989 [6], to date numerous compositional developments have been conducted aiming at the achievement of maximum GFA and therefore, the maximum critical diameter for solidification into fully amorphous states. Those Mg-based BMGs exhibit typically extremely high specific strength values, which recommend them as new types of high-strength light-metal–base structural materials. However, the applicability of these new alloys also depends significantly on their corrosion stability. The high reactivity of Mg and the rare earths components makes this a critical issue despite their particular amorphous structure. So far, only for selected Mg-based BMGs have corrosion properties been fundamentally assessed.

This chapter starts with a brief overview about Mg-based BMGs and describes principal strategies and challenges for the achievement of maximum amorphous sample thickness. Characteristic thermal data and mechanical properties are summarized. Some fundamental considerations about the effect of microstructural refinements up to the adjustment of single-phase amorphous states on the corrosion reactivity of Mg-based alloys are made.

The current state of knowledge regarding the free corrosion and passivation behaviour of selected Mg-based BMGs is described, before the susceptibility to chloride-induced local corrosion is evaluated. Finally, the effect of cathodically generated hydrogen on the degradation of the amorphous structure is concisely discussed.

5.2 Magnesium (Mg)-based bulk metallic glasses (BMGs)

In the early 1990s Inoue and co-workers [2,7] explored the amorphous phase formation in binary and ternary Mg-based alloy systems by rapid solidification from the melt and stated the Mg–TM–RE systems (TM = transition metal, RE = rare earth) to be most prospective. Only these showed during heating a distinct glass transition T_g and a wide undercooled liquid regime ΔT_x, which was regarded to be a decisive criterion for high glass-forming ability. Moreover, highest tensile fracture strength of ≥600 MPa was predicted to be achievable [8]. In particular the Mg–Cu–Y system was most promising with $Mg_{65}Cu_{25}Y_{10}$ for which a maximum amorphous sample diameter of 4 mm was obtained by copper mould injection casting [9] and 7 mm by high-pressure die casting [8]. The amorphous structure of $Mg_{65}Cu_{25}Y_{10}$ comprises a near densely packed icosahedral type of short-range order and an icosahedral medium-range order [10]. This particular alloy composition was then the starting point for many compositional modifications in order to further increase the glass-forming ability.

While partial substitution of the main component Mg with other light metals such as Al was not successful [11], substitutions of Cu and Y appeared to be quite promising. Table 5.1 summarizes typical Mg-based bulk metallic glass compositions, maximum amorphous sample diameters, characteristic thermal data and some reported mechanical data. For $Mg_{65}Cu_{25}Y_{10}$, partial substitutions of Cu with Ag [12], Ag and Pd [13], Zn [14] or Ag and Zn and Ni [15] led to a gradual increase of the amorphous sample diameter to 9 mm achievable by Cu mould casting under Ar atmosphere. Additional partial substitution of Y with Gd [16] resulted in $Mg_{65}Cu_{7.5}Ni_{7.5}Ag_5Zn_5Y_5Gd_5$ with 14 mm of maximum diameter which, remarkably, was obtained in an air atmosphere. The compositional modifications yielded also an improvement of the mechanical properties. For the $Mg_{65}Cu_{7.5}Ni_{7.5}Ag_5Zn_5Y_5Gd_5$ bulk metallic glass ultimate compressive fracture strength of 928 MPa was reported corresponding to a specific strength of ~260 MPa cm^3/g, which is over 20% higher than that of many conventional crystalline alloys. Moreover, unlike earlier Mg-based glasses this multi-component alloy exhibits a certain plastic strain.

Furthermore, a complete substitution of the rare earth component Y with Gd led to $Mg_{65}Cu_{25}Gd_{10}$ with increased GFA, but no further improvement

Table 5.1 Mg-based alloys with high glass-forming ability prepared by Cu mould injection casting (maximum sample thickness, thermal data from differential scanning calorimetry (DSC): 20 K/min, mechanical properties, publication year)

Composition	$D_{max.}$ [mm]	T_g [K]	T_x [K]	ΔT_x [K]	Mechanical properties	Year	Ref.
$Mg_{65}Cu_{25}Y_{10}$	4	–	–	60	σ_{cf} = 822 MPa	1991	9
	7*	420	490	70		1993	8
$Mg_{65}Cu_{15}Ag_{10}Y_{10}$	6(10**)	428	469	41	–	2000	12
$Mg_{65}Cu_{15}Ag_5Pd_5Y_{10}$	7	437	472	35	σ_{cf} = 770 MPa	2000	13
$Mg_{65}Cu_{20}Zn_5Y_{10}$	6	404	456	52	–	2002	14
$Mg_{65}Cu_{7.5}Ni_{7.5}Ag_5Zn_5Y_{10}$	9	430	459	29	σ_{cf} = 825 MPa, ε_{el} = 1.91%	2003	15
$Mg_{65}Cu_{7.5}Ni_{7.5}Ag_5Zn_5Y_5Gd_5$	14_{air}	434	472	38	σ_{cf} = 928 MPa	2005	16
					ε_{el} = 1.80%, ε_{pl} = 0.57%		
$Mg_{54}Cu_{28}Ag_7Y_{11}$	16	433	501	68	–	2006	24
$Mg_{65}Cu_{25}Gd_{10}$	8	408	478	70	σ_{cf} = 834 MPa, E = 56 GPa	2003	17
					ε_{el} = 1.5%		
$Mg_{65}Cu_{20}Ni_5Gd_{10}$	5	423	484	61	σ_{cf} = 904 MPa, E = 59 GPa	2005	18
					ε_{el} = 1.5%, ε_{pl} = 0.15%		
$Mg_{65}Cu_{20}Ag_5Gd_{10}$	11_{air}	427	465	38	σ_{cf} = 909 MPa	2005	19
					ε_{el} = 1.71%, ε_{pl} = 0.5%		
$Mg_{65}Cu_{15}Ag_5Pd_5Gd_{10}$	10	430	472	42	σ_{cf} = 817 MPa, ε_{total} = 1.6%	2003	20
$Mg_{61}Cu_{28}Gd_{11}$	12	422	483	61	σ_{cf} = 461–732 MPa	2007	25
					ε_{pl} = 0.4%		
$Mg_{54}Cu_{26.5}Ag_{8.5}Gd_{11}$	25	–	–	–	σ_{cf} ~ 1000 MPa, ε_{pl} ~ 3%	2005	26
$Mg_{59.5}Cu_{22.9}Ag_{6.6}Gd_{11}$	27	425	472	47	–	2007	27
$Mg_{65}Cu_{25}Tb_{10}$	5	426	488	62***	σ_{cf} = 776 MPa, ε_{total} = 1.47%	2004	21
$Mg_{65}Cu_{20}Zn_5Tb_{10}$	7	428	475	48	σ_{cf} = 992 MPa, ε_{total} = 1.89%	2008	22
$Mg_{65}Cu_{20}Ag_5Tb_{10}$	10	429	471	42***	σ_{cf} = 971 MPa, ε_{total} = 1.86%	2009	23

*High-pressure die-casting, **squeeze casting, ***40 K/min.

of mechanical properties was observed [17]. Also for this alloy fractional Cu substitutions [18–20] were in part successful, resulting in an amorphous sample diameter of 11 mm for $Mg_{65}Cu_{20}Ag_5Gd_{10}$ by casting in air [19]. In the following substitutions with other rare earths were investigated, among them Tb [21–23], but results have not so far been as promising as those obtained with Gd.

Besides this strategy of partially substituting the components of $Mg_{65}Cu_{25}Y_{10}$ with other transition metals or rare earths in order to obtain multicomponent alloys with increased GFA, more recent studies pursued another strategy. This is based on the conviction that not the deepest eutectic composition, e.g. $Mg_{65}Cu_{25}Y_{10}$, but an off-eutectic composition close to it would reveal maximum GFA. Moreover, high GFA is regarded to be related to a strong liquid behaviour of an alloy which requires the consideration of fragility parameters of the glass-forming liquid. This new strategy led to a sharpening of glass-forming ternary and quaternary M–(Cu,Ag)–Y/Gd alloy compositions resulting in much larger amorphous sample diameters [24–27]. To date maximum reported diameters are 16 mm for $Mg_{54}Cu_{28}Ag_7Y_{11}$ [24] as well as 25 mm and 27 mm for $Mg_{54}Cu_{26.5}Ag_{8.5}Gd_{11}$ and $Mg_{59.5}Cu_{22.9}Ag_{6.6}Gd_{11}$ [26,27], respectively. For the latter compressive fracture strengths in the order of ~1000 MPa and plastic strains of ~3% are reported.

In parallel to those alloy developments, Mg–Ni–RE systems have also been explored. Starting from Mg–Ni–Y glasses prepared by melt-spinning [7], replacements of Y with Pr [28], Gd [29,30] and Nd [30–33] were investigated. So far, only low GFA corresponding to sample diameters of only a few millimetres has been detected, e.g. 5 mm for $Mg_{75}Ni_{15}Gd_5Nd_5$ [30] or 3.5 mm for $Mg_{65}Ni_{20}Nd_{15}$ [31]. Moreover, mechanical properties of those alloys are not improved compared with those of the Cu-containing Mg-based glasses.

Table 5.1 also summarizes the characteristic thermal data of the Mg-based bulk metallic glasses. Compared with other BMG types [1–4] their glass transition temperatures T_g are relatively low, ranging from 404 K (131 °C) [14] to 437 K (164 °C) [13]. This strongly limits their temperature range for applications, since above T_g the glasses start to soften and crystallization sets in, as reflected by the crystallization temperature T_x. Furthermore, it is obvious that the undercooled liquid regime $\Delta T_x = T_x - T_g$ which reaches values between 29 K [15] and 70 K [17] does not systematically vary with the component substitutions and the related maximum sample diameters. Consequently, the earlier predicted relation that a high GFA requires a large ΔT_x [1] does not hold within the Mg-based systems. Meanwhile other important criteria have been identified [26,27,34].

Besides the search for suitable alloy compositions with high GFA, another practical challenge for the preparation of Mg-based bulk metallic glasses is the strict control of the casting process. Main problems during Cu mould casting are losses of Mg due to evaporation in the melting regime, which leads to deviations of the real from the optimized nominal compositions and therefore, reduction of GFA. Overheating far above the principal melting temperature is often necessary to melt all intermetallic phases which formed in the pre-alloy and to obtain a homogeneous liquid phase for subsequent solidification of a single amorphous phase. Also, the impurity level of the melt, the atmosphere and the mould surface must be kept very low to avoid heterogeneous nucleation and primary crystallization processes. Therefore, Cu mould casting is typically performed in highly purified Ar atmosphere and with the use of highly purified starting elements. Moreover, other casting techniques such as high-pressure die casting [8] or squeeze casting [12] are applied for better control of the partial pressure of Mg and for improvements of the cooling conditions.

Alternatively, first attempts following the powder metallurgical route are reported to prepare Mg-based bulk metallic glasses. For Mg–Cu–Y alloys, e.g. $Mg_{55}Cu_{30}Y_{15}$ [35–39] or $Mg_{49}Cu_{36}Y_{15}$ [40], mechanical alloying of elemental powders by ball milling or intensive milling of melt-spun material was applied to obtain mainly amorphous powders with small fractions of nanocrystalline phases. Subsequently, those powders were consolidated by vacuum hot pressing [37,38,40] or spark plasma sintering [39] in the temperature region above the glass transition and below the onset of crystallization. The low viscous state of the undercooled liquid was utilized to deform the powder particles during hot pressing and, thus, to prepare nearly fully dense bulk samples of up to 20 mm diameter. Moreover, bulk glassy $Mg_{85}Cu_5Y_{10}$ samples with up to 6 mm diameter containing fine dispersions of a hexagonal close packed (hcp) Mg-phase were obtained by hot extrusion of gas atomized alloy powders at temperatures below the onset of crystallization [8,41]. Main challenges of these powder metallurgical routes are for example: (i) the strict control of the phase formation reactions during powder preparation, e.g. via milling intensity, milling duration, cooling rate (gas atomization) and impurity levels, (ii) the determination of suitable temperature–time windows for hot consolidation without the initiation of crystallization, and (iii) the avoidance of powder oxidation during the whole process. Therefore, for Mg-based bulk metallic glasses these techniques are so far only rarely applied compared with casting techniques.

5.3 Effect of micro-structural refinement on the corrosion of magnesium (Mg)-based alloys

The limited corrosion resistance of commercial Mg and microcrystalline Mg-based alloys origins from the high reactivity of Mg, which is only in alkaline media inhibited by the formation of protective hydroxide layers, and from the existence of impurities and second phases, which act as local cathodes and, thus, cause galvanic corrosion processes [42–44]. The application of non-equilibrium preparation techniques such as melt- or splat-quenching, sputter deposition, laser melting and ball milling is expected to lead to improved corrosion properties. This can be due to (i) the production of finer and more homogeneous microstructures, (ii) the reduction of detrimental effects of impurities and precipitates by extension of solid solubility ranges, (iii) the possible formation of new corrosion-resistant phases, and (iv) the formation of homogeneous glassy oxide films [45–47].

High-energy ball milling of Mg powders was applied to reduce the grain size to several tens of nanometres and in consequence, an improvement of the corrosion resistance of Mg was observed in alkaline solutions [48]. This was attributed to the high defect density and large fraction of grain boundaries in milled Mg powders, which increases the number of nucleation sites for Mg hydroxide $Mg(OH)_2$ formation. The effect of rapid quenching of binary Mg-based alloys on the behaviour in alkaline buffer solutions with up to 3.5% NaCl was studied and only Al was found to be beneficial for improving the corrosion resistance, while Ca, Li, Si and Zn were detrimental [45]. Rapidly quenched Mg–Al alloys form the same α-Mg and β-$Al_{12}Mg_{17}$ phases like slowly cooled materials, but a significant amount of Al is dissolved in Mg forming supersaturated solid solutions, which reduces the fraction of the fine-dispersed second phase [49]. With sputter deposition complete solid solutions of Mg–4Al–4Zr and Mg–3Al–15Zr with nanometre grain sizes were obtained, which showed increased corrosion resistance in 3.5% NaCl solution when compared with their Zr-free counterparts [50]. Also for sputter-deposited nanocrystalline/amorphous Mg–Y films improved corrosion behaviour in chloride electrolytes was detected compared with Mg and commercial Mg-based alloys as long as Y was in solid solution. Participation of Y in the formation of homogeneous passive films and significant changes in the film morphology were identified as decisive factors [51–53]. Rapidly quenched Mg–Y and Mg–Nd alloys with refined two-phase microstructures and increased solubility of the rare earths in the Mg phase exhibited enhanced passivation ability in 0.01 M NaCl solution in terms of pseudo-passive potential regions which did not occur for the coarse-grained cast material [54]. The passive films formed during anodization of ribbons were found to comprise, in addition to $Mg(OH)_2$, fractions of Y_2O_3 [55].

These principal considerations and research examples for materials processed under non-equilibrium conditions point out that, ideally, highly corrosion-resistant Mg-based alloys must be complete solid solutions, but should at the same time comprise passivating alloying elements in sufficient concentrations. Theoretically, fully glassy or amorphous alloys can to a certain degree meet those requirements [5,56–58]. Their ideally single-phase and defect-free nature is expected to enable the growth of laterally very homogeneous passive layers and to minimize the number of higher energetic, electrochemically active surface sites which can induce local corrosion processes. On the other hand, the glassy state is metastable, which can give rise to enhanced reactivity due to the lower activation barrier for charge-transfer processes than that needed for equilibrium crystalline states. In addition, in particular for bulk metallic glasses the kind and concentration of constituents are strictly defined by the GFA (see previous section), i.e. cannot be easily modified to improve corrosion properties. Those opposing structural, thermodynamic and chemical effects act superimposed on the corrosion activity of glassy alloys and thus, can hardly be analysed separately. Altogether, one cannot per se predict a high corrosion resistance for metallic glasses, particularly for those with a highly reactive main constituent such as Mg.

5.4 General corrosion and passivation behaviour of magnesium (Mg)-based bulk metallic glasses (BMGs)

Following the developments in alloy design (section 5.2), for selected Mg-based BMGs the corrosion behaviour in electrolytes with different pH values was fundamentally assessed. Due to their unique compositions which strongly deviate from those of conventional Mg-based alloys, a new quality of the corrosion stability under open circuit conditions and of the anodic passivation ability is attained in particular in near-neutral solutions.

Figure 5.1 summarizes typical potentiodynamic polarization curves for three prominent Mg-based BMGs of the Mg–Cu–Y-based system in comparison with Mg and AZ31, which were recorded in a limited potential range close to the corrosion potential E_{corr} in borate buffer solution with pH = 8.4 [59–61]. For Mg and AZ31 corrosion potentials are very negative and corrosion current densities i_{corr} are rather high, indicating their strongly reactive state. Meanwhile, for all Mg-based glassy alloys corrosion potentials are at least 1000 mV more positive and corrosion current densities are reduced by two orders of magnitude. This comparatively low corrosion reactivity is mainly explained by the effect of the alloying elements of the metallic glasses, rather than by their particular amorphous structure. While the rare earth elements with standard electrode potentials of –2290 mV (Gd/Gd^{3+}) and –2370 mV (Y/Y^{3+}) are similarly reactive like Mg (–2370 mV for Mg/Mg^{2+}) [62], the presence of the noble elements Cu and Ag, but also of Ni and Zn, leads to this significant improvement of the corrosion stability.

For $Mg_{65}Cu_{25}Y_{10}$ the comparison between amorphous and crystalline counterpart states is also demonstrated [59]. The corrosion activity of the crystalline alloy is slightly increased. A similar trend was also detected for $Mg_{65}Cu_{25}Gd_{10}$ under comparable electrolyte conditions [63]. These small differences cannot be attributed only to the particularities of the atomic ordering states and grain boundary effects, but are mostly related with the

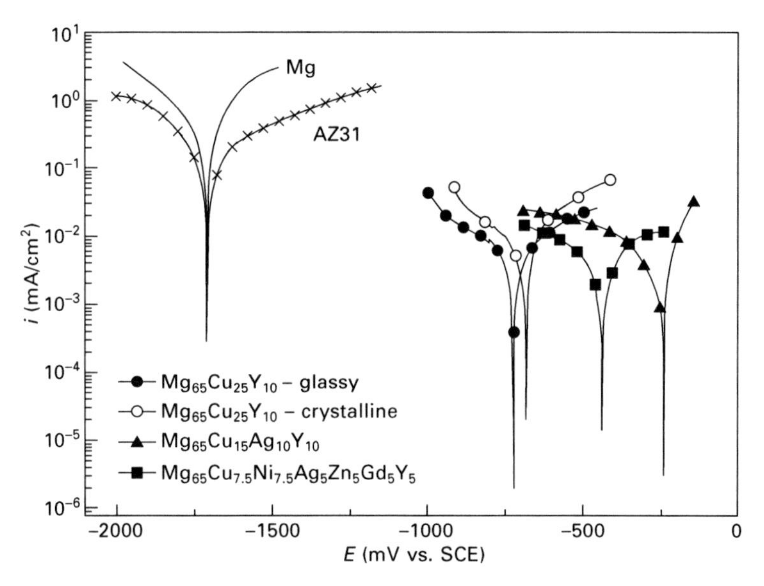

5.1 Potentiodynamic polarization curves close to E_{corr} of Mg-based bulk metallic glasses in comparison to Mg and AZ31 recorded in 0.3 M $H_3BO_3/Na_2B_4O_7$ buffer solution with pH=8.4 (scan-rate 0.5 mV/s) (adapted from [61]).

degree of chemical homogeneity of the two alloy states. While the rapidly quenched states are single-phase amorphous typically with some few rare earth oxide inclusions, the pre-alloyed or annealed crystalline states comprise various coexisting phases of very different compositions [59]. Galvanic coupling between those crystalline phases must be considered as a decisive reason for the observed enhanced corrosion activity.

However, with increasing pH of the electrolyte, differences in the corrosion reactivity of glassy and various kinds of crystalline Mg-based alloys increasingly diminish due to the reduced reactivity of the main alloying component Mg which is spontaneously passive in alkaline media. Thus, at pH≥13 very low corrosion rates are generally noticed [59,61,64,65].

Furthermore, the analysis of the anodic passivation ability of Mg-based glasses was in the focus of various studies. Figure 5.2 shows cyclic potentiodynamic polarization curves for the glassy $Mg_{65}Cu_{25}Y_{10}$ alloy and its crystalline counterpart in comparison to Mg and AZ31 recorded in borate buffer solution with pH = 8.4 [59,61]. In the case of Mg and AZ31, starting from very negative corrosion potentials during anodization a pronounced active state and an active/passive transition followed by a pseudo-passive region with a relatively high current density level of ~2.5 mA/cm² occur. During reverse potential scanning a similar behaviour is observed. This indicates the formation of highly permeable Mg hydroxide surface films with low stability providing very limited protection against general corrosion.

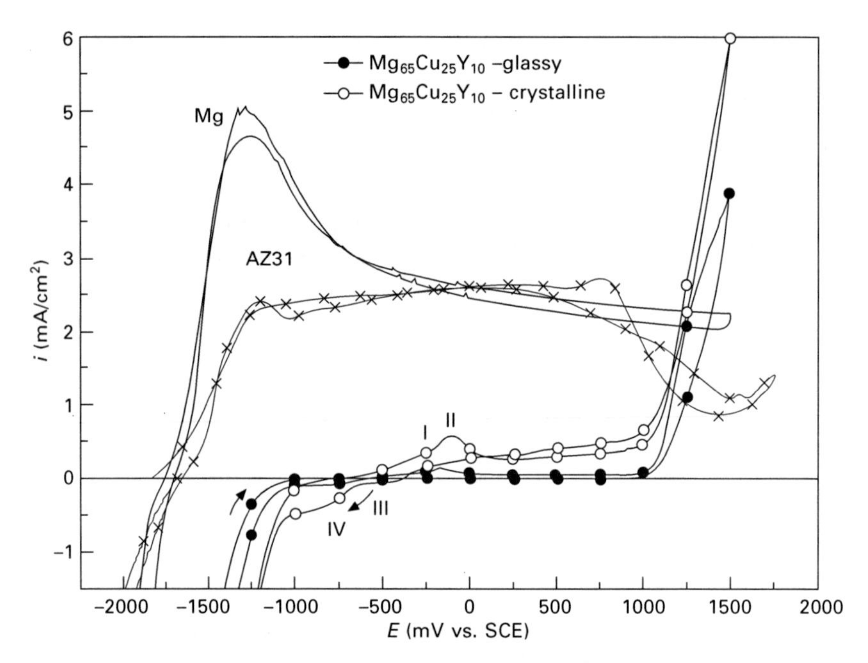

5.2 Cyclic potentiodynamic polarization curves for the $Mg_{65}Cu_{25}Y_{10}$ bulk metallic glass in comparison to its crystalline counterpart and Mg and AZ31 recorded in 0.3 M $H_3BO_3/Na_2B_4O_7$ buffer solution with pH=8.4 (scan-rate 5 mV/s) (adapted from [59]).

However, the amorphous $Mg_{65}Y_{10}Cu_{25}$ alloy and its crystalline counterpart exhibit very different anodic behaviour. The polarization curve of the crystalline alloy sample reveals a characteristic double peak (I,II) in the potential range of –500 mV to 0 mV vs. SCE with corresponding peaks (III,IV) in the reverse scan. These characteristic peaks are mainly due to oxidation of Cu(0) to Cu(I), i.e. Cu_2O, and further oxidation of Cu(I) to Cu(II), i.e. CuO, and the corresponding reverse reactions. Peak IV must also be attributed to the reduction of oxygen dissolved in the electrolyte. As derived from Pourbaix diagrams [62], during the anodic polarization the stability ranges of magnesium hydroxide and yttrium hydroxide are attained. Current densities in the passive region of the multiphase crystalline alloy are ~300 μA/cm² and thus, one order of magnitude lower than those measured for Mg and AZ31, which indicates a remarkable improvement of the protective properties of the surface layers. Further reduction of the anodic reactivity is achieved when establishing an amorphous alloy state. Its polarization curve shows the same characteristic features as observed for the multiphase crystalline alloy. However, the surface oxidation peaks are less pronounced and the current densities in the passive region were found to be only ~50 μA/cm². This remarkably improved passivity is attributed to a higher lateral homogeneity of the surface layers which grow on the nearly single-phase amorphous alloy compared with those which grow unevenly on the heterogeneous crystalline material. Additional potentiostatic anodic polarization tests combined with surface analytical studies confirmed that in weakly alkaline solutions all components of the $Mg_{65}Y_{10}Cu_{25}$ alloy participate in the formation of passive films [59]. Figure 5.3 shows Auger electron spectroscopy (AES) atomic concentration depth profiles of passivated surfaces of an amorphous $Mg_{65}Y_{10}Cu_{25}$ sample (a) and of a crystalline counterpart sample (b). For both the distribution of O, Mg, Cu, Y and B species in the layer region is in principle similar, indicating comparable in-depth compositions of the passive layers. But from comparison of the oxygen profiles it can be concluded that on the crystalline sample much thicker and more heterogeneous layers grow. A higher porosity and, therefore, permeability of these layers could explain the detected higher passive current densities.

Partial substitution of Cu by Ag to obtain bulk glassy $Mg_{65}Cu_{15}Ag_{10}Y_{10}$ only slightly modifies the anodic passivation ability at pH = 8.4. The presence of oxidized Ag species in the passive layer was analytically confirmed [66]. Based on these findings, the role of the components in the passive layer growth on the glassy $Mg_{65}Cu_{7.5}Ni_{7.5}Ag_5Zn_5Y_5Gd_5$ alloy was discussed. Due to its multicomponent nature with alloying elements in minor fractions, an analysis of passivated alloy surfaces is extremely difficult. Theoretically [62] it can be predicted that at pH = 8.4 and at anodic potentials all elements, i.e. also Ni, Zn and Gd, can form stable oxides. Thus, participation of all alloying elements in the passive layer formation is assumed. This would explain the further reduced anodic passive current density level of ~20 μA/cm², compared with that of the ternary or quaternary glassy alloys [61].

As for conventional crystalline alloys uniform corrosion rates and anodic passivity of glassy Mg-based alloys are strongly dependent on electrolyte parameters, e.g. the pH value and the kind and concentration of anions. These factors are especially critical in *neutral and acidic* regimes.

Detailed studies on the glassy $Mg_{65}Cu_{15}Ag_{10}Y_{10}$ alloy including partial comparison with the glassy $Mg_{65}Cu_{25}Y_{10}$ [67] and $Mg_{65}Cu_{7.5}Ni_{7.5}Ag_5Zn_5Y_5Gd_5$ [61] alloys revealed, that with gradual decrease of the pH value of the borate electrolyte from pH = 8.4 to pH = 5 an increase of the corrosion current density and therefore, a detectable enhancement of the corrosion rate occurs. The i_{corr} values for the quaternary and the multicomponent alloys were generally lower than those of the ternary alloy. However, compared with Mg and AZ31 the reactivity level of these glassy alloys is strongly reduced in neutral and weakly acidic solutions [61]. Moreover, while for the conventional materials a stable anodic passive state is usually hardly attainable in these environments, this was instantaneously established for the quaternary glassy alloy up to pH≥6, though with reduced protective effect [60,67]. The passive layers which anodically form on the amorphous $Mg_{65}Cu_{15}Ag_{10}Y_{10}$

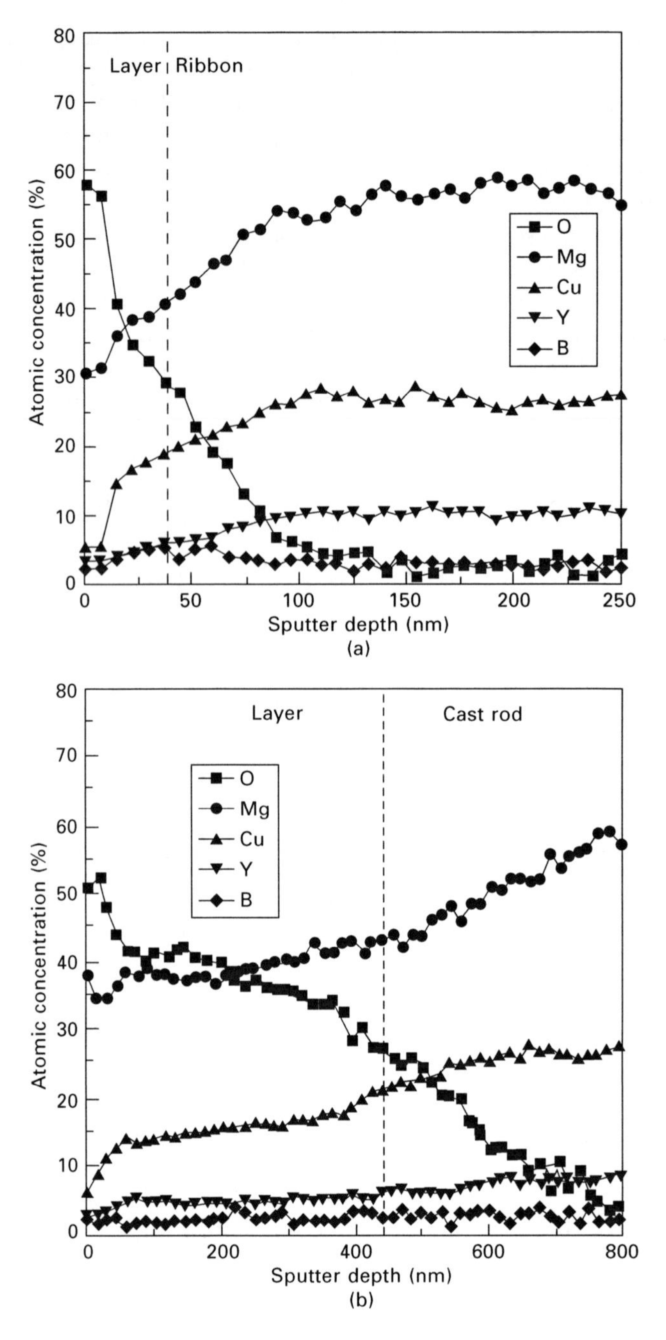

5.3 Auger electron spectroscopy (AES) depth profiles of $Mg_{65}Cu_{25}Y_{10}$ sample surfaces recorded after potentiostatic polarization tests in 0.3 M $H_3BO_3/Na_2B_4O_7$ buffer solution with pH=8.4: (a) amorphous alloy state; (b) crystalline alloy state [59].

alloy surface at pH = 6–7 consist mainly of Y and Mg oxide species in the outermost regions. At the interface between the passive layer and the bulk alloy slight enrichment of Ag-oxide species is noticed. This is in contrast to compositions of passive layers formed at pH = 8.4, where Cu oxide species were mainly detected in the outermost regions, followed by Y and Ag oxide species in an intermediate region and Mg oxide species at the interfacial region. This pH dependence of passive layer compositions is attributed to the higher solubility of the Cu and Ag species at pH = 6–7. In comparison, the glassy $Mg_{65}Cu_{7.5}Ni_{7.5}Ag_5Zn_5Y_5Gd_5$ alloy exhibits stable passivity in borate solutions only at pH = 8.4 and pH = 7, while at pH ≤ 6 anodic alloy dissolution dominates [61]. A reason for this relative destabilization of the multi-component alloy can be the partial replacement of Ag with Zn (Zn becomes quite reactive in the presence of noble elements) and possibly also the substitution of Cu with the less noble Ni.

In addition to this pH value effect the kind of anions in solution plays a decisive role. For the amorphous $Mg_{65}Cu_{15}Ag_{10}Y_{10}$ alloy it was demonstrated that, although at pH = 7 in borate solution a stable anodic passive state is easily established, severe alloy dissolution occurs in electrolytes containing phthalate or sulphate ions [67]. This must be considered as an expression of the weakness of the ability of the glassy alloys to passivate in neutral and weakly acidic environments.

Various studies have been devoted to the analysis of the anodic passivation behaviour of Mg-based metallic glasses in *strong alkaline environments*. Under those conditions the main alloying element Mg is spontaneously passive forming a dense and stable hydroxide layer. This is also determining for the behaviour of Mg-based glasses [59,61,64,66].

Figure 5.4 shows cyclic potentiodynamic polarization curves for Mg, glassy $Mg_{65}Cu_{25}Y_{10}$ and glassy and crystalline $Mg_{65}Cu_{15}Ag_{10}Y_{10}$ recorded in sodium hydroxide solution with pH = 13. Both glassy alloys exhibit Mg-like anodic behaviour with a wide stable passive range; only the oxygen evolution reaction at high anodic potentials occurs less inhibited. Their anodic passive current densities are in the order of $\leq 6\,\mu A/cm^2$ and, thus, lower than those for Mg with $\geq 16\,\mu A/cm^2$ [59,64,66], which may be indicative for the participation of other alloying components in the passive film formation. Under similar polarization conditions, for the glassy $Mg_{65}Cu_{7.5}Ni_{7.5}Ag_5Zn_5Y_5Gd_5$ alloy slightly enhanced passive current densities of $\sim 12\,\mu A/cm^2$ and a more limited passive potential range were detected [61].

Surface analyses by means of AES depth profiling of glassy ternary and quaternary alloy samples which were anodically passivated in an electrolyte with pH = 13 confirmed that passive layers are very thin and dense and are mainly composed of Mg hydroxide, while Y (and Ag) oxide species were identified at the inner metal-near region and Cu species were not found to be incorporated in the layer [59,66]. Comparative X-ray photoelectron

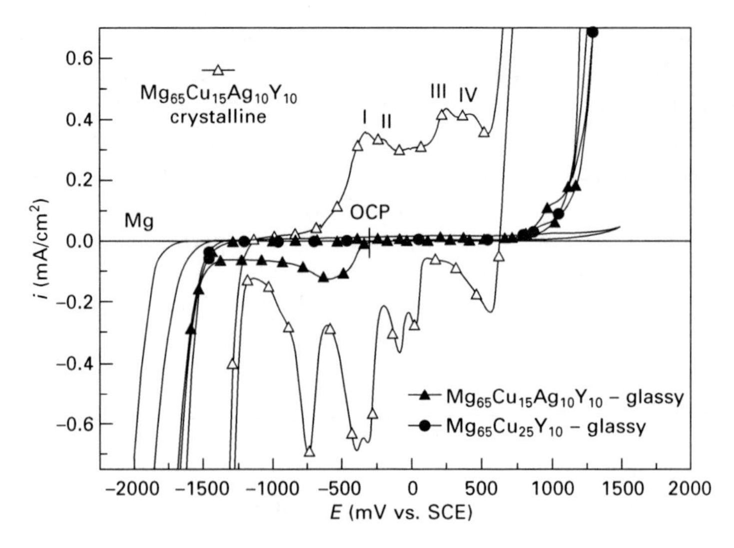

5.4 Cyclic potentiodynamic polarization curves for the $Mg_{65}Cu_{25}Y_{10}$ bulk metallic glass, the $Mg_{65}Cu_{15}Ag_{10}Y_{10}$ bulk metallic glass and its crystalline counterpart and for Mg recorded in 0.1 M NaOH solution with pH=13 (scan-rate 5 mV/s) (adapted from [66]).

spectroscopy (XPS) studies on glassy $Mg_{65}Cu_{25}Y_{10}$ and $Mg_{65}Ni_{20}Nd_{15}$ samples which were passivated in a sodium chloride solution with pH=12 revealed similar surface layer compositions, i.e. a dominance of Mg oxides/hydroxides, an enrichment of Y or Nd oxides in the inner regions and a depletion of Cu or Ni oxides throughout the whole layer [68]. However, in a very detailed approach it was also demonstrated for passivated glassy Mg–Cu(–Ag)–Y and Mg–Ni–Y alloys that surface analytical methods comprising a sputter depth profiling have to be considered critically, since quantification of the results is impeded by preferential sputtering of Mg [69].

Furthermore, from Fig. 5.4 it is obvious that in a strong alkaline environment the polarization behaviour of a crystalline alloy, e.g. $Mg_{65}Cu_{15}Ag_{10}Y_{10}$, is significantly different from that of its amorphous counterpart which demonstrates the beneficial effect of the single-phase state. Enhanced reactivity of the crystalline alloy is obvious from the occurrence of pronounced Cu oxidation peaks (I and II) and Ag oxidation peaks (III and IV) along with their reduction counterparts in the reverse scan as well as from lower overpotentials for cathodic hydrogen and anodic oxygen evolution reactions [66]. This has been similarly observed for the crystalline and amorphous $Mg_{65}Cu_{25}Y_{10}$ [59,64] and $Mg_{65}Cu_{25}Gd_{10}$ [63] alloys. The increased anodic reactivity of the crystalline counterpart alloys must be mainly attributed to their heterogeneous nature. While the single-phase amorphous alloys enable a laterally homogeneous passive layer growth under dominance of Mg oxidation, the crystalline counterpart alloys consist of various intermetallic phases of very different composition. This causes a local enrichment of alloying components, e.g. Cu or Ag, which in consequence leads to a higher local reaction rate. Galvanic coupling between adjacent phases and facilitated diffusion of constituent atoms along grain boundaries may also be supporting effects. However, it is very difficult to identify those local reaction processes by means of surface analytical methods due to the very fine-grained microstructure of the crystalline states [59].

Selected investigations have been conducted on the stability of glassy $Mg_{65}Cu_{25}Y_{10}$ $Mg_{50}Ni_{30}Y_{20}$ and $Mg_{63}Ni_{30}Y_7$ alloys in *extreme alkaline solutions*, i.e. 6 M KOH with pH≥14 [65]. While these alloys exhibit the lowest corrosion rate and the best passivation ability in electrolytes with pH = 13 (due to the high thermodynamic stability of Mg and Y hydroxides/oxides), at pH≥14 gradual oxidation and dissolution of Cu or Ni from surface-near regions governs the initial anodic polarization behaviour before at more positive potentials a stable passive state is attained. The dissolution of Ni from Mg–Ni–Y alloys is much more inhibited than that of Cu from Mg–Cu–Y due to the lower tendency of Ni to form soluble oxidized ions and to the effect of higher fractions of Y in the alloy in supporting a more rapid stable passive film growth. Cathodic hydrogen charging leads to enrichment of Ni or Cu species at the alloy surface. This gives rise to preferential oxidation and dissolution processes during the subsequent exposure under free corrosion and anodic polarization conditions. Chemical pre-treatments with HF solutions or YCl_3/H_2O_2 mixtures can significantly alter the surface reactivity in extreme alkaline environments [70].

5.5 Chloride-induced local corrosion behaviour of magnesium (Mg)-based metallic glasses

Of particular importance for the evaluation of the corrosion stability of glassy Mg-based alloys is the analysis of their behaviour in chloride-containing media. The occurrence of local corrosion phenomena is expected to be in strong relation with the defect density of the glassy samples, which depends on the glass-forming ability of the alloy and on the preparation conditions (see Section 5.2). Therefore, the dimensions and micro-structural states of the samples play a very decisive role.

The first studies were performed on Mg–Ni–Nd alloys [71–73]. Due to their low GFA, only samples of thin melt-spun ribbons or small tips of wedge-shaped cast bodies samples have been investigated. Those were exposed to 3–3.5% NaCl solutions saturated with $Mg(OH)_2$ and the corrosion rate was evaluated in terms of the hydrogen evolution rate. This way, for glassy $Mg_{66}Ni_{20}Nd_{14,}$ $Mg_{68.3}Ni_{17.5}Nd_{14.2}$ and $Mg_{69.8}Ni_{15.7}Nd_{14.5}$ alloys corrosion rates which were three or four times lower than for Mg and AZ91E were determined [71]. The lowest corrosion rate was detected for $Mg_{65}Ni_{20}Nd_{15}$

(with the highest GFA in the Mg–Ni–Nd system [31]) and was attributed to the elimination of micro-galvanic reactions which typically occurred between Mg-rich matrix and Ni-rich particles in not fully amorphous samples [72]. For annealed Mg–Ni–Nd samples with fully crystalline microstructure strongly enhanced corrosion rates were noticed [71,73]. Furthermore as shown in Fig. 5.5, anodic polarization studies performed on glassy $Mg_{65}Ni_{20}Nd_{15}$ and $Mg_{65}Cu_{25}Y_{10}$ ribbon samples in 0.01 M NaCl with pH = 12 revealed in comparison with $Mg_{82}Ni_{18}$ and $Mg_{79}Cu_{21}$ samples a higher passivation ability and lower pitting susceptibility [68,74,75]. By means of XPS analysis this was attributed to the beneficial presence of the rare earth oxides Nd_2O_3 and Y_2O_3 in the passive films besides the dominating Mg hydroxide $Mg(OH)_2$. Hydroxidation of the rare earth oxides led to passivity breakdown.

Cast bulk samples with 1 mm thickness of the glassy $Mg_{65}Cu_{25}Gd_{10}$ alloy and its crystalline counterpart exhibited during anodic polarization in 0.02 M NaCl solution with pH = 11 (adjusted with NaOH) severe active dissolution at high current densities [63]. A somewhat lower dissolution rate of amorphous samples was attributed to the absence of heterogeneities. More uniform morphologies of the corroded amorphous sample surfaces were observed comprising voluminous corrosion products composed of Mg oxide and Cu oxides and a large number of small cracks. Film breakdown occurred randomly with a uniform distribution of pits. Also in neutral 0.01 M

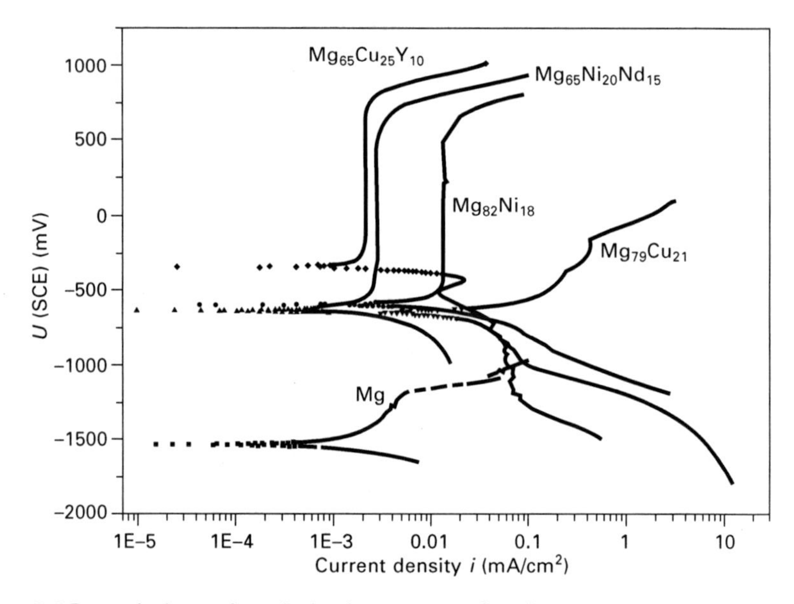

5.5 Potentiodynamic polarization curves of melt-spun ternary $Mg_{65}Ni_{20}Nd_{15}$ and $Mg_{65}Cu_{25}Y_{10}$ and binary $Mg_{82}Ni_{18}$ and $Mg_{79}Cu_{21}$ alloys and Mg recorded in 0.01 M NaCl solution (pH=12) (scan-rate 1 mV/s) [68].

NaCl solution a low corrosion resistance of bulk glassy $Mg_{65}Cu_{25}Gd_{10}$ was observed and explained with the formation of more porous and less compact passive films than in chloride-free environments [76]. Partial substitution of Cu with Zn was found to result in a significant improvement of the corrosion resistance in 5% NaCl solution under open circuit conditions [77]. Moreover, partial replacement of Cu by Ni, e.g. $Mg_{65}Cu_{20}Ni_5Gd_{10}$, led to an increase of the GFA and was quite effective in reducing the mass loss during exposure in 1% NaCl for 168 h [78].

For bulk glassy $Mg_{65}Cu_{7.5}Ni_{7.5}Ag_5Zn_5Gd_5Y_5$ alloy samples with 3 mm diameter systematic polarization studies were conducted in 0.0001 and 0.001 M NaCl solutions with pH values varying between pH = 6 and pH = 9 and corroded sample surfaces were examined with scanning electron microscopy (SEM) in order to understand local corrosion phenomena [61]. Figure 5.6 shows that in all test solutions an anodic passive layer breakdown occurred, corresponding to a sharp increase of the current density. In the backward potential scan the current density level remained high in all cases indicating that there was no significant re-passivation of the glassy alloy surface, but rather continuous corrosion. As a measure for the local corrosion resistance the difference between the passivity breakdown potential and the corrosion potential $\Delta E = (E_{breakdown} - E_{corr})$ was determined. Obviously the pH value

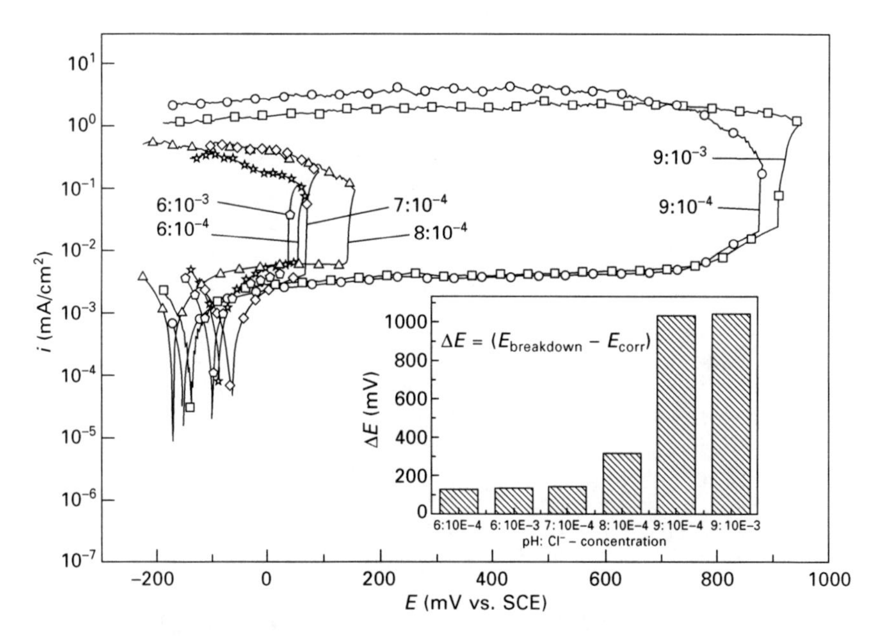

5.6 Potentiodynamic polarization curves with reverse scans of the bulk metallic glass $Mg_{65}Cu_{7.5}Ni_{7.5}Ag_5Zn_5Gd_5Y_5$ alloy backward recorded in NaCl solutions with different pH values (pH value: chloride concentration (M)) (scan-rate 0.5 mV/s) (adapted from [61]).

of the electrolyte, which determined the quality of the growing passive layers, was most decisive for the local corrosion resistance. A high chloride sensitivity was given at pH = 6 and pH = 7 and it was slightly reduced at pH = 8, while a high resistance was noticed at pH = 9, where the breakdown occurred only in the potential range of beginning oxygen evolution. Moreover, a direct comparison of the bulk glassy $Mg_{65}Cu_{7.5}Ni_{7.5}Ag_5Zn_5Gd_5Y_5$ alloy with the conventional AZ31 alloy revealed significant differences [61]. AZ31 established in a 0.0001 M NaCl solution with pH = 7 a very negative corrosion potential $E_{corr}\sim -1500$ mV and showed during anodization an active dissolution behaviour and only low hysteresis during backward scanning. In contrast, the corrosion potential of the glassy alloy was much more positive, i.e. $E_{corr}\sim -100$ mV and with initial anodic polarization a passive state was attained before a sudden breakdown occurred. During backward scanning the current density level remained high and dropped only at rather negative potentials of $E_{cathodic}\sim -750$ mV, indicating a high reactivity of the glassy alloy even under cathodic conditions.

Filiform corrosion was identified as the dominating corrosion mode on the glassy $Mg_{65}Cu_{7.5}Ni_{7.5}Ag_5Zn_5Gd_5Y_5$ alloy sample surfaces in chloride-containing near-neutral media [61]. Figure 5.7 shows an SEM image of a typical corroded area. Starting from single pits filaments formed and propagated on the alloy surface. At pH = 9 they appeared to be more convoluted and thicker than those formed at lower pH values. Filaments were mainly enriched in Mg and O species indicating $Mg(OH)_2$ formation. Filiform corrosion was earlier

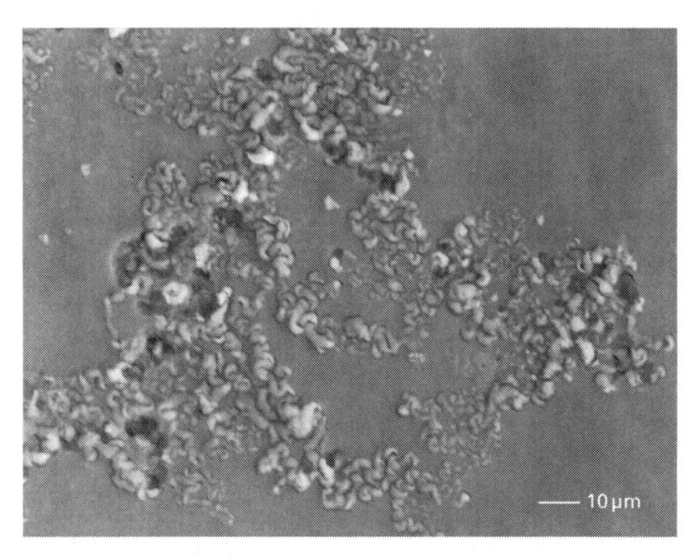

5.7 SEM image of a corroded surface of a bulk glassy $Mg_{65}Cu_{7.5}Ni_{7.5}Ag_5Zn_5Gd_5Y_5$ alloy sample after anodic polarization in a 0.0001 M NaCl solution with pH=9 pitting and filiform corrosion.

described for conventional Mg-based alloys, e.g. AZ31 and AZ91 [79]. For those the morphology and propagation direction of filaments was found to be determined by the alloy microstructure and crystallographic orientation of the grains. In contrast, on a nearly single-phase glassy alloy surface, the filaments propagated quite randomly. Several Y-rich dendrite inclusions were present in the cast bulk glassy sample, but those were obviously not involved in the corrosion process. This suggests that other heterogeneities or defects of the bulk glassy alloy surface, which are not yet identified are responsible for the initiation of the observed chloride-induced corrosion phenomena.

In summary, these first reported studies already clearly indicate that the corrosion stability of Mg-based metallic glassy alloys is generally low in chloride-containing aqueous solutions. Further significant modifications in alloy design and optimization of preparation conditions will be necessary to overcome this critical problem.

5.6 Effect of hydrogen on the stability of magnesium (Mg)-based glassy alloys

Mg-based alloys with amorphous and/or nanocrystalline microstructures exhibit a high capacity to absorb hydrogen even at room temperature. Therefore, the stability of their microstructural state under cathodic polarization conditions, which are dominated by the hydrogen evolution reactions, is of critical importance.

First studies on melt-spun single-phase amorphous Mg–Ni–La alloys with up to 25 at.% La [80,81] revealed a maximum hydrogen uptake of 2.4 wt% during galvanostatic cathodic charging. This was related to gradual alloy decomposition into La hydrides and various intermetallic phases. The thermal crystallization behaviour of those alloys is strongly affected by the presence of hydrogen [82].

Also melt-spun $Mg_{65}Cu_{25}Y_{10}$ ribbon samples exhibited during galvanostatic cathodic polarization at −1 or −10 mA/cm^2 in 0.1 M NaOH solution a high tendency to transfer from a single-phase amorphous state into a very fine nanocrystalline state under hydrogen uptake of up to 3.5 wt% [83]. Obviously, small absorbed hydrogen atoms can easily induce rearrangement processes in the short-range ordered amorphous structure of $Mg_{65}Cu_{25}Y_{10}$ leading to nanocrystallization. Mg–Cu-based compounds are generally known to disproportionate under the influence of hydrogen, e.g. Mg_2Cu decomposes into MgH_2 and $MgCu_2$ [84]. For amorphous Mg–Cu-based alloys those disproportion reactions are particularly eased due to the higher hydrogen diffusivity [83].

Mg–Ni-based compounds exhibit typically a much lower tendency to disproportionate under influence of hydrogen [85]. This may principally explain the higher stability of amorphous Mg–Ni–Y alloys, which mostly maintained their as-quenched microstructure during cathodic polarization [86–89]. The hydrogen uptake during electrolytic charging was found to strongly depend on the microstructure and on the composition of an alloy. Fully amorphous Mg–Ni–Y samples showed in comparison to partially or fully crystallized samples higher initial hydriding kinetics and a strong dependence of the absorbed hydrogen concentration from the applied cathodic current density. This is mainly explained by the fact that in the short-range ordered amorphous structures there are interstitial sites with broad energy distribution, which are successively filled with hydrogen atoms [90]. As a result of those studies it was proposed to use amorphous/nanocrystalline Mg–Ni–Y alloys for electrochemical hydrogen storage in Ni/metal hydride batteries. Electrodes made from mechanically milled Mg–Ni–Y alloys, e.g. $Mg_{61}Ni_{30}Y_9$, exhibited very high maximum discharge capacities, but suffered from low cycling stability mainly due to progressing corrosion [91]. During cathodic charging in extreme alkaline environment a change of the surface chemistry occurred, i.e. an enrichment of metallic Ni-rich species whereby Mg and Y transformed to their hydroxides and oxides. During subsequent anodic discharging Ni-rich species oxidized and were dissolved from the alloy surface [65].

Altogether, under cathodic polarization conditions, which are related to the generation of hydrogen, a de-stabilization of Mg-based glassy alloys must be generally considered. This is particularly the case for Cu-containing alloys due to favoured disproportion reactions, but may be also critical for Ni-containing alloys in extreme alkaline solutions. Therefore, for example galvanic coupling with less noble materials, e.g. conventional Mg alloys, must be avoided. Nevertheless, more studies are necessary to analyze the effect of hydrogen on particular amorphous Mg-based alloys in more detail.

5.7 Future trends

Corrosion studies on selected Mg-based bulk metallic glasses have shown that indeed, owing to their unusual alloy compositions and unique single-phase nature based on a short-range ordered amorphous structure, lower corrosion rates and an improved passivation ability in particular in near-neutral media are attained. On the other hand, their high susceptibility to chloride-induced corrosion processes and their strong tendency for hydrogen absorption under cathodic polarization conditions and the related possible de-stabilization of the amorphous structural state are critical aspects. More detailed studies will be necessary to reach a deeper understanding of these degradation phenomena and to clarify the role of each constituent element and of the particular microstructural state.

There is also a need for a more systematic approach in the corrosion studies.

So far various research groups have investigated Mg-based metallic glass corrosion, but each used samples with only one to three specific compositions and different qualities of their structural state, depending on the preparation conditions, and those samples were tested in one or two specific electrolytes. This makes a general comparison of the corrosion performance of the Mg-based glassy alloys at present very difficult. In this context, it is not clear, yet, to what degree results of frequently performed corrosion studies on very thin melt-spun ribbon samples are transferable and relevant for the behaviour of cast bulk samples, which are most interesting for engineering applications. It must be considered that differences between the structural relaxation states (related with the degree of free volume), the defect densities (e.g. crystalline inclusions) as well as the degree of chemical fluctuations in rapidly quenched and slowly cooled samples may lead to non-neglectable deviations in their electrochemical behaviour. Those questions will have to be clarified before also more complex, not yet considered, corrosion phenomena can be reliably studied, e.g. the effect of temperature on the passivity or aspects of stress-corrosion cracking.

Based on the corrosion studies on Mg-based metallic glasses done so far it must be predicted that a high long-term stability under environmental conditions, which is indispensable for various anticipated applications as new engineering materials, is not available for the presently known alloy compositions. Further alloy developments considering constituent elements in sufficient concentrations which lead to improved corrosion performance are necessary. This will have to be balanced with the need to further increase the glass-forming ability of the alloys in order to achieve a higher resistance against crystallization during cooling from the melt and to further increase the sample dimensions.

However, presently the main focus in Mg-based bulk metallic glass research is directed towards the achievement of maximum sample thickness in combination with excellent mechanical performance. Therefore, so-called *glass matrix composites* are under development, which combine a high-strength bulk glassy Mg-based matrix with finely-dispersed second phase particles resulting in further strengthening or increase of ductility. Those composites can be achieved *in situ* by additional alloying elements triggering phase separation or nanocrystal precipitation under suitable casting conditions, e.g. Nb [92], Be [93,94], Ag [94], Zn [95], Ti [96] or Li [97]. They can also be prepared *ex situ*, e.g. by mixing and hot pressing of Mg-based bulk metallic glass powder with fine oxide particles, e.g. ZrO_2 [98] or Y_2O_3 [99,100]. In any case, with respect to the achievement of high corrosion resistance, these Mg-based glass matrix composites will be extremely challenging.

5.8 Acknowledgements

A. Gebert gratefully acknowledges the contribution of colleagues from the IFW Dresden, Germany to corrosion research on Mg-based bulk metallic glasses: U. Wolff, V. Hähnel, B. Khorkounov, M. Johne, K. Hennig, M. Uhlemann, A. Teresiak, Ch. Mickel, S. Baunack, M. Frey, S. Donath, J. Eckert and L. Schultz. Part of this work has been conducted with financial support of the German Research Foundation (DFG) under grant Ge1106/3 '*Reactivity of Mg-based glasses*'. Fruitful collaboration with various researchers is gratefully acknowledged: R.V. Subba Rao (Indira Gandhi Centre for Atomic Research Kalpakkam, India), E.S. Park and D.H. Kim (Center for Non-crystalline Materials, Yonsei University Seoul, Korea) and M. Savyak (Academy of Science Kiev, Ukraine).

5.9 References

[1] Inoue A, 'Stabilization of metallic supercooled liquid and bulk amorphous alloys', *Acta Materialia*, 2000 **48** 277–304.

[2] Inoue A, Takeuchi A, 'Recent progress in bulk glassy alloys', *Materials Transactions JIM*, 2002 **43** 1892–1906.

[3] Inoue A, Wada T, Wang XM, Greer AL, 'Bulk non-equilibrium alloys and porous glassy alloys with unique mechanical characteristics', *Materials Science and Engineering A*, 2006 **442** 233–242.

[4] Inoue A, Wang XM, Zhang W, 'Developments and applications of bulk metallic glasses', *Reviews in Advanced Materials Science*, 2008 **18** 1–9.

[5] Scully JR, Gebert A, Payer JH, 'Corrosion and related mechanical properties of bulk metallic glasses', *Journal of Materials Research*, 2007 **22** 302–313.

[6] Inoue A, Kohinata M, Tsai A, Masumoto T, 'Mg–Ni–La Amorphous alloys with a wide supercooled liquid region', *Materials Transactions*, 1989 **30** 378–381.

[7] Kim SG, Inoue A, Masumoto T, 'High mechanical strengths of Mg–Ni–Y and Mg–Cu–Y amorphous alloys with significant supercooled liquid region', *Materials Transactions JIM*, 1990 **31** 929–934.

[8] Inoue A, Masumoto T, 'Mg-based amorphous alloys', *Materials Science and Engineering A*, 1993 **173** 1–8.

[9] Inoue A, Kato A, Zhang T, Kim SG, Masumoto T, 'Mg–Cu–Y amorphous alloys with high mechanical strengths produced by a metallic mold casting method', *Materials Transactions*, 1991 **32** 609–616.

[10] Hui X, Gao R, Chen GL, Shang SL, Wang Y, Liu ZK, 'Short-to-medium-range order in $Mg_{65}Cu_{25}Y_{10}$ metallic glass', *Physics Letters A*, 2008 **372** 3078–3084.

[11] Pryds NH, Eldrup M, Ohnuma M, Pedersen AS, Hattel J, Linderoth S, 'Preparation and properties of Mg–Cu–Y–Al bulk amorphous alloys', *Materials Transactions JIM*, 2000 **41** 1435–1442.

[12] Kang GK, Park ES, Kim WT, Kim DH, Cho HK, 'Fabrication of bulk Mg–Cu–Ag–Y glassy alloy by squeeze casting', *Materials Transactions*, 2000 **41** 846–849.

[13] Amiya K, Inoue A, 'Thermal stability and mechanical properties of Mg–Y–Cu–M (M = Ag,Pd) bulk amorphous alloys', *Materials Transactions*, 2000 **41** 1460–1462.

[14] Men H, Hu ZQ, Xu J, 'Bulk metallic glass formation in the Mg–Cu–Zn–Y system', *Scripta Materialia,* 2002 **46** 699–703.

[15] Ma H, Ma E, Xu J, 'A new $Mg_{65}Cu_{7.5}Ni_{7.5}Ag_5 Zn_5Y_{10}$ bulk metallic glass with strong glass-forming ability', *Journal of Materials Research,* 2003 **18** 2288–2291.

[16] Park ES, Kim DH, 'Formation of Mg–Cu–Ni–Ag–Zn–Y–Gd bulk glassy alloy by casting into cone-shaped copper mold in air atmosphere', *Journal of Materials Research,* 2005 **20** 1465–1469.

[17] Men H, Kim DH, 'Fabrication of ternary Mg–Cu–Gd bulk metallic glass with high glass-forming ability under air atmosphere', *Journal of Materials Research,* 2003 **18** 1502–1504.

[18] Yuan G, Inoue A, 'The effect of Ni substitution on the glass-forming ability and mechanical properties of Mg–Cu–Gd metallic glass alloy', *Journal of Alloys and Compounds,* 2005 **387** 134–138.

[19] Park ES, Lee JY, Kim DH, 'Effect of Ag addition on the improvement of glass-forming ability and plasticity of Mg–Cu–Gd bulk metallic glass', *Journal of Materials Research,* 2005 **20** 2379–2385.

[20] Men H, Kim WT, Kim DH, 'Fabrication and mechanical properties of $Mg_{65}Cu_{15}Ag_5 Pd_5 Gd_{10}$ bulk metallic glass', *Materials Transactions,* 2003 **44** 2141–2144.

[21] Xi KK, Zhao DQ, Pan MX, Wang WH, 'Glass-forming Mg–Cu–RE (RE = Gd, Pr, Nd, Tb, Y, and Dy) alloys with strong oxygen resistance in manufacturability', *Journal of Non-crystalline Solids,* 2004 **344** 105–109.

[22] Qin WD, Li J, Kou H, Gu X, Kecskes L, Chang H, Zhou L, 'Effects of Zn addition on the improvement of glass-forming ability and plasticity of Mg–Cu–Tb bulk metallic glass', *Journal of Non-crystalline Solids,* 2008 **354** 5368–5371.

[23] Qin W, Li J, Kou H, Gu X, Xue X, Zhou L, 'Effects of alloy addition on the improvement of glass forming ability and plasticity of Mg–Cu–Tb bulk metallic glass', *Intermetallics,* 2009 **17** 253–255.

[24] Ma H, Shi LL, Xu J, Li Y, Ma E, 'Improving glass-forming ability of Mg–Cu–Y via substitutional alloying: effects of Ag versus Ni', *Journal of Materials Research,* 2006 **21** 2204–2214.

[25] Zheng Q, Cheng S, Strader JH, Ma E, Xu J, 'Critical size and strength of the best bulk metallic glass former in the Mg–Cu–Gd ternary system', *Scripta Materialia,* 2007 **56** 161–164.

[26] Ma H, Shi LL, Xu J, Li Y, Ma E, 'Discovering inch-diameter metallic glasses in three-dimensional composition space', *Applied Physics Letters,* 2005 **87** 181915.

[27] Zheng Q, Xu J, Ma E, 'High glass-forming ability correlated with fragility of Mg–Cu(Ag)–Gd alloys', *Journal of Applied Physics,* 2007 **102** 113519.

[28] Wei YX, Xi XK, Zhao DQ, Pan MX, Wang WH, 'Formation of MgNiPr bulk metallic glasses in air', *Materials Letters*, 2005 **59** 945–947.

[29] Park ES, Chang HJ, Kim DH, 'Mg–rich Mg–Ni–Gd ternary bulk metallic glasses with high compressive specific strength and ductility', *Journal of Materials Research,* 2007 **22** 334–338.

[30] Yin J, Yuan GY, Chu ZH, Zhang J, Ding WJ, 'Mg–Ni–(Gd,Nd) bulk metallic glasses with improved glass-forming ability and mechanical properties', *Journal of Materials Research,* 2009 **24** 2130–2139.

[31] Li Y, Ng SC, Ong CK, Hng HH, Jones H, 'Critical cooling rates of glass formation in Mg-based Mg–Ni–Nd alloys', *Journal of Materials Science,* 1995 **14** 988–990.

[32] Li Y, Liu HY, Jones H, 'Easy glass formation in magnesium-based Mg–Ni–Nd alloys', *Journal of Materials Science,* 1996 **31** 1857–1863.

[33] Li Y, Jones H, Davies HA, 'Determination of critical thickness for glass-formation in new easy glass-forming magnesium-base alloys by the wedge chill casting technique', *Scripta Metallurgica et Materialia,* 1992 **26** 1371–1375.

[34] Kim D, Lee BJ, Kim NJ, 'Prediction of composition dependency of glass-forming ability of Mg–Cu–Y alloys by thermodynamic approach', *Scripta Materialia,* 2005 **52** 969–972.

[35] Seidel M, Eckert J, Zueco-Rodrigo E, Schultz L, 'Mg-based amorphous alloys with extended supercooled liquid region produced by mechanical alloying', *Journal of Non-Crystalline Solids,* 1996 **205–207** 514–517.

[36] Schlorke N, Eckert J, Schultz L, 'Formation and stability of bulk metallic glass forming Mg–Y–Cu alloys produced by mechanical alloying and rapid quenching', *Materials Science Forum*, 1998 **269–272** 761–766.

[37] Schlorke N, Weiss B, Eckert J, Schultz L, 'Properties of Mg–Y–Cu glasses with nanocrystalline particles', *NanoStructured Materials*, 1999 **12** 127–130.

[38] Lee PY, Hsu CF, Wang CC, 'Mg–Y–Cu bulk metallic glasses obtained by mechanical alloying and powder consolidation', *Materials Science Forum,* 2007 **534–536** 205–208.

[39] Mear FO, Xie GQ, Louzguine-Luzgin DV, Inoue A, 'Spark plasma sintering of Mg-based amorphous ball-milled powders', *Materials Transactions,* 2009 **50** 588–591.

[40] Lee PY, Kao MC, Lin CK, Huang JC, 'Mg–Y–Cu bulk metallic glass prepared by mechanical alloying and vacuum hot pressing', *Intermetallics*, 2006 **14** 994–999.

[41] Kato A, Suganuma T, Horikiri H, Kawamura Y, Inoue A, Masumoto T, 'Consolidation and mechanical properties of atomized Mg-based amorphous powders', *Materials Science and Engineering A,* 1994 **179–180** 112–117.

[42] Makar GL, Kruger J, 'Corrosion of magnesium', *International Materials Reviews,* 1993 **38** 138–153.

[43] Song GL, Atrens A, 'Corrosion mechanisms of magnesium alloys', *Advanced Engineering Materials*, 1999 **1** 11–33.

[44] Song GL, 'Recent progress in corrosion and protection of magnesium alloys', *Advanced Engineering Materials*, 2005 **7** 563–586.

[45] Makar GL, Kruger J, 'Corrosion studies of rapidly solidified magnesium alloys', *Journal of the Electrochemical Society,* 1990 **137** 414–421.

[46] Polymear IJ, 'Magnesium alloys and applications', *Materials Science and Technology*, 1994 **10** 1–15.

[47] Nakatsugawa I, Kamado S, Kojima Y, Ninomiya R, Kubota K, 'Corrosion of magnesium alloys containing rare earth elements', *Corrosion Reviews*, 1998 **16** 139–157.

[48] Grosjean MH, Zidoune M, Roue L, Huot J, Schulz R, 'Effect of ball milling on the corrosion resistance of magnesium in aqueous media', *Electrochimica Acta,* 2004 **49** 2461–2470.

[49] Cho SS, Chun BS, Won CW, Kim SD, Lee BS, Baek H, Suryanarayana C, 'Structure and properties of rapidly solidified Mg–Al alloys', *Journal of Materials Science,* 1999 **34** 4311–4320.

[50] Grigucevicience A, Leinartas K, Juskenas R, Juzeliunas E, 'Structure and initial corrosion resistance of sputter deposited nanocrystalline Mg–Al–Zr alloys', *Materials Science and Engineering A,* 2005 **394** 411–416.

[51] Miller PL, Shaw BA, Wendt RG, Moshier WC, 'Improving the corrosion resistance of magnesium by nonequilibrium alloying with Y', *Corrosion,* 1993 **49** 947–950.

[52] Miller PL, Shaw BA, Wendt RG, Moshier WC, 'Assessing the corrosion resistance of nonequilibrium magnesium–yttrium alloys', *Corrosion,* 1995 **51** 922–931.

[53] Heidersbach KL, Shaw BA, 'Study of the corrosion behaviour of Mg-Y nonequilibrium thin films', *Electrochemical Society Series,* 1998 **97** 858–869.

[54] Krishnamurthy S, Khobaib M, Robertson E, Froes FH, 'Corrosion behavior of rapidly solidified Mg–Nd and Mg–Y alloys', *Materials Science and Engineering,* 1988 **99** 507–511.

[55] Yao HB, Li Y, Wee ATS, 'Passivity behaviour of melt-spun Mg-Y alloys', *Electrochimica Acta,* 2003 **48** 4197–4204.

[56] Masumoto T, Hashimoto K, 'Chemical properties of amorphous metals', *Annual Reviews of Materials Science,* 1978 **8** 215–233.

[57] Walmsley RG, Lee YS, Marshall AF, Stevenson DA, 'Electrochemical characterization of amorphous and microcrystalline metals', *Journal of Non-Crystalline Solids*, 1984 **61/62** 625–630.

[58] Latanision RM, Turn JC, Compeau CR, 'The corrosion resistance of metallic glasses' in *Mechanical Behaviour of Materials, Proceedings of ICM 3,* Eds. K.J. Miller, R.F. Smith, Pergamon Press, Oxford, 1980 **2** 475–483.

[59] Gebert A, Wolff U, John A, Eckert J, Schultz L, 'Stability of the bulk glass-forming $Mg_{65}Cu_{25}Y_{10}$ alloy in aqueous electrolytes', *Materials Science and Engineering A,* 2001 **299** 125–135.

[60] Gebert A, Subba Rao RV, Wolff U, Baunack S, Eckert J, Schultz L, 'Corrosion behaviour of the $Mg_{65}Y_{10}Cu_{25}Ag_{10}$ bulk metallic glass', *Materials Science and Engineering A*, 2004 **375–377** 280–284.

[61] Gebert A, Haehnel V, Park ES, Kim DH, Schultz L, 'Corrosion behaviour of $Mg_{65}Cu_{7.5}Ni_{7.5}Ag_5 Zn_5 Gd_5Y_5$ bulk metallic glass in aqueous environments', *Electrochimica Acta,* 2008 **53** 3403–3411.

[62] Pourbaix M, *Atlas of Electrochemical Equilibria in Aqueous Solutions*, Pergamon Press Oxford, 1966.

[63] Qin FX, Bae GT, Dan ZH, Lee H, Kim NJ, 'Corrosion behaviour of the $Mg_{65}Cu_{25}Gd_{10}$ bulk amorphous alloy', *Materials Science and Engineering A,* 2007 **449–451** 636–639.

[64] Gebert A, Wolff U, John A, Eckert J, 'Corrosion behaviour of $Mg_{65}Cu_{25}Y_{10}$ metallic glass', *Scripta Materialia,* 2000 **43** 279–283.

[65] Gebert A, Khorkounov B, Wolff U, Mickel Ch, Uhlemann M, Schultz L, 'Stability of rapidly quenched and hydrogenated Mg–Ni–Y and Mg–Cu–Y alloys in extreme alkaline medium', *Journal of Alloys and Compounds,* 2006 **419** 319–327.

[66] Subba Rao RV, Wolff U, Baunack S, Eckert J, Gebert A, 'Corrosion behaviour of the amorphous $Mg_{65}Y_{10}Cu_{25}Ag_{10}$ alloy', *Corrosion Science,* 2003 **45** 817–832.

[67] Subba Rao RV, Wolff U, Baunack S, Eckert J, Gebert A, 'Stability of the $Mg_{65}Y_{10}Cu_{25}Ag_{10}$ metallic glass in neutral and weakly acidic media', *Journal of Materials Research,* 2003 **18** 97–105.

[68] Yao HB, Li Y, Wee ATS, 'Corrosion behaviour of melt-spun $Mg_{65}Ni_{20}Nd_{15}$ and $Mg_{65}Cu_{25}Y_{10}$ metallic glasses', *Electrochimica Acta,* 2003 **48** 2641–2650.

[69] Baunack S, Subba Rao RV, Wolff U, 'Characterization of oxide layers on amorphous Mg-based alloys by Auger electron spectroscopy with sputter depth profiling', *Analytical and Bioanalytical Chemistry,* 2003 **375** 896–901.

[70] Gebert A, Khorkounov B, Schultz L, 'Effect of chemical pre-treatments on the stability of melt-spun $Mg_{50}Ni_{20}Y_{20}$ in extreme alkaline electrolyte', *Reviews in Advanced Materials Science*, 2008 **18** 639–643.

[71] Dobson SJ, Whitaker I, Jones H, Davies HA, 'Some characteristics of Mg–Ni–Nd bulk glass forming alloys', in *Proceedings of the 3rd International Magnesium Conference*, ed. Lorimer GW, The Institute of Materials, Manchester UK, 1996 507–516.

[72] Li Y, Ng SC, Kam CH, 'Dissolution of Mg-based Mg–Ni–Nd amorphous alloys in 3% NaCl solution', *Materials Letters*, 1998 **36** 214–217.

[73] Kam CH, Li Y, Ng SC, Wee A, Pan JS, Jones H, 'The effect of heat treatment on the corrosion behaviour of amorphous Mg–Ni–Nd alloys', *Journal of Materials Research*, 1999 **14** 1638–1644.

[74] Ong MS, Li Y, Blackwood DJ, Ng SC, 'The influence of heat treatment on the corrosion behaviour of amorphous melt-spun binary Mg-18 at.% Ni and Mg-21 at.% Cu alloy', *Materials Science and Engineering A*, 2001 **304–306** 510–514.

[75] Yao HB, Li Y, Wee ATS, Chai JW, Pan JS, 'The alloying effect of Ni on the corrosion behaviour of melt-spun Mg-Ni ribbons', *Electrochimica Acta*, 2001 **46** 2649–2657.

[76] Li GQ, Zheng LJ, Li HX, 'Corrosion behaviour of the bulk amorphous $Mg_{65}Cu_{25}Gd_{10}$ alloy', *Rare Metal Materials and Engineering*, 2009 **38** 110–114.

[77] Li GQ, Huang W, Li HX, Zheng LJ, Hashmi MF, 'Corrosion behaviour of $Mg_{65}Cu_{25-x}Zn_xGd_{10}$ (x = 0,5) metallic glass', *Journal of Wuhan University of Technology – Materials Science Edition*, 2008 **23** 678–682.

[78] Yuan G, Qin C, Inoue A, 'Mg-based bulk glassy alloys with high strength above 900 MPa and plastic strain', *Journal of Materials Research*, 2005 **20** 394–401.

[79] Revie RW, *Uhlig's Corrosion Handbook*, John Wiley & Sons, New York, 2000, 812–820.

[80] Yamaura S, Kim HY, Kimura H, Inoue A, Arata Y, 'Thermal stabilities and discharge capacities of melt-spun Mg–Ni-based amorphous alloys', *Journal of Alloys and Compounds*, 2002 **339** 230–235.

[81] Liu FJ, Suda S, 'F-treatment effect on the initial activation characteristics of Mg–La–Ni amorphous alloys', *Journal of Alloys and Compounds*, 1995 **231** 696–701.

[82] Teresiak A, Uhlemann M, Gebert A, Thomas J, Eckert J, Schultz L, 'Formation of nanostructured $LaMg_2Ni$ by rapid quenching and intensive milling and its hydrogen reactivity', *Journal of Alloys and Compounds*, 2009 **481** 144–151.

[83] Savyak M, Hirnyj S, Bauer HD, Uhlemann M, Eckert J, Schultz L, Gebert A, 'Electrochemical hydrogenation of $Mg_{65}Cu_{25}Y_{10}$ metallic glass', *Journal of Alloys and Compounds*, 2004 **364** 229–237.

[84] Seiler A, Schlapbach L, Von Waldkirch T, Shaltiel D, Stucki F, 'Surface analysis of Mg_2Ni–Mg, Mg_2Ni and Mg_2Cu', *Journal of Less-Common Metals*, 1980 **73** 193–199.

[85] Reilly JJ, 'Metal hydride technology', *Zeitschrift für Physikalische Chemie Neue Folge*, 1979 **117** 155–184.

[86] Spassov T, Köster U, 'Thermal stability and hydriding properties of nanocrystalline melt-spun $Mg_{63}Ni_{30}Y_7$ alloy', *Journal of Alloys and Compounds*, 1998 **279** 279–286.

[87] Spassov T, Köster U, 'Hydrogenation of amorphous and nanocrystalline Mg-based alloys', *Journal of Alloys and Compounds*, 1999 **287** 243–250.

[88] Spassov T, Köster U, 'Nanocrystalline Mg–Ni–based hydrogen storage alloys produced by nanocrystallization', *Materials Science Forum*, 1999 **307** 197–202.

[89] Teresiak A, Gebert A, Savyak M, Uhlemann M, Mickel C, Mattern N, '*In situ* high temperature XRD studies of the thermal behaviour of the rapidly quenched $Mg_{77}Ni_{18}Y_5$ alloy under hydrogen', *Journal of Alloys and Compounds*, 2005 **398** 156–164.

[90] Orimo S, Züttel A, Ikeda K, Saruki S, Fukunaga T, Fujii H, Schlapbach L, 'Hydriding properties of the MgNi-based systems', *Journal of Alloys and Compounds*, 1999 **293–295** 437–442.

[91] Khorkounov B, Gebert A, Mickel C, Schultz L, 'Improving the performance of hydrogen storage electrodes based on mechanically alloyed $Mg_{63}Ni_{30}Y_9$', *Journal of Alloys and Compounds*, 2008 **458** 479–486.

[92] Zhang CM, Hui X, Li ZG, Chen GL, 'Improving the strength and the toughness of Mg–Cu–(Y,Gd) bulk metallic glass by minor addition of Nb', *Journal of Alloys and Compounds*, 2009 **467** 241–245.

[93] Li ZG, Hui X, Zhang CM, Wang ML, Chen GL, 'Strengthening and toughening of Mg–Cu–(Y,Gd) bulk metallic glass by minor addition of Be', *Materials Letters*, 2007 **61** 5018–5021.

[94] Park ES, Kyeong JS, Kim DH, 'Enhanced glass forming ability and plasticity in Mg-based bulk metallic glasses', *Materials Science and Engineering A*, 2007 **449–451** 225–229.

[95] Hui X, Dong W, Chen GL, Yao KF, 'Formation, microstructure and properties of long-period order structure reinforced Mg-based bulk metallic glass composites', *Acta Materialia*, 2007 **55** 907–920.

[96] Li F, Guan S, Shen B, Makino A, Inoue A, 'High specific strength and improved ductility of bulk (Mg0.65Cu0.25Gd0.1)(100–*x*) Ti–*x* metallic glass composites', *Materials Transactions*, 2007 **48** 3193–3196.

[97] Liu W, Johnson WL, 'Precipitation of bcc nanocrystals in bulk Mg–Cu–Y amorphous alloys', *Journal of Materials Research*, 1996 **11** 2388–2392.

[98] Chang LJ, Fang GR, Jang JSC, Lee IS, Huang JC, Tsa CYA, Jou JL, 'Mechanical properties of the hot-pressed amorphous Mg65Cu20Y10Ag5/nanoZrO(2) composite alloy', *Materials Transactions*, 2007 **48** 1797–1801.

[99] Eckert J, Schlorke-de Boer N, Weiss B, Schultz L, 'Mechanically alloyed Mg-based metallic glass composites containing nanocrystalline particles', *Zeitschrift für Metallkunde*, 1999 **90** 908–913.

[100] Eckert J, Bartusch B, Gebert A, 'The effect of nanosized Y_2O_3 as a second phase in mechanically alloyed Mg–Y–Cu glass matrix composites', *Journal of Metastable and Nanocrystalline Materials*, 2003 **15–16** 37–42.

6
Corrosion of innovative magnesium (Mg) alloys

P. B. SRINIVASAN, C. BLAWERT, D. HÖCHE,
GKSS-Helmholtz-Zentrum Geesthacht, Germany

Abstract: This chapter deals with the corrosion performance of innovative or unconventional magnesium (Mg) alloys. It considers, on the one hand, all Mg alloys that have an unusual microstructure and/or composition mostly as a result of extreme processing conditions and, on the other hand, the innovative concepts that are used to improve the corrosion performance of conventional alloys. The available literature is reviewed and the specific influences of the microstructure and/or composition on corrosion behaviour is discussed and general possibilities to improve corrosion resistance of these alloys are highlighted.

Key words: recycling alloys, amorphous alloys, alloy coatings, laser surfacing, laser alloying.

6.1 Recycled alloys

Recycled alloys are more or less standard alloys and neither the alloying elements nor the processing is innovative. Innovation lies only in the domain of alloy design, seeking for corrosion resistance without following the concept of high-purity alloys. Magnesium (Mg) secondary alloys are of common interest because their production process requires less energy compared to the primary magnesium production, thus helping to reduce the CO_2 emission further (Ditze and Scharf, 2004; Ehrenberger *et al.*, 2008). In spite of this, no post-consumer scrap is used for recycling of magnesium and furthermore no defined secondary Mg alloys comparable to aluminium secondary alloys such as A380 are available. The main reason for this is the possible uptake of heavy metal impurities during the recycling process with extremely detrimental effects on the corrosion resistance (Ditze and Scharf, 2008). This problem has to be solved and a recycling system including secondary alloys must be developed to promote the general use of Mg alloys.

However, the development of secondary Mg alloys requires a completely different concept compared with standard alloys, which obtain their corrosion resistance by reducing the levels of impurities below certain alloy-specific and process-dependent limits. Although control of impurities is technically possible, this would make the secondary alloys too expensive. A cheaper solution is the modification of the microstructures of the secondary alloys to make them more tolerant against impurities without too many negative effects on the corrosion and mechanical properties. It is this concept of fewer reduced impurities that is innovative. The resulting alloys are not much different from standard alloys and can be processed with all standard industrial processes and equipment.

So far two different approaches for modifying the microstructure have been followed. The first quite simple approach, which works for cast alloys with higher Al contents, e.g. AZ91 and the related secondary alloy AZC1231, is to improve the barrier function of the ß-phase ($Mg_{17}Al_{12}$), forming a dense corrosion barrier along the grain boundaries. However, if ductility is required, the Al content should be restricted to a maximum of about 5% and thus the previously protecting ß-phase network is present only in the form of isolated localised precipitates with detrimental effects on the corrosion resistance. Therefore, the second approach suitable for wrought alloys, e.g. AM50 and the related secondary alloy AZC531, uses a new concept trying to replace the ß-phase by τ-phase (which is able to incorporate more impurities while being less detrimental to the matrix from the corrosion point of view). The chemical compositions of the standard, contaminated standard and the secondary alloys are given in Table 6.1.

Several different corrosion tests revealed that the secondary alloy AZC1231 tolerates up to 1% copper and 40 ppm Ni under various casting conditions (gravity die-casting, semi-solid processing as well as high-pressure die-casting) having corrosion rates still comparable to the standard alloy AZ91D (Ditze *et al.*, 2005, 2006; Scharf *et al.*, 2005, 2007; Blawert *et al.*, 2006, 2008). The salt spray performance of the alloys is presented as an example in Table 6.2. For comparison, the performance of the standard alloys contaminated with the same amount of copper as in the secondary alloys is also given. The correlation of microstructure with corrosion rate suggests that for this alloy different mechanisms are responsible for improving the impurity tolerance limits (Fig. 6.1). The secondary alloy forms the intended ß-phase network around the grains. A grain refinement caused by the increase of Al, Zn and Mn makes it easier to form a dense network, improving the barrier properties.

Table 6.1 Chemical composition of the standard, contaminated standard and secondary alloys (wt%)

Alloy	Al	Zn	Mn	Si	Cu
AZ91D	8.75	0.67	0.2	0.054	0.008
AZC911	7.79	0.74	0.23	0.13	0.5
AZC1231	11.7	3.04	0.48	0.39	0.47
AM50	4.9	0.02	0.26	0.026	0.008
AMC501	4.84	0.023	0.26	0.028	0.5
AZC531	5.1	3.06	0.27	0.027	0.49

Table 6.2 Corrosion rates of the standard and modified alloys after 48 h of salt spray as per ASTM B117

Alloy	Corrosion rate (mm/year)
AZ91D	1.07 ± 0.23
AZC911	8.53 ± 0.06
AZC1231	0.99 ± 0.58
AM50	0.71 ± 0.20
AMC501	8.99 ± 0.48
AZC531	1.11 ± 0.03

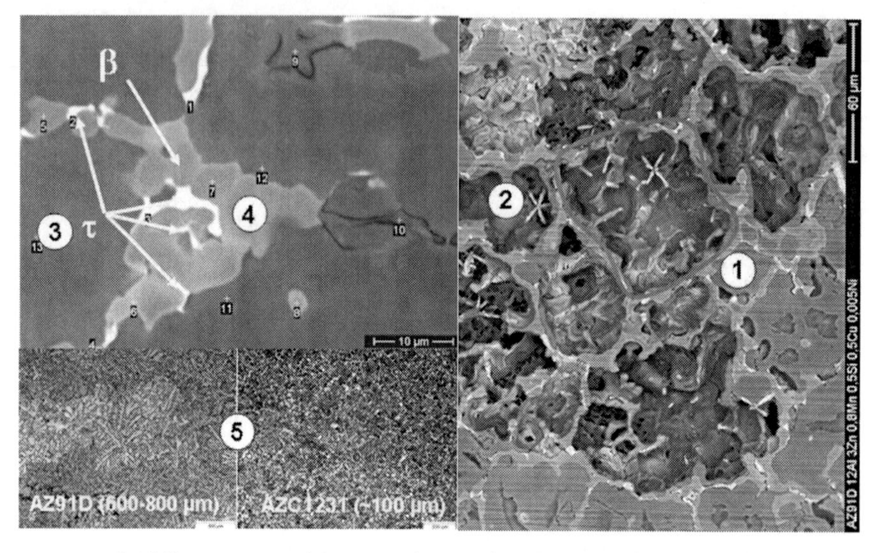

6.1 Microstructural features improving the corrosion resistance of an AZC1231 alloy: 1, barrier function of ß-phase; 2, Al_8Mn_5 precipitates embedding Fe and Ni; 3, τ-phase incorporating Cu and preventing the formation of binary Mg-Cu phases; 4, more noble phases isolated from the matrix by the ß-phase; 5, grain refinement.

The setting of impurities in intermetallic phases is effective as well. A detailed analysis of the phases present in real castings of AZ91 and AZ123 alloys at the same impurity level of 0.5% Cu, 0.008% Fe and 0.0035% Ni reveals the advantage of the new alloy (Table 6.3). Copper is largely incorporated into the τ-phase and is no longer available to form binary Mg–Cu phases. Thus, the formations of those phases that are detrimental to the corrosion resistance are suppressed. Furthermore, the dominating ß-phase also embeds the τ-phase and the other nobler critical precipitates, thus isolating them from a direct contact with the matrix. This reduces the galvanic effects in the alloy to a large extent. The question whether the increasing Al and Zn contents are stabilising or contributing to the passive film formation needs

Table 6.3 Phases detected in pure AZ91D and AZM1231 and in their Cu (0.5 wt%) contaminated counterparts AZ91 and AZC1231

Phase	AZ91D	AZ91	AZM1231	AZC1231
α-phase	×	×	×	×
β-phase	×	×	×	×
Mg_2Si	×	×	×	×
Al_8Mn_5	×	×	×	×
$Mg_6Cu_3Al_7$	–	×	–	×
Mg_2Cu	–	×	–	–
$MgCu_2$	–	×	–	–
MgZn	–	–	×	–
τ-phase	–	–	×	×

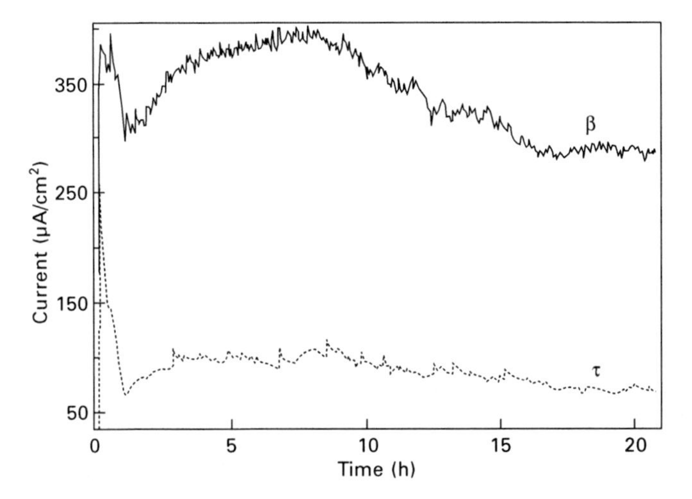

6.2 Comparison of the galvanic current measured between either ß or τ and pure magnesium in 5% NaCl solution (pH 11).

further detailed investigations. Presently it seems as if they are enhancing short-term protection, while in the long run increasing film thickness results in flaking-off of protective surface oxide films.

In spite of good castability, strength and corrosion behaviour of the AZC1231 secondary alloy, suggesting that this alloy may be a suitable alternative for many applications, it cannot be used in places where ductility is important. For wrought processing or more ductile cast applications an additional secondary Mg alloy group is required.

The essential step in the development of the ductile secondary alloy AZC531 was the finding that the τ-phase causes less galvanic corrosion impact to an Mg matrix than the ß-phase (Fig. 6.2). Consequently the alloy composition was modified to suppress the ß-phase and enhance the formation of the τ-phase (Blawert *et al.*, 2009b). The additional benefit for the recycled secondary alloys would be that the τ-phase is capable of incorporating a large amount of Cu impurities, without too many negative effects on the corrosion performance. The tolerable copper content is largely extended, reaching 0.7 wt% and the tolerable nickel content is also slightly enhanced, but still has to be kept lower than 20 ppm (0.002 wt%) if the salt spray corrosion rate does not exceed 1 mm/y. The maximum content of impurities in the standard alloys according to ASTM B94-07 is presented in Table 6.4. The changes in the composition and microstructure of the secondary alloy AZC531 do not affect the ductility much. Unlike the case of the secondary alloy AZC1231, which suffers from a significant loss of ductility compared with its standard counterpart AZ91D, the AZC531 secondary alloy possesses a ductility level similar to that of the standard AM50.

To summarise, one can state that the overall experimental effort in the alloy development correlating composition, microstructure and corrosion resistance was reduced using thermodynamic calculations (Scheil) to optimise the alloy composition. The outcome is a new, more impurity-tolerant alloy class with the two main secondary alloys AZC1231 and AZC531 having a composition between standard AZ and ZC systems, covering cast and wrought applications. While the AZ91 alloy scrap is suitable for the secondary alloy AZC1231, the AM50 scrap can be used for the making of secondary alloy AZC531. Further, there exists the potential to extend the same concept to obtain AZC321 or AZC211 secondary alloys using AZ31 scrap.

6.2 Amorphous alloys

In the late 1980s, new classes of magnesium-based ternary metallic glasses with a wide range of compositions represented by Mg–TM–Ln (TM (transition metal) = Ni, Cu or Zn; Ln (lanthanides) = Y, Ce or Nd) were developed (Inoue and masumoto, 1991). The so-called bulk amorphous Mg alloys are produced by solidification techniques involving relatively very high cooling rates, namely, copper mould casting. The requirement of a very high cooling rate to avoid nucleation and growth of a crystalline phase in order to obtain the amorphous structure limits the section thickness of components, which is a drawback in the context of structural applications.

Table 6.4 Tolerance levels for the standard alloys and for the secondary alloys

	Impurity content (wt%)		
Alloy	Cu	Fe	Ni
AM50A[1]	0.010	0.004	0.002
AZ91D[1]	0.030	0.005	0.002

[1]ASTM B94-07.

Amorphous magnesium alloys can also be produced in the form of foils, ribbons or strips of thin sections by melt spinning technique, and by mechanical alloying/high-energy ball-milling (particulate processing) followed by warm compaction. Mg–Ni-based amorphous alloys produced by mechanical alloying are candidate materials for hydrogen storage (Yuan *et al.*, 2003; Xie *et al.*, 2009; Anik, 2010). Furthermore, amorphous Mg alloys can be obtained as coatings on metallic substrates through physical vapour deposition and ion-irradiation techniques. The corrosion behaviours of amorphous alloys are discussed in detail in a different chapter of this book, and a brief account of some of the aspects is given in this section.

One of the early works on Mg–Ni–Y and Mg–Cu–Y amorphous alloys with 0–40% Ni or Cu and 0–40%Y produced by melt spinning reported fracture stress values of 830 and 800 MPa for the $Mg_{80}Ni_{15}Y_5$ and $Mg_{85}Cu_5Y_{10}$ alloys, respectively. The alloys were found to have good bend-ductility as well (Kim *et al.*, 1990). Amorphous Mg–Cu–Y alloy cylinders of 4 mm diameter produced by Inoue *et al.* (1991) contained a disordered structure typical for that of a super-cooled alloy and were reported to possess mechanical strength and deformation characteristics close to that of melt-spun ribbons, indicating similarities in the disordered structures. The possibility of producing up to 7 mm and up to 10 mm in thickness or diameter by high-pressure die-casting and squeeze-casting, respectively, has also been demonstrated (Inoue *et al.*, 1992; Kang *et al.*, 2000). Li *et al.* (1995) investigated 30–100 μm thick magnesium-based amorphous alloy ribbons, containing up to 20% Ni and 15% Nd, produced by rapid solidification using a chill-block melt spinning facility and reported tensile strength and Young's modulus values in the range 330–640 MPa and 3.0–5.1 GPa, respectively. Amorphous Mg–23.5Ni melt spun ribbons possess high strength and low ductility in the temperature range of 25–125 °C (Pérez *et al.*, 2004). With increasing temperature, due to the evolution of nanocrystalline structure with stable phases of Mg_2Ni and $Mg_{5.5}Ni$ phases, the ductility of these amorphous alloys increases substantially.

Mg alloys produced by conventional ingot metallurgy route, in general, exhibit a poor corrosion resistance. The corrosion resistance of such alloys is governed by the level of impurities, alloying elements, chemical homogeneity, heat-treatment condition and the amount/distribution of secondary phases in the matrix. The differences in the electrochemical potentials of the grains, grain boundaries and the secondary phases in the polycrystalline alloys also greatly influence the corrosion behaviour. Production of Mg alloys through non-equilibrium processing can improve their corrosion resistance due to (a) control over the microstructure/phases and (b) extension of solubilities of elements and thereby the minimisation of the negative effects of impurities. In addition, the stability of passive films is also influenced by the incorporation of specific alloying elements in the glassy amorphous alloys.

Yao *et al.* (2001a) produced Mg–Ni alloy ribbons by melt spinning. The alloys with low concentration of nickel contained crystalline Mg and Mg_2Ni intermetallic phases in the concentration below 5 at% Ni. X-ray diffraction (XRD) spectra revealed that the alloys that contained above 11 at.% Ni were amorphous in nature (Fig. 6.3). The alloys with higher nickel content exhibited more noble corrosion potentials in 0.01 M NaCl solution of pH 12. The corrosion resistance of the amorphous Mg–11.2% Ni and Mg–18.3% Ni alloys was better than the crystalline pure magnesium, while the crystalline alloys with lower nickel contents showed an accelerated corrosion rate, which is attributed to the galvanic drive on the Mg matrix induced by Mg_2Ni phase. In very dilute alkaline chloride electrolyte pure magnesium exhibits a slight passivation, and does not pit at the corrosion potential. However, the crystalline alloys containing 1.2 and 4.8% Ni undergo active dissolution without any appreciable passivity in the anodic region due to the galvanic effect induced by the Mg_2Ni precipitates. The polarisation plots (Fig. 6.4) show that the amorphous alloys with 11.2 and 18.3% Ni exhibit an enhanced passive range in this alkaline chloride solution.

The introduction of Nd significantly enhances the corrosion resistance of the binary Mg–Ni amorphous alloys. Based on the immersion tests with measurement of hydrogen in 3% NaCl solutions, Li *et al.* (1998) reported

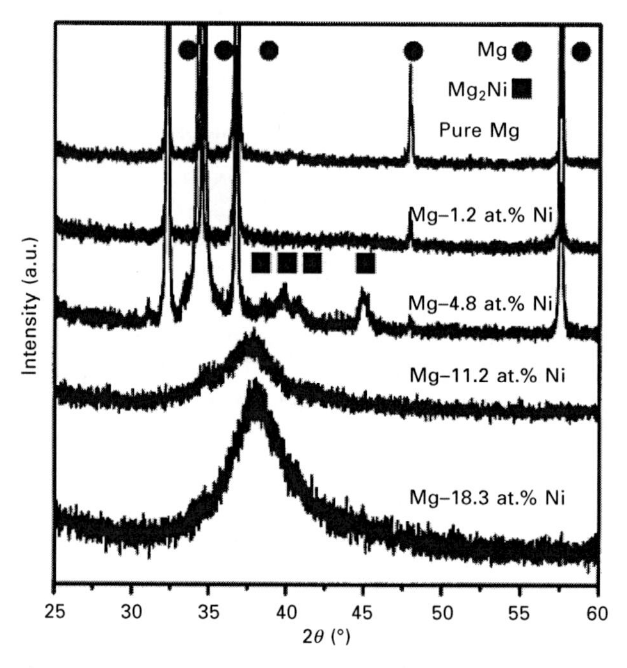

6.3 XRD spectra of melt spun binary Mg–Ni alloys ribbons showing the transition with increasing Ni content, from single α-Mg to α-Mg plus Mg_2Ni, and then to an amorphous structure (Yao *et al.*, 2001a).

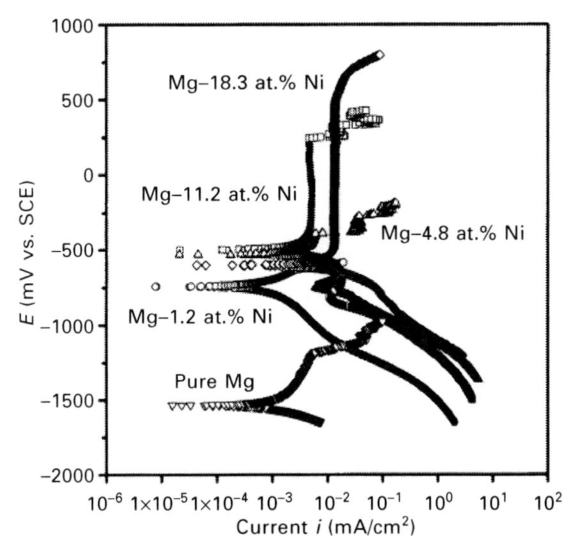

6.4 Polarisation behaviour of pure magnesium and Mg–Ni ribbons in 0.01 M NaCl (pH 12) electrolyte (Yao *et al.*, 2001a).

a corrosion rate of 38 mpy for the Mg–Ni–Nd alloys containing ~35 at.% of Ni+Nd. Yao *et al.* (2001b) observed that the corrosion rate of the $Mg_{67}Ni_{18}Nd_{15}$ amorphous alloys was much lower than $Mg_{77}Ni_{18}Nd_5$ alloy (Fig. 6.5) suggesting the beneficial influence of higher concentration of Nd for a better resistance. X-ray photoelectron spectroscopy (XPS) studies revealed very low concentrations of Cl^- and CO_3^{2-} on the sputtered surface of the corroded amorphous alloys after different sputtering durations, and the high contents of oxidised Nd in the corrosion film seemed to inhibit the ingress of damage causing aggressive anions. Studies on the corrosion behaviour by Gebert *et al.* (2001, 2004) and Yao *et al.* (2003) claimed that the introduction of rare earth elements to the amorphous alloys improves the corrosion resistance by providing a better stability to the passive films owing to the formation of oxides of Ni, Nd, Cu, Y and Ag, depending on the constitution of alloy.

Qin *et al.* (2007) reported a slightly better corrosion resistance for the $Mg_{65}Cu_{25}Gd_{10}$ amorphous alloy produced by copper mould injection casting technique over its crystalline counterpart in alkaline borate buffer and neutral dilute chloride solutions, and attributed the performance to the structural homogeneity of the amorphous alloy. Uhlenhaut *et al.* (2009) attempted to develop Mg amorphous alloys containing aluminium and gallium. As the Pourbaix diagrams of Ga and Al are similar, the improved corrosion behaviour of the alloy was anticipated. The melt spun $Mg_{70}Al_{15}Ga_{15}$ bulk amorphous alloy assessed by potentiodynamic polarisation in 0.1 M NaCl solution was found to exhibit a more active corrosion potential than that of pure 99.5%

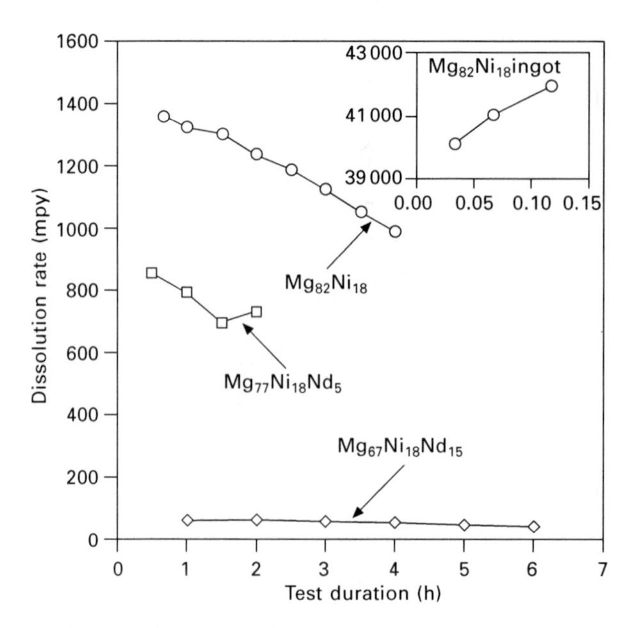

6.5 Dissolution rates of $Mg_{82}Ni_{18}$ crystalline ingot and $Mg_{82-x}Ni_{18}Nd_x$ (x = 0, 5, 15) amorphous alloys in 3% NaCl solution saturated with $Mg(OH)_2$ (Yao *et al.*, 2001b).

Mg and possessed a superior corrosion resistance than a couple of Mg alloys and another amorphous alloy $Mg_{65}Cu_{25}Y_{10}$ (Fig. 6.6). Interestingly, the Auger electron spectroscopy (AES) analysis showed that there was no trace of Ga compounds in the passive layer and that it was enriched only with aluminium oxide. However, AES depth profiles suggested the deposition of metallic Ga below the corrosion layer, which was further confirmed by the XRD results. It appears that the enhanced corrosion resistance was only due to the aluminium oxide enrichment at the surface of this alloy. These research findings opened up avenues for the development of amorphous alloys with higher aluminium content that could provide not only an improved electrochemical behaviour but superior mechanical properties as well.

Mg alloy-based hydrogen storage alloys are produced from the mechanical alloyed/high-energy balled powders and subsequently compacted into pellets (Han *et al.*, 1999; Xiao *et al.*, 2006; Okonska and Jurczyk, 2009). These alloys are reported to possess several advantages, namely, higher discharge capacity, lighter mass and low cost, but their rapid degradation in KOH solution is a major disadvantage. MgNi–CoB composite alloys produced by mechanical alloying possessed a better corrosion resistance than the MgNi amorphous alloy (Feng *et al.*, 2007). Element substitution has been an accepted and effective method for improving the properties of magnesium-based hydrogen storage alloys. The elements that come as partial substitution to

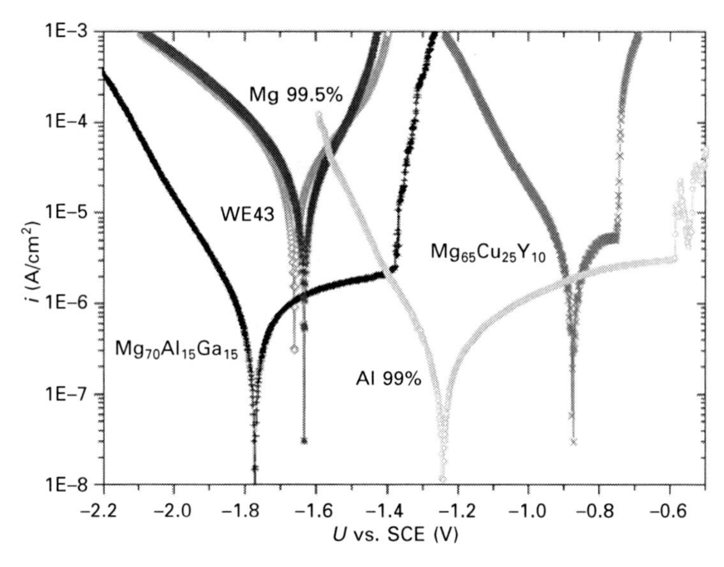

6.6 Potentiodynamic polarisation behaviour of $Mg_{70}Al_{15}Ga_{15}$, $Mg_{65}Cu_{25}Y_{10}$ amorphous alloys in neutral 0.1 M NaCl solution. Polarisation behaviour of the conventional pure magnesium, pure Al and WE43 magnesium alloys are shown for comparison (Uhlenhaut *et al.*, 2009).

magnesium in the hydrogen storage alloys include Al, Ti, V and Y (Nohara *et al.*, 1998; Lenain *et al.*, 1999; Xue *et al.*, 2000; Zhang *et al.*, 2001). The discharge characteristics and the anti-corrosion properties of the electrodes are influenced greatly by the partial substitution. Further to the above, effects of substitution of Ni with Co, Mn and Pd in the $Mg_{0.9}Ti_{0.1}Ni_{1-x}M_x$ (M = Co, Mn, Pd) hydrogen storage amorphous alloys seem to influence the electrochemical charge/discharge cycle and corrosion characteristics appreciably (Tian *et al.*, 2006; Huang *et al.*, 2010).

Even though the amorphous alloys produced by melt spinning possess a low ductility, the high strength and better corrosion behaviour make them attractive. Attempts to improve their ductility by controlled heat treatments would help in enhancing application potentials. From the viewpoint of hydrogen storage applications, improving the electrochemical stability with small additives would be highly beneficial, and incidentally the contemporary research is directed towards this.

6.3 Alloy coatings

The innovative character of these types of Mg alloys (coatings) lies mainly in the non-equilibrium composition as a result of the extreme conditions during their formation, which is a rapid transition from the gas or plasma phase to the solid state. Most magnesium-based coatings are produced by physical vapour deposition (PVD) processes, such as simple vacuum evaporation and deposition (Yamamoto *et al.*, 2001a,b; Yu and Uan, 2006), ion plating (Tsubakino *et al.*, 2003; Lee *et al.*, 2008), vacuum arc deposition (Bohne *et al.*, 2007), ion beam sputtering (Mitsuo and Aizawa, 2003; Blawert *et al.*, 2007a) and magnetron sputter deposition (Lee *et al.*, 2003; Grigucevicienе *et al.*, 2004, 2005; Blawert *et al.*, 2007a). In general the PVD processes were considered to be good tools for alloy design as basically all elements are available for alloying with magnesium independent of their solubility or melting point. Even metals with markedly different melting points, such as Mg and Ti, can be easily transferred into the vapour phase and subsequently deposited together offering the possibility to create new alloys. The ability to produce solid solution alloys far beyond the equilibrium solubility is especially mentioned (Adeva-Ramos *et al.*, 2001). The solubility limits known from conventional casting are no longer valid and all PVD techniques can produce supersaturated, precipitation free and microcrystalline magnesium layers. The only requirement appears to be a source to produce the metal vapour (Blawert *et al.*, 2007a). The advantage of PVD processes is the simultaneous deposition of energetic particles, neutral atoms and charged ions (Nastasi *et al.*, 1996). With increasing ion energy (in the range of 0.05 to 500 eV) more compact films with higher density can form because of an increasing surface mobility (Thornton, 1974). At the same time, due to the non-thermal energy input in combination with high cooling rates (at and beyond 10^6 K/s), the formation of non-equilibrium phases is promoted. The segregation of phases or the spinodal decomposition of non-solid solution systems can be completely suppressed depending on the film growth rate and ion energy used (Nastasi *et al.*, 1996).

The microstructure of the coating can range from micro- or nanocrystalline to amorphous morphologies (Song and Haddad, 2010). It is influenced not only by the alloy composition but also by the process parameters (temperature, pressure, deposition angle, deposition rate, average kinetic energy per particle, etc.) and the substrate material. An enhanced surface mobility, e.g. by higher particle energies, can result in a transition from columnar to layer-by-layer growth (Grigucevicienе *et al.*, 2004), thus resulting in coatings with fewer defects such as pinholes. However, different amounts of internal stress may be incorporated as a secondary effect while changing the energy balance (Lee *et al.*, 2003). All this can affect the performance of such coating systems and has to be considered if new innovative Mg alloys with tailored corrosion properties should be designed by PVD processes.

Despite the availability of the knowledge-base on such processing technologies, coatings based on magnesium alloys were rarely studied and the full potential of multi-element systems is still to be exploited. Most of the work concentrated on pure magnesium (Yamamoto *et al.*, 2001a,b), binary systems (Diplas *et al.*, 1999; Mathieu *et al.*, 1999; Adeva-Ramos *et al.*, 2001; Gray and Luan, 2002) and ternary systems (Griguceviciene *et al.*, 2005; Blawert *et al.*, 2007a). Generally these layers are highly textured and can have corrosion properties that make them possible candidates for cathodic corrosion protection of conventional magnesium alloy components (Blawert *et al.*, 2007a). However, it should be noted that the shift of the free corrosion potential compared with conventional Mg alloys is reported for the same systems by some researchers to be positive (Lee *et al.*, 2003; Griguceviciene *et al.*, 2004) and by others to be negative (Mathieu *et al.*, 1999; Bohne *et al.*, 2007). An explanation for this is hard to find and might be related to the processing parameters. Even for magnetron sputtered pure megnesium, different observations were made. Lee *et al.* (2003) found a finer structure of the coating with increasing working pressure (0.13 to 1.3 Pa) and at the same time the corrosion resistance increased with an associated shift of the free corrosion potentials to more noble values. In contrast, Störmer *et al.* (2007) reported that the films produced at low pressure (0.2 Pa) and low deposition angle (0°) are smoother, more compact and exhibit a stronger fibre texture, since the densely packed basal planes of the hexagonal close packed (HCP) structure grow preferentially parallel to the substrate surface. The film growth is generally columnar with voided boundaries, but with increasing pressure and/or angle, the morphology and the microstructure (texture) of the film was more disturbed.

It is further reported that disturbing the layer growth can be used to modify the electrochemical properties of the coatings without changing the composition. All pure Mg coatings exhibited a lower free corrosion potential than cast or extruded magnesium, but variation between −1735 and −1880 mV (vs. Ag/AgCl) is possible depending on the degree of disturbance. The denser and less disturbed films had a better corrosion resistance. The only common finding in the two studies was the fact that the PVD coatings were more corrosion resistant than the target material (Lee *et al.*, 2003; Störmer *et al.*, 2007).

Information on the corrosion properties of such coatings is even more limited. Only a few studies reported some details and moreover the findings are sometimes contradictory, which might be a result of employment of different deposition systems and processing parameters. However, as mentioned above, the sputter deposited magnesium-based coatings can have some interesting corrosion properties, especially showing a shift of the free corrosion potential (Mathieu *et al.*, 1999; Seegar *et al.*, 2005; Blawert *et al.*, 2007a,b; Bohne *et al.*, 2007; Störmer *et al.*, 2007) which can offer cathodic protection to conventional Mg alloys. Alloying is a possible way to influence the shift of the free corrosion potential as well as the overall corrosion resistance. The simplest way is to deposit conventional alloys, and an overview about the observed changes in the corrosion resistance and the free corrosion potential is given in Fig. 4.7. It can be observed that the potentials of the deposited coatings were all negatively shifted relative to the potential of the sputter targets (cast alloys). The improvements in the corrosion resistance are mainly related to the obtained super-saturation preventing the formation of more noble precipitates, which are detrimental for a dense passive film formation. Mostly general corrosion attack without micro-galvanic and/or localised corrosion caused by the second phase precipitates was reported for the PVD coatings (Blawert *et al.*, 2009a).

However, a long-term resistance in neutral or even acidic surroundings requires a more selective alloying. Higher levels of super-saturation are required

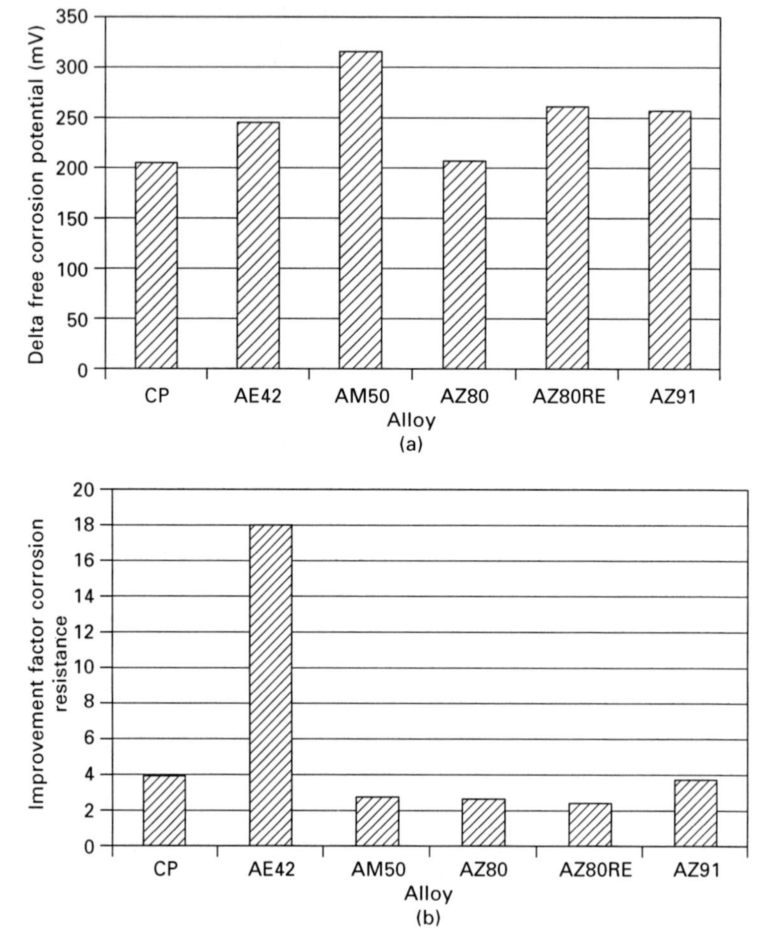

6.7 Difference between the free corrosion potential of cast alloy and coating alloy (a) and improvement factor of the corrosion resistance by sputter deposition of the alloys (b). All properties were determined in 0.5% NaCl solution at pH 11.

and additional alloying elements that improve the resistance especially in acidic surroundings. Every alloying element has a specific effect on the free corrosion potential and the corrosion rate of binary Mg–X sputter coatings. The studied combinations are still limited and positive influences on the corrosion resistance were reported for the systems Mg–Al, Mg–Si, Mg–RE and Mg–Y (Griguceviciene *et al.*, 2005; Blawert *et al.*, 2007a) as well as negative effects for the systems Mg–Ti and Mg–Sn (Mitchel *et al.*, 2005; Blawert *et al.*, 2007a,b; Bohne *et al.*, 2007). Nevertheless, the combination of various alloying elements with their specific corrosion properties allows tailoring of the corrosion performance of the coatings over a wide range. Such an example is the Mg–Al system whose good corrosion resistance can be further improved by adding zirconium without the problem of a negative interaction of aluminium and zirconium. Improved corrosion resistance is reported for the MgAlZr system compared with the binary Mg–Al alloy (Griguceviciene *et al.*, 2005; Blawert *et al.*, 2007a). In this way a large number of new alloys can be designed involving multi-element systems, which may offer further improvements.

The substrate on which the Mg coatings are deposited can have two major influences, which are not independent from each other. On the one hand the substrate can affect the microstructure of the coating and on the other hand the overall corrosion performance is always a function of the coating/substrate system. This is especially a problem for coatings on Mg, as almost all commercial metallic coatings have a more noble corrosion potential than the substrate. Such coatings will have a negative influence on the corrosion resistance of the whole system as galvanic corrosion can cause an enhanced attack of the substrate in the case of remaining defects within the coating. This is demonstrated in Fig. 6.8, which compares the corrosion rates and corrosion potentials measured for the nobler MgTi/AM50 and MgTi/Si and the less noble MgAl/AM50 and MgAl/Si systems with the uncoated AM50 Mg alloy substrate. None of the coatings is free from defects as indicted by the measured mixed potentials. As a consequence, the MgTi coatings with micropores led to an accelerated corrosion of the substrate, indicated by the much higher corrosion rates of the coated AM50 compared with the uncoated AM50. In contrast, Mg–Al alloys showed persistent low corrosion rates on either Si or AM50 substrates despite the presence of similar pores in the coatings. The corrosion rates of the above systems are lower than the corrosion rate of the uncoated AM50 which shows that there is a positive effect due to the coating. This is related to the cathodic protection by MgAl layers. As shown in Fig. 6.8 the substrate is protected by the coating and only the coating is corroding with its inherent lower corrosion rate. However, PVD coatings made from pure magnesium (Tsubakino *et al.*, 2003; Yu and Uan 2006), standard Mg alloys as well as special magnesium-based alloys (Blawert *et al.*, 2007a; Bohne *et al.*, 2007) have the potential to offer cathodic

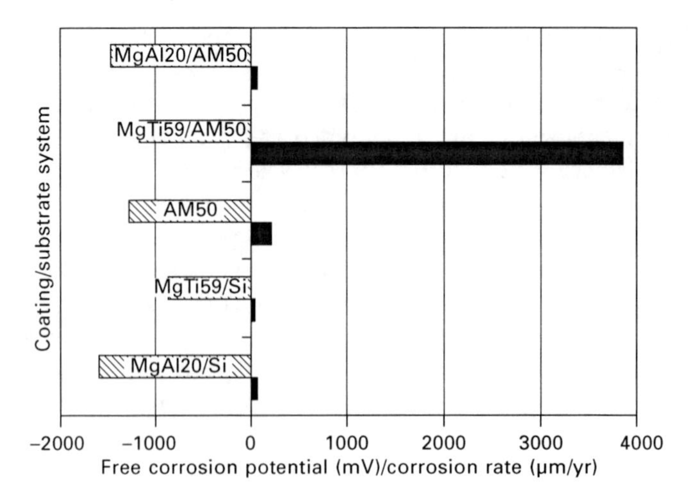

6.8 Influence of inert Si and active AM50 substrate on the corrosion performance of various coating/substrate systems.

protection, thus they can be used to protect cast or wrought Mg components from corrosion. The challenge is to design coatings that have a sufficiently high corrosion resistance in the defect-free condition, a high enough potential difference compared with the standard Mg substrate as the driving force for the galvanic current, a high enough galvanic current to guarantee sufficient cathodic protection and are thick enough to provide the cathodic protection for a reasonable time. Further potential in improving the corrosion performance is expected if glassy-like magnesium-based coatings are produced, which should be available by combining several alloying elements.

6.4 Ion implantation

Ion implantation is another possibility to alloy magnesium with almost any other element independent of the thermodynamic constraints; thus concentrations far from equilibrium can be reached. Whether super-saturation or formation of precipitates is reached depends mainly on the concentration/type of the implanted ions and the treatment temperature. The disadvantage in comparison with other surface treatment techniques is the shallow depth of modification (normally less than 1 μm), thus the effect on the corrosion resistance may be available for only a limited time. A few metal or gas atoms were implanted into magnesium or Mg alloys with different success with regard to the corrosion performance (please note that just a few examples are given and that the selection is far from being complete). Positive effects on the aqueous electrochemical corrosion resistance in chloride-containing solutions are reported from implantation of Al (Lei *et al.*, 2007), Ta (Wang *et al.*, 2007), C/H (Li *et al.*, 2008) and negative effects from Cr (Vilarigues *et al.*, 2008) and O (Wan *et al.*, 2007). The oxidation resistance was reported to be increased by the implantation of Ce (Wang *et al.*, 2008). The results indicated that there is a strong influence of the implanted dose on the corrosion resistance, independent of whether the effect of the implantation is positive or negative for the corrosion resistance (Fig. 6.9). An interesting approach

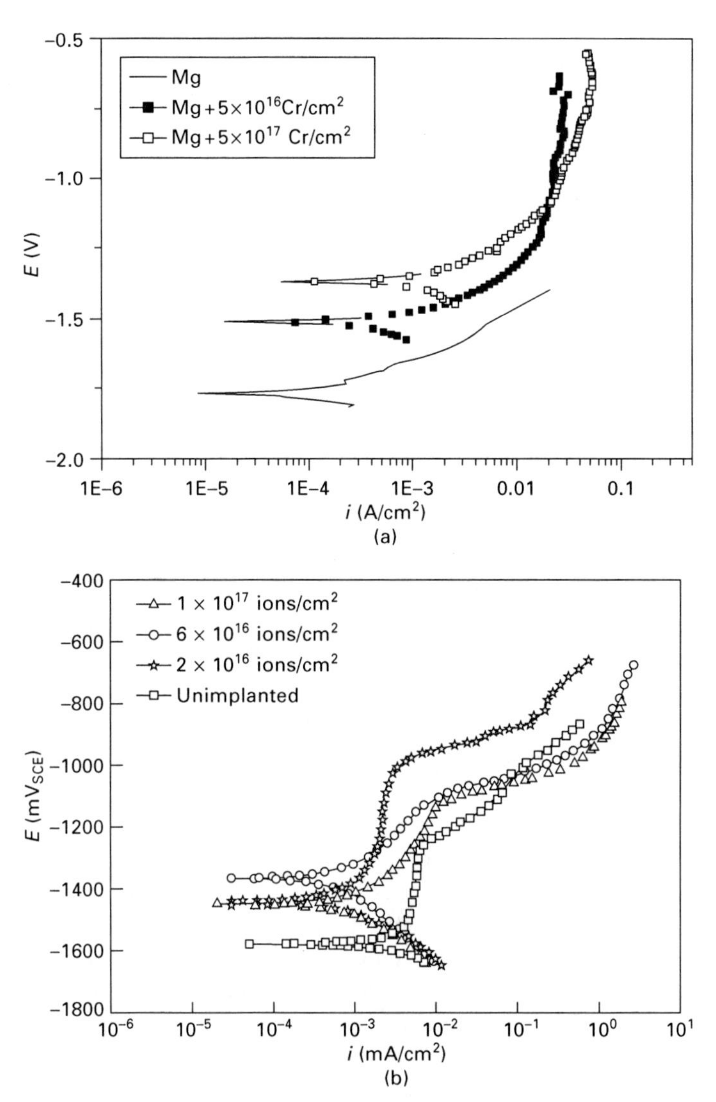

6.9 Influence of the implanted dose on the corrosion performance of Mg alloys: (a) Cr implantation into pure magnesium, polarisation curves determined in 0.5 M Na_2SO_4 solution (Vilarigues *et al.*, 2008) and (b) Al implantation in AZ31, polarisation curves measured in 0.01 M NaCl solution at pH 12 (Lei *et al.*, 2007).

to increase the modified treatment depth is the combination of implantation and coating using hydrocarbon plasma immersion ion implantation. A diamond-like carbon (DLC) film was deposited after carbon implantation by switching the process gas from CH_4 to C_2H_2. The film had a good adhesion to the substrate and very small microporosity was produced with good corrosion resistance (Yekehtaz *et al.*, 2009). However, it should be noted that the improvement is no longer related to alloying of the Mg substrate, but simply to the formation of the DLC film.

Summarising, the ion implantation technique offers interesting possibilities for alloying of magnesium, but with the limitation on the effective corrosion protection due to shallow treatment depths. A solution might be the combination of this with other coating processes, where the implantation is used to modify the structure and composition of the growing film (e.g. to produce denser PVD magnesium-based coatings) to improve the corrosion resistance.

6.5 Laser processed magnesium (Mg) alloys

Laser processing of Mg alloys is performed to obtain surfaces with controlled microstructures and chemical compositions, possessing improved tribological characteristics and corrosion resistance. It can be broadly grouped into three categories, namely, (a) laser surface melting, (b) laser alloying and (c) laser cladding. Techniques (a) and (b) either modify the surface of the alloys only with use of the lasing source without the addition of any materials or involve the use of additional materials in the form of powder, wire or paste to alter the composition of the surface of the alloy. Technique (c), laser cladding, is similar to laser alloying in some respect, but employs noble metal/alloy wire, powder/paste to form overlay coatings (Subramanian *et al.*, 1991; Yue and Li, 2008). The first two categories can be called innovative from the perspective of alloy design at the surface, and some aspects of laser processing are addressed in this section.

The effectiveness of the laser processing is dictated by the energy density, beam dimension, the degree of overlap in tracks and the beam–surface interaction time (or in other words, the scanning speed). Considering the size of the bulk substrate, the depth/thickness of the laser surface-modified layers is generally very small. The large temperature gradients between the surface and the underneath substrate facilitate a rapid quenching and solidification, that could result in microstructural refinement, extension of solubility limits and the formation of non-equilibrium crystalline or amorphous phases depending on the other processing conditions.

The technique of laser surface melting (LSM) is well known and has been in use for many years for modifying the surfaces of both ferrous and non-ferrous substrates (Liu *et al.*, 2006). LSM of the engineering alloys, in general, is aimed at improving the corrosion resistance by modifying the microstructural features and the consequent electrochemical characteristics of the surface. The conditions experienced by the surface of alloys during LSM could be similar or close to those realised in the rapid solidification techniques, namely, melt spinning or splat quenching. While the above-mentioned rapid solidification techniques are for the production of amorphous melt-spun ribbons/bulk alloys, the LSM is intended to alter the conditions of the alloy at the surface only to a shallow depth of a few hundred microns. The surfacing is done as multiple scans with a small degree of overlap. The scan layer width is governed by the dimensions of the laser beam as can be seen in the schematic diagram of the laser surfacing set up in Fig. 6.10 (Majumdar *et al.*, 2003). The electrochemical characteristics of the overlap regions could be influenced by the heat-induced microstructural variations.

Evaporation of magnesium from the surface during LSM leads to the enrichment of the re-melted layers with other major elements in the alloy. For example, in the case of AZ91D and AM60B alloys, the LSM layer was found to contain about 11–12% and 8–9% Al, respectively (Dubé *et al.*, 2001). The enrichment of aluminium at the surface may either increase the ß-phase volume fraction in these regions or may promote an enhanced solid solubility of aluminium in the matrix at the surface, depending on the alloy composition and the laser processing conditions. It is believed that the oxide films formed on the rapidly solidified alloys containing higher amounts of aluminium in solid solution may provide a better passivation behaviour/corrosion resistance. Abbas *et al.* (2005) reported that the corrosion resistance of AZ31, AZ61 and WE43 alloys could be enhanced significantly by LSM

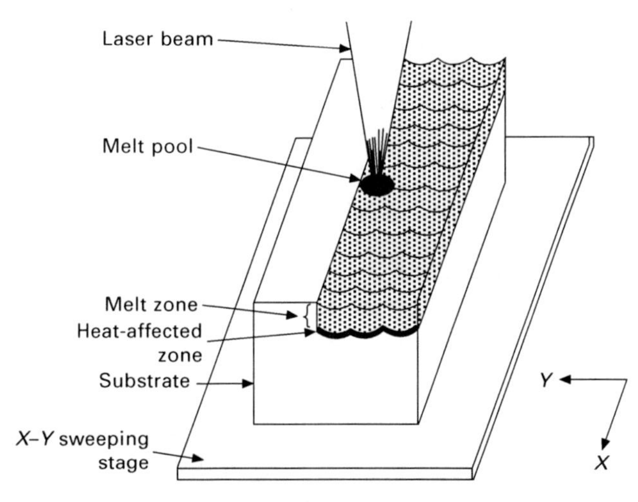

6.10 Schematic representation of laser surface melting (Majumdar *et al.*, 2003).

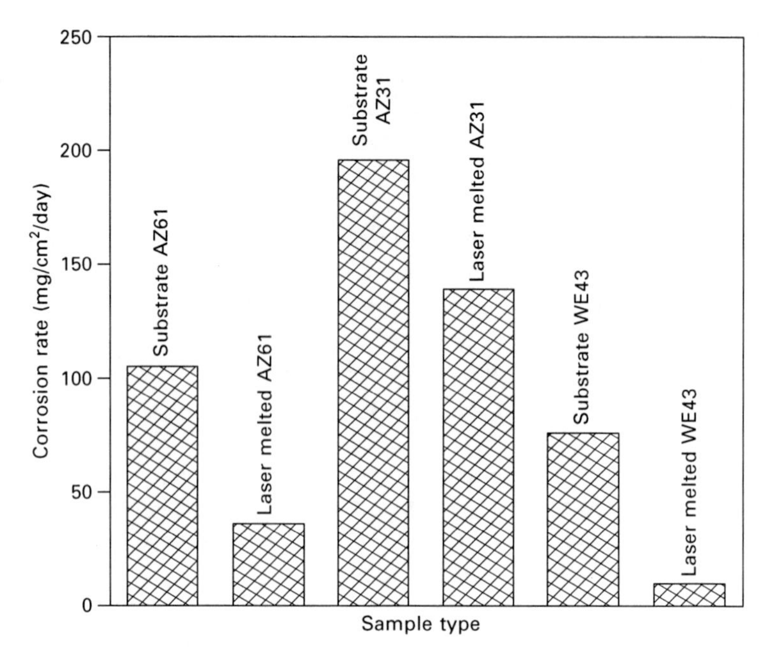

6.11 Average corrosion rates of Mg alloys before and after LSM (immersion tests in 5% NaCl solution of pH 10.5) (Abbas *et al.*, 2005).

(Fig. 6.11), which was attributed to the refinement of grains and the more uniform distribution of secondary phase(s). Similarly, LSM of a commercial Mg alloy containing small amounts of Zn, Mn, Zr and rare earth elements using a continuous wave CO_2 laser with different laser powers and scan speeds resulted in an extremely fine-grained surface (Fig. 6.12), leading to an increased micro-hardness and corrosion resistance (Majumdar *et al.*, 2003). Gao *et al.* (2007)also reported the development of a fine-grained structure by LSM on an AZ91HP alloy using CO_2 laser, and observed a significant enhancement in the corrosion resistance. The corrosion damage on the untreated surface was more uniform and severe, while the LSM surface had small regions of localised damage. The scanning electron micrographs of the untreated and LS melted specimens before and after the corrosion tests are shown in Fig. 6.13. LSM with high laser powers or slow scanning speeds may provide a large case depth. However, as the above conditions are likely to reduce the cooling rates, they would hence be congenial for the formation of more amounts of secondary phase particles, namely, ß-phase, which in turn could greatly influence the corrosion resistance.

Dubé *et al.* (2001) reported that the LSM of AZ91D and AM60 alloys using a pulsed Nd-YAG laser did not improve the corrosion resistance and instead it resulted in an enhanced corrosion rate. A non-uniform dendritic

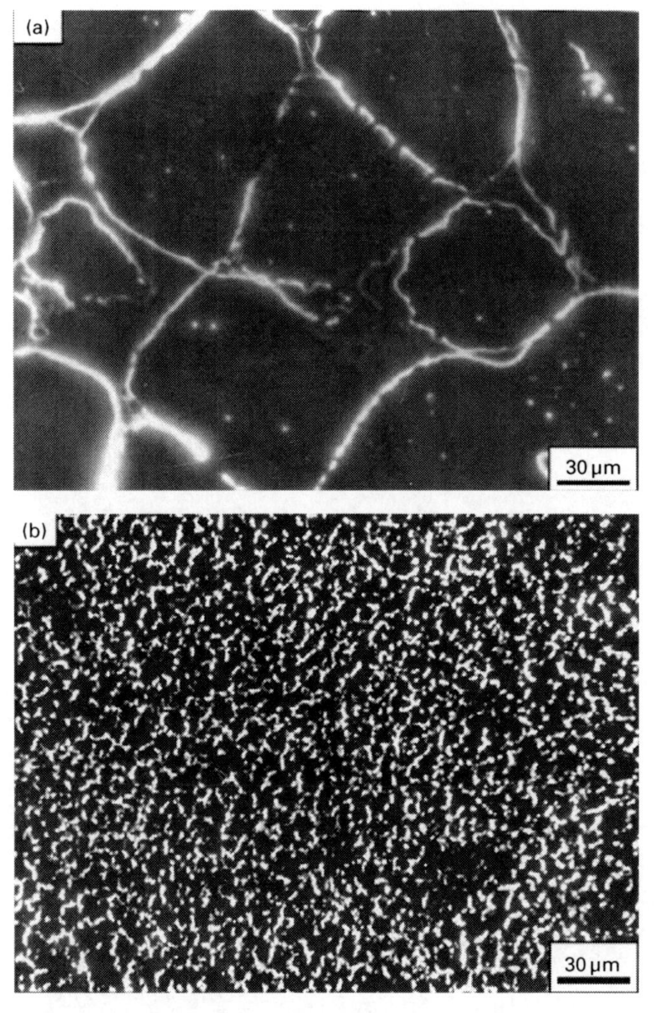

6.12 Scanning electron micrographs of (a) untreated and (b) laser surface melted MEZ alloy specimens (Majumdar *et al.*, 2003).

microstructure and the formation of aluminium-rich curled bands on the treated surface were believed to be the cause for the inferior corrosion behaviour. Elimination of the Al_2Ca phase from the grain boundaries of an ACM720 Mg alloy and also an increase in aluminium concentration at the surface by LSM could increase the corrosion resistance of the alloy appreciably (Mondal *et al.*, 2008).

Short pulse irradiation of Mg alloys using XeCl excimer laser produced a wavy structure on the surface and the development of about 3–6 μm thick amorphous layers on the surface. The corrosion resistance was reported to be enhanced (Schippman *et al.*, 1999). Control of laser scanning speed plays an important role in providing the necessary heat input for the dissolution of secondary phases in the alloys subjected to LSM and to obtain a homogenised surface. However, excessive heat may promote formation of secondary phases again, due to slow cooling, and also may increase the grain size. The effect of scanning speed on the dissolution of $Mg_{12}Nd$ phase in a WE43 Mg alloy treated by excimer laser is shown in Fig. 6.14. The effective dissolution of this cathodic phase was found to improve the corrosion resistance of the alloy

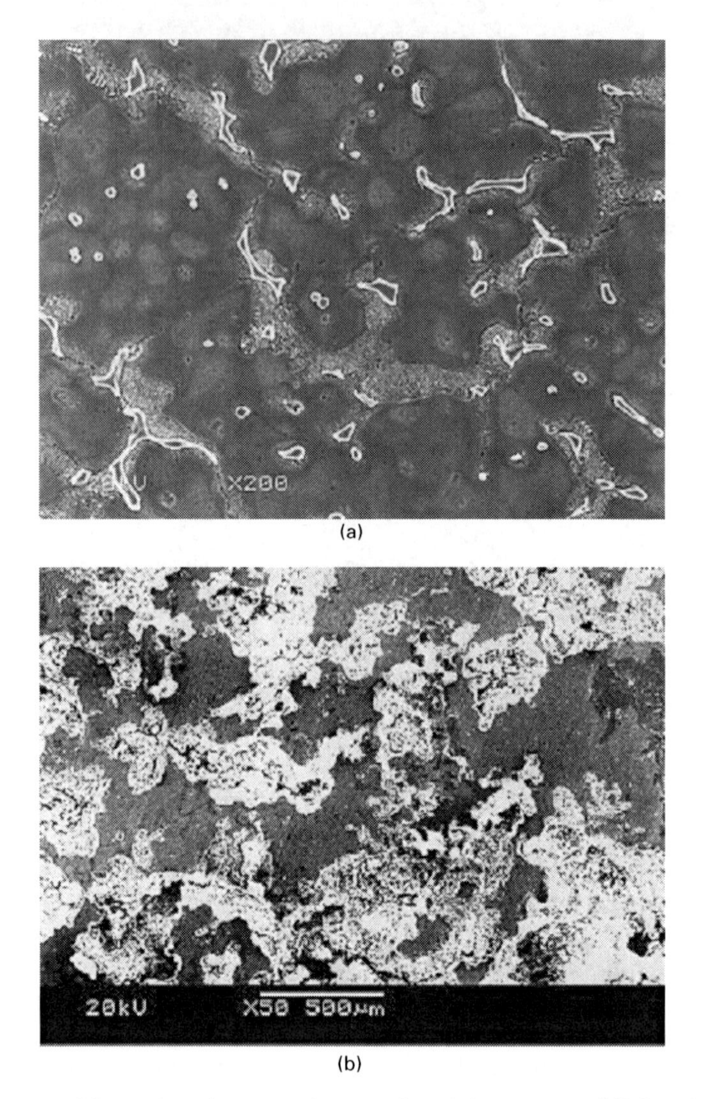

6.13 Scanning electron micrographs of the untreated ((a) and (b)) and laser surface melted ((c) and (d)) AZ91HP alloy before and after the corrosion tests (Gao *et al.*, 2007).

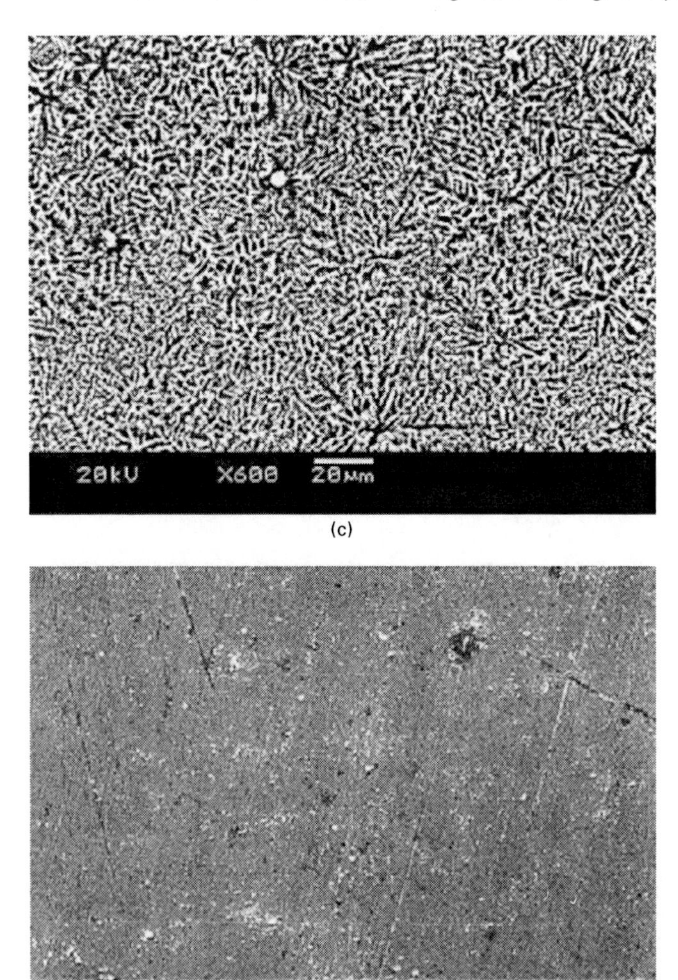

(c)

(d)

6.13 Continued

significantly. A lower scan speed of 2 mm/s yielded a surface that possessed a better corrosion resistance than the surface obtained with 10 mm/s (Fig. 6.15) (Guo *et al.*, 2005). The number of pulses in the excimer LSM seems to be critical in dictating the thickness and quality of re-melted layers. Coy *et al.* (2010) showed that an increase in the number of pulses per unit area from 10 to 50 could increase the thickness of the re-melted layer from 4 to 13 μm (Fig. 6.16). However, the LSM at higher pulse conditions lead to the formation of pores and micro-cracks, and thus result in poor corrosion

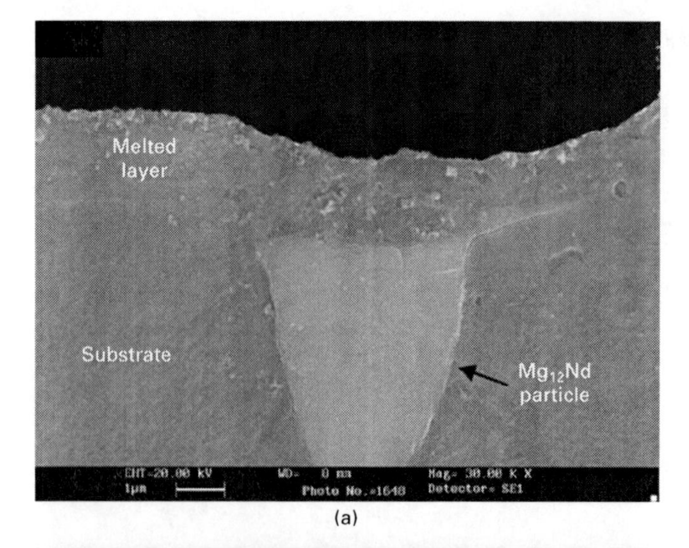

(a)

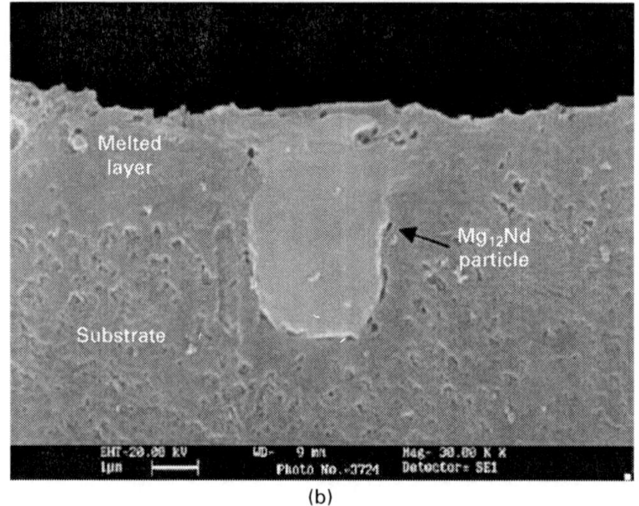

(b)

6.14 Cross-sections of laser-treated WE43 Mg alloy specimens at two different laser scanning speeds: (a) 2 mm/s; (b) 10 mm/s, with a constant energy density (6 J/cm^2) (Guo *et al.*, 2005).

resistance of the treated surface. Excimer LSM has successfully been applied to improve the corrosion resistance of magnesium-based metal matrix composites as well (Yue *et al.*, 1997).

Wang *et al.* (1993) showed the feasibility of alloying/cladding magnesium- and aluminium-rich Mg alloys by using a CO_2 laser facility. The formation of eutectic phase at the surface could completely be suppressed in an alloy with a composition $Mg_{27}Al_{73}$ by the rapid cooling rates of the laser processing.

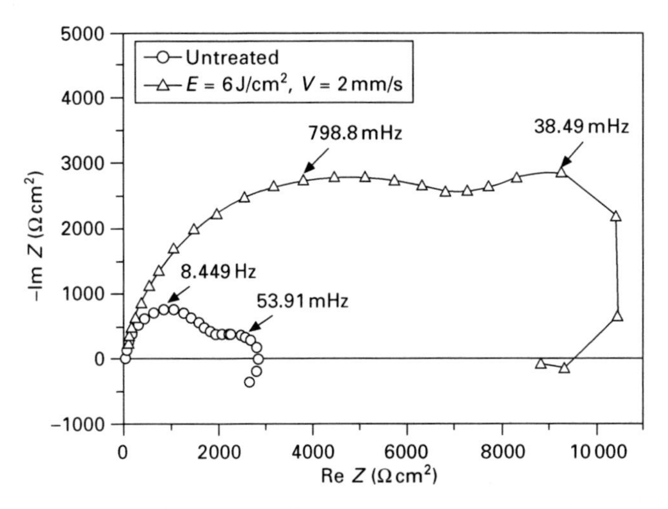

6.15 Impedance spectra (Nyquist plot) of the untreated and laser-treated WE43 alloy in 3.5% NaCl solution (Guo *et al.*, 2005).

In the aluminium-lean alloys the formation of eutectic phase was noticed throughout the modified layer. The laser-clad alloy $Mg_{27}Al_{73}$ showed a nobler potential and a better corrosion resistance than the cast AZ91B and cast magnesium (Fig. 6.17). The feasibility of cladding and alloying of WE43 and ZE41 Mg alloys with aluminium injection using Nd–YAG laser was also demonstrated by Ignat *et al.* (2004), in which layers with $Mg_{17}Al_{12}$ with a hardness of around 200 HV could be achieved. However, the modified layers contained defects in the form on pores and cracks. Cladding of the AZ91HP Mg alloy with an Al–Si alloy using CO_2 laser could constitute a clad layer, bond zone and heat-affected zone (Fig. 6.18(a)). The analysis of the surface using scanning electron microscopy/energy dispersive spectroscopy (SEM/EDS) showed that the surface was constituted with Mg_2Si dendrites in a matrix of $Mg_{17}Al_{12}$ (Fig. 6.18b). Superior wear resistance and corrosion behaviour were claimed on account of the modified phase composition of these alloys, and especially the formation of the $Mg_{17}Al_{12}$ at the surface (Gao *et al.*, 2006). Higher laser scanning speeds during cladding of Al–Si over ZE41 magnesium alloy could result in smaller volume fraction of Mg_2Si phase (Volovitch *et al.*, 2008) and hence reported to provide better corrosion resistance.

The existing knowledge-base on the laser processing of Mg alloys for improving the corrosion resistance suggests there is ample scope for expanding it further. In particular, the processing with use of aluminium powders to enrich the surface composition to obtain modified structures for imparting wear and corrosion resistance looks promising. However, tackling issues such as the formation micro-cracks during laser processing

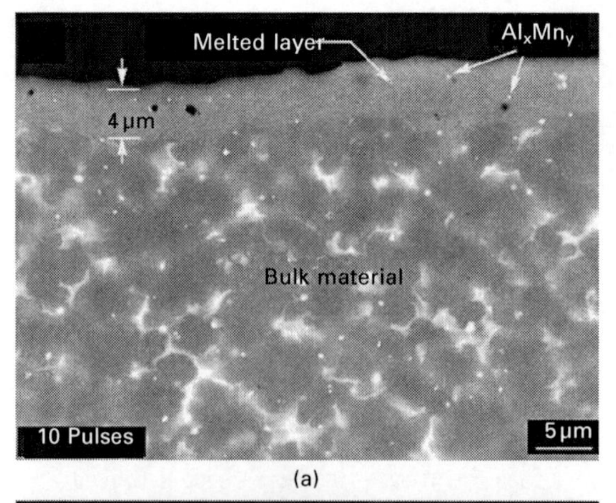

(a)

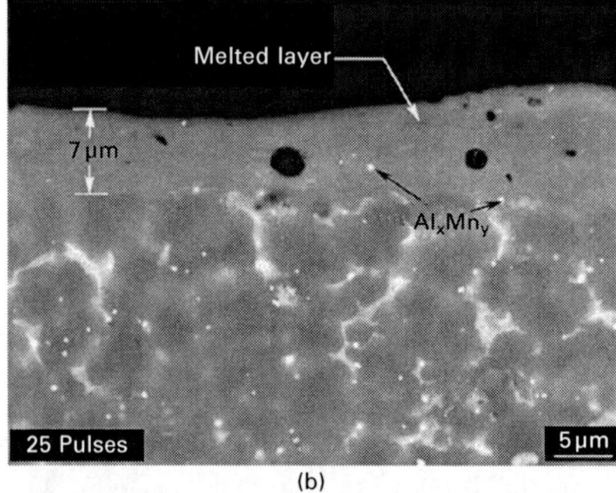

(b)

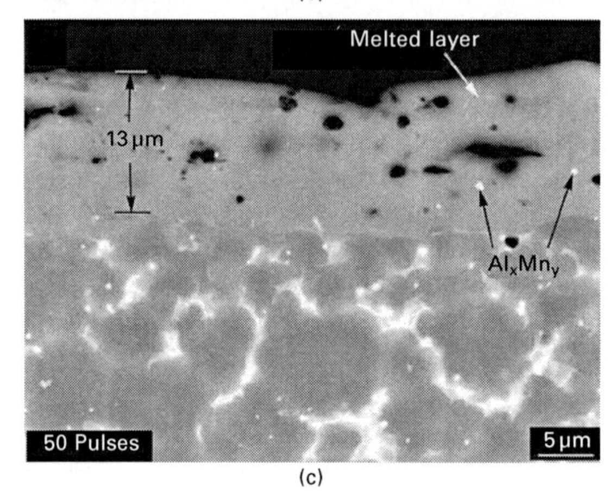

(c)

6.16 Back-scattered scanning electron micrographs of cross-section of the specimens after laser treatment with different number of laser pulses: (a) 10, (b) 25 and (c) 50 pulses (Coy *et al.*, 2010).

is still a challenging task. The laser treatment of magnesium alloy surfaces for long-term corrosion protection is another important aspect that needs to be addressed in the days to come. Finally, it should be mentioned that the laser can also be replaced by other higher energy beams, namely, electron beam, to produce similar effects on the Mg alloys.

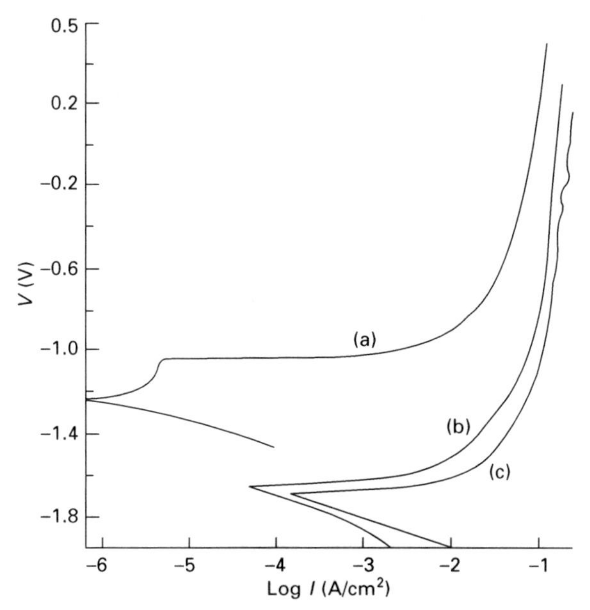

6.17 Polarisation behaviour of (a) laser-clad $Mg_{27}Al_{73}$ alloy, (b) AZ91B alloy and (c) pure magnesium (Wang *et al.*, 1993).

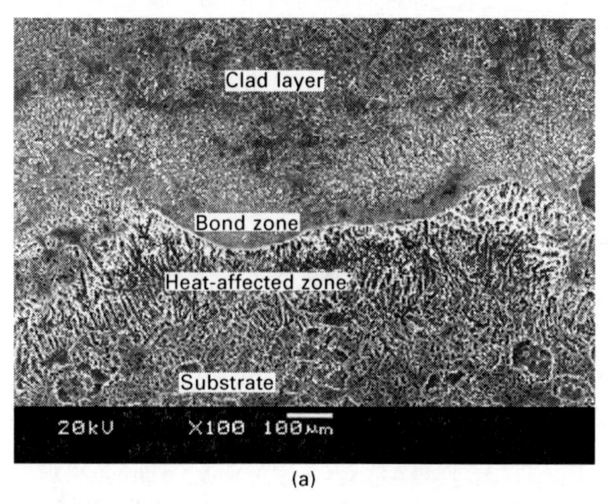

(a)

6.18 Scanning electron macro/micrographs of the laser-clad AZ91HP alloy: (a) cross-section; (b) surface morphology (Gao *et al.*, 2006).

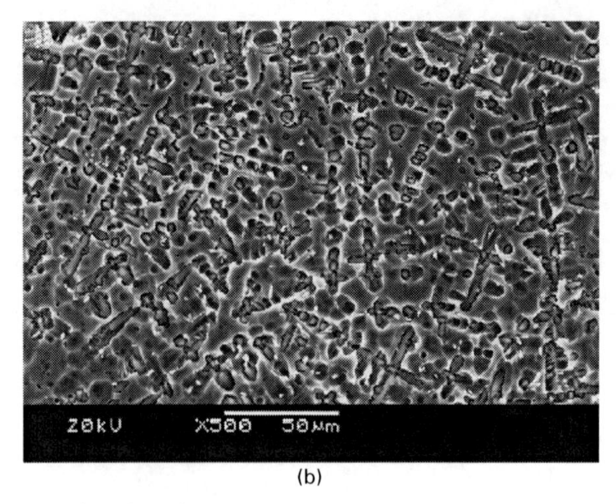

(b)

6.18 Continued

6.6 References

Abbas G, Liu Z, Skeldon P (2005), 'Corrosion behaviour of laser-melted magnesium alloys', *Applied Surface Science*, **247**, 347–353.

Adeva-Ramos P, Dodd SB, Morgan P, Hehmann F, Steinmetz P, Sommer F, (2001), 'Autopassive wrought magnesium alloys', *Advanced Engineering Materials*, **3**, 147–152.

Anik M, (2010), 'Electrochemical hydrogen storage capacities of Mg_2Ni and MgNi alloys synthesized by mechanical alloying', *Journal of Alloys and Compounds*, **491**, 565–570.

Blawert C, Morales GED, Dietzel W, Kainer KU, Scharf C, Ditze A, (2006), 'Influence of the copper content on microstructure and corrosion resistance of AZ91 based secondary magnesium alloys', SAE paper 2006–01–0254.

Blawert C, Heitmann V, Dietzel W, Stoermer M, Bohne Y, Mändl S, Rauschenbach B, (2007a), 'Corrosion properties of supersaturated magnesium alloy systems', *Materials Science Forum*, **539–543**, 1679–1684.

Blawert C, Heitmann V, Dietzel W, Störmer M, Bohne Y, Gerlach JW, Manowa D, Mändl S, (2007b), 'Design and deposition of corrosion protective supersaturated magnesium alloy systems by pvd processes', *Proceedings of the Light Metals Technology Conference 2007*, Saint-Sauveur (CDN), 24–26 September 2007, 135–139.

Blawert C, Morales GED, Kainer KU, Scharf C, Živanović P, Ditze A, (2008), 'Development of a new AZ based secondary magnesium alloy', *Proceedings 65th Annual World Magnesium Conference*, Warsaw, Poland, 18–20 May 2008, 57–64.

Blawert C, Heitmann V, Scharnagl N, Störmer M, Lutz J, Prager-Duschke A, Manova D, Mändl S, (2009a), Different underlying corrosion mechanism for Mg bulk alloys and Mg thin films, *Plasma Processes and Polymers*, **6**, S690–S694.

Blawert C, Kainer KU, Dietzel W, Ditze A, Scharf C, Živanović P, (2009b), Duktile Magnesiumlegierung, DE 10 2008 020 523 A1, 2008; EP 2 116 622 A1.

Bohne Y, Manova D, Blawert C, Störmer M, Dietzel W, Mändl S, (2007), 'Deposition and properties of novel microcrystalline Mg alloy coatings', *Surface Engineering*, **23**, 5, 339–343.

Coy AE, Viejo F, Garcia GFJ, Liu Z, Skeldon P, Thompson GE, (2010), 'Effect of excimer laser surface melting on the microstructure and corrosion performance of the die cast AZ91D magnesium alloy', *Corrosion Science*, **52**, 387–397.

Diplas S, Tsakiropoulos P, Brydson RMD, (1999), 'Development of physical vapour deposited Mg–Zr alloys. Part 3 – Comparison of alloying and corrosion behaviour in Mg–V and Mg–Zr physical vapour deposited alloys', *Materials Science and Technology*, **15**, 1373–1378.

Ditze A, Scharf C, (2004), 'Destillationsprozesse zur Gewinnung von Magnesiummetall und zur Aufbereitung von Magnesiumschrott', *World of Metallurgy– Erzmetall.*, **57**, 251–257.

Ditze A, Scharf C, (2008), *Recycling of Magnesium*, Papierflieger Verlag, Clausthal-Zellerfeld, Germany, 2008.

Ditze A, Scharf C, Blawert C, Kainer KU, Morales GED, (2005), Magnesiumsekundärlegierung, DE 10 2005 033835 A1, 2005.

Ditze A, Scharf C, Blawert C, Kainer KU, Morales GED, (2006) Magnesium Alloy, WO 2007/009435 A1, 2006.

Dubé D, Fiset M, Couture A, Nakatsugawa I, (2001),'Characterization and performance of laser melted AZ91D and AM60B', *Materials Science and Engineering a – Structural Materials Properties Microstructure and Processing*, **299**, 38–45.

Ehrenberger SI, Schmid SA, Friedrich HE, (2008), Magnesium production and automotive applications: Life-cycle analysis focussing on green house gases, 16th Magnesium Automotive and User Seminar, Europäische Forschungsgemeinschaft Magnesium, 2008, Aalen, Germany.

Feng Y, Jiao L, Yuan H, Zhao M, (2007), 'Study on the preparation and electrochemical characteristics of MgNi–CoB alloys', *Journal of Alloys and Compounds*, **440**, 304–308.

Gao Y, Wang, C, Lin Q, Liu H, Yao M, (2006), 'Broad-beam laser cladding of Al–Si alloy coating on AZ91HP magnesium alloy', *Surface and Coatings Technology*, **201**, 2701–2706.

Gao Y, Wang C, Yao M, Liu H, (2007), 'Corrosion behavior of laser melted AZ91HP magnesium alloy', *Materials and Corrosion*, **58**, 463–466.

Gebert A, Wolff U, John A, Eckert J, Schultz L, (2001), 'Stability of the bulk glass-forming $Mg_{65}Cu_{25}Y_{10}$ alloy in aqueous electrolytes', *Materials Science and Engineering*, **A299**, 125–135.

Gebert A, Subba Rao RV, Wolff U, Baunack S, Eckert J, Schultz L, (2004), 'Corrosion behavoiur of $Mg_{65}Y_{10}Cu_{15}Ag_{10}$ bulk metallic glass', *Materials Science and Engineering*, **A375–377**, 280–284.

Gray JE, Luan B, (2002), 'Protective coatings on magnesium and its alloys – a critical review', *Journal of Alloys and Compounds*, **336**, 88–113.

Griguceviciene A, Leinartas K, Juskenas R, Juzeliunas E, (2004), 'Voltammetric and structural characterization of sputter deposited Al-Mg films', *Journal of Electroanalytical Chemistry*, **565**, 203–209.

Griguceviciene A, Leinartas K, Juskenas R, Juzeliunas E, (2005), 'Structure and initial corrosion resistance of sputter deposited nano-crystalline Mg–Al–Zr alloys', *Materials Science and Engineering*, **A 394**, 411–416.

Guo LF, Yue TM, Man HC, (2005), 'Excimer laser surface treatment of magnesium alloy WE43 for corrosion resistance improvement', *Journal of Materials Science*, **40**, 3531–3533.

Han SC, Jiang, JJ, Park JG, Jang, KJ, Chin YE, Lee YJ, (1999), 'The electrochemical evaluation of ball milled MgNi-based hydrogen storage alloys', *Journal of Alloys and Compounds*, **285**, L8–L11.

Huang H, Huang K, Liu S, Chen D, (2010), 'Microstructure and electrochemical properties of $Mg_{0.9}Ti_{0.1}Ni_{1-x}M_x$ (M=Co, Mn; x = 0, 0.1, 0.2) hydrogen storage alloys', *Powder Technology*, **198**, 144–148.

Ignat S, Sallamand P, Grevey D, Lambertin M, (2004), 'Magnesium alloys laser (Nd:YAG) cladding and alloying with side injection of aluminium powder', *Applied Surface Science*, **225**, 124–134.

Inoue A, Masumoto T, (1991), 'Production and properties of light-metal-based amorphous alloys', *Materials Science and Engineering*, **A133**, 6–9.

Inoue A, Kato A, Zhang T, Kim SG, Masumoto T, (1991), 'Mg–Cu–Y amorphous alloys with high mechanical strengths produced by a metallic mold casting method', *Materials Transactions, JIM*, **32**, 609–616.

Inoue A, Nakamura T, Nishiyama N, Masumoto T, (1992), 'Mg–Cu–Y bulk amorphous alloys with high tensile strength produced by a high-pressure die casting method', *Materials Transactions, JIM*, **33**, 937–945.

Kang HG, Park ES, Kim WT, Kim DH, Cho HK, (2000), 'Fabrication of bulk Mg–Cu–Ag–Y glassy alloy by squeeze casting', *Materials Transactions, JIM*, **41**, 846–849.

Kim SG, Inoue A, Masumoto T, (1990), 'High mechanical strengths of Mg–Ni–Y and Mg–Cu–Y amorphous alloys with significant super cooled liquid region', *Materials Transactions, JIM*, **31**, 929–934.

Lee MH, Bae IY, Kim KJ, Moon KM, Oki T, (2003), 'Formation mechanism of new corrosion resistance magnesium thin films by PVD method', *Surface and Coatings Technology*, **169–170**, 670–674.

Lee MH, Moon KM, Kim KJ, Bae Y, Baek SM, (2008), 'Influence of inter-layers on corrosion resistance of ion-plated Mg thin films', *Surface and Coatings Technology*, **202**, 5603–5606.

Lei MK, Li P, Yang HG, Zhu XM, (2007), 'Wear and corrosion resistance of Al ion implanted AZ31 magnesium alloy', *Surface and Coatings Technology*, **201**, 5182–5185.

Lenain C, Aymard L Dupont L, Tarascon JM, (1999), 'A new $Mg_{0.9}Y_{0.1}Ni$ hydride forming composition obtained by mechanical grinding', *Journal of Alloys and Compounds*, **299**, 84–89.

Li P, Han XG, Xin JP, Zhu XP, Lei MK, (2008), 'Wear and corrosion resistance of AZ31 magnesium alloy irradiated by high-intensity pulsed ion beam', *Nuclear Instruments and Methods in Physics Research*, **B266**, 3945–3952.

Li Y, Ng SC, Ong CK, (1995), 'New amorphous alloys with high strength and good bend ductility in the Mg–Ni–Nd system', *Journal of Materials Processing Technology*, **48**, 489–493.

Li Y, Ng SC, Kam CH, (1998), 'Dissolution of Mg-based Mg–Ni–Nd amorphous alloys in 3% NaCl solution', *Materials Letters*, **36**, 214–217.

Liu Z, Chong PH, Skeldon P, Hilton PA, Spencer JT, Quayle B, (2006), 'Fundamental understanding of the corrosion performance of laser-melted metallic alloys', *Surface and Coatings Technology*, **200**, 5514–5525.

Majumdar JD, Galun R, Mordike BL, Manna I, (2003), 'Effect of laser surface melting on corrosion and wear resistance of a commercial magnesium alloy', *Materials Science and Engineering*, **A361**, 119–129.

Mathieu S, Hazan J, Rapin C, Steinmetz P, (1999), 'Corrosion of cast and non equilibrium magnesium alloys', *Proceedings COM1999, Environmental Degradation of Materials and Corrosion Control in Metals*, Ed. by M. Elboujdaini and E. Ghali, MetSoc, 69–83.

Mitchell T, Diplas S, Tsakiropoulos P, (2005), 'Characterisation of corrosion products formed on PVD in-situ mechanically worked Mg–Ti alloys', *Journal of Alloys and Compounds*, **392**, 127–141.

Mitsuo A, Aizawa T, (2003), 'Cold coating of magnesium base alloy films by ion beam sputtering', *Materials Science Forum*, **419–422**, 927–930.

Mondal AK, Kumar S, Blawert C, Dahotre NB, (2008), 'Effect of laser surface treatment on corrosion and wear resistance of ACM720 Mg alloy', *Surface and Coatings Technology*, **202**, 3187–3198.

Nastasi M, Mayer JW, Hirvonen JK, (1996), *Ion-Solid Interactions: Fundamentals and Applications*, Cambridge University Press, Cambridge.

Nohara S, Hamasaki K, Zhang SG, Inoue H, Iwakura C, (1998), 'Electrochemical characteristics of an amorphous $Mg_{0.9}V_{0.1}Ni$ alloy prepared by mechanical alloying', *Journal of Alloys and Compounds*, **280**, 104–106.

Okonska I, Jurczyk M, (2009), 'Electrochemical properties of an amorphous 2Mg+ 3d alloys doped by nickel atoms (3d = Fe, Co, Ni, Cu)', *Journal of Alloys and Compounds*, **475**, 289–293.

Pérez P, Garcés G, Adeva P, (2004), 'Mechanical behaviour of amorphous Mg–23.5Ni ribbons', *Journal of Alloys and Compounds*, **381**, 114–123.

Qin FX, Bae GT, Dan ZH, Lee H, Kim NJ, (2007), 'Corrosion behaviour of the $Mg_{65}Cu_{25}Gd_{10}$ bulk amorphous alloys', *Materials Science and Engineering*, **A449–451**, 636–639.

Scharf C, Ditze A, Shkurankov A, Morales GED, Blawert C, Dietzel W, Kainer KU, (2005), 'Corrosion of AZ 91 secondary magnesium alloy', *Advanced Engineering Materials*, **7**, 1134–1142.

Scharf C, Živanović P, Ditze A, Horny K, Franke G, Blawert C, Kainer KU, Morales GED, (2007), 'Untersuchungen zum Einsatz einer neuen Magnesium-Sekundärlegierung in der industriellen Praxis', *Gießerei*, **94** (11), 38–50.

Schippman D, Weisheit A, Mordike BL, (1999), 'Short pulse irradiation of magnesium based alloys to improve surface properties', *Surface Engineering*, **15**, 23–26.

Seeger DM, Blawert C, Dietzel W, Bohne Y, Mändl S, Rauschenbach B, (2005), 'Comparison of as-cast and plasma deposited commercial magnesium alloys', Powell, *Magnesium Technology 2005*, Ed by N.R. Neelameggham, H.I. Kaplan, B.R. TMS 2005, 134th Annual Meeting & Exhibition. San Francisco, CA, 13–17 February 2005, 323–328.

Song, Haddad, (2010), Private communication.

Störmer M, Blawert C, Hagen H, Heitmann V, Dietzel W, (2007), 'Structure and corrosion of magnetron sputtered pure Mg films on silicon substrates', *Plasma Processes and Polymers*, **4**, S557–S561.

Subramanian R, Sircar S, Mazumder J, (1991), 'Laser cladding of zirconium on magnesium for improved corrosion properties', *Journal of Materials Science*, **26**, 951–956.

Thornton JA, (1974), 'Influence of apparatus geometry and deposition conditions on the structure and topography of thick sputtered coatings', *Journal of Vacuum Science and Technology*, **11**, 666–670.

Tian Q, Zhang Y, Tan Z, Xu F, Sun L, Zhang T, Yuan H, (2006), 'Effects of Pd substitution for Ni on corrosion performances of $Mg_{0.9}Ti_{0.1}Ni_{1-x}Pd_x$ hydrogen storage alloys', *Transactions of Nonferrous Metals Society of China*, **16**, 497–501.

Tsubakino H, Yamamoto A, Sugahara K, Fukumoto S, (2003), 'Corrosion resistance in magnesium alloys and deposition coated magnesium alloy', *Materials Science Forum*, **419–422**, 915–920.

Uhlenhaut DI, Furrera A, Uggowitzera PJ, Löffler JF, (2009), 'Corrosion properties of glassy $Mg_{70}Al_{15}Ga_{15}$ in 0.1 M NaCl solution', *Intermetallics*, **17**, 811–817.

Vilarigues M, Fernandes JCS, Alves LC, da Silva RC, (2008), 'Electrochemical behaviour of chromium-implanted magnesium in hydroxide, chloride and sulphate solutions', *Surface and Coatings Technology*, **202**, 4086–4093.

Volovitch P, Masse JE, Fabre A, Barrallier L, Saikaly W, (2008), 'Microstructure and corrosion resistance of magnesium alloy ZE41 with laser surface cladding by Al–Si powder', *Surface and Coatings Technology,* **202**, 4901–4914.

Wan GJ, Maitz MF, Sun H, Li PP, Huang N, (2007), 'Corrosion properties of oxygen plasma immersion ion implantation treated magnesium', *Surface and Coatings Technology*, **201**, 8267–8272.

Wang AA, Sircar S, Mazumder J, (1993), 'Laser cladding of Mg–Al alloys', *Journal of Materials Science,* **28**, 5113–5122.

Wang X, Zeng X, Wu G, Yao S, Lai Y, (2007), 'Effects of tantalum ion implantation on the corrosion behaviour of AZ31 magnesium alloy', *Journal of Alloys and Compounds*, **437**, 87–92.

Wang XM, Zeng XQ, Wu GS, Yao SS, Lai Y J, (2008), 'The effects of cerium implantation on the oxidation behaviour of AZ31 magnesium alloys', *Journal of Alloys and Compounds*, **456**, 384–389.

Xiao X, Wang X, Gao L, Wang W, Chen C, (2006), 'Electrochemical properties of amorphous Mg–Fe alloys mixed with Ni prepared by ball-milling', *Journal of Alloys and Compounds*, **413**, 312–318.

Xie DH, Li P, Zeng CX, Sun JW, Qu XH, (2009), 'Effect of substitution of Nd for Mg on the hydrogen storage properties of Mg_2Ni alloy', *Journal of Alloys and Compounds*, **478**, 96–102.

Xue J, Li G, Hu Y, Du J, Wang C, Hu G, (2000), 'Electrochemical characteristics of Al-substituted Mg_2Ni As negative electrode', *Journal of Alloys and Compounds*, **307**, 204–244.

Yamamoto A, Watanabe A, Sugahara K, Fukumoto S, Tsubakino H, (2001a) 'Applying a vapor deposition technique to improve corrosion resistance in magnesium alloys' in *Proceedings of the Second International Conference on Environment Sensitive Cracking and Corrosion Damage*, Ed by M. Matsumura, H. Nagano, K. Nakasa and Y. Isomoto, Nishiki Printing Ltd, ESCCD Hiroshima, Japan, 160–167.

Yamamoto A, Watanabe A, Sugahara K, Tsubakino H, Fukumoto S, (2001b), 'Improvement of corrosion resistance of magnesium alloys by vapour deposition', *Scripta Materialia*, **44**, 1039–1042.

Yao HB, Li Y, Wee ATS, Chai JW, Pan JS, (2001a), 'The alloying effect of Ni on the corrosion behavior of melt-spun Mg–Ni ribbons', *Electrochimica Acta*, **46**, 2649–2657.

Yao, HB, Li A, Wee ATS, Pan JS, Chai JW, (2001b), 'Correlation between the corrosion behavior and corrosion films formed on the surfaces of $Mg_{82-x}Ni_{18}Nd_x$ ($x = 0, 5, 15$) amorphous alloys', *Applied Surface Science*, **173**, 54–61.

Yao HB, Li Y, Wee ATS, (2003), 'Corrosion behaviour of melt-spun $Mg_{65}Ni_{20}Nd_{15}$ and $Mg_{65}Cu_{25}Y_{10}$ metallic glasses', *Electrochimica Acta*, **48**, 2641–2650.

Yekehtaz M, Baba K, Hatada R, Flege S, Sittner F, Ensinger W, (2009), 'Corrosion resistance of magnesium treated by hydrocarbon plasma immersion ion implantation', *Nuclear Instruments and Methods in Physics Research*, **B267**, 1666–1669.

Yu BL, Uan JY, (2006), 'Sacrificial Mg film anode for cathodic protection of die cast Mg–9 wt.%Al–1 wt.%Zn alloy in NaCl aqueous solution', *Scripta Materialia*, **54**, 1253–1257.

Yuan Q, Song H, Wang Y, Liu J, (2003), 'Electrochemical characteristics of Mg_2Ni-

type alloys prepared by mechanical alloying', *Journal of Alloys and Compounds*, **353**, 322–326.

Yue TM, Wang AH, Man HC, (1997), 'Improvement in the corrosion resistance of magnesium ZK60/SiC composite by excimer laser surface treatment', *Scripta Materialia*, **38**, 191–198.

Yue TM, Li T, (2008), 'Laser cladding of Ni/Cu/Al functionally graded coating on magnesium substrate', *Surface and Coatings Technology*, **202**, 3043–3049.

Zhang Y, Liao B, Chen L, Lei YQ, Wang QD, (2001), 'The effect of Ni content on the electrochemical and surface characteristics of $Mg_{90-x}Ti_{10}Ni_x$ (x = 50, 55, 60) ternary hydrogen storage electrode alloys', *Journal of Alloys and Compounds*, **327**, 195–200.

Part III

Environmental influences

7
Atmospheric corrosion of magnesium (Mg) alloys

M. JÖNSSON and D. PERSSON, SWEREA KIMAB, Sweden

Abstract: Depending on the atmospheric condition, and the deposition of particles and gases, this thin aqueous layer varies in thickness. In marine atmospheres and other chloride containing environments, corrosion attacks are triggered by the formation of chloride-containing aqueous films on the surface. Gaseous constituents, such as CO_2 are more important in atmospheric conditions, compared to the corrosion of magnesium alloys in bulk solutions. The microstructure of magnesium alloys is also of vital importance to the corrosion behaviour in atmospheric conditions. However, in atmospheric conditions the electrolyte layer is thin, which decreases the possibility of galvanic coupling of alloy constituents located at larger distances from each other.

Key words: magnesium, atmospheric corrosion, field test, corrosion products, microstructure.

7.1 Introduction

The atmospheric corrosion of magnesium (Mg) alloys is of considerable interest in several technically important areas, such as the automotive and aerospace industries and in portable electronics. While the corrosion of Mg alloys in bulk solutions has been the subject of investigation in numerous academic papers, studies of atmospheric corrosion of Mg alloys are rarer, even though magnesium as a structural metal is most commonly used in atmospheric environment. Magnesium has very low nobility and a reputation for ready corrosion; nevertheless the corrosion rate of bare Mg panels is in both marine and mobile field exposures considerably lower than the corrosion rate of carbon steel. Atmospheric corrosion is triggered by atmospheric humidity, which forms a thin layer of liquid on the surface. Depending on the atmospheric condition and the different particles and gases, this thin water layer has different thickness. Owing to poor conductivity of the thin electrolyte layer present during atmospheric corrosion, the microstructure has a different influence on the corrosion properties of Mg alloys from that in bulk solutions. The aim of this chapter is to look more closely at the fundamental processes that govern the atmospheric corrosion behaviour of Mg alloys.

7.2 The atmospheric environment

The atmospheric corrosion of Mg alloys is a complex process which results from the interaction between a metal and its atmospheric environment. A prerequisite for atmospheric corrosion is the presence of a water layer on the surface. The thickness of the water layer varies considerably with the climatic conditions and may range from monomolecular thickness to clearly visible water films. The formation of an aqueous layer occurs in humid air by adsorption on the hydroxylated oxide present on most metal surfaces exposed to ambient conditions. The thickness of the reversible adsorbed water film varies with the relative humidity (RH). Table 7.1 shows the approximate number of water monolayers on a metal surface at 25 °C and steady state conditions (1). Thicker aqueous films can also form in the atmospheric environment by condensation, precipitation or water absorption by hygroscopic substances on the surface.

The thin aqueous layer on a metal surface in atmospheric environments acts as a medium for atmospheric gaseous constituents or particles that occur in the surrounding atmosphere and interact with common gaseous pollutants that affect the atmospheric corrosion of metals. These constituents include, for example, sulphur dioxide (SO_2), nitrogen dioxide (NO_2), ozone (O_3), carbon dioxide (CO_2) and nitric acid (HNO_3) (1). In some cases the species are incorporated in the thin electrolyte as a component of precipitation, while in other cases they are directly deposited in the thin electrolyte present on the surface. Particles such as sea salt and ammonium sulphate ($(NH_4)_2SO_4$), are important, as they are particles which dissolve in the water film and form ionic species that increase the conductivity of the aqueous layer and facilitate electrochemical corrosion processes on the surface. ($(NH_4)_2SO_4$) particles are formed by combination of NH_3 and oxidised SO_2 in the atmosphere where NH_3 originates mainly from anthropogenic sources while SO_2 is emitted in large amounts by combustion of fossil fuels.

Water-soluble species and particles also play an important role in the formation of the aqueous film on the surface. This is due to the tendency of a soluble salt to take up water and form an electrolyte solution (deliquescence) above a certain RH. For instance, NaCl has a deliquescence point of 78% of the relative humidity at 25 °C. The resulting thickness of the film is

Table 7.1 The approximate number of water monolayers on a metal surface at 25 °C and steady state conditions

Relative humidity (%)	Number of water monolayers
20	1
40	1.5–2
60	2–5
80	5–10

dependent on the RH and the amount of soluble species on the surface. Since the salt solution strives to be in equilibrium with the surrounding atmosphere, more salt on the surface results in increased thickness of the electrolyte layer and not in a higher concentration of Cl^- in the electrolyte layer. In marine environments chloride-containing aerosols have a large impact on the corrosion rate. The main chloride contribution comes from NaCl, but $MgCl_2$ and other components in the sea salt also contribute. In addition, significant amounts of other species such as sulphate are present in the sea salt aerosol. The deliquescence process of sea salt aerosols is more complicated than that of pure NaCl and may pass through partially dissolved stages before finally becoming a homogeneous electrolyte solution. Since atmospheric corrosion is strongly dependent on pollutants present in the atmosphere, different atmospheres are classified according to exposure conditions. The major categories based on potential corrosion rates are Rural, Urban, Industrial, Marine and Indoor. All of these atmospheres can be further subdivided, for example into wet, dry and arctic temperate zone.

A rural atmosphere is that of an inland rural countryside with little or no heavy manufacturing industries. Hence the corrosion rate is generally low in these areas. In urban atmospheres, even in areas free from heavy industries, there are pollution contributions from road traffic and emissions from fossil fuels. Road traffic is responsible for oxides of nitrogen which may oxidise to nitric acid. Fossil fuels have the potential of generating sulphur dioxide, which is converted to sulphuric and sulphurous acids when in contact with moisture. In an industrial environment there are often high emission pollutants stemming from the use of fossil fuels. The atmosphere here often contains contributions from all types of contamination, for example sulphur dioxide. However, in western countries the emission of sulphur dioxide has decreased significantly. Hence the corrosion rate of metals due to these emissions has also generally decreased.

7.3 Electrochemical reactions

The atmospheric corrosion of metals is largely dependent on the electrochemical reactions occurring in the thin aqueous layer on the surface and at the interface between the solid substrate and the thin electrolyte layer. The thin aqueous layer on the surface also acts as a conductive medium which can support electrochemical processes on the surface. Due to the presence of different phases with different electrochemical properties in magnesium alloys the anodic and cathodic reactions are often localised in different areas on the magnesium surface. The microelectrodes may consist of different phases present in the microstructure of the alloys. The influence of the microstructure on the atmospheric corrosion behaviour of magnesium alloys will be discussed in more detail further on. In atmospheric corrosion the thin electrolyte reduces the conducting path between the anodes and cathodes present on the surface. Hence, the area fraction between the anode and cathode does not play such an important role as in the case of corrosion in solution (2). Further, with the thin electrolyte layer the oxygen and carbon dioxide in the air can easily be transported through the thin electrolyte layer to the metal interface. It was shown in a fairly old work by Tomashov and Matveeva (3) that part of the cathodic reaction is due to oxygen reduction. This is different from a bulk electrolyte solution, where hydrogen evolution is the main cathodic process (4). The cathodic and anodic processes during atmospheric corrosion can then be written as follows:

Cathodic reaction

Water reduction:

$$2H_2O + 2e^- \rightarrow H_2 + 2OH^- \qquad 7.1$$

Oxygen reduction:

$$O_2 + 2H_2O + 4e^- \rightarrow 4OH^- \qquad 7.2$$

Anodic reaction:

$$Mg \rightarrow Mg^{++} + 2e^- \qquad 7.3$$

Reaction 7.4 describes the overall reaction in the case of water reduction:

$$Mg + 2H_2O \rightarrow Mg(OH)_2 + H_2 \qquad 7.4$$

7.4 The oxide film

Magnesium is a reactive metal, and after exposure to air an oxide layer will quickly form on a magnesium surface. This oxide layer will give some protection against atmospheric corrosion (5). It is generally believed that in damp conditions a thin MgO layer is rapidly formed on the pure magnesium metal. In experiments using transmission electron microscopy (TEM), performed by Nordlien *et al.* (6), magnesium was exposed to ambient atmosphere for 60 min and scratched. After scratching, a film was rapidly formed. This film consisted of a dense mixture of amorphous MgO and $Mg(OH)_2$.

The thickness of the oxide layer formed on pure magnesium in ambient conditions was calculated by McIntyre and Chen (7) using X-ray photoelectron spectroscopy (XPS). After exposure for only 10 s the oxide layer was measured to be 2.2 ± 0.3 nm (*c.* seven monolayers of MgO). The oxide layer was found to increase in thickness slowly and linearly with the logarithm of exposure time during a test period of 10 month's exposure to laboratory atmosphere. Hydration by absorbed water molecules contributed to the measured thickening of the air-formed film through hydroxide formation.

Santamaria *et al.* (8) suggested a model where the oxide of pure magnesium, in an aqueous electrolyte, has a bilayer structure, consisting of a thin MgO layer (~2.5 nm) and an $Mg(OH)_2$ external layer. The thickness of the $Mg(OH)_2$ was influenced by the immersion time and solution temperature. Hence, Nordlien *et al.* (6, 9) reported thicker oxide films that are formed on the surface after immersion in distilled water. Aluminium is a frequent alloying element in Mg alloys. It is well known that small additions of Al to an Mg–base metal improve the corrosion resistance. Such additions cause formation of a mixture of magnesium and aluminium oxides with increased stability compared with pure magnesium (10). This can explain the improved corrosion of single-phase Mg alloys with small additions of aluminium (e.g. AZ31) compared with commercially pure magnesium (11).

It has also been reported that an increase in aluminium in the bulk causes an enrichment of aluminium in the oxide film formed on the surface of Mg–AL alloys (12–14), thus indicating that the surface content of Al is related to the bulk content. The rate of oxide nucleation and growth is enhanced on Mg–Al surfaces compared with pure Mg surfaces especially at higher Al contents (13, 14). The film becomes more compact and protective when the aluminium content in the bulk alloy increases (9). The thickness of the subsurface layer is in the order of 4–6 nm, the depth at which the bulk composition is reached increases with increasing bulk Al-alloying content (12). Investigations using Auger electron spectroscopy (AES) (15) on the actual constituents in an AZ91D magnesium alloy show that there is an enrichment of aluminium in the oxide layer. This enrichment depends on the aluminium content in the different constituents of the alloy: the higher the aluminium content in the alloy constituents is, the higher is the enrichment of aluminium in the oxide layer. Another interesting fact is that XPS analysis reveals that an increasing aluminium concentration in Mg–Al alloys results in greater magnesium carbonate formation and lower amounts of MgO and $Mg(OH)_2$ (16).

7.5 The effect of atmospheric gases and particles

There are several gaseous constituents that might affect the atmospheric corrosion of magnesium and magnesium alloys. SO_2 is an atmospheric gas which is mainly emitted during combustion of fossil fuels and has historically been the most important pollutant for the atmospheric corrosion of metals in urban and industrialised environments. The level of SO_2 has decreased in industrialised countries and the atmospheric corrosion of metals, such as zinc in a corresponding way. The situation has moved from a single gas, SO_2, as the main pollutant affecting the corrosion behaviour to a multipollutant situation where several gases influence the corrosion process. This may not be valid in the developing countries where environments with very high SO_2

levels now can be found. However, little information is available about the influence of gaseous pollutants on the atmospheric corrosion of Mg alloys, especially for NO_2 and O_3, which are known to affect the corrosion process on metals such as zinc and copper. Substantial amounts of magnesium sulphate have sometimes been found on Mg alloys after field exposure in polluted industrial atmospheres (17). It was found in several older investigations that the atmospheric corrosion of magnesium was only weakly dependent on the pollutant levels but rather dependent on the time of presence of an aqueous film on the surface (18, 19). Laboratory investigation (20) of the effect of SO_2 on the corrosion of magnesium showed that the corrosion increased strongly with ppb levels of SO_2 and that the addition of NO_2 and O_3 increased the deposition rate of SO_2. The corrosion products were mainly magnesium sulphite after the laboratory exposure to SO_2.

The effect of CO_2 is of great importance for the corrosion mechanism of magnesium in the atmosphere. As can be seen from reactions 1 and 2, OH^- ions are formed on the surface during the corrosion process, and a high pH often forms quite rapidly in thin water films. This high pH tends to further facilitate the formation of brucite ($Mg(OH)_2$). As the corrosion attack proceeds, the metal surface experiences a local pH increase because of the formation of $Mg(OH)_2$, which results in a pH of about 11 in saturated solutions of $Mg(OH)_2$. However, a high pH in the thin electrolyte layer will to a larger extent attract CO_2 from the ambient air. The dissolution of CO_2 in the surface electrolyte causes a decrease in pH according to (21):

$$CO_2(aq) + H_2O \rightarrow HCO_3^- + H^+ \qquad 7.5$$

$$HCO_3^- + OH^- \rightarrow CO_3^{2-} + H_2O \qquad 7.6$$

The decreasing pH and the presence of carbonate cause $Mg(OH)_2$ to transform into magnesium hydroxy carbonates. The presence of ambient levels of CO_2 inhibits the atmospheric corrosion rate of Mg alloys in humid air by a factor of four (21). It is argued that both of these gaseous constituents influence the corrosion of Mg–Al alloys by decreasing the pH on the surface, resulting in a greater stability of aluminium-containing passive films. When the concentrations of the ions in the liquid eventually become supersaturated, the ions will precipitate to a solid phase with prolonged exposure. The number of precipitated nuclei will increase, and eventually they will cover the metallic surface and form corrosion products. Hence, through investigation of the corrosion products the mechanisms of the corrosion process can be understood, as well as the influence of atmospheric pollutants and other environmental parameters.

Other important factors for the atmospheric corrosion of magnesium are soluble atmospheric particles, such as sea salt and corrodant-containing precipitation. Chloride is of special interest as it is the major corrodant in sea salt in marine environments and in road environments where de-icing salt is used. It should be noted that the corrosion products on magnesium seldom contain less-soluble chloride-containing phases, as in the case for other metals such as zinc. In fact a large part of the NaCl deposited initially on a magnesium surface remains in soluble form after exposure, showing that the surfaces were covered with a surface electrolyte throughout the experiments (22). In another study (23) it was shown that the corrosion of magnesium was proportional to the amount of NaCl deposited on the surface. The corrosion also increased with the humidity. Likewise in this study substantial amounts of chloride in soluble form could be found after exposure. The important role of RH in atmospheric corrosion has been clearly illustrated in a simulated ocean exposure corrosion test; the corrosion damage of pure Mg, AZ91 and AM50 is significantly reduced by simply decreasing the relative humidity of the simulated chlorides containing ocean environment [24]. While chloride ions probably play a role in the breakdown of the passive films on magnesium, these investigations indicate that the main effect of chlorides is probably related to the formation of a phase film of electrolyte on chloride-contaminated surfaces. The conductive electrolytes support electrochemical reactions on the surface and facilitate the corrosion processes in microgalvanic cells set up between alloy constituents with different nobility.

7.6 Corrosion of magnesium (Mg) alloys during field exposure

Some corrosion rates of Mg alloys at different field exposure sites given in the literature (25–27) are displayed in Table 7.2. The highest corrosion rate, 8.8 µm/year, is shown for the AM50 alloy exposed in marine environment for one year, i.e. 2005-05-31 to 2006-05-23. The corrosion rate of AZ91D, measured in µm/year, is after 12 months of exposure 4.2, 2.2 and 1.8 for the marine, rural and urban exposures, respectively. The weight loss of the Mg alloys is linear with time (25). This was also seen in the laboratory (22) and has been reported in the literature (26). The weight of the field-exposed

Table 7.2 Corrosion rate, given in µm/year, of Mg alloy AZ91D obtained from three different field stations (25). Also included in the table are corrosion rates, found in the literature (26, 27), of field-exposed Mg alloys

	Rural	Industrial	Marine	Urban
Jönsson *et al.* (22)	2.18		4.15	1.75
Godard *et al.*[1] (26)	4.3	15.7	22	
Southwell *et al.*[2] (27)	12.6		19.2	

[1] Magnesium alloy AZ91A-T6 ; [2] Magnesium alloy AZ61X.

panels was measured before the pickling procedure (25) and was in many cases below the initial weight, i.e. the corrosion product had been removed by rainfall, wind and sea spray from the waves. All in all the results suggest that the formation of corrosion products works poorly as a physical barrier against corrosion attacks. The microstructure was found to play a vital role for the initiation of the corrosion attack in the field (25), which was initiated in the less noble α-phase, with the more noble eutectic α/β-constituent close to the α-phase acting as the cathodic site. The Al content of different phases was closely related to the extent of the corrosion attack. Further, the intermetallic Al–Mn particles did not play an important role in the corrosion process of either AZ91D or AM50 in the field. Hence the influence of the microstructure on the corrosion behaviour in the field was similar to findings in the laboratory, which will be described further on.

As a comparison reference corrosion panels of carbon steel, zinc, copper and aluminium can be seen in Table 7.3. These panels have been exposed during 7 October, 2005 to 21 October, 2006 (i.e. a slightly different time period) at the same field station. The field station is located at Sainte Anne du Portzic, Brest, France, at the facilities of Institut de la Corrosion. The measurements have been made in accordance with ISO standard 9223. A comparison between Table 7.2 and 7.3 reveals that the corrosion rate of Mg alloys AZ91D is much lower than the corrosion rate for steel, comparable to the corrosion rate of copper but four times as high as aluminium. The comparison shows that the corrosion rate of magnesium is not as high as one would initially believe, bearing the low nobility of the magnesium metal in mind.

In a project called 'Assessment of corrosivity of global vehicle environment' (28) corrosion coupons of different materials as well as different material combinations have been exposed for three years on trailers operating in different parts of the world. The exposure setup on a trailer in Dubai can be seen in Fig. 7.1. In the road environment the de-icing salts as well as the road dirt have a strong effect on the corrosion rate of the different material. A layer of dirt from the road covers the exposed materials, prolonging the wetting period; the dirt also contains different corrosive species, such as de-icing salt and copper particles from braking pads. One of the exposed materials in the project has been Mg alloy AZ91D. Some of the corrosion values for Mg alloy AZ91D are given in Table 7.4. As a reference some

Table 7.3 Corrosivity, given in μm/year, from Ste Anne field site for steel, zinc, copper and aluminium – 2006

	Steel	Zinc	Copper	Aluminium
Corrosion rate	103 ± 4	1.7 ± 0.1	4.0 ± 0.2	0.7 ± 0.1

7.1 A matrix of corrosion panels exposed under a truck in Dubai.

Table 7.4 Corrosion rate, given in μm/year, of Mg AZ91D, carbon steel and zinc obtained after mobile exposures at different locations around the world

	AZ91D	Carbon steel	Zinc
France	3.9 ± 0.3	20.7 ± 2.0	5.2 ± 0.3
Dubai	4.0 ± 0.3	6.4 ± 1.1	0.7 ± 0.3
Thailand	3.1 ± 0.1	11.1 ± 1.0	0.3 ± 0,0
Canada	8.5 ± 1.3	57.3 ± 2.4	4.8 ± 0.4
Sweden	14.7 ± 2.3	50.1 ± 2.7	3.6 ± 0.2
England	10.7 ± 2.1	66.8 ± 2.3	8.9 ±1.1

corrosion rates for carbon steel and zinc, obtained during the same exposure, are included in the table. It can be seen that the corrosion rates for magnesium are between, 3.1 μm/year in Thailand and 10.7 and 14.7 μm/year for England and Sweden, respectively. This is not surprising since England and Sweden use large amounts of de-icing salt during the winter periods. However, the corrosion rates are not that different from the ones obtained during the marine field exposure. Comparing the corrosion rates for magnesium with the ones obtained for carbon steel, it can be inferred that magnesium has a lower corrosion rate than carbon steel at all of the exposure sites; in some of the cases the corrosion rates is about six times as low.

7.7 Corrosion products

7.7.1 Initial formation of corrosion products in the laboratory environment

The composition of the corrosion products formed is a result of ion pairing reactions of the metal ions and anions present in the aqueous layer. The ionic species present in the thin aqueous layer originate from deposited gases and particles, dissolved metal ions and electrochemical reaction products. By analysing the corrosion products, it is possible to gain information about the reactions taking place on the surface. In most investigations into the corrosion of magnesium alloys the corrosion products are analysed with X-ray diffraction (XRD), which gives such compositions of crystalline corrosion products as $Mg(OH_2)$ Al_2O_3 and MgO (29). However, magnesium carbonate is often poorly crystalline and can be problematic to detect using XRD (21, 22, 29). The use of other techniques shows that magnesium carbonates are a main corrosion product in the atmospheric corrosion of Mg alloys. Analysis with Fourier transform infrared (FTIR) spectroscopy clearly reveals the presence of large amounts of carbonates under atmospheric conditions (22, 29, 30). Feliu *et al.* (16) analysed the surface chemistry of a number of Mg alloys using XPS under atmospheric conditions. The results indicate that there are large amounts of carbonates on the surface of the Mg alloys tested, and that there is a relationship between increasing carbonate formation and increasing aluminium content in the alloys tested.

The reaction sequence of the formation of corrosion products on NaCl-contaminated AZ91 under laboratory conditions was studied by the authors of this chapter (22). As seen from reaction 4.4 above, $MgOH_2$ is the overall corrosion product and is also the main corrosion product in bulk solution (31). However, brucite is stable only at low CO_2 partial pressures (32). At the levels of CO_2 found in ordinary atmosphere, brucite will react directly with CO_2 to form magnesite. The reaction then becomes:

$$\underbrace{Mg(OH)_2}_{\text{Brucite}} + CO_2 \rightarrow \underbrace{MgCO_3}_{\text{Magnesite}} + H_2O \qquad 7.7$$

Magnesite is subsequently transformed to nesquehonite after 2–3 days of exposure according to:

$$\underbrace{MgCO_3}_{\text{Magnesite}} + 3H_2O \rightarrow \underbrace{MgCO_3 \cdot 3H_2O}_{\text{Nequehonite}} \qquad 7.8$$

After longer exposures hydromagnesite is formed:

$$\underbrace{5MgCO_3 \cdot 3H_2O}_{\text{Nequehonite}} \rightarrow \underbrace{4MgCO_3 \cdot Mg(OH)_2 \cdot 4H_2O}_{\text{Hydromagnesite}} + CO_2 + 10H_2O \qquad 7.9$$

Closer investigation revealed that beneath the crust of hydromagnesite, which was formed through the transformation of nesquehonite, brucite ($MgOH_2$) was formed. Beneath the crust of hydromagnesite there is a limited transport of CO_2. Hence brucite can be formed in the pits, if the corrosion process is rapid and the hydromagnesite lid covers the pit, as was the case in this exposure. Brucite has a pH of 11.2 in saturated solutions. Therefore, as the brucite formation increases, the pH within the pit also increases. When the pit reaches the saturation pH value of brucite, the corrosion reaction slows down radically. Similar results were also obtained by Liu *et al.* (29) in the investigation of Mg alloy AM60 under thin electrolyte layers, where corrosion products $Mg(OH)_2$ and compounds containing CO_3^{2-} could be found in heavily corroded areas. But only carbonate-containing products could be found in areas that were not affected by corrosion to the same extent.

7.7.2 Corrosion products in the field

For evaluation and comparison of Mg alloys, accelerated corrosion tests are commonly used in industrial research to simulate real conditions. However, in order to design corrosion tests that mimic the working environment of a Mg component, further understanding of the corrosion mechanisms of Mg alloys in the field is needed. Few studies (17, 19, 25–27, 30, 33–36) have been dedicated to the atmospheric corrosion of Mg alloys in the field. Most of the studies are older. In a recent investigation Mg alloys AZ91D and AM50 were exposed for times ranging from 3 months up to 2 years (25) in urban and rural field stations in Sweden and a marine field station Sainte Anne du Portzic, Brest, France. At all the exposure sites the main corrosion product turned out to be hydromagnesite. Older studies report that similar corrosion products are formed on Mg alloys in the field. Godard *et al.* (26) refer to field studies where hydromagnesite ($Mg_5(CO_3)_4(OH)_2 \cdot 4H_2O$) and nesquehonite ($MgCO_3 \cdot 3H_2O$) were formed. In a more recent study by Takigawa *et al.* (30) magnesite was found to be the main corrosion product in marine atmosphere for the AZ91D alloy. For the AZ31B alloy with a lower aluminium content the corrosion products consisted of brucite magnesite and hydromagnesite. In addition, it was reported that crystalline Mg(OH)Cl was detected in marine environments (37). In some older studies, sulphur-containing corrosion products, such as $MgSO_4 \cdot 6H_2O$, have been reported to form on magnesium at urban industrial sites (26) as well as during exposure to SO_2 in laboratory investigations (21, 38). In the above-mentioned study (25) two differences can be noted between the corrosion products formed in the field and the corrosion product formed in the laboratory (22):

- *Firstly*: On the field-exposed panels there were small contributions from sulphate ions. Since under the current field conditions only minor bands caused by sulphate could be found, the results suggest that SO_2-induced corrosion did not have any large impact on the exposed panels, probably due to too low SO_2 levels in the atmosphere at the exposure sites.

- *Secondly:* Brucite ($Mg(OH_2)$) could not be found on any of the field-exposed samples. In the laboratory, however, brucite could be detected after six days of exposure (22). In the field the corrosion products consisted exclusively of magnesium carbonates. However, Takigawa *et al.* (30) reported findings of brucite as a corrosion product of AZ31B in the marine environment. One explanation for the deviating results of the two studies can be a high corrosion rate of the AZ31B alloy. In that case brucite can be formed under a lid of carbonates (22) (as explained above). Another explanation can be that increasing aluminium concentration in Mg–Al alloys results in greater magnesium carbonate formation and lower amounts of Mg, MgO and $Mg(OH)_2$, as reported by Feliu *et al.* (16).

With the exception of the two differences mentioned above, there are considerable similarities between the corrosion products formed on magnesium samples in the laboratory and those formed in the field. This indicates that the sequence found in the laboratory (22), according to which $Mg(OH)_2$ will react with CO_2 in the surrounding air to form magnesite ($MgCO_3$), and magnesite together with water will form hydromagnesite or nesquehonite, also applies to the field-exposed samples.

7.8 Influence of microstructure on the atmospheric corrosion behaviour

In engineering applications magnesium is seldom used in its pure form. Instead it is alloyed with other elements and these alloying elements often form various secondary phases and intermetallic particles in the alloy. The various phases in the microstructure have different chemical compositions and may also have different electrochemical properties. The phases may form anodic and cathodic areas during the electrochemical corrosion processes of the alloy. The different phases, acting as microelectrodes, can have a large impact on the corrosion behaviour of magnesium alloys, especially in bulk electrolyte. However, in the atmosphere the electrolyte layer is thin and hence the conducting path is narrower. The microelectrodes play a somewhat different role in this case, which can result in differences in the corrosion mechanisms. AZ91D is one of the most commonly used magnesium alloys and will in this chapter work as a model to illustrate the influence of the microstructure on the atmospheric corrosion of magnesium alloys. The microstructure of AZ91D typically consists of a matrix of primary α-phase grains (Fig. 7.2) (39). The grains have an aluminium content of around 4

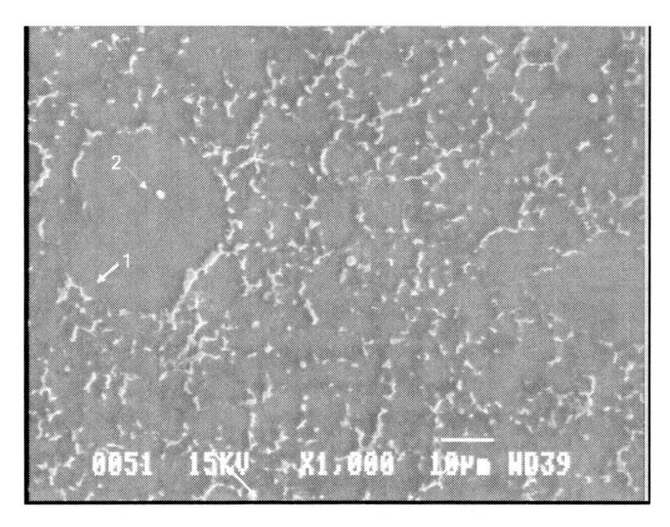

7.2 SEM back-scattered image of AZ91D Mg alloy showing α-phase grains. In the grain boundaries the β-$Mg_{17}Al_{12}$ phase can be found (1). Between the α-phase and the α-phase a eutectic area of β-phase and α-phase precipitates. Al_8Mn_5 inclusions are also present in the alloy (2).

wt% in their middle. In the grain boundaries the β-phase ($Mg_{17}Al_{12}$) has been formed. The β-phase has an aluminium content of around 30 wt%. Between the β-phase and the α-phase a eutectic area of β-phase and α-phase can be found. The aluminium content in this phase varies between that of the α-phase and that of the β-phase. The AZ91D alloy also contains different intermetallic phases of Al–Mn type. These Al–Mn intermetallics are often of the type η-Al_8Mn_5 (39). Al–Mn has a small content of iron, in the range of ~1–3 at.% (39). It is well known that manganese has a high tendency to bind to iron to form various intermetallic particles (40, 41). Iron is often a result of impurities in the casting process (42).

7.8.1 Characterisation of potential distribution and corrosion attack morphology in atmosphere conditions

The corrosion processes of Mg alloys in the atmosphere as well as in solutions lead to corrosion attacks which are inhomogeneously distributed over the surface with a localised and preferential attack of some alloy constituents. The difference in the electrochemical properties of different phases and intermetallic particles in the alloys is of great importance for the corrosion mechanisms. It is therefore important to be able to measure the potential differences between the phases. During recent decades there has been a rapid development of techniques for local electrochemical measurements, such as the scanning Kelvin probe force microscope (SKPFM) as well as methods for visualisation and quantification of corrosion attacks, such as atomic force microscopy, confocal microscopy and laser beam profilometry. The pit initiation, growth of the attack and the development of the corrosion attack morphology have been studied in detail by optical techniques such as confocal microscopy (15, 25, 43) and laser beam profilometry (44). Focused ion beam scanning electron microscopy (FIB/SEM) is another technique with great potential for detailed studies of the sub-surface micro- and nanostructures of corroded Mg alloys. Studies performed by FIB/SEM of atmospheric corrosion of Mg alloy AZ91 revealed corrosion morphologies related to the evolution of gas bubbles during the corrosion process (45).

Investigations have been made (46–52) on the electrochemical properties of synthetically prepared phases in solution. However, these results are not directly applicable in the atmospheric environment, since the state of the surface of the different phases is altered rapidly in a bulk electrolyte. In an ambient atmosphere the surface state is different owing to the thin electrolyte layer, which leads to a different local chemistry at the aqueous/metal interface compared with a bulk electrolyte. Measurements of the open circuit potential in atmospheric environment are difficult due to the thin aqueous layers which prevent the use of ordinary electrochemical techniques. However, it is possible to use techniques such as the scanning Kelvin probe (53–55) and SKPFM (56, 57) for investigation of the nobility, i.e. the Volta potential, of these phases in the atmosphere. The SKPFM can measure the Volta potential with a high resolution, making it possible to measure the Volta potential of the microconstituents of the alloy investigated.

Using the calibrated SKPFM instrument (39), a typical grain within the microstructure of the AZ91D alloy was studied. The same grain was also examined using scanning electron microscopy/energy dispersive X-ray spectroscopy (SEM-EDx). The results can be seen in Fig. 7.3. A darker colour in this figure indicates a lower Volta potential. The β-$Mg_{17}Al_{12}$ phase is present at the grain boundaries, and the intermetallic particle Al_8Mn_5 can be seen as a white particle in the middle of the grain. From Table 7.5, where the Volta potential of the different phases is shown, it can be inferred that both the Al_8Mn_5 particle and the β-$Mg_{17}Al_{12}$ phase in AZ91D have a more noble potential than the α-Mg phase. It can also be seen that the variation of aluminium in the eutectic zone results in changes in the Volta potential that are large enough to be measured. Hence, since the potential difference between the Al_8Mn_5 particle and the α-Mg phase is large, close to 400 mV, it can be assumed that the corrosion would be initiated in the α-Mg phase in the vicinity of the Al–Mn particles. It should be borne in mind that the Volta potential is to a large extent dependent on the top layer of the surface (i.e. the oxides), as discussed in Section 7.4 on oxides. An increase in aluminium

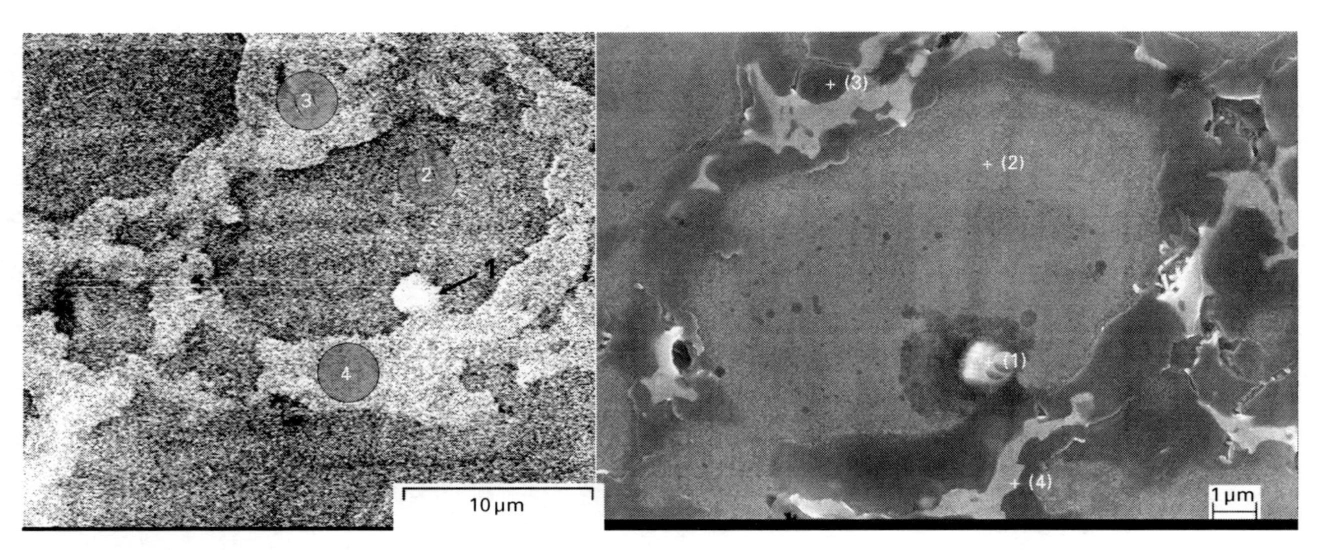

7.3 Left: SKPFM image of AZ91D alloy. Right: field emission gun scanning electron microscopy (FEG-SEM) image of the same area. The compositions and Volta potentials of points 1–4 are given in Table 7.5.

Table 7.5 Compositions and Volta potentials measured by SKPFM (at 40% RH and at 20 °C) at points 1–4 in Fig. 7.3.

Location	Concentration (atom %)			Phase	Volta potential (mV vs. SHE)
	Mg	Al	Mn		
1	15.8	53.8	30.4	Al_8Mn_5	–640
2	93.8	6.1	0.1	α-Mg	–1025
3	89.2	10.6	0.2	Eutectical α/β-phase	–805
4	71.2	28.7	0.1	β-$Mg_{17}Al_{12}$	–760

in the bulk causes an enrichment of aluminium in the oxide film formed on the surface of Mg–Al alloys (12–14). Auger measurements indicated that the aluminium content of the oxides on the surface was related to the Al content in different constituents of the AZ91D alloy (15).

7.8.2 The initial corrosion attack

It is important to remember that the potential of the different constituents only shows the driving force of the process that is to take place, whereas nothing is said about the kinetics, which, as will be shown, are dependent on a number of factors. Usually potentiodynamic measurements are used to investigate the kinetics of an alloy and of synthetically prepared phases (51, 52, 58). However, as is the case when measuring the corrosion potential, potentiodynamic measurements also require a bulk electrolyte. To get a more comprehensive idea about the atmospheric corrosion behaviour, the surface itself needs to be studied and other techniques have to be utilised.

Confocal microscopy is an optical imaging technique that makes it possible to obtain a three-dimensional image of the surface of a sample. Using confocal microscopy, the initial stages of a corrosion attack can be seen. Figure 7.4 shows an SEM image of an interesting part of the AZ91D magnesium surface. Marked by arrows, two intermetallic particles can be observed. The same areas were identified again, using confocal microscopy, after the samples had been exposed to humid air (95% RH) for 4 days (Fig. 7.5). After the corrosion products had been removed, the corrosion attack was clearly visible. No corrosion attack could be detected in the vicinity of either of the two Al_8Mn_5 particles. Furthermore, the initiation of the corrosion attack was initiated in the largest α-Mg grain. However, both the Al–Mn particles are embedded in areas where the aluminium content is high, i.e. the eutectic α-/β-phase or the β-phase, and the Al–Mn particles are not located in the vicinity of the α-Mg phase. In fact, due to the solidification process the majority of the Al–Mn particles can be found in the areas of a eutectic α-/β-phase (59, 60). Thus in atmospheric environments, where the electrolyte layer is thin, the driving force between the eutectic α-/ β-phase and the intermetallic Al–Mn particles

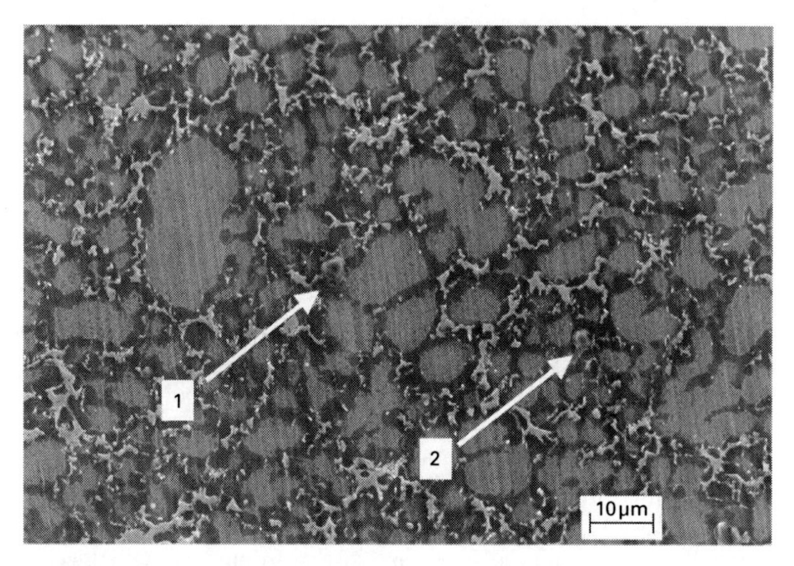

7.4 SEM image of the microstructure of AZ91D Mg alloy. Arrows 1 and 2 point at intermetallic particles of the Al_8Mn_5 type.

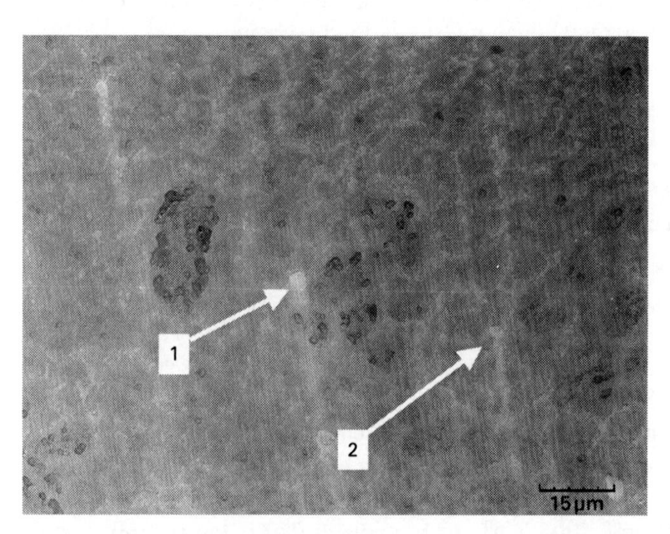

7.5 Confocal image of the same area as seen in Fig. 7.4. The area was exposed for 4 days in 95% RH, and the corrosion products were removed through pickling. Arrows 1 and 2 point at intermetallic particles of the Al_8Mn_5 type.

is not strong enough for the initiation of a corrosion attack in the vicinity of the Al–Mn particles due to the high aluminium content in the eutectic α-/β-constituent. Instead, as can be inferred from Fig. 7.5, the aluminium-containing phases, i.e. the eutectic α-/β-constituent and the α-phase, play a more important role in the initiation process of a corrosion attack. A closer investigation of a sample exposed for 3 months in marine environment (Fig. 7.6) reveals more about the influence of aluminium on the corrosion behaviour of the Mg alloy. The confocal image shows trenches that were formed at the boundary between the α-phase and the eutectic α-/β-constituent. The lines for the depth profile and the aluminium content exhibit a close similarity. It appears that the aluminium content in these trenches and in the rest of the grain is below 6% (Fig. 7.7). In the eutectic area the aluminium content increases and so does the corresponding depth profile line, indicating that the corrosion attack is not very severe in eutectic areas. The formation of the trenches suggests that a galvanic couple was formed between areas with a high and a low aluminium content and that the anodic process takes place in areas with an aluminium content below 6%. The initial corrosion attack in the largest grain seen in both Figs 7.5 and 7.6 can hence be explained by the lower aluminium content in these larger grains.

These results were obtained with relatively low amounts of chloride contamination on the surfaces and in environments with larger chloride deposition on the surfaces this behaviour may not be observed readily since thicker electrolyte layers lead to an increased corrosion rate and rapid attack of the surface. Galvanic coupling of the grains with more noble intermetallic particles would then also be possible. The situation will be more similar to the corrosion process observed during immersion or in salt spray. Academic papers on the influence of the microstructure on the corrosion behaviour in atmospheric conditions are scarce (15, 25, 39, 43). But there is literature to be found on the influence of the microstructure in bulk solutions. In the case of corrosion in bulk solutions the role of the aluminium content in the α-phase is somewhat disputed. It has been reported that in solutions with a high pH the anodic activity of the α-Mg phase is increased when the aluminium content is increased to a certain level (49, 52, 61). Lunder *et al.* (52) measured current vs. time curves for binary alloys containing magnesium and up to 8% aluminium. They showed that a small amount of aluminium content (up to 8%) in the α-phase leads to higher anodic activity. Similar results have been reported by Song *et al.* (49) and Lee *et al.* (61). Song *et al.* (49) made electrochemical measurements of binary Mg–Al alloys containing Mg–2%Al and Mg–9%Al. They showed that if the α-grain is only weakly polarised anodically, the dissolution rate of Mg–2%Al is higher than that of Mg–9%Al, but lower if the anodic polarisation is strong. Hence, the eutectic α-/β-area would corrode to a greater extent in solution.

In order to understand the initial steps of the atmospheric corrosion process, the Volta potential of a magnesium grain was examined by means of SKPFM. Figure 7.8 shows an SEM image of part of the AZ91D surface together with the corresponding Volta potential image. Comparing the EDX measurement from Fig. 7.9 with the Volta potential measurement along the

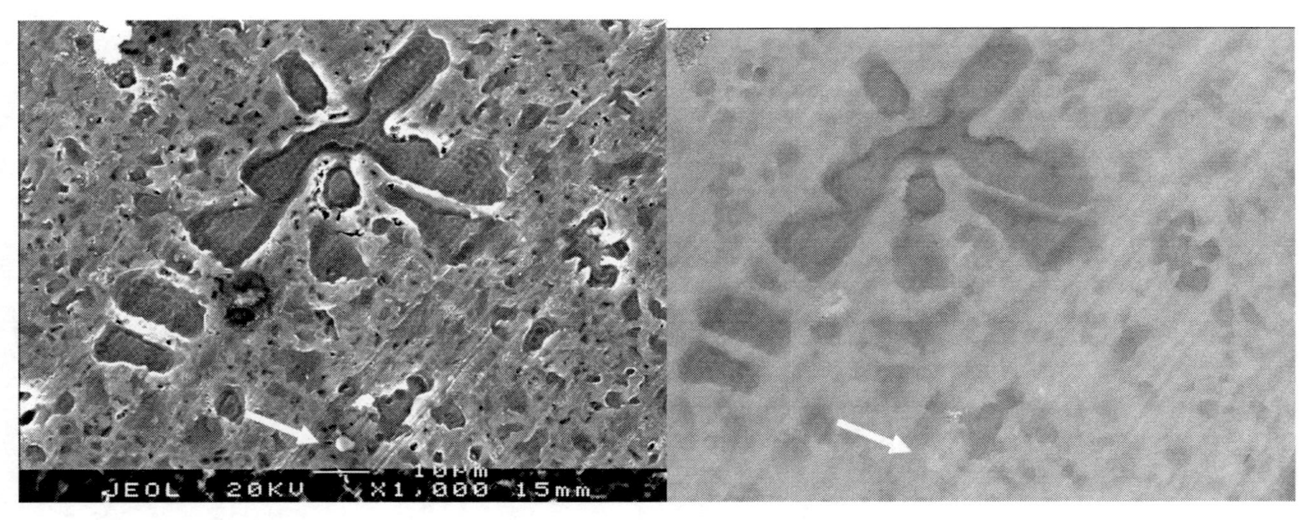

7.6 Left: SEM image of the corroded AZ91D surface exposed for 3 months in marine environment. Right: Confocal image of the same area. Marked with an arrow is an intermetallic particle of the Al–Mn type.

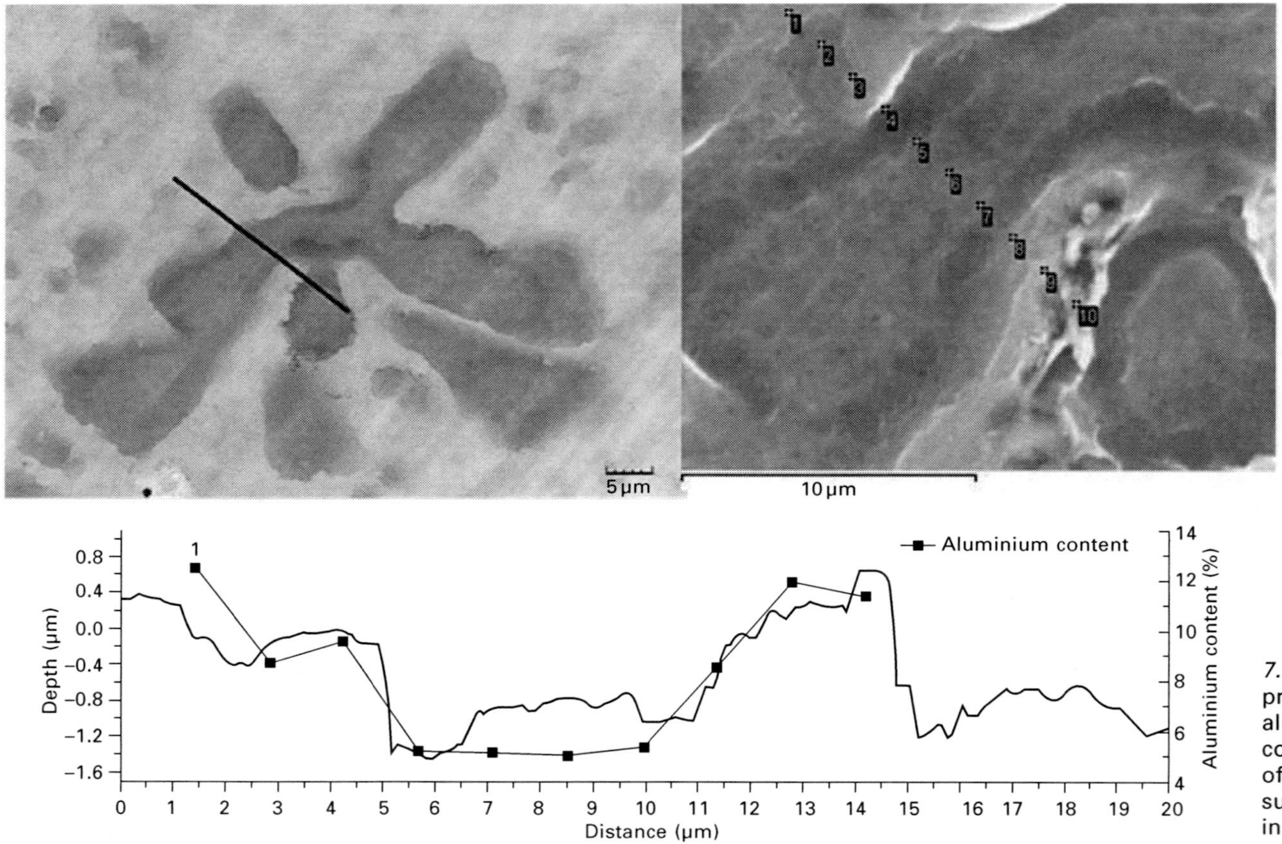

7.7 Surface profile and aluminium content of part of the AZ91D surface seen in Fig. 7.6.

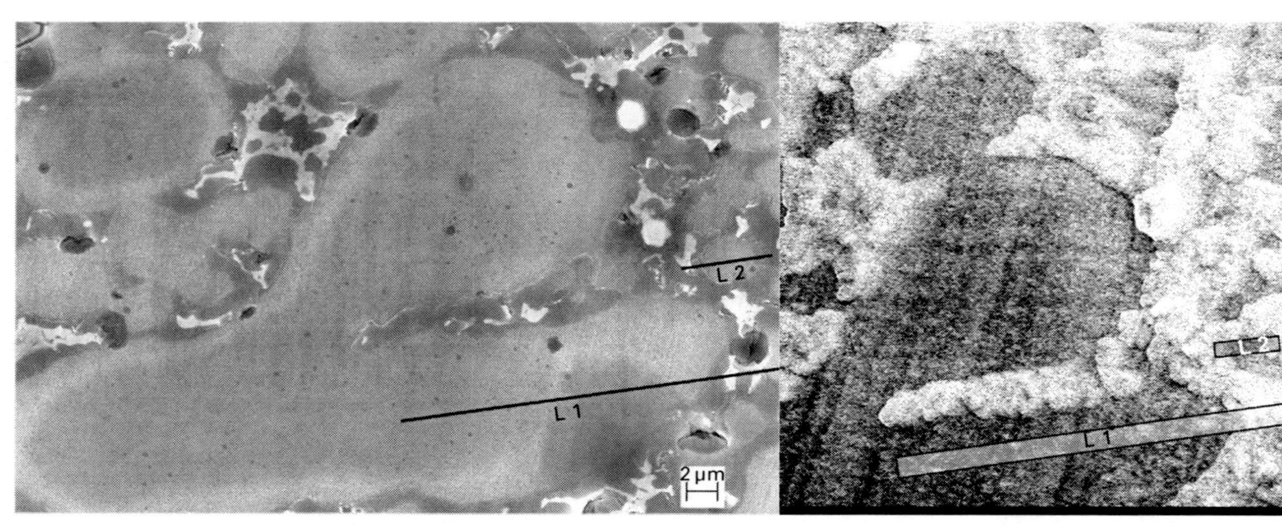

7.8 Left: SEM image of a grain in the AZ91D alloy. Right: the corresponding Volta potential image. The aluminium content and Volta potential are given in Figs 7.9 and 7.10.

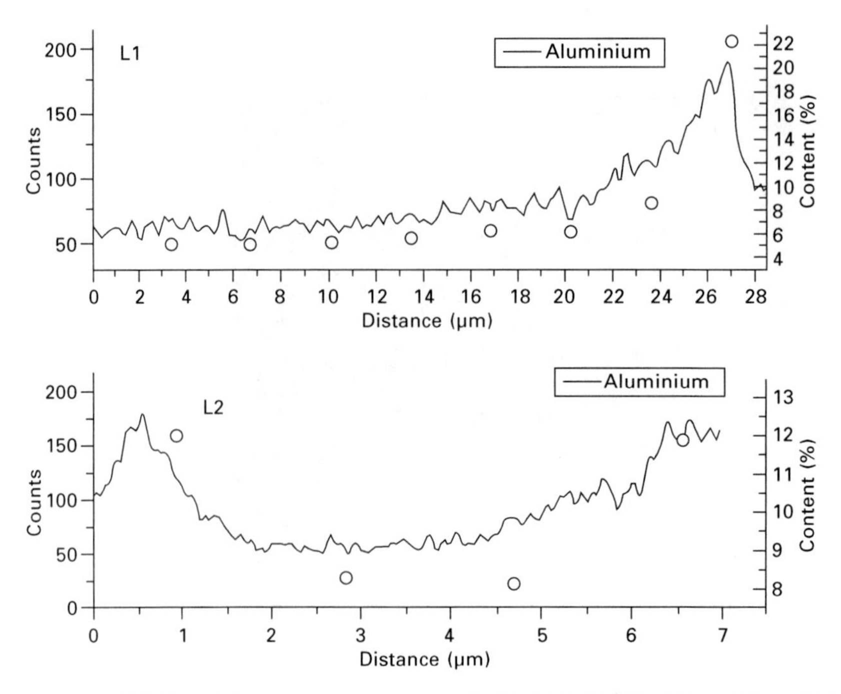

7.9 Aluminium content measured with SEM-EDX of lines L1 and L2 in Fig. 7.8.

same lines (Fig. 7.10), it can be seen that the Volta potential follows the aluminium content in the grains and that areas with a higher aluminium content give a higher Volta potential. Further the aluminium content in the different grains is dependent on the sizes of the grains. More accurately, the aluminium content in the grains is dependent on the cooling rate, and a high cooling rate results in small grains with a high aluminium content. In the same way the SKPFM measurement shows that the Volta potential for the same areas is also dependent on the grain size, the smallest grains giving the highest Volta potential and vice versa. Hence the lower potential of the larger grains explains the primary corrosion attack on these grains. The results indicate that a higher aluminium content in the grains is beneficial for the corrosion behaviour under atmospheric conditions.

7.8.3 Model of the atmospheric corrosion process over time

Based on the results of the present work a schematic model was produced, describing the influence of the microstructure on corrosion behaviour and the development and formation of corrosion products during atmospheric corrosion. This model is schematically presented in Fig. 7.11.

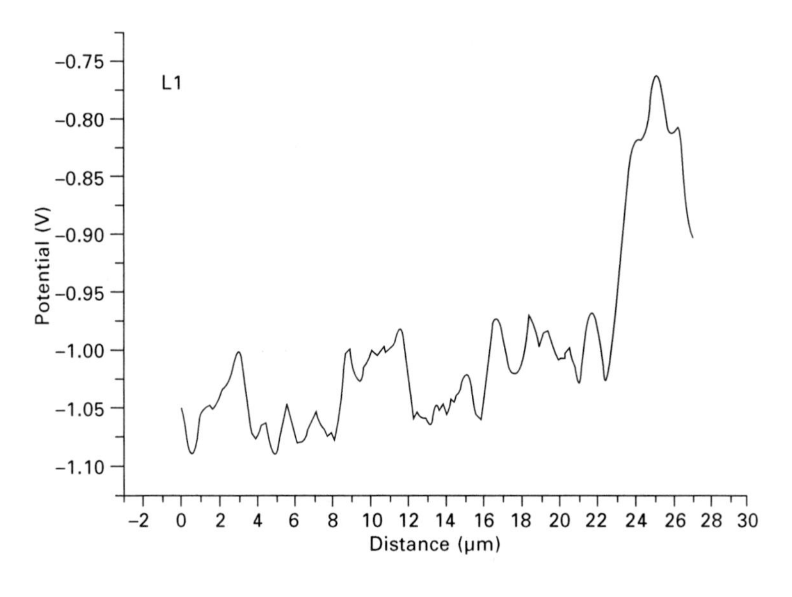

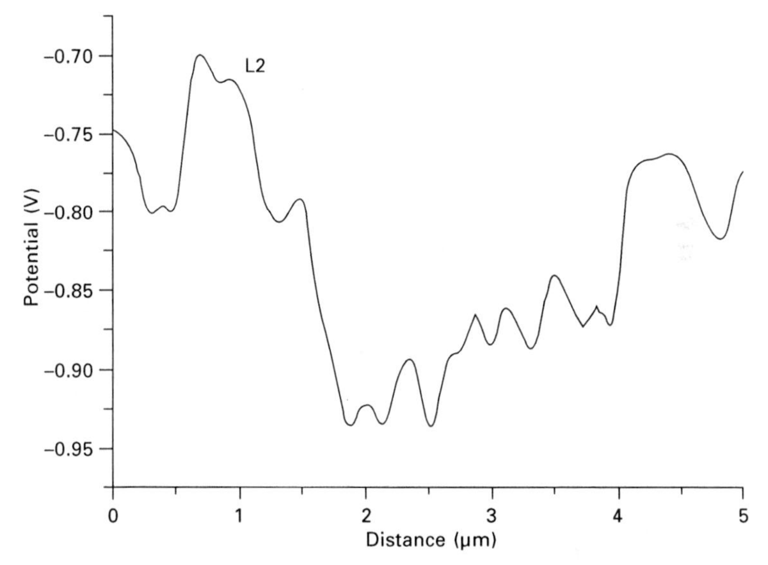

7.10 Volta potential of lines L1 and L2 in Fig. 7.8.

Initially (Fig. 7.11(a)), the anodic dissolution occurs at the location where electrolyte droplets are present on the surface and where the Cl^- content is high. CO_2 diffuses through the water droplets, and reacts with the $Mg(OH)_2$ formed by the anodic dissolution of Mg. This results in rapid formation of magnesium carbonate magnesite. As long as sufficient oxygen is present at the reaction sites on the surface, the cathodic reaction is probably to a large

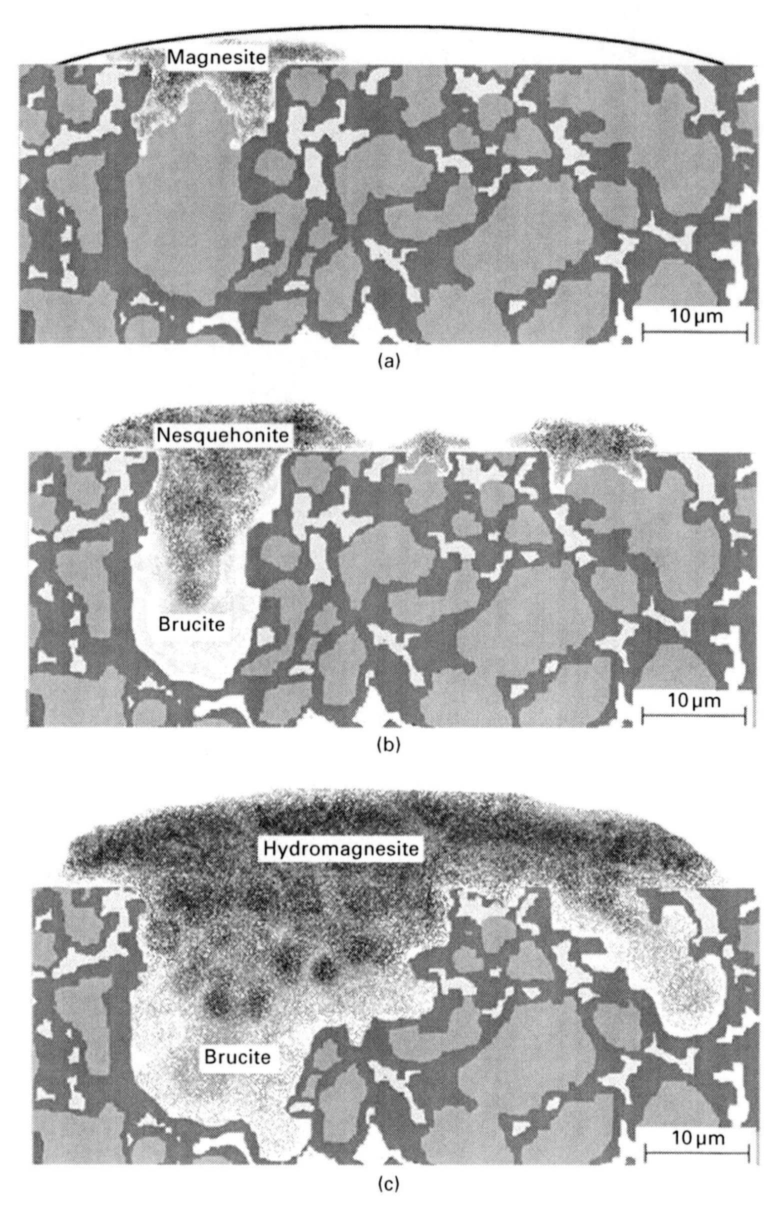

7.11 A schematic image of the corrosion process of magnesium alloy AZ91D. A discussion of steps (a) to (c) follows in the text.

extent due to oxygen reduction. In the cathodic areas, the pH increases due to the formation of OH^- by the cathodic reaction. The dissolution of CO_2 in the surface electrolyte causes a pH decrease, which is favourable for the aluminium-containing surface film. Looking at the microstructure of the alloy, the large grains are mainly attacked due to the lower aluminium content of these grains. A microgalvanic element is formed between the α-phase and the eutectic α-/β-phase, and this results in trenches in the α-phase, where the aluminium content is low.

After some days of exposure (Fig. 7.11(b)) the whole grains are corroded away, mainly due to the galvanic coupling between the α-phase and the eutectic α-/β phase, which results in the formation of deep pits which can be seen after a few days of exposure (15). The corrosion product magnesite is hydrated into nesquehonite. It is also possible that brucite starts to form due to a low CO_2 content at the bottom of the pit at this stage.

After longer exposure periods the corrosion (Fig. 7.11(c)) also starts to affect the grains in the vicinity, even the grains with a smaller size and a higher aluminium content and hence with higher corrosion resistance. The corrosion product nesquehonite is transformed into magnesium carbonate hydromagnesite, which forms a lid on the surface that hinders the transport of species such as CO_2 and O_2 through the crust. The low CO_2 content in the pit explains why brucite is not transformed into magnesite, which would be the case in ambient atmospheres. Further, the absence of CO_2 and the formation of brucite indicate that the pH is higher in the pits compared to the surface. The higher pH in the pits is unfavourable for the aluminium-containing phases, although aluminium is passive in neutral solutions. It has, however, an increased corrosion rate in alkaline solutions (48, 52). Hence the aluminium-containing phases, i.e. the eutectic α/β constituents and the β-phase, are affected by corrosion. More severe corrosion attacks can be seen when these phases are corroded away. Since water is present beneath the crust, and the oxygen is probably depleted, the cathodic reaction is most likely dominated by water reduction. Closer to the bottom of the pit all components for the corrosion process are present, and hence the corrosion process can proceed with little exchange with the surrounding atmosphere. At this stage the corrosion attack is more of a general nature, with hemispherically shaped corrosion patterns.

7.9 Differences between field-exposed magnesium (Mg) and accelerated tests

To evaluate magnesium alloys both immersion and salt spray tests are often used. Table 7.6 shows corrosion rates obtained after different types of accelerated corrosion tests taken from the literature (52, 62, 63). The corrosion rates show a wide scatter. For some of the accelerated corrosion tests such as the immersion in NaCl solution the corrosion rates are extremely high.

Two major factors differentiate the laboratory from the field.

Table 7.6 Corrosion rate of Mg AZ91D, given in μm/year, in different accelerated corrosion tests taken from the literature (52, 62, 63)

	Corrosion rate
Lunder *et al.*[1] (52)	2307.47
Wei *et al.*[2] (62)	62.9
Isacsson *et al.*[3] (63)	49.95

[1] 3 days' exposure in 5%NaCl saturated with Mg(OH$_2$).
[2] 10 days' exposure NaCl spray test in accordance with ASTM B117.
[3] 6 weeks' exposure in climatic chamber with cyclic RH immersion in 1% NaCl 5 min twice a week.

- *Firstly*: The pH of the electrolyte is a factor of great importance for the corrosion properties of Mg–Al alloys. It has been shown that the pH has a profound effect on the activity of the Mg–Al phases (48, 52, 64). Corrosion potential data show that Mg is stable in alkaline solutions and exhibits an increasing corrosion rate in decreasing pH, while aluminium is passive in neutral solutions but active and suffers a high corrosion rate in alkaline solutions (48, 52). Ambat *et al.* (64) tested the corrosion rate of AZ91 in three different pH values, 2.00, 7.25 and 12.00, respectively, the corrosion rate increases in the following order, pH 7.25 < pH 12.00 < pH 2.00. On the other hand it has been reported that pure magnesium has a corrosion rate 26 times lower at pH 12 than at pH 8.5 (52), due to the passivation of magnesium in alkaline environments. Hence, pH influences the corrosion rate in completely different ways depending on the proportion of magnesium and aluminium in the material tested. Measuring the corrosion properties of Mg alloys in solutions in a closed vessel the pH will increase to 10.5 within hours (46). This is due to the release of $Mg(OH)_2$ that is the main corrosion product in the absence of CO_2 (21). The high pH is detrimental to the protective aluminium oxide containing surface layer resulting in a selective dissolution of aluminium in the outermost layers of the surface (52, 61). In contrast, during atmospheric conditions the electrolyte layer is thin and the corrosion products consist of magnesium carbonates (22). The dissolution of CO_2 on the surface electrolyte causes a decrease in the pH (21). The surface film on Mg–Al alloys, that contains a high Al content (9, 10, 15), will stabilise due to the more neutral pH, thus increasing the corrosion protective properties of the oxide film.
- *Secondly*: The electrolyte layer is another aspect that differentiates the laboratory from the field. To accelerate the corrosion process in the laboratory, high relative humidity or bulk electrolyte is used, often together with high amounts of NaCl. A higher relative humidity results in

a thicker electrolyte layer and similarly higher amounts of NaCl applied on the surface increase the thickness and the hence the conductivity of the surface electrolyte, which of course will have an influence on the corrosion behaviour. Rozenfeld (17) has shown that with a decreasing thickness of the electrolyte film both magnesium and aluminium are able to polarise towards more positive values before the metals go into an active state. Not surprisingly aluminium can polarise towards more noble values than magnesium before going into an active state. Hence the degree of passivity of magnesium and aluminium increases with a thinner electrolyte layer, due to the easy access to oxygen, and consequently the ability to withstand corrosion increases under thin electrolyte layer. Also the cathodic ability is affected by a thinner electrolyte layer (17). If there is an easy access to oxygen, as is the case in under thin electrolyte layers, oxygen reduction will be the preferred cathodic reaction. The thicker electrolyte also results in a higher conductivity on the surface. A better conductivity of the electrolyte layer allows a greater separation between the local anode and cathode on the surface. In a bulk electrolyte, the conductivity between the different micro constituencies is high. Thus, galvanic coupling is possible without any larger ohmic potential drop in the solution. Hence the micro-constituents, such as the β-phase and the Al–Mn phases, will play a more important role as cathodes in the corrosion process. In contrast during field conditions a larger ohmic potential drop will be formed in the thin water layer on the surface. This decreases the possibility of galvanic coupling of alloy constituents located at larger distances from each other. In conclusion, during accelerated tests, such as immersion tests in NaCl solution or salt spray test, the protective abilities of aluminium-rich phases are reduced and the importance of the micro-constituencies on the surface is enhanced.

7.10 Concluding remarks

Even though magnesium has very low nobility and a reputation to readily corrode, the corrosion rate of bare magnesium panels is, in both marine and mobile field exposures, considerably lower than the corrosion rate of carbon steel. Atmospheric corrosion is triggered by a thin aqueous layer formed on metal surfaces in the presence of humid air. This thin aqueous layer varies in thickness depending on the atmospheric condition, and the deposition of particles and gases, In marine atmospheres and other chloride-containing environments, corrosion attacks are triggered by the formation of chloride-containing aqueous films on the surface. Owing to the presence of alloy constituents with different nobility and the presence of the conductive electrolyte layer, microgalvanic elements are present on the surface during corrosion. Gaseous constituents, such as CO_2, are more important in atmospheric conditions, compared with corrosion of Mg alloys in bulk solutions. The corrosion products formed in atmospheric conditions are mostly magnesium carbonate, indicating that CO_2 is readily solvated in the thin electrolyte layer. The CO_2 lowers the pH in areas on the surface which are alkaline due to the cathodic reaction and stabilises the aluminium-containing corrosion product film that is present on the aluminium-rich phases and results in increased corrosion protection. Corrosion product composition was similar on field exposed samples and dominated by magnesium carbonates and only small contributions from sulphate-containing corrosion products. This reflects the decreasing levels of SO_2 resulting in smaller impact of SO_2-induced corrosion on magnesium, at least in western countries. The microstructure of the Mg alloys is also of vital importance to the corrosion behaviour in atmospheric conditions. However, in atmospheric conditions the electrolyte layer is thin, which decreases the possibility of galvanic coupling of alloy constituents located at larger distances from each other. This means that intermetallic constituents such as Al–Mn phases are less important under atmospheric conditions, even though these have a high potential compared with the surrounding matrix. Further, the potential of other aluminium-rich phases is related to the aluminium content. A high aluminium content in the grains leads to a higher protective ability of the aluminium-containing surface film. The corrosion attack, both in the laboratory and under field conditions, is initiated in large α-phase grains due to the lower aluminium content in these grains.

7.11 References

1. C. Leygraf, in *Corrosion Mechanisms in Theory and Practice*, P. Marcus Editor, p. 529, Marcel Dekker, New York (2002).
2. R. Francis, *Galvanic Corrosion: A practical guide for engineers*, NACE International, Houston (2001).
3. N. D. Tomashov and T. V. Matveeva, *Trudy Institut. Physicalnaya Chemia AN SSSR*, **2**, 163 (1951).
4. E. Ghali, in *Uhlig's Corrosion Handbook*, W. R. Revie Editor, Wiley and Sons (2000).
5. J. H. Nordlien, S. Ono, N. Masuko and K. Nisancioglu, *Corrosion Science*, **39**, 1397 (1997).
6. J. H. Nordlien, O. Sachiko, N. Masuko and K. Nisancioglu, *Journal of Electrochemical Society*, **142**, 3320 (1995).
7. N. S. McIntyre and C. Chen, *Corrosion Science*, **40**, 1697 (1998).
8. M. Santamaria, F. Di Quarto, S. Zanna and P. Marcus, *Electrochimica Acta*, **53**, 1315 (2007).
9. J. H. Nordlien, K. Nisancioglu, S. Ono and N. Masuko, *Journal of Electrochemical Society*, **144**, 461 (1997).
10. J. H. Nordlien, O. Sachiko, M. Noburo and K. Nisancioglu, *Journal of Electrochemical Society*, **143**, 2564 (1996).

11. S. Feliy Jr, M. C. Merino, R. Arrabal, A. E. Coy and E. Matykina, *Surface and Interface Analysis*, **41**, 143 (2009).
12. L. P. H. Jeurgens, M. S. Vinodh and E. J. Mittemeijer, *Acta Materialia*, **56**, 4621 (2008).
13. S. J. Splinter and N. S. McIntyre, *Surface Science*, **314**, 157 (1994).
14. S. J. Splinter, N. S. McIntyre, P. A. W. van der Heide and T. Do, *Surface Science*, **317**, 194 (1994).
15. M. Jönsson, D. Persson and R. Gubner, *Journal of the Electrochemical Society*, **154**, C684–C691 (2007).
16. S. Feliu Jr, A. Pardo, M. C. Merino, A. E. Coy, F. Viejo and R. Arrabal, *Applied Surface Science*, **255**, 4102 (2009).
17. I. L. Rozenfeld, *Atmospheric Corrosion of Metals*, p. 221, National Association of Corrosion Engineers, Huston (1972).
18. G. K. Berukshtis and G. B. Klark, *Corrosion Resistance of Metals and Metallic Coatings in Atmospheric Conditions,* Nauka, Moscow (1971).
19. Y. Mikhailovskii, A. Skurikhin, M. Czerny, R. Wellesz and M. Zaydel, *Protection of Metals*, **15**, 419 (1979).
20. D. Bengtsson Blücher, Carbon dioxide: the unknown factor in the atmosperic corrosion of light metals, a laboratory study, Doctoral thesis, Department of chemical and biological engineering, Chalmers, Göteborg (2005).
21. R. Lindström, J. E. Svensson and L. G. Johansson, *Journal of the Electrochemical Society*, **149**, B103–B107, (2002).
22. M. Jönsson, D. Persson and D. Thierry, *Corrosion Science*, **49**, 1540 (2007).
23. N. LeBozec, M. Jönsson and D. Thierry, *Corrosion*, **60**, 356 (2004).
24. G. Song, S. Hapugoda and D. StJohn, *Corrosion Science*, **49** 1245 (2006).
25. M. Jönsson, D. Persson and C. Leygraf, *Corrosion Science*, **50**, 1406 (2008).
26. H. P. Godard, W. B. Jepson, M. R. Bothwell and R. L. Lane, *The Corrosion of Light Metals*, Wiley and Sons, New York (1967).
27. C. R. Southwell, A. L. Alexander and C. W. Hummer, *Materials Protection*, **4**, 30 (1965).
28. B. Rendahl and N. LeBozec, Assesment of corrosivity of glaobal vehicle enviroment, Research report, Swerea KIMAB Stockholm (2009).
29. W. Liu, F. Cao, B. Jia, L. Zheng, J. Zhang, C. Cao and X. Li, *Corrosion Science*, **52**, 639 (2010).
30. S. Takigawa, I. Muto and N. Hara, in *ECS Transactions*, p. 71, Corrosion in Marine and Saltwater Environments 3 – 214th ECS Meeting, Honolulu, HI (2009).
31. M. Jönsson and D. Persson, Accelerated corrosion tests for magnesium alloys: Do they really simulate field conditions? *The Annual World Magnesium Conference*, Warsaw, Poland (2008).
32. W. B. White, *Environmental Geology*, **30**, 46 (1997).
33. J. E. Hillis and R. W. Murray, in *14th International Die Casting Congress and Exposition*, p. paper no. G, Toronto, Canada (1987).
34. F. Pearlstein and L. Teitell, *Materials Performance*, **13**, 22 (1974).
35. P. V. Strekalov, *Protection of Metals*, **29**, 673 (1993).
36. D. V. Vy, A. A. Mikhajlov, Y. N. Mikhajlovskij, P. V. Strekalov and T. B. Do, *Zashchita Metallov*, **30**, 578 (1994).
37. H. D. Hui, P. V. Strekalov, M. Yu., D. T. Bin and A. Mikhailov, *Protection of Metals*, **437** (1994).
38. D. B. Blücher, PhD thesis, Carbon dioxide: The unknown factor in the atmospheric

corrosion of light metals: A laboratory study, Chalmers University of Technology, Göteborg, Sweden, (2005).

39. M. Jönsson, D. Thierry and N. LeBozec, *Corrosion Science*, **48**, 1193 (2006).
40. O. Lunder, T. K. Aune and K. Nisancioglu, *Corrosion*, **43**, 291 (1987).
41. I. J. Polmear, *Light Alloys Metallurgy of the Light Metals*, Butterworth Heinemann, Oxford, p. 362 (1995).
42. O. Lunder, M. Videm and K. Nisancioglu, *Journal of Materials and Manufacturing*, **No 950428**, 352 (1995).
43. M. Jönsson and D. Persson, *Corrosion Science*, **52**, 1677–1085 (2010).
44. R. B. Alvarez, H. J. Martin, M. F. Horstemeyer, M. Q. Chandler, N. Williams, P. T. Wang and A. Ruiz, *Corrosion Science*, **52**, 1635–1648 (2010).
45. Y. Wan, J. Tan, G. Song and C. Yan, *Metallurgical and Materials Transactions A: Physical Metallurgy and Materials Science*, **37**, 2313 (2006).
46. O. Lunder, K. Nisancioglu and R. S. Hansen, *International Congress and Exposition*, 1–5 March 1993, p. 117, SAE Special Publications, Detroit, MI, USA (1993).
47. O. Lunder, J. H. Nordlien and K. Nisangliou, *Corrosion Reviews*, **15**, 439 (1997).
48. S. Mathieu, C. Rapin, J. Steinmetz and P. Steinmetz, *Corrosion Science*, **45**, 2741 (2003).
49. G. Song, A. Atrens and M. Dargusch, *Corrosion Science*, **41**, 249 (1999).
50. L.-Y. Wei, H. Westengen, T. K. Aune and D. Albright, in *Magnesium Technology 2000*, p. 153, TMS Annual Meeting, Nashville, TN, United States (2000).
51. O. Lunder and K. Nisangliou, *The Understanding and Prevention of Corrosion*, p. 1249, Barcelona, Spain (1993).
52. O. Lunder, J. E. Lein, T. K. Aune and K. Nisancioglu, *Corrosion*, **45**, 741 (1989).
53. M. Stratmann and H. Streckel, *Corrosion Science*, **30**, 681 (1990).
54. M. Stratmann and H. Streckel, *Corrosion Science*, **30**, 697 (1990).
55. M. Stratmann, H. Streckel, K. T. Kim and S. Crockett, *Corrosion Science*, **30**, 715 (1990).
56. V. Guillaumin, P. Schmutz and G. S. Frankel, *Journal of Electrochemical Society*, **148**, B163 (2001).
57. P. Schmutz and G. S. Frankel, *Journal of the Electrochemical Society*, **145**, 2295 (1998).
58. G. Song, A. Atrens, X. Wu and B. Zhang, *Corrosion Science*, **40**, 1769 (1998).
59. V. Y. Gertsman, J. Li, S. Xu, J. P. Thomson and M. Sahoo, *Metallurgical and Materials Transactions A: Physical Metallurgy and Materials Science*, **36**, 1989 (2005).
60. C. J. Simensen, B. C. Oberlander, J. Svalestuen and A. Thorvaldsen, *Zeitschrift fuer Metallkunde/Materials Research and Advanced Techniques*, **79**, 696 (1988).
61. C. D. Lee, C. S. Kang and K. S. Shin, *Metals and Materials*, **6**, 441 (2000).
62. L.-Y. Wei, Development of microstructure in cast magnesium alloys, Chalmers University of Technology, Göteborg, Sweden, in, p. 217 (1990).
63. M. Isacsson, M. Strom, H. Rootzen and O. Lunder, *Proceedings of the 1997 International Congress and Exposition*, 24–27 Feb 1997, p. 43, SAE Special Publications, Detroit, MI, USA (1997).
64. R. Ambat, N. N. Aung and W. Zhou, *Journal of Applied Electrochemistry*, **30**, 865 (2000).

8
Stress corrosion cracking (SCC) of magnesium (Mg) alloys

A. ATRENS, The University of Queensland, Australia, N. WINZER, Fraunhofer Institute for Mechanics of Materials IWM, Germany, W. DIETZEL and P. B. SRINIVASAN, Helmholtz-Zentrum Geesthacht, Germany and G.-L. SONG, General Motors Corporation, USA

Abstract: This chapter reviews current research into stress corrosion cracking (SCC) in magnesium (Mg) alloys, particularly intergranular stress corrosion cracking (IGSCC) and transgranular stress corrosion cracking (TGSCC).

A nearly continuous second phase along grain boundaries causes IGSCC. The second phase accelerates corrosion of the adjacent matrix by micro-galvanic corrosion; the applied stress opens the crack and allows propagation through the alloy. IGSCC can be avoided by appropriate Mg alloy design. TGSCC is the intrinsic form of SCC. This is caused by an interaction of hydrogen (H) with the microstructure, so a study of H-trap interactions is needed to understand this damage mechanism, and in order to design alloys more resistant to TGSCC. This understanding is urgently needed if Mg alloys are to be used safely in service because prior research indicates that many Mg alloys have a threshold stress for SCC of the order of half the yield stress in common environments, including high-purity water.

Key words: magnesium, corrosion, stress corrosion cracking.

8.1 Introduction

Magnesium (Mg) alloys are starting to be used in structural applications and new classes of wrought Mg alloys are emerging. Their stress corrosion cracking (SCC) behaviour has not been characterised. Clearly, significant risk of SCC would negate much of the potential for application of stressed Mg–alloy auto-components exposed to road spray or stressed medical parts in contact with body fluids. This chapter provides a surprising indication that SCC could occur at a stress as low as 30% of the yield stress for a high Al-containing alloy under certain environmental conditions. Some common alloys (such as AZ91, AZ31 and AM30) are susceptible to SCC failure even in a mild environment such as distilled water. This implies that the SCC threshold can be easily met in automotive and medical applications. The most important implication is that SCC will be an issue for Mg–alloy structure parts in automotive and medical applications. Urgently required are more detailed investigations in this area with existing and new Mg alloys that may be used in the auto industry and for stressed medical applications. This chapter provides a basis for future research work in this area. Moreover, in automotive and medical applications, another big concern for Mg parts in service is their corrosion fatigue performance. There is a shortage of this knowledge for Mg alloys. Owing to the close relationship between SCC and corrosion fatigue, this chapter should provide some insight in this unknown area as well

The chapter builds on our critical reviews on Mg corrosion [1–4] and Mg SCC [5,6]. SCC [5–8] involves: (1) a stress, (2) a susceptible alloy and (3) an environment. SCC is related to hydrogen embrittlement (HE). HE is SCC that is caused by hydrogen (H), which can be gaseous, can come from corrosion, or can be internal from prior processing. HE is often postulated as the SCC mechanism. SCC can be extremely dangerous. Under safe loading conditions, SCC causes slow crack growth. Fast fracture occurs when the crack reaches a critical size. SCC, for any alloy + environment combination, can be characterised by [7,8] the threshold stress, σ_{SCC}, the threshold stress intensity factor, K_{1SCC}, and the stress corrosion crack velocity.

SCC is typically attributed to mechanisms related to dissolution or embrittlement. Figure 8.1 illustrates an embrittlement mechanism: (1) an embrittled region forms ahead of the crack tip, (2) there is a crack propagation, (3) the crack stops as it enters the parent material, and (4) the process recurs when an embrittled region has re-formed. Corrosion is an important issue for Mg [1–3,5,10–13], and much research has documented corrosion of Mg alloys in common environments such as 3% NaCl [14–58].

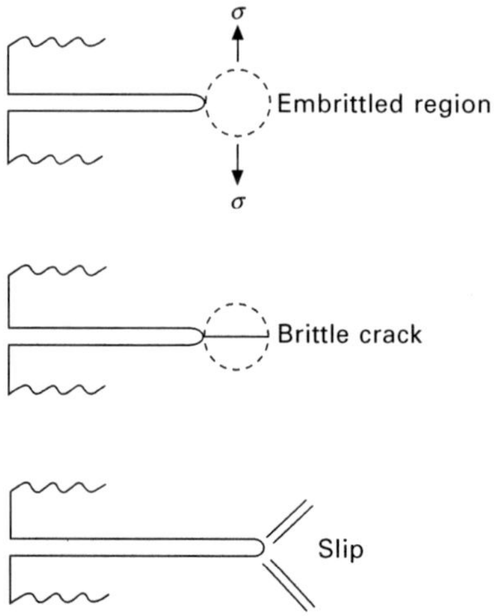

8.1 Model for transgranular stress corrosion cracking (TGSCC) [9].

Mg alloys are susceptible to SCC, but it was proposed that pre-1980s service failures were rare [59,60]. This includes (1) airplane panels examined by Dow, (2) failures of a number of forged AZ80 French aircraft components, apparently resulting from excessive assembly and residual stresses, and (3) service failures of cast and forged South African Mg aircraft wheels. Speidel and Fourt [61] estimated that 10 to 60 aerospace Mg component SCC service failures occurred each year from 1960 to 1970. More than 70% involved either cast AZ91-T6 or wrought AZ80-F. Unpublished work by MEL, and confidential industrial reports from the USA, indicate some incidents of SCC for Zr-containing alloys [62]. The section on Mg SCC in the *ASM Handbook* [63] can be summarised as follows. Mg alloys containing more than 1.5% Al are susceptible to SCC. Wrought alloys appear more susceptible than cast alloys. While there is little documentation of service SCC of castings, laboratory tests can cause SCC at tensile loads less than 50% of the yield stress in environments causing negligible corrosion. The low incidence of service SCC failures is attributable to low service stresses [63].

The incidence of SCC may increase because Mg parts are increasingly used in structural applications compared with the previous non-structural applications. Service conditions are becoming more severe, particularly in the automobile industry, where cast Mg components are increasingly used in load-bearing applications. Automobile examples include: engine blocks, transmission housings, engine oil pans, wheels, and structural body castings such as engine cradles, instrument panels, doorframes, body connectors and cross-members. Humid air is no longer the only environment that is of concern. Road splash of aqueous chloride solutions must also be considered. Light weighting produces a significant drive for designers to decrease section sizes to reduce weight. Moreover, loaded Mg components are increasing in complexity. Galvanic, localised, or intergranular corrosion, and thinning by uniform corrosion, could initiate SCC.

Our critical reviews [5,6] and our research [36,37,64–70] elucidated critical aspects of Mg SCC. This chapter provides an update on Mg SCC, based on our understanding of Mg corrosion [1–3], SCC [7,8,71–85], HE [86–89], corrosion fatigue (CF) [90–93], diffusion [88,89,94] and passivity [95–97].

8.2 Alloy influences

8.2.1 Pure Mg

Pure Mg is susceptible to SCC [60,98–102]. Winzer *et al.* [102] found transgranular stress corrosion cracking (TGSCC) for pure Mg in 5 g/L NaCl. Meletis and Hochman [99] reported crystallographic TGSCC for 99.9% pure Mg in a chloride-chromate solution. Fracture was cleavage-like, consisting of flat, parallel facets on $\{2\bar{2}03\}$ planes separated by steps also on $\{2\bar{2}03\}$ planes. The cleavage was attributed to a reduction in the surface energy due to atomic H or to Mg hydride. Stampella *et al.* [101] reported SCC for commercial purity (CP) Mg (99.5% Mg) and high purity (HP) Mg (99.95% Mg) in deaerated, pH = 10, 10^{-3} M Na_2SO_4 solution (Fig. 8.2). They proposed that SCC occurred by cleavage facilitated by atomic H in solid solution based on:

- the load–elongation curves for specimens tested in the solution were similar to those of pre-exposed specimens;
- embrittlement of pre-exposed specimens was partly reversible by storing for 24 h in a desiccator;
- SCC was invariably accompanied by pitting;
- SCC was prevented by cathodic polarisation, which inhibited pitting;
- SCC fractures were cleavage-like.

Stampella *et al.* [101] found that the crack morphology was exclusively TGSCC for fine grain (0.025 mm) CP Mg, whereas larger grain (0.075 mm) HP Mg produced mixed TGSCC and intergranular stress corrosion cracking (IGSCC).

Lynch and Trevena [98] studied SCC of cast 99.99% pure Mg in aqueous

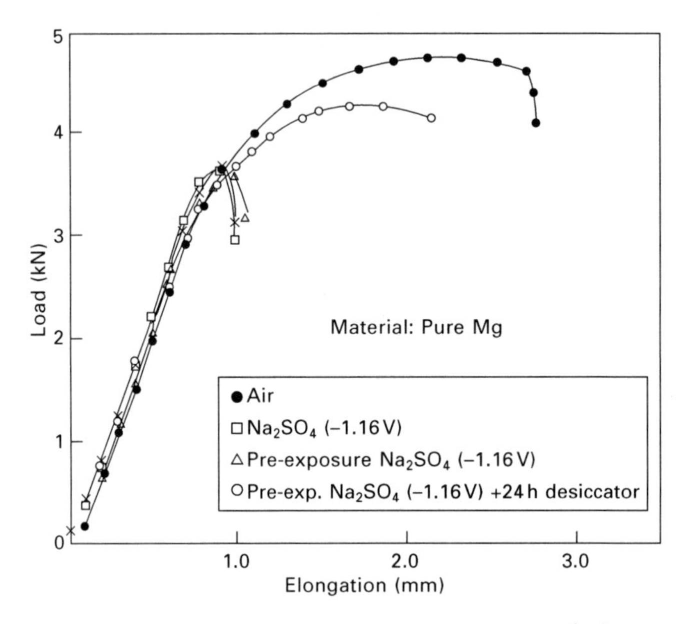

8.2 Load vs. elongation at a strain rate of $5.7 \times 10^{-6} s^{-1}$ for commercial purity Mg, with anodic polarisation to –1.16 V: ● stressed in air; □, X stressed in Na_2SO_4 solution; △ pre-exposed to Na_2SO_4 solution, stored for 24 h and stressed in air [101].

3.3% NaCl + 2% K_2CrO_4 solution, by cantilever bending of specimens at various deflection rates up to high rates. They proposed that SCC occurred by localised micro-void coalescence resulting from dislocations at crack tips, due to weakening of inter-atomic bonds by adsorbed H atoms. They observed concave features on opposing fracture surfaces, so that the opposing fracture surfaces were not matching. They suggested that tubular voids were nucleated at intersections of {0001} and $\{10\bar{1}X\}$ slip bands resulting in a fluted fracture surface parallel to $\{10\bar{1}X\}$ planes (Fig. 8.3). There was fractographic evidence of SCC for crack velocities between 10^{-8} and 5×10^{-2} m/s. They proposed that SCC was induced by adsorption of H since insufficient time was available for H diffusion or localised dissolution at the fastest crack velocities. At the lower velocities, it was conceded that there

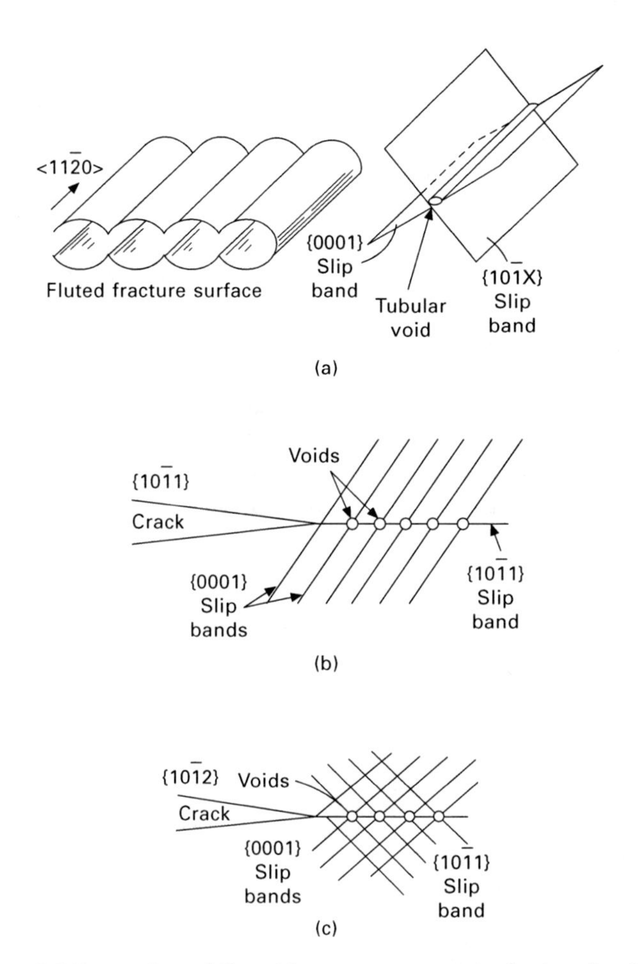

8.3 Formation of fluted fracture topography by localised micro-void coalescence.

might be some H diffusion ahead of the crack tip. There was no fractographic evidence of Mg hydride.

8.2.2 Alloy composition

Table 8.1 presents values of the SCC threshold stress for common Mg alloys [5].

Influence of Al

Al in Mg alloys increases strength and fluidity in casting. Mg–Al alloys are susceptible to SCC [63,100,101,103–109] in air, distilled water and chloride-containing solutions. SCC-induced fractures may occur at stresses as low as 50% of the yield strength. Figure 8.4(a) shows that SCC susceptibility increased as the Al content increased from 1% to 8% [106]. If the failure stress at 100 000 s is compared with values of tensile yield stress [110,111], the (ratio of failure stress)/(tensile yield stress) decreases to values somewhat lower than those in Table 8.1. The tensile stress values are for comparable material but may not be exactly the same as those in [106]. Nevertheless, Fig. 8.4 does indicate the high SCC susceptibility of some alloys.

Miller [107] showed SCC in distilled water for the popular alloys AZ91, AM60 and AS41. The time to failure data, for experiments lasting up to 500 days, indicated an SCC threshold at ~ 40–50% of the yield strength for all three alloys. Miller found the same SCC behaviour in HP AZ91 and AZ91–0.01% Fe (AZ91 containing a high iron content of 0.01%). Arnon and Aghion [108] found SCC in AZ31 in 0.9% NaCl. Chen *et al.* [109] observed TGSCC for AZ91 in 0.5 M $MgCl_2$, with the threshold stress ~ 1/3 yield stress. There was also TGSCC in more dilute solutions to 0.005 M.

Table 8.1 SCC threshold stress for common Mg alloys [5]

Alloy, environment	σ_{SCC}
HP Mg, 0.5% KHF_2	60% YS
Mg2Mn, 0.5% KHF_2	50% YS
MgMnCe, air, 0.001N NaCl, 0.01Na_2SO_4	85% YS
ZK60A-T5, rural atmosphere	50% YS
QE22, rural atmosphere	70–80% YS
HK31, rural atmosphere	70–80% YS
HM21, rural atmosphere	70–80% YS
HP AM60, distilled water	40–50% YS
HP AS41, distilled water	40–50% YS
AZ31, rural atmosphere	40% YS
AZ61, costal atmosphere	50% YS
AZ63-T6, rural atmosphere	60% YS
HP AZ91, distilled water	40–50% YS

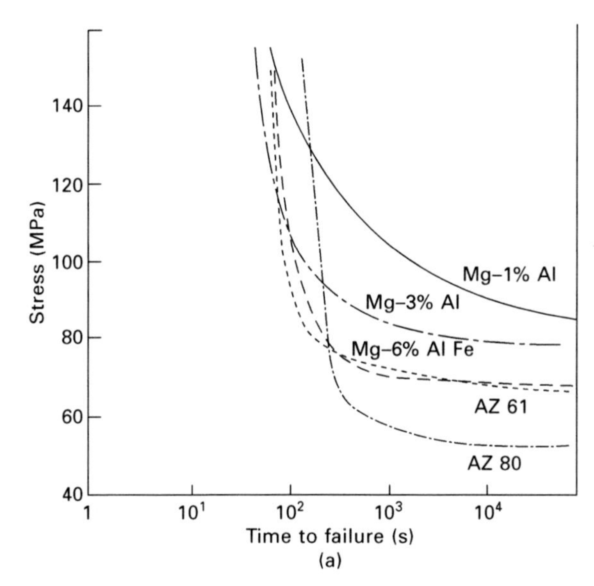

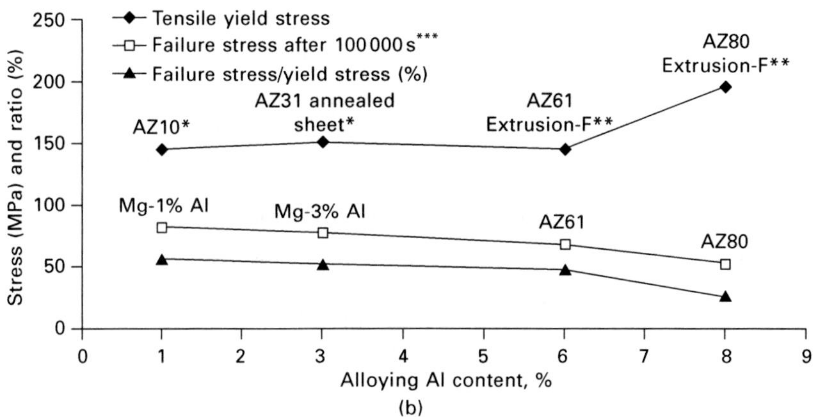

8.4 (a) Initial applied stress vs. time to failure for various Mg–Al alloys in chloride–chromate solutions [106]. (b) Dependence of Al content on (i) failure stress at 100 000 s from (a) [104], (ii) tensile yield strength for AZ10, AZ31 [110], AZ61 and AZ80 [111], and (iii) ratio of (failure stress)/(tensile yield strength). (Data from **ASM Handbook* online, ASM International, Materials Park, OH, 2002; **Magnesium Electron (Private communication); *** L. Fairman, H.J. Bray, 'Transgranular SCC in magnesium alloys; *Corrosion Science*, **11** (1971) 5330541.)

An HE mechanism was postulated. Kannan and Raman [112] found small decreases of strength and ductility for AZ91 in a simulated body fluid using the constant extension rate test (CERT); these decreases usually indicate SCC although the decreases were claimed to be insignificant.

Makar *et al.* [113] showed that (i) increasing Al from 1 to 9% increased the repassivation rate of Mg alloys, attributed to the superior passivation of Al, and (ii) Al extended the pH range over which Mg alloys form a protective film. By analysing the $Mg(OH)_2$ surface film formed in aqueous environments using X-ray and electron diffraction, Fairman and Bray [106] proposed that two Al^{3+} ions replace three Mg^{2+} ions in the tetrahedral $Mg(OH)_2$ lattice, resulting in vacant lattice sites. The resulting film was thicker and was considered more protective. This was used to explain the perceived increased general corrosion resistance of Mg–Al alloys.

Influence of Zn

Zn was reported to increase SCC susceptibility [60], although this was disputed by Fairman and Bray [106]. Mg–Zn alloys containing rare earths, such as ZExx, are considered to have moderate SCC susceptibility relative to Mg–Al–Zn alloys. AZxx alloys contain both Al and Zn and are considered particularly susceptible in air and chloride-containing solutions. They are also the most common, which is one reason why many studies have focused on Mg–Al alloys [114–120], Mg–Al–Zn alloys [104, 106,121–125] and pure Mg [98, 99,101].

Influence of Mn

Mg–Mn alloys were considered to be immune in the atmosphere, chloride solutions and chloride–chromate solutions [60], but are susceptible (i) in the atmosphere and in distilled water [126], (ii) in solutions containing chloride and sulphate ions [126], (iii) and Mn–2%Mn–0.5%Ce showed SCC in distilled water and 0.5% KHF [119]. Timonova [127] stated that addition of Mn or Zn to Mg–8% Al decreased susceptibility; however, addition of both elements increased susceptibility.

Influence of rare earth (RE) elements

Mg alloys incorporate rare earth (RE) elements [128] to improve (i) creep resistance, which is primarily achieved by RE-containing phases along grain boundaries [129,130], (ii) castability, (iii) age hardening [131], and corrosion resistance [132]. Chang *et al.* [132] reported that Mg–3Nd–0.2Zn–0.4Zr had a corrosion rate lower than AZ91D. Nordlien *et al.* [133] reported that RE elements improved passivation. Krishnamurthy *et al.* [134] suggested that pseudo-passivation in rapidly solidified Mg–Nd was due to Nd enrichment at the surface.

The literature on SCC of RE containing Mg alloys is limited [5,68,128]. Winzer *et al.* [5] indicated that Nd had little or no influence on SCC susceptibility, whereas Rokhlin [128] reported that Nd addition to Mg–Zn–Zr increased SCC resistance. Kannan *et al.* [68] found SCC for three RE alloys (ZE41, QE22 and EV31A) and AZ80 in 0.5 wt% NaCl and in distilled water. TGSCC in AZ80, ZE41 and QE22 in distilled water and in AZ80 in 0.5 wt% NaCl was consistent with HE, whereas IGSCC in ZE41, QE22 and EV31A in 0.5 wt% NaCl was caused by micro-galvanic corrosion associated with the second phase along grain boundaries (Fig. 8.5).

Influence of Fe

Fe is present only as an impurity, and significantly increases the corrosion rate [1,2,15,100] above the tolerance limit. Perryman [119] reported that Mg–5Al + 0.13%Fe had SCC susceptibility in distilled water higher than Mg–5Al + 0.0019% Fe. Similarly, Pelensky and Gallaccio [105] reported that SCC susceptibility for Mg–5Al alloys increased with Fe concentration and Pardue *et al.* [125] reported that the fraction of TGSCC in AZ61 increased with Fe concentration. In contrast, Fairman and Bray [106] stated that Fe contributes to general corrosion but has minimal effect on SCC. Similarly, Miller [107] found equivalent SCC susceptibility in distilled water for low purity (0.010% Fe) and HP AZ91 and Timonova [127] reported that Fe had no effect on the SCC of Mg–Al–Zn–Mn alloys. Intermetallic FeAl could increase SCC susceptibility [104,119,135]. Priest *et al.* [104] postulated that FeAl occurs on the basal plane, and that FeAl causes SCC by preferential corrosion of the adjacent matrix; however, there was no evidence of FeAl on basal planes.

Other elements

Reports are varied on other elements. Li, Ag, Nd, Pb, Cu, Ni, Sn and Th have little or no influence on SCC susceptibility [60,100]. Busk [103] stated that cast Mg–Zr alloys have negligible SCC whereas the *ASM Handbook* [100] states that alloys containing Zr and rare earths have intermediate SCC susceptibility in atmospheric environments. Cd, Ce and Sn may increase susceptibility in certain alloys [100, 127]. Conversely, Rokhlin [136] reported that addition of Cd and Nd to Mg–Zn–Zr increased SCC resistance.

8.2.3 AZ91, AZ31 and AM30

Winzer *et al.* [37] characterised SCC of AZ91, AZ31 and AM30 in distilled water using the CERT and the linearly increasing stress test (LIST) [69,82]. AZ91 consists of an α-matrix with a significant amount of β-phase, whereas AZ31 and AM30 consisted essentially only of an α-matrix with an Al-concentration similar to that in the α-phase of AZ91. AZ91, AZ31 and AM30

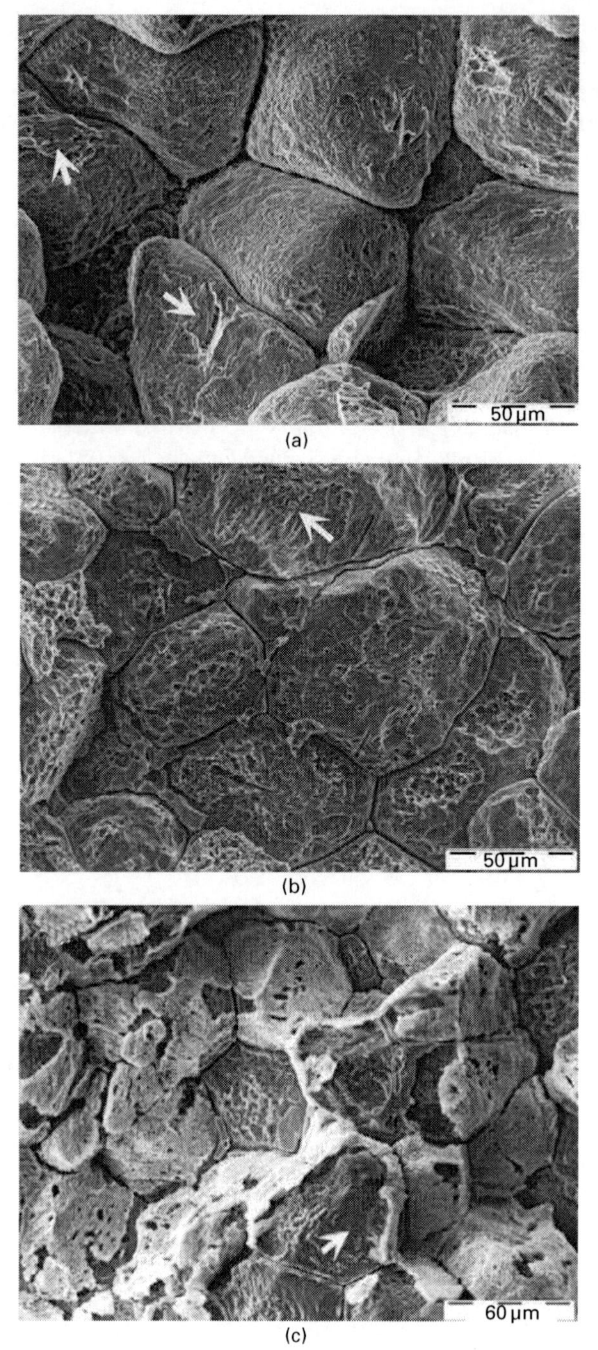

8.5 SCC of RE-containing Mg alloys in 3.5% NaCl. (a) ZE41: predominantly IGSCC with isolated TGSCC, arrows; (b) EV31: IGSCC with some TGSCC, arrow; and (c) QE22: IGSCC and TGSCC, arrow [68].

were susceptible to SCC in distilled water (Fig. 8.6), so water itself is the key environment factor. A TGSCC tendency is expected for all aqueous solutions unless there is clear contrary evidence. The mechanism for SCC initiation in AZ31 and AM30, under CERT conditions, involved localised dissolution at stresses significantly higher than for AZ91. A mechanism involving mechanical film rupture at emerging slip steps seems likely. The mechanism for crack initiation in AZ91 involves fracture of β-particles close to the surface, with H ingress facilitated by mechanical rupture of the surface film; a similar mechanism is proposed for crack propagation.

The threshold stress, σ_{SCC}, was measured using a DC potential drop (DCPD) technique (Fig. 8.7). The threshold stress was 55–75 MPa for AZ91, 105–170 MPa for AZ31 and 130–140 MPa for AM30. SCC susceptibility increased with decreasing strain rate. The low σ_{SCC} of AZ91 is attributed to: (i) the tendency for the β-particles to fracture; and (ii) their behaviour as reversible H traps, which enhance H transport within the matrix. This implies that the increasing susceptibility of Mg alloys to SCC with increasing Al concentration is related to β-particles. Generalisation implies low SCC resistance for two-phase Mg alloys if there is an H influence on the fracture of the second phase.

The stress corrosion crack velocities for AM30 ($V_c = 3.6 \times 10^{-10}$–$9.3 \times 10^{-10}$ m/s) were slower than for AZ91 ($V_c = 1.6 \times 10^{-9}$–1.2×10^{-8} m/s) and AZ31 ($V_c = 1.2 \times 10^{-9}$–6.7×10^{-9} m/s). This is consistent with a lower

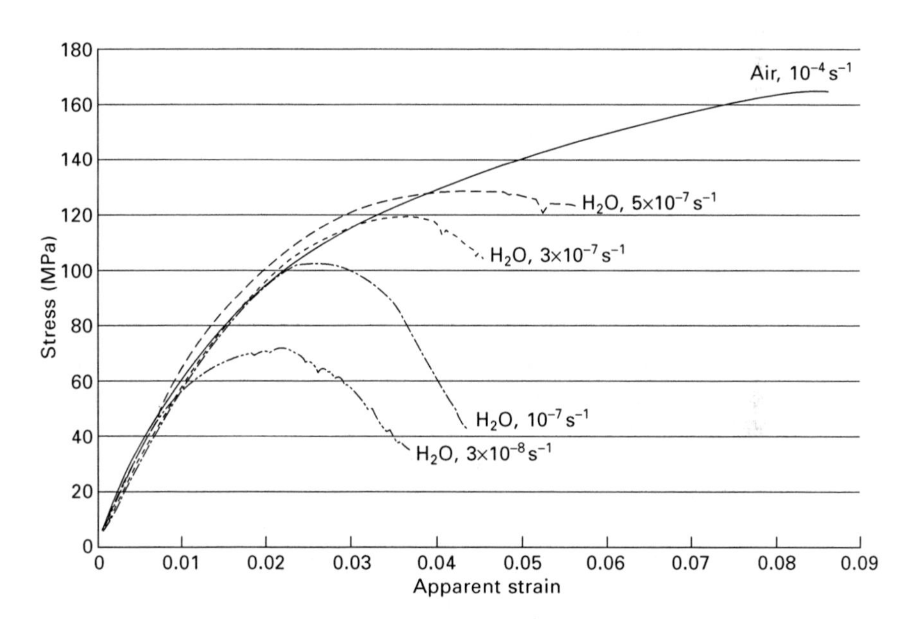

8.6 Stress vs. apparent strain curves for AZ91 in distilled water and air under CERT conditions.

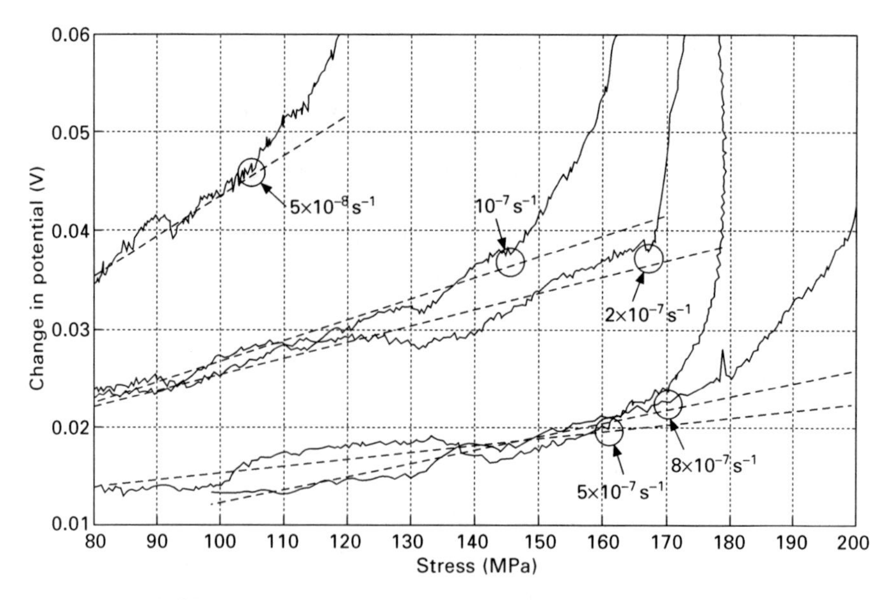

8.7 DCPD vs. stress curves for AZ31 in distilled water and air under CERT conditions showing threshold stress values. This technique works equally well for LIST and similar threshold values are measured.

H diffusivity in the α-phase in the absence of Zn. AM30 has a similar Al concentration as AZ30, but there is no Zn.

AZ91, pre-charged in gaseous H_2 at 3 MPa for 14 h at 300 °C, fractured just above the yield stress, σ_Y, without apparent plastic strain or reduction in load. This is consistent with a mechanism involving the formation and fracture of a brittle hydride phase.

8.2.4 Wrought vs. cast

Influences of microstructure, composition and strain rate for SCC of cast alloys are similar to those of wrought alloys. SCC susceptibility is influenced by (i) the presence and distribution of the β-phase, (ii) grain size and (iii) residual stresses. These factors are related to processing (casting or mechanical working) and heat treatment.

Figure 8.8 shows that the SCC susceptibility for sand-cast AZ63 in a rural atmosphere was lower than for extruded or rolled AZ61 [137]. This may be due to the residual stresses resulting from extrusion or rolling or may be due to surface pick up of Fe during rolling [138]. Stephens *et al.* [139] reported that cast AZ91E-T6 was highly susceptible to SCC in 3.5% NaCl. Lynch and Trevena [98] reported SCC in as-cast 99.99% pure Mg in 3.3% NaCl + 2% K_2CrO_4. The casting process can influence SCC susceptibility. Makar

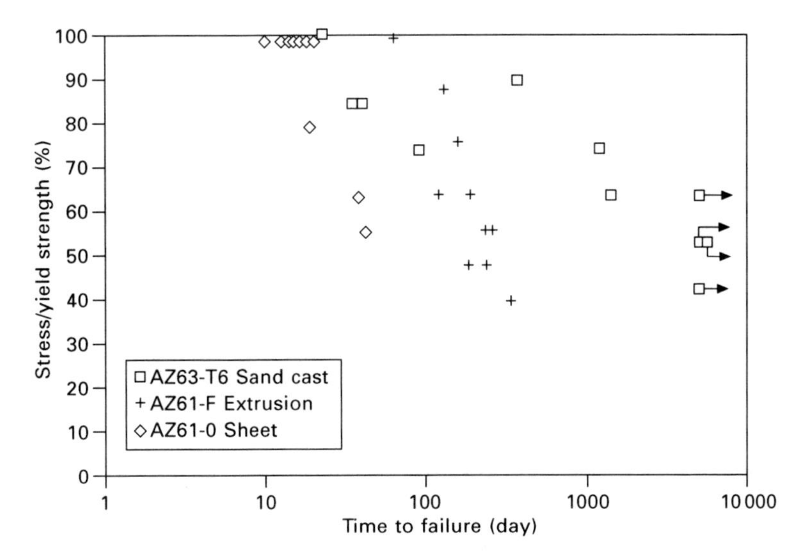

8.8 Long-term rural atmosphere SCC susceptibility, as characterised by normalised applied stress as a percentage of yield strength and time to failure and wrought Mg alloys with similar composition.

et al. [113] compared the TGSCC of HP rapidly solidified (RS) and as-cast (AC) Mg–1Al and Mg–9Al in 3.5% NaCl + 4.0% K_2CrO_4, pH 9.0. They found that, although the overall current densities for RS and AC alloys were similar, repassivation occurred more quickly in the RS alloys, while corrosion in AC alloys was more severe and localised. The greater localised corrosion resistance and higher repassivation rate of the RS alloys was attributed to greater compositional and microstructure homogenisation. Rapid solidification increased the solubility of Al in the Mg matrix from around 2% (for as-cast alloys) to between 5 and 9%. Therefore AC alloys were more likely to contain $Mg_{17}Al_{12}$ grain–boundary precipitates. Similarly, Mathieu *et al.* [140] stated that the concentration of Al in the primary α matrix for AZ91D is 1.8% for die-castings and 3% for semi-solid castings. The uniformity and corrosion protection provided by the surface film depends on the composition. AC alloys tended to be less uniform and have more localised corrosion due to galvanic interactions with the second phase. Localised corrosion can assist SCC by providing film-free surfaces for easy H ingress [120].

SCC resistance is reduced by residual tensile stress [100]. Timonova *et al.* [141] found reduced SCC resistance for Mg–Y–Zn alloys in NaCl following hot deformation, and that resistance was partly recovered by annealing, attributed to the removal of residual tensile stresses. Marichev and Shipilov [142] observed anisotropic SCC resistance of hot rolled Mg–Y–Zn. Chakrapani and Pugh [115, 117] proposed that SCC occurred by cleavage on $\{31\bar{4}0\}$ planes for hot-rolled Mg–7.5Al in 4% NaCl + 4% K_2CrO_4. The cleavage planes were orientated perpendicular to the rolling plane, so, for specimens loaded in bending, the direction of crack propagation was defined by the orientation of the tensile face with respect to the rolling plane. Timonova *et al.* [141] stated that for Mg–Y–Zn alloys formed by hot deformation, all anisotropy was removed by annealing.

8.2.5 TGSCC vs. IGSCC

Many workers [60] have stated that most SCC occurs by TGSCC. Others [99,101,104,124] claim that IGSCC is dominant. There are different mechanisms for IGSCC and TGSCC. The mechanism is determined by the microstructure and the environment. Pardue *et al.* [125] proposed that TGSCC is discontinuous and involves alternating fracture and dissolution whereas IGSCC is continuous and completely electrochemical. It also appears that H mechanisms may produce both TGSCC and IGSCC, depending particularly on the grain size. Stampella *et al.* [101] observed TGSCC for fine grained (0.025 mm) 99.5% CP Mg whereas mixed TGSCC–IGSCC occurred for large-grained (0.075 mm) 99.95% HP Mg. IGSCC was attributed to the large grains being subject to higher stresses across grain boundaries, due to more concentrated dislocation pile-ups. In contrast, Meletis and Hochman [99] reported TGSCC for 99.9% purity Mg, heat-treated and furnace cooled to produce large grains. Stampella *et al.* [101] used a dilute 10^{-3} Na_2SO_4 solution while Meletis and Hochman [99] used a chloride–chromate solution, suggesting that crack morphology is influenced by the environment. This was supported by Perryman [119] who showed that Mg–5Al evinced TGSCC in saturated $MgCO_3$ solution, 0.5% KF solution, 0.5% KHF solution and 0.5% HF solution, but IGSCC in 0.05% potassium chromate solution. If HE is the mechanism tending to produce TGSCC, then the critical parameter characterising the solution–material interaction might be the effective H activity produced by each solution; it might be expected that a higher H activity might be linked to a greater tendency to produce TGSCC, even for pure Mg with a large grain size.

Priest *et al.* [104] and Fairman and Bray [124] found IGSCC for Mg–Al–Zn and Mg–Al alloys, heat treated to produce fine grains (~5 μm) and with heavy $Mg_{17}Al_{12}$ precipitation at grain boundaries. Fairman and Bray [124] stated that Mg–Al alloys are more inclined to IGSCC if the Al content is greater than 6%. This was based on the suggestion that IGSCC occurs by preferential corrosion of the Mg matrix adjacent to the continuous $Mg_{17}Al_{12}$ precipitate at grain boundaries (Fig. 8.9). Miller [60] and Pardue *et al.* [125] attribute all IGSCC to this micro-galvanic corrosion, although Priest *et al.* also noted that some limit of the grain size was required to induce IGSCC regardless of the presence of $Mg_{17}Al_{12}$. The maximum grain size for IGSCC was found to be around 28 μm. Priest *et al.* proposed that, where both IGSCC

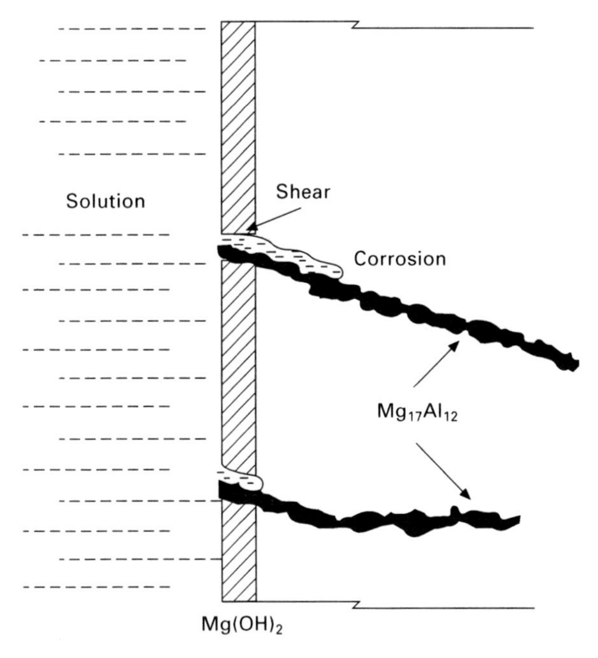

8.9 Preferential corrosion of metal matrix adjacent to $Mg_{12}Al_{12}$.

and TGSCC are possible, due to the presence of $Mg_{17}Al_{12}$ and a source of H, increasing the grain size causes a transition to TGSCC because the crack may propagate along a more direct path, or alternatively HE might be easier with large grains.

Mears *et al.* [143] found IGSCC for Mg–6.5Al–1Zn in the pH 5.0 chloride–bichromate solution (3.5% NaCl + 2.0% $K_2Cr_2O_7$) and TGSCC in the pH 8.1 chloride–chromate solution (3.5% NaCl + 4.0% K_2CrO_4.). However, that crack morphology was influenced by pH was rejected on the basis of experimental evidence by Fairman and West [122] and Pardue *et al.* [125]. Fairman and West tested single-phase Mg–7Al–1Zn in chloride–chromate solutions. The popular model for IGSCC cracking indicates that, in single-phase alloys, TGSCC is due to the absence of $Mg_{17}Al_{12}$. Also, Pardue *et al.* tested large-grained (~80 μm) Mg–6Al–1Zn in a chloride–chromate solution, which according to Priest *et al.* [104] would show TGSCC due to the large grain size.

8.2.6 Heat treatment

Speidel *et al.* [144] stated that the threshold stress intensity factor and stage 2 crack velocity for all high-strength Mg alloys (including ZKxx) do not vary significantly with heat treatment. This would imply that SCC of high-strength Mg alloys does not depend on ageing and also does not depend on slip morphology as is the case for high-strength steels and high-strength aluminium alloys.

However, the previous discussion on the roles of $Mg_{17}Al_{12}$ precipitates and grain size suggests that heat treatment does influence SCC of Mg–Al alloys with grain size close to the IG–TG transition. There may be $Mg_{17}Al_{12}$ in Mg alloys with more than 2% Al [113] and $Mg_{17}Al_{12}$ is precipitated by slow cooling from the solution-treatment temperature [60] and slow cooling of castings [12].

Similarly, Priest *et al.* [104] found TGSCC in chloride–chromate solutions for Mg–6Al–1Zn heat-treated for 24 h at 345 °C and water quenched, whereas there was IGSCC for furnace-cooled samples. $Mg_{17}Al_{12}$ was only precipitated in the furnace-cooled samples. Moccari and Shastry [123] compared SCC for rolled AZ61, as-received and after heat treatment (i) annealed at 475 °C for 90 min and quenched in boiling water or iced brine (A&Q), and (ii) aged at 200 °C for 48 h. There was IGSCC for the as-received specimens whereas there was some TGSCC for the heat-treated specimens. This suggests that, there was discontinuous grain boundary precipitation in the heat-treated specimens. The heat treatment did not significantly improve the SCC resistance.

Perryman [119] studied the influence of heat treatment on SCC susceptibility of Mg–5Al in distilled water. The grain boundary precipitate was completely dissolved by solution treating at 360 °C for 3 h. Subsequent ageing at 150 °C for 2 h to produce a continuous film of $Mg_{17}Al_{12}$ at grain boundaries had no significant effect on TGSCC. It was inferred that the $Mg_{17}Al_{12}$ at the grain boundaries was not significant for the observed TGSCC. However, precipitation within grains, by ageing at 200 °C for 4 and 14 days, caused a reduction in localised corrosion and improved SCC resistance.

Pardue *et al.* [124] investigated the influence of grain size and cooling rate on SCC of wrought AZ61 in a chloride–chromate solution. The specimens were heat treated at 345, 425 or 480 °C for 24 h and furnace cooled or quenched. The furnace-cooled specimens had high concentrations of $Mg_{17}Al_{12}$ at grain boundaries, and consequently had primarily IGSCC, whereas water quenched specimens had higher percentages of TGSCC. Specimens that were heat treated at 425 and 480 °C had larger grains than those treated at 345 °C and this correlated with a higher percentage of TGSCC.

Residual tensile stresses decrease the minimum applied stress required for SCC [127]. Busk [104] noted that residual tensile stresses reduce the threshold stress but can be relieved by T4 and T5 heat treatment: T4 consists of solution heat treatment and ageing to a stable condition, whereas T5 consists of artificially ageing at an elevated temperature. Timonova [127] stated that relieving residual stresses in MA5 by annealing significantly increased SCC resistance, with resistance increasing with annealing temperature.

Kiszko [145] found IGSCC in humid air for Mg–14Li with more than 1%

Al, rapidly cooled after hot-working at 370 °C. The IGSCC susceptibility was associated with the Al content and the formation of a second phase during rapid cooling. SCC resistance was restored by heating 24 h at 150 °C.

8.2.7 Welding

Welding is used to fabricate engineering structures. Fusion welding changes the microstructure due to the thermal cycle. The weld metal can have a composition significantly unlike the parent metal, depending on the welding consumable used. These changes in microstructure and chemical composition may influence SCC. There is extensive research on Mg welding [146–148]. Laser beam (LB) welding of Mg alloys has the following advantages: (i) high welding speed; (ii) very narrow joints with reduced heat-affected zone (HAZ) and low distortion; and (iii) a high joint efficiency (joint strength/base metal strength) approaching 100% [146], whereas conventional fusion welding, such as tungsten inert gas (TIG) welding, has joint efficiencies of ~ 70–90% [147]. Friction stir welding (FSW) [149] achieves a metallic bond at temperatures below the melting point of the base material, thereby avoiding the issues associated with melting of the alloy [150].

Only a few researchers [151–153] have studied SCC of welded Mg. Winzer *et al.* [153] reported high SCC susceptibility of TIG welds of continuous-cast AZ31 sheet. SCC initiated at the interface between the weld metal and the HAZ. Kannan *et al.* [151] also reported high SCC susceptibility of LB welded AZ31. SCC occurred in the fusion boundary (Fig. 8.10(a)). It was suggested that SCC was caused by galvanic corrosion between the fusion zone and the base metal, because of the difference in Al concentration. The fracture was mixed IGSCC and TGSCC (Fig. 8.10(b)). The high SCC susceptibility for fusion welded Mg alloys appears to be caused by the ease of producing significant galvanic corrosion. Kannan *et al.* [151] also reported high SCC susceptibility of FSW AZ31. SCC occurred in the stir zone (SZ) (Fig. 8.11). The authors suggested that hydrogen-assisted cracking could cause SCC, due to the high dislocation density in the stir zone.

Surface treatment is a popular method for corrosion prevention. Bala Srinivasan *et al.* [154,155] studied plasma electrolytic oxidation (PEO) coated LB welded AZ31 and FSW AZ61. The PEO coating increased the corrosion resistance, but did not improve SCC resistance. The failure to improve SCC resistance may be because the PEO coating is brittle.

8.3 Loading

8.3.1 Fracture mechanics

A given material/environment combination has a simple relation between the stress intensity factor, K_I, and the growth rate of the stress corrosion crack,

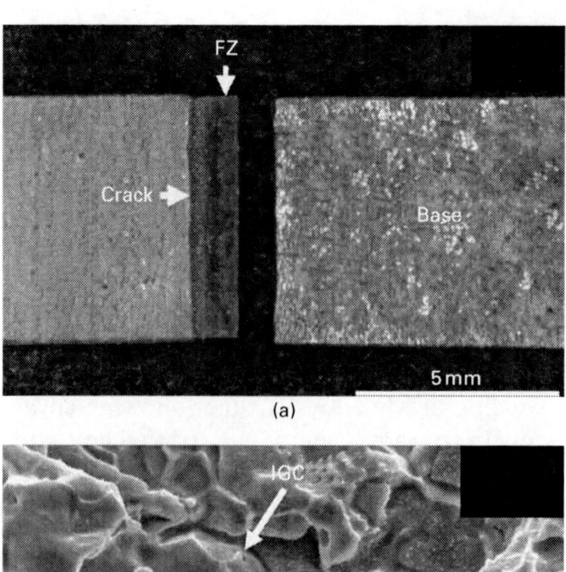

(a)

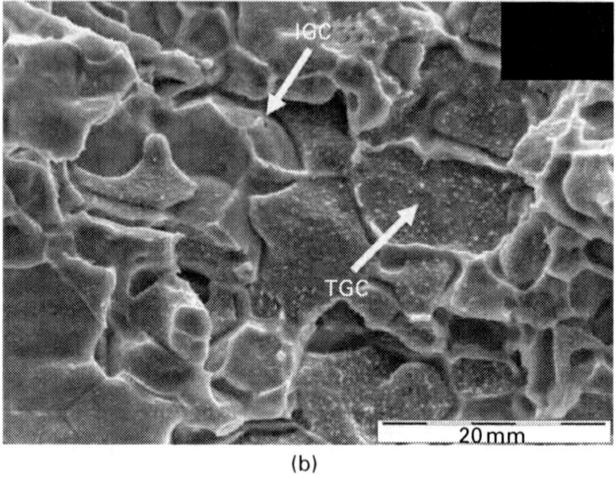

(b)

8.10 (a) Overview of LB AZ31 after SCC test, and (b) the fracture surface showed IGSCC and TGSCC.

d*a*/d*t* or *v*. Figure 8.12 provides typical data [144]. The threshold for SCC, K_{ISCC}, is the stress intensity factor above which occurs the first measurable crack extension. This does not imply that there is, under all circumstances, a threshold in the sense of a real cut-off. On the contrary, for many material/environment combinations, the existence of a true threshold appears rather doubtful. The practical meaning of K_{ISCC} lies in the fact that below this stress intensity factor, the growth rates of stress corrosion cracks fall below a lower limit of e.g. 10^{-10} m/s, corresponding to a crack increment of roughly 3 mm per year. The typical curve has three stages: initiation and stage 1 propagation; steady state or stage 2; and final failure or stage 3. Region 3 is not always observed; for example, region 3 is absent in Fig. 8.12 [144] for ZK60 in distilled water, 1.4 m Na_2SO_4 and 5 m NaBr. During stage 1, the crack velocity increases rapidly as the stress intensity factor increases above K_{ISCC}. During stage 2, the crack velocity is essentially constant. As

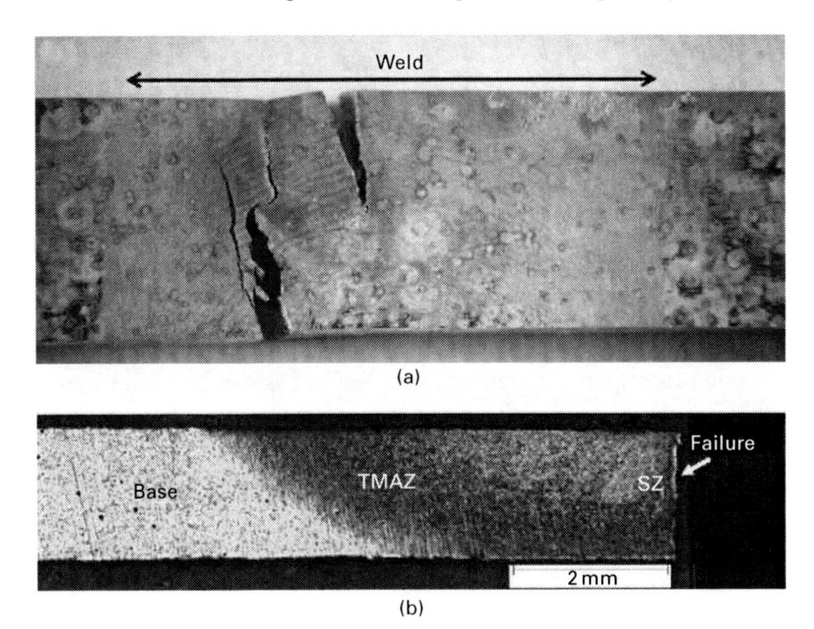

8.11 (a) Overview of FSW AZ31 after SCC test, and (b) SCC was in the stir zone.

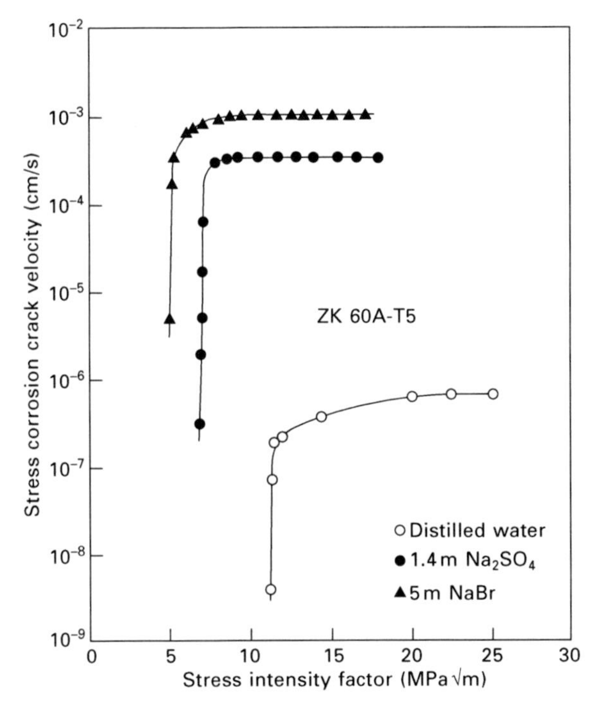

8.12 Effect of the stress intensity factor on crack velocity for ZK60 alloy in various environments [144].

the stress intensity factor approaches K_{1c}, the crack propagation mechanism becomes dominated by ductile tearing and rapid crack growth ensues until failure occurs. In regions 1 and 3, crack velocity is strongly dependent on the stress intensity factor, whereas the steady state crack velocity in region 2 is largely independent of the stress intensity factor [7,8,144]. Quantitative prediction of crack velocities has not been particularly successful. Figure 8.13 shows that environmental conditions can influence the steady-state crack velocity and the threshold values [120].

Results from the different types of SCC tests, and on various specimen configurations, are identical if all requirements are carefully met. The data obtained from part through-cracked specimens can be used to predict crack initiation and growth in plates containing small semi-elliptical surface cracks, which simulate defects in large structures. This underlines the capability of the fracture mechanics approach to transfer SCC data from small-scale laboratory specimens to real components and structures.

The advantage of using the parameter K_{ISCC} lies in its ability to predict the combinations of stress, flaw size and shape that lead to SCC. K_{ISCC} may be used as a design criterion for ensuring no SCC growth in service, provided that the stress, minimum detectable flaw size and environmental conditions are well defined, and that the service loads are essentially sustained, i.e. that cyclic loading is not significant. Figure 8.14 illustrates how K_{ISCC} values can be used in an assessment of structural integrity. In an inert environment,

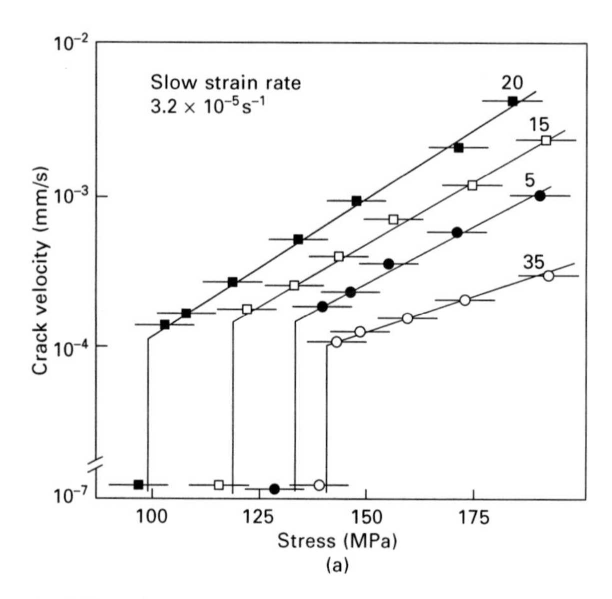

8.13 Threshold stresses and intensity factors for Mg–9Al under various loading conditions in aqueous solutions with 35 g/L NaCl and various K_2CrO_4 concentrations (given in g/L) [120].

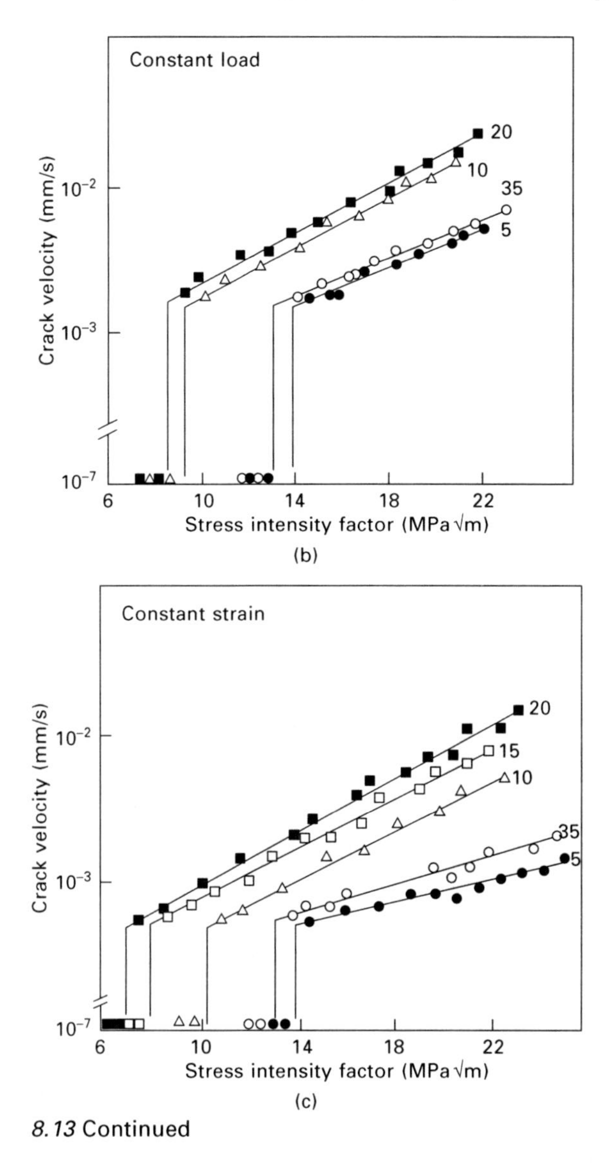

8.13 Continued

the mechanical limits are defined by the yield stress, σ_y, and the fracture toughness, K_{1C}. In a SCC environment, the mechanical limits are reduced to the threshold stress intensity factor, K_{1SCC}, and the threshold stress, σ_{SCC}, determined on smooth tensile specimens of the same material and in the same environment. These two parameters define an acceptable region immune to SCC. This region can be significantly smaller than the original acceptable region derived from tests in air.

For small cracks, the use of the K concept, and hence of K_{ISCC}, is non-conservative (Fig. 8.14). A combination of the SCC threshold stress, σ_{SCC},

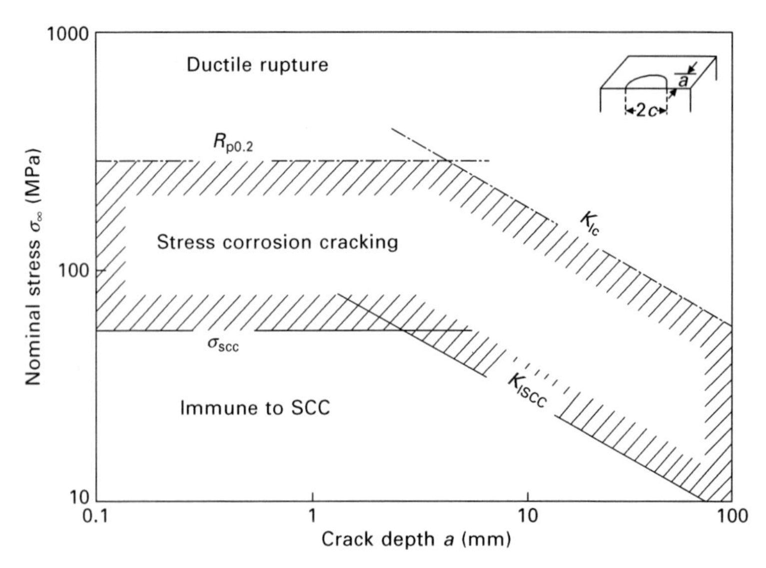

8.14 Mechanical limits of a typical system involving an inert and an SCC environment.

determined on smooth specimens, and K_{ISCC} may serve as a short crack design criterion defining a minimum crack size for the use of the K concept.

8.3.2 Stress and strain rate

The probability of SCC increases with increasing applied and internal tensile stresses [7,127]. Typically σ_{SCC} is ~ ½ yield strength [5,100,114,120]. However, attempts to predict σ_{SCC} have not been successful. K_{1SCC} may be determined from data as in Figs 8.12 and 8.13. σ_{SCC} may be determined from time to failure data as in Figs 8.4 or 8.8; or by a potential drop technique associated with LIST or CERT as in Fig. 8.7.

Wearmouth *et al.* [114] found, for Mg–7Al in 3.5% NaCl + 2% K_2CrO_4, σ_{SCC} correlated with K_{1SCC}. Ebtehaj *et al.* [120] observed a similar relationship for Mg–9Al in aqueous solutions with 35 g/L NaCl and various K_2CrO_4 concentrations, and suggested that this correlation was because, for pre-cracked or plain specimens, cracking was controlled by the strain rate, or alternatively cracks in plain specimens were initiated at corrosion pits such that they were essentially notched specimens. Both K_{1SCC} and σ_{SCC} represent the minimum values for which a crack would continue to propagate to total failure. Both thresholds can be dependent on the environment as shown by Figs 8.12 and 8.13.

Constant load tests can be more severe than constant displacement tests [60, 125] in that SCC has occurred in constant load tests whereas the constant displacement tests have indicated no SCC [84]. For constant load tests, the stress increases as the cross-sectional area reduces with crack propagation, whereas for constant displacement tests, the stress decreases as the crack lengthens. Creep also decreases the stress by relaxation of residual stresses and those caused by applied displacements.

Wei *et al.* [156], in constant displacement tests of Mg–9Al (and Mg–9Al alloys containing various additions of RE, Zn and Ca) found that Mg–9Al had a threshold strain of 0.35% in a solution of 3.5% NaCl + 2% K_2CrO_4, and higher thresholds for the alloys containing the various additions of RE, Zn and Ca. These results may indicate that plastic deformation is required for SCC, or alternatively may simply reflect that constant deflection tests give threshold values higher than constant load tests.

The role of strain in SCC has been attributed to the rupture of a surface film, to allow H ingress, or to allow localised dissolution. Wearmouth *et al.* [114] investigated the influence of strain rate for high-purity extruded Mg–7Al in an aqueous 2% K_2CrO_4 + 3.5% NaCl. They proposed that the SCC crack propagation velocity was defined by the balance between corrosion inhibition by film growth and bare metal exposure by the crack tip strain rate. Consequently, SCC occurred in a limited range of strain rates, where the strain at the crack tip was sufficient to overcome repassivation.

Chakrapani and Pugh [118] reported that HP Mg–7.5Al exhibited reduced ductility, characteristic of SCC/HE, when exposed to gaseous H or cathodically generated H in 4% NaCl + 4% K_2CrO_4 under load. Figure 8.15 shows that the difference in tensile properties for pre-exposed and vacuum annealed specimens was dependent on strain rate. At low strain rates there was a large difference in the ultimate tensile strength (UTS) and elongation of the annealed and pre-exposed specimens. The fracture surfaces at the higher strain rates were completely dimpled, suggesting that ductile tearing was the dominant failure mechanism. The authors proposed that, at the higher strain rates, insufficient time was available for H diffusion. For the specimens tested at the slower strain rates, the difference in UTS and elongation to failure of the annealed and pre-exposed specimens was attributed to loss of H during annealing.

Ebtehaj *et al.* [120] investigated the influence of strain rate on SCC susceptibility for cast Mg–9Al. They proposed that the SCC mechanism involved diffusion of cathodically generated H, and that the SCC susceptibility is defined by opposing effects relating to H ingress as suggested by Wearmouth *et al.* [114]. At slow strain rates, film integrity was maintained, preventing H ingress, and ductile fracture ensued. As the strain rate was increased, the effectiveness of repassivation was reduced, allowing H to ingress more freely. At high strain rates, ductile tearing occurred before embrittlement, because insufficient time was available for H ingress. This agreed with the work by Chakrapani and Pugh [118]. Maximum susceptibility occurred at an intermediate strain rate (Fig. 8.16). Furthermore, the strain rate corresponding

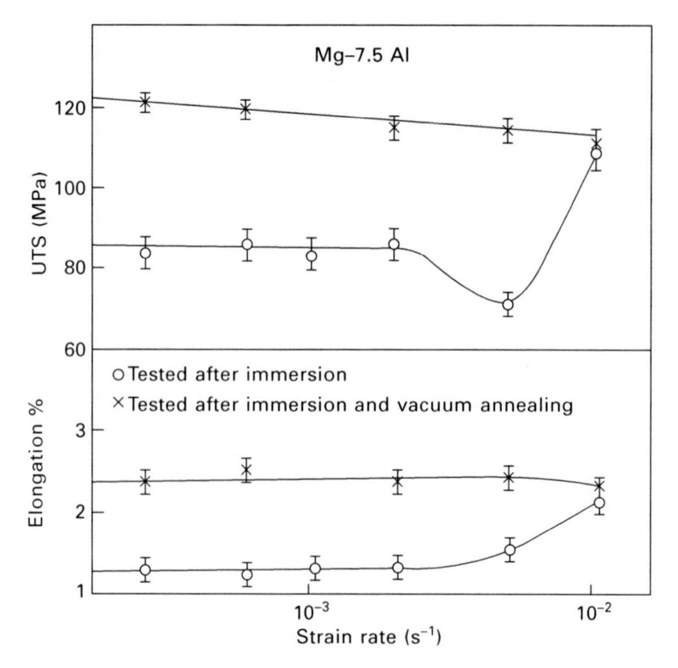

8.15 Effect of strain rate on tensile properties for pre-exposed and vacuum annealed Mg–7.5Al specimens [118].

to maximum susceptibility was dependent on the balance between active and passive corrosion (as defined by the ratio of chloride and chromate ion concentrations) with this effect becoming less significant at high strain rates. As the chromate concentration increased, the tendency for repassivation increased, so higher strain rates were required to overcome repassivation and the overall resistance to cracking increased. This explains the locations of the mimima with respect to strain rate for the curves representing 5, 20 and 35 g/L K_2CrO_4.

The influence of strain rate on the susceptibility of Mg alloys to SCC in the work of Ebtehaj *et al.* [120] was ascribed to, at low strain rates, the balance between repassivation and mechanical film rupture at the crack tip and, at high strain rates, the propensity for the inert fracture mechanism to overwhelm the SCC fracture mechanism. In contrast is SCC in distilled water [37], in which decreasing strain rate caused a continuous increase in SCC susceptibility (characterised by: (i) an increasing difference between σ_{SCC} and the UTS; (ii) a decreasing elongation-to-failure; and (iii) a decrease in σ_{SCC} for AZ91 and AZ31). At these strain rates, the influence of repassivation at the crack tip is negligible. In contrast the data of Fig. 8.16 used strongly passivating solutions containing K_2CrO_4. This indicates that the occurrence of maximum SCC susceptibility at intermediate strain rates for Mg alloys in strongly passivating solutions (Fig. 8.16) is a characteristic of these environments rather than a characteristic of Mg alloys.

In agreement, Nozaki *et al.* [121] also found that SCC susceptibility increased as the strain rate decreased from $8.3 \times 10^{-4}\,s^{-1}$ to $8.3 \times 10^{-7}\,s^{-1}$ for AZ31B in distilled water (Fig. 8.17). SCC susceptibility was characterised by a SCC susceptibility index, I_{SCC}, given by

$$I_{SCC} = [(E_{oil} - E_{SCC})/E_{oil}] \times 100\% \qquad 8.1$$

where E_{SCC} and E_{oil} are the area under the stress–strain curve in the SCC solution and in oil, respectively, the latter being inert. I_{SCC} is high for a system with high SCC susceptibility whereas I_{SCC} tends to zero for a system with low susceptibility. Figure 8.17 shows that susceptibility in distilled water increased from 0 to 85% as the strain rate decreased from $8.3 \times 10^{-4}\,s^{-1}$ to $8.3 \times 10^{-7}\,s^{-1}$. For NaCl concentrations of 4 and 8%, susceptibility was around 90%, independent of strain rate. This again indicates no maximum of SCC susceptibility with decreasing strain rate and contrasts with the work of Ebtehaj *et al.* [120].

Makar *et al.* [113] also found that the SCC velocity increased continuously with increasing loading rate for RS Mg–1Al and RS Mg–9Al in 0.21 M

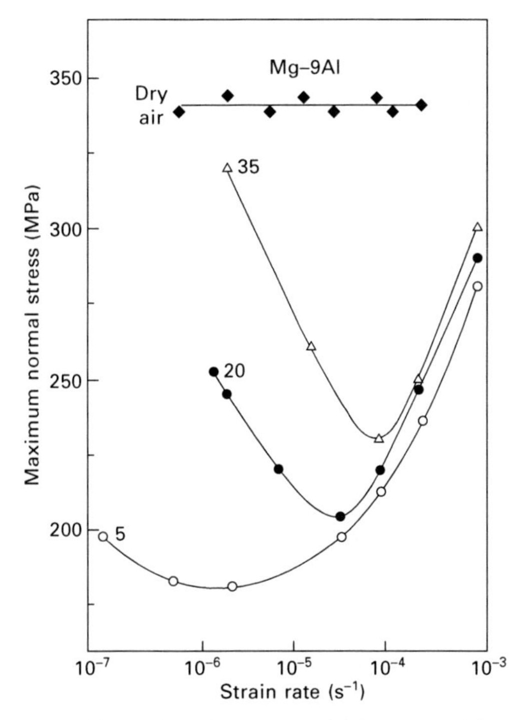

8.16 Effect of strain rate on SCC susceptibility for 5, 20 and 35 g/L K_2CrO_4 and 5 g/L NaCl [120].

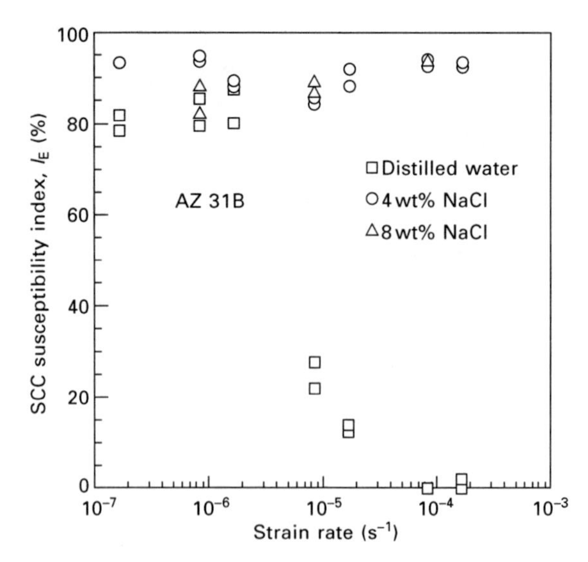

8.17 Effect of strain rate on SCC susceptibility index for AZ31B in distilled water and NaCl solutions [121].

K_2CrO_4 + 0.6 M NaCl. Markar *et al.* indicated that ductile tearing became increasingly dominant, as the loading rate increased from 4.8×10^{-3} mm s^{-1} to 8.9×10^{-3} mm s^{-1}, above which the fracture surfaces were mostly ductile. This could explain the high rates and ductile tearing mechanism observed by Lynch and Trevena [98].

8.3.3 CERT vs. LIST

Winzer *et al.* [69] compared the LIST and the CERT in the evaluation of TGSCC of AZ91 in distilled water and 5 g/L NaCl. The LIST apparatus [82], illustrated in Fig. 8.18, is based on the lever principle. The specimen is attached to one end of the lever arm. A known mass is attached to the other end. The tensile load applied to the specimen increases linearly as the distance between the fulcrum and the mass is increased by means of a screw thread and synchronous motor. LIST is load controlled whereas CERT is extension controlled. They are essentially identical until SCC initiation. Thereafter, LIST ends as soon as a critical crack size is reached whereas CERT can take much longer as typically CERT only ends when the final ligament suffers ductile rupture.

Figure 8.19 shows a comparison between a LIST at 7.3×10^{-4} MPa/s and a CERT at 10^{-7} s^{-1}, under identical environmental conditions. These loading rates are equivalent for purely elastic behaviour. A LIST was typically complete soon after crack initiation; fast fracture occurred when the crack reached a critical size. In contrast stress corrosion cracks during

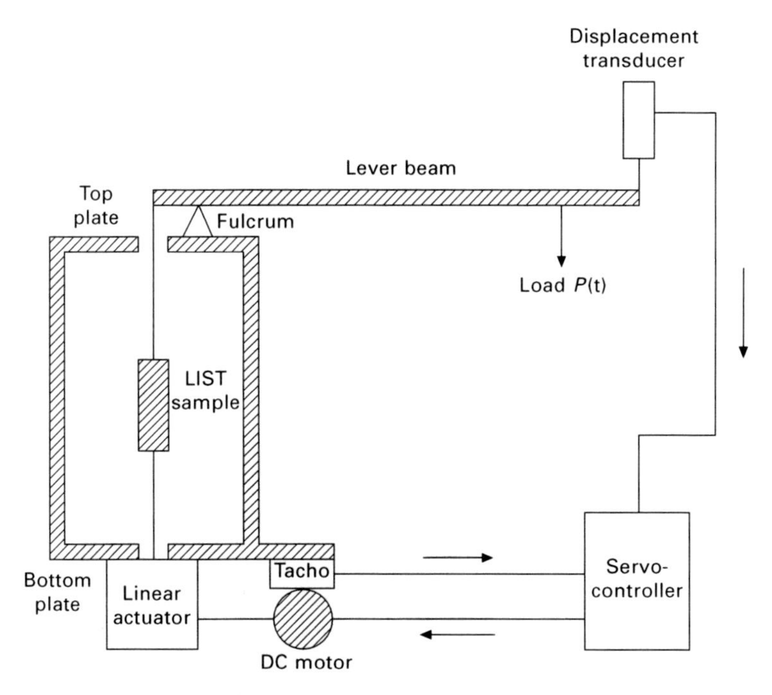

8.18 Schematic of LIST apparatus.

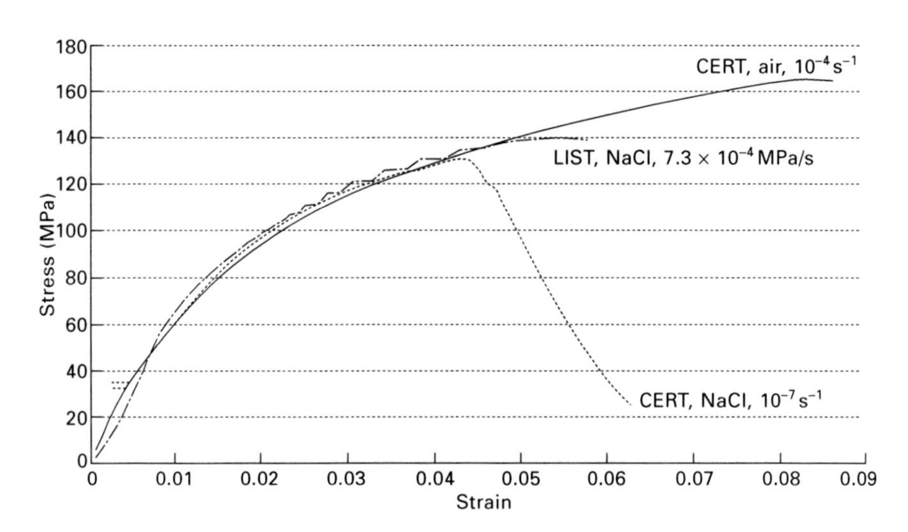

8.19 Comparison of stress–strain results for LIST and CERT.

a CERT propagated for a long period of time. A CERT was typically two or three times longer in duration. Figure 8.19 also shows that considerable crack propagation occurs even after the nominal stress is reduced to a value below the threshold stress; this effect is primarily due to the reduction in cross sectional area of the specimen by crack propagation.

The LIST and CERT techniques are both useful in identifying the occurrence of SCC. When coupled with a technique for identifying crack initiation, both LIST and CERT can measure the threshold stress (see Fig. 8.7) and crack velocity from the final crack size divided by the time for cracking. The increased crack propagation time under CERT conditions may be important in determining the mechanism for SCC in Mg alloys. However, the stress corrosion crack growth occurs under conditions of decreasing load, and the fractography towards the end of a CERT may not be typical of that in the earlier stages of SCC.

8.3.4 C(T) experiments

Winzer *et al.* [36] studied SCC of the Mg–Al alloys AZ91 and AZ31 using compact tension specimens, designated C(T) specimens. For AZ31, the stress corrosion crack velocity, V_c, was $2.5 \times 10^{-9} - 8 \times 10^{-9}$ m/s, and the threshold stress intensity factor, K_{ISCC}, was $14\,\text{MPa}\sqrt{\text{m}}$. The fracture surfaces for AZ31 contained cleavage-like facets similar to those on SCC fracture surfaces for AZ31 and AZ91 cylindrical tensile specimens; however, the cleavage-like facets did not appear to contain micro-dimples as for the cylindrical tensile specimens. SCC in AZ91 C(T) specimens was characterised by superficial stress corrosion crack branches extending from the fatigue crack tip along the maximum principal stress contours in the region where the maximum principal stress exceeded σ_{SCC}. The absence of a primary stress corrosion crack through the entire width of the specimen prevented V_c and K_{ISCC} from being measured for AZ91.

8.4 Environmental influences

8.4.1 Hydrogen

Mg alloys have shown SCC in gaseous H [37,118,120], these studies have been related to the SCC mechanism. Figure 8.20 shows the influence of H_2 pre-charging. The specimen fractured just above the yield stress, σ_Y, and without the apparent plastic strain or reduction in load that typically characterises SCC of AZ91 in aqueous solutions under CERT conditions (see Fig. 8.6). The control sample, which was exposed to gaseous Ar at 3 MPa for 15 h at 300 °C, had a UTS and ductility comparable to that of the sample tested in air. It follows that HE was responsible for the large reductions in UTS and elongation-to-failure for the specimen pre-charged in gaseous H_2.

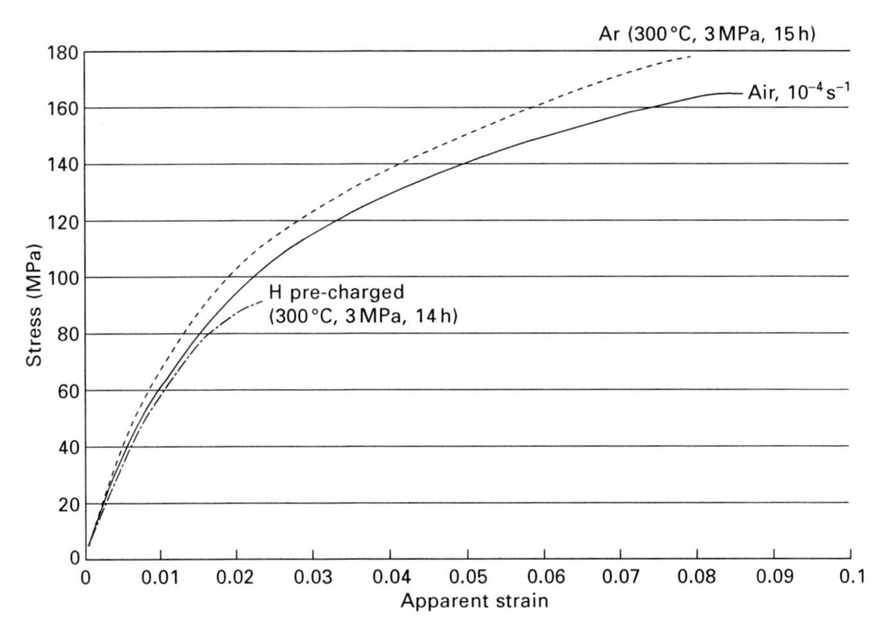

8.20 Stress vs. apparent strain curves for AZ91 pre-charged in gaseous H_2, and Ar at 3 MPa and 300 °C and fractured in air, compared with a non-charged specimen fractured in air.

8.4.2 Atmosphere

Dry air is generally considered inert for pure Mg [98,101,106], albeit SCC has been reported in damp and outdoor air atmospheres. Pelensky and Gallaccio [105] reported that all AZxx alloys failed during outdoor exposure, including marine and rural atmospheres, with susceptibility increasing with increasing periods of rain, humidity or high temperature. For indoor air at ambient temperatures, AZ61 was susceptible to SCC at 98 to 100% relative humidity. Increasing O_2 or CO_2 concentration decreased the critical relative humidity for AZ61 to 95%. If the relative humidity was near 100%, increasing the concentration of SO_2 and CO_2 decreased the time to cracking. Pelensky and Gallacio reported that M1 (Mg–1.2Mn) was immune to SCC in marine environments. Kiszka [145] reported SCC for rapidly cooled Mg–14Li alloys containing 1% or 1.5% Al in humid air. Loose and Barbian [59] reported SCC of AZ61 in annealed and as-rolled conditions in rural and coastal atmospheres.

Marrow *et al.* [157] induced SCC in air in cast WE43-T6. Crack initiation occurred at relatively high stresses and was attributed to the cracking of an intergranular intermetallic due to its inability to accommodate the locally high plastic deformation of the α-matrix. Propagation occurred as TGSCC when the crack was sufficiently long. There were non-propagating cracks at lower stresses on the smooth specimens, reminiscent of those reported by Wearmouth *et al.* [114]. Generalisation implies some SCC susceptibility to all multi-phase Mg alloys at high stress.

8.4.3 Solution composition

A vast literature has dealt with SCC in aqueous solutions containing chloride and chromate ions. Chromate ions inhibit corrosion by promoting protective film growth. Chloride ions tend to cause film breakdown and H evolution, which facilitate SCC [113,114,120]. Ebtehaj *et al.* [120] investigated the influence of chloride and chromate ion concentration on TGSCC susceptibility of homogenised fine-grained cast Mg–9Al. The specimens were tested under monotonically increasing strain, constant strain and constant load. The threshold stress or threshold stress intensity factor was a minimum when the ratio of chloride ions to chromate ions was approximately 1–2. The authors proposed that the role of chromate was to increase the open circuit potential, and therefore the pitting potential, for a given chloride concentration. SCC required a balance between active and passive corrosion behaviour: chromate-only solutions result in complete passivity preventing localised corrosion which was essential for H ingress, whereas chloride-only solutions result in excessive general corrosion which outran crack growth. Maximum susceptibility occurred at intermediate chloride/chromate ratios. This was contradicted by Pelensky and Gallaccio [105] who stated that susceptibility of AZ61 increased as the concentration of K_2CrO_4 was increased from 3 to 200 g/L in 35 g/L NaCl and, similarly, susceptibility increased as the concentration of NaCl was increased from 40 to 200 g/L in 5 g/L K_2CrO_4 solution.

Makar *et al.* [113] found that SCC of rapidly solidified and as-cast Mg–1Al and Mg–9Al occurred in a limited range of relatively high strain rates in NaCl-only solutions and not at all in K_2CrO_4-only solutions. The requirement for a relatively high strain rate coupled with the trends outlined by Ebtehaj *et al.* [120] suggested that HE is dependent on the localised corrosion and H-evolution reactions caused by appropriate chloride–chromate ratios. Varying the chloride–chromate ratio varied the strain rate at which localisation occurred (Fig. 8.16).

However, the relationship between strain rate and environment varies with system. Winzer *et al.* [37] found that SCC susceptibility of AZ91, AZ31 and AM30 in distilled water increased continuously with decreasing strain rate. Nozaki *et al.* [121] reported no significant variation in the SCC susceptibility, index, Eq (8.1), for strain rates between $8.3 \times 10^{-5}\,s^{-1}$ and $8.3 \times 10^{-7}\,s^{-1}$ for NaCl-only solutions with concentrations between 2 and 8% and also found an increase in susceptibility index from 0 to 85% for distilled water (Fig. 8.17).

Fairman and West [122] found that the threshold stress of single-phase Mg–6.5Al in NaCl + $K_2Cr_2O_7$ was lower than in NaCl + K_2CrO_4. Speidel *et al.* [144] reported that crack growth in distilled water was accelerated by additions of sulphate or bromide ions as shown in Fig. 8.12. Fairman and Bray [106] tested pure Mg and various Mg–Al, Mg–Al–Fe and Mg–Al–Zn alloys, heat treated to ensure TGSCC, in 4% NaCl + 4% Na_2CrO_4. They found that additions of $NaNO_3$ or Na_2CO_3 to the solution improved the ability to repair defects in the surface film. Such vulnerable sites would otherwise contribute to SCC. This also increased the thickness and stability of the film and made the corrosion potential less negative. The inhibition of SCC for Mg–6Al in chloride–chromate solutions by NO_3^- ions was also investigated by Frankenthal [158] who proposed that NO_3^- ions prevent breakdown of the chromate-produced film by chloride ions. Frankenthal observed that the unstressed specimens in 4% NaCl + 4% K_2CrO_4 were subject to profuse localised corrosion; however, with the addition of 3% $NaNO_3$, there was only slight localised corrosion. Furthermore, the addition of nitrate to the chloride–chromate solution resulted in a more positive corrosion potential, similar to that for a chromate-only solution.

Timonova [126] stated that MA3 was susceptible to SCC in sodium carbonate solution, with a susceptibility constant for concentrations between 0.005 and 0.15 M, and increasing with concentrations between 0.15 and 1.0 M.

Pelensky and Gallaccio [105] carried out an extensive assessment of environmental influences for SCC of pure Mg and Mg–Mn, Mg–Al, Mg–Al–Zn, Mg–Zn–Zr, Mg–Al–Mn and Mg–Li alloys in air, water and numerous aqueous solutions. Most alloy–environment combinations resulted in SCC. The following general trends were noted:

- High-purity Mg and Mg–2Mn failed in KHF_2.
- Mg–5Al failed in distilled water.
- ZK60A-T5 (Mg–5.5Zn–0.45Zr) failed rapidly in distilled water and seawater.
- ZK60A-T5 stressed to 90% of its yield strength failed rapidly in KCl, CsCl, NaBr, NaCl and NaI solutions.
- AZ61 failed in Na_2CO_3 solution with susceptibility constant for concentrations between 0.265 and 15.9 g/L and increasing with concentration between 15.9 and 53 g/L.
- AZ61 did not fail in K_2CrO_4 + $NaCH_3COO$, K_2CrO_4 + Na_2CO_3 or K_2CrO_4 + $NaNO_3$ solutions.
- AZ61 was susceptible in NaCl + K_2CrO_4, K_2CrO_4 + Na_2SO_4, Na_2SO_4, $NaNO_3$, Na_2CO_3, NaCl, $NaCH_3COO$ (in order of decreasing susceptibility).
- The susceptibility of AZ61 and Al–1.5Mn in NaCl solutions increased with solution concentration.

Little explanation of these trends was offered by the authors. Many of the environments were selected to represent common service conditions, but it was noted that accelerated SCC tests should not be used to predict service lifetimes, because of the large number of metallurgical, environmental and mechanical variables, and that accurate lifetime prediction requires accurate simulation of service conditions.

Yakovlev *et al.* [159] found that MA2 exhibited similar SCC kinetics in Na_2SO_4 + NaOH solution as in NaCl + $K_2Cr_2O_7$ solution. Tomashov and Modestova [126] also stated that, for Mg–Mn alloys, there were similar corrosion characteristics and SCC failure times for Na_2SO_4 and NaCl solutions of equal concentration. Marichev and Shipilov [142] stated that NaOH was associated with passivation of Mg alloys. Timonova [126] showed that MA3 was susceptible to SCC in H_2SO_4 solutions, with susceptibility increasing with solution concentration, to some maximum, and then decreasing with further increase in concentration as general corrosion became more profuse.

Timonova [126] examined the resistance of Mg–8Al to SCC and general corrosion in 0.01 M solutions of NaCl, HCl, HNO_3, NaOH, NaF and HF. NaF produced limited general corrosion, limited by the formation of a dense surface film, which also inhibited SCC. In NaOH, general corrosion was also limited although SCC occurred, which was attributed to a less dense surface film. For HCl and NaCl, the general corrosion rate was high, preventing SCC. Perryman [119] observed SCC for Mg–5Al in saturated magnesium carbonate, 0.5% potassium fluoride, 0.5% potassium hydrogen fluoride and 0.5% hydrofluoric acid (in order of decreasing time to failure).

KHF_2 solutions appear to cause SCC in most alloys including high-purity Mg and Mg–2% Mn [119,160]. Since the F^- ion inhibits Mg corrosion by the formation of a fluoride film, at least part of the electrochemical explanation may lie in film breakdown and repair kinetics.

8.4.4 Electrochemical potential

It was generally believed that SCC in Mg is inhibited by cathodic polarisation, and accelerated by anodic polarisation [60]. Stampella *et al.* [101] found that SCC was invariably preceded by pitting for pure Mg in a 10^{-3} M Na_2SO_4 solution at the free corrosion potential, and that the corrosion potential (–1.36 V) coincided with the pitting potential. Anodic polarisation to –1.16 V resulted in more severe pitting and greater SCC susceptibility. Cathodic polarisation to –1.5 V and –2.5 V resulted in a more stable film and prevented pitting. At the more cathodic potentials, repassivation was sufficient to overcome mechanical film rupture by strain rates as high as 5.7×10^{-6} s^{-1}. Priest *et al.* [104] and Logan [161] also showed that SCC in Mg is prevented by cathodic polarisation. Ebtehaj *et al.* [120] demonstrated that SCC susceptibility, characterised in Fig. 8.21 by reduction in area, decreased

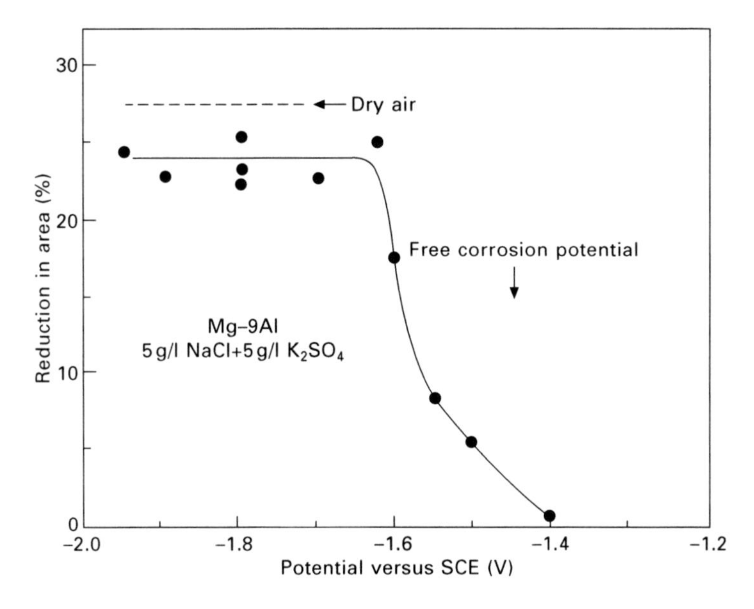

8.21 Influence of applied potential on reduction area for Mg–9Al subject to a strain rate of $2 \times 10^{-6}\,s^{-1}$ in 5 g/L NaCl + 5 g/L K_2CrO_4 solution [120].

rapidly with decreasing applied potential. Below approximately –1.6 V, the reduction in area was constant and equal to that for dry air; the lack of SCC was attributed to the rapid formation of a surface film that prevented hydrogen ingress and SCC. In contrast, Kannan *et al.* [162] found SCC under CERT conditions for AZ80 cathodically polarised in distilled water; it was proposed that these testing conditions did not allow surface film formation sufficiently rapidly to prevent SCC. Chen *et al.* [49] similarly measured an influence of cathodic precharging for AZ91, indicating that H entry under cathodic conditions for two-phase alloys like AZ91 and AZ80 may be easier than for pure Mg.

Marichev and Shipilov [142] showed that, for various Mg alloys containing Al, Zn, Mn and RE in NaCl solutions, cathodic polarisation decreased and anodic polarisation increased the crack velocity. These trends were attributed to the alkalisation and acidification of the crack tip solutions leading to film stabilisation and breakdown respectively. In contrast, cathodic polarisation resulted in rapid acceleration of cracking for the high-strength Mg–7.6Y–1.7Zn–1.5Cd–0.3Zr in the passivating solutions: 1 M NaOH and 2 M CrO_3. Ebtehaj *et al.* [120] showed that the strain rate corresponding to maximum SCC susceptibility (the maximum nominal stress below which there was no SCC) was increased as the corrosion potential became more cathodic (Fig. 8.22). This trend was attributed to the promotion of film growth by cathodic

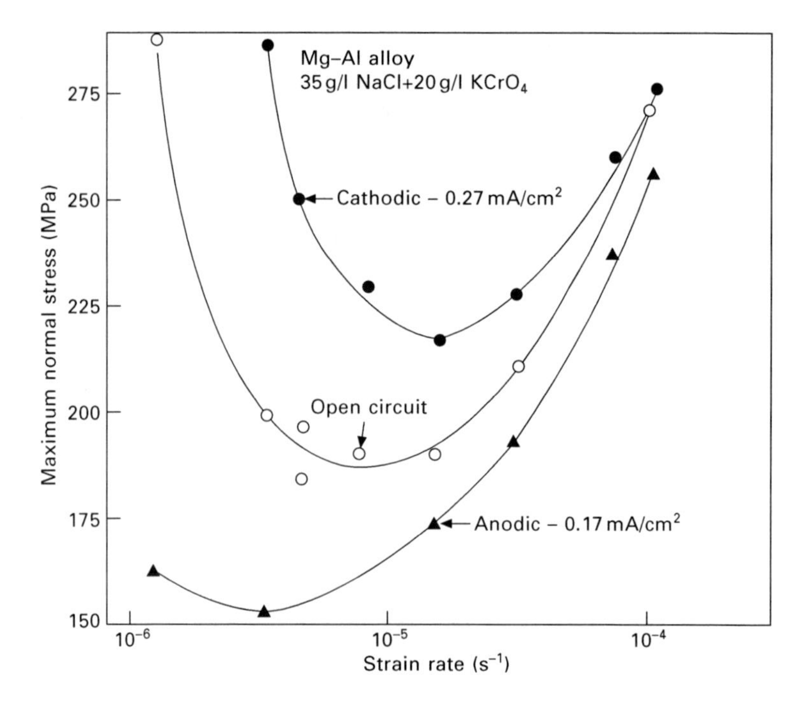

8.22 Influence of strain rate on SCC susceptibility for various applied potentials.

polarisation such that higher strain rates were required to induce film rupture and localised corrosion.

8.4.5 pH

Reports on the influence of pH on SCC are few but they agree that SCC does not occur in chloride–chromate solutions with pH values greater than 12. Sager *et al.* [163] reported that the time to failure for Mg–6.5Al–1Zn in a chloride–chromate solution was constant for pH values between 5 and 12 (Fig. 8.23). At lower pH values SCC susceptibility increased sharply, whereas SCC was inhibited at higher pH values. Pelensky and Gallacio [105] found SCC for AZ61 in a chloride–chromate solution for pH values between 2 and 12 but not at higher pH values. Several other workers [104,124,125] have established that pH has no influence on crack morphology. Loose [164] and Scully [160] report that, in non-fluoride solutions, SCC is inhibited at pH values greater than 10.2. This is probably related to the greater ease of film formation that occurs in highly alkaline solution. However, Timonava [127] reported SCC in 0.01N NaOH.

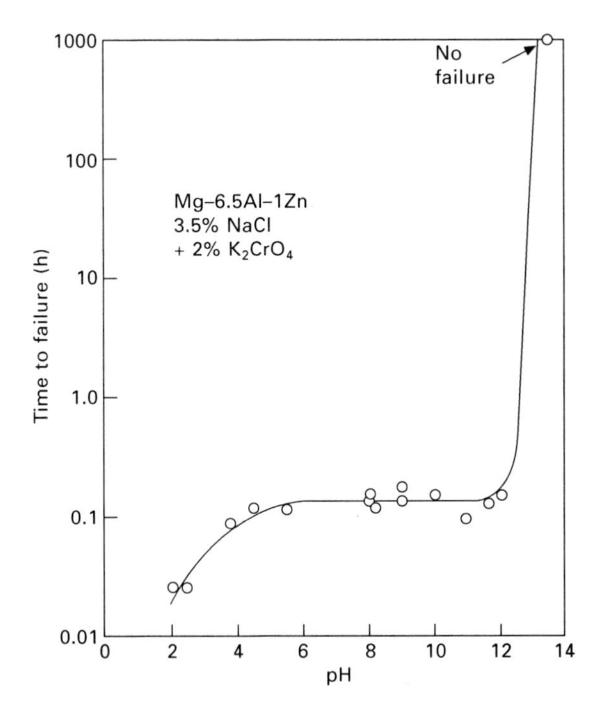

8.23 Variation in time to failure with pH for Mg–6.5Al–1Zn alloy in 3.5% NaCl + 2% K_2CrO_4 solution [163].

8.4.6 Temperature

Miller [60] stated that Mg SCC susceptibility in air and water increases with temperature. Pelensky and Gallaccio [105] reported that SCC susceptibility for AZ61, in a chloride–chromate solution, was a maximum at ~ 40 °C whereas susceptibility of MA2 and AZ80 in a H_2SO_4 + NaCl solution increased with temperature to ~70 °C. Romanov [165] also showed that SCC susceptibility increased with temperature for MA2 in a H_2SO_4 + NaCl solution, whereas for passivating environments, increased temperature decreased SCC susceptibility due to improved passivation. Increasing temperature also increases the amount of creep, which may cause stress relaxation resulting in improved SCC resistance or may contribute to film rupture resulting in reduced SCC resistance.

8.4.7 Coatings

The prime function of coatings, anodising and surface treatments is to prevent Mg contacting the environment and to thereby prevent SCC. Breaks or holidays in the coating can lead to SCC [127, 159]. Scully [160] reported that anodising increased SCC life. Srinivasan and co-workers [166–168] found that PEO coating did not prevent SCC of AM50 and AZ61 in a dilute aqueous solution.

8.5 Fractography

The fracture surfaces in air for pure Mg and Mg alloys are ductile. In contrast, Meletis and Hochman [99] reported cleavage for SCC of pure Mg in a NaCl + K_2CrO_4 solution. Cleavage consisted of long, parallel facets separated by perpendicular steps with jogs at the edges. The facets and steps corresponded to $\{2\bar{2}03\}$ planes, with the ledges of each step oriented in the $<10\bar{1}1>$ direction. Similar fractography was reported by Chakrapani and Pugh [115,117,118] for Mg–7.5Al in NaCl + K_2CrO_4 solution, but the cleavage facets corresponded to $\{31\bar{4}0\}$ planes. Chakrapani and Pugh also observed (i) opposite fracture surfaces were interlocking, and (ii) fine parallel markings ~1.5 μm apart, interpreted as crack arrest markings, within some cleavage-like facets. Similar markings were also reported by Fairman and West [122]. Quasi-cleavage fracture for SCC of Mg alloys has also been reported by [114,121,159].

In marked contrast, Lynch and Trevena [98] for pure Mg in a NaCl + K_2CrO_4 solution, reported that the fracture surfaces consisted of flutes containing equiaxed dimples on $\{10\bar{1}X\}$ planes and grain boundaries as well as cleavage-like features on {0001} planes, with the flutes and dimples being smaller and shallower than those occurring in air.

Winzer *et al.* [70] found that the SCC fracture morphologies for Mg alloys vary widely for different combinations of alloy and environment, and even between different regions of the same fracture surface, i.e. with crack depth. The diversity of SCC fracture morphologies is attributed to the number of mechanisms that are involved in stress corrosion crack initiation and propagation. For example, Winzer *et al.* [70] showed there were significant differences in SCC fracture surfaces for AM30 and AZ31, which had generally similar microstructures except that AM30 had a lower concentration of Zn and second phase particles. The fracture surfaces for AZ31 consisted of relatively smooth regions containing small, elongated dimples whereas those for AM30 consisted of cleavage-like markings.

Fractography has contributed to the discussion of the involvement of hydrides in Mg SCC. Early works [9,99,115,117,118] proposed that Mg SCC involved delayed hydrogen cracking (DHC). The plausibility of this mechanism was reinforced by the cleavage-like fracture surfaces, as such features were also related to hydrides in other metals. More recent fractography [70] has indicated that hydrides may indeed play a role in SCC of Mg alloys at very slow crack velocities; however, the formation of MgH_2 during SCC was related to a 'quasi-porous' rather than cleavage-like fracture.

8.6 Stress corrosion cracking (SCC) mechanisms

Mg SCC has been generally attributed to one of two groups of mechanisms: continuous crack propagation by anodic dissolution at the crack tip, or discontinuous crack propagation by a series of mechanical fractures at the crack tip (Fig. 8.1), [60,101]. Dissolution mechanisms include preferential attack, film rupture and tunnelling while mechanical mechanisms include cleavage and HE. It appears that there are different mechanisms for IGSCC and TGSCC. IGSCC occurs by microgalvanic acceleration of the Mg matrix adjacent to cathodic grain boundary precipitates (Fig. 8.9). For TGSCC, it appears that dissolution cannot explain the crack propagation rates and fracture surfaces, particularly the interlocking fracture surfaces found by Pugh and co-workers [9,115,117,118]. There is considerable evidence to support HE-induced cleavage. H-induced plasticity may also play a role, particularly to produce the fracture surfaces reported by Lynch and Trevena [98] at high stressing rates which had concave features on opposing fracture surfaces (Fig. 8.6).

It has been suggested [114] that the threshold stress, σ_{SCC}, is related to the yield stress. In single crystals, it has been reported [169] that the yield strength must be exceeded to produce plastic deformation and this is necessary for SCC. However, Table 8.1 [5] indicates that σ_{SCC} is significantly less than the yield stress for many alloy–environment combinations.

8.6.1 Preferential corrosion

The preferred model for IGSCC in Mg–Al alloys (Fig. 8.9), is accelerated dissolution of the metal matrix adjacent to $Mg_{17}Al_{12}$ grain boundary precipitates, which precipitate during slow cooling of Mg alloys with Al concentrations greater than 2.1% [124]. Cracking initiates and grows by preferential dissolution of the Mg matrix [124]. Cracking is driven by the potential difference between $Mg_{17}Al_{12}$ and the Mg–Al matrix, which is ~ 300 mV [2,124]. Stress pulls apart opposite crack surfaces and allows solution access to the crack tip. Pardue *et al.* [125] showed continuous cracking, for a continuous grain boundary precipitate, as indicated by a lack of noise signals in acoustic emission studies. Fairman and Bray [124] reported some discontinuities in IGSCC extension, attributing growth discontinuities to crack intersection with unfavourably orientated grains. A similar mechanism would be expected for other Mg alloy systems, because all second phases tend to accelerate the corrosion of the adjacent Mg matrix [2].

8.6.2 Galvanic corrosion and film rupture

Logan [161,170] proposed an alternative electrochemical mechanism. He postulated that, when the surface film is ruptured by an applied strain rate, an electrochemical cell occurs between the anodic film-free area and the cathodic filmed area. The potential difference between the anode and cathode was estimated to be 0.2 V so that rapid dissolution could occur. Stress concentration at the crack tip was postulated to prevent the film from reforming such that crack propagation was continuous. Miller [60] stated that, for film rupture models, continuous crack growth is possible only where repassivation is insignificant, otherwise there is discontinuous crack growth by alternating phases of film rupture and repassivation. Crack growth, defined by the competing rates of film rupture and growth, was also proposed by Ebtehaj *et al.* [120] and Wearmouth *et al.* [114].

Logan's hypothesis [161,170] that the SCC was entirely electrochemical was based on the observation that cathodic polarisation could prevent SCC. However, HE models generally require the exposure of film-free surfaces, which may be prevented by cathodic polarisation. Furthermore, Logan calculated from Faraday's law that the observed crack propagation rates (10×10^{-6} m/s) required an effective current density of 14 A/cm^2. Similarly Pugh *et al.* [171] observed crack velocities between 6×10^{-3} and 40×10^{-6} m/s for Mg–7.6Al in a chloride–chromate solution, which correspond to current densities between 8 and 60 A/cm^2. Such current densities were considered by Pugh *et al.* to be prohibitively high.

8.6.3 Tunnelling

The tunnelling model, introduced by Pickering and Swan [172], proposed that film rupture at emerging slip steps causes localised electrochemical cells, resulting in tubular pits. The direction of these pits was thought to be initially defined by the electrochemical potential difference between the metal matrix and the fine precipitates (assumed to be $Mg_{17}Al_{12}$). Subsequently, once there was a local electrochemical cell capable of preventing film repair and sustaining enhanced anodic dissolution, propagation was thought to continue independent of the precipitate. Crack extension occurred by ductile tearing of the narrow ligaments of metal between parallel tunnels. Ductile tearing destroyed the local cells at the crack tips so new pits were formed in order for the process to repeat. Unlike the film rupture model, Pickering and Swan claimed that the current density required to achieve the observed crack penetration rates was realistic. However, this model cannot explain the interlocking fracture surfaces often observed for TGSCC. Furthermore, Pickering and Swan proposed this mechanism for Mg–1Al and Mg–7Al, but $Mg_{17}Al_{12}$ precipitates are unlikely in Mg–1Al.

8.6.4 Cleavage

The most commonly proposed mode of TGSCC of Mg alloys is discontinuous cleavage. Most Mg alloys have hexagonal close packed (HCP) crystal structures, which are susceptible to cleavage due to the lack of slip systems [173]. Cleavage is usually evidenced by fluctuating acoustic emission and corrosion potential, and by faceted, interlocking fracture topography.

Pardue *et al.* [125] suggested that TGSCC in AZ61 is initiated by highly localised pitting; that the high stress concentration at a corrosion pit induces a cleavage crack, which propagates within the grain until it intercepts some obstruction, such as a grain boundary. The obstruction is removed by further electrochemical corrosion and the process is repeated. The alternating electrochemical and mechanical stages were evidenced by discrete acoustic emissions and corrosion current fluctuations, respectively. The occurrence of brittle fracture was attributed to dislocation blocking, and thus inhibition of plastic deformation, by FeAl precipitates within grains or to lattice distortion by Fe ions providing the activation energy for Mg dissolution. Fairman and West [121] elaborated this model by proposing that pitting is initiated by film rupture due to basal slip and that continued film rupture at the base of the pit causes it to deepen by ~0.1–0.2 μm. Stress concentration at the base of the pit initiates cleavage on whichever cleavage plane, (0001), $(10\bar{1}0)$ or $(10\bar{1}1)$, is most normal to the tensile axis, as evidenced by a stepped fracture surface with wide forward steps in the cracking plane. The width of the steps was correlated with changes in longitudinal extension of the specimens. After limited propagation, the crack arrests due to stress relief, cross-slip of dislocations, or interference by precipitates or flaws. The process is repeated when slip occurs again at the crack tip resulting in the formation of another pit. The crack changes orientation at grain boundaries; however, since there are multiple cleavage planes it still propagates roughly perpendicular to the tensile axis. Ductile failure finally occurs when the cross-section of the specimen is reduced such that the ultimate tensile stress is reached. Fairman and West also acknowledged the electrochemical stage in crack propagation, based on observations that cracking could be stopped at any time by cathodic polarisation. They also concede that film rupture is unnecessary where there is severe localised corrosion, due to $Mg_{17}Al_{12}$ grain boundary precipitates.

Pugh *et al.* [171] criticised Fairman and West's model for inadequately explaining why stress intensification at the immerging pit initiates cleavage, rather than ductile fracture, or why the cleavage crack stops, in spite of the fact that the stress intensity factor increases with crack length. Instead, Pugh *et al.* proposed a variation of the film-rupture model for transgranular stress corrosion cracking by proposing that embrittlement of the metal immediately ahead of the crack tip occurs by the formation of a porous surface layer of an oxide (other than MgO) or selective dissolution of Mg or Al, and that crack arrest is due to the blunting of the crack tip as it enters the ductile substrate ahead of the embrittled layer. Thus, characteristic alternating stages were proposed: (i) embrittlement of a thin surface film, (ii) film rupture and crack propagation, and (iii) crack arrest. This model was also supported by crack propagation velocities, measured using a travelling microscope to be ~5.8 to 42 × 10^{-6} m/s (0.35–2.5 mm/min), which Pugh *et al.* proposed could only be explained by a mechanism involving a fast mechanical stage. The embrittling surface film was speculated to be an oxide or a de-alloyed later. HE was rejected based on observations that cracking is retarded and accelerated by cathodic and anodic polarisation, respectively; however, this effect of polarisation has since been commonly attributed to the negative difference effect.

The cleavage model proposed by Pugh *et al.* [171] was disputed by Fairman and Bray [106] on two counts. Firstly, they observed no fractographic evidence of brittle fracture for Mg–Al alloys in chloride–chromate solutions. Secondly, they detected no oxide layer on the fracture surfaces. Fairman and Bray instead proposed a ductile fracture model as evidenced by striations perpendicular to the crack propagation direction, which were attributed to positions of successive crack arrest. This ductile fracture interpretation conflicts with the majority of fractographic evidence provided by other workers. Fairman and Bray stated that for Mg–Al alloys, Al^{3+} ions displace Mg^{2+} ions in the $Mg(OH)_2$ surface film, requiring vacancies in the film lattice to preserve electroneutrality. This increased the susceptibility of the surface film to localised rupture. Film rupture was thought to occur by a flow of dislocations, causing slip in the metal and failure on the film basal plane. Chloride ions migrate to regions of high tensile strain; therefore tunnelling occurs at the rupture site if the dislocation concentration is sufficient to overcome repassivation. Stress concentration at the tip of the tunnel would then lead to ductile tearing until the crack is relaxed by plastic flow. Since Mg–Al has multiple slip systems a new tunnel may then be created at the crack tip and the process is repeated.

Pardue *et al.* [125] and Chakrapani and Pugh [115] showed that stages of crack propagation coincided with discrete spikes in acoustic emission measurements. Unstressed specimens in solution emitted a steady acoustic signal, which was associated with H evolution from pitted areas while specimens undergoing plastic deformation in air emitted no signal. Stressed specimens in solution emitted discrete acoustic signals superimposed onto the continuous 'H' signal, indicating discontinuous crack advance. Pardue *et al.* also observed that the electrochemical step in the cracking process corresponded with fluctuations in corrosion currents, due to the repeated exposure of fresh anodic metal.

Chakrapani and Pugh [115] determined from measurements of the distance between acoustic emission spikes that the average crack velocities in Mg–7.6Al were between 5 and 30 × 10^{-6} m/s. Similarly, using a travelling microscope Pugh *et al.* [171] determined that, for the same alloy, average crack velocities were between 6 and 40 × 10^{-6} m/s. Such velocities could not be reconciled with dissolution models according to Faraday's law, adding further support to a mechanical fracture model.

Various workers [9,99,115,117] have reported that TGSCC in Mg results in fracture surfaces consisting of flat, parallel facets separated by perpendicular steps, consistent with a cleavage mechanism. The steps and facets are generally parallel to the direction of crack propagation and change direction at grain boundaries [99]. Opposite fracture surfaces are matching and interlocking, which is also consistent with the occurrence of cleavage and is difficult to explain by a dissolution model [9]. The fracture surfaces show numerous jogs resulting from overlap of parallel cleavage cracks (Fig. 8.24), further evidencing discontinuous propagation. Reports on the crystallography of these surfaces are generally inconsistent and Meletis *et al.* [99] attribute this to the common use of two surface analysis methods rather than more accurate photogrammetric methods. That there can be a ductile component of the fracture mechanism was shown by Lynch and Trevena [98] who observed concave features on opposing fracture surfaces for higher crack propagation velocities, such that the opposing fracture surfaces were not interlocking.

8.6.5 Hydrogen embrittlement

The most commonly cited mechanism for SCC in Mg alloys is HE. On first blush, HE might seem an unlikely mechanism: cathodic polarisation might

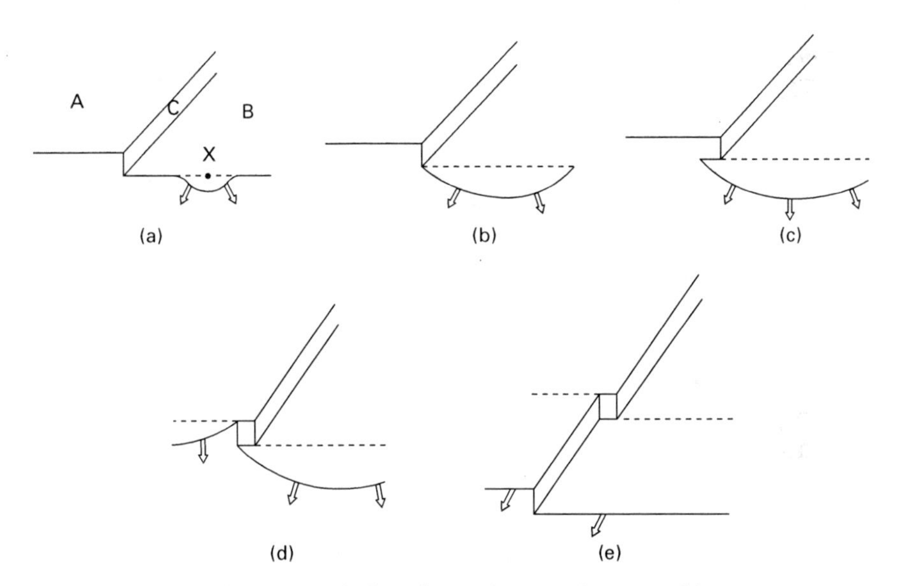

8.24 Jog formation during discontinuous cleavage. After arrest, the crack reinitiates at X (a) and advances on facet B (b). The crack overshoots the cleavage step (c), resulting in jog formation by secondary cleavage (d) and finally advances uniformaly (e) [115].

be expected to increase H evolution but has been shown to retard cracking whereas anodic polarisation might be expected to decrease H evolution and has been shown to accelerate cracking [60,101,120]. This contradiction was explained in terms of the negative difference effect by Ebtehaj *et al.* [120], who showed that large amounts of H were evolved within a discrete range of anodic polarisation. Ebtehaj *et al.* supported the H-ingress mechanism for SCC in Mg–Al alloys by correlating test results for cast Mg–9Al stressed in dry gaseous H and in chloride–chromate solutions. They also reported that cracking, in initially plain specimens, invariably began at regions of localised pitting and that, for slow applied strain rates, the crack initiation potential coincided with the pitting potential. Thus it was inferred that pitting allows H ingress through exposed metal surfaces.

Chakrapani and Pugh originally proposed that transgranular SCC for hot-rolled Mg–7.5Al in NaCl–K_2CrO_4 solution occurred by discontinuous cleavage evidenced by the stepped fracture surface topography; however, they also observed that H evolution invariably occurred at corrosion pits [115,117]. Subsequent work [118] proposed that the cleavage-like fracture surfaces could be due to HE. Specimens, stressed in gaseous H or exposed to aqueous solution prior to the application of stress, exhibited a loss of ductility and cleavage-like fracture surfaces. This proposal was further evidenced by inert gas fusion methods, which showed that the H concentration of the specimens progressively increased with time, and vacuum annealing, which partially reversed the effects of H exposure (Fig. 8.15).

Chakrapani and Pugh [118] inferred that the mechanism for crack growth was repeated cycles of H diffusion to the crack tip region followed by brittle fracture. This was supported by discontinuous acoustic emission signals. It was also suggested that the role of H in the brittle fracture could be in the formation of brittle hydrides or to produce decohesion. They inferred that since SCC fractures tend to occur on $\{31\bar{4}0\}$ planes, these may correspond to the habit or cleavage planes of a hydride. Fracture surfaces for SCC and pure-HE systems were different: the latter tended to be flatter and without the pleated/stepped structure. It was speculated that this could be related to H fugacity and H entry kinetics, and the fact that, for SCC conditions, dissolution occurs at the crack tip.

H effects were also identified by Makar *et al.* [113] by correlating the results of pre-exposed and *in situ* tests. There were differences in the fracture surfaces for the two tests, suggesting that the mechanisms for samples subject to HE in the presence and absence of stress were somewhat different. This was explained in terms of the crack tip area available to H entry and/or the different distributions of hydrided regions. Some samples were vacuum annealed prior to testing and the environment effects were mostly reversible, as also reported by Chakrapani and Pugh [118].

Makar *et al.* modelled the active mechanisms at various strain rates for rapidly solidified Mg–9Al in chloride–chromate solution. Crack propagation velocities for anodic dissolution, ductile tearing and HE were modelled according to Faraday's law, fracture mechanics and the solubility of H in the Mg lattice, respectively. By plotting crack velocity with respect to crack tip strain rate it was proposed that HE (which was proposed to occur by the formation of MgH_2 ahead of the crack tip) was the dominant propagation mechanism for values of crack tip strain rate less than 10^{-2}–$10\,s^{-1}$, depending on the H diffusivity. At higher strain rates, crack velocity due to ductile tearing overtakes and becomes the dominant mechanism. The crack velocity due to anodic dissolution is relatively insignificant for all strain rates and at some discernible strain rate repassivation prevents any dissolution.

Stampella *et al.* [101] studied the connection between pitting and cracking in pure Mg in 10^{-3} M Na_2SO_4, with the pH adjusted to 10.0 by NaOH. They proposed that for SCC to occur, H must be evolved from film-free pit walls. Figure 8.2 shows that specimens anodically polarised and stressed in solution exhibited embrittlement and a reduction in tensile strength; moreover, these effects were mostly reversible by exposure to air at room temperature. This contradicted the suggestion of HE by the formation of high-pressure molecular H bubbles or by the formation of brittle hydrides, since MgH_2 is not stable at room temperature. Instead, Stampella *et al.* proposed that the embrittlement mechanism involves atomic H in solid solution lowering the cleavage strength of the Mg matrix.

Lynch and Trevena [98] identified the mechanism for SCC in pure Mg by comparing the fracture surfaces for slow and rapid crack growth in an aqueous environment, rapid crack growth in liquid alkali metals (Na, Cs and Rb) and overload crack growth in dry air. Liquid metal embrittlement (LME) was thought to occur by the adsorption of metal atoms at the crack tip at high crack velocities since insufficient time was available for other reactions and the solubilities of the tested alkali metals in Mg were negligible. The authors proposed that, for stress corrosion crack velocities ~5×10^{-2} m/s in aqueous environments, HE also occurs by adsorption since insufficient time is available for localised dissolution or H diffusion; this is now called the adsorption-induced dislocation emission (AIDE) mechanism [174]. They also suggested that dislocation transport of H is unlikely at these high crack velocities.

At low crack velocities, Lynch and Trevena proposed that H diffusion is likely to occur; however, adsorption was thought to still be the dominant mechanism for embrittlement. This was evidenced by the correlation between the fracture planes, crack propagation directions and fracture surfaces for LME and slow SCC fracture. It was also noted that no evidence of hydride formation was apparent on the fracture surfaces. At low crack velocities, adsorption at external crack tips could be inhibited by film formation, necessitating diffusion to and adsorption at internal cracks. The authors suggest that this could explain why HE characteristics are apparent for pre-exposed specimens.

It is important to note that the calculations by Lynch and Trevena for H diffusion were based on an approximate value for H diffusivity of $\sim 10^{-9}$ cm^2/s. This was based on hydride formation kinetics in Mg–2Ce since values for H diffusivity in pure Mg were unavailable. It was stated that for the given crack velocity insignificant H diffusion would occur for $D < 2 \times 10^{-7}$ cm^2/s. However, Makar *et al.* [113] postulated that the H diffusivity might be as large as 10^{-6} cm^2/s.

Meletis and Hochman [99] reported that crack initiation and propagation for pure Mg in a chloride-chromate solution was accompanied by H evolution. Using electron channelling pattern analysis and scanning electron microscopy (SEM) photogrammetry, they showed that cracks propagated primarily by cleavage on $\{2\bar{2}03\}$ planes as evidenced by parallel facets. These facets were separated by steps also on $\{2\bar{2}03\}$ planes. The authors proposed that cleavage occurs by reduction in surface energy of $\{2\bar{2}03\}$ planes by HE resulting from preferential H accumulation or hydride formation on these planes. H may be absorbed from the solution at the crack tip or transported to the region by dislocation motion.

Bursle and Pugh [9] also concluded that cracking in Mg–Al occurs by discontinuous cleavage induced by HE. They rejected the adsorption model (reduction of metal inter-atomic bond strength at the crack tip by the interacting of absorbed ions) on the basis that adsorption would only affect a few atomic layers ahead of the crack tip, and therefore it would not cause discontinuous crack propagation involving the distances between crack arrest markings observed by SEM. Adsorption-induced propagation would be macroscopically continuous at a rate determined by the transport of ions to the crack tip. Dealloying models were also rejected on the basis that solutions that cause dealloying are not usually associated with TGSCC and there has been no correlation reported between dealloying and cleavage.

Bursle and Pugh observed a 1 μm layer of brittle Mg hydride (MgH_2) on the cleavage fracture surfaces of Mg–Al specimens, contradicting Lynch and Trevena [98]. That this compound was not apparent on external or ductile fracture surfaces suggests that hydride formation was stress-induced. The authors also proposed that the $\{31\bar{4}0\}$ planes, determined by the two surface trace technique and a photogrammetric method, to be the orientation of the cleavage facets, may also be the habit or cleavage plane of the hydride. These factors lead to the proposal that hydrogen embrittlement as illustrated in Fig. 8.1 was the dominant mechanism for crack propagation.

8.6.6 Delayed hydride cracking

The critical review by Winzer *et al.* [5] indicated that TGSCC is the inherent mode of SCC for Mg alloys and that the mechanism for TGSCC is still equivocal. It is generally accepted that TGSCC of Mg involves H [5] and thus TGSCC can be considered an example of hydrogen environment assisted cracking (HEAC) [175].

The most recent mechanistic studies [5] have established the important elements in the propagation of TGSCC. Pugh and co-workers [9,115,117,118,171] proposed that TGSCC occurred by a brittle cleavage mechanism involving H, which resulted in stepped and faceted interlocking fracture surfaces. In contrast, the fractography of Lynch and Trevena [98] indicated some plasticity, particularly at higher strain rates and crack velocities. Slow strain rate testing by Ebtehaj *et al.* [120] and Stampella *et al.* [101] suggested a mechanism involving strain-induced film rupture leading to corrosion and H production, with crack advance due to H absorption. Makar *et al.* [113] confirmed the stepped and faceted interlocking fracture surface morphology and the importance of strain rate and H absorption, but proposed a mechanism involving formation and fracture of brittle hydrides. Thus, there is agreement that H is part of the SCC propagation mechanism but disagreement on the role that H plays. The prior review [5] also indicates that SCC is associated with environmental conditions leading to the local breakdown of a partially protective surface film, allowing absorption of H produced by the Mg corrosion reaction. Film breakdown can be caused by the environment (e.g. pitting due to chloride ions). Film breakdown can also be caused by the mechanical loading, because SCC occurs in non-pitting environments, e.g. (i) pure Mg in distilled water and a dilute sulphate solution; and (ii) Mg alloys (AZ91, AM60, AS41, ZK60A-T5) in distilled water.

Important characteristics of TGSCC of Mg alloys that must be rationalised by a TGSCC mechanism include: (i) that the stress corrosion crack velocity, V_c, has been measured by Speidel *et al.* [144] to be independent of the applied stress intensity factor, K_1, and by Bursle and Pugh [9] to be proportional to K_I^2 above a critical stress intensity factor for SCC, K_{1SCC}; (ii) that crack propagation is discontinuous (as evidenced by discrete high-amplitude acoustic emissions [99,101,115,117,118,125]); (iii) the development of a stepped and interlocking fracture surface [99,101,114,115,117,118]; (iv) that the fracture process zone, l_{fpz}, is typically 0.1–0.8 μm (as evidenced by parallel markings within cleavage steps associated with consecutive crack arrest fronts [9,115,117,118]), and (v) that measured values for K_{1SCC} lie within the range 4–14 MPa m$^{1/2}$ [144,120].

A possible mechanism for TGSCC of Mg alloys is delayed hydride cracking (DHC) [5,9,99,113,114,118]. Previous workers have proposed various models relevant to DHC in hydride-forming metals, particularly Zr alloys. Liu [176] and van Leeuwen [177] derived analytical solutions for the steady state and transient distribution, respectively, of lattice H ahead of a sharp crack; however, these models did not include hydride precipitation. Dutton *et al.* [178] derived an expression for the rate of hydride growth in the elastic field ahead of a crack tip in Zr by assuming: (i) a steady state flux through the crack tip; and (ii) that the hydride grows in line with the crack with a uniform thickness of twice the crack tip radius. More recent models [179–183] used finite element analysis (FEA) solutions to the diffusion equations. The FEA approach had not previously been applied to DHC in Mg alloys. The FEA method facilitates consideration of the influence of hydride precipitation on H diffusion and stress distribution. These FEA models assumed isotropic conditions (except for that proposed by Varias and Massih [181]) and considered the interrelation of diffusion/precipitation and lattice deformation.

Winzer *et al.* [67] provided a critical evaluation of the DHC mechanism in TGSCC of Mg alloys. The DHC model was critically evaluated to determine the maximum predicted values for V_c. A DHC model was formulated with the following components: (i) transient H diffusion towards the crack tip driven by stress and H concentration gradients; (ii) hydride precipitation when the H solvus is exceeded; and (iii) crack propagation through the extent of the hydride when it reaches a critical size of ~0.8 μm. The stress corrosion crack velocity, V_c, was calculated from the time for the hydride to reach the critical size. The model was implemented using a finite element script developed in MATLAB. The input parameters were chosen, based on the information available, to determine the highest possible value for V_c. Values for V_c of ~10^{-7} m/s were predicted by this DHC model. These predictions are consistent with measured values for V_c for Mg alloys in distilled water but cannot explain values for V_c of ~10^{-4} m/s measured in other aqueous environments. Insights for understanding Mg TGSCC were drawn. A key outcome is that the assumed initial condition for the DHC models is unlikely to be correct. During steady state stress corrosion crack propagation of magnesium in aqueous solutions, a high dynamic hydrogen concentration would be expected to build up immediately behind the crack tip. Stress corrosion crack velocities ~ 10^{-4} m/s, typical for Mg alloys in aqueous solutions, might be predicted using a DHC model for magnesium based on the time to reach a critical hydride size in steady state, with a significant residual hydrogen concentration from the previous crack advance step.

8.6.7 Hydrogen diffusion

Our critical review of Mg SCC [5], and the above discussion, indicate that the mechanism for Mg SCC most likely involves hydrogen (H) [101,120]. Although not all details have been established, it appears that Mg SCC involves the following. Mechanical or chemical rupture of a partially protective film allows aqueous solution access to the Mg metal and H is liberated as the cathodic partial reaction of the Mg corrosion. For the subsequent step(s), the speed of H diffusion in Mg is critical for several of the mechanisms proposed for the SCC of magnesium [9,98,113]. Available literature data [184–186] are presented in Fig. 8.25 [64]. Renner and Grabke [184] hydrided samples of an Mg–Ce alloy; the samples were of significant size, ~cms. The Ce in the alloy reacted with H to form hydrides in a surface layer of thickness ξ, which was related to the diffusion coefficient of H in the alloy. The data [184], when extrapolated to ambient temperature, indicated a low diffusion coefficient, $\sim 10^{-13}\,m^2/s$, so that doubts had been expressed [98] about the viability of any H mechanism for Mg SCC, in which H transport is required some distance ahead of the crack tip. However, this view may need to be changed by the more recent data [185,186] for the diffusion of H in Mg.

Available literature data [184–186] are presented in Fig. 8.25 for the diffusion coefficient of H in Mg. This compares the data obtained by Renner and Grabke [184] in 1978 with the more recent data of Nishimura *et al.* [185] and of Schimmel [186]. Nishimura *et al.* [185] carried out permeation experiments through 99.9% pure Mg sheets of thickness 0.6–2.9 mm. The Mg was annealed in vacuum at 573 K for 3.6 ks and mechanically polished to 0.05 μm alumina abrasive. A 0.1 μm thick Pd over-layer was deposited on both sides as an H catalyst. Pure H_2 gas was the charging medium. The H permeation flux was measured with a calibrated vacuum gauge on the exit side of the membrane that was maintained at a pressure of $\sim 10^{-7}\,Pa$. Schimmel [186] estimated the H diffusion coefficient from molecular dynamics simulations.

There is a reasonable agreement among these three estimates of the H diffusion coefficient in Mg in Fig. 8.25, so that it is possible to put a line through all these data, and extrapolate to 23 °C. This yields an estimate of the diffusion coefficient at ambient temperature to be $\sim 10^{-9}\,m^2/s$–$10^{-5}\,cm^2/s$. This value of the H diffusion coefficient is sufficient to allow significant H transport ahead of a stress corrosion crack in Mg at ambient temperature. A least squares regression through the data [184–186] of Fig. 8.25 yielded the following equation for the diffusion coefficient of H in Mg, $D_{H\,in\,Mg}$:

$$D_{H\,in\,Mg} = 1.02 \times 10^{-7} \exp(-1620/T) \qquad 8.2$$

where $D_{H\,in\,Mg}$ is in m^2/s and T is in K. A better equation to describe the H diffusion coefficient would be possible if there were data at ambient temperatures.

Dietzel *et al.* [187] estimated the H diffusion coefficient in AZ91 to be $D_{H\,in\,AZ91} = 2 \times 10^{-13}\,m^2/s$ at RT based on modelling the CERT results of Fig. 8.6 for AZ91 in distilled water [37]. The stress–strain behaviour of the cylindrical specimen was represented by a bundle of fibres orientated parallel to the direction of the applied strain. To simulate tensile tests, the fibre-bundle model assumed that each individual fibre followed the same stress–strain curve as the bulk material. For an environment leading to HE,

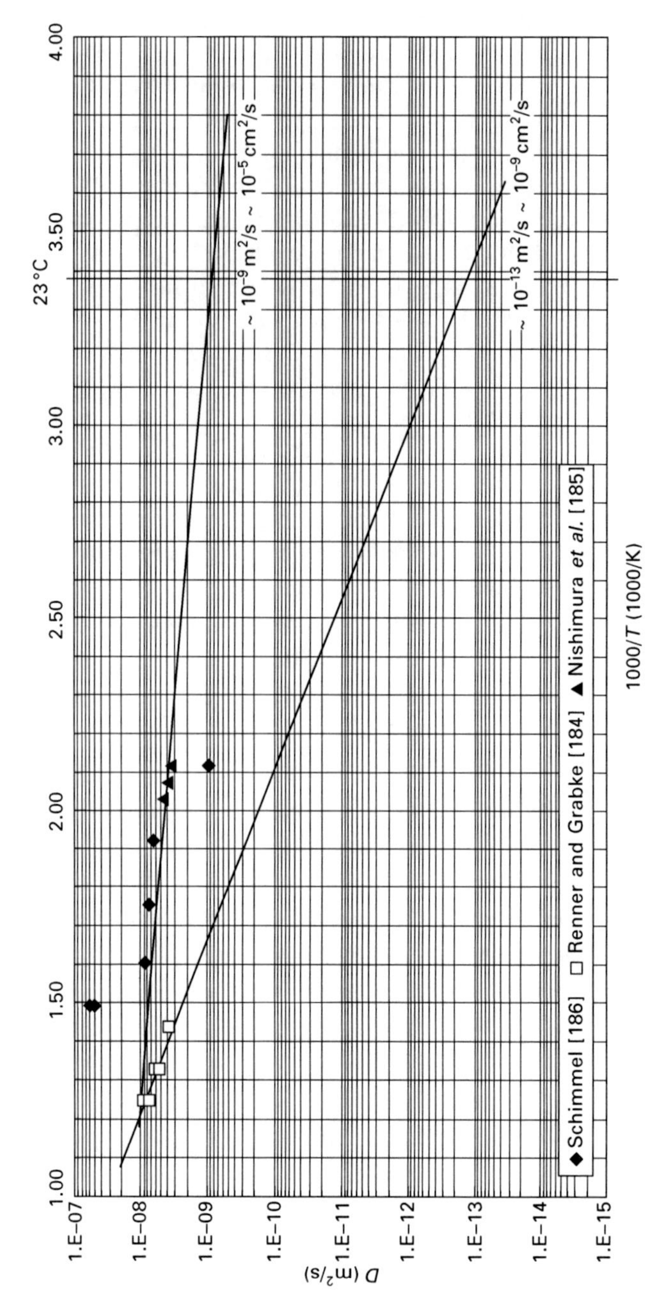

8.25 Literature data for the diffusion coefficient of H in Mg, $D_{\text{H in Mg}}$.

it was assumed that H diffused into the material and reduced the strain-to-failure of individual fibres and that fracture of a particular fibre occurred once the combination of applied strain and local H concentration reached a critical value. Crack initiation and growth were treated as a sequence of failure events at individual fibres. This model simulated the results of the SCC tests presented in Fig. 8.6. It was assumed that H was generated by the corrosion reaction inside pits at the specimen surface. These pits were created by mechanically straining the specimen and thus rupturing the hydroxide layer on the surface, where part of the hydrogen thus generated diffused into the bulk of the material. The model produced fracture surfaces macroscopically similar in extent to those produced by CERT, and stress–strain curves were generated which reflected the influence of the applied strain rate on the H induced fracture of the Mg alloy. The model had two unknown parameters: $D_{\mathrm{H\ in\ AZ91}}$ and x_{H} (which relates to the fracture strain in the presence of H). By systematically varying the values of these two parameters, best fit was obtained $D_{\mathrm{H\ in\ AZ91}} = 2 \times 10^{-13}\ \mathrm{m^2/s}$.

An alternative approach to the available H diffusion data is presented in Fig. 8.26 [188]. Separate best-fit lines are drawn through the diffusion data for Mg–2%Ce [184] and the data for pure Mg [185,186]. These lines have slopes similar to each other and to data for other close packed lattices (e.g. HCP, Zr and Ti, and face centred cubic, (FCC), Ni, Cu, Au). Included on the Fig. 8.26 from [188] as 'AZ91 (sim)' is the modelling estimate from the work of Dietzel *et al.* [187] of $D_{\mathrm{H\ in\ AZ91}} = 2 \times 10^{-13}\ \mathrm{m^2/s}$. This value is in good agreement with lower shelf values predicted for H diffusion in Mg at room temperature (Fig. 8.25) by extrapolating the elevated temperatures measurements of Renner and Grabke [184].

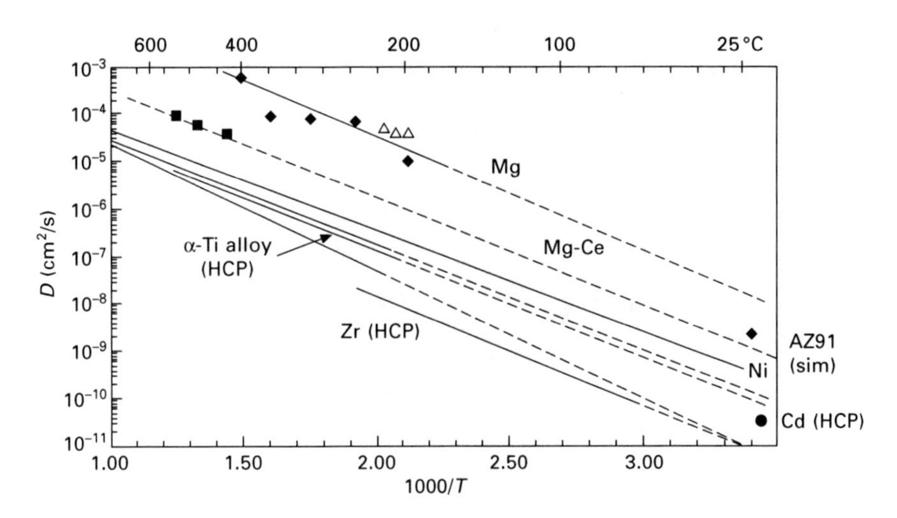

8.26 Diffusion coefficient of H in Mg, $D_{\mathrm{H\ in\ Mg}}$.

8.7 Recent insights

Recent work has resulted in a number of important insights. Multiple mechanisms can occur, and these mechanisms behave synergistically. Winzer *et al.* [37,69,70,153] found that, for AZ31, AM30 and AZ91 in water, tested with CERT, TGSCC initiated by highly localised dissolution, with a transition to HE at some critical crack length (Fig. 8.27). The HE mechanism is dependent on microstructure. TGSCC initiation in LIST is directly by HE.

8.7.1 β particles

Another key insight is that β particles play a critical role in TGSCC in Mg–Al alloys such as AZ91, whose microstructure consists of β-phase particles in an α-phase matrix. The β-phase is the $Mg_{17}Al_{12}$ intermetallic. Winzer *et al.* [37,70] found that, for AZ91 containing large β-phase particles, TGSCC propagation involved crack nucleation within β particles ahead of the principal crack tip (Fig. 8.28). Crack nucleation at β particles could be due to: (i) the inherently low fracture toughness of the β particles; (ii) a reduction in the fracture toughness of β particles by internal H by hydrogen enhanced decohesion (HEDE) or by transformation to MgH_2; or (iii) the synergistic effects of dislocation pile-ups at the α–β interface, as observed by Wang *et al.* [189], and H-enhanced dislocation mobility. Using the same AZ91 as Winzer *et al.* [37,70], Chen *et al.* [49,54] showed that cathodic polarisation in Na_2SO_4, even in the absence of an external stress, resulted in fracture of

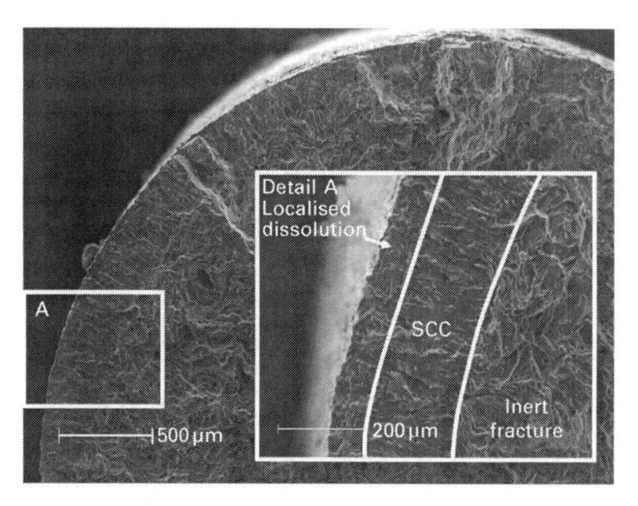

8.27 Fracture surface for AM30 in distilled water under CERT conditions showing zones corresponding to localised dissolution, SCC and inert fracture [70].

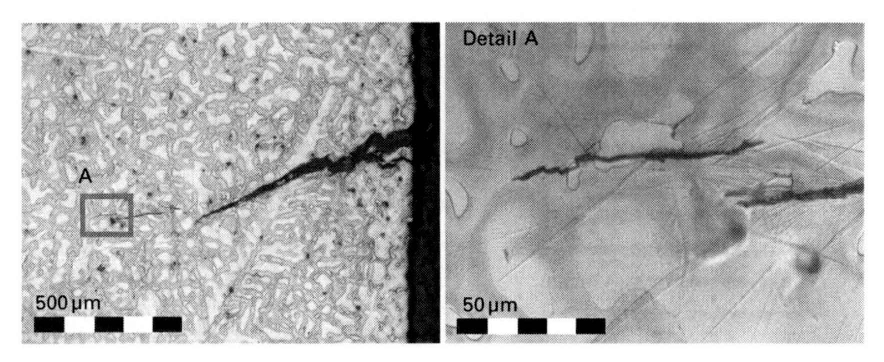

8.28 Micrograph of secondary crack in AZ91 tested distilled water at $3 \times 10^{-7} s^{-1}$ showing crack nucleation within a β particle.

the β at the surface, indicating that β particle fracture is indeed due to H and not to their inherently low fracture toughness. The factors contributing to the behaviour of β particles as H-trapping and crack nucleation sites were evaluated in Winzer *et al.* [190]; however, further work is required to validate this model.

The role of β particles implies an increase in SCC susceptibility with increasing β particle concentration, and typically increasing Al content consistent with Fig. 8.4. However, Winzer *et al.* [153] recently showed that, for weld metal containing relatively low concentrations of β particles, increasing bulk Al concentration decreased SCC and corrosion susceptibility. The positive influence of Al was attributed to the strengthening of the $Mg(OH)_2$ surface film by Al enrichment and the reduction of the galvanic current output of β particles (due to increased Al concentration of the α matrix). These factors have been largely overlooked by prior researchers, but may be relevant to the SCC and corrosion of alloys containing β refining elements (e.g. rare earths or alkaline metals).

8.7.2 H diffusion and trapping

It is generally accepted that the mechanism for TGSCC propagation in Mg alloys is a form of HE. This implies (i) that the durability of Mg alloys, subject to TGSCC, is dependent on H diffusivity, and (ii) that the microstructure influences can be understood in terms of H transport and trapping. Winzer *et al.* [70] showed that small changes in microstructure can result in significant variations in the mechanisms or rate-limiting processes. They compared extruded AZ31 and AM30, which have generally similar microstructures except that AZ31 contains higher concentrations of Zn and second phase particles. There were significant differences in fracture morphologies: the fracture for AZ31 was characterised by relatively smooth regions containing small, elongated dimples whereas those for AM30 were cleavage-like (Fig. 8.29). Moreover, the crack velocities for AM30 were lower than those for AZ31. These differences may be related to the influences of Zn and second phase particles on H diffusivity such that: (i) different HE mechanisms occur in AZ31 and AM30; or (ii) the same HE mechanism occurs, with the difference in H diffusivity changing the crack velocity and fracture surface morphology. There is a complete absence of measured data on H diffusion in Mg alloys at room temperature, and there is no understanding of H trapping by microstructure features in Mg alloys. This represents a substantial gap in understanding the SCC/HE mechanisms, particularly since there is evidence to suggest that key microstructure features (i.e. β particles) can act as H trapping and crack nucleation sites.

8.7.3 Hydrides

A mechanism commonly proposed by early work on TGSCC of Mg was DHC [99,113,118]. This mechanism involves repeated stages of: (i) stress-assisted diffusion of H to the maximum stress area ahead of the crack tip; (ii) precipitation of hydride as the H concentration exceeds the local solvus; and (iii) fracture of the brittle hydride. The mechanism is supported by (i) the cleavage-like appearance of fracture surfaces (it was assumed that the cleavage planes corresponded to the habit planes of hydrides or cleavage of hydrides) and (ii) the similarity of the Mg fracture surfaces to those associated with DHC in other metals, e.g. Zr, Nb and Ti alloys. However, hydrides were never observed in Mg by metallography or fractography. Winzer *et al.* [67] developed a numerical DHC model for Mg using an approach similar to that of Lufrano *et al.* [191] for Zr. The model predicts crack velocities similar to those measured for Mg TGSCC in distilled water. However,

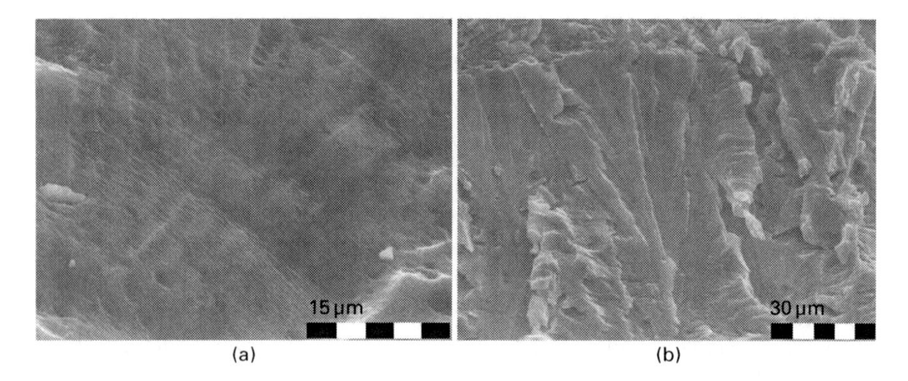

8.29 Morphology of SCC fracture surface for AZ31 (a) and AM30 (b) [70].

there is uncertainty regarding the hydride morphology and key parameters, particularly the H diffusion coefficient and H solubility. Winzer *et al.* [70] proposed that hydride precipitation indeed did occur at very slow crack velocities, based on the quasi-porous appearance of fracture surfaces for AZ91 tested in distilled water (Fig. 8.30). These quasi-porous features were identified as associated with hydride because there were similar features on fracture surfaces for AZ91 after tensile testing specimens after charging for 14h at 300°C and 30 bar gaseous H_2. Numerical analysis of the charging indicated that MgH_2 precipitation was likely. Schober [192] also reported precipitation of large hydride agglomerates in Mg for charging in H_2 gas under similar conditions. These MgH_2 hydride agglomerates dissolved on exposure to water, leaving large, pore-like cavities consistent with the quasi-porous fracture morphology of Winzer *et al.* [70], but in stark contrast to that for hydrides in other metals.

8.7.4 Texture

Another key outcome of recent work is that IGSCC is possible in Mg alloys not containing grain boundary precipitates. Previously IGSCC of Mg alloys was attributed exclusively to grain boundary β, causing micro-galvanic localised dissolution of the α matrix adjacent to grain boundary β precipitates [106]. This mechanism is expected in all Mg alloys containing continuous or nearly continuous second phase along the grain boundary. In contrast, Winzer *et al.* [153] found IGSCC in rolled AZ31 without grain boundary precipitates. The IGSCC fracture surfaces were characterised by slip traces on intergranular

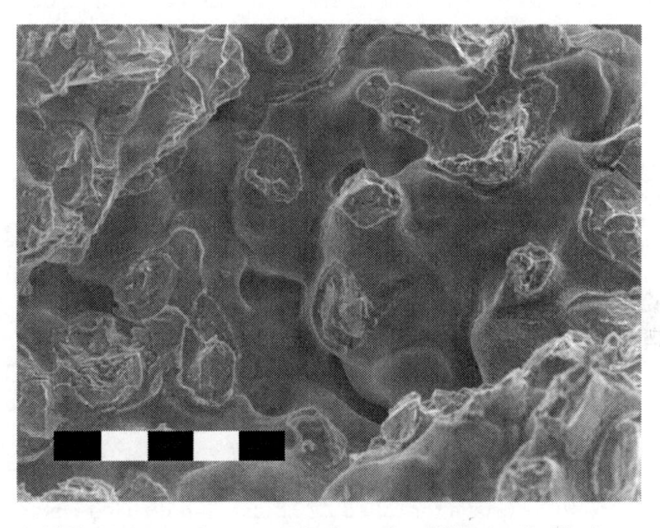

8.30 Fracture surface morphology for AZ91 in distilled water at $3 \times 10^{-8} s^{-1}$ showing quasi-porous features [70].

facets (Fig. 8.31), indicative of large-scale crack-tip plasticity [98]. This case is the first of IGSCC propagation by HE. Mixed-mode IGSCC and TGSCC attributed to H have been reported for pure Mg [98,101]. Kannan *et al.* [151] reported IGSCC in AZ31 adjacent to laser-beam welds attributed to galvanic corrosion between the fusion zone and the base metal. Winzer *et al.* [153] proposed that (i) the IGSCC of AZ31 was associated with the texture imparted by rolling, the grains were orientated with their *c*-axis normal to the rolling plane; (ii) this texture caused high stresses at grain boundaries and (iii) these stresses, in combination with solute H, caused grain boundary separation. This model is supported by (i) the work of Agnew *et al.* [193] who showed composite-like load sharing between differently orientated grains during plastic deformation of a similar material, and (ii) the work of Prakash *et al.* [194] who showed, using numerical models, that such load sharing causes elevated stress at grain boundaries.

8.8 Open issues

IGSCC of Mg alloys is reasonably well understood. IGSCC in Mg alloys, illustrated in Fig. 8.5, occurs typically by micro-galvanic corrosion, caused if there is a continuous or nearly continuous second phase along the grain boundaries, a microstructure typical of many cast creep-resistant Mg alloys. In contrast, TGSCC occurs through the α-phase Mg matrix. TGSCC is the intrinsic form of SCC and TGSCC can occur in alloys resistant to IGSCC. TGSCC occurs for common Mg alloys in environments such as distilled water and dilute chloride solutions.

The most recent mechanistic studies have established the main components important for propagation of TGSCC. The work of Pugh and co-workers [9,115,117,118,171] has provided convincing evidence for a brittle cleavage-

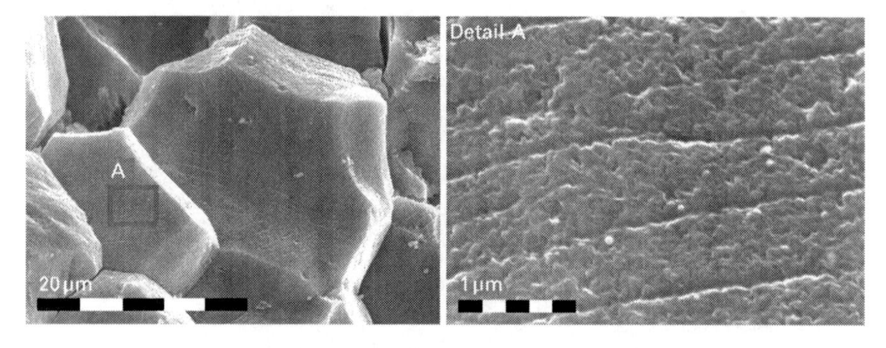

8.31 Fracture surface for rolled AZ31 tested in distilled water under CERT conditions showing fine parallel markings within intergranular facets [153].

type mechanism involving H. Particularly noteworthy are the stepped and faceted interlocking fracture surfaces. In contrast, the fractography of Lynch and Trevena [98] indicates some plasticity, particularly at higher loading rates and faster crack velocities. The slow strain rate testing of both Ebtehaj *et al.* [120] and Stampella *et al.* [101] indicates a mechanism involving strain-induced film rupture leading to corrosion and H production, with crack advance due to H. Makar *et al.* [113] confirmed the fractography, the importance of strain rate and H, but they propose a brittle hydride model. Thus, there is agreement that H is part of the SCC propagation mechanism but disagreement on the role of H. Moreover, the diffusion coefficient for H in Mg has not been measured at room temperature. Thus, there are no good data on H transport that are needed for an evaluation of the role of H.

Concerning the key aspects of the environment, the mechanistic hypothesis that emerges from the prior work is that SCC is associated with environmental conditions leading to the local breakdown of a partially protective surface film. Film breakdown can be caused by the environment, e.g. localised corrosion by chloride ions. However, SCC does occur for pure Mg and Mg alloys (AZ91, AM60, AS41, ZK60A-T5) in distilled water. This indicates that film breakdown may also be caused by the loading. The role of the film is also consistent with it acting as a barrier preventing hydrogen ingress.

All two-phase alloys, like AZ91, are expected to have poor TGSCC properties if there is H facilitated fracture of the large second-phase particles. In contrast, at present there is no information regarding the H-influenced fracture behaviour of the small precipitates that are effective for precipitation hardening, although there are promising indications, extrapolating from their positive influence for some heat treatment conditions in Al alloys. SCC resistance is much higher for Al alloys in the over-aged condition and for particular double ageing conditions [195].

AM30 has an SCC propagation velocity significantly lower than that of AZ31 and AZ91 [37,38,69,70], consistent with a lower effective H diffusivity, probably due to the absence of Zn in AM30. This again demonstrates the necessity to understand the influence of solid-solution alloying on SCC propagation velocity and H diffusivity.

The next major advance in understanding Mg TGSCC requires an understanding of the microstructure influences on SCC. Included needs to be an understanding of the effective H diffusion rate to the fracture process zone. This requires an understanding of H-trap interactions in Mg alloys.

Mg alloys are starting to be used in structural applications and new classes of wrought Mg alloys are emerging. These include high-performance high-strength heat-treatable wrought alloys. Promising alloy systems include Mg–Zn-Φ, Mg–Sn-Λ and Mg–RE-Σ. Particular mention may be made of the Mg–Y–Zn alloy ZW21 [196] and Mg–10Gd [197]. Their SCC behaviour has not been characterised. Clearly, significant risk of SCC would negate much of the potential for application of Mg alloys in critical service components, such as stents (catastrophic fracture due to SCC in a heart artery would probably be fatal) or stressed Mg-alloy auto components exposed to road spray.

The prior work on pure Mg, Mg–Al alloys and Zr-containing alloys showed that SCC is a significant issue, and that SCC can occur for a load condition equivalent to 50% of the yield stress for many combinations of alloy + environment. Since the prior research indicates that all Mg alloys are likely to be more or less susceptible, guidelines are needed to ensure safe application of Mg alloys in service. It would seem that an urgent task would be to delineate safe operational limits for common alloys in likely service environments.

Table 8.1 indicates that it would be conservative to assume that the threshold stress was ~ 50% YS, unless there is other convincing data for the particular alloy + environment combination. It would be prudent to apply this recommendation for the common Mg alloys in the environments enunciated above.

8.9 Acknowledgements

This research was supported by an Australian Research Council (ARC) Linkage grant in collaboration with General Motors Corporation USA. Atrens wishes to thank GKSS-Forschungszentrum Geesthacht GmbH for their support that allowed him to work at GKSS as a visiting scientist.

8.10 References

1 G Song, A Atrens, Corrosion mechanisms of magnesium alloys, *Advanced Engineering Materials*, **1999**, *1*, 11.
2 GL Song, A Atrens, Understanding magnesium corrosion mechanism: a framework for improved alloy performance, *Advanced Engineering Materials*, **2003**, *5*, 837.
3 G Song, A Atrens, Recent insights into the mechanism of magnesium corrosion and research suggestions, *Advanced Engineering Materials*, **2007**, *9*, 177–183.
4 Z Shi, M Liu, A Atrens, Measurement of the corrosion rate of magnesium alloys using Tafel extrapolation, *Corrosion Science*, **2010**, *52*, 579–588.
5 N Winzer, A Atrens, G Song, E Ghali, W Dietzel, KU Kainer, N Hort, C Blawert, A critical review of the stress corrosion cracking (SCC) of magnesium alloys, *Advanced Engineering Materials*, **2005**, *7*, 659–693.
6 A Atrens, N Winzer, W Dietzel, Stress corrosion cracking of magnesium alloys, *Advanced Engineering Materials*, **2009**, doi: 10:1002/adem.200900287.
7 A Atrens, ZF Wang, Stress corrosion cracking, *Materials Forum*, **1995**, *19*, 9.
8 W Dietzel, in Eds KHJ Buschow, RW Cahn, MC Flemings, B Ilschner, EJ Kramer, S. Mahajan, Encyclopedia of Materials: Science and Technology, Elsevier Science Ltd., Amsterdam, **2001**, 8883.
9 AJ Bursle, EN Pugh, in Eds PR Swann, FP Ford and ARC Westwood, *Mechanisms*

of Environment Sensitive Cracking of Materials, Materials Society, London, **1977**, 471.

10 MC Zhao, P Schmutz, S Brunner, M Liu, G Song, A Atrens, An exploratory study of the corrosion of Mg alloys during interrupted salt spray testing, *Corrosion Science*, **2009**, *51*, 1277–1292.

11 MC Zhao, M Liu, G Song, A Atrens, Influence of pH and chloride ion concentration on the corrosion of Mg alloy ZE41, *Corrosion Science*, **2008**, *50*, 3168.

12 MC Zhao, M Liu, G Song, A Atrens, Influence of the β-phase morphology on the Corrosion of the Mg Alloy AZ91, *Corrosion Science*, **2008**, *50*, 1939.

13 M Liu, S Zanna, H Ardelean, I Frateur, P Schmutz, G Song, A Atrens, P Marcus, Influence of the β-phase morphology on the corrosion of the Mg alloy AZ91, *Corrosion Science*, **2009**, *51*, 1115–1127.

14 A Atrens, W Dietzel, The negative difference effect and unipositive Mg^+, *Advanced Engineering Materials*, **2007**, *9*, 292–297.

15 M Liu, PJ Uggowitzer, AV Nagasekhar, P Schmutz, M Easton, G Song, A Atrens, Calculated phase diagrams and the corrosion of die-cast Mg–Al alloys, *Corrosion Science*, **2009**, *51*, 602.

16 M Liu, D Qiu, MC Zhao, G Song, A Atrens, The effect of crystallographic orientation on the active corrosion of pure magnesium, *Scripta Materialia*, **2008**, *58*, 421–424.

17 JX Jia, GL Song, A Atrens, Influence of geometry on galvanic corrosion of AZ91D coupled to steel, *Corrosion Science*, **2006**, *48*, 2133–2153.

18 JX Jia, A Atrens, G Song, T Muster, Simulation of galvanic corrosion of magnesium coupled to a steel fastener in NaCl solution, *Materials and Corrosion*, **2005**, *56*, 468–474.

19 GL Song, A Atrens, M Dargusch, Influence of microstructure on the corrosion of diecast AZ91D, *Corrosion Science*, **1999**, *41*, 249–273.

20 MC Zhao, M Liu, G Song, A Atrens, Influence of homogenization annealing of AZ91 on mechanical properties and corrosion behavior, *Advanced Engineering Materials*, **2008**, *10*, 93–103.

21 MC Zhao, M Liu, G Song, A Atrens, Influence of microstructure on corrosion of as-cast ZE41, *Advanced Engineering Materials*, **2008**, *10*, 104.

22 LM Peng, JW Chang, XW Guo, A Atrens, WJ Ding, YH Peng, Influence of heat treatment and microstructure on the corrosion of magnesium alloy Mg-10Gd-3Y-0.4Zr, *Journal of Applied Electrochemistry*, **2009**, *39*, 913.

23 GL Song, A Atrens, X Wu and B Zhang, Corrosion behaviour of AZ21, AZ501 and AZ91 in sodium chloride, *Corrosion Science*, **1998**, *40*, 1769–1791.

24 GL Song, A Atrens, DH StJohn, J Nairn, Y Lang, Electrochemical corrosion of pure magnesium in 1N NaCl, *Corrosion Science*, **1997**, *39*, 855–875.

25 A Seyeux, M Liu, P Schmutz, G Song, A Atrens, P Marcus, ToF-SIMS depth profile of the surface film on pure magnesium formed by immersion in pure water and the identification of magnesium hydride, *Corrosion Science*, **2009**, *51*, 1883–1886.

26 G Song, A Atrens, DH St. John in ed J Hryn, *Magnesium Technology*, TMS, New Orleans, **2001**, 255.

27 Z Shi, G Song, A Atrens, Influence of the β phase on the corrosion performance of anodised coatings on magnesium–aluminium alloys, *Corrosion Science*, **2005**, *47*, 2760.

28 MB Haroush, CB Hamu, D Eliezer, L Wagner, The relation between microstructure and corrosion behavior of AZ80 Mg alloy following different extrusion temperatures, *Corrosion Science*, **2008**, *50*, 1766–1778.

29 S Bender, J Goellner, A Atrens, A study of the corrosion of AZ91 in 1N NaCl and the mechanism of magnesium corrosion, *Advanced Engineering Materials*, **2008**, *10*, 583.
30 G Wu, Y Fan, A Atrens, C Zhai, W Ding, Electrochemical behavior of magnesium alloys AZ91D, AZCe2, and AZLa1 in chloride and sulfate solutions, *Journal of Applied Electrochemistry*, **2008**, *38*, 251–257.
31 JW Chang, XW Guo, PH Fu, A Atrens, LM Peng, WJ Ding, XS Wang, A comparison of the corrosion behaviour in 5% NaCl solution of Mg alloys NZ30K and AZ91D, *Journal of Applied Electrochemistry*, **2008**, *38*, 207.
32 G Williams, HN McMurray, Localized corrosion of magnesium in chloride-containing electrolyte studied by a scanning vibrating electrode technique, *Journal of the Electrochemical Society*, **2008**, *155*, C340–C349.
33 W Zhou, NN Aug, Y Sun, Effect of antimony, bismuth and calcium addition on corrosion and electrochemical behaviour of AZ91 magnesium alloy, *Corrosion Science*, **2009**, *51*, 403–408.
34 WC Neil, M Forsyth, PC Howlett, CR Hutchinson, BRW Hinton, Corrosion of magnesium alloy ZE41 – the role of microstructural features, *Corrosion Science*, **2009**, *51*, 387–394.
35 MP Staiger, AM Pietak, J Huadmai, G Dias, Magnesium and its alloys as orthopedic biomaterials: a review, *Biomaterials*, **2006**, *27*, 1728–1734.
36 N Winzer, A Atrens, W Dietzel, G Song, KU Kainer, Stress corrosion cracking (SCC) in Mg–Al alloys studied using compact specimens, *Advanced Engineering Materials*, **2008**, *10*, 453–458.
37 N Winzer, A Atrens, W Dietzel, VS Raja, G Song, KU Kainer, Characterisation of stress corrosion cracking (SCC) of Mg–Al Alloys, *Materials Science and Engineering A*, **2008**, *488*, 339–351.
38 N Winzer, A Atrens, W Dietzel, G Song, KU Kainer, Comparison of the linearly increasing stress test and the constant extension rate test in the evaluation of transgranular stress corrosion cracking of magnesium, *Materials Science and Engineering A*, **2008**, *472*, 97–106.
39 JX Jia, G Song, A Atrens, Experimental measurement and computer simmulation of galvanic corrosion of magnesium coupled to steel, *Advanced Engineering Materials*, 2007, **9**, 65–74.
40 Z Shi, G Song, A Atrens, The corrosion performance of anodised magnesium alloys, *Corrosion Science*, **2006**, *48*, 3531–3546.
41 Z Shi, GL Song, A Atrens, Influence of anodising current on the corrosion resistance of anodised AZ91D magnesium alloy, *Corrosion Science*, **2006**, *48*, 1939–1959.
42 Z Shi, GL Song, A Atrens, Corrosion resistance of anodised single-phase Mg alloys, *Surface and Coatings Technology*, **2006**, *201*, 492–503.
43 RG Song, C Blawert, W Dietzel, A Atrens, A study of the stress corrosion cracking and hydrogen embrittlement of AZ31 magnesium alloy, *Materials Science and Engineering*, **2005**, 399, 308–317.
44 JX Jia, G Song, A Atrens, Boundary element predictions of the influence of the electrolyte on the galvanic corrosion of AZ91D coupled to steel, *Materials and Corrosion*, **2005**, *56*, 259–270.
45 A Atrens, Suggestions for research directions in magnesium corrosion arising from the Wolfsburg Conference, *Advanced Engineering Materials*, **2004**, *6*, 83–84.
46 SB Abhijeet, R Balasubramaniam, M Gupta, Corrosion behaviour of Mg–Cu and Mg–Mo composites in 3.5% NaCl, *Corrosion Science*, **2008**, *50*, 2423–2428.

47 J Zhang, D Zhang, Z Tian, J Wang, K Liu, H Lu, D Tang, J Meng, Microstructures, tensile properties and corrosion behavior of die-cast Mg–4Al-based alloys containing La and/or Ce, *Materials Science and Engineering: A*, **2008**, *489*, 113–119.
48 E Zhang, W He, H Du, K Yang, Microstructure, mechanical properties and corrosion properties of Mg–Zn–Y alloys with low Zn content, *Materials Science and Engineering*, **2008**, *A488*, 102–111.
49 J Chen, J Wang, E Han, W Ke, Effect of hydrogen on stress corrosion cracking of magnesium alloy in 0.1 M Na_2SO_4 solution, *Materials Science and Engineering*, **2008**, *A488*, 428–434.
50 M Jönsson, D Persson, C Leygraf, Atmospheric corrosion of field-exposed magnesium alloy AZ91D, *Corrosion Science*, **2008**, *50*, 1406–1413.
51 MB Kannan, RK Singh Raman, *In vitro* degradation and mechanical integrity of calcium-containing magnesium alloys in modified-simulated body fluid, *Biomaterials*, **2008**, *29*, 2306–2314.
52 J Chen, J Dong, J Wang, E Han, W Kev, Effect of magnesium hydride on the corrosion behavior of an AZ91 magnesium alloy in sodium chloride solution, *Corrosion Science*, **2008**, *50*, 3610–3614.
53 X Zhou, Y Huang, Z Wei, Q Chen, F Gan, Improvement of corrosion resistance of AZ91D magnesium alloy by holmium addition *Corrosion Science*, **2006**, *48*, 4223–4233.
54 J Chen, J Wang, E Han, J Dong, W Ke, States and transport of hydrogen in the corrosion process of an AZ91 magnesium alloy in aqueous solution, *Corrosion Science*, **2008**, *50*, 1292–1305.
55 LJ Liu, M Schlesinger, Corrosion of magnesium and its alloys, *Corrosion Science*, **2009**, *51*, 1733–1737.
56 N Birbilis, MA Easton, AD Sudholz, SM Zhu, MA Gibson, On the corrosion of binary rare earth alloys, *Corrosion Science*, **2009**, *51*, 683–689.
57 G Song, Effect of tin modification on corrosion of AM70 magnesium alloy, *Corrosion Science*, **2009**, *51*, 2063–2070.
58 W Liu, F Cao, L Chang, Z Zhang, J Zhang, Effect of rare earth element Ce and La on corrosion behavior of AM60 magnesium alloy, *Corrosion Science*, **2009**, *51*, 1334–1343.
59 WS Loose, HA Barbian, in *Symposium on Stress Corrosion Cracking of Metals*, American Society for Testing Materials, USA, **1945**, 273.
60 WK Miller, in Ed RH Jones, *Stress Corrosion Cracking: Materials Performance and Evaluation*, ASM International, Metals Park, Ohio, **1992**, 251.
61 MO Speidel, PM Fourt, in *Stress Corrosion Cracking and Hydrogen Embrittlement of Iron Base Alloys*, NACE, Houston, TX, **1977**, 57.
62 P Lyon, Magnesium Electron UK, private communication Feb **2001**.
63 *ASM Handbook*, Vol 13, '*Corrosion*' ASM International, fourth printing J.R. Davised, Metals Park, Ohio, **1992**.
64 A Atrens, N Winzer, GL Song, W Dietzel, C Blawert, Stress corrosion cracking and hydrogen diffusion in magnesium, *Advanced Engineering Materials*, **2006**, *8*, 749–751.
65 N Winzer, A Atrens, W Dietzel, G Song, KU Kainer, Stress corrosion cracking in magnesium alloys: characterisation and Prevention, *Journal of Metals*, **2007**, *59*(8), 49–53.
66 N Winzer, A Atrens, W Dietzel, G Song, KU Kainer, Magnesium stress corrosion cracking, *Transactions of Nonferrous Metals Society of China*, **2007**, *17*, S150–S155 Part A Sp. Iss. 1, Nov.

67 N Winzer, A Atrens, W Dietzel, G Song, KU Kainer, Evaluation of the delayed hydride cracking mechanism for transgranular stress corrosion cracking of magnesium alloys, *Materials Science and Engineering A*, **2007**, *466*, 18–31.

68 MB Kannan, W Dietzel, C Blawert, A Atrens, P Lyon, Stress corrosion cracking of rare-earth-containing magnesium alloys ZE41, QE22, and Elektron 21 (EV31A) compared with AZ80, *Materials Science and Engineering A*, **2008**, *480*, 529–539.

69 N Winzer, A Atrens, W Dietzel, G Song, KU Kainer, Comparison of the linearly increasing stress test and the constant extension rate test in the evaluation of transgranular stress corrosion cracking of magnesium, *Materials Science and Engineering A*, **2008**, *472*, 97–106.

70 N Winzer, A Atrens, W Dietzel, G Song, KU Kainer, The fractography of stress corrosion cracking (SCC) of Mg–Al alloys, *Metallurgical and Materials Transactions* A, **2008**, *39*, 1157.

71 A Atrens, JQ Wang, K Stiller and HO Andren, Atom probe field ion microscope measurements of carbon segregation at an α:α grain boundary and service failures by intergranular stress corrosion cracking, *Corrosion Science*, **2006**, *48*, 79–92.

72 JQ Wang, A Atrens, Analysis of service stress corrosion cracking in a natural gas transmission pipeline, active or dormant?, *Engineering Failure Analysis*, **2004**, *11*, 3.

73 E Gamboa, A Atrens, Material influence on the stress corrosion cracking of rock bolts, *Engineering Failure Analysis*, **2005**, *12*, 201.

74 J Wang, A Atrens, SCC initiation for X65 pipeline steel in 'high' pH carbonate/bicarbonate solution, *Corrosion Science*, **2003**, *45*, 2199.

75 E Gamboa, A Atrens, Environmental influence on the stress corrosion cracking of rock bolts, *Engineering Failure Analysis*, **2003**, *10*, 521.

76 JQ Wang, A Atrens, DR Cousens, PM Kelly, C Nockolds, S Bulcock, Measurement of grain boundary composition for X52 pipeline steel, *Acta Materialia*, **1998**, *46*, 5677.

77 A Oehlert, A Atrens, SCC Propagation in Aermet 100, *Journal of Materials Science*, **1998**, *33*, 775–781.

78 A Oehlert, A Atrens, Environmental assisted fracture for 4340 steel in water and air of various humidities, *Journal of Materials Science*, **1997**, *32*, 6519.

79 A Oehlert, A Atrens, Initiation and propagation of stress corrosion cracking in AISI 4340 and 3.5NiCrMoV rotor steel in constant load tests, *Corrosion Science*, **1996**, *38*, 1159.

80 ZF Wang A Atrens, Initiation of stress corrosion cracking for pipeline steels in a carbonate–bicarbonate solution, *Metallurgical and Materials Transactions*, **1996**, *27A*, 2686.

81 A Oehlert, A Atrens, Room temperature creep of high strength steels, *Acta Metallurgica et Materialia*, **1994**, *42*, 1493.

82 A Atrens, CC Brosnan, S Ramamurthy, A Oehlert, IO Smith, Linearly increasing stress test (LIST) for SCC research, *Measurement Science and Technology*, **1993**, *4*, 1281.

83 S Ramamurthy, A Atrens, The stress corrosion cracking of as-quenched 4340 and 3.5NiCrMoV steels under stress rate control in distilled water at 90°C, *Corrosion Science*, **1993**, *34*, 1385.

84 RM Rieck, A Atrens, IO Smith, The role of crack tip strain rate in the stress corrosion cracking of high strength steels in water, *Metallurgical Transactions*, **1989**, *20A*, 889.

85 RG Song, W Dietzel, BJ Zhang, WJ Liu, MK Tseng and A Atrens, Stress corrosion cracking and hydrogen embrittlement of an Al–Zn–Mg–Cu alloy, *Acta Materialia*, **2004**, *52*, 4727–4743.
86 RM Rieck, A Atrens, IO Smith, Stress corrosion cracking and hydrogen embrittlement of cold worked AISI type 304 austenitic stainless steel in mode I and mode III, *Materials Science and Technology*, **1986**, *2*, 1066–1073.
87 CD Cann, A Atrens, A metallographic study of the terminal solubility of hydrogen in zirconium at low hydrogen concentrations, *Journal of Nuclear Materials*, **1980**, *88*, 42.
88 A Atrens, D Mezzanotte, NF Fiore, MA Genshaw, Electrochemical studies of hydrogen diffusion and permeability in Ni, *Corrosion Science* **1980**, *20*, 673.
89 A Atrens, JJ Bellina, NF Fiore, RJ Coyle, *The Metal Science of Stainless Steels,* WE Collings and HW King, Eds, TMS-AIME, **1978**, 54.
90 R Zeng, E Han, W Ke, W Dietzel, KU Kainer, A Atrens, Influence of microstructure on tensile properties and fatigue crack growth in extruded magnesium alloy AM60, *International Journal of Fatigue*, **2010**, *32*, 411–419.
91 A Atrens, H Meyer, G Faber, K Schneider, in Eds MO Speidel and A Atrens, *Corrosion in Power Generating Equipment*, Plenum, New York, **1984**, 299.
92 A Atrens, W Hoffelner, TW Duering, J Allison, Subsurface crack initiation in high cycle fatigue in Ti–6Al–4V and a typical martensitic stainless steel, *Scripta Metallurgica*, **1983**, *17*, 601.
93 R Zeng, E Han, W Ke, W Dietzel, KU Kainer, A Atrens, Influence of microstructure on tensile properties and fatigue crack growth in extruded magnesium alloy AM60, *International Journal of Fatigue*, **2010**, *32*, 411–419.
94 IG Ritchie, A Atrens, The Diffusion of oxygen in alpha-zirconium, *Journal of Nuclear Materials*, **1977**, *67*, 254.
95 S Jin, A Atrens, ESCA – Studies of the structure and composition of the passive film formed on stainless steels by various immersion times in 0.1M NaCl solution, *Applied Physics A*, **1987**, *42,* 149.
96 AS Lim, A Atrens, ESCA Studies of Fe–Ti, *Applied Physics A*, **1992**, *54*, 500.
97 AS Lim, A Atrens, ESCA studies of nitrogen containing stainless steels, *Applied Physics A*, **1990**, *51*, 411.
98 SP Lynch, P Trevena, Stress corrosion cracking and liquid metal embrittlement in pure magnesium, *Corrosion*, **1988**, *44*, 113–123.
99 EI Meletis, RF Hochman, Crystallography of stress corrosion cracking in pure magnesium, *Corrosion*, **1984**, *40*, 39–48.
100 *ASM Specialty Handbook: Magnesium and Magnesium Alloys*, ASM International, Metals Park Ohio, **1999**, 211.
101 RS Stampella, RPM Procter, V Ashworth, Environmentally induced cracking of magnesium, *Corrosion Science*, **1984**, *24*, 325–341.
102 N Winzer, G Song, A Atrens, W Dietzel, C Blawert, Corros Preven 2005, ACA, **2005**, 37.
103 RS Busk, *Magnesium Products Design* Marcel Dekker, New York, **1986**, 256.
104 DK Priest, FH Beck, MG Fontana, Stress corrosion mechanism in a magnesium–base alloy, *Transactions of the American Society for Metals*, **1955**, *47*, 473–492.
105 MA Pelensky, A Gallaccio, *Stress Corrosion Testing*, *STP425*, ASTM, West Conshohocken, Pennsylvania, **1967**, 107.
106 L Fairman, HJ Bray, Transgranular SCC in magnesium alloys, *Corrosion Science*, **1971**, *11*, 533–541.

107 WK Miller, *Materials Research Society Symposium Proceedings*, **1988**, *125*, 253.

108 A Arnon, E Aghion, Stress corrosion cracking of nano/sub-micron E906 magnesium alloy, *Advanced Engineering Materials*, **2008**, *10*, 742–745.

109 J Chen, M Ai, J Wang, E Han, W Ke, Stress corrosion cracking behaviours of AZ91 magnesium alloy in deicer solution using constant load, *Materials Science and Engineering*, **2009**, *A515*, 79–84.

110 ASM, ASM Handbook on line, ASM International, Materials Park, OH, **2009**.

111 Minimum values, Magnesium Electron, private communication, **2009**.

112 MB Kannan, RKS Raman, *In vitro* degradation and mechanical integrity of calcium-containing magnesium alloys in modified-simulated body fluid, *Biomaterials*, **2008**, *29*, 2306–2314.

113 GL Makar, J Kruger, K Sieradzki, Stress corrosion cracking of rapidly solidified magnesium alloys, *Corrosion Science*, **1993**, *34*, 1311–1342.

114 WR Wearmouth, GP Dean, RN Parkins, Role of stress in stress corrosion cracking of a Mg–Al alloy, *Corrosion*, **1973**, *29*, 251–260.

115 DG Chakrapani, EN Pugh, The transgranular SCC of a Mg–Al alloy: Crystallographic, fractographic and acoustic-emission studies, *Metallurgical Transactions*, **1975**, *6A*, 1155–1163.

116 G Oryall, D Tromans, Transgranular stress corrosion cracking of solution treated and quenched Mg–86 Al alloy, *Corrosion*, **1971**, *27*, 334–341.

117 DG Chakrapani, EN Pugh, On the fractography of transgranular stress corrosion failures in a Mg–Al alloy, *Corrosion*, **1975**, *31*, 247–251.

118 DG Chakrapani, EN Pugh, Hydrogen embrittlement in a Mg–Al Alloy, *Metallurgical Transaction*, **1976**, *7A*, 173–178.

119 ECW Perryman, Stress corrosion of magnesium alloys, *Journal of the Institute of Metals*, **1951**, *79*, 621–642.

120 K Ebtehaj, D Hardie, RN Parkins, The influence of chloride–chromate solution composition on the stress corrosion cracking of a Mg–Al alloy, *Corrosion Science*, **1993**, *28*, 811–821.

121 T Nozaki, S Hanaki, M Yamashita, H Uchida, *Proceedings of the 13th Asian-Pacific Corrosion Control Conference*, Osaka, **2003**, K–15.

122 L Fairman, JM West, Stress corrosion cracking of a magnesium alloy, *Corrosion Science* **1965**, *5*, 711–716.

123 A Moccari, CR Shastry, An investigation of stress corrosion cracking in Mg AZ61 alloy in 3.5% NaCl + 2% K_2CrO_4 aqueous solution at room temperature, *Journal of Materials Technology (Zeitschrift fur Werkstofftechnik)*, **1979**, *10*, 119–123.

124 L Fairman, HJ Bray, *British Corrosion Journal*, **1971**, *6*, 170–174.

125 WM Pardue, FH Beck, MG Fontana, Propagation of stress-corrosion cracking in a magnesium–base alloy as determined by several techniques, *Transactions of the American Society for Metals*, **1961**, *54*, 539–548.

126 ND Tomashov, VN Modestova, in Ed. IA Levin, *Intercrystalline Corrosion and Corrosion of Metals Under Stress*, London, **1962**, 251–262.

127 MA Timonova, in Ed. I.A. Levin, *Intercrystalline Corrosion and Corrosion of Metals Under Stress*, London, **1962**, 263.

128 LL Rokhlin, *Magnesium Alloys Containing Rare Earth Metals*, Taylor and Francis, London, **2003**.

129 JF Nie, X Gao, SM Zhu, Enhanced age hardening response and creep resistance of Mg–Gd alloys containing Zn, *Scripta Materialia*, **2005**, *53*, 1049–1053.

130 C Sanchez, C Nussbaum, P Azavant, H Octor, Elevated temperature behaviour of rapidly solidified magnesium alloys containing rare earths, *Materials Science Engineering*, **1996**, *A221*, 48–57.

131 P Lyon, T Wilks, I Syed, in: N.R. Neelameggham, in Eds (H.I. Kaplan, B.R. Powell), *The Influence of Alloying Elements and Heat Treatment upon the Properties of Elektron 21 (EV31A) Alloy*, Magnesium Technology 2005, Warrendale, PA, **2005**, p. 303.

132 J Chang, X Guo, P Fu, L Peng, W Ding, Effect of heat treatment on corrosion and electrochemical behaviour of Mg–3Nd–0.2Zn–0.4Zr (wt.%) alloy, *Electrochimica Acta*, **2007**, *52*, 3160–3167.

133 JH Nordlien, K Nisancioglu, S Ono, N Masuko, Morphology and structure of water-formed oxides on ternary MgAl alloys, *Journal of the Electrochemical Society*, **1997**, *144*, 461–466.

134 S Krishnamurthy, M Khobaib, E Robertson, FH Froes, Corrosion behaviour of rapidly solidified Mg–Nd and Mg–Y alloys, *Materials Science Engineering*, **1988**, *99*, 507–511.

135 R Heidenrich, CH Gerould, FE McNulty, *Transactions of the AIME*, **1946**, *166*, 15.

136 LL Rokhlin, *Magnesium Alloys Containing Rare Earth Metals*, Taylor and Francis, London, **2003**, 221.

137 *Exterior Stress Corrosion Resistance of Commercial Magnesium Alloys, Report Mt 19622*, Dow Chemical USA, **1966**.

138 G Song, Z Xu, The surface, microstructure and corrosion of magnesium alloy AZ31 sheet, *Electrochimica Acta*, **2010**, *55*, 4148.

139 RI Stephens, CD Schrader, DL Goodenberger, KB Lease, VV Ogarevic, SN Perov, Society of Automotive Engineers, No. 930752, USA, **1993**.

140 S Mathieu, C Rapin, J Hazan, P Steinmetz, Corrosion behaviour of high pressure die-cast and semi-solid cast AZ91D alloys, *Corrosion Science*, **2002**, *44*, 2737–2756.

141 MA Timonova, LI D'yalchenko, YM Dolzhanskii, MB Al'tman, NV Sakharova, AA Blyablin, *Protection of Metals*, **1983**, *19*, 99–102.

142 VA Marichev, SA Shipilov, Influence of electrochemical polarisation on crack-growth in corrosion cracking and corrosion fatigue of magnesium alloys, *Soviet Materials Science*, **1986**, *33*, 240–244.

143 RB Mears, RH Brown, EH Dix, A Generalized Theory of Corrosion in Alloys, in *Symposium on Stress-Corrosion Cracking of Metals*, American Society for Testing Materials, West-Conshohocken, Pennsylvania, **1945**, 323.

144 MO Speidel, MJ Blackburn, TR Beck, JA Feeney, *Corrosion Fatigue: Chemistry, Mechanics and Microstructure*, NACE-2, 1**972**, 324.

145 JC Kiszka, Stress corrosion tests of some wrought magnesium–lithium base alloys, *Materials Protection*, **1965**, *4*, 28–29.

146 A Munitz, C Cotler, H Shaham, G Kohn, Electron beam welding of magnesium AZ91D plates, *Welding Journal*, **2000**, *79*, 202s–208s.

147 J Matsumoto, M Kobayashi, M Hotta, Arc welding of magnesium casting alloy AM60, *Welding International*, **1990**, *4*, 23–28.

148 X Cao, M Jahazi, JP Immarigeon, W Wallace, A review of laser welding techniques for magnesium alloys, *Journal of Materials Processing Technology*, **2006**, *171*, 188–204.

149 WM Thomas, US Patent No. 5,460,317.

150 R Zettler, AC Blanco, JF dos Santos, S Marya, The effect of process parameters and tool geometry on thermal field development and weld formation in friction stir welding of the alloy AZ31 and AZ61 in: Ed. N.R. Neelameggham, H.I. Daplan, B.R. Powell, *Magnesium Technology*, TMS, The Minerals, Metals & Materials Society, 2005, 409–423.
151 MB Kannan, W Dietzel, C Blawert, S Riekehr, M Kocak, Stress corrosion cracking behavior of Nd:YAG laser butt welded AZ31 Mg sheet, *Materials Science and Engineering*, **2007**, *A444*, 220–226.
152 MB Kannan, W Dietzel, R Zeng, R Zettler, JF dos Santos, A study on the SCC susceptibility of friction stir welded AZ31 Mg sheet, *Materials Science and Engineering*, **2007**, *A460–461*, 243–250.
153 N Winzer, P Xu, S Bender, T Gross, WES Unger, CE Cross, Stress corrosion cracking of gas-tungsten arc welds in continuous-cast AZ31 Mg alloy sheet, *Corrosion Science*, 2009, *51*, 1950–1963.
154 PB Srinivasan, R Zettler, C Blawert, W Dietzel, A study on the effect of plasma electrolytic oxidation on the stress corrosion cracking behaviour of a wrought AZ61 magnesium alloy and its friction stir weldment, *Materials Characterization, 2009*, **60**, 389–396.
155 PB Srinivasan, S Riekehr, C Blawert, W Dietzel, M Kocak, Slow strain rate stress corrosion cracking behaviour of as-welded and plasma electrolytic oxidation treated AZ31HP magnesium alloy autogenous laser beam weldment, *Materials Science and Engineering A*, **2009**, *A517*, 197–203.
156 ZL Wei, QR Chen, ZC Guo, L Yang, NX Xiu and YW Huang, in Ed. KU Kainer, *Magnesium Proceedings of the 6th International Conference Magnesium Alloys and Their Applications*, Wiley-VCH, Weinheim, Berlin, Darmstadt und Zürich, **2004**, 649.
157 TJ Marrow, AB Ahmad, IN Kahn, SMA Sim, S Torkamani, Environment-assisted cracking of cast WE43-T6 magnesium, *Materials Science and Engineering*, **2004**, *A387–389*, 419–423.
158 RP Frankenthal, The inhibition of pitting and stress corrosion cracking of Mg–Al Alloys by NO_3^-, *Corrosion Science*, **1967**, 7, 61–62.
159 VB Yakovlev, LP Trutneva, NI Isaev, G Nemetch, Influence of protective films on kinetics of stress corrosion cracking of magnesium alloy MA2-1, *Protection of Metals*, **1984**, *20*, 300–306.
160 JC Scully, in Ed LL Shrier, RA Jarman and GT Burstein, *Corrosion*, Butterworth Heinemann, Oxford, **1993**, *8*, 115–142.
161 HL Logan, Mechanism of stress-corrosion cracking in the AZ31B magnesium alloy, *Journal of Research of the National Bureau of Standards,* **1958**, *61*, 503–508.
162 MB Kannan, W Dietzel, RKS Raman, P Lyon, Hydrogen induced cracking in magnesium alloy under cathodic polarisation, *Scripta Materialia*, **2007**, *57*, 579–581.
163 GF Sager, RH Brown, RB Mears, *Symposium on Stress-Corrosion Cracking of Metals*, ASTM, **1945**, 267.
164 WS Loose, *Magnesium*, ASM, Cleveland, OH, **1956**, 173.
165 VV Romanov, Ed. IA Levin, *Intercrystalline Corrosion and Corrosion of Metals Under Stress*, London **1962**, 283.
166 PB Srinivasan, C Blawert, W Dietzel, Effect of plasma electrolytic oxidation treatment on the corrosion and stress corrosion cracking behaviour of AM50 magnesium alloy, *Materials Science and Engineering*, **2008**, *A494*, 401–406.

167 PB Srinivasan, C Blawert, W Dietzel, Effect of plasma electrolytic oxidation coating on the corrosion and stress corrosion cracking behaviour of wrought AZ61 magnesium alloy, *Corrosion Science*, **2008**, *50*, 2415–2418.
168 PB Srinivasan, C Blawert, W Dietzel, KU Kainer, Stress corrosion cracking behaviour of a surface-modified magnesium alloy, *Scripta Materialia*, **2008**, *59*, 43–46.
169 F Meller, M Metzger, *U.S.N.A.C.A. Tech Note No 4019 (1985).*
170 HL Logan, Film rupture mechanism of stress corrosion, *Journal of Research of the National Bureau of Standards*, **1952**, *48*, 99–105.
171 EH Pugh, JAS Green, PW Slattery, in, Ed P.L. Pratt, *Fracture 1969: The Proceedings of the Second International Conference on Fracture*, Chapman and Hall Ltd, London, **1969**, 387.
172 HW Pickering, PR Swann, Electron metallography of chemical attack upon some alloys susceptible to stress corrosion cracking, *Corrosion*, **1963**, *19*, 373–389.
173 TL Anderson, *Fracture Mechanics: Fundamentals and Applications*, 2nd Edition, CRC Press, Boca Raton, FL, **1992**.
174 SP Lynch, 11th International Conference on Fracture, **2005**, paper p 3885.
175 RP Gangloff, Hydrogen-assisted cracking, in Eds I Milne, RO Ritchie and B Karihaloo *Comprehensive Structural Integrity*, Vol 6 *Environmentally assisted failure*, Amsterdam, San Diego, CA: Elsevier Pergamon, **2003**, 31–101.
176 HW Liu, Stress-corrosion cracking and the interaction between crack-tip stress field and solute Atoms, *Transactions of the ASME*: *Journal of Basic Engineering*, **1970**, *92*, 633–638.
177 HP Van Leeuwen, The kinetics of hydrogen embrittlement: a quantitative diffusion model, *Engineering Fracture Mechanics*, **1974**, *6*, 141–161.
178 R Dutton, K Nuttall, MP Puls, LA Simpson, Mechanisms of hydrogen induced delayed cracking in hydride forming materials, *Metallurgical Transactions*, **1977**, *8A*, 1553–1562.
179 J Lufrano, P Sofronis, HK Birnbaum, Modelling of hydrogen transport of elastically accommodated hydride formation near a crack tip, *Journal of Mechanics and Physics of Solids*, **1996**, *44*, 179–205.
180 P Sofronis, RM McMeeking, Numerical analysis of hydrogen transport near a blunting crack tip, *Journal of Mechanics and Physics of Solids*, **1989**, *37*, 317–350.
181 AG Varias, AR Massih, Hydride-induced embrittlement and fracture in metals – effect of stress and temperature distribution, *Journal of Mechanics and Physics of Solids*, **2002**, *50*, 1469–1510.
182 AG Varias, JL Feng, Simulation of hydride induced steady-state crack growth in metals – Part I: Growth near hydrogen chemical equilibrium, *Computational Mechanics*, **2004**, *34*, 339–356.
183 AG Varias, JL Feng, Simulation of hydride induced steady-state crack growth in metals – Part II: General near tip field, *Computational Mechanics*, **2004**, *34*, 357–376.
184 J Renner, HJ Grabke, Bestimmung von Diffusionskoeffizienten bei der Hydrierung von Legierungen, *Zeitshrist Metallkunde*, **1978**, *69*, 639.
185 C Nishimura, M Komaki, M Amano, Hydrogen permeation through magnesium, *Journal of Alloys and Compounds*, **1999**, *293–295*, 329.
186 HG Schimmel, Towards a hydrogen-driven society, PhD Thesis, **2004**, Technical Univ Delft.
187 W Dietzel, M Pfuff, N Winzer, Testing and mesoscale modelling of hydrogen

assisted cracking of magnesium, *Engineering Fracture Mechanics*, **2010**, *77*, 257–263.

188 SP Lynch, unpublished work, Defence Science and Technology Organisation, Australia, 2009, presented at an International Symposium on Stress Corrosion Cracking in Structural Materials at ambient temperatures, Padova, Italy, Aug 30–Sept 4, 2009.

189 RM Wang, A Eliezer, EM Gutman, An investigation on the microstructure of an AM50 magnesium alloy, *Materials Science Engineering* **2003**, *355A*, 201–207.

190 N Winzer, CE Cross, On the role of ß particles in stress corrosion cracking of Mg–Al Alloys, *Metallurgical and Materials Transactions*, **2009**, *40A*, 273–274.

191 J Lufrano, P Sofronis, HK Birnbaum, Modeling of hydrogen transport and elastically accommodated hydride formation near a crack tip, *Journal of Mechanics and Physics of Solids*, **1996**, *44*, 179–205.

192 T Schober, The magnesium-hydrogen system: Transmission electron microscopy, *Metallurgical and Materials Transactions*, **1981**, *12A*, 951–957.

193 SR Agnew, CN Tomé, DW Brown, TM Holden, SC Vogel, Study of slip mechanisms in a magnesium alloy by neutron diffraction and modelling, *Scripta Materialia*, **2003**, *48*, 1003–1008.

194 A Prakash, SM Weygand, H Riedel, Modelling the evolution of texture and grain shape in Mg alloy AZ31 using the crystal plasticity finite element method, *Computational Materials Science*, **2009**, *45*, 744–750.

195 RG Song, W Dietzel, BJ Zhang, WJ Liu, M Tseng, A Atrens, Stress corrosion cracking and hydrogen embrittlement of an Al-Zn-Mg-Cu alloy. *Acta Materialia*, **2004**, *52*, 4727–4743.

196 AC Haenzi, I Gerber, M Schinhammer, JF Loeffler, PJ Uggowitzer, On the *in vitro* and *in vivo* degradation performance and biological response of new biodegradable Mg–Y–Zn alloys, *Acta Biomaterialia*, **2009**, *6*, 182–1833.

197 N Hort, Y Huang, D Fletcher, M Stormer, C Blawert, F Witte, H Drucker, R Willumeit, KU Kainer, F Feyerabend, Magnesium alloys as implant materials – Principles of property design for Mg–RE alloys, *Acta Biomaterialia*, **2009**, *6*, 1714–1725.

9

Corrosion creep and fatigue behavior of magnesium (Mg) alloys

Y. B. UNIGOVSKI and E. M. GUTMAN, Ben-Gurion University of the Negev, Israel

Abstract: Environment-enhanced creep, which we have called 'corrosion creep', and high cycle corrosion fatigue were investigated in pure magnesium (Mg) and Mg alloys in different corrosive solutions in comparison to those in air. Extruded alloys, especially AZ80 and ZK60, show a significantly higher sensitivity to the action of corrosive solutions under stress in comparison with diecast alloys. The strong chemomechanical effect of increased plasticity of diecast and extruded alloys due to surface electrochemical reactions was found in borate buffer solution (pH 9.3).

Key words: magnesium alloys, corrosion creep, corrosion fatigue, buffer solution, NaCl, Na_2SO_4.

9.1 Introduction

The chapter includes four sections:

- Historical review of environmentally enhanced creep and fatigue of metals.
- Mechanoelectrochemical behavior of Mg alloys.
- Corrosion creep of magnesium and diecast Mg alloys.
- Corrosion fatigue of magnesium alloys.

Section 9.2 includes an extended review on creep and fatigue of metals accompanied by anodic dissolution. It is shown that in real service, mechanical and corrosion processes develop together and inevitably lead to the appearance of synergistic mechanochemical effects that significantly reduce the lifetime of Mg alloys due to static and dynamic corrosion fatigue. Section 9.3 includes, by way of example, the effect of preliminary plastic deformation and the second phase content on the corrosion rate of two diecast Mg alloys. Section 9.4 presents extended results on the creep behavior of magnesium and its alloys in various electrolytic media. Section 9.5 shows high-cycle corrosion fatigue behavior of Mg alloys produced by extrusion, common diecasting and rheoformed diecastings. At a reversible cyclic loading, extruded alloys show a significantly higher sensitivity to the action of corrosive solution in comparison with diecast alloys. In correlation with mechanoelectrochemical behavior of Mg alloys, the highest sensitivity to a corrosive medium during creep and fatigue is observed in the alloy with the highest Al content or in alloys with a higher micro- and macrostructure heterogeneity.

9.2 Historical review of environmentally enhanced creep and fatigue of metals

The investigation of environmentally assisted creep is very important to reveal mechanisms of stress corrosion processes in the tip of a crack. The environments around creeping metals can be divided into several large groups: (a) gaseous (oxidizing, reducing or neutral gases); (b) vacuum; (c) solutions of some polar surface-active agents such as organic acids; (d) non-conductive liquids containing some reagents which react with the solid (organic solutions, melted polymers); (e) water-based and organic electrolytes (acid, base and salt solutions), mineral melts, e.g. melted slag; and (f) liquid metals, e.g. steels in liquid sodium.

Rehbinder and co-workers first found an adsorption effect of surface-active additions on the mechanical behavior of mica and calcite in 1928 [1,2]. The resistance of single- and polycrystalline materials to plastic deformation was found to be reduced when they were surrounded by solutions of surface-active agents [1–6]. For example, creep deformation of lead, tin and copper under constant load in solutions of small amounts of certain polar surface-active substances, e.g. oleic acid in paraffin oil, is greatly facilitated compared with that in air or in a liquid without surface-active reagents. The effect reaches the maximum at complete saturation of the monomolecular adsorption layer by those polar substances [1]. For instance, creep strain and creep rate of single tin crystals increase several times in octane containing 0.005 mol/L oleic acid [2,3]. The effect of the surface-active acid is attributed to its penetration into the microcracks formed in the process of elongation along glide planes. The Rehbinder effect becomes more pronounced with an increase in the deformation rate of the metal [4,5]. However, it is small even at a fast deformation rate *V*; e.g. with an increase in *V* by two orders of magnitude – from 10^{-4} to $10^{-2} s^{-1}$ – a decrease in tensile yield stress (TYS) of cadmium single-crystal in 1% butyl alcohol in water equaled around 30% for samples both with and without an oxide film [5].

The effect of surface oxide films on an increase of the strength of metals was, probably, discovered for the first time by Roscoe in 1936 on cadmium single-crystal wires [7]. The effect was attributed to the healing of submicroscopic cracks by the oxide film, but now there is no evidence to support this idea. Harper and Cottrell confirmed the hardening effect of an oxide film on Zn crystals crept in air and softening of crystals by immersion in paraffin and oleic acid [8]. They supposed that the liquid containing surface-active additions acts by penetrating the oxide film, presumably through cracks, and weakens adherence of oxide to the surface of the metal, thereby reducing its hardening effect.

Adsorption of one or more reactants on surfaces is only one of the steps of the reaction at the metal/liquid interface. Therefore, creeping metals surrounded by electrolytes undergo electrochemical and mechanical interactions (corrosion attack, stress corrosion cracking, etc.) and show a pronounced difference in plastic flow compared with that in a neutral medium. For example, the multifold increase in the creep rate of zinc crystal after the removal of oxide film by hydrochloric acid during a creep test [8] or cadmium single crystal at the replacement of water by sulfuric acid [9] can be explained not only by adsorption phenomena but also by electrochemical and mechanochemical interactions of liquid with stressed solid.

The effect of electrolytes on the creep of metals at a constant load was discovered by Andrade and Randall using the apparatus presented in Fig. 9.1(a) [10,11]. They observed an immediate increase in the secondary creep rate of a cadmium single-crystal wire from about 0.01 h^{-1} to 0.02–0.04 h^{-1}

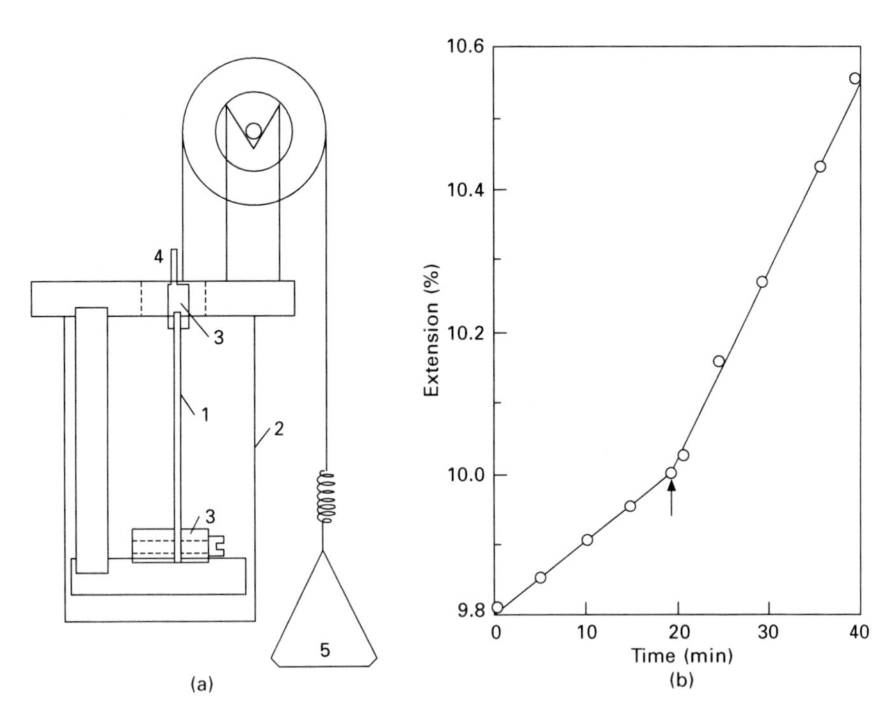

9.1 Andrade's apparatus for straining metals in electrolytes (a) and a creep curve for cadmium single crystal immersed in cadmium sulfate solution at the moment indicated by an arrow (b). 1 – wire sample of about 6 cm long, 2 – glass beaker; 3 – grips; 4 – a fiducial mark allowing a cathetometer to read the extension of the sample; 5 – pan with weights [11].

when the surrounding of the wire was replaced with $CdSO_4$ solution (Fig. 9.1(b)). An analogous effect occurs in zinc for Zn ions. However, Cd sulfate and chloride solutions have no effect on the creep flow of Cd single-crystal wires that had been heated in vacuum with the formation of a really clean surface without oxides [12]. On the contrary, solutions of cadmium and zinc nitrates stop the flow and raise the critical shear stress owing, probably, to hydroxide film formation [11]. A surface effect on polycrystalline metals has been detected only with polycrystalline lead, but it is appreciable only when the grain size is comparable with the wire diameter [13].

The effect of polarization on creep was initially studied by Venstrem and Rhebinder on single crystals of lead and tin in 0.1N sodium sulfate solution [14]. The polarization and surface-active compounds affect the metal surface itself rather than oxide layers on it. Some contradictory results obtained by Andrade and other researchers [8,9,10,13,14] during creep tests in electrolytes can be explained by the action of surface reactions that form different films. Creep rate in this case depends on the rate of specific physical, adsorptional and chemical stages which are inherent to those interactions. Another origin for contradictions in the results of, for example, Andrade, was, probably, a small value of adsorption effect in his experiments performed at a very low flow rate of crystals (less than $10^{-4} s^{-1}$). Probably, as Kramer and Demer suggested [15], when products form adherent films which block the egress of dislocations from the surface, the creep strength is enhanced, e.g. at the introduction of Cd nitrate into creeping cadmium crystal [10].

Various mechanisms offered throughout the 1930s–1950s for the effects of oxide and metal films on tensile plastic flow, creep and fatigue in different media were divided by Kramer and Demer into four main groups: (a) penetration of oxide into surface cracks; (b) load-carrying capacity of the film; (c) locking of surface dislocation sources and (d) blocking of dislocations at the surface resulting in pile-ups [15].

How do films formed on the surface of metal as a reaction product between stressed solid and a surrounding liquid influence creep and creep-rupture of metals? There is no evidence to support the suggestion of surface cracks healing-up by an oxide film. If the relation of the film thickness to the specimen diameter is relatively large (thin wires) compared with common samples, load capacity of the film is a valid hypothesis. For example, a decrease in TYS of single cadmium crystal covered by thick oxide film from 25% to zero was found with an increase in the sample diameter from 0.3 to 2.8 mm [6].

Plastic deformation of metals in electrolytes is accompanied by charge transfer between the solid and electrolyte. There is a lot of data in literature concerning electrochemically controlled stress corrosion, stress corrosion cracking and creep tests performed under a constant load or stress in water solutions on tin and lead single crystals [9]; copper and its alloys [16–21];

nickel [22,23]; gold [24] and steels, mainly stainless steels [25–37]. Moreover, as reported by Feld *et al.* [38], melted sulfate-based slag accelerates creep of Ni-based alloy under potentiostatic conditions at 800 °C. The viscoelastic and fatigue behaviors of magnesium and Mg alloys in borate buffer solutions and in aqueous solutions of different salts have been investigated by Gutman and co-authors as 'corrosion creep' (CC) [39–45], 'corrosion stress relaxation' (CSR) [46,47] and high-cycle corrosion fatigue (HCCF) [48–51].

The explanations suggested by different authors concerning the acceleration of plastic deformation of metals in solutions are based on the electrocapillary effect [14], adsorption-induced loss of strength and a surface crack idea e.g., [1,2]; soap formation [15] or debris removal [15,23]; divacancies diffusion and surface energy lowering [16], etc. These hypotheses are very contradictory because the results depend on different factors, first of all, on experimental conditions. For example, Revie and Uhlig [16] support the idea of the generation of divacancies by the dissolution process during anodic polarization of oxide-free copper, and of a corresponding reduction in the surface energy by cathodic polarization. These divacancies diffuse into the metal, causing the climb of sessile dislocations, thereby facilitating slip. Only *simultaneous* application of stress and anodic polarization markedly accelerates the creep in copper wire in de-aerated acetate buffer solution (pH 3.7). This effect increases with anodic current density. The cathodic polarization under the same conditions has confirmed the Rehbinder effect, as well as the observed increased creep rate in acetate buffer solution with an additive of surface-active agent [16]. However, Van Der Wekken concludes that the accelerated creep rates of copper single-crystal specimens during anodic dissolution in acetate buffer solutions under a constant load in experiments of Revie and Uhlig are essentially due to an increasing stress. Under the conditions of constant stress, the effect largely disappeared in single crystals, as well as in polycrystalline wire specimens [17]. The creep rate of brass, e.g. in 3.5% NaCl, increased linearly with the increase in the anodic current density as a result of dislocation climb induced by vacancy supersaturation during the anodic polarization or by a mechanism related to hydrogen-facilitated local plastic deformation during the cathodic polarization [18]. A metal–electrolyte interface usually leads to charge separation. Sircar and Thakur reported that negatively charged high-purity copper wire creeps at a higher rate in air or in ammonium acetate buffer solution compared with that in air or the solution in uncharged state. A positively charged specimen exhibits a decelerated creep rate of similar nature [21].

In case of the interaction of a solid with active medium, it may lead both to plasticity increase and to its reduction (accompanied by hardening), depending on the results of surface chemical (electrochemical) reactions. E. M. Gutman has discovered in 1967 that (electro)chemical reactions on the surface of a stressed solid cause additional dislocation flux, which changes mechanical properties and fine microstructure. This effect of increasing the plasticity of a solid under the influence of chemical reactions was named the *chemomechanical effect* [52]. This additional dislocation flux is generated as a result of rapid surface layer saturation with dislocations that gives favorable conditions for multiple slip, and, hence, for unhardening and microstress relieving.

For example, in experiments on chromium–nickel steel plasticization by means of anodic polarization within the region of passive state potentials, it obviously prevails over possible manifestations of the barrier effect [53]. A linear dependence of hardness loss on the logarithm current density has been established over all the ranges of active and passive state potentials (Fig. 9.2). This points to the predominant role of chemomechanical effect despite the formation of passive film which is transparent for dislocations.

The influence of various media on strength characteristics of 0.17 mm thick stainless steel of 18–10 type was realized by recording the process of steel strain hardening in different solutions at a steady potential using a tensile test machine with strain rate of 8%/min. Na_2SO_4 solution for obtaining a steady passive state and H_2SO_4 solution for the steel to be in the active state were used as media, [53]. As given in Fig. 9.2(b), the same strain values lead, in comparison with metal deformation in the air, to higher stresses in the metal in passivating medium (sodium sulfate) and lower ones – in active medium (sulfuric acid). Hardening of 18–10 steel, especially, of vacuum-annealed samples with coarse austenite grains of about 0.4 mm, under deformation in sodium sulfate can be attributed to barrier mechanism action. The probability of dislocation passage through a fast passive film is sharply reduced. This causes metal surface layer hardening, and in the case of especially thin metal plates, this has an essential effect on the behavior of the metal on the whole. This effect is especially marked in annealed steel, since annealed steel contains fewer obstacles blocking dislocation motion in the bulk metal. As for sulfuric acid, in such a case surface films are absent and, in contrast, plasticization takes place due to chemomechanical effect.

Taking into account the fact that stress field created by a dislocation propagates within one crystal, in case of corrosion under stress, specific conditions for active development of chemomechanical effect arise in the crack tip. Further propagation is determined by the properties of one crystal (transcrystalline failure) or two adjacent crystals (intercrystalline failure). Then the chemomechanical effect, contributing to an increase in chemical potential of surface atoms (dislocations exit), promotes the mechanochemical effect. The latter, in its turn, promotes an exit of dislocations. On the basis of such a synergistic interaction, the *autocatalytic mechanism* of mechanochemical failure in the crack tip during stress corrosion cracking and corrosion fatigue was proposed [53].

Thus, the literature data on the environment-assisted creep or *corrosion*

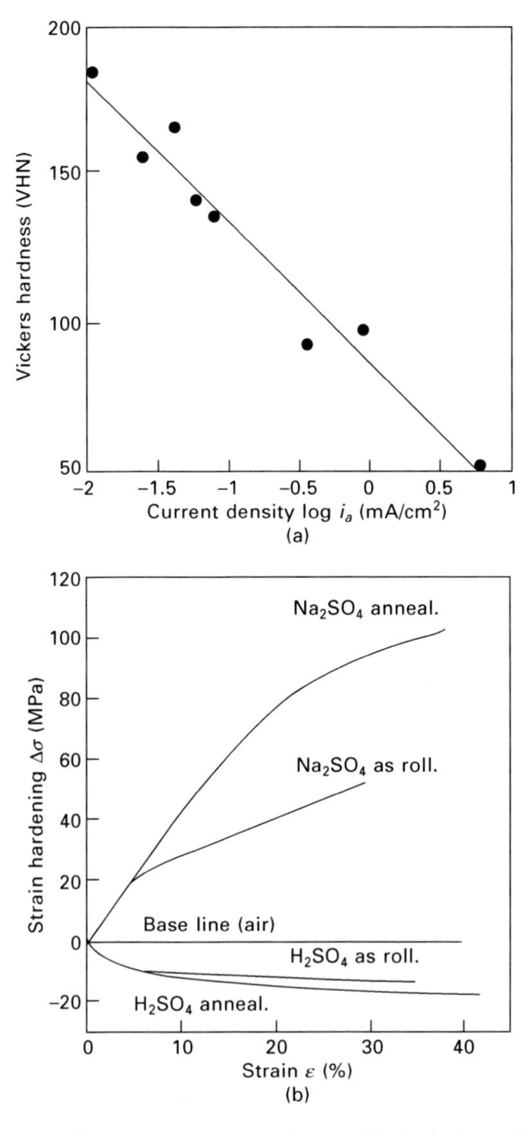

9.2 Effect of current density (a), kind of electrolytes and heat treatment (b) on hardening parameters of 18–10 type stainless steel [53].

creep and *corrosion fatigue* are contradictory. In the present work, the effect of corrosive medium and processing on the creep, stress relaxation and high-cycle fatigue behavior of pure magnesium and diecast Mg alloys has been investigated. The chapter contains a critical review including earlier publications of the authors.

9.3 Mechanoelectrochemical behavior of magnesium (Mg) alloys

For two diecast Mg alloys AZ91D (Mg–9Al–1Zn) and AM50 (Mg–5Al) a significant acceleration of their corrosion was found in sodium tetraborate buffer solution 0.1N $Na_2B_4O_7$ (further in the text named as 'buffer solution') saturated with NaOH caused by plastic deformation in comparison with unstressed alloys [45,49]. Moreover, mechanoelectrochemical measurements demonstrated a higher corrosion rate of AZ91D in a deformed state than that of AM50, while in non-deformed state AZ91D has a lower corrosion rate than AM50 (Fig. 9.3).

Under the strain increasing from 0 to 4%, the average dissolution rate derived from electrochemical polarization data and represented by dotted lines in Fig. 9.3 increases for AZ91D and AM50 alloys 7 and 1.5 times, respectively [49]. Thus, the smallest mechanochemical effect is observed in AM50 alloy, and the greatest one in AZ91 alloy. It can be explained by an increasing strain-hardening coefficient due to the growth of Al content and hard β-phase ($Mg_{17}Al_{12}$) precipitates. The maximums on the curves appear at work hardening stages due to the creation and destruction of dislocation pile-ups according to the theory [53]. This is confirmed by transmission electron microscopy (TEM) observation (Fig. 9.4): the maximum of the curve for AM50 in Fig. 9.3 lies near 7% deformation, which exactly corresponds to intensive pile-ups formation presented in Fig. 9.4(b).

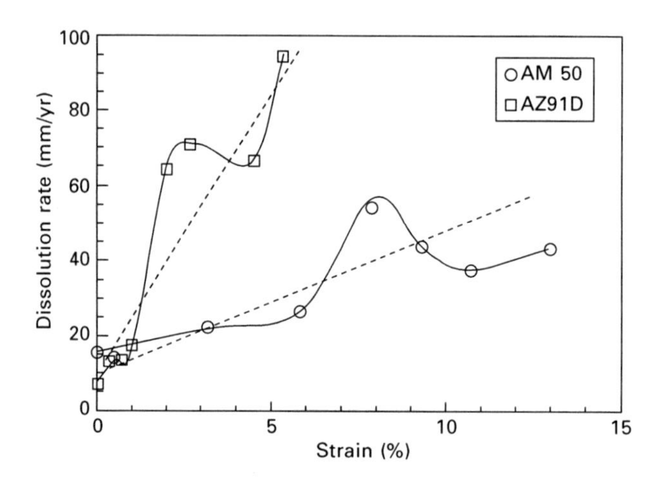

9.3 Dissolution rate of two diecast Mg alloys in 0.1 N $Na_2B_4O_7$ + NaOH solution (25 °C, pH 10.5) derived from electrochemical measurements vs. tensile strain (ZRA = zero resistance ammeter) [49].

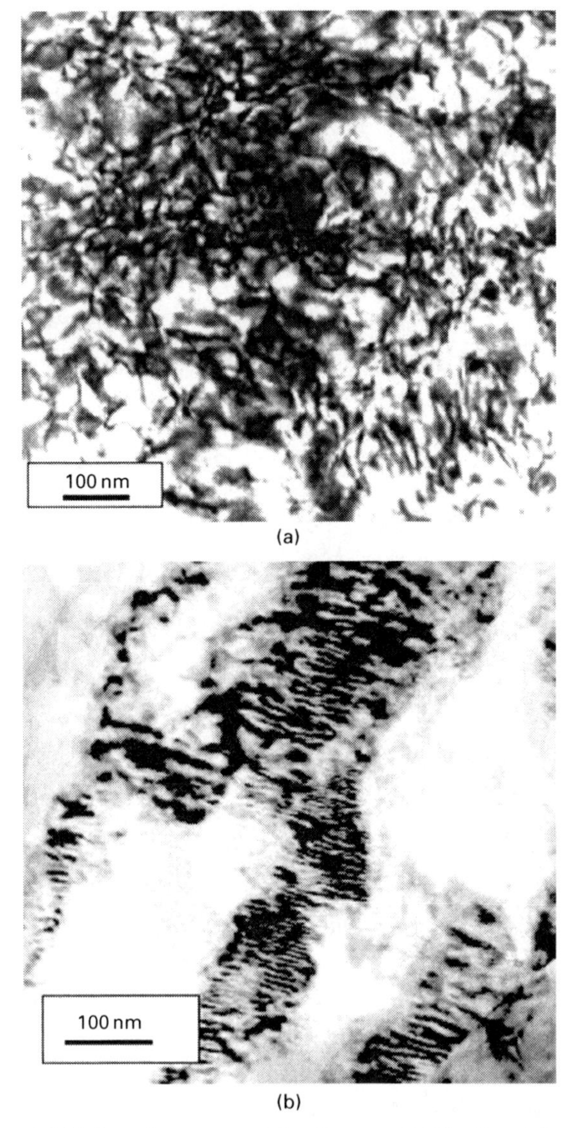

9.4 Dislocation network in diecast AM50 alloy (a) and dislocation pile-ups (b) after 6.8% deformation [49].

9.4 Corrosion creep of magnesium (Mg) and diecast magnesium (Mg) alloys

9.4.1 Experimental procedure

Flat specimens of polycrystalline 99.9653% Mg with gauge sizes of 10 mm wide, 4 mm thick and 42 mm long were cut out in the transverse direction from commercial ingots produced by Dead Sea Magnesium Works (DSM). Round specimens (5.9 mm in diameter, gauge length 75 mm) of Mg alloys AZ91D, AM50 and AS21 were produced on diecast machines with the locking forces of 2000 kN (Israel Institute of Metals, Technion) and 3450 kN (DSM). Rheoformed diecast AZ91D alloy was produced by stirring the melt during the solidification in a spinning crucible at the casting temperature 580 °C, where the solid fraction amounted to about 40–60% [44,54]. Chemical composition and standard mechanical properties of pure magnesium and its alloys are presented in Tables 9.1 and 9.2, including the data for extruded alloys used in Section 9.5 in experiments on corrosion fatigue. AZ91D and AM50 samples were studied both as-cast and coated with an anodic non-chromate coating ALGAN-2 (Algat Co., Israel).

The grain size of Mg–Al solid solution in the studied Mg alloys is 1–10 and 10–15 μm in the surface layer and in the bulk of diecastings, respectively. Since diecasting involves rapid cooling rates, even alloys with relatively low Al content contain a certain volume fraction of divorced eutectic β-phase $Mg_{17}Al_{12}$. Such secondary phases as $Mg_{17}Al_{12}$ (β-phase), etc., are mainly distributed at grain boundaries (Fig. 9.5(a), (b) and (c)) or within the grains like silicide Mg_2Si (Fig. 9.5(d)). Rheoformed AZ91D alloy has a globular microstructure (Fig. 9.5(b)) as compared with a dendritic microstructure

Table 9.1 Chemical composition (in mass percent) of diecast (DC) and extruded (E) alloys, Mg – the rest

Alloy	Al	Mn	Zn	Zr	Ni	Cu	Fe	Be
AZ91D (DC)	9.3	0.21	0.71	0.02 (Si)	0.0004	0.0008	0.0028	0.0006
AM50 (DC)	5.1	0.57	0.15	0.02 (Si)	0.0006	0.0007	0.0040	0.0013
AS21 (DC)	2.3	0.23	–	1.10 (Si)	0.0007	0.0005	–	–
AZ80 (E)	7.8	0.19	0.40	–	0.0009	0.0008	0.0033	<0.00001
AM50 (E)	4.4	0.36	0.001	–	0.0008	0.0005	0.0155	0.0006
AZ31 (E)	2.8	0.28	0.96	–	0.0007	0.0017	0.0111	< 0.00001
ZK60 (E)	0.04	0.008	5.30	0.44	0.0003	0.0021	0.0367	< 0.00001

Table 9.2 Tensile properties of pure Mg, diecast (DC) and extruded (E) magnesium alloys

	Pure Mg	AZ91D (DC)	AM50 (DC)	AS21 (DC)	AZ80 (E)	AM50 (E)	AZ31 (E)	ZK60 (E)
Ultimate tensile strength (UTS), MPa	60	225	229	221	278	291	273	336
$TYP_{0.2\%}$, MPa	24	170	136	134	194	179	155	227
Elongation to fracture, %	5.6	2.8	11.7	9.7	14.2	18.1	19.4	16.6

of a conventional alloy (Fig. 9.5(a)). Samples of pure magnesium cut out from the columnar zone of an ingot consist of relatively large, long grains crystallographically oriented with their dendrite directions parallel to the heat flux direction. The size of these crystallites is equal to about 2–3 mm in diameter and more than 10 mm (gauge width) in length (Fig. 9.5(e)).

Creep tests were carried out on Model 3 SATEC creep tester (Satec Inc., USA) in air at room temperature (25 ± 2 °C; 150 °C ± 0.5 °C and 175 °C ± 0.5 °C) and in solutions at room temperature. As active environments, buffer

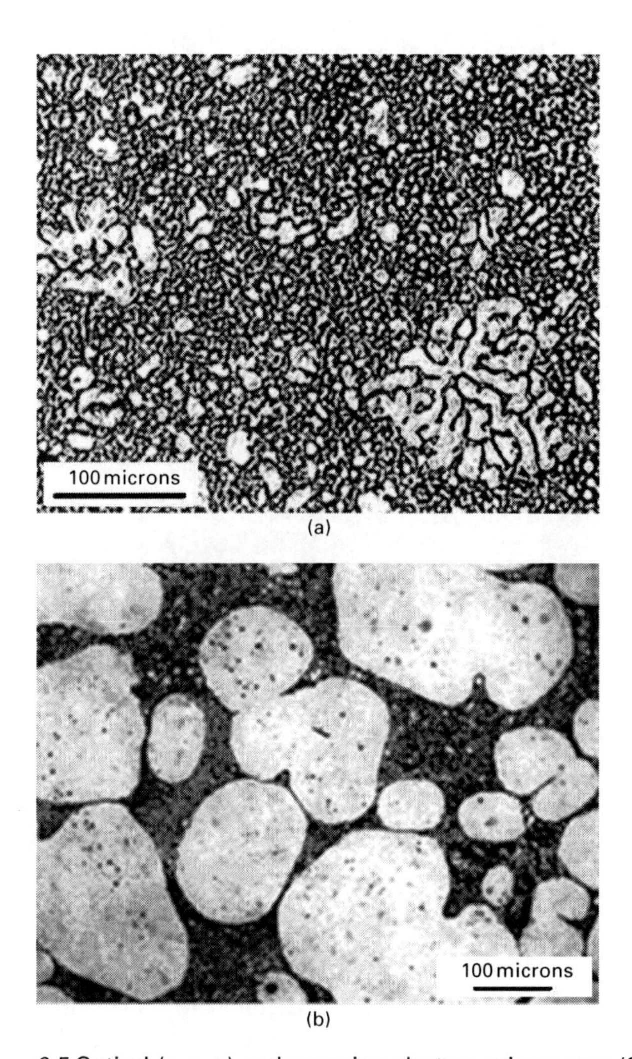

9.5 Optical (a–c, e) and scanning electron microscopy (SEM) (d) micrographs of diecast common (a) and rheoformed (b) AZ91D; diecast AM50 (c) and AS21 (d) alloys and pure Mg sample with a large transgranular crystallite (e).

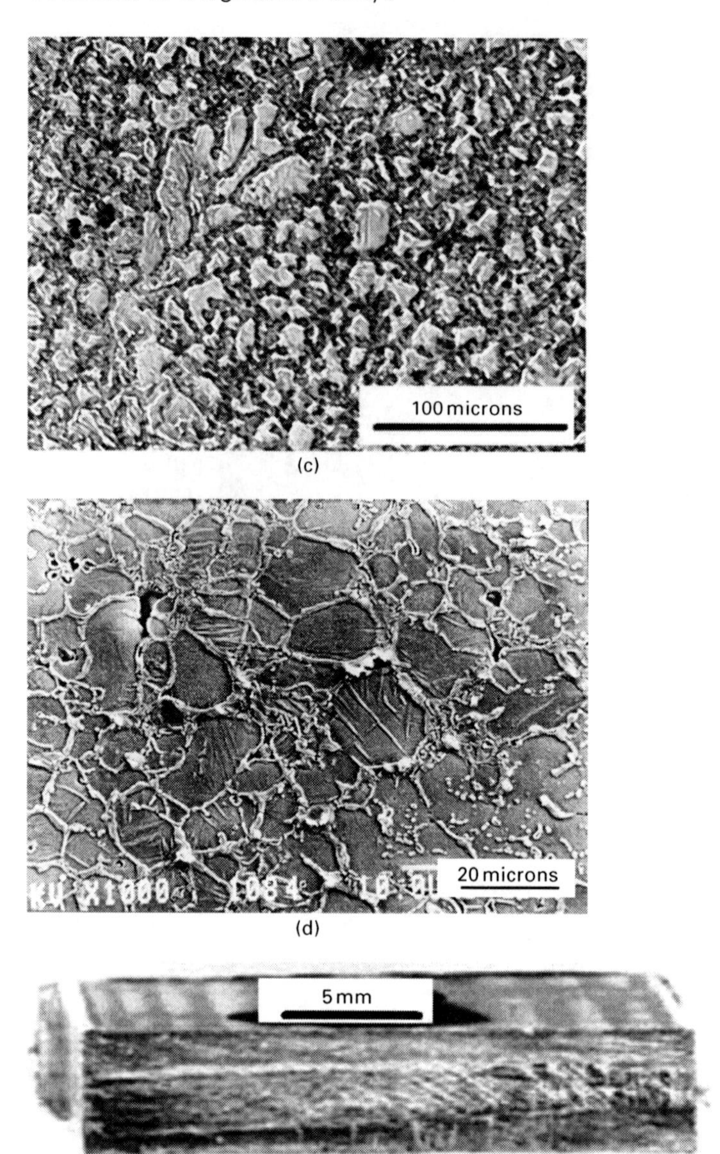

(c)

(d)

(e)

9.5 Continued

solution 0.1N $Na_2B_4O_7$ saturated with magnesium hydroxide with pH = 9.3, 3.5% NaCl and sodium sulfate solution 3% Na_2SO_4 were used. All solutions were prepared from analytical grade chemicals and distilled water. The small cell for corrosion creep (CC) tests was made from glass, transparent plastic (poly(methylmethacrylate), PMMA) or lightly powdered natural rubber latex

(Fig. 9.6) [39,42]. Specimens were studied after rinsing in distilled water and alcohol and wiping with acetone.

Al specimens were cleaned after CC tests by dipping into concentrated 70% nitric acid for several minutes (ASTM G34-90). The oxide layer was scrubbed lightly in a stream of water with a rubber stopper or a bristly brush so as not to abrade mechanically this soft material. Cleaning of Mg-based alloys from the surface oxide layer was performed by the oxide dissolution in 15% CrO_3 solution at 80 °C for 0.5–1 min (ASTM G1-88). Microstructure and pitting studies were carried out using an optical microscope 'Nicon' and a scanning electron microscope (SEM) JEOL JSM-5600 with 'NORON' energy-dispersive analysis system (EDS).

9.4.2 Results and discussion

At room temperature in air, pure magnesium and diecast Mg alloys show a decrease in the creep rate pointing to the primary stage typical of low-temperature creep without rupture lasting for 600 h in the case of magnesium and for at least 800 h in case of its alloys (Figs 9.7–9.9), while the high-temperature creep that occurs at temperatures exceeding a half of the melting point (in kelvin) is predominantly a steady state creep. Thus, it is assumed that high-temperature creep corresponds, e.g. for AZ91D alloy, to temperatures more than 93 °C. In Fig. 9.10 two typical curves as examples of high-temperature creep in air at 150 °C and 175 °C are presented for AZ91D alloy.

However, creep in solutions is characterized, as a rule, by all the three stages of creep: the initial stage with decreasing creep rate (the primary or

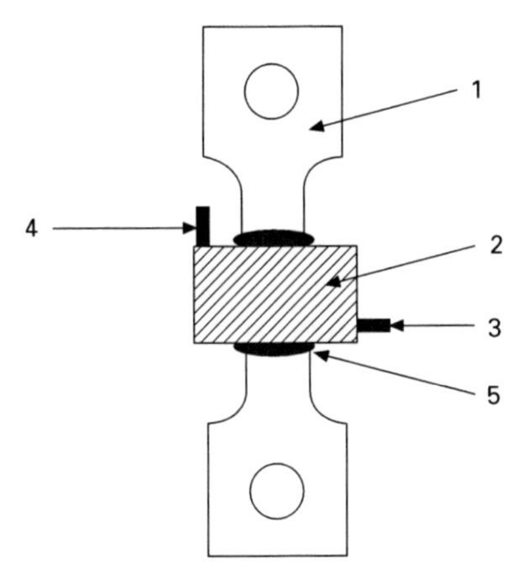

9.6 The scheme of setup for corrosion creep tests including a specimen 1; a glass or rubber cell 2; solution inlet 3 and outlet 4; and sealing 5.

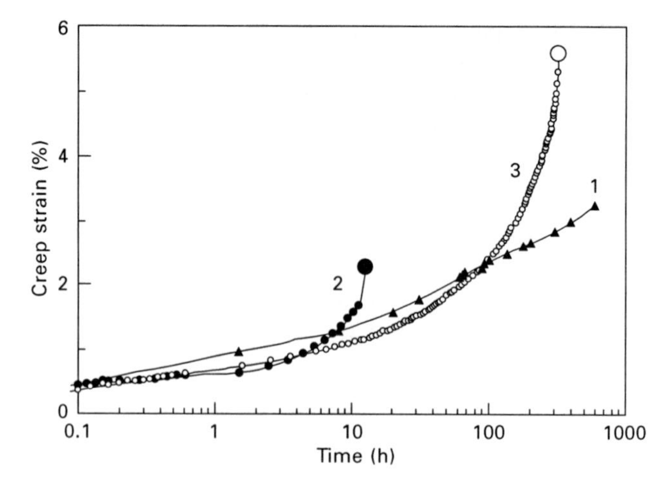

9.7 Creep behavior of pure magnesium in air (1), 3.5% NaCl (2) and borate buffer solution saturated with magnesium hydroxide (3). The large points show the creep rupture; 25 °C, 28 MPa.

transient creep), the secondary or steady state creep with approximately constant creep rate and tertiary creep with final rupture of the sample (Fig. 9.11(a)). Two main factors, besides the test temperature and the stress value, affect creep strain and lifetime (time to creep rupture) of magnesium and its alloys in corrosive solutions. The first is the kind of the environment, and the second is aluminum content in the alloy [39–42].

For pure magnesium and its alloys, an opposite effect of borate buffer solution and sodium chloride solutions on the lifetime and strain-to-fracture during corrosion creep was found. The lifetime of pure magnesium in 3.5% NaCl was more than an order of magnitude shorter than in buffer solution and varied from 12±5 h to about 240±70 h, respectively (Fig. 9.7). On the contrary, as shown in Fig. 9.9, the creep life of AZ91D alloy in buffer solution under stress of 120 MPa was more than threefold shorter than in NaCl solution (250 h and 816 h, respectively). The same creep behavior in these solutions was also found for AM50 alloy [42]. In sodium sulfate solution, stressed AZ91D alloys experienced very fast stress corrosion cracking at very low strain values with very short creep life compared with that in other studied solutions (Fig. 9.9).

The surface electrochemical heterogeneity of a magnesium sample due to the local activating effect of chlorine ions and a friable oxide-based film causes embrittlement caused by microcracking with a much shorter lifetime in NaCl solution than that in buffer solution. However, in the latter medium there is no embrittlement effect and, in contrast, plasticization is observed, i.e. a strong chemomechanical effect is found. The depth of pits in magnesium after corrosion creep is much larger in sodium chloride

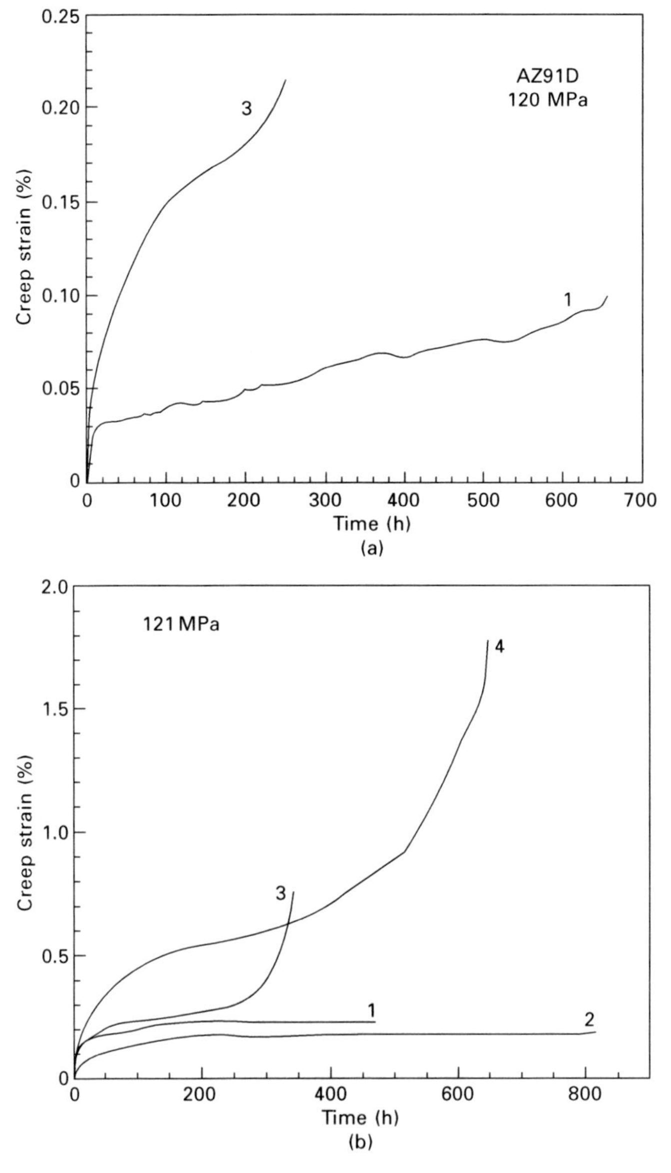

9.8 Typical low-temperature creep in air (1, 2) and corrosion creep in buffer solution (3, 4) of diecast Mg alloys AZ91D (a); AM50 (b, curves 1, 3) and AS21 (b, curves 2, 4) at applied stress of 120.5±0.5 MPa and room temperature. (1, 2 – no rupture and 3, 4 – rupture takes place) [42].

than in buffer solution, and in typical creep-rupture tests it equals around 800 μm and 70 μm, respectively (Fig. 9.12). Therefore, its lifetime in NaCl solution was four times shorter (12 h against 48 h in Na_2BO_4 buffer solution).

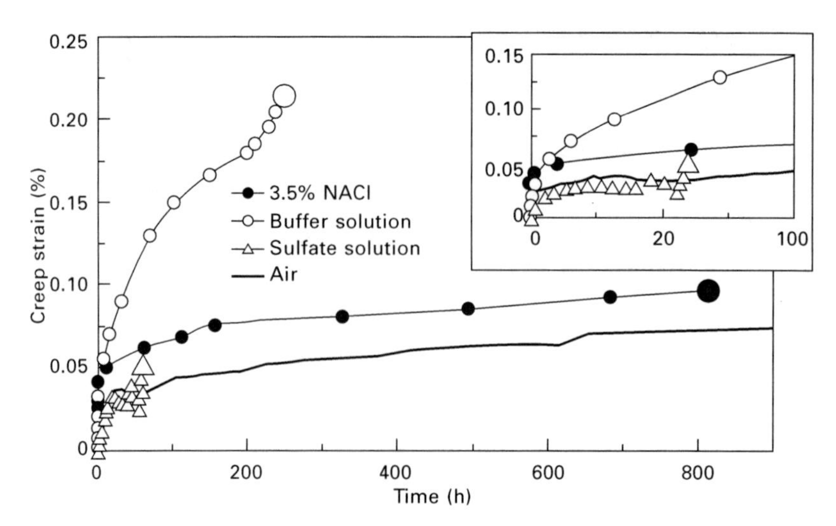

9.9 Effect of the environment on the typical creep behavior of AZ91D alloys in air, sodium borate buffer (0.1N $Na_2B_4O_7$), sodium sulfate (3.5% Na_2SO_4) and sodium chloride (3.5% NaCl) solutions. 120.5±0.5 MPa; room temperature; large points show creep rupture.

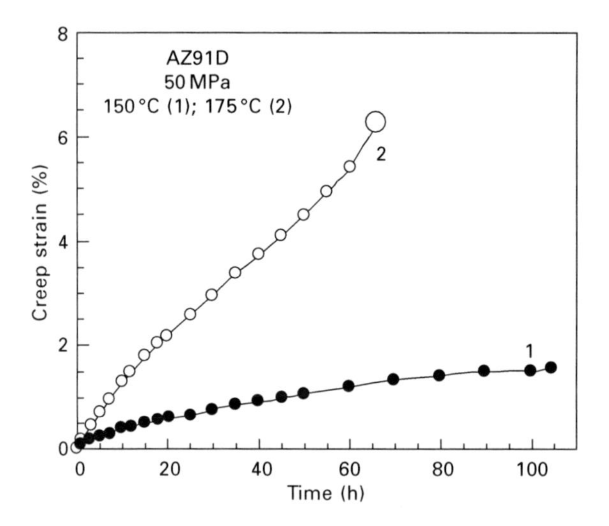

9.10 Typical high-temperature creep of AZ91D alloy in air at applied stress of 50 MPa and temperature of 150 °C (1) and 175 °C (2) (1 – no rupture, 2 – rupture takes place).

Fracture patterns observed after corrosion creep show the crack area with pitting and final ductile fracture in NaCl solution (Fig. 9.13(a)) and ductile fracture channels in two Mg grains in a sample tested in buffer solution (Fig. 9.13(b)). Pure magnesium shows transgranular fracture in corrosion

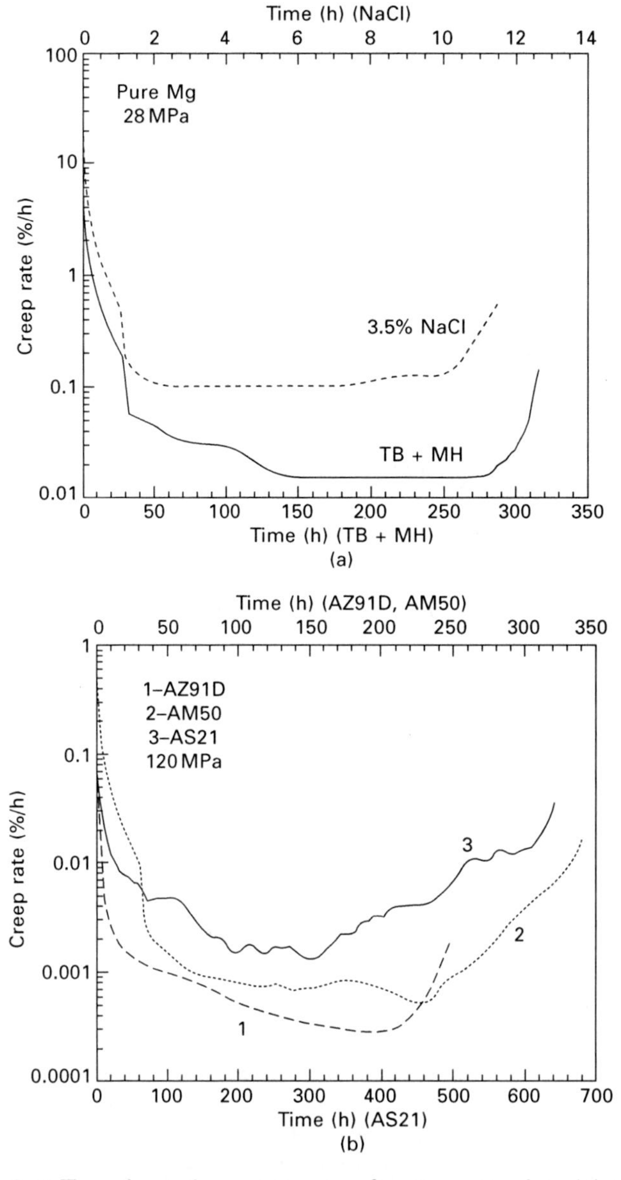

9.11 Time-dependent creep rate of pure magnesium (a) and diecast Mg alloys (b) AZ91D (1), AM50 (2) and AS21 (3) in 3.5% NaCl (a) and in 0.1 N $Na_2B_4O_7$ + $Mg(OH)_2$ (a, b); Abbreviation 'TB + MH' means sodium tetraborate and magnesium hydroxide (buffer solution) [42].

creep tests against, as shown in Fig. 9.14(a), intergranular fracture of alloys in corrosive media [42].

Thus, the effect of the corrosive medium on stressed metals was related to

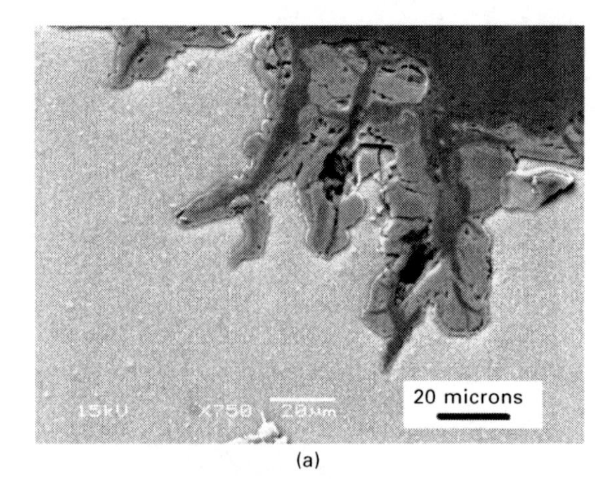

(a)

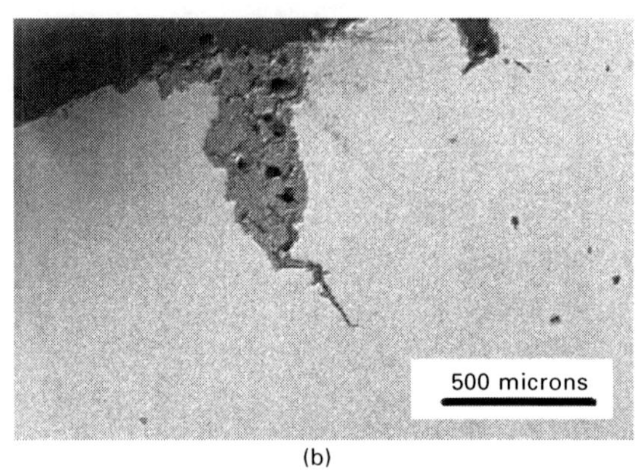

(b)

9.12 Relief of corrosion pits in pure magnesium after creep-rupture CC tests in buffer (a) and NaCl (b) solutions. Creep life, hours: 48 (a) and 12 (b); Depth of pits, μm: 70 (a) and around 800 (b).

the autocatalytic interaction of chemomechanical effect or plasticization and softening of surface layer as a result of anodic dissolution [53]. In all cases, a marked plasticization of magnesium and its alloys (a strong chemomechanical effect) was found in buffer solution, namely, creep strain-to-rupture of pure magnesium and its alloys in buffer solution were approximately twice as high as those in NaCl (Figs 9.7 and 9.9). However, much longer creep life of Mg alloys in NaCl solution compared with that in buffer solution can be explained by weakening of local activating effect of chlorine ions owing to a much more protective oxide film, and blunting of the tip of a crack, as shown by way of example in Fig. 9.14(b) for AM50 alloy.

A decrease in creep life of Mg alloys was found with increasing aluminum content from 2.3 to 8.4% in AS21, AM50 and AZ91D alloys [39,40–42]. Under the same stress of 120.5 ± 0.5 MPa, the creep life of these alloys in

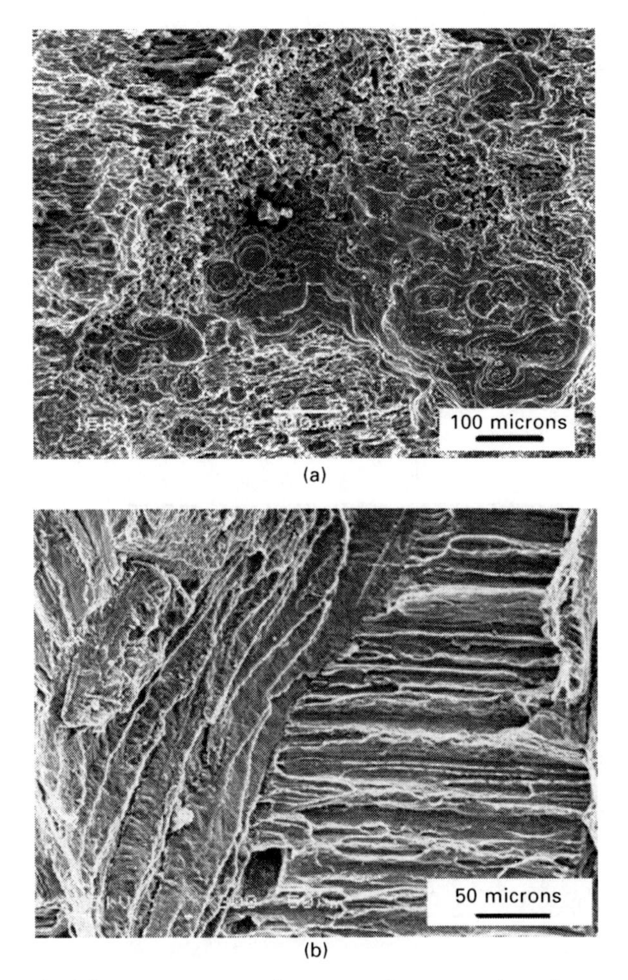

9.13 Fracture patterns in magnesium samples after corrosion creep tests in 3.5% NaCl (a) and in buffer solution [42] (b). (a) The crack area with pitting and final ductile fracture; (b) ductile fracture channels in two Mg grains.

buffer solution decreases and amounts to 500, 340 and 250 h, respectively, whereas elongation-to-fracture ε_{fr} amounts to 2.1, 0.8 and 0.2%, respectively (Figs 9.8 and Fig. 9.11(b)). Low values of ε_{fr} are in the range of the primary creep, but due to the plasticization effect of the solution, we also observed secondary and tertiary creep.

Corrosion creep of diecast Mg–5Al and Mg–9Al–1Zn alloys anodized by a ~8 μm thick coating showed that anodic coating delays metal dissolution facilitating plastic deformation of the stressed metal, and for this reason alone increases the corrosion creep life of metals (Fig. 9.15). For instance, time-to-fracture of anodized Mg–5Al alloy increases from 340 to 637 hours

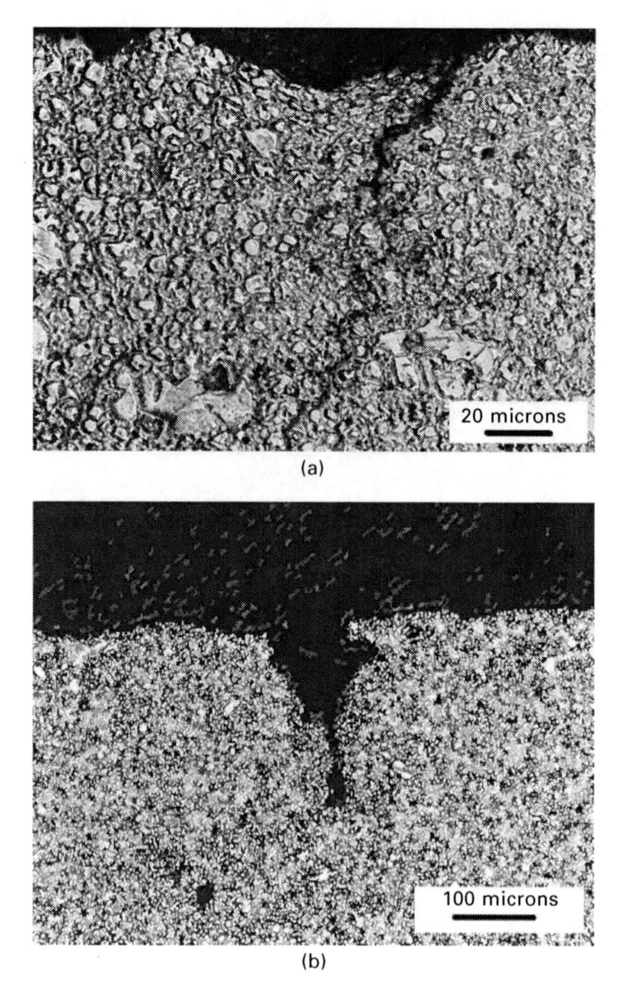

9.14 Intergranular cracking [42] (a) and a blunted crack emanating from the pit (b) in AM50 alloy in 3.5% NaCl solution during an interrupted creep test. 120 MPa, duration of the test is 1108 h, no rupture.

in comparison with uncoated specimens in the borate buffer solution under the stress of 121 MPa [43]. Anodizing leads to a certain increase in the creep strain of Mg–5Al in air during long-term tests without rupture (Fig. 9.15). An increase in the plasticity of coated alloys in air is connected, probably, with a decrease in the alloy surface energy during the anodizing process in basic electrolytes. After the creep test of alloys, one can observe the intergranular fracture mode, as presented in Fig. 9.16 for anodized AZ91D with a crack that extends beyond the coating and propagates into the surface layer of metal.

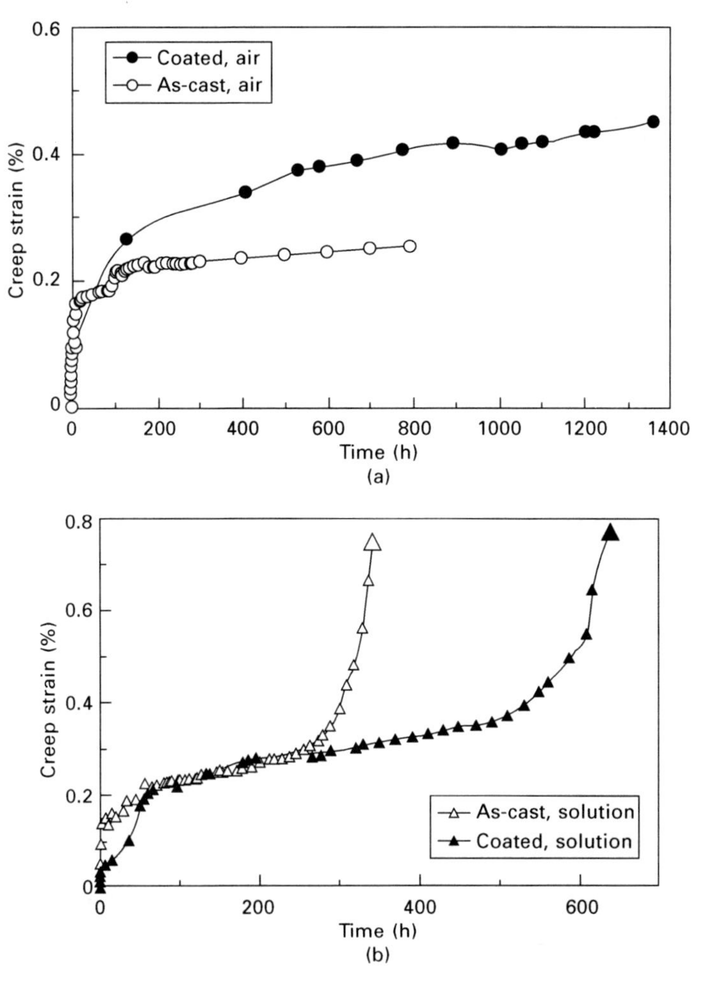

9.15 Effect of anodized coating on the creep behavior of dicast Mg–5%Al alloy in air (a, no rupture) and in buffer solution (b, rupture occurs). 121 MPa, 25 °C, large points indicate the fracture [43].

In the sodium sulfate solution, stressed Mg alloys experienced fast stress-corrosion cracking. Creep life of conventional and rheoformed AZ91D alloy drops in 3% Na_2SO_4 solution compared with buffer solution by more than two orders of magnitude (Table 9.3). Creep life of anodized rheoformed alloy increases compared with that of an uncoated alloy from 4 to 116 h, i.e. more than an order of magnitude, due to the delay of metal dissolution (Table 9.3). As reported earlier [44], a conventional diecast AZ91 alloy shows a more significant pitting corrosion than diecast rheoformed alloy during the corrosion creep tests in buffer solution. Pitting corrosion, intergranular cracks and final fracture of rheoformed alloy were observed just in the region between large

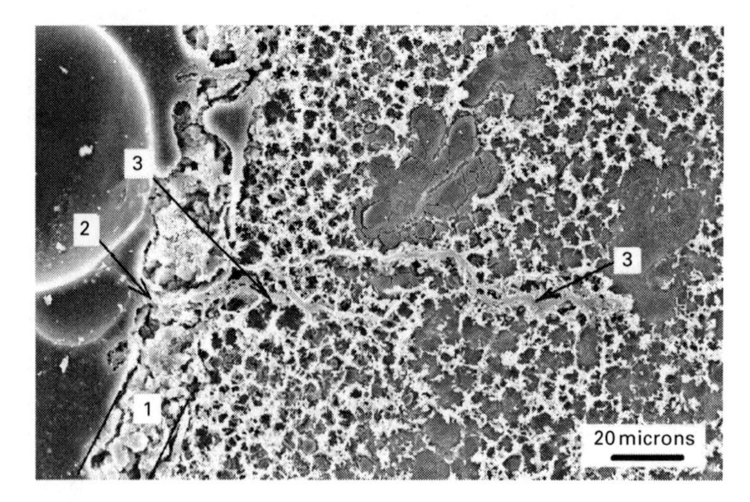

9.16 Cracks filled by phenolic resin in the vicinity of the coated surface of AZ91D alloy after a corrosion creep test in buffer solution (dark area is α-Mg solid solution; light area is $Mg_{17}Al_{12}$ phase). 1 – coating; 2 – the crack in coating; 3 – the crack in the bulk of the alloy.

Table 9.3 Creep life of conventional and rheoformed diecast AZ91D alloy in buffer and sodium sulfate solutions at room temperature under the stress of 58 MPa [44]

Alloy	σ/TYS	Solution	Surface conditions	Creep life (h)
Conventional AZ91D	0.34	Buffer solution	Machined	>430
Conventional AZ91D	0.34	3% Na_2SO_4	Machined	4
Rheoformed AZ91	0.39	Buffer solution	Machined	898
Rheoformed AZ91	0.39	3% Na_2SO_4	Machined	5
Rheoformed AZ91	0.39	3% Na_2SO_4	Anodized	116

Mg(α) grains. The reduction in macro-segregation and surface cracking due to a lower cast temperature is the main reason of a better corrosion and creep resistance of rheoformed alloys [54].

A catastrophic failure of stressed metals in corrosive solutions in comparison with their stable behavior in air can be explained by the synergetic interaction between mechanical and chemical processes described as mechanochemical phenomena [53]. Chemical (electrochemical) reactions proceeding on the metal surface and causing additional dislocation flux and localized enhanced plasticity, affect the fine microstructure and creep properties of a solid.

The synergistic effect of corrosion and stress on the creep behavior of magnesium and diecast Mg alloys named *corrosion creep* has been investigated. We can conclude that at room temperature marked plasticization and fracture are observed in sodium tetraborate buffer solution even under stresses less than 40% of TYS as compared with common low-temperature creep in air almost without any plastic deformation. This fact points to a strong chemomechanical effect in this corrosive solution manifested in plasticization and softening of surface layer as a result of anodic dissolution [53]. However, this effect is much less pronounced in sodium chloride solution. The much longer creep life of Mg alloys in this solution than in buffer solution can be explained by weakening of local activating effect of chlorine ions owing to a much more protective oxide film and blunting of the tip of a crack. Mechanically enhanced anodic dissolution of metal causes crack initiation and propagation, especially in sodium sulfate solution.

Anodized Mg alloys reveal a significantly longer creep life than uncoated diecast alloys. Pure magnesium shows transgranular fracture in corrosion creep tests against the ordinary intergranular fracture of alloys in corrosive media.

9.5 Corrosion fatigue of magnesium (Mg) alloys

9.5.1 State-of-the art of fatigue and corrosion fatigue behavior of Mg alloys

It is of significant scientific and practical interest to study corrosion fatigue of diecast and extruded Mg alloys, because many mechanically loaded parts of, say, automobiles are often subjected to prolonged cyclic stresses in an active medium. Usually, the fatigue life of Mg–Al–(Zn)–Mn alloys increases with Al percentage growth both in air [48,55–57] and in corrosive solutions [48,50,57] due to the solid solution hardening and an increase in ultimate tensile strength at the addition of Al. However, it was found that an increase in Al content from 5–6% in AM50 and AM60 alloys up to 8–9% in AZ81 and AZ91D alloys can lead to a certain decrease in fatigue life both in air [57,58] and in distilled water or NaCl solutions [51,55–57,59–62]. It is of interest that such a processing variable as diecasting temperature affects the corrosion fatigue of AM50 and AZ81 alloys to a larger extent than their fatigue in air, which is attributed to the structural sensitivity of stress corrosion [57].

The fatigue life of Mg alloys in such corrosive solutions as, for example, NaCl, is always less than in air [55,56,59–62]. The corrosion fatigue life of Mg alloys depends significantly on the solution's acidity. For example, in NaCl–$KCrO_4$ solution, a decrease in pH below 5 results in a considerable increase in the rate of stress-corrosion cracking of a strained Mg–6.5A1–1Zn alloy. In the pH interval from 5 to 12, the rate remains stable, and at pH exceeding 12 it begins to decrease rapidly [63]. In the solution containing 0.1N Cl^-, the maximal loss of endurance to cyclic stress applied by reversible bending at 8 Hz was in the pH range varying from 1.3 to 4 compared with a much higher lifetime of samples tested at pH = 4.6–14.0 [64].

Owing to the anodic dissolution of magnesium and instability of pH in basic electrolytes, the borate 0.1N $Na_2B_4O_7$ buffer solution saturated by magnesium hydroxide (pH 9.3) was used in high-cyclic fatigue tests [48,49,62]. In this solution, an unusual result was observed: the fatigue life of AM50 and AZ91D alloys was longer than in air. Apparently, such a behavior of magnesium alloys means that the inhibiting action of buffer solution during fatigue tests is dominant in comparison with the alloy degradation due to stress corrosion in this medium. Besides, borate anions $B_4O_7^{2-}$ can suppress, for example, stress corrosion cracking of stainless steel by delaying the crack initiation time and reducing the crack initiation frequency [65].

At the lowest Al content in diecast and wrought Mg alloys, the weakest environmental effect in static fatigue (creep) [39,40] and dynamic (cyclic) fatigue was observed [48,55,57,62]. These results correlate with mechanoelectrochemical and immersion corrosion tests, which demonstrates a higher corrosion rate of AZ91D in a deformed state than that of AM50, while in non-deformed state AZ91D has a lower corrosion rate than AM50 (Fig. 9.3). Besides, it is well known that the stress-corrosion cracking rate in Mg alloys essentially increases with the growth of aluminum content [66–68]. Stress accelerates several times the corrosion of the β-phase, Mg –1.6% Al, Mg–8.0% Al and Mg–3.5Al–0.6Zn–0.3Mn alloys [67]. Such additions to aqueous solutions as J^-, SO_4^{2-}, Cl^- and Br^-, which can accelerate stress-corrosion crack growth, also accelerate corrosion fatigue crack growth in Mg alloy ZK60A (5.2% Zn, 0.45% Zr, 0.22% Mn, Mg – the rest) [69]. Under static (stress-corrosion cracking) and cyclic (corrosion fatigue) loading, the subcritical crack growth rate in this alloy is slower in air, argon and distilled water [69]. The degradation in fatigue strength for the high-strength AZ80 or AZ91D alloys due to NaCl is more pronounced than that of lower-strength alloys AZ31 [55] and AM20–AM40 [57] due to a higher percentage of the second phase in the former. It also agrees with the known fact of a higher sensitivity of AZ91D alloy to high cyclic fatigue in 3.5% NaCl solution at stresses below 130 MPa compared with AM50 alloy [48].

It is known that the relative fatigue life N_{sol}/N_{air} of Mg alloys containing about 9% Al in (3–5)% NaCl solutions varies from 0.01 to about 0.9–1 [48,55,59,60,62]. For example, at a given stress amplitude, the fatigue life of sand-cast AZ91E-T6 alloy in 3.5% NaCl solution was less than in air by a factor of between 2 and 500 [59]. At longer fatigue lives ($N = 10^5$–10^6), N_{sol}/N_{air} amounts to 0.01, but at shorter fatigue lives ($N = 10$–100) $N_{sol}/N_{air} \sim$ 0.9–1 [59]. According to high cyclic fatigue tests of a diecast AZ91D alloy, N_{sol}/N_{air} ratio was about 0.1 for samples cut out from diecast components [60] and slightly decreased from 0.74 to 0.64 with stress increasing from 115 to 165 MPa in diecast specimens [48]. However, unlike Gutman *et al.* [48], where a continuous solution feed flow was performed, in the tests carried out by Witt *et al.* [60], permanent sprinkling of 5% NaCl was used. It is known that due to spraying, i.e. under the conditions of increased aeration, cracking begins more rapidly [65]. The acceleration of the process of stress-corrosion cracking might also be caused by an increase in salt concentration on the surface of the specimen during its alternating moistening and drying.

In air, Mg alloys have a marked fatigue limit varying in a very broad range [56,58,62]. For example, diecast AZ91D showed a fatigue limit at 38 MPa [58], 80 MPa [56], 130 MPa [48,62]. Probably, the lowest fatigue limit of 38 MPa [58] results from the experimental method in which specimens taken from the inner section of diecast ingots with a higher porosity level were used. Besides, these data were obtained at ultrasonic frequencies (20 kHz) in a resonance vibration regime that can lead to the sample heating due to internal friction. As-cast AZ91D specimens have smaller pores and better fatigue resistance than machined samples [48,61,62].

Thus, as follows from mentioned above, the data on corrosion fatigue of Mg alloys are very contradictory. The present research is devoted to the correlation analysis of corrosion fatigue behavior of diecast and extruded Mg alloys in 3.5% NaCl (pH ≈ 5) and 0.1N $Na_2B_4O_7$ solution with a stable pH of 9.3 (buffer solution) and to the investigation of fatigue fracture mechanisms of these alloys depending on chemical composition, e.g, Al content, environment and microstructure.

9.5.2 Experimental setup for corrosion fatigue

Diecast specimens of Mg alloys AZ91D, AM50 with the gauge diameter of 5.9 mm and gauge length of 75 mm were produced on cold-chamber machines with the locking force of 2000 kN (Israel Institute of Metals, Technion) and 3450 kN (Dead Sea Magnesium Works, Israel). Diecast specimens were used without any mechanical treatment of the gauge [49]. Extruded hourglass-shaped specimens (the minimum gauge diameter and length of 8.0 and 49 mm, respectively, ASTM E466-82) were cut from extruded rods of AZ31, AZ80, AM50 and ZK60 alloys produced from raw material 30 mm in diameter (Rotem Industries Ltd, Israel) and machined by final turnery (roughness R_c below 3.2 μm). Fatigue tests were performed on a rotating beam type fatigue machine (Satec System, Inc., USA) equipped with a special electrolytic cell at room temperature (25 ± 2 °C) under reversible load (stress ratio $R = -1$) and the frequency of about 30 Hz [51]. The tests were performed in 3.5% NaCl (pH ≈ 5) and 0.1 N $Na_2B_4O_7$ saturated with magnesium hydroxide (pH 9.3). All tests were carried out under open circuit conditions. The detailed experimental conditions and setup have been described earlier [49,51]. The chemical composition and standard mechanical properties of alloys are given in Tables 9.1 and 9.2.

9.5.3 Results and discussion

Figures 9.17 and 9.18 represent stress–fatigue life (*S–N*) diagrams describing fatigue and corrosion fatigue behavior of diecast AM50, AZ91D and extruded AM50 and AZ80 Mg alloys in air, 3.5% NaCl and 0.1N $Na_2B_4O_7$ buffer solution. Diecast AM50 is supposed to have a fatigue limit in air approximately between 100 and 110 MPa, because, on the one hand, at 110–115 MPa its fracture was observed, and on the other hand, at the stress of 95 MPa one of specimens remained unbroken up to 4×10^7 cycles. Extruded AM50 shows the fatigue limit in air at approximately 120 MPa (Fig. 9.17(a)). In air, diecast AZ91D has a marked fatigue limit at 130 MPa (Fig. 9.18(a)). In NaCl solution, diecast AM50 and AZ91D alloys demonstrate approximately the same fatigue limits as in air, i.e. 110 and 130 MPa, respectively (Figs 9.17 and 9.18).

Fatigue life of all extruded alloys (AM50, AZ31, AZ80 and ZK60) in air was significantly longer than in corrosive solutions (Figs 9.17–9.20). In comparison with diecast alloys, they demonstrate significantly longer fatigue life both in air and in NaCl-containing solutions (Figs 9.17–9.20). Among extruded alloys, the longest fatigue life in air is observed for ZK60 alloy and the shortest one for AM50 and AZ31 alloys. AM50 has a somewhat lower fatigue life under low stresses in air than that of AZ31 (Fig. 9.20).

However, in a buffer solution, extruded AM50 alloy shows the longest durability among all extruded alloys, especially at low stress values, in spite of its relatively short lifetime in air (Fig. 9.20). A very interesting fact was discovered, namely: extruded alloys, especially AZ80 and ZK60, showed a significantly greater difference between the lifetime in buffer solution N_{sol} and in air N_{air} (a shorter relative fatigue life N_{sol}/N_{air} or a stronger embrittlement) under cyclic loading with respect to that for diecast alloys (Figs 9.18–9.20). In this solution, the lifetime of extruded AZ80 was even shorter than that of diecast AZ91D with approximately the same composition (Figs 9.18(c) and 9.20). Moreover, under low stresses all extruded alloys, excluding AM50, demonstrated a shorter lifetime in buffer solution with respect to that for diecast alloys (Fig. 9.20).

In the borate buffer solution, as reported earlier [49], the fatigue life of diecast AZ91D and AM50 was even longer than in air, probably due to the action of borate anions $B_4O_7^{2-}$ delaying the crack initiation time, as was reported for stainless steel [65]. Additionally, unlike extruded alloys, the presence of fine-grained surface layer in 'as-cast' samples [70] with increased corrosion resistance (e.g in AZ91D alloy this layer contains up to 30 vol.% of β-phase $Mg_{17}Al_{11.7}Zn_{0.3}$ [39]) can extend the lifetime of the sample. Besides, 'crusts' of β-phase on the 'as-cast' surface up to 500 μm long can also retard the crack initiation [71]. However, the protective action of borate buffer solution under cycle loading of diecast alloys compared with air (Figs

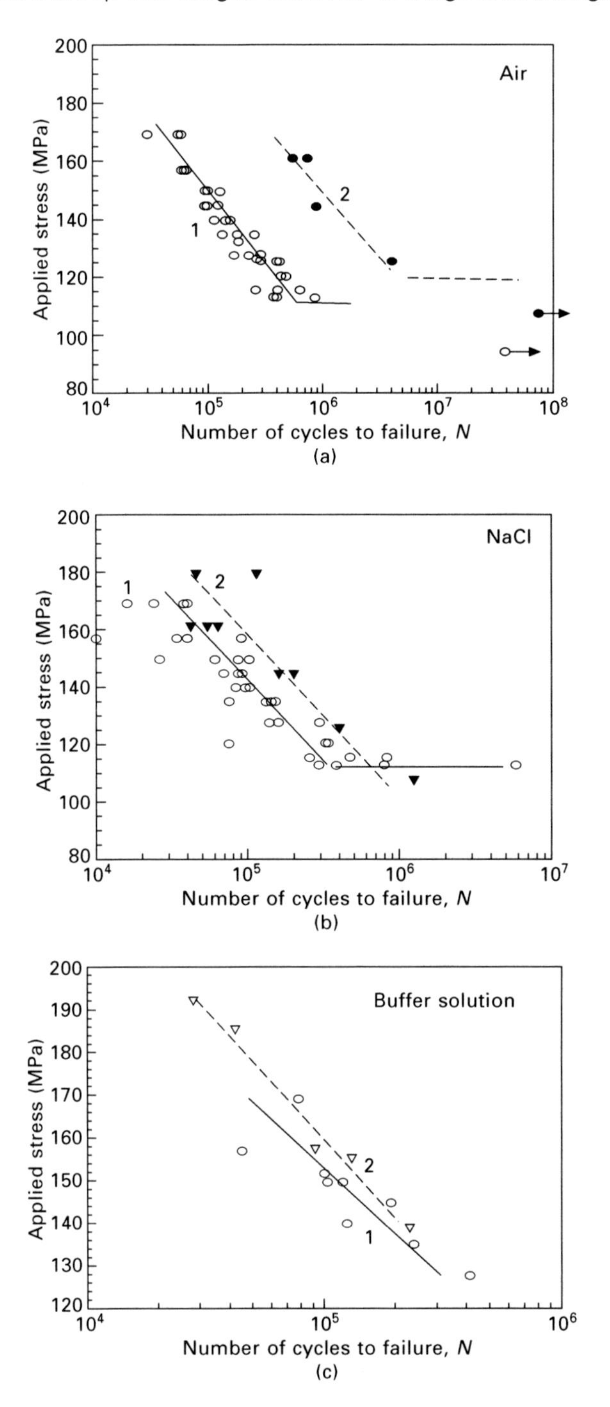

9.17 S–N diagrams for diecast (1) and extruded (2) AM50 alloys tested in air (a), 3.5% NaCl (b) and buffer solution (c).

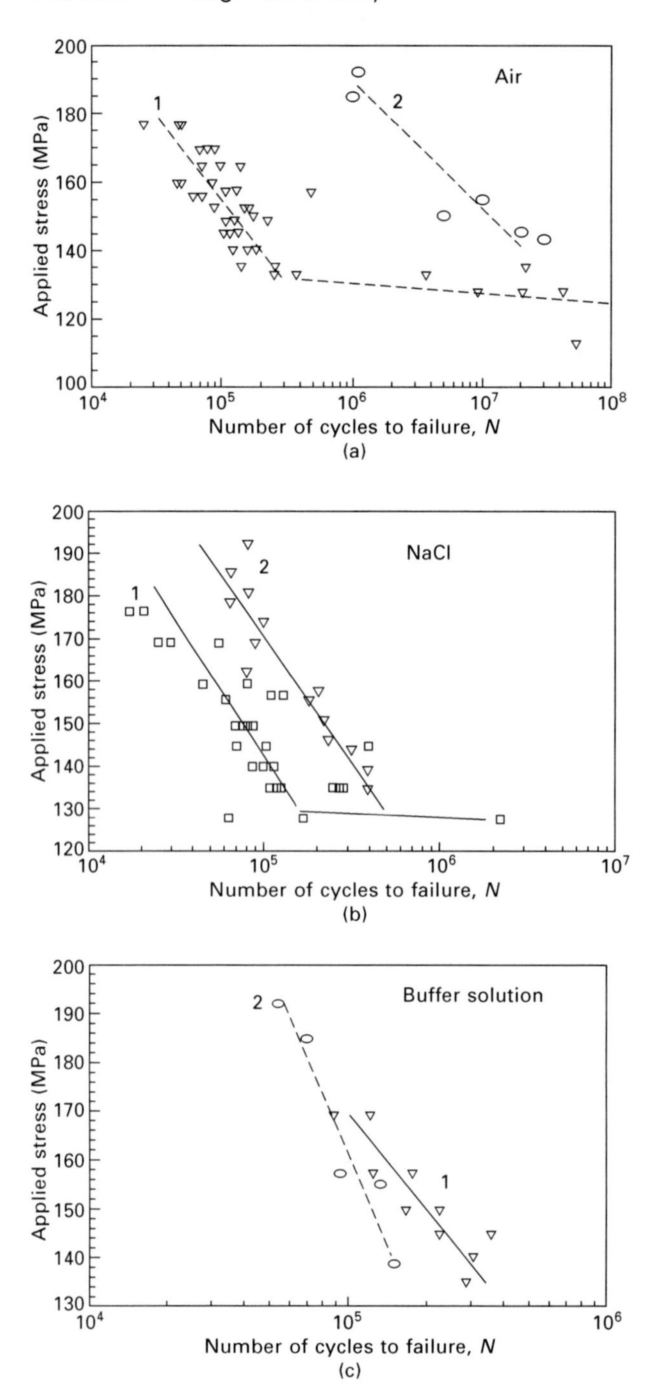

9.18 S–N diagrams for diecast AZ91D (1) and extruded AZ80 (2) alloys tested in air (a), 3.5% NaCl (b) and in buffer solution (c).

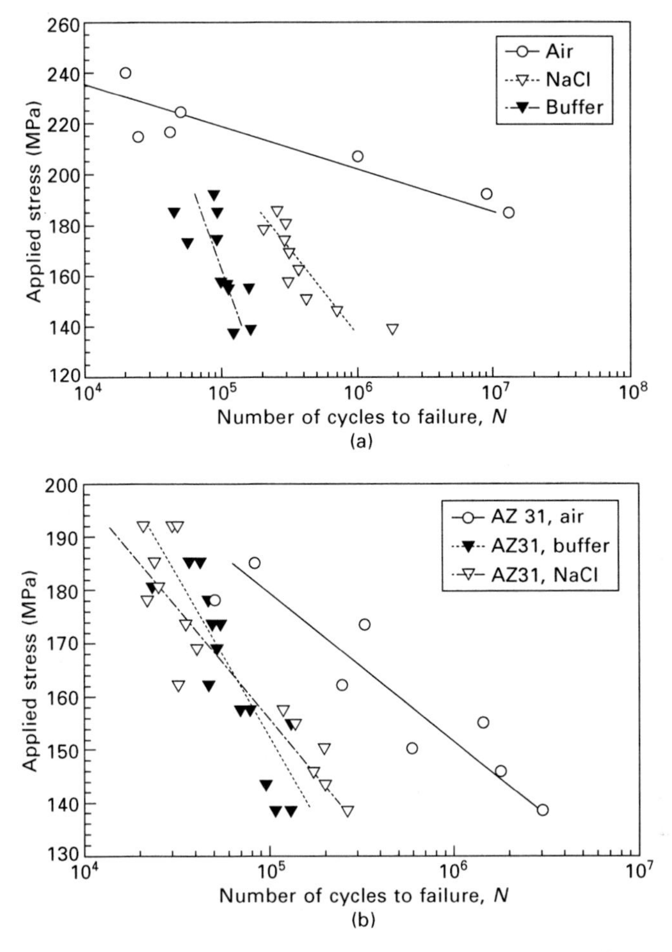

9.19 S–N diagrams for extruded alloys ZK60 (a) and AZ31 (b) in air, buffer solution and 3.5% NaCl.

9.17 and 9.18) was not confirmed at static loading in corrosion creep tests: the fracture of all diecast alloys was observed in buffer solution and did not occur in air (Fig. 9.9). To explain this phenomenon, further studies are necessary.

As noted above, under static loading, a marked plasticization of Mg alloys (a strong chemomechanical effect) in buffer solution was found, namely, the strain-to-rupture in creep tests approximately twice as high as that in NaCl. The data presented in Figs 9.9, 9.17 and 9.18 confirm the same effect of sodium borate and sodium chloride solutions on the lifetime of Mg alloys both under static and cyclic loading. A higher plastic deformation in the tip of a crack in an initially elastically loaded sample can accelerate the crack propagation and shorten the lifetime of the alloy. Meanwhile, much longer

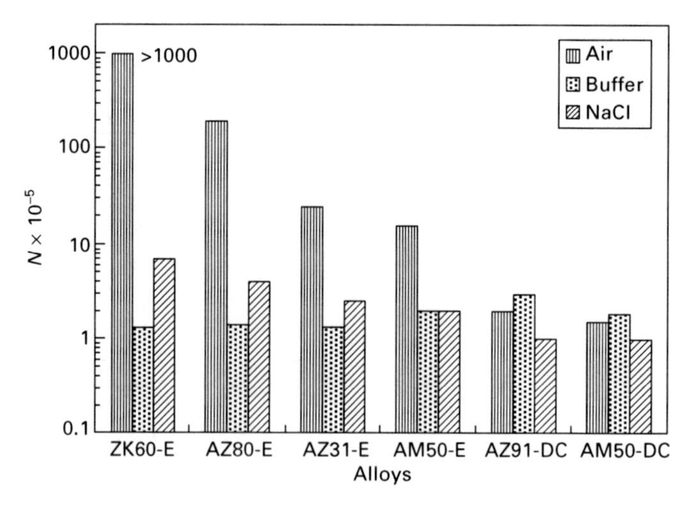

9.20 Comparison of lifetime *N* of extruded and diecast alloys in air, borate buffer solution and 3.5% NaCl at the same applied stress of 140 MPa. The *N* value for ZK60 alloy in air was obtained by extrapolation, *N* >> 1000.

creep and fatigue lives of Mg alloys in NaCl solution compared with that in buffer solution can be explained also by blunting of the tip of a crack as shown, for example in Fig. 9.14 for corrosion creep of AM50 alloy.

The correlation analysis of the corrosion fatigue behavior of Mg alloys at high stresses was carried out using so-called Basquin's equation [72]:

$$N\sigma^{p} = C \text{ or } N = C\sigma^{-p} \qquad 9.1$$

where N is the number of cycles to fracture; σ is the maximum nominal applied stress; p and C are coefficients. Correlation coefficients r in Eq. (9.1) varied for all alloys, as a rule, from 0.85 to 0.97, except the fatigue data for AZ91 in air, where $r = 0.76$. This analysis was carried out using fitted N_{sol}/N_{air} ratios (the relative lifetime) calculated at given stresses for samples examined in air and in NaCl solution (Fig. 9.21).

Extruded alloys show a significantly higher sensitivity to the action of 3.5% NaCl solution compared with diecast alloys. For instance, N_{sol}/N_{air} ratios corresponding to the applied stress change from 100 to 220 MPa vary in the range of $\sim 10^{-1}$–10^{0} for extruded AM50 and AZ31 alloys and 0.6–1 for diecast AM50 and AZ91D alloys. Being the most sensitive to corrosive environment, ZK60 alloy shows a relative lifetime in NaCl solution several orders of magnitude shorter than in air (Fig. 9.20). Among extruded alloys, the highest sensitivity to the action of NaCl-based solutions was observed in ZK60 ($N_{sol}/N_{air} \sim 10^{-4}$–$10^{0}$) and the lowest one – in AM50 and AZ31 alloys ($N_{sol}/N_{air} \sim 10^{-1}$–$10^{0}$).

In contrast to our data, according to Hilpert and Wagner [55], extruded

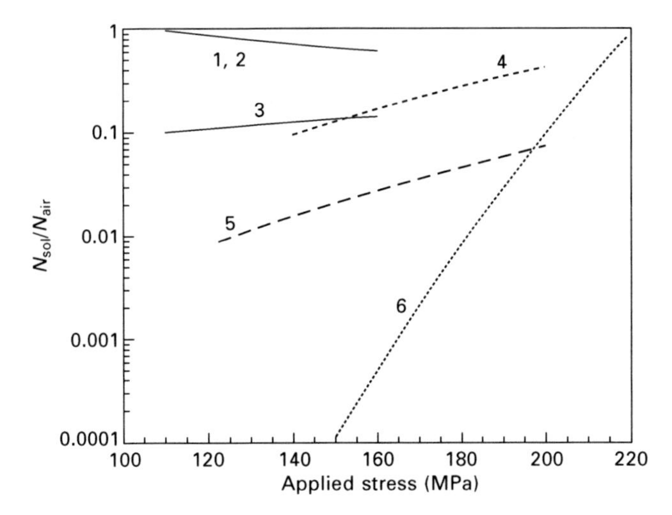

9.21 Effect of applied stress on the relative fatigue life N_{sol}/N_{air} of diecast alloys AZ91D (1) and AM50 (2) and extruded alloys AM50 (3); AZ31 (4); AZ80 (5) and ZK60 (6) in 3.5% NaCl solution.

AZ31 and AZ80 alloys have almost the same fatigue life both in air and in NaCl solutions (N_{sol}/N_{air} ratio close to 1). Apparently, it is connected with a marked decrease in sensitivity of alloys to the action of a corrosive medium on electrolytically polished specimens in comparison with lathe-formed specimens in our experiments.

Typical microstructure of extruded alloys is characterized by the presence of extended deformation bands of the solid solutions Mg–Al in AM50, AZ31 and AZ80 alloys or Mg–Zn in ZK60, and second phase stringers. In AM50 and AZ31 alloys, the second phases are represented by β-phase ($Mg_{17}Al_{12}$), Al_mMn_n and Mn_3Mg_2. Very high sensitivity of ZK60 to corrosion fatigue is connected, probably, with the presence of a significant content of very small (as a rule, less than 0.5 μm) Zn_2Zr precipitates in Mg–Zn solid solution [51]. Such intermetallides promote strain hardening and, thus, increase chemical potential of metal atoms and their mechanochemical dissolution [53]. In spite of a higher sensitivity of extruded alloys to the action of NaCl solution, their fatigue life in this medium is significantly longer than that of diecast alloys. For instance, fatigue life of extruded AM50 in NaCl is five to ten times longer than that of diecast AM50 alloy under the same applied stress (Fig. 9.21).

As a result of stress-corrosion attack during fatigue tests, pits and cracks are observed on the lateral surface of samples and in the area of crack propagation, as shown, for example, in Figs. 9.22 and 9.23. The fracture surface in the crack origin area of diecast AZ91D and extruded AZ31 alloys after a corrosion fatigue test in 3.5% NaCl shows a faceted appearance of

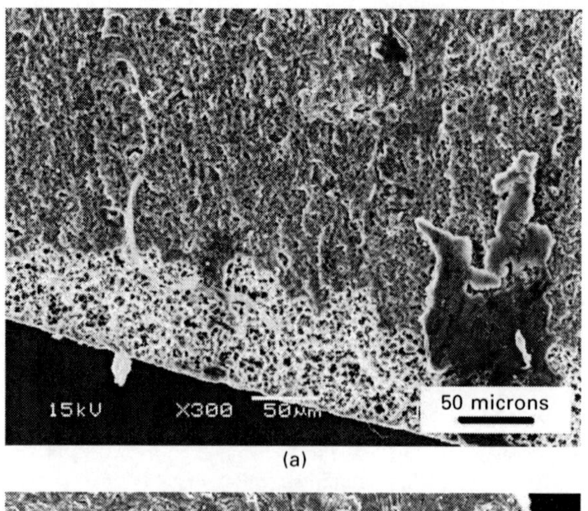

(a)

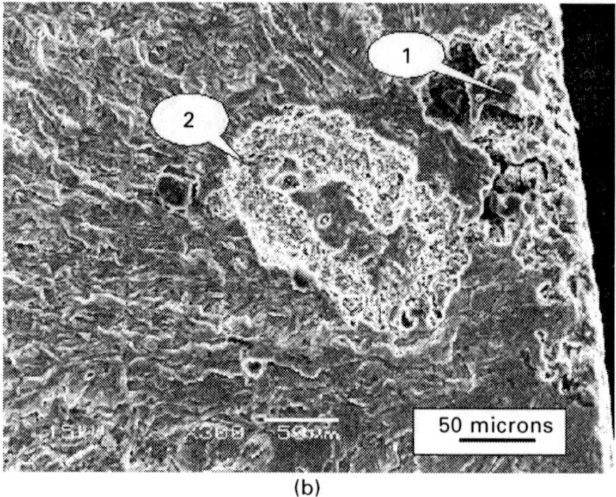

(b)

9.22 Fracture surface in the crack origin area of diecast AZ91D (a) and extruded AZ31 (b) [51] alloys with pitting on the specimen surface (1) and on the fracture surface in crack origin area (2) after a fatigue test in 3.5% NaCl.

fatigue fracture in the crack origin area (Fig. 9.22). A fatigue crack initiates from a pit, which grows up to the critical size, at which the stress intensity factor reaches the threshold for fatigue cracking. During a fatigue test, solution propagates into a crack, and this leads to corrosion pit appearance in the crack origin area (Fig. 9.22). Extruded ZK60 alloy is significantly more sensitive to the action of corrosive environment than extruded AM50 alloy, as mentioned above, and shows a much more developed surface degradation in NaCl (Figs 9.21 and 9.23).

We can conclude that diecast AM50 and AZ91D alloys have a fatigue limit in air of around 110 and 130 MPa, respectively. The same levels of

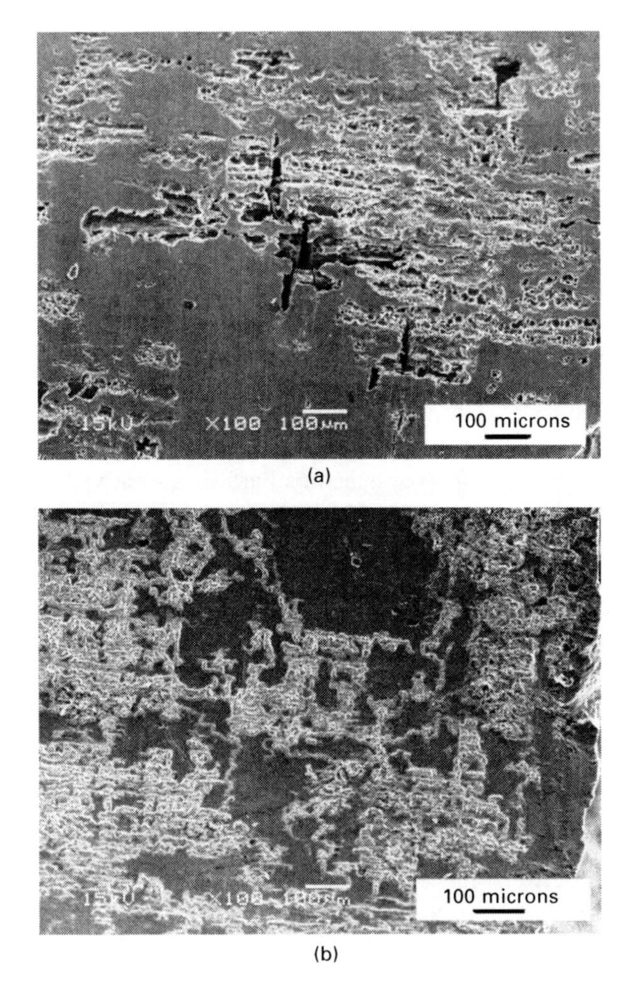

9.23 Surface pits in extruded AM50 (a) and ZK60 (b) alloys after fatigue in 3.5% NaCl solution.

fatigue limit for these alloys were found in 3.5% NaCl solution. Extruded AM50 shows fatigue limit in air at approximately 120 MPa. Fatigue limits in other examined alloys were not observed.

Strongly textured extruded alloys show a significantly longer fatigue life than diecast alloys both in air and in NaCl-containing solutions. The longest fatigue life in air is observed in ZK60 alloy and the shortest one in extruded AM50 and AZ31 alloys. However, in buffer solution, extruded AM50 alloy shows the longest durability among all extruded alloys, especially at low stress values, in spite of its relatively short lifetime in air and in sodium chloride solution.

A very interesting fact was discovered, namely: extruded alloys, especially

AZ80 and ZK60, showed a significantly stronger embrittlement in borate buffer solution under cyclic loading with respect to that for diecast alloys. In this solution, the lifetime of extruded AZ80 was even shorter than that for diecast AZ91D with approximately the same composition. Moreover, under low stresses all extruded alloys, excluding AM50, demonstrated a shorter lifetime in buffer solution with respect to that for diecast alloys.

9.6 Summary

The creep and fatigue behavior of Mg alloys produced by extrusion, common diecasting and rheoformed die-casting depend very strongly on processing and a kind of environment. For pure magnesium and its alloys the significant plasticization is observed in buffer solution, i.e. a strong chemomechanical effect is found. The greatest mechanochemical effect is observed in AZ91D and AS21 diecast alloys and in ZK60 and AZ80 extruded alloys. It can be explained by an increasing strain-hardening coefficient due to the growth of Al content and hard precipitates of the second phases ($Mg_{17}Al_{12}$; Mg_2Si or Zn_2Zr). TEM observation confirms the creation and destruction of dislocation pile-ups appeared at work-hardening stages. Corrosion creep of coated common diecast and rheoformed Mg alloys shows that anodic coating delays metal dissolution facilitating plastic deformation of the stressed metal, and for this reason alone increases the corrosion creep life of metals. Under cyclic loading extruded alloys show a significantly stronger embrittlement in borate buffer solution with respect to that for diecast alloys. Moreover, under low stresses all extruded alloys, excluding AM50, demonstrated a shorter lifetime in buffer solution than did diecast alloys.

9.7 References

1. P. A. Rebinder, E. K. Venstrem, The effect of the medium and of adsorption layers on the plastic flow of metals, *Bulleten Acad. Sci. U. R. S. S., Classe Sci. Mat. Nat., Ser. Phys.*, (1937) 531–548 (in German 548–550).
2. P. A. Rehbinder, E. D. Shchukin, Surface phenomena in solids during deformation and fracture processes, in: *Progress in Surface Science*, 3(Pt. 2) (1972) 97–188.
3. V. I. Likhtman, E. D. Shchukin and P. A. Rebinder, *Physicochemical mechanics of metals, Adsorbtion phenomena in the process of deformation and failure of metals* (Israel Program for Scientific Translations, Ierusalem, 1964).
4. P. A. Rebinder, V. I. Likhtman, V. M. Maslennikov, Deformation of single crystals of metals as facilitated by adsorption of surface-active substances, *Doklady Akademii Nauk SSSR, Seriya A* **32** (1941) 125–129.
5. V. S. Ostrovsky, V. I. Likhtman, Influence of surface-active materials and oxide films on deformation of cadmium single crystals, *Doklady Akademii Nauk SSSR*, **XCVI**, No. 2 (1954) 319–321.
6. V. I. Likhtman, V. S. Ostrovsky, Influence of oxide films on mechanical properties of cadmium single crystals, *Doklady AN SSSR*, **XCIII**, No. 1 (1953) 105–107.

7. R. Roscoe, Plastic deformation of cadmium single crystals, *Philosophical Magazine*, **21** (1936) 399–406.
8. S. Harper, A. H. Cottrell, Surface effects and the plasticity of zinc crystals, *Proc. Phys. Soc., London* **63B** (1950) 331–338.
9. D.J. Philips, N. Tompson, Surface effects in creep of cadmium crystals, *Proc. Phys., Soc. London (B)* **63B** (1950) 839–847.
10. E. N. da C. Andrade, R. F. Y. Randall, Surface effects with single crystal wires of cadmium, *Nature*, **162** (1948) 890–891.
11. E. N. da C. Andrade, R. F. Y. Randall, The influence of electrolytes on the mechanical properties of certain metal single crystals, *Proceedings of the Physical Society, London*, **65B** (1952) 445–454.
12. E. N. da C. Andrade, Surface effect and structure of single crystal wires, *Nature*, **164** (1949) 536–537.
13. E. N. da C. Andrade, The effect of surface conditions on the mechanical properties of metals, mainly single crystals, *Symp. on Properties of Metallic Surfaces* (Inst. Metals, London) (1952) 133–143.
14. E. K.Venstrem, P. A. Rebinder, The electrocapillary effect of the lowering of the hardness of metals, *Doklady Akademii Nauk SSSR*, **68** (1949) 329–332.
15. I. R. Kramer, L. J. Demer, Effects of Environment on Mechanical Properties of Metals, in: *Progress in Materials Science*, Vol. 09, No.3, New York, Pergamon, 1961, pp. 131–199.
16. R. W. Revie, H. Uhlig, Effect of applied potential and surface dissolution on the creep behavior of copper, *Acta Metallurgica*, **22** (1974) 619–627.
17. C. J. Van Der Wekken, The effect of surface dissolution on the creep rate of copper, *Acta Metallurgica*, **25** Issue 10 (1977) 1201–1207.
18. B. Gu, W. Y. Chu, W. Chu, L. J. Qiao, C. M. Hsiao, *Corrosion Science*, **36**, Issue 8 (1994) 1437–1445.
19. Y. Suzuki, Y. Hisamatsu, Stress corrosion cracking of pure copper in dilute ammoniacal solutions, *Corrosion Sci.*, **21** Issue 5 (1981) 353–368.
20. R. Nishimura, T. Yoshida, Stress corrosion cracking of Cu–30% Zn alloy in Mattsson's solutions at pH 7.0 and 10.0 using constant load method – A proposal of SCC mechanism, *Corrosion Sci.*, **50** Issue 4 (2008) 1205–1213.
21. S. C. Sircar, S. Thakur, Charge effect on creep, *J. Materials Sci. Lett.*, **6**, No. 11 (1987) 1323–1324.
22. C. J. Van Der Wekken, Adsorbed species on nickel in electrolyte solutions, *J. Electrochem. Soc.*, **133**, 11 (1986) 2293–2295.
23. R. M. Latanision, R. W. Staehle, Plastic deformation of electrochemically polarized nickel single crystals, *Acta Metallurgica*, **17** Issue 3 (1969) 307–319.
24. C. J. Van Der Wekken, Surface dislocation pinning by adsorbed ions and water molecules, *J. Electrochem. Soc.*, **131**, 11 (1984) 2481–2483.
25. T. P. Hoar, J. C. Scally, Mechanochemical anodic dissolution of austenitic stainless steel in hot chloride solution at controlled electrode potential, *J. Electrochem. Society*, **111** (1964) 348–352.
26. H. Graefen, D. Kuron, Intercrystalline stress corrosion of mild steel in alkaline solutions, *Archiv fuer das Eisenhuettenwesen* **36**(4) (1965) 285–291.
27. R. Muenster, H. Graefen, The effects of carbon, nitrogen, aluminum, and titanium contents on intercrystalline stress corrosion in plain carbon steels, *Archiv fuer das Eisenhuettenwesen*, **36**(4) (1965) 277–284.
28. M. C. Petit, D. Desjardins, Elongation measurements of stainless steel during SCC

tests, *International Corrosion Conference Series* (1977), NACE-5 (Stress Corros. Cracking Hydrogen Embrittlement Iron Base Alloys), 1205–1210.
29. M. Smialowski, J. Kostanski, Creep and stress corrosion cracking of austenitic stainless steel in boiling 35% magnesium chloride solution, *Corrosion Sci.*, **19**(12) (1979) 1019–1029.
30. H. Iwanaga, T. Oki, Creep behavior of pitting stainless steels under controlled potential, *Zairyo*, **31**(342) (1982) 288–294 [Japanese].
31. R. Nishimura, K. Kudo, Stress corrosion cracking of AISI 304 and AISI 316 austenitic stainless steels in HCl and H_2SO_4 solutions – prediction of time-to-failure and criterion for assessment of SCC susceptibility, *Corrosion*, (1989) 308–316.
32. M. C. Petit, M. Cid, M. Puiggali, Z. Amor, An impedance study of the passivity breakdown during stress corrosion cracking phenomena, *Corrosion Sci.*, **31** (1990) 491–496.
33. M. Touzet, M. Cid, M. Puiggali, M. C. Petit, An EIS study and Auger analysis on 304L stainless steel in hot chloride media before and after a sample straining, *Corrosion Sci.* **34**(7) (1993) 1187–1196.
34. H. Leinonen, Stress corrosion cracking and life prediction evaluation of austenitic stainless steels in calcium chloride solution, *Corrosion*, **52**(5) (1996) 337–346.
35. H. Leinonen, I. Virkkunen, H. Hanninen, Stress corrosion cracking and life prediction of austenitic stainless steels in calcium chloride solution, in: *Hydrogen Effects on Material Behavior and Corrosion Deformation Interactions*, Proc. Int. Conf., Moran, WY, USA, 22–26 Sept. 2002 (2003) 673–682.
36 K. S. Raja, S. A. Namjoshi, D. A. Jones, Corrosion-creep interaction of stainless alloys in acid chloride solutions, *Metallurgical and Materials Transactions A: Physical Metallurgy and Materials Science*, 36A(5) (2005) 1107–1120.
37. H. S. da Costa-Mattos, I. N. Bastos, J. A. C. P. Gomes, A simple model for slow strain rate and constant load corrosion tests of austenitic stainless steel in acid aqueous solution containing sodium chloride, *Corrosion Sci.*, **50** (2008) 2858–2866.
38. U. Feld, A. Rahmel, M. Schmidt, Investigation on the interactions between creep and corrosion of nickel-based alloys in sulfate melts, in: *Corrosion and Mechanical Stress at High Temperatures*, Applied Sci. Publishers Ltd, Barking, 1981, pp. 171–194, discuss. 195–196.
39. E. M. Gutman, Ya. Unigovski, A. Eliezer, E. Abramov, Mechanoelectrochemical behavior of magnesium alloys stressed in aqueous solutions, *J. Materials Synthesis Processing*, **8**, Nos.3/4 (2000) 133–138.
40. E. M. Gutman, A. Eliezer, Ya. Unigovski, E. Abramov, Mechanoelectrochemical behavior and creep corrosion of magnesium alloys, *Material Sci. Eng. A*, **A302** (2001) 63–67.
41. E. M. Gutman, Ya. Unigovski, A. Eliezer, E. Abramov, Corrosion creep of magnesium and diecast magnesium alloys, *J. Mater. Sci. Letters*, **20** (2001) 1541–1543.
42. Ya. Unigovski, Z. Keren, A. Eliezer and E.M. Gutman, *Mater. Sci. Eng.*, **A398**, Issues 1–2 (2005) 188–197.
43. Ya. B. Unigovski, E M Gutman, Z. Koren, T. Poryadkov, H. Rosenson, Corrosion creep of metals, *J. Physics: Conference Series*, **98** (2008) 072004.
44. Ya. Unigovski, E. M. Gutman, Z. Koren, H. Rosenson, Y. Hao, T. Chen, Effect of processing on stress-corrosion behavior of diecast Mg–Al alloy, *J. Mater. Processing Technology*, **208** Issues 1–3 (2008) 395–399.
45. A. Eliezer, G. Ben-Hamu, E. Abramov, Ya. Unigovski, E. M. Gutman, Mechanoelectrochemical behavior of magnesium alloys, in: Intern. Corrosion

Congress: *Frontiers in Corrosion Science and Technology*, 15th, Granada, Spain, 22–27 Sept. 2002 (2002), 435/1–435/8.
46. Ya. Unigovski, A. Eliezer, L. Riber, E. M. Gutman, in: *Proceed. Int. Conf. on Mg Alloys and Their Applications*, Wiley-VCH, Wolfsburg, Germany, 19–23 Nov. 2003, pp. 632–637.
47. Ya. B. Unigovski, L. Riber, E. M. Gutman, Corrosion stress relaxation in pure magnesium and diecast Mg alloys, *J. Metals, Materials Minerals*, **17**(1) (2007) 1–7.
48. E. M. Gutman, Ya. Unigovski, A. Eliezer, E. Abramov, L. Riber, Effect of processing and environment on mechanical properties of die cast magnesium alloys, *Light Metal Age*, (December 2000), 14, 15, 16–20.
49. A. Eliezer, E. M. Gutman, E. Abramov, Ya. Unigovski, Corrosion fatigue of diecast and extruded magnesium alloys, *J. Light Metals*, **1**, Issue 3 (2001) 179–186.
50. E. M. Gutman, A. Eliezer, Ya. Unigovski, E. Abramov, Corrosion fatigue of magnesium alloys, *Materials Science Forum*, **419–422** (2003) 115–120.
51. Ya. Unigovski, A. Eliezer, E. Abramov, Y. Snir, E. M. Gutman, Corrosion fatigue of extruded magnesium alloys, *Mater. Eng.*, **A360** (2003) 132–139.
52. E. M. Gutman, Thermodynamics of the mechanochemical effect. 1. Derivation of basic equations nature of the effect, *Sov. Mater. Sci.* **3** (1967) 190–196; Thermodynamics of the mechanochemical effect. 2. Nonlinear relations, ibid., 304–310; Interdependence of corrosion phenomena and mechanical factors acting on metal, ibid., 401–409.
53. E. M. Gutman, *Mechanochemistry of Solid Surfaces*, World Scientific, Singapore, New Jersey, London, 1994, p. 322.
54. Z. Koren, H. Rosenson, E. M. Gutman, Ya. B. Unigovski, A. Eliezer, Development of semisolid casting for AZ91 and AM50 magnesium alloys, *J. Light Metals*, **2** (2002) 81–87.
55. M. Hilpert, L. Wagner, Effect of mechanical surface treatment and environment on fatigue behavior of wrought magnesium alloys, in: *Proceedings of the International Congress: Magnesium 2000, Magnesium Alloys and Their Applications*, Munich, 2000, pp. 463–468.
56. C. M. Sonsino, K. Dieterich, L.Wenk, A. S. Till, Fatigue design with cast magnesium alloys, in: *Proceedings of the International Congress: Magnesium 2000, Magnesium Alloys and Their Applications*, Munich, 2000, pp. 304–311.
57. A. Eliezer, E. M. Gutman, E. Abramov, E. Aghion, Corrosion fatigue and mechanochemical behavior of magnesium alloys, in: Prof. B.L. Mordike (Guest Ed.), *Corrosion Reviews*, Special Issue on Corrosion Resistance of Magnesium Alloys, **16**, Nos. 1–2, Freund Publ. House Ltd, London, 1998, pp. 1–26.
58. H. R. Mayer, H. Lipovsky, M. Papakyriacou, R. Rosch, A. Stich, S. Stanzl-Tschegg, Applications of ultrasound for fatigue testing of lightweight alloys, *Fatigue Fracture Eng. Materials Structures*, **22** No.7 (1999) 591–599.
59. R. I. Stephens, C. D. Schrader, K. B. Lease, Corrosion fatigue of AZ91E-T6 cast magnesium alloy in a 3.5 percent NaCl aqueous environment, *J. Eng. Mat. Techn.*, **117** (1995) 293–298.
60. M. Witt, K. Poetter, H. Zenner, K. Sponheim, P. Heuler, Fatigue strength of cast aluminum and magnesium chassis parts, in: *Magnesium 2000, Proceedings of the 2nd Int. Conf. on Magnesium Science and Technology*, Eds. E. Aghion and D. Eliezer, Pub. by Magnesium Research Institute (MRI) Ltd, Beer-Sheva, Israel, 2000, pp. 263–275.
61. C. Müller, R. Koch, G. H. Deinzer, Corrosion of the magnesium alloy AZ91D and

its influence on fatigue properties, in: *Proceedings of the International Congress: Magnesium 2000, Magnesium Alloys and Their Applications*, Munich, 2000, pp. 457–462.
62. A. Eliezer, E. M. Gutman, E. Abramov, Y. Unigovski, E. Aghion, Corrosion fatigue and corrosion creep of magnesium alloys, in: *Proceedings of the International Congress: Magnesium 2000, Magnesium Alloys and Their Applications*, Munich, 2000, pp. 498–505.
63. G. F. Sager, R. H. Brown, R. B. Mears, Tests for determining susceptibility to stress-corrosion cracking, in: *Symposium on Stress-corrosion Cracking of Metals*, American Society for Testing and Materials and the American Institute of Mechanical Engineers, 1945, pp. 255–272.
64. V. E. Belyakov, S. V. Pushkina, A. K. Prokin, V. V. Romanov, Effect of pH on the corrosion fatigue of the magnesium alloy MA-2-1, *Fiziko-Khimichna Mekhanika Materialiv*, **6**(1) (1970) 38–41 [Russian].
65. S. Zhang, T. Shibata, T. Haruna, Inhibition effect of the borate ion on intergranular stress corrosion cracking on sensitized type 304 stainless steel, *Corrosion*, **54**(6) (1998) 428–434.
66. V. V. Romanov, *Stress Corrosion Cracking of Metals* (Israeli program for scientific translations), Israel, Jerusalem, 1961.
67. M. A. Timonova, The connection between the magnesium alloy structure and its stress corrosion tendency, *Doklady Akademii Nauk SSSR*, **117** (1957) 848–851 [in Russian].
68. J. J. Harwood, The influence of stress on corrossion in: *Stress Corrosion Cracking and Embrittlement*, WD Roberton, ed., John Wiley & Sons, 1956, pp. 1–8.
69. M. O. Speidel, M. J. Blackburn, T. R. Beck, J. A. Feeney, Corrosion fatigue and stress corrosion crack growth in high strength aluminum alloys, magnesium alloys, and titanium alloys exposed to aqueous solutions, in: *Proceed. Inter. Conf.: Corrosion Fatigue: Chemistry, mechanics and microstructure*, National Association of Corrosion Engineers, 1972, pp. 324–343, discussion pp. 343–345.
70. E. F. Emley, *Principles of Magnesium Technology* Pergamon Press, Oxford, 1966, p. 1813.
71. Ya. Unigovski, E. Gutman, Surface morphology of a diecast Mg-alloy, *Applied Surface Sci.*, **153** (1999) 47–52.
72. O. H. Basquin, The exponential law of endurance tests, *American Society of Testing and Materials Proceedings*, **10** (1910) 625–630.

10 Magnesium (Mg) corrosion: a challenging concept for degradable implants

F. WITTE, Hannover Medical School, Germany, N. HORT and F. FEYERABEND, Helmholtz-Zentrum Geesthacht, Germany, and C. VOGT, Leibniz Universität Hannover, Germany

Abstract: Degradable metals are breaking the current paradigm in biomaterial science to develop only corrosion-resistant metals. In particular, metals which consist of trace elements existing in the human body are promising candidates for temporary implant materials. Such implants would be needed for a short time to provide mechanical support during the healing process of the injured or pathological tissue. Magnesium (Mg) and its alloys have been investigated recently by many authors as a suitable degradable biomaterial. The degradation of Mg alloys *in vivo* has been investigated in several animal studies. The findings of these studies will be critically discussed and related to Mg corrosion principles. In contrast to slow corroding metals the designated complete degradation or corrosion of Mg alloys is conjunct with the limited use of the standard procedure for biocompatibility testing (ISO 10993). In particular, established test systems for biocompatibility and cytotoxicity of long-term biomaterials have limited use and reliability when used to investigate degradable Mg alloys. Additionally, the results obtained *in vitro* are substantially different from *in vivo* observations. The physiological background and possible hypotheses will be elucidated and possible mechanism of *in vivo* corrosion of Mg alloys will be discussed. Several approaches to simulate the *in vivo* conditions on the laboratory scale have been investigated in the literature so far. These approaches will be presented and critically reviewed. This chapter will summarize the latest achievements and comment on the selection and use, test methods and the approaches to develop and produce Mg alloys that are intended to perform clinically with an appropriate host response.

Key words: degradable metals, biocompatibility, *in vivo* corrosion, *in vitro* corrosion, magnesium implant.

10.1 An introduction to degradable magnesium (Mg) implants

More than 200 years after the first production of elemental magnesium (Mg) by Sir Humphrey Davy, the attention to magnesium-based alloys as biomaterials has increased during the past decade with biomaterial scientists and medical device developers (Mantovani and Witte, 2010). Surprisingly, the concept of using magnesium as an implantable material is relatively old. In the first decades of the twentieth century magnesium was investigated extensively for several medical applications, but with the advent and adoption of stainless steels as metallic implants this interest waned in the post-World War II period to little, if any, research. The attraction of a lightweight metal with mechanical properties suitable for many applications brought a renewed focus on Mg alloys in the automotive and aerospace industries. This interest spread to the current, rapidly growing interest in magnesium-based alloys for medical applications. In the words of the author of Ecclesiastes, 'there is nothing new under the sun' (Mantovani and Witte, 2010).

The late 1990s marked the beginning of notable research into the idea that degradable metallic materials might be utilized in implants. Although the concept of temporary implants has been widely accepted and adopted for polymeric materials in many application areas, it breaks the current paradigm for metallic materials, which have been selected for, among other properties, their corrosion resistance. The growing literature in degradable metallic biomaterials has come from a variety of perspectives, including traditional materials science, tissue engineering, surface modification and characterization, toxicology, cell and molecular biology, medical imaging and a variety of clinical disciplines, most notably cardiology and orthopaedic surgery. In particular, the clinical testing of coronary stents made from a degradable Mg alloy has served as a prime example of how advances in materials design and processing have translated all the way to interventional cardiology and patient care with interdisciplinary support along the way.

10.1.1 Temporary implants

In biomaterial sciences, temporary biomaterials support the healing tissue until full regeneration or scarring healing is completed. During this tissue healing process the temporary biomaterial can gradually lose its mechanical property by its continuing degradation process (Fig. 10.1). Thereby degradable implants prevent the need for a second operation for patients to remove the implant after completed tissue healing. Biodegradable metals have an advantage over existing biodegradable materials such as polymers, ceramics or bioactive glasses in load-bearing applications that require higher initial tensile strength and Young's modulus closer to that of bone (Witte *et al.*, 2008b).

10.1.2 The basic concept of biodegradable magnesium implants

This degradation process can also be used to release drugs supporting or modulating the implant surrounding tissue response. The control and adaptation

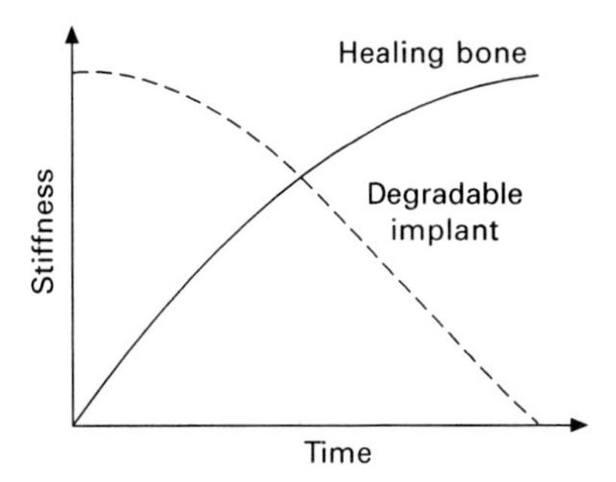

10.1 The gradual loss of mechanical integrity of a temporary or biodegradable implant while the surrounding tissue is regaining mechanical resistance.

of the degradation rate are crucial, since the resorption capacity of the tissue is limited. Thus, the local physiology of the implant environment determines the maximal degradation rate of a temporary implant. In this sense, an option to create biocompatible Mg alloys is to design slow corroding implants by appropriate alloying, microstructure design and/or an appropriate coating. All these corrosion protective strategies may be planned carefully while keeping in mind that temporary implants need to degrade completely – otherwise parts of the implant may persist locally and will act as long-term biomaterials.

A basic concept in magnesium corrosion is the reduction of corrosion rate; thus the evolving hydrogen gas can be removed by local fluid flow or diffusion without the appearance of clinically visible gas cavities. An effective reduced corrosion rate can be obtained by alloy composition design, processing or coating. However, it has to be kept in mind that the corrosion rate will be further reduced *in vivo* after implantation due to adherent proteins and inorganic salts such as calcium phosphates which stabilize the corrosion layer (Witte *et al.*, 2005). Based on this theory, an Mg implant with an initially reduced corrosion rate could lead finally to an arresting corrosion process *in vivo*. Thus, the right balance of a reduced initial corrosion rate and an assured complete corrosion *in vivo* will create a useful biodegradable Mg implant.

10.1.3 Major recent advances in the field

The major recent advances in Mg alloys as temporary biomaterials have been in understanding the interface and interaction of Mg alloys and their biological environment. In contrast to previous technical alloy developments aiming at the improvement of mechanical properties, corrosion resistance and production costs, in biomedical application the main focus is shifting to the influence of the alloying elements on the formation of the corrosion protective interfaces and on the surrounding biological environment *in vitro* and *in vivo*. However, currently available Mg alloys were investigated in different biomedical applications. Probably the most advanced clinical applications are biodegradable cardiovascular Mg stents which have been successfully investigated in animals (Di Mario *et al.*, 2004; Waksman *et al.*, 2006a,b); the first clinical human trials have been conducted (Di Mario *et al.*, 2004; Erbel *et al.*, 2007; Zhang *et al.*, 2007). Mg alloys have also been investigated as bone implants (Witte *et al.*, 2005; Kaya *et al.*, 2007; Xu *et al.*, 2007) and can be applied in various designs, e.g. as screws, plates or other fixture devices. Mg chips have been investigated for vertebral fusion in spinal surgery of sheep (Wen *et al.*, 2004) and open-porous scaffolds made of Mg alloys have been introduced as load-bearing biomaterials for tissue engineering (Feyerabend *et al.*, 2006; Witte *et al.*, 2006a, 2007a,b). However, dissolved ions from metal implants are always a concern to induce hypersensitivity and allergy. Mg alloys AZ31, AZ91, WE43 and LAE442 have been shown to be non-allergenic in an epicutaneous patch test in accordance with the ISO standard (Witte *et al.*, 2008a). Moreover, high extracellular Mg concentrations have been found beneficial for cartilage tissue engineering (Feyerabend *et al.*, 2006). Even though rare earth-containing Mg alloys (Witte *et al.*, 2010) as well as metallic glasses (Zberg *et al.*, 2009) can corrode *in vivo* without the clinical appearance of subcutaneous gas cavities, ongoing research will elucidate more suitable Mg alloys and coatings in the near future which will accelerate the development of biodegradable metal implants.

10.2 The appropriate selection and use of biodegradable magnesium (Mg) alloys

Today Mg alloys are mainly used in transportation and 3C industries (computer, communication and consumer electronics). Most applications are castings and only a minority of Mg alloys are used as wrought materials. In any case mechanical strength, appropriate corrosion behaviour and a good processability are major requirements. Additionally, deformability also plays a role when wrought materials are used. In general, a property profile has to be considered rather than a single property such as tensile yield strength. This is important since changing a single property also has an impact on the other properties from the property profile. Figure 10.2 illustrates this in a scheme for the development of implants.

It is extremely important to know the requirements of the implant regarding its strength, whether it has to be degradable or not, what is the required degradation rate and under which environmental conditions the implant will be used. In fact, the application in the cardiovascular or musculoskeletal system of the human body determines the entire property profile for the intended implant. These requirements are the benchmarks which will be used to decide if a development is ready for application or if additional work is necessary.

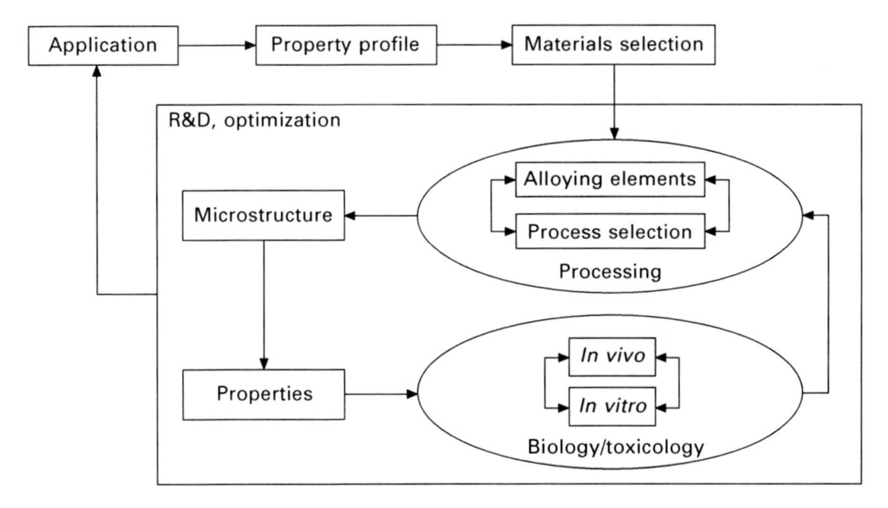

10.2 The cycle of development to achieve a suitable biodegradable Mg implant for the required properties of the implant.

After determining the boundary conditions the material and the according manufacturing processes can be selected. Special care has to be given to the selection of the alloying elements which must be neither toxic nor harmful. All intermetallic phases and reaction products that occur during processing must also be neither harmful nor toxic.

For some elements such as calcium (Ca) and a number of rare earth elements (REE) their suitability in binary Mg alloys has already been proven. But further work still needs to be undertaken to assure that the intermetallic compound of Mg–Ca and Mg–REE are also not harmful to the human body. In the case that ternary or even more complex alloys will be developed the same tests have to be applied to make sure that all compounds that may form are safe. The experimental work in this area has to be accompanied by thermodynamic calculations. While binary phase diagrams are available for most elemental combinations, there are also reliable phase diagrams for a number of ternary systems. A major challenge is still the calculation of phase formation in more complex systems. Moreover, it might be the case that Mg alloy systems which contain more than three alloy components will be better solutions for biodegradable implants than those ones which are actually presented in literature.

Alloys such as AM50, AZ91, WE43, LAE442 which are proposed as suitable implant materials can only be seen as some examples to prove some concepts regarding strength, degradability and design of implants. In fact, these alloys normally contain many more elements as shown in their designation: while AM is a ternary Mg–Al–Mn alloy, AZ91 is actually Mg–Al–Zn–Mn, WE43 is Mg–Y–Nd–REE–Zr and LAE is Mg–Li–Al–Mn–REE. A master alloy is normally used to add REE. This master alloy contains one REE in larger quantities but also almost all of the REEs. It has to be noted that currently silicon (Si) can be added to the Mg alloy in the range of up to 0.3 wt% without any declaration while impurities need to be declared if they exceed 0.3 wt% in total. These uncertainties in the alloy composition may be acceptable for applications in transportation or 3C industries but cannot be tolerated for biodegradable Mg implants.

It is not only the toxicity of alloying elements that is of importance: their influence on the property profile and the processability of the alloy is also relevant. Not all biocompatible alloying elements can be used due to restrictions in solubility, interactions with the processes and process parameters, etc. In fact, this is a complex system which, at the moment, has to be investigated in detail with a special regard to any alloying element and their influence on the chosen processing route.

Both alloy composition and processing influence the microstructure and phase formation. Composition and microstructure are the important features that determine the property profile. A major part of the work in the development of an implant is based on research and development, followed by the optimization of the process to achieve a defined property profile. As long as the benchmarks of the previously selected application are not reached, this work has to be revised as often as necessary.

The designation of different heat treatments of Mg alloys follows the same system as for aluminium alloys (Avedesian and Baker, 1999). In cast components T4 tempering is used to level the differences in the distribution of alloying elements in grains as well as to remove the eutectics and intermetallic phases. Normally the grain size should remain stable, but often this cannot be ensured. A T4 heat treatment requires high temperatures in combination with long heat treatment times. For cast components it is recommended that a temperature below the lowest eutectic temperature in the alloy system is selected to avoid a first melting of the eutectic phase. The time for this type of temper is quite often in the range of several hours up to a day (or more). If a wrought process follows, a grain growth could be negligible in this case. Otherwise the tempering has to be designed in a way that the process is stopped when all precipitates on grain boundaries have been dissolved. In general a water quench follows the heat treatment to freeze the microstructure and to avoid diffusion and re-precipitation of intermetallic phases. In general, the mechanical strength is lowered by this method to a certain extent. But owing to the fact that the distribution of alloying elements is more equal after a T4 heat treatment this temper normally has a positive influence on the corrosion behaviour. A T6 treatment is built upon the aforementioned T4 temper. The purpose of this temper is to precipitate intermetallic phases in the grain and on grain boundaries. With respect to the chosen tempering time and temperature it is possible to adjust the size of precipitates as well as their distribution.

The presence of alpha grains and an eutectic phase has a direct influence on mechanical properties as well as on the corrosion behaviour (Kainer *et al.*, 2009). Especially for Mg alloys all intermetallic phases are nobler than the matrix. This will lead to an increase of the corrosion rate along the grain boundaries. While the segregation of alloying elements does not influence mechanical properties too much it definitely influences the corrosion behaviour as well owing to differences in the potential of the centres of the grains and the areas close to the grain boundary.

The influence of different heat treatments on microstructure could be effectively shown in the binary Mg–Gd alloy system where the amount of gadolinium (Gd) was varied in combination with a heat treatment (Kainer *et al.*, 2009). As stated before, a T4 treatment could improve the corrosion rate drastically compared with the as-cast state (F). In contrast to the F state which consists of alpha grains and eutectics on the grain boundaries, the T4 state consists only of alpha grains and the alloying element (Gd in this case) is homogeneously distributed in the sample. A T6 treatment reprecipitates intermetallic phases which are nobler than the matrix and again have a poor effect on the corrosion behaviour. The investigations also showed that the T6 state creates a material that is normally more corrosion resistant than the as-cast material. In both cases intermetallics are present. In the T6 state, however, they can be regarded as being in a thermo-dynamic equilibrium. In the best case these intermetallics are also finely distributed over the matrix and not only present at the grain boundaries. Intermetallic precipitates on the grain boundaries could be regarded as the worst case scenario with respect to the corrosion behaviour. This is mainly due to the fact that grain boundaries are weakened areas with a high internal energy. It is most likely that a chemical attack would start at the grain boundary or even close to it. If, as in the case of as-cast materials, the grain boundaries are additionally decorated with noble intermetallics and would promote a severe corrosive attack in these areas which would explain why the highest corrosion rates can be observed in as-cast binary Mg–Gd alloys.

10.3 *In vivo* corrosion of magnesium (Mg) alloys: what happens in living tissue?

Magnesium and its alloys are generally known to degrade in aqueous environments via an electrochemical reaction (corrosion) which produces magnesium hydroxide and hydrogen gas. Thus, magnesium corrosion is relatively insensitive to various oxygen concentrations in aqueous solutions which occur around implants in different anatomical locations. The overall corrosion reaction of magnesium in aqueous environments is given below:

$$Mg\ (s) + 2H_2O\ (aq) \rightleftharpoons Mg(OH)_2\ (s) + H_2\ (g) \qquad 10.1$$

This overall reaction may include the following partial reactions:

$$Mg\ (s) \rightleftharpoons Mg^{2+}\ (aq) + 2\ e^- \quad \text{(anodic reaction)} \quad 10.2$$

$$2H_2O\ (aq) + 2e^- \rightleftharpoons H_2\ (g) + 2OH^-\ (aq) \quad \text{(cathodic reaction)} \quad 10.3$$

$$Mg^{2+}\ (aq) + 2OH^-\ (aq) \rightleftharpoons Mg(OH)_2\ (s) \quad \text{(product formation)} \quad 10.4$$

Magnesium hydroxide accumulates on the underlying Mg matrix as a corrosion protective layer in water, but when the chloride concentration in the corrosive environment rises above 30 mmol/l (Shaw, 2003), it starts to convert into highly soluble magnesium chloride. Therefore, severe pitting corrosion can be observed on Mg alloys *in vivo* where the chloride content of the body fluid is about 150 mmol/l (Witte *et al.*, 2005, 2006c; Xu *et al.*, 2007). In magnesium and its alloys, elements (impurities) and cathodic sites with a low hydrogen overpotential facilitate hydrogen evolution (Song and Atrens, 1999), thus causing substantial galvanic corrosion and potential local gas cavities *in vivo*. Synchrotron radiation-based infrared spectroscopies of explanted magnesium–bone interfaces have shown that proteins partly cover the corroding Mg surface *in vivo* (Witte *et al.*, 2006a). Thus, partial protein adhesion could initiate or provoke local Mg implant corrosion. Apparently, serum proteins interact more likely with the cathodic and alkaline sites of the magnesium surface. Therefore, the corrosion morphology of magnesium and its alloys depends on the alloy chemistry and the environmental conditions (Song and Atrens, 1999; Witte *et al.*, 2005). In particular, the local accumulation of calcium phosphates in the corrosion layer has been observed *in vivo* in bone and intravascular (Witte *et al.*, 2005; Erbel *et al.*, 2007), which are both quite different biological environments. However, as discussed in the field of biodegradable materials, there is at least a two-way relationship between the material and the biological host response, i.e. the degradation process or the corrosion products can induce local inflammation and the products of inflammation can enhance the degradation process. The complexity of this relationship is generally unknown for biodegradable metals, even though first results have shown that fast-corroding Mg alloys respond with a mild foreign body reaction (Witte *et al.*, 2007b). *In vivo* studies were predominantly performed in small animals, i.e. rats (subcutaneously), guinea pigs and rabbits (Witte *et al.*, 2005, 2006c, 2007a,b, Xu *et al.*, 2007; Zhang *et al.*, 2007). While magnesium-based devices were tested in large animals (sheep, pigs) reporting on the corrosion of magnesium chips in spinal applications (Kaya *et al.*, 2007) and preclinical experiments for cardiovascular stent applications (Waksman *et al.*, 2005, 2006a,b). Since the local blood flow and the water content of the different tissues (local chloride content, hydrogen diffusion coefficient) can be assumed to be different in various animal models (Table 10.1), the obtained *in vivo* corrosion rates are not directly comparable. Currently, the local *in vivo* corrosion pattern of Mg alloys in various anatomical locations

Table 10.1 The table demonstrates the consistency of the tissue water content in various species, while significant differences can be observed in the local blood flow in different species. Thus, it is influencing the local hydrogen diffusion coefficient

	Mouse		Rabbit		Human	
	Water content (%)	Blood flow (ml/min/100 g)	Water content (%)	Blood flow (ml/min/100 g)	Water content (%)	Blood flow (ml/min/100 g)
Heart	79.0 ± 0.2	39	78.2 – 79.0	50.0 ± 0.8	71.2–80.3	1000
Skin	65.1 ± 0.7	18.9 ± 1.4	54.0 – 67.8	12.7 ± 1.7	67.8–75.8	120
Bone	44.6 ± 1.7	2.3 ± 2.0	39.2 – 58.1	19.1 ± 1.7	43.9	120

Table 10.2 Analytical methods used *in vitro* and *in vivo*

Analytical methods used *in vitro*	Analytical methods used *in vivo*
Hydrogen evolution method	Atomic absorption spectroscopy (AAS)
Electrochemical measurements (linear polarization, EIS)	Atomic emission spectroscopy (AES)
Volume change of the metallic volume of the remaining sample, microtomography	Mass spectrometry with inductively coupled plasma (ICP-MS)
	Laser ablation for solid sampling
	X-ray fluorescence analysis with synchrotron source (XRF)
	Electron beam (SEM-EDX)
	X-ray diffraction (XRD)
	Microtomography
	Neutron activation analysis (NAA)
	Glow-discharge optical emission spectroscopy (GD-OES)

and different mechanical loading situations are under investigation. These results will shed light on the underlying complex corrosion processes of the investigated Mg alloys *in vivo*.

10.4 Methods to characterize *in vivo* corrosion

10.4.1 Element distribution around corroding implants *in vivo*

Various analytical methods have been used to determine the elemental components of biodegradable magnesium alloys (Mg, Al, Li, Zn, REE) in histological sections, bone, tissue and body fluids (Witte *et al.*, 2008b) (see Table 10.2). The application of these methods for trace and ultra-trace analysis in small sample volumes is hampered by several problems. The typical concentrations of the elements mentioned above range from < 1 μg/L to about 1 mg/L in serum and from < 1 mg/kg up to about 500 mg/kg for example in liver and bone. Thus, the sensitivity of several analytical methods is not sufficient (AES, GD-OES, XRF, SEM-EDX) (Witte *et al.*, 2008b). Further limitations are caused by time-consuming sample preparation (AES, OES, ICP-MS), the access to the appropriate method (NAA, synchrotron-based methods), the lack of sufficient lateral resolution for solid sample analysis (GD-OES) or challenging interferences during the measuring process (AAS, AES, ICP-MS, XRF) (Witte *et al.*, 2008b). In detail, phosphate ions in dissolved bone samples may hamper the accurate determination of trace metal concentrations by AAS due to the formation of very stable phosphate compounds. High concentrations of alkaline and earth alkaline elements cause problems in AES measurements due to their strong influence on the line intensities of other elements.

In ICP-MS measurements, signal distortion can occur due to contributions of other ions or molecular ions with the same mass-to-charge (m/z) ratios as the elements which are currently analysed. These effects are known for calcium, phosphorus, iron and zinc. Additionally, oxide formation of REEs could be observed leading to decreased signals and shifts in ICP-MS. Furthermore, X-ray spectra of REE mixtures are characterized by strong signal overlaps when using energy dispersive measurements. Sensitivity problems could be minimized if the samples are completely dissolved. In this case, ICP-MS, with or without pre-concentration of the analyte, will provide good results for most elements of the periodic system. However, locally resolved multi-element analysis of solid samples is still a challenging task. At present, micro-XRF and laser ablation ICP-MS are the most promising methods (Witte *et al.*, 2008b), even though their sensitivity is limited.

10.4.2 Corrosion rate determined from area measurements of the remaining metallic implant

Histological cutting-grinding techniques are currently used to investigate the histological response to biomaterials at the bone–implant interphase. The biological response can be accurately quantified by the level of cellular activity and the amount of existing bone mass around the implant using a method called histomorphometry. The aim of histomorphometry is to understand the 3D reality of the 2D histological sections (An, 1999). The 3D expression from 2D sections can be obtained only if a series of histological section is examined (An, 1999). This technique can also be exploited to analyse magnesium corrosion in bone, even though this method detects only significant differences in the corrosion rate *in vivo* (Witte *et al.*, 2010). The use of histomorphology for Mg corrosion analysis is limited by various factors such as overlapping boundaries of tissue and rough corrosion layer, an imperfect observation angle or brittle and water-soluble corrosion products (Fig. 10.3). These histological techniques are less favourable if magnesium corrosion needs to be determined.

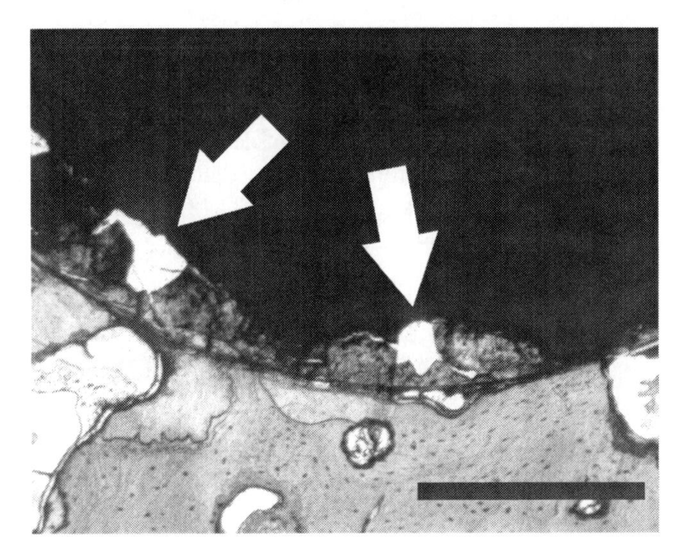

10.3 Uncalcified histological section of an implanted Mg rod in bone tissue. White arrows indicate drop outs in the bone–implant interphase due to technical limitations inherently coupled to the cutting-grinding technique. Scale bar = 400 μm.

10.4.3 General and local corrosion rate determined non-destructively in 3D by SRμCT

Mg corrosion can be determined *in vivo* using microtomography. Synchrotron-based microtomography (SRμCT) is a non-destructive method with a high density and high spatial resolution. Previous studies reported that SRμCT is a non-destructive tool for investigating the interface of materials and bone providing a high resolution and a high accuracy (Bonse *et al.*, 1994; Bonse and Busch, 1996; Bernhardt *et al.*, 2005). In a comparison of different conventional X-ray tubes, SRμCT and conventional histomorphometry, only SRμCT showed a reasonable fit of the bony morphology with the classical histological sections (Bernhardt *et al.*, 2004). Therefore, SRμCT has been investigated in attenuation mode to determine the volume decrease of the implanted metal as well as element-specific SRμCT to determine the spatial distribution of the alloying elements during *in vivo* corrosion (Witte *et al.*, 2006b).

The non-destructive determination of the *in vivo* corrosion rate using SRμCT was performed at beamline HARWI I (W2) and BW2 at HASYLAB at Deutsches Elektronensynchrotron in Hamburg (Germany) (Witte *et al.*, 2006c). The residual implant volume was analysed using VGStudio Max 1.2® Software (Volume Graphics GmbH, Germany). After segmentation of the grey values followed by a 3D region growing method the remaining metallic Mg alloy was separated from the surrounding bone matrix and the corrosion layer (Fig. 10.4).

Thus, the remaining non-corroded metal volume as well as the sample surface was determined in three dimensions non-destructively on a micrometer scale. The reduction of metal implant volume could be converted into a corrosion rate by using a modification of the ASTM G31-72, 2004 equation (10.5) for weight loss measurements:

$$\mathrm{CR} = \frac{W}{A \cdot t \cdot \rho} \qquad 10.5$$

where CR is the corrosion rate (mm/year), W is the weight loss of the metal or alloy, A is the initial surface area exposed to corrosion, ρ is the standard density of the metal or alloy and t is the time of immersion. Herein the weight loss was substituted by the reduction in volume (ΔV) multiplied by the standard density (ρ), resulting in equation (10.6):

$$\mathrm{CR} = \frac{\Delta V}{A \cdot t} \qquad 10.6$$

where ΔV is the reduction in volume that is equal to the remaining metal volume subtracted from the initial implant volume.

This method provides a non-destructively general corrosion rate of the implanted metal. A more local corrosion analysis is possible if the corrosion rates are calculated based on the pitting depth. It was shown that even slowly

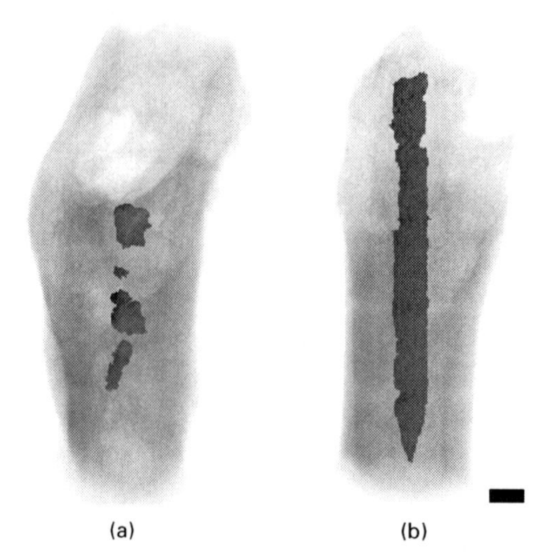

10.4 3D reconstruction of remaining Mg alloy (black, after 18 weeks of implantation) segmented from the bone matrix (grey) by a voxel growing method: (a) AZ91D, (b) LAE442; bar = 1.5 mm. With kind permission from Elsevier (Witte *et al.*, 2006).

corroding Mg alloys as well as the effect of coating *in vivo* could be analysed using a segmented data set of *in vivo* corroded Mg alloy LAE442 (Witte *et al.*, 2010). Measurements of the residual implant volume indicated that the uncoated Mg alloy LAE442 degraded by 10% (3.7 mm^3) of its initial volume after 12 weeks postoperatively, while the MgF_2 coated LAE442 lost about 4% (1.4 mm^3) of its initial volume. SRμCT enables the detection of corrosion *in vivo* for LAE442 and LAE442+MgF_2 as early as 2 weeks postoperatively (Fig 10.5) while differences between the coated and uncoated Mg alloy could be detected as early as 4 weeks postoperatively (Witte *et al.*, 2010). The MgF_2 coated Mg alloy degraded significantly slower than the non-coated Mg alloy (Table 10.3). The corrosion morphology could be determined non-destructively in 3D by SRμCT (Fig. 10.6). Localized corrosion attack occurred on the coated and non-coated Mg alloy (Fig. 10.6). Corrosion occurred as irregular shallow or deep pits that were locally spread over the whole implant surface. The pits were observed to be more frequently aligned to the extrusion direction, while areas which were covered by bone exhibited no or less corrosion attack. In this case, the direct bone contact or bone apposition acted as a protective layer. Even though the initial corrosion rate determined by SRμCT for the MgF_2-coated Mg alloy was low, the corrosion rate of the MgF_2 coated Mg alloy decreased faster to lower corrosion rates than the uncoated Mg alloy during the implantation period (Table 10.4). Both coated and uncoated Mg alloy LAE442 showed low corrosion rates after 12 weeks of implantation (Table 10.4).

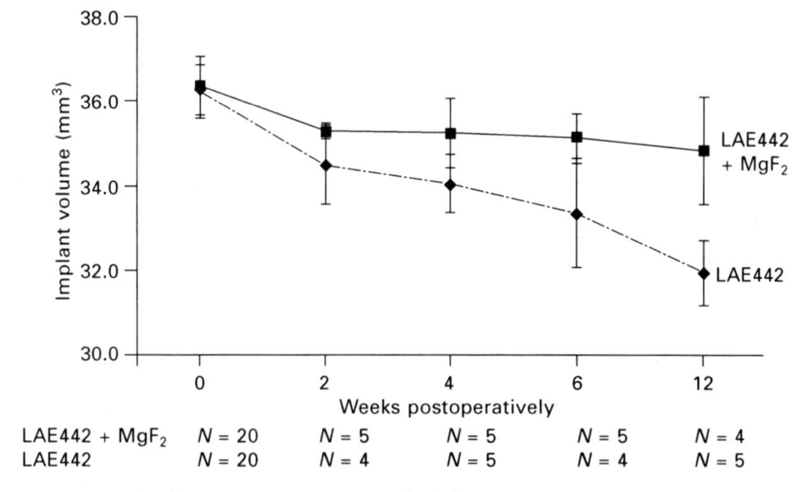

10.5 The implant volume of LAE442 and magnesium fluoride (MgF_2)-coated LAE442 at different postoperative intervals. Results were obtained from SRμCT analysis. The number of analysed samples (*n*) at each time point is stated below the diagram. With kind permission from Elsevier (Witte *et al.*, 2010).

Table 10.3 In vivo corrosion rates (mm/yr) calculated from the volume reduction of the corroding metal implant according to equation 10.6[a]. With kind permission from Elsevier (Witte *et al.*, 2010).

	2 weeks (mm/yr)	4 weeks (mm/yr)	6 weeks (mm/yr)	12 weeks (mm/yr)
LAE442	0.58 ± 0.06	0.46 ± 0.11	0.43 ± 0.10	0.31 ± 0.06
LAE442 + MgF_2	0.40 ± 0.03*	0.29 ± 0.04*	0.14 ± 0.02*	0.13 ± 0.03*

[a] Values presented as mean of all samples of one time interval ± standard deviation.
* Indicates sig. differences ($p < 0.05$) between LAE442 and LAE442 + MgF_2 coating.

10.6 Reconstructed and visualized SRμCT data showing the morphology of *in vivo* corroded magnesium alloys LAE442 (a) and magnesium fluoride-coated LAE442 (b) 12 weeks postoperatively. The 3D reconstruction shows the surface of the residual metallic Mg alloy exhibiting a regular pattern of pitting corrosion on LAE442 samples (a) and more uniform corrosion with singular deep pits with the LAE442 + MgF2 sample. The dominant corrosion morphology appeared as pitting corrosion (a and b). With kind permission from Elsevier (Witte *et al.*, 2010).

The weight loss measurements are usually limited to materials exhibiting uniform corrosion, but might give a more complete picture if additionally reported to local corrosion rates. In case of SRμCT-based *in vivo* corrosion measurements of Mg alloys the weight loss measurements as well as measurements of pit depths can be compared in different biological environments. The pitting factor is the ratio of the deepest pits resulting from corrosion divided by the average pit depth as calculated from weight loss (ASTM G46), with higher values indicating a greater susceptibility to pitting while a pitting factor of one represents uniform corrosion. In a previous study, the pitting factor at 6 weeks was significantly higher with MgF_2-coated LAE442 implants than with uncoated implants (Witte *et al.*, 2009). Interestingly, the 2D measurements of the remaining metallic area also confirmed that MgF_2-coated LAE442 has the tendency to corrode more slowly *in vivo* than the uncoated Mg alloy LAE442. However, significant

Table 10.4 Corrosion rate (mm/yr) based on the maximum pit depth and based on the mean of the pit depths that were calculated from pit measurements on SRμCT tomograms[a]. The pitting factors demonstrate that pitting is the prominent corrosion form in MgF_2 coated LAE442 after 6 and 12 weeks of implantation. With kind permission from Elsevier (Witte *et al.*, 2010).

	2 weeks (mm/yr)	4 weeks (mm/yr)	6 weeks (mm/yr)	12 weeks (mm/yr)
LAE442				
CR $_{(max\ pit\ depth)}$ [a]	2.28 ± 0.40	1.89 ± 0.20	1.61 ± 0.06	1.32 ± 0.12
CR $_{(mean\ pit\ depth)}$ [a]	1.75 ± 0.20	1.58 ± 0.20	1.31 ± 0.06	0.96 ± 0.08
Pitting factor[b]	3.91 ± 0.71	4.54 ± 0.96	3.85 ± 0.84	4.33 ± 0.75
LAE442 + MgF_2				
CR $_{(max\ pit\ depth)}$ [a]	1.76 ± 0.34	1.46 ± 0.24*	1.35 ± 0.29	1.10 ± 0.18
CR $_{(mean\ pit\ depth)}$ [a]	1.29 ± 0.26	0.98 ± 0.10*	1.00 ± 0.22	0.77 ± 0.13
Pitting factor[b]	4.45 ± 1.11	5.26 ± 0.85	9.96 ± 2.29*	8.55 ± 0.52*

[a] Values presented as mean of all samples of one time interval ± standard deviation.
[b] Factors were calculated individually according to ASTM G46 and presented as mean values.
* Indicates sig. differences ($p < 0.05$) between LAE442 and LAE442 + MgF_2 coating.

differences in the remaining metallic implant area between the coated and uncoated Mg alloys could be determined by this 2D measurement as early as 12 weeks postoperatively (Witte *et al.*, 2010). The high standard deviation expresses the limitations with 2D area measurements. Therefore, corrosion rates based on SRμCT are closer to reality than corrosion rates based on 2D measurements on histological serial sections. However, the advantage of histological serial sections is that the tissue and cells can be evaluated at a high lateral resolution. In general, SRμCT seems to be a superior method for the non-destructive evaluation of the corrosion morphology and the determination of the corrosion rate of biodegradable metal implants. The corrosion morphology can be evaluated three-dimensionally, which seems to be an advantage over standard metallurgical and histomorphometrical methods.

10.4.4 Local gas cavity formation accompanying Mg implant corrosion

A common observation from *in vivo* experiments in biodegradable Mg research is the local formation of gas cavities which accompanies the implant corrosion. However, there are contradictory reports on the occurrence of gas cavities subcutaneously while intravasal application showed no local gas accumulation. An explanation for this observation might be based on the diffusion and solubility coefficient of hydrogen in biological tissues which has been widely reviewed (Lango *et al.*, 1996). The solubility of hydrogen in tissues is influenced by the content of lipids, proteins and salinity, but in fat and oils, the solubility seems to be approximately independent of temperature in the physiological range (Piiper *et al.*, 1962; Lango *et al.*, 1996). Not only viscosity but also different tissue components and structures such as lipids, proteins and glycosaminoglycans influence the numeric value of the hydrogen diffusion coefficient (Vaupel, 1976; Lango *et al.*, 1996). Depending on experimental configuration, the diffusion coefficient may be underestimated in both stagnant and flowing media due to a boundary layer formation, which increases the effective diffusion distance (Lango *et al.*, 1996). This finding might be important for intravascular Mg applications. Correlating the hydrogen diffusion coefficients from various biological media having fractional water contents from about 68% to 100% demonstrated that the diffusion coefficient of hydrogen increases exponentially with the increasing water fraction of the tissue (Vaupel, 1976). Table 10.1 demonstrates that the tissue water content increases from adipose tissue to skin to bone and to muscles in animals and humans, but is similar for the same tissue regardless of the species. This might explain why different corrosion rates and gas cavities were observed for Mg alloys in different anatomical implantation sites (Wen *et al.*, 2004; Witte *et al.*, 2005, 2007b; Xu *et al.*, 2007).

In an animal study with rats, it was shown that the adsorption of hydrogen gas from subcutaneous gas pockets was limited by the diffusion coefficient of hydrogen in the tissue; the overall hydrogen adsorption rate was determined as 0.954 ml per hour (Piiper *et al.*, 1962). Thus, the local blood flow and the water content of the tissue surrounding the implant are the most important parameters which should be considered in designing biodegradable Mg alloys with an appropriate corrosion rate. Concomitantly, it can be assumed that local hydrogen cavities occur when more hydrogen is produced per time interval than can be dissolved in the surrounding tissue or diffuse from the implant surface into the extracellular medium which is renewed depending on the local blood flow. This means that Mg alloys are corroding *in vivo* with an appropriate corrosion rate when no local gas cavities are observed during the implantation period in a specific anatomical site.

10.5 *In vitro* corrosion test methods

When dealing with materials for biomedical applications one has to deal with a more complex environment than for technical applications. Although this includes a quite defined surrounding for a specific tissue, the problem is that many of the parameters found in such tissues are not well described. Additionally, the biological reactions to a degrading material are poorly understood. This can be observed, for example, for degradable polymers, of which only polylactid acid (PLA) and polyglycolic acid (PGA) have reached the state of certification by ASTM and are approved by the US Food and Drug Administration (FDA), although many others have been in clinical application for some decades.

10.5.1 Basic and technical tests for *in vitro* corrosion

The basic *in vitro* tests derived from technical applications are the salt spray test (ASTM B-117) and the submersion test, performed in saline solutions (3.5% NaCl, ASTM G31-72 (2004) Standard Practice for Laboratory Immersion Corrosion Testing of Metals). Both tests are conducted at room temperature. By such tests the mass loss can be determined according to equation (10.5). The mass loss is calculated after removing the corrosion products with chromic acid, which at the same time removes the corrosion products and inhibits further corrosion (Lorking, 1964), which is suitable for technical applications but may raise problems in biological environments.

An alternative test is the hydrogen evolution method (Kray, 1934). This test is based on the fact that during Mg dissolution an equal part of hydrogen is produced (equation 10.7). By collecting the evolving hydrogen gas the corrosion rate can be determined easily and rapidly (Song and Atrens, 2003) with material available in most laboratories (Fig. 10.7).

$$2Mg + 2H^+ + 2H_2O \rightarrow 2Mg^{2+} + 2OH^- + 2H_2 \qquad 10.7$$

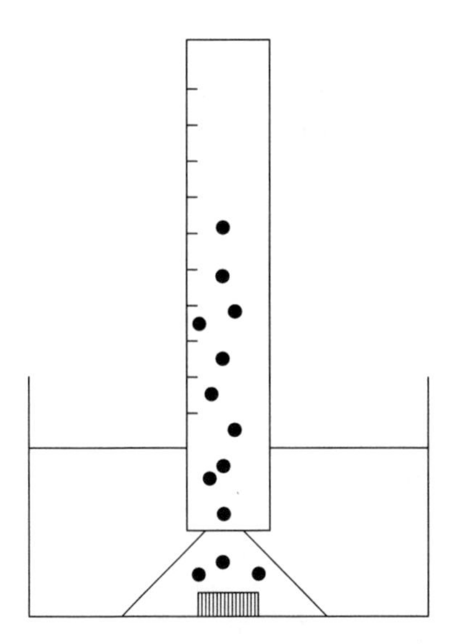

10.7 Schematic drawing of the setup for measuring corrosion rates by the hydrogen evolution method. The Mg sample is immersed below a funnel which is directing the gas bubbles into a burette which is filled with a solution.

The calculation of the corrosion rate in mm per year is done by converting the total amount of collected hydrogen into material loss (1 ml H_2 gas = 0.001083 g dissolved Mg) and using equation (10.8) (weight change Δg in g, surface area A in cm^2, time t in h, density of the alloy ρ in g/cm^{-3}). Corrosion rates determined by hydrogen evolution and mass loss methods show a very good correlation (Song and Atrens, 2003; Hort *et al.*, 2010), independently of the use of pure magnesium or Mg alloys.

$$CR = \frac{8.76 \cdot 10^4 \cdot \Delta g}{A \cdot t \cdot \rho} \qquad 10.8$$

One advantage of this method is the option for time-resolved measurements. By measuring at different time points, variations in the corrosion rate can be determined. As the mass loss method is an endpoint measurement, it is not possible to use it for time-resolved experiments. For the measurement of the evolving hydrogen a more sophisticated instrumentation is necessary. Eudiometers (as recommended by DIN 38414-8) are less prone to experimental errors as well as being highly standardized and calibrated. They also offer the option for automation which increases the sample number of parallel truly online measurements.

10.5.2 Measurement of element and ion release in the corrosion solution

The characterization of ion and element release may be crucial for some applications mainly in the biomedical field. There are various methods available and the most common ones are listed in this section. The easiest setup is the corrosion of pure Mg samples, but there are also manifold methods to determine the release kinetics of alloying elements even though they are more challenging to analyse.

10.5.3 Osmolality measurements

Osmolality measurements are very basic measurements without the possibility of quantifying elements in solution. This method determines the amount of ions in solutions by changing the freezing point. In clinical applications, osmolality measurements are normally used to determine blood plasma osmolality. However, for complex solutions with undetermined components, osmolality measurements are a very fast, simple and reliable method. If a calibration is performed by increasing concentrations of MgC_{l2}, the osmolality measurements can be used to quantify pure magnesium corrosion. The measurement range is from 0 to about 3000 mOsm/kg.

10.5.4 Titration

The standard method to quantify element release is titration. There are standard procedures for most commonly used elements. Therefore titration can be recommended for pure magnesium or binary alloys with alloying elements for which established titration methods are available. The advantage is the relative low cost of the instrumentation (beaker and burette). The disadvantages are the use of special chemicals and the time-consuming procedure. For the titration of Mg ions calcium ions must be excluded. If calcium ions are present a further titration step has to be included.

10.5.5 Ion-selective electrodes (ISE)

The determination of magnesium can also be performed by using ion-selective electrodes (ISE). Although the electrodes are called 'ion specific' there is interference with other ions. This is especially true for magnesium and calcium. Therefore the method includes two electrodes: one for magnesium and calcium (water hardness) and one for calcium. The Mg concentration can be determined by the subtraction of the calcium measurement from the water hardness value. The measurement is fast and easy, the major cost for this method is related to the electrodes. The basic principle of the measurement is the conversion of ion activity to an electric potential (Bobacka *et al.*, 2008). Even though the measurement range is very broad, both electrodes are able to measure concentrations between 5×10^{-6} M (0.02 ppm) and 1 M (40 000 ppm). This technique is very promising because of the development of microelectrodes for cellular measurements and for *in vivo* experiments (Günzel and Schlue, 2002; Bobacka *et al.*, 2008).

10.5.6 Sophisticated methods: availability and challenges

There are various methods derived from trace or ultratrace analysis of elements available. However, the instruments are expensive; sample preparation may be difficult or has to be done very carefully to avoid even minute contamination, or the method is only available at neutron or synchrotron sources with limited access.

10.5.7 Applicability and modification of submersion test: solutions, proteins and environment

The corrosion resistance of magnesium and its alloys is highly dependent on the purity of the material. A high-purity alloy can exhibit 10–100 times higher corrosion resistance than standard purity alloy in salt solutions (Song and Atrens, 1999). However, salt solutions do not represent the physiological or biological environment found at the implant site. Moreover, anatomical implant sites differ highly in their properties, e.g. the hydrogen diffusion coefficient or the water fraction (Witte *et al.*, 2008b). Therefore, more physiological solutions must be used to approach the physiological conditions. A standardized mixture of simulated body fluid (SBF) was recommended by Bohner and Lemaitre (2009), because there are several SBF solutions with slight variations in compositions (Kokubo *et al.*, 1987; Tas, 2000; Oyane *et al.*, 2003), leading to non-standardized conditions and incompatibility of the results. The approach to simulate the inorganic compound of blood plasma already is a step in the right direction; however, in the physiological surrounding proteins are also widely abundant.

Only a few studies on the influence of proteins are available. Liu *et al.* (2007) found a significant increase in corrosion resistance just by adding albumin to SBF. A clear influence and dose-dependence of albumin was also reported by Mueller *et al.* (2009). In another study the corrosion resistance of magnesium in physiological solutions was not measured and the composition of the corrosion layer was analysed (Rettig and Virtanen, 2009). No differences in the corrosion resistance were observed after extensive exposure (days) to physiological solutions with or without proteins (Rettig and Virtanen, 2009). Our own studies indicate a clear delay of corrosion by proteins and the evaluation of the corrosion mechanism under the influence of protein mixtures is an important topic in current research. Therefore, a reliable *in vitro* setup should contain a certain amount of proteins. For standardization purposes undefined mixtures such as fetal bovine serum (FBS) should not be used, defined protein mixtures with the most important proteins in the targeted tissues being preferred.

10.6 Future trends

The first hurdle to developing Mg alloys that corrode *in vivo* without clinically visible hydrogen gas cavities has been already taken. The solution is based on a proper selection of alloying elements combined with an extrusion process and can be further combined by coating options. Another approach to obtain very slow corroding Mg alloys *in vivo* is based on metallic Mg glasses which contain a zinc content higher than 30 at.%. However, it has to be kept in mind that the *in vivo* corrosion rate will decrease after implantation while a more complex and stable corrosion layer is formed on the corroding implant. Thus, previous developments in polymer and calcium phosphate ceramics reveal that very slow degrading biomaterials have a high risk of remaining at least partly in the tissue, hence contradicting the initial idea to create temporary implant materials. Therefore, the Mg corrosion rate still needs to be adapted to the special requirements of the specific biomedical application. In the near future, more advanced *in vitro* tests systems will be developed which resemble more accurately the *in vivo* conditions and are also suitable for high content screening. This test system development is necessary, since many different cell types will be in contact with the corroding implant and its coating or its corrosion products.

10.7 References

An, Y. H., Friedman, R. J. (1999) *Animal Models in Orthopeaedic Research*, Boca Raton, FL, CRC Press.

Avedesian, M. M. & Baker, H. (1999) *ASM Specialty Handbook – Magnesium and Magnesium alloys*, ASM International, Materials Park, OH, USA.

Bernhardt, R., Scharnweber, D., Muller, B., Thurner, P., Schliephake, H., Wyss, P., Beckmann, F., Goebbels, J. & Worch, H. (2004) Comparison of microfocus- and synchrotron X-ray tomography for the analysis of osteointegration around Ti6Al4V implants. *Eur Cell Mater*, **7**, 42–51; discussion 51.

Bernhardt, R., van den Dolder, J., Bierbaum, S., Beutner, R., Scharnweber, D., Jansen, J., Beckmann, F. & Worch, H. (2005) Osteoconductive modifications of Ti-implants in a goat defect model: characterization of bone growth with SR muCT and histology. *Biomaterials*, **26**, 3009–19.

Bobacka, J., Ivaska, A. & Lewenstam, A. (2008) Potentiometric ion sensors. *Chem Rev*, **108**, 329–51.

Bohner, M. & Lemaitre, J. (2009) Can bioactivity be tested *in vitro* with SBF solution? *Biomaterials*, **30**, 2175–9.

Bonse, U. & Busch, F. (1996) X-ray computed microtomography (microCT) using synchrotron radiation (SR). *Prog Biophys Mol Biol*, **65**, 133–69.

Bonse, U., Busch, F., Gunnewig, O., Beckmann, F., Pahl, R., Delling, G., Hahn, M. & Graeff, W. (1994) 3D computed X-ray tomography of human cancellous bone at 8 microns spatial and 10(-4) energy resolution. *Bone Miner*, **25**, 25–38.

Di Mario, C., Griffiths, H., Goktekin, O., Peeters, N., Verbist, J., Bosiers, M., Deloose, K., Heublein, B., Rohde, R., Kasese, V., Ilsley, C. & Erbel, R. (2004) Drug-eluting bioabsorbable magnesium stent. *J Interv Cardiol*, **17**, 391–395.

Erbel, R., Di Mario, C., Bartunek, J., Bonnier, J., de Bruyne, B., Eberli, F. R., Erne, P., Haude, M., Heublein, B., Horrigan, M., Ilsley, C., Bose, D., Koolen, J., Luscher, T. F., Weissman, N. & Waksman, R. (2007) Temporary scaffolding of coronary arteries with bioabsorbable magnesium stents: a prospective, non-randomised multicentre trial. *Lancet*, **369**, 1869–75.

Feyerabend, F., Witte, F., Kammal, M. & Willumeit, R. (2006) Unphysiologically high magnesium concentrations support chondrocyte proliferation and redifferentiation. *Tissue Eng*, **12**, 3545–556.

Günzel, D. & Schlue, W.-R. (2002) Determination of [Mg2+]i – an update on the use of Mg^{2+}-selective electrodes. *BioMetals*, **15**, 237–249.

Hort, N., Huang, Y., Fechner, D., Stormer, M., Blawert, C., Witte, F., Vogt, C., Drucker, H., Willumeit, R., Kainer, K. U. & Feyerabend, F. (2010) Magnesium alloys as implant materials – principles of property design for Mg-RE alloys. *Acta Biomater*, **6**, 1714–1725.

Kainer, K., Hort, N., Willumeit, R., Feyerabend, F. & Witte, F. (2009) Magnesium alloys for the design of medical implants. In Niinomi, M., Morinaga, M., Nakai, M., Bhatnagar, N. & Srivatsan, T. (Eds.) *Proc. PFAM XVIII – Processing and Fabrication of Advanced Materials.* Sendai, Japan.

Kaya, R. A., Cavusoglu, H., Tanik, C., Kaya, A. A., Duygulu, Ö., Mutlu, Z., Zengin, E. & Aydin, Y. (2007) The effects of magnesium particles in posterolateral spinal fusion: an experimental *in vivo* study in a sheep model. *J Neurosurgery: Spine*, **6**, 141–9.

Kokubo, T., Ito, S., Shigematsu, M., Sanka, S. & Yamamuro, T. (1987) Fatigue and life-time of bioactive glass-ceramic A-W containing apatite and wollastonite. *J Mater Sci*, **22**, 4067–70.

Kray, R. H. (1934) Modified hydrogen evolution method for metallic magnesium, aluminum, and zinc. *Ind & Engi Chem Anall Edn*, **6**, 250–51.

Lango, T., Morland, T. & Brubakk, A. O. (1996) Diffusion coefficients and solubility coefficients for gases in biological fluids and tissues: a review. *Undersea & Hyperbaric Medicine*, **23**, 247–72.

Liu, C., Xin, Y., Tian, X. & Chu, P. K. (2007) Degradation susceptibility of surgical magnesium alloy in artificial biological fluid containing albumin. *J Mater Res*, **22**, 1806–814.

Lorking, K. F. (1964) Inhibition of corrosion of magnesium in chromic acid. *Nature*, **201**, 75.

Mantovani, D. & Witte, F. (2010) Editorial to Special Issue on Biodegradable Metals. *Acta Biomater*, **6**, 1.

Mueller, W.-D., Mele, M. F. L. D., Nascimento, M. L. & Zeddies, M. (2009) Degradation of magnesium and its alloys: dependence on the composition of the synthetic biological media. *J Biomed Mater Res Part A*, **90A**, 487–95.

Oyane, A., Kim, H. M., Furuya, T., Kokubo, T., Miyazaki, T. & Nakamura, T. (2003) Preparation and assessment of revised simulated body fluids. *J Biomed Mater Res A*, **65**, 188–95.

Piiper, J., Canfield, R. E. & Rahn, H. (1962) Absorption of various inert gases from subcutaneous gas pockets in rats. *J Appl Physiol*, **17**, 268–274.

Rettig, R. & Virtanen, S. (2009) Composition of corrosion layers on a magnesium rare-earth alloy in simulated body fluids. *J Biomed Mat Res Part A*, **88A**, 359–69.

Shaw, B. A. (2003) Corrosion resistance of magnesium alloys. In Cramer, S. D. & Covino, B. S., Jr (Eds.) *ASM Handbook volume 13a: corrosion: fundamentals, testing and protection*. ASM International, Materials Park, OH, USA.

Song, G. L. & Atrens, A. (1999) Corrosion mechanisms of magnesium alloys. *Advanced Engig Mater*, **1**, 11–33.

Song, G. & Atrens, A. (2003) Understanding magnesium corrosion – a framework for improved alloy performance. *Adv Engig Mater*, **5**, 837–858.

Tas, A. C. (2000) Synthesis of biomimetic Ca-hydroxyapatite powders at 37 degrees C in synthetic body fluids. *Biomaterials*, **21**, 1429–38.

Vaupel, P. (1976) Effect of percentual water-content in tissues and liquids on diffusion-coefficients of O_2, CO_2, N_2, and H_2. *Pflugers Archiv–European J Physiol*, **361**, 201–204.

Waksman, R., Pakala, R, Baffour, R., Hellinga, D, Seabron, R., Tio, F. O. (2005) Bioabsorbable magnesium alloy stents attenuate neointimal formation in; a porcine coronary model. *Circulation*, **112**, U539.

Waksman, R., Pakala, R, Wittchow, E., Hartwig, S, Harder, C., Rohde, R, Heublein, B., Haverich, A, Andreae, A., Waldman, K. H. (2006a) Effect of magnesium alloy stents in porcine coronary arteries: morphometric analysis of a long-term study. *JACC*, **47**, 23B.

Waksman, R., Pakala, R., Kuchulakanti, P. K., Baffour, R., Hellinga, D., Seabron, R., Tio, F. O., Wittchow, E., Hartwig, S., Harder, C., Rohde, R., Heublein, B., Andreae,

A., Waldmann, K.-H. & Haverich, A. (2006b) Safety and efficacy of bioabsorbable magnesium alloy stents in porcine coronary arteries. *Catheterization and Cardiovascular Interventions*, **68**, 607–17.

Wen, C. E., Yamada, Y., Shimojima, K., Chino, Y., Hosokawa, H. & Mabuchi, M. (2004) Compressibility of porous magnesium foam: dependency on porosity and pore size. *Mater Letts*, **58**, 357–360.

Witte, F., Kaese, V., Haferkamp, H., Switzer, E., Meyer-Lindenberg, A., Wirth, C. J. & Windhagen, H. (2005) *In vivo* corrosion of four magnesium alloys and the associated bone response. *Biomaterials*, **26**, 3557–63.

Witte, F., Dargel, R., Bechstein, K., Schade, U. & Vogt, C. (2006a) Synchrotron-based IR-microspectroscopy – a tool for the detection of organic structures in the vicintiy of implanted magnesium-alloys. *ORS Trans*, **52**, 939.

Witte, F., Fischer, J., Nellesen, J. & Beckmann, F. (2006b) Microtomography of magnesium implants in bone and their degradation. *Progr Biomedi Optics Imaging – Proc SPIE*. San Diego, CA.

Witte, F., Fischer, J., Nellesen, J., Crostack, H. A., Kaese, V., Pisch, A., Beckmann, F. & Windhagen, H. (2006c) *In vitro* and *in vivo* corrosion measurements of magnesium alloys. *Biomaterials*, **27**, 1013–18.

Witte, F., Reifenrath, J., Müller, P. P., Crostack, H.-A., Nellesen, J., Bach, F. W., Bormann, D. & Rudert, M. (2006d) Cartilage repair on magnesium scaffolds used as a subchondral bone replacement. *Mat.-wiss. u. Werkstofftech*, **37**, 504–8.

Witte, F., Ulrich, H., Palm, C. & Willbold, E. (2007a) Biodegradable magnesium scaffolds: Part II: peri-implant bone remodeling. *J Biomed Mater Res A*, **81**, 757–65.

Witte, F., Ulrich, H., Rudert, M. & Willbold, E. (2007b) Biodegradable magnesium scaffolds: Part 1: appropriate inflammatory response. *J Biomed Mater Res A*, **81**, 748–56.

Witte, F., Abeln, I., Switzer, E., Kaese, V., Meyer-Lindenberg, A. & Windhagen, H. (2008a) Evaluation of the skin sensitizing potential of biodegradable magnesium alloys. *J Biomedi Mater Res Part A*, **86A**, 1041–7.

Witte, F., Hort, N., Vogt, C., Cohen, S., Willumeit, R., Kainer, K. U. & Feyerabend, F. (2008b) Degradable biomaterials based on magnesium corrosion. *Current Opin Solid State Mater Sci*, **12**, 63–72.

Witte, F., Fischer, J., Nellesen, J., Vogt, C., Vogt, J., Donath, T. & Beckmann, F. (2010) *In vivo* corrosion and corrosion protection of magnesium alloy LAE442. *Acta Biomater*, **6**, 1792–1799.

Xu, L., Yu, G., Zhang, E., Pan, F. & Yang, K. (2007) *In vivo* corrosion behavior of Mg–Mn–Zn alloy for bone implant application. *J Biomed Mater Res A*, **83**, 703–11.

Zberg, B., Uggowitzer, P. J. & Loffler, J. F. (2009) MgZnCa glasses without clinically observable hydrogen evolution for biodegradable implants. *Nature Mater*, **8**, 887–891.

Zhang, G. D., Huang, J. J., Yang, K., Zhang, B. C. & Ai, H. J. (2007) Experimental study of *in vivo* implantation of a magnesium alloy at early stage. *Acta Metallurgica Sinica*, **43**, 1186–90.

11
Corrosion of magnesium (Mg) alloys in engine coolants

G.-L. SONG, General Motors Corporation, USA and
D. H. StJOHN, The University of Queensland, Australia

Abstract: The corrosion of engine blocks by engine coolant is a critical issue for the automotive industry. In this chapter, after a summary of some fundamental aspects of the corrosion of pure magnesium in ethylene glycol solutions, a review of the corrosion performance of magnesium alloys in ethylene glycol solutions is presented. Based on the knowledge gained, the corrosion behaviour of AZ91D and some recently developed magnesium engine block alloys in several selected commercial coolants is assessed by means of immersion testing, hydrogen evolution measurement, galvanic current monitoring and the well-accepted ASTM D1384 standard test. Finally, a corrosion inhibition strategy is presented and inhibitors suitable for magnesium alloys in coolants are identified.

Key words: magnesium alloys, corrosion, coolant.

11.1 Introduction

Magnesium (Mg) alloys are promising structural materials for the automotive, aerospace and electronic industries. Particularly in automotive applications, a magnesium engine block can significantly reduce the weight of an automobile and therefore its fuel consumption and environmental impact. In the near future, magnesium alloys may become important engine materials for the automotive industry. For potential applications, new magnesium alloys have recently been developed [1,2]. For example, a three-cylinder turbo diesel engine with a lightweight AM-SC1 engine block was successfully developed by AVL List using a proprietary coolant additive that was supplied by BASF [3].

Since magnesium alloys are more susceptible to corrosion attack in aqueous solutions than iron and aluminum-based alloys, corrosion caused by contact with the engine coolant is a major concern if magnesium alloys are to be used for engine blocks. To ensure successful commercialization of magnesium engine blocks it is important to understand the corrosion performance of magnesium alloys in coolants. Unfortunately, even though the corrosion mechanisms of magnesium alloys in aqueous media have been widely investigated recently [4–7], relevant studies on the corrosion mechanisms of magnesium in coolants are quite rare [8–10]. Currently, there is little confidence in answering some critical questions with regard to a magnesium alloy's ability to survive in conventional coolants that have been developed for non-magnesium alloys. It is unclear whether a suitable coolant is already available in the market that works for magnesium alloys. Therefore, it is important to understand the corrosion of magnesium alloys when in contact with engine coolant and then, based on that knowledge, develop or search for a suitable coolant for magnesium alloy engine blocks.

11.2 Magnesium (Mg) alloys and coolants

Magnesium alloys are not currently used as engine block materials containing coolant channels. Existing standards and regulations for cast iron and aluminium engine systems cannot be directly applied to a magnesium engine system. There are no particular regulations or specifications to follow regarding the selection of suitable alloys and coolants for a magnesium engine block.

11.2.1 Magnesium engine block alloys

So far, there is limited power-train experience with magnesium alloys. Owing to critical mechanical and thermal requirements, such as high-strength stability, creep resistance, fatigue performance and thermal conductivity not many commercial magnesium alloys can be used as an engine block material. Table 11.1 lists some basic laboratory property specifications. Before sufficient experience has been gained by the auto industry regarding the use of magnesium engine blocks, Table 11.1 is a good starting point.

Currently, only a few magnesium alloys have been evaluated or tested in the laboratory for application in a magnesium engine block. These are EZ33, WE43, AM-SC1, AS21X, AS31, AXJ530, AJ52X, AJ62, MRI153M, MRI230D, ML10 and MRI202S. Generally speaking, those magnesium alloys that contain a high concentration of rare earth elements and relatively easily meet the property specifications. The cheaper aluminium-containing

Table 11.1 Critical property specifications for a sand-cast Mg engine block alloy over a range of operating temperatures [2]

	Operating temperature		
	Room temp	150 °C	177 °C
0.2% proof stress (MPa)	120		110
Creep strength (MPa)		110	90
Fatigue limit (MPa)	50		
Thermal conductivity (W/km)	115		

magnesium alloys normally do not have good high-temperature properties and have been limited to some low-temperature power-train applications.

11.2.2 Engine coolants

Commercial coolants are solutions of ethylene glycol (30~70 vol.%) in water combined with additives. The additives include corrosion inhibitors, buffers, lubricants and anti-foaming agents. Many brands of coolants have been developed and are commercially available for aluminium and cast iron engine blocks. Although there was effectively no change in coolants when the automotive industry went from iron to aluminium engine blocks, innovation in coolants is required this time for the industry to be able to move into a magnesium engine block era, as magnesium is 'special'.

The main composition of a conventional coolant is ethylene glycol ($C_2H_6O_2$). Its molar mass is 62.068 g/mol, density 1.1132 g/cm^3, melting point –12.9 °C and boiling point 197.3 °C. It can mix with water in all proportions and the boiling point decreases with increasing water content. The freezing point of an ethylene glycol solution can reach a minimum –51 °C when the water content is 60 vol.%. Therefore, ethylene glycol solutions have a wide operating range as a coolant.

The critical component of a commercial coolant is its inhibitor package. It is well known that different inhibitors can lead to different inhibition performance. Considerable research has been carried out in the area of coolant corrosion and inhibition [11–16].

Traditional coolants usually contain molybdate, phosphate, borate, nitrate, nitrite, tolyltriazole, benzoate silicate, etc., as inhibitors. In addition, some organic acid-based long-life coolants have recently emerged in the market, which contain organic acid carboxylate ions as the main inhibitors [17–19]. In summary [20,21] most current coolants include silicate for high temperature corrosion protection of Al; European coolants more likely contain a borate–benzoate plus nitrite, nitrate, silicate, triethanol amine, phosphoric acid, sodium mercaptobenzothiazole; and the triazole and the nitrite are currently being replaced by salts of organic aliphatic acids. In the USA, coolants contain a large amount of phosphate and varying amounts of nitrate, silicate and triazole. Currently, other inhibitors are being investigated, such as disodium sebacate, the sodium salt of decandioic acid, monobasic and dibasic carboxylic acids, etc., and these may have recently emerged in the market.

The introduction of formulations that use organic acid salts as the major inhibitors opened new prospects for long-life coolants. In the organic acid-based long-life coolants, carboxylate ions determine the inhibition performance. An organic species is likely to be adsorbed on a metal surface. The adsorbed organic species more or less act as inhibitors and their degree of inhibition depends on the coverage of the adsorptive film on the metal. It has been reported that an inhibitor package free of silicate, nitrate, borate and phosphate, containing aliphatic mono- and di-carboxylic acids and tolyltriazole that deplete much more slowly than conventional inhibitors, can significantly extend the service life of a coolant [22].

From a long-term perspective, a coolant should have at least the following properties in addition to corrosion inhibition: (1) be insensitive to hard water (phosphate-free); (2) be resistant to silica gelation for long-term storage and water pump sealing (silicate free); (3) have reduced toxicity and environmental hazard. It is clear that the development of organic-based long-life coolants will continue to be an important research direction in order to further reduce coolant corrosion.

11.3 Laboratory evaluation methodology

There are many standard testing methods for evaluating the corrosion performance of a material in coolants, including large-scale pilot tests to laboratory-scale simulation tests. As magnesium alloys for engine block applications are still in an early stage of development, laboratory screening of suitable coolants is an essential step in the successful selection of coolants. In this chapter, only laboratory evaluation methodologies are discussed.

11.3.1 Coolant solutions for laboratory tests

Many coolant relevant solutions can be tested in the laboratory. Generally speaking, these solutions can be classified into three groups: (1) ethylene glycol solutions; (2) commercial coolants; and (3) modified and newly developed coolants.

The 'basic' ethylene glycol solutions can be made from AR grade ethylene glycol and demineralized water. Corrosive (aggressive) water (ASTM D1384-96 [23]) can also be used to make corrosive (aggressive) ethylene glycol solutions. The aggressive water solution can be prepared by dissolving 148 mg Na_2SO_4, 165 mg NaCl and 138 mg $NaHCO_3$ into ASTM type II demineralized water, and the resulting solution is then made up to 1 L with the demineralized water. NaCl, Na_2SO_4, $NaHCO_3$, $Mg(OH)_2$, $Mg(NO_3)_2$, KF, etc. can also be added into the basic ethylene glycol solutions for the investigation of the effects of these chemicals on the corrosion performance of magnesium.

Commercial coolants are those currently available in the market. Solutions are prepared by simply following the product instructions for use in a vehicle radiator. Commercial coolants can be modified by the addition of new inhibitors to make them suitable for magnesium alloys. New coolants can also be developed specifically for magnesium alloys and can also be evaluated in the laboratory.

11.3.2 Immersion, hydrogen collection and weight loss measurement

The combination of immersion, hydrogen collection and weight loss measurement is an easy corrosion evaluation method, particularly for magnesium alloys. The method, first established and used by Song *et al.* [5] to estimate and monitor the corrosion rate of magnesium in a NaCl solution, has been widely adopted as a common corrosion rate measurement technique for magnesium alloys in various aqueous solutions. The reliability of the method has been theoretically and experimentally demonstrated [24,25] and the details will not be repeated here.

Before immersion, the specimens are polished with SiC paper, cleaned with distilled water and acetone, dried and weighed for the original weight (w_0), and then immersed in beakers that contain 500 ml of coolant at controlled temperatures. The duration of immersion is normally 2 weeks. At the same time, hydrogen evolved from the magnesium specimens is collected. After immersion, the loose corrosion products on the specimen surfaces are removed with a brush. The corrosion products that could not be brushed off are removed by immersing these magnesium specimens in a chromic solution (200 g/L CrO_3 + 10 g/L $AgNO_3$) for 5–10 minutes. The specimens are subsequently washed in demineralized water, dried and weighed again to obtain the final weight (w_1). The weight loss caused by corrosion in the coolant was then calculated (w_0–w_1).

11.3.3 Galvanic current measurement

Galvanic corrosion within the engine cooling system is one of the biggest concerns. It is quite possible that a magnesium engine block has an Al alloy head. The galvanic effect between magnesium alloys and the aluminium alloy should be evaluated in the laboratory. The setup shown in Fig. 11.1 is quite simple. Magnesium and aluminium alloy electrodes are immersed in a beaker of the coolant, face to face with a fixed distance (2 cm) between them. The galvanic current between these two electrodes is then monitored.

Recently, a sandwich-like assembly consisting of aluminium, stainless steel and magnesium coupons, which simulates a magnesium engine block with an aluminium head and a steel gasket inserted between the two alloys, has also been used by Zhang *et al.* to evaluate the galvanic corrosion of magnesium in a coolant [9].

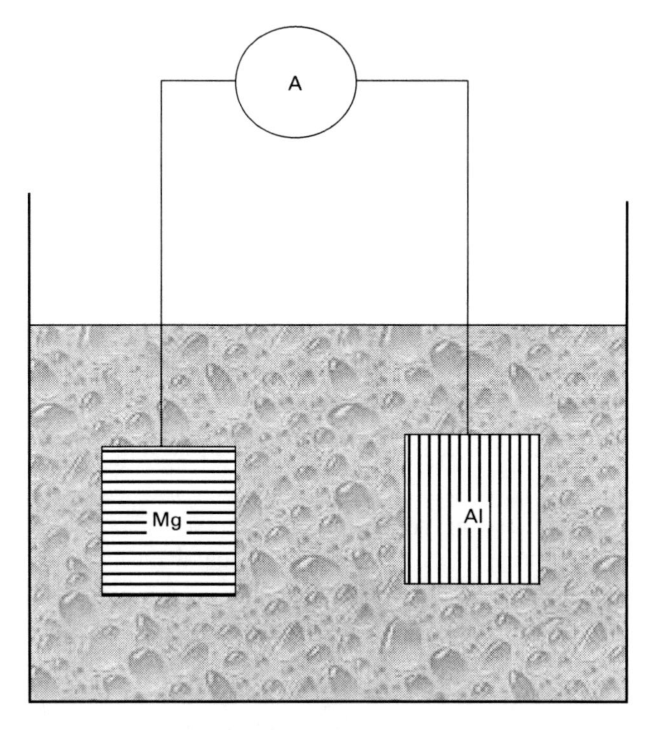

11.1 Schematic illustration of a setup for galvanic current measurement (based on Song and StJohn [26]).

11.3.4 Polarization and electrochemical impedance spectrum (EIS)

In the laboratory, electrochemical techniques can quickly compare corrosion rates of different magnesium alloys in different coolant solutions. More importantly, electrochemical results are useful for elucidating the corrosion mechanism of a magnesium–coolant system. Polarization curves and electrochemical impedance spectra (EIS) of a magnesium alloy can be measured in an electrolytic cell containing about 500 mL ethylene glycol solution using an electrochemical measurement system. The polishing and cleaning of electrodes should be the same as specified in the above hydrogen evolution and weight loss measurements. The polarization starts from a cathodic potential of about –700 to –800 mV relative to the corrosion potential and stops at an anodic potential of +1000 to +1500 mV relative to the corrosion potential. The scanning rate is set at 10 mV/min. AC impedance measurements can be conducted at the open-circuit potentials (OCP). The amplitude of AC signals is normally 5 mV, and the measured frequency range is from 1 mHz to 1 kHz.

11.3.5 ASTM standards (D1384 and D4340)

ASTM D1384 and D4340 are two of the most popular laboratory standards. ASTM D1384 was designed to simulate the coolant circuit in an engine assembly. This test mainly evaluates the galvanic effect on induced corrosion damage in the cooling system. Commercially made coupon bundles as described in ASTM D1384 can be purchased. It needs to be slightly modified with a magnesium plate inserted into the standard bundle for laboratory use.

ASTM D4340 was originally established to evaluate the effectiveness of engine coolants in combating corrosion of Al alloys with the coolant heat-transfer conditions set at 135°C under a pressure of 193 kPa. Crevice and cavitation corrosion damage can also be assessed by carefully analysing the sample surface and weight loss measurement after a 1 week test. Certainly, the standard can be very easily adapted to evaluate the corrosion of magnesium alloys under the same testing conditions by simply replacing the Al alloy disk with a magnesium alloy one in the experimental set-up.

11.3.6 Acceptable corrosion threshold for laboratory evaluation

There is no specific criterion for an acceptable laboratory corrosion rate for magnesium alloys in coolants. However, according to ASTM D4340, a rate of weight loss greater than 1 mg/cm^2/week for an aluminium alloy is considered to be unacceptable. Considering that the same volume loss caused by corrosion should be used as the criterion for the acceptance of a magnesium alloy, 0.67 mg/cm^2/week is therefore regarded as an acceptable threshold for magnesium alloys [26,27].

For galvanic corrosion, 0.67 mg/cm^2/week corresponds to a current density of 8.8 μA/cm^2. Therefore, a galvanic current density of 8.8 μA/cm^2 is regarded as an acceptable threshold for magnesium when tested in the current laboratory galvanic measurement test.

11.4 Corrosion of magnesium (Mg) in ethylene glycol solution

Since the primary component of an engine coolant is ethylene glycol and magnesium is the basis of magnesium alloys, pure magnesium in ethylene glycol can be regarded as a simplified representative corrosion system. Results from this system will reveal the corrosion behaviour of magnesium alloys in coolants. Moreover, an engine block is not operating all the time. While the engine is not running, the alloy is simply exposed to static coolant at room temperature. The corrosion performance of an engine material immersed in a coolant at room temperature should also have a vital contribution to the service life. Therefore, fully understanding the corrosion behaviour and mechanism of magnesium immersed in ethylene glycol solutions is an essential step prior to identifying successful coolants for magnesium alloys. The results obtained at ambient temperature will provide a baseline for the study of the corrosion of magnesium alloys in hot, flowing coolants.

11.4.1 Effect of the concentration of ethylene glycol on corrosion rate

There is very limited information on the corrosivity of magnesium in ethylene glycol. It has been indicated [28] that magnesium is suitable for exposure to any concentration of ethylene glycol, but inhibitors may be required for warranted service. Basically, pure ethylene glycol is inert to magnesium. However, the presence of water in an ethylene glycol solution may cause magnesium to corrode to some degree [27]. Figure 11.2 shows that the corrosion rate of pure magnesium slightly decreases as the concentration of ethylene glycol increases. When the concentration of ethylene glycol is less than 33 vol.%, the corrosion rate is higher than 1 mg/week/cm^2 (~0.14 mg/cm^2/day), which is an unacceptable threshold for an aluminium alloy in a coolant [29].

These results indicate that pure ethylene glycol is almost inert to magnesium and the corrosion of magnesium in an ethylene glycol solution is closely related to the water content of the solution.

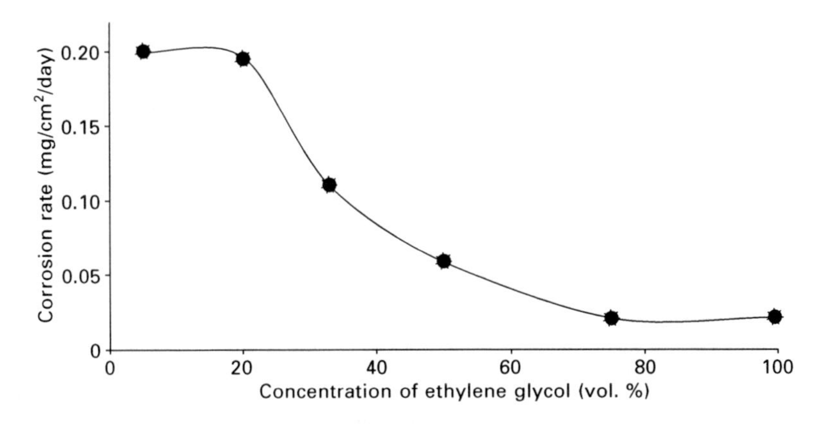

11.2 Corrosion rate of pure magnesium in various concentrations of ethylene glycol solution for 2 weeks [27].

11.4.2 Effect of water impurities on corrosion rate

Water quality can influence the corrosion of magnesium in an ethylene glycol solution. If an ethylene glycol solution is made from corrosive water, then the corrosivity of the solution is expected to be higher. It was found [27] that the corrosion rate of magnesium in an ethylene glycol solution made from the aggressive water is much higher than that in the solution made from an ASTM type II demineralized water. The increased corrosivity can be attributed to the contaminants in the aggressive water, i.e. NaCl, Na_2SO_4 and $NaHCO_3$.

NaCl is known to be detrimental to the corrosion of magnesium in most aqueous solutions and Na_2SO_4 has been reported [5] to have an inhibitive effect on magnesium in a NaCl solution. The individual influences of NaCl, Na_2SO_4 and $NaHCO_3$ on the corrosion rate of magnesium in 33 vol.% basic ethylene glycol solution have been reported by Song and StJohn [27]; the corrosion rate increases with increasing concentration of any one of the contaminants; NaCl has the most detrimental influence on the corrosion of magnesium, while $NaHCO_3$ exhibits a less adverse effect and the effect of Na_2SO_4 is relatively mild in comparison.

Similarly, adverse effects on the corrosion of magnesium in ethylene glycol solutions can also be observed for the impurities of $Mg(NO_3)_2$ and $Mg(OH)_2$ (see Fig. 11.3).

An interesting finding regarding impurity effects is the complicated interaction between these two contaminants in ethylene glycol (see Fig. 11.4). The corrosion rate of magnesium increases with increasing concentration of Na_2SO_4 or $NaHCO_3$ in a chloride-containing ethylene glycol solution when the concentration of Na_2SO_4 or $NaHCO_3$ is relatively high. However, a small amount of Na_2SO_4 or $NaHCO_3$ addition appears to inhibit the corrosion

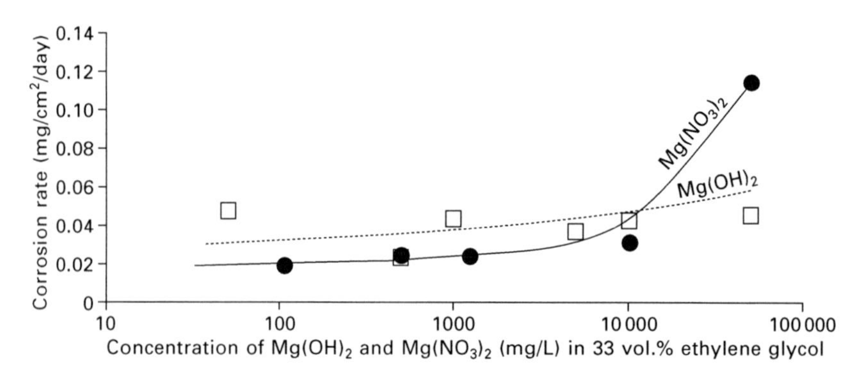

11.3 The 2 week average corrosion rates of pure magnesium in 33 vol.% of ethylene glycol solution with various additions of $Mg(NO_3)_2$ and $Mg(OH)_2$.

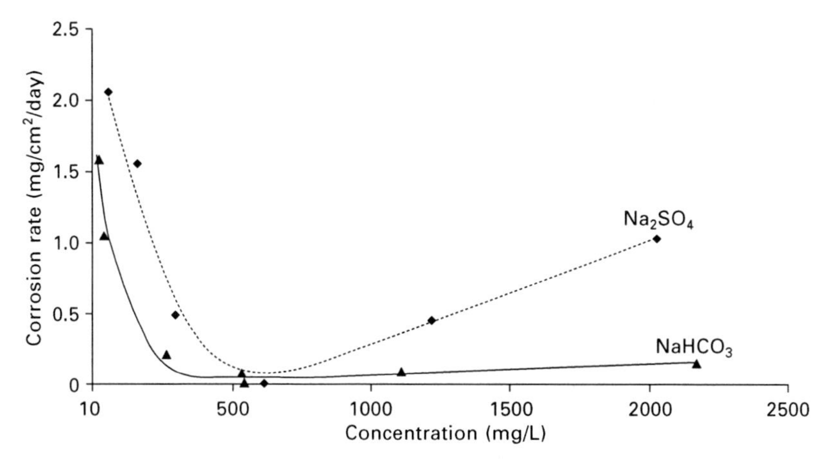

11.4 The 2 week average corrosion rates of pure magnesium in 33 vol.% of ethylene glycol solution contaminated by 336.16 mg/L NaCl with various additions of Na_2SO_4 and $NaHCO_3$ [27].

of magnesium to some degree in the chloride-containing ethylene glycol solution.

11.4.3 Corrosion mechanism

Pure ethylene glycol has very poor electrical conductivity and is almost an insulator. It has been measured in the laboratory that the resistivity of pure ethylene glycol is about 40 times higher than ASTM type II water [27]. Therefore, the resistivity of ethylene glycol solution would decrease with an increase in the water content. Moreover, dilution by water tends to facilitate the hydrolysis of the hydroxyl groups of ethylene glycol, leading to increased conductivity as well. The resistivity of a 33 vol.% ethylene glycol solution in the laboratory has been measured to be about $10^4\,\Omega\,\text{cm}$ [27], much higher than that of a normal aqueous solution (usually less than $10^2\,\Omega\,\text{cm}$). Therefore, the high solution resistivity plays an important role in corrosion.

In an ethylene glycol solution, the corrosion damage to magnesium is localized. Pitting can be clearly seen on the surface [27], and the surrounding region where it has not been corroded is relatively shiny. The pit is an active anode and the surrounding shiny region is a cathode protected from corrosion by pitting. Figure 11.5 schematically illustrates the localized pitting corrosion and the corresponding equivalent circuit for magnesium in an ethylene glycol solution. In the equivalent circuit, C_d is the capacitance of the non-corroded area of the magnesium specimen, R_{pt} and C_{pt} are the resistance and capacitance at the bottom of the corroding (pitting) area, R_{ps} is the resistance of the solution in the pit, and R_s is the solution resistance between the reference electrode and magnesium specimen. Therefore, the

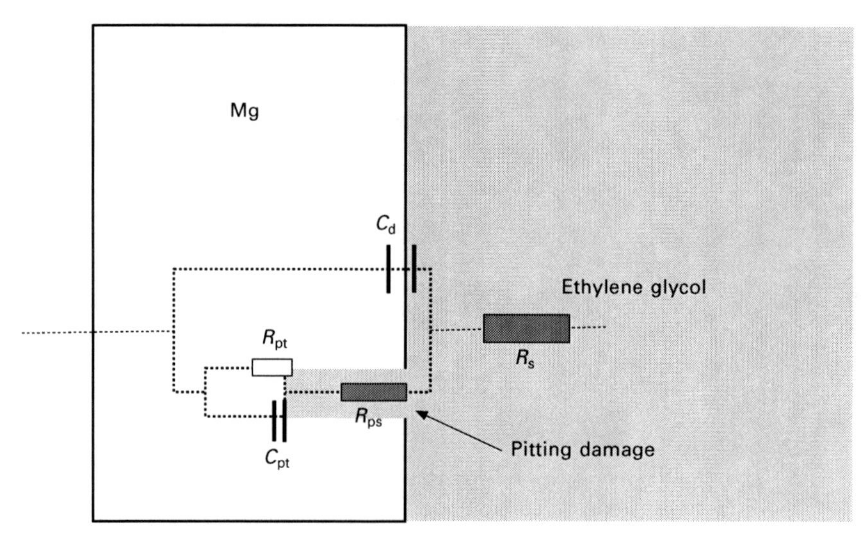

11.5 Pitting model of magnesium in an ethylene glycol solution and the corresponding equivalent circuit.

overall polarisation resistance R_p can be expressed as: $R_p = R_{ps} + R_{pt}$. Because of the high solution resistivity, the ratio of R_{ps} over R_{pt} is very high ($R_{ps} >> R_{pt}$), and thus $R_p \approx R_{ps}$.

The localized corrosion damage implies that the pitting corrosion rate is limited by the solution resistance in the pit, and thus the solution resistivity can significantly affect the corrosion resistance or polarization resistance of magnesium in the ethylene glycol solution. The inference has been supported by a measured linear relationship between polarization resistance and solution resistivity [27]; as the solution resistance increases with increase in concentration of ethylene glycol, the polarization resistance also increases. This explains the decreasing corrosion rate of magnesium with increasing concentration of ethylene glycol (Fig. 11.2).

In a contaminated ethylene glycol solution, the solution resistance decreases with increasing concentration of each contaminant NaCl, Na_2SO_4 or $NaHCO_3$. The decreased polarization resistance of magnesium can be attributed to the increased total content of ions in the solution. When the contents of the contaminants are low, the conductivity of the solution is proportional to the concentrations of the electrolytes. The decrease in solution resistance is particularly evident with increasing concentration of NaCl. This explains the most detrimental effect of NaCl on the corrosion performance of magnesium in the contaminated ethylene glycol solution. The decrease in solution resistance caused by contaminants can also account for the slightly increased corrosion rate of magnesium in ethylene glycol by addition of $Mg(OH)_2$ and $Mg(NO_3)_2$ (Fig. 11.3).

Apart from the solution resistance, the most fundamental explanation for the corrosion behaviour of magnesium in ethylene glycol may come from the magnesium/ethylene glycol interface. Like most other organics, ethylene glycol can be adsorbed onto an electrode surface. The capacitance C_d is a good indication of the adsorption of ethylene glycol on the magnesium surface. It has been measured [27] that the capacitance decreases as ethylene glycol concentration increases. A decreasing interface capacitance can be caused by high dielectric water at the interface being replaced by larger long dielectric molecules. The ethylene glycol molecule is larger and less polar than water. Its adsorption on the surface of magnesium can certainly result in a lower C_d. When the concentration of ethylene glycol increases, more ethylene glycol will be adsorbed on the surface, leading to a lower C_d. In other words, the magnesium surface is more completely covered by ethylene glycol molecules in a more concentrated ethylene glycol solution, which more effectively protects magnesium from attack by water. This explains the decreasing corrosion rate of magnesium with increasing concentration of ethylene glycol (Fig. 11.2).

In a contaminated solution, NaCl, Na_2SO_4 and $NaHCO_3$ are smaller in size than ethylene glycol. The increase in capacitance should be due to the replacement of ethylene glycol on the magnesium surface by the contaminants. SO_4^{2-} and HCO_3^- could be more strongly (easily) adsorbed on the magnesium surface than Cl^-. Additional adsorbed ethylene glycol could be replaced by Na_2SO_4 or $NaHCO_3$ and a higher surface capacitance results from this exchange than from NaCl at the same weight/volume concentration of these contaminants. Nevertheless, chlorides are well known to be much more aggressive than sulphates and carbonates. Only a few adsorbed chloride ions can be enough to significantly accelerate the dissolution of magnesium at the adsorbed sites. Thus, the corrosion rate of magnesium is higher in the chloride-containing ethylene glycol solution than in the ethylene glycol solution containing Na_2SO_4 or $NaHCO_3$.

The stronger adsorption of Na_2SO_4 or $NaHCO_3$ than chloride implies that, if Na_2SO_4 or $NaHCO_3$ ions are added into a chloride-containing ethylene glycol solution, the adsorbed Cl^- ions on the magnesium surface will be replaced by sulphate or carbonate. Since sulphate or carbonate is much less corrosive to magnesium than chloride, the replacement of the adsorbed chlorides with sulphates or carbonates in effect passivates the active sites on the magnesium surface. Hence, the dissolution of magnesium is retarded. Therefore, in a chloride-containing solution, Na_2SO_4 or $NaHCO_3$ would have a dual role. On one hand, the inhibition effect of Na_2SO_4 or $NaHCO_3$ leads to passivation. On the other hand, the addition of Na_2SO_4 or $NaHCO_3$ into a chloride-containing ethylene glycol solution reduces solution resistance, resulting in lower corrosion resistance. After most of the adsorbed chloride ions on the magnesium surface are repelled, further addition of Na_2SO_4 or

$NaHCO_3$ could only reduce R_s of the solution. Hence, the corrosion rate of magnesium decreases first and then increases with further addition of Na_2SO_4 or $NaHCO_3$.

11.5 Magnesium (Mg) alloys in ethylene glycol solution

Magnesium alloys should have a similar corrosion behaviour to that of pure magnesium in an ethylene glycol solution. This has been confirmed by Fekry and Fatayerji's systematic study [30], which is briefly described as follows.

Firstly, the corrosivity of ethylene glycol solution to AZ91D also decreases as the concentration of ethylene glycol increases (see Fig. 11.6). This can again be interpreted by the replacement of water by ethylene glycol molecules on the AZ91D surface as the concentration of ethylene glycol increases. This interpretation is supported by the increasing value of $1/C_T$ with ethylene glycol concentration where C_T is the total capacitance (Fig. 11.6).

Secondly, chlorides in the ethylene glycol solution have a detrimental effect on the corrosion of AZ91D. When the concentration of chlorides is over 0.05 M, an evident corrosion acceleration effect is measured (see Fig. 11.7). The anodic polarization current density increases dramatically with increasing chloride concentration in the solution.

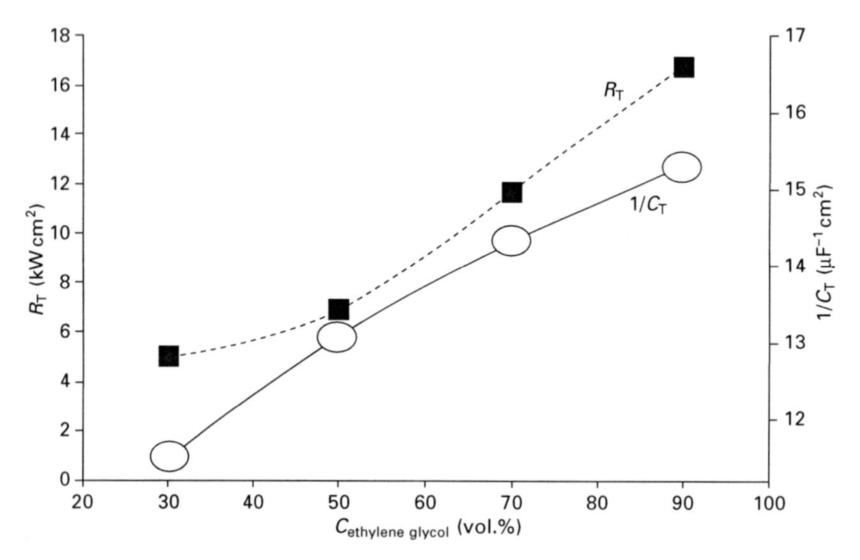

11.6 Total resistance (R_T) and relative thickness indicated by the measured total capacitance (C_T) for AZ91D alloy at various concentrations of ethylene glycol, measured after 2 h immersion [30].

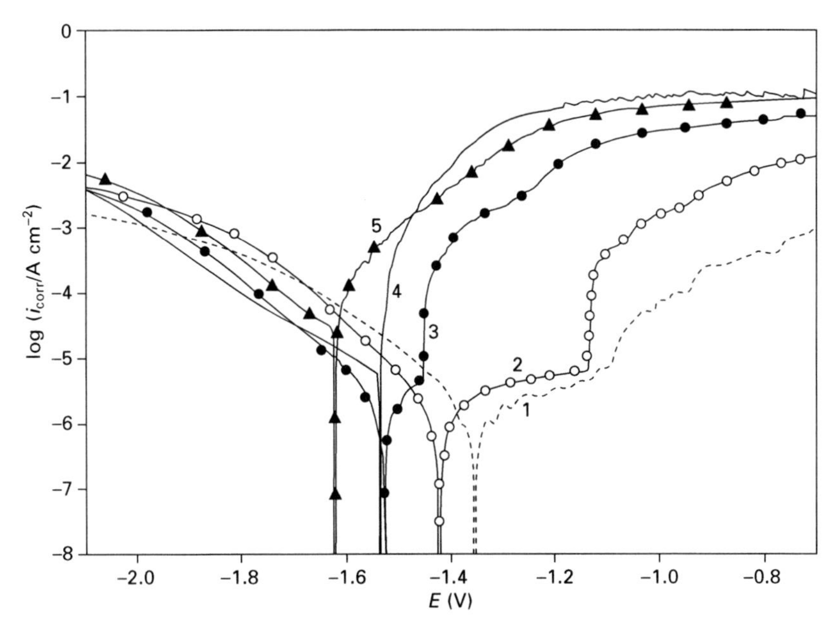

11.7 Polarization curves of AZ91D after 2 h immersion in 30% ethylene glycol solution with chloride additions (1) 0.01 M, (2) 0.05 M, (3) 0.1 M, (4) 0.3 M and (5) 0.6 M [30].

11.6 Magnesium (Mg) alloys in commercial coolants

It has been illustrated above that ethylene glycol is not corrosive to magnesium and its alloys. The water and additives reduce the resistivity and activate the magnesium surface, resulting in corrosion damage. Unfortunately, commercial coolant contains a large number of additives and hence the solution resistance is far below that of ethylene glycol. In this case, the corrosion performance should not be determined by the solution resistance. Instead, the additives will be playing a critical role in corrosion. Therefore, the corrosion behaviour of a magnesium alloy in a commercial coolant is not exactly the same as in an ethylene glycol solution.

Although the corrosion of pure magnesium in commercial coolants has been reported [9,31], in practice, an engine block cannot be made of pure magnesium, and the coolant cannot be pure ethylene glycol solution only. Therefore, it is important to understand the corrosion behaviour of magnesium alloys in commercial coolants.

11.6.1 Corrosion performance of AM-SC1 in selected coolants

AM-SC1 is a promising magnesium engine block material which has been selected as an engine block material for evaluation in the USCAR programme

Table 11.2 Corrosion rates of AM-SC1 in a few ethylene glycol-based commercial coolants [data based on ref. 26]

Coolants	General corrosion at 25 °C (mg/cm²/week)	General corrosion at 95 °C (mg/cm²/week)	Galvanic corrosion at 25 °C (μA/cm²)	Galvanic corrosion at 95 °C (μA/cm²)
Mobil 33 plus (MBL)	12~13	~0.5	~800	~60
Castrol (CTL)	8~9	~0.4		~30
Tectaloy green long life (CTL)	8~9	~0.4		~30

[32]. It is used here to represent an example of rare earth-containing magnesium alloys. Its corrosion behaviour in coolants should be representative of the group of creep-resistant magnesium alloys that do not contain aluminium. The corrosion rates of AM-SC1 in a few commercial ethylene glycol-based coolants are summarised in Table 11.2. The general corrosion rates were measured with a magnesium alloy coupon simply immersed into coolants for two weeks. The galvanic current was determined between a coupon of magnesium and a coupon of a typical aluminium cylinder head alloy AlSi9Cu3 immersed in coolant [26].

None of these commercial coolants has acceptable corrosivity to AM-SC1 when the general and galvanic corrosion rates at room and high temperatures are considered. This indicates that an engine block made of AM-SC1 alloy will suffer from either general or galvanic or combined corrosion attack no matter whether the vehicle is running or not when these selected commercial coolants are used.

It is noted that the corrosion rates of AM-SC1 alloy in these commercial coolants are much higher than pure magnesium in ethylene glycol solution. This can be attributed to the significantly decreased solution resistance. For example, the resistivity of the MBL coolant was measured to be in the 10 Ωcm order of magnitude, nearly three orders of magnitude lower than that of the pure ethylene glycol. The lower resistivity of commercial coolants is due to the large number of additives that are strong conductive electrolytes in the coolants.

The ASTM D1384 test also confirmed that the weight loss rate of AM-SC1 in MBL is higher than 0.67 mg/cm^2/week, indicating that AM-SC1 will not be able to survive in the MBL coolant.

11.6.2 Corrosivity of long-life coolants

In aqueous solutions, organic compounds normally have some inhibition effect on metals. For example, it has been found [33] that simple sodium linear-saturated carboxylates can effectively inhibit the corrosion of a magnesium

Table 11.3 Corrosion rates of AM-SC1 in two commercial organic-based long-life coolants [data based on ref. 26]

Coolants	General corrosion at 25 °C (mg/cm²/week)	General corrosion at 95 °C (mg/cm²/week)	Galvanic corrosion at 25 °C (μA/cm²)	Galvanic corrosion at 95 °C (μA/cm²)
Long-life red genuine Toyota (LLC-F)	1.2~1.7	0.8~1.1	~1	~100
Long-life red genuine Ford (LLC-T)	0.12~0.17	0.15~0.25	~10	~40

alloy in a corrosive water; the inhibition efficiency increases with immersion time and the aliphatic chain length up to C12. The formation of magnesium carboxylate on the surface should be responsible for the corrosion inhibition effect. It has been reported that [34] some carboxylate ions are effective inhibitors for magnesium alloys in ethylene glycol solutions.

Organic-based long-life coolants contain various carboxylates. It is expected that these coolants have a better inhibition effect on magnesium alloys. In experiments, it does appear that the organic-based long-life coolants perform better for AM-SC1 alloy. Table 11.3 summarizes the corrosion rates of AM-SC1 alloy in two commercial organic-based long-life coolants.

According to Table 11.3, the corrosivity of coolants LLC-T and LLC-F is very close to the acceptable threshold. With further modification they could meet the corrosivity requirements for the AM-SC1 alloy.

In addition to AM-SC1, AZ91D may be used to make some parts of a magnesium engine, e.g. a water pump housing. Hence, it is important to examine whether this metal can survive in a proposed coolant formulation.

Figure 11.8 displays the general corrosion rates of AZ91D alloy in the two most promising coolants at room and high temperatures. The general corrosion rates of AZ91D ingot in these two coolants are lower than the acceptable threshold of 0.67 mg/cm^2/week.

AM-SC1, ingot AZ91D and diecast AZ91D were also tested according to ASTM D1384 in the commercial coolants (see Table 11.4). These results show the corrosion rates of AM-SC1, ingot AZ91D and diecast AZ91D are all acceptable in LLC-T, while the corrosion rate of AM-SC1 is not acceptable in MBL.

11.6.3 Other alloy–coolant systems

A few magnesium alloy–coolant systems have been tested under the USCAR programme and it is found that systems behave quite differently from one

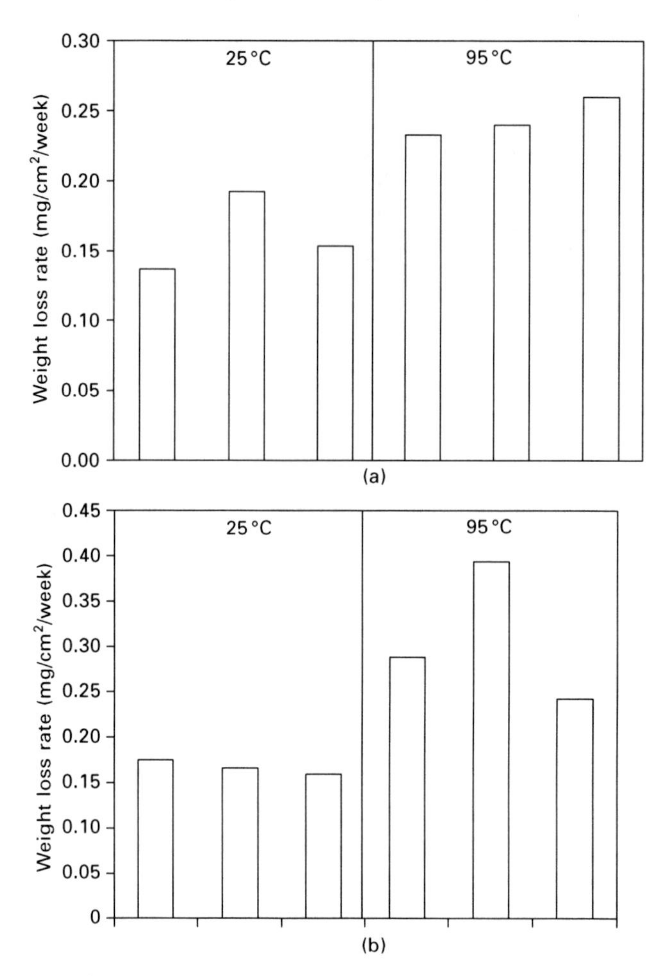

11.8 General corrosion rates of AZ91D in LLC-F (a) and LCC-T (b) coolants at room temperature and 95 °C (based on Song and StJohn [26]).

another [32]. Recently, AZ91D and magnesium engine block alloys AM-SC1 and NZK have been investigated in a few different commercial coolants, such as Total, Castrol are Caltex. It was found that the NZK alloy has a similar corrosion behaviour to AM-SC1 alloy in these coolants, but AZ91 behaves differently. This is understandable as AZ91 contains a significantly large amount of aluminium while NZK and AM-SC1 both contain rare earth elements as critical alloying elements and do not contain aluminium. The corrosion performance of these alloys also varies in different coolants. The organic carboxylate-based long-life coolant (Total) shows a relatively better inhibition effect than Castrol and Caltex in terms of general corrosion during the simple immersion test. However, the ASTM D1384 test indicates

Table 11.4 ASTM D 1384 testing results for alloys AM-SC1, ingot AZ91D and diecast AZ91D immersed in coolants MBL and LLC-T [26]

#	Coolant	Magnesium alloy	Weight loss rate (mg/cm^2/week)
1	MBL	AM-SC1	3.767
2	MBL	Ingot AZ91D	0.129
3	LLC-T	AM-SC1	0.845
4	LLC-T	Ingot AZ91D	0.277
5	LLC-T	Diecast AZ91D	0.565

that the Total coolant is actually not so good. The galvanic effect could be responsible for the worse corrosion performance of Total coolant during the ASTM D1384 test. Desorption of organic inhibitors from the magnesium alloy surfaces may occur because of the strong anodic polarization due to the galvanic effect between the magnesium alloys and other metals in the standard test.

11.6.4 Alloy effect

Different magnesium alloys exhibit different corrosion behaviours. Generally, AZ91D, particularly diecast, is more resistant to corrosion by coolants than AM-SC1 [26]. Magnesium alloys can normally be classified into two groups. The first group of alloys has aluminium as a primary alloying element and the alloys in the second group usually contain rare earth elements and zinc as well as zirconium as a grain refiner (and do not contain aluminium). These two groups of alloys have been demonstrated to behave differently in a salt solution [35]. AZ91D is a typical aluminium-containing alloy of the first group, and AM-SC1 and NZK belong to the second group. Thus, it is unlikely that their corrosion performance would be the same in a coolant. The higher corrosion resistance of AZ91D in the coolants could be due to its aluminium content. These commercial coolants were originally designed for aluminium and cast iron. They should be non-corrosive to aluminium alloys, and could also have a certain degree of inhibition effect on an aluminium-containing alloy such as AZ91D. The corrosion of magnesium might not have been seriously considered in the design of coolant formulations. Thus, it is not surprising that they do not show a strong inhibition effect on NZK and AM-SC1.

11.6.5 Influence of temperature

The response of magnesium alloys to temperature in terms of corrosion behaviour may also be different from the traditional engine materials. For example, MBL is more corrosive to AM-SC1 at room temperature than at the high temperature of 95 °C. The average general corrosion rate of AM-SC1 immersed in MBL decreases as temperature increased [26]. When the temperature is lower than 85 °C, the general corrosion rate is unacceptable. The corrosion rate of AM-SC1 at a higher temperature increases dramatically at the very beginning and then quickly slows down with time. However, at room temperature, the corroded amount of AM-SC1 slowly increased, and the increasing trend continues for a couple of hours until it achieves a certain stable high level. These results suggest that AM-SC1 becomes passive with time, and at a higher temperature the 'passivating' process is much faster. This could be due to the formation of a certain type of film on the surface and the film formation process is faster at higher temperatures. X-ray photoelectron spectroscopy (XPS) analysis of the specimen after immersion in the MBL coolant reveals that the corrosion products on the magnesium surface are mainly magnesium silicates. Silicate is one of the important inhibitors in the MBL coolant.

In organic acid-based long-life coolants, such as LLC-T and LLC-F, the influence of temperature on corrosion is different. A higher temperature appears to increase the general and galvanic corrosivity of the coolant to AM-SC1. The different temperature-dependent corrosivities of the organic acid-based long-life coolants and the traditional coolants could result from their different corrosion inhibition mechanisms of the inhibitors in the coolants. It seems that the inhibition effect of the organic acid in the long-life coolants becomes less significant at an elevated temperature.

11.7 Corrosion inhibition

Theoretically, most inhibitors only selectively inhibit the corrosion of certain metals. Coolants developed for aluminium and cast iron may not always be suitable for magnesium alloys. Nevertheless, the possibility cannot be excluded that some inhibitors used in particular commercial coolants for aluminium and iron-based alloys may fortuitously also have some inhibition effect on magnesium.

11.7.1 Strategy for the development of suitable coolants for magnesium alloys

Since protection of magnesium alloys has not been considered in developing these commercial coolants before, some companies have started developing new coolants with new inhibitors particularly for magnesium alloys [36]. To develop new coolants and inhibitors for magnesium alloys, a normal approach [36,37] is by starting with screening possible inhibitors in ethylene glycol. This approach will lead to numerous standard tests for various coolant properties and performance, from an initial ethylene glycol and inhibitor recipe to a final coolant product.

Another short-term approach is directly searching for a suitable commercial coolant and an inhibitor for magnesium alloys through assessing the corrosion performance of magnesium alloys in the existing commercial coolants. The reason is that all the commercial coolants have passed various performance and property tests, including their corrosivity to all the traditional engine materials, such as aluminium alloys, cast iron, copper, brass and solder. These performance and property requirements, including corrosivity to the traditional materials, are also critical to a new coolant for magnesium engine blocks, as some of these traditional materials will be used in the construction of magnesium engines. Therefore, as long as a coolant selected from those existing commercial brands is measured to be non-corrosive to magnesium alloys, it is likely to meet the other essential requirements of a normal engine coolant, and does not need to be tested again for those essential performance and properties. In selection of inhibitors for magnesium alloys, only those inhibitors that are effective for magnesium alloys should be considered, as there are already inhibitors in commercial coolants for the other engine block materials.

11.7.2 Inhibitors in coolants for magnesium alloys

Organic inhibitors used in coolants include non-silicate antifreeze formulations containing alkali metal salts of benzoic acid, dicarboxylic acid and nitrate, alkali metal aromatic trazoles, aliphatic monobasic acids or salts, hydrocarbyl dibasic acids or salts and hydrocarbonyl triazole, hydrocarbyl dibasic acids or salts, hydrocarbyl azoles, specific hydrocarbyl alkali metal sulphonates, cyclohexane acid, sebacic acid and tolytriazole, cyclohexane hexacarboxylic acid, etc. Some organic compounds have been found to have an inhibition effect on magnesium alloys in an ethylene glycol solution. For example, aliphatic mono- or di-basic acids or aromatic carboxylate acids, or the alkali metal, ammonium or amine salts can provide corrosion protection to magnesium when combined with a fluoride or a fluorocarboxylic acid or a salt [36]. It is important that these organic compounds also offer corrosion protection to aluminium, iron, copper and solder that are normally used in a cooling system. It has also been claimed [36] that a combination of the above acids or salts with fluoro compounds can provide synergistic corrosion protection for magnesium, and an optional addition of a hydrocarbyl triazole and/or a thiazole to these combinations further improves the inhibition effect on copper and aluminium. The presence of fluoride and/or fluorocarboxylate can significantly improve the high-temperature magnesium corrosion inhibition efficiency.

Slavcheva and Schmitt [38] evaluated the corrosion inhibition effect of two groups of organic inhibitors on AZ91D in 50 vol.% ethylene glycol solution at 80 °C. The first group includes derivatives of lactobionic acid: lactobiono-tallowamide (LBTA), lactobiono-oleylamid (LNOA), lactobionao-cocosamid (LBCA) and potassium lactobionate (KLB). The second group is six-ring organic compounds containing N-heteroatom: α-(prridil)-1,3-indan-dione (α-PP), γ-(pyridil)-1,3-indan-dione (γ-PP) and 8-hydroxyquinoline (HQ). Their test results show that the first group of inhibitors is more effective than the second group. Based on this, they further investigated the inhibition efficiency of LBTA in an ethylene glycol solution containing chlorides [39], and confirmed that the efficiency could reach 77% at a relatively low concentration of 0.2 g/L. The inhibition mechanism was identified as mixed type, hindering both the cathodic and the anodic partial reactions. An interesting finding is that paracetamol at a certain concentration has also been shown to significantly inhibit the corrosion of AZ91D in an ethylene glycol solution [30].

11.7.3 Inhibitive effect of KF in ethylene glycol solutions

It is well known that magnesium can react with F^- ions and form an MgF_2 film which is relatively insoluble [40]. The film can also inhibit the corrosion of magnesium in an ethylene glycol solution [27]. Figure 11.9 displays a clear decrease in the corrosion rate of magnesium in 33 vol.% ethylene glycol solution resulting from the addition of KF. The inhibition effect of KF is further confirmed by reduced corrosivity of an ethylene glycol solution made from the aggressive water after addition of 1 wt% KF. This signifies the practical effectiveness of KF as an inhibitor for magnesium in aqueous ethylene glycol.

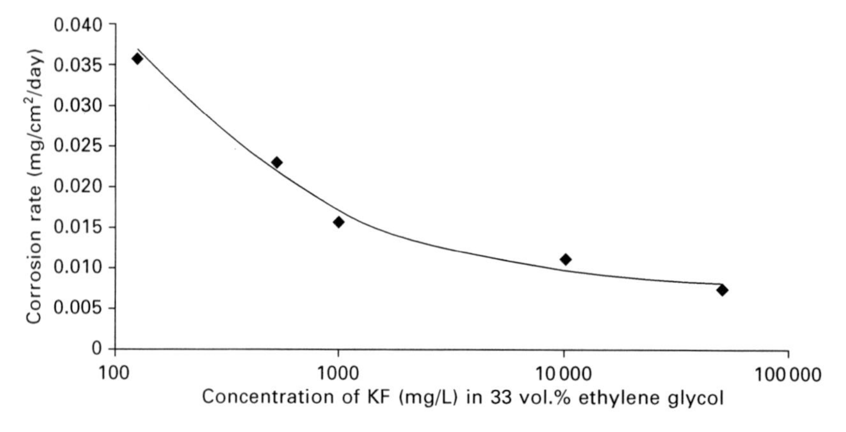

11.9 Corrosion rates of magnesium in 33 vol.% basic ethylene glycol solution with various additions of KF for 14 days [27].

The formation of the fluoride film on magnesium can be confirmed by XPS analysis. About 5~9 at.% fluorine can be detected on the specimen surface after immersion in an ethylene glycol solution with 1 wt% KF addition. Such a strong XPS signal can only come from a phase layer rather than adsorbed ions. The result strongly suggests that a fluoride-containing film formed on the magnesium surface and it should be a three-dimensional film with an appreciable thickness.

EIS measurements further show that the addition of KF into ethylene glycol leads to a strikingly enhanced R_p but decreased C_d. The reduced C_d and improved R_p can be associated with the formation of a three-dimensional film on the magnesium surface. Polarization curve measurements also confirm that the phase film formed on the magnesium surface is responsible for the reduced corrosion rate of magnesium. After addition of KF, the anodic polarization current is significantly reduced to a very low value and a low 'current-plateau' is seen where the anodic current is almost independent of the polarization potential.

The inhibition effect of fluorides has also recently been observed on AZ91 in an ethylene glycol solution [30]. However, the effect is not as evident as on pure magnesium. This could be due to the content of aluminium in the alloy. Fluorides cannot inhibit the dissolution of aluminium.

11.7.4 Inhibition by KF in commercial coolants

The inhibition mechanism of fluorides described above also operates on magnesium alloys in commercial coolants. For example, the inhibition effect of KF on the general corrosion of AM-SC1 in MBL coolant at room temperature is illustrated in Fig. 11.10. The general corrosion rate of AM-SC1 decreases as the concentration of KF increases. When the concentration of KF is greater than 1 wt%, the corrosion rate becomes lower than the corrosion rate threshold of 0.67 mg/cm^2/week [26]. However, for galvanic corrosion at high temperature, KF does not work very well in the MBL coolant.

Similar general and galvanic corrosion tests were carried out for some other coolants with KF as an inhibitor. The results are summarized in Table 11.5. A comparison of Table 11.5 with Tables 11.2 and 11.3 leads to a conclusion that KF has an inhibition effect in all of these coolants. LLC-F + 1 wt% KF has acceptable corrosivity to AM-SC1 alloy and LLC-T + 1 wt% KF is also close to the acceptable level.

The likely mechanism of the inhibitive effect of KF on AM-SC1 is similar to that for pure magnesium and AZ91D in a fluoride-containing ethylene glycol solution, i.e. fluorides react with magnesium in the matrix of AM-SC1 alloy, forming a low-solubility magnesium fluoride product which deposits on the alloy surface and prevents further corrosion attack to the alloy.

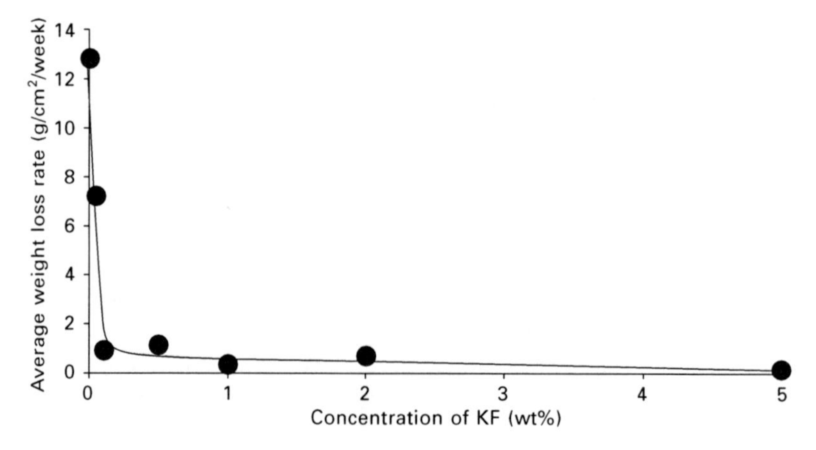

11.10 Dependence of the general corrosion rate of AM-SC1 on the concentration of KF in MBL coolant for two weeks at room temperature (based on Song and StJohn [26]).

Table 11.5 Corrosion performance of AM-SC1 in various coolants with 1 wt% KF as inhibitor [26]

1 wt% KF in coolants	General corrosion at 25 °C (mg/cm^2/week)	General corrosion at 95 °C (mg/cm^2/week)	Galvanic corrosion at 25 °C (μA/cm^2)	Galvanic corrosion at 95 °C (μA/cm^2)
MBL	~0.5	~0.5	2~3	~20
CTL	0.08~0.19	0.18~1		~25
LLG	~0.16	~1.9		~30
LLC-F	0.14~0.17	0.07~0.1	~1	~4
LLC-T	0.06~0.11	0.06~0.09	~5	~18

The galvanic corrosion rates of AZ91D in these coolants are slightly higher than the acceptable threshold at 95 °C. The addition of 1 wt% KF at this temperature reduces the galvanic current very close to 8.8 μA/cm^2 in LLC-T, but in LLC-F it is still much higher than the threshold. Therefore, LLC-T appears to be slightly better than LLC-F.

To evaluate whether KF can effectively inhibit the corrosion of a magnesium alloy in a more realistic cooling system at a high temperature, KF is added, at various concentrations, to coolant LLC-T [41] using the ASTM D1384 test. Significant reductions in the corrosion rate are found with an addition of 0.2 wt%, 0.5 wt% and beyond. The dramatic effect of KF additions on the corrosion of AM-SC1 in LLC-T is highlighted in Fig. 11.11.

The ASTM D1384 test has been conducted for AM-SC1, ingot AZ91D and diecast AZ91D in the other coolants with 1 wt% KF addition. The results are listed in Table 11.6.

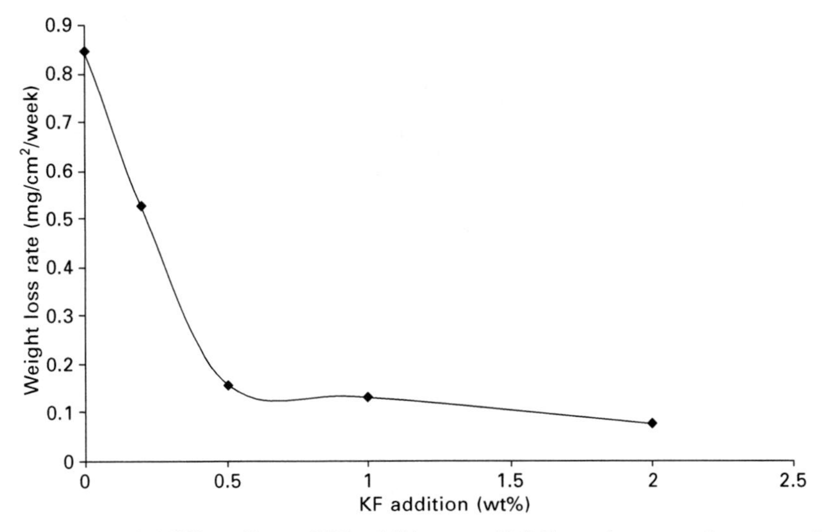

11.11 The effect of KF additions to LLC-T on the corrosion rate of AM-SC1 under ASTM D1384 test [41].

Table 11.6 ASTM D 1384 testing results for AM-SC1, ingot AZ91D and diecast AZ91D immersed in LLC-T coolant with a range of KF contents [26]

#	Coolant + KF	Magnesium alloy	Weight loss rate ($mg/cm^2/week$)
1	LLC-T+0.2 wt% KF	AM-SC1	0.525
2	LLC-T+0.5 wt% KF	AM-SC1	0.157
3	LLC-T+1 wt% KF	AM-SC1	0.131
4	LLC-T+2 wt% KF	AM-SC1	0.078
5	LLC-T+0.2 wt% KF	Ingot AZ91D	0.798
6	LLC-T+0.5 wt% KF	Ingot AZ91D	–0.046*
7	LLC-T+1 wt% KF	Ingot AZ91D	–0.091*
8	LLC-T+0.5 wt% KF	Diecast AZ91D	0.151
9	LLC-T+1 wt% KF	Diecast AZ91D	0.139

*The weight gains (#6 and 7) could be experimental errors caused by some reaction products deposited from the corrosion product removal solution or a passive film/layer formed in the coolants.

The corrosion rates of AM-SC1, ingot AZ91D and diecast AZ91D are all acceptable in LLC-T with 1 wt% KF as an inhibitor. The other metals such as copper, brass, cast iron, solder and aluminium alloys, have all been measured to have very low corrosion rates in these coolants, which means that the addition of KF does not affect the corrosion rates of these metals. Figure 11.12 shows the surfaces of the ASTN D1384 tested coupons in LLC-T + 1 wt% KF. It is clearly shown that all the plates are nearly intact after the corrosion test. The significance of these findings is that LLC-T may be a suitable coolant for magnesium engine blocks, and KF can be added to further reduce the corrosivity of the coolant.

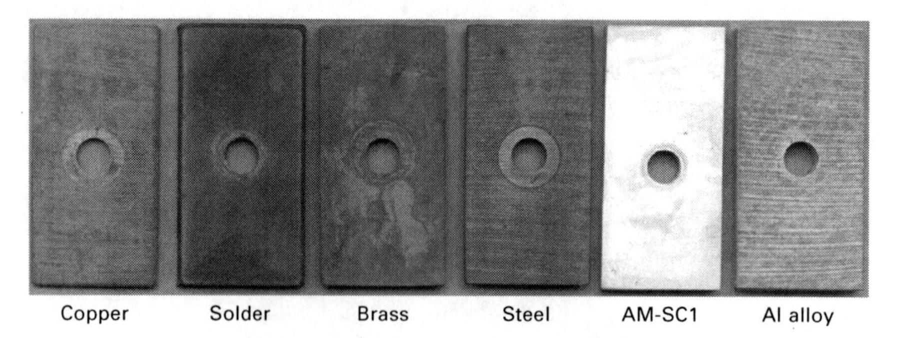

11.12 Corrosion morphologies of metals after ASTM D1384 testing in LLC-T + 1 wt% KF [26].

The above results indicate that an organic acid-based coolant might be more suitable for magnesium alloys. The difference in inhibition mechanism between a traditional coolant (e.g. MBL) and an organic acid-based long-life coolant (e.g. LLC-T and LLC-F) should be responsible for the different degrees of the inhibition effect of KF observed in these two groups of coolants. As MgF_2 has a much lower solubility than other magnesium salts resulting from the reactions between magnesium with the inhibitors in the traditional coolants, fluoride may preferentially form a protective MgF_2 film replacing other relatively soluble salt films (e.g. magnesium silicate) on the magnesium surface. Hence, the corrosion rate of AM-SC1 in MBL is likely to be significantly reduced by the addition of KF. A high temperature is favourable to the formation and deposition of some inorganic magnesium films due to the change of solubility of the salts with temperature. Therefore, a better inhibitive effect may be achieved at a higher temperature due to an inorganic magnesium salt film formed on the magnesium surface in addition to the MgF_2. In contrast, in an organic acid-based long-life coolant, it is likely that a good adsorptive carboxylate film might have already formed on the magnesium surface, which prevents the reaction of fluorides with magnesium. Therefore, the inhibitive effect on AM-SC1 by KF in LLC-T or LLC-F is less significant. At a higher temperature the organic additives desorbed from the magnesium surface, if no other magnesium salt deposition occurs, the inhibition efficiency should become lower.

11.8 Health and environmental concerns

Ethylene glycol is moderately toxic. On ingestion, ethylene glycol is oxidized to glycolic acid which is, in turn, oxidized to oxalic acid which is toxic. It is critically important that additives do not significantly increase the toxicity or make the coolant hazardous to the environment. This to a great extent limits the selection of inhibitors.

In the case of KF, although fluorides are present throughout the environment at low levels and are not harmful (in fact, small amounts of sodium fluoride can help prevent tooth decay) high levels of fluoride or hydrogen fluoride gas are hazardous to health [34]. Concentrated hydrogen fluoride is very corrosive and would badly burn exposed plants, birds or land animals.

In a coolant system, the health and environmental hazards of fluoride are relatively low. This is because of the following:

- The cooling system is a closed cyclic device and coolant will not be released into the environment under normal operating conditions. The chance of humans being directly exposed to the fluoride in a coolant is extremely low.
- Current commercial coolants need to be recycled and are not allowed to be disposed into the environment directly after use. Thus the fluoride will also be recycled together with the coolants.
- Even though fluoride has slightly higher health and environmental hazard ratings than ethylene glycol, its concentration in a coolant as an inhibitor is much lower than the concentration of ethylene glycol in the coolant. Thus the hazard due to the fluoride may not necessarily be higher than ethylene glycol in this case.
- Most coolants are slightly alkaline, thus it is unlikely that a significant amount of hydrogen fluoride will be generated from KF in these solutions. Therefore, the formation of hydrogen fluoride in coolants should not be a significant hazard.

Therefore, if KF is added to coolants to prevent the corrosion of magnesium alloys, the health and environmental issues may not be a major concern.

11.9 Summary

This chapter presents the results and knowledge generated from research undertaken to date on the susceptibility of magnesium alloys to corrosion when in contact with engine coolants. A commercial solution that satisfies all the property requirements for a coolant designed for magnesium alloys for engine block applications has not been conclusively developed or successfully applied commercially. However, our understanding has progressed so that pathways for the development of such a coolant have been identified. It is apparent that KF is an effective inhibitor in ethylene glycol solution and in some commercial coolants. Further work needs to be undertaken to show in which commercial coolant the best inhibition effect can be achieved without any side effects on other materials. It is also clear that much additional research is required to identify and develop more effective inhibitors before a reliable cost-effective commercial coolant is developed.

11.10 References

1. C.J. Bettles, C.T. Forwood, D. StJohn, *et al.*, 'AMC-SC1: An elevated temperature magnesium alloy suitable for precision sand casting of powertrain components', in H. Kaplan (ed), *Magnesium Technology*, TMS, San Diego (2003) pp. 223–226.
2. C.J. Bettles, C.T. Forwood, J.R. Griffiths, 'AMC-SC1: A new magnesium alloy suitable for powertrain applications', SAE Technical paper # 2003-01-1365, SAE World Congress, Detroit, March 2003.
3. W. Schoffmann *et al.*, 'Magnesium crankcase on a lightweight diesel engine', (presented at) Magdeberg Conference, Magdeburg, Germany, Feb. 2005.
4. G. Song, A. Atrens, D. StJohn, J. Nairn, Y. Li, 'The electrochemical corrosion of pure magnesium in 1N NaCl', *Corr. Sci.*, **39**(5) (1997) 855.
5. G. Song, A. Atrens, D. StJohn, X. Wu, J. Nairn, 'The anodic dissolution of magnesium in chloride and sulphate solutions', *Corr. Sci.*, **39**(10–11) (1997) 1981.
6. G. Song, A. Atrens, X. Wu, B. Zhang, 'Corrosion behaviour of AZ21 AZ501, and AZ91 in sodium chloride', *Corr. Sci.*, **40**(10) (1998) 1769.
7. G. Song, A. Atrens, M. Dargush, 'Influence of microstructure on the corrosion of diecast AZ91D', *Corr. Sci.*, **41** (1999) 249.
8. Z. Shi, P.K. Mallick, R.C. McCune, S. Simko, F. Naab, 'A study of corrosion film growth on pure Mg and a creep-resistant magnesium alloy in an automotive engine coolant', in S.R. Agnew, N.R. Neelameggham, E.A. Nyberg and W.H. Sillebens (eds), *Magnesium Technology 2010*, TMS2010, pp. 173–179.
9. P. Zhang, X. Nie, D.O. Northwood, 'Influence of coating thickness on the galvanic corrosion properties of Mg oxide in an engine coolant', *Surface Coatings Technol.*, **203** (2009) 3271–3277.
10. L.H. Han, X.Y. Nie, H. Hu, 'Electrochemical behaviour of squeeze cast AJ62 magnesium alloy in salt solution and engine coolant', *Mater. Technol.* **24** (2009) 170–173.
11. B. Sales, G. Delgadillo, 'Corrosion control in cooling systems of heavy-duty diesel engines', *Corro. Rev.*, **3**(2–4) (1995) 245–259.
12. A.D. Mercer, *Corrosion Inhibitors in Internal Combustion Engine Cooling Systems*, The Institute of Materials, UK (1994) 58–63.
13. S.M. Woodward, A.V. Gershun, 'Characterisation of used engine coolant by statistical analysis', in R.E. Beal (ed), *Engine Coolant Testing*: 3rd Volume, ASTM STP 1192 American Society for Testing and Materials, Philadelphia (1993) pp. 234–247.
14. B.D. Oakes, 'Cavitation corrosion', in W.H. Ailor (ed), *Engine Coolant Testing: State of the Art*, ASTM STP 705, American Society for Testing and Materials, Philadelphia (1980), pp. 284–294.
15. R.L. Chance, 'Electrochemical corrosion of an aluminium alloy in cavitating ethylene glycol solutions', in W.H. Ailor (ed), *Engine Coolant Testing: State of the Art*, ASTM STP 705, American Society for Testing and Materials, Philadelphia (1980) pp. 270–283.
16. M.S. Vukasovich, F.J. Sullivan, 'Inhibitors and coolant corrosivity', in R.E. Beal (ed), *Engine Coolant Testing: 2nd Symposium*, ASTM STP 887, American Society for Testing and Materials, Philadelphia (1986) pp. 86–98.
17. R. Pellet, P. Van de Ven, D. Amaez, P. Fritz, L. Bartley, D. Hunsicker, 'The role of nitrite and carboxylate ions in repressing diesel engine cylinder liner cavitation corrosion' in *Corrosion 98, 53rd Annual Conference and Exhibition, California, Conference Proceedings*, Vol. 4, Paper No. 545 March 1998.

18. T.W. Weir, 'Testing of organic acids in engine coolants', in R. E. Beal (ed), *Engine Coolant Testing*: 4th Volume, ASTM STP 1335, American Society for Testing and Materials, Philadelphia (1999) pp. 7–22.
19. F.T. Wagner, T.E. Moylan, S.J. Simko, M.C. Militello, 'Composition of incipient passivating layers on heat-rejecting aluminum in carboxylate-and silicate-inhibited coolants: correlation with ASTM D 4340 weight losses', in R.E. Beal (ed), *Engine Coolant Testing*: 4th Volume, ASTM STP 1335, American Society for Testing and Materials, Philadelphia (1999) pp. 23–42.
20. P. Van De Ven, J.P. Maes, 'A compatibility study of mixtures of a monoacid/dibasic acid coolant and a traditional nitrite-free coolant', SAE technical paper series # 940769, SAE International, Detroit, 1994.
21. P. Van De Ven, J.-P. Maes, 'The effect of silicate content in engine coolants on the corrosion protection of aluminum heat-rejecting surface', SAE technical paper series # 940498, SAE International, Detroit, 1994.
22. D.A. Washington, D.L. Miller, J. Maes, P.V. de Ven, J.E. Orth, 'Long life performance of carboxylic acid based coolants', SAE technical # 940500, SAE International, Detroit, 1994.
23. ASTM D1384-97, 'Standard Test Method for Corrosion Test for Engine Coolants in Glassware', 1999 annual book of ASTM Standards, Vol.15.05, ASTM, Philadelphia, p. 9.
24. G. Song, A. Atrens, 'Understanding magnesium corrosion, a framework for improved alloy performance', *Adv. Eng. Mater.*, **5**(12) (2003) 837–858.
25. G. Song, A. Atrens, D. StJohn, 'An hydrogen evolution method for the estimation of the corrosion rate of magnesium alloys', in J. N. Hryn (ed), *Magnesium Technology 2001*, TMS (2001) pp. 255–262.
26. G. Song, D.H. StJohn, 'Corrosion of magnesium alloys in commercial engine coolants', *Materials and Corrosion*, **56** (2005) 15–23.
27. G. Song, D. StJohn, 'Corrosion behaviour of magnesium in ethylene glycol', *Corr. Sci.*, **46** (2004) 1381–1399.
28. M.M. Avedesian, H. Baker, (ed), *ASM Specialty Handbook, Magnesium and Magnesium Alloys*, ASM International, The Materials Information Society (1999) pp. 194–210.
29. ASTM D4340-96, 'Standard test method for corrosion of CAST aluminium alloys in engine coolants under heat-rejecting conditions', ASTM International (2007) Vol. 15.05.
30. A.M. Fekry, M.Z. Fatayerji, 'Electrochemical corrosion behaviour of AZ91D alloy in ethylene glycol', *Electrochim. Acta* **54** (2009) 6522–6528.
31. Z. Shi, P.K. Mallick, R.C. McCune, 'Characterization of film formation on magnesium alloys due to corrosion in engine coolants', SAE technical paper 2008-01-1155, SAE International, Detroit. 2008.
32. B.R. Powell, L.J. Ouiment, J.E. Allison, J.A. Hines, R.S. Beals, L. Kopka, P.P. Ried, 'The magnesium powertrain cast components project: part I – accomplishments of phase I and the objective and plans for the magnesium engine in phase II', in A. Luo (ed), *Magnesium Technology 2004*, TMS (2004) pp. 3–10.
33. D. Daloz, C. Rapin, P. Steinmetz G. Michot, 'Corrosion inhibition of rapidly solidified Mg–3%Zn–15%Al magnesium alloy with sodium carboxylates', *Corrosion*, **54** (1998) 444–450.
34. http://www.npi.gov.au/database/substance-info/profiles/44.html
35. G. Song, D. StJohn, 'Corrosion performance of magnesium alloys MEZ and AZ91', *Int. J. Cast Metals Res.*, **12** (2000) 327–334.

36. J.P. Meas, S. Lievens, 'Corrosion inhibitors and synergistic inhibitor combinations for the protection of light metals in heat-transfer fluids and engine coolants', PCT/IB99/01659 (1999); WO 00/22189.
37. M. Starostin, S. Tamir, 'New engine coolant for corrosion protection of magnesium alloys', *Materials and Corrosion*, **57** (2006) 345–349.
38. E. Slavcheva, G. Schmitt, 'Screening of new corrosion inhibitors via electrochemical noise analysis', *Materials and Corrosion*, **53** (2002) 647–655.
39. E. Slavcheva, G. Petkova, P. Andreev, 'Inhibition of corrosion of AZ91 magnesium alloy in ethylene glycol solution in presence of chloride anions', *Materials and Corrosion* **56** (2005) 83–87.
40. R.C. Weast, M.J. Astle, W.H. Beyer (Eds.), *CRC Handbook of Chemistry and Physics*, CRC Press, 67th edition (1986–1987) B104.
41. G. Song, D. StJohn, C. Bettles, G. Dunlop, 'The corrosion performance of magnesium alloy AM-SC1 in automotive engine block applications', *JOM*, **57**(5) (2005) 54–56.

12
Numerical modelling of galvanic corrosion of magnesium (Mg) alloys

A. ATRENS and Z. SHI, The University of Queensland, Australia and G.-L. SONG, General Motors Corporation, USA

Abstract: This chapter reviews numerical modelling of magnesium (Mg) galvanic corrosion, in particular, the prediction of the galvanic current density distribution of a typical Mg alloy such as AZ91 in contact with steel in a typical corrosive solution such as 5% NaCl. The galvanic current density distribution predicted by boundary elemental modelling was in good agreement with experimental measurements. The galvanic current density distribution caused by the interaction of two independent galvanic couples was equal to the sum of the galvanic current density caused by each individual galvanic couple. However, experimental measurements indicate that the measured corrosion rate was significantly higher than the galvanic corrosion rate, and this was interpreted as self-corrosion of Mg, corresponding to a penetration rate of ~230 mm/yr. However, does the necessity to postulate self-corrosion indicate some fundamental flaw in the methodology? Issues and future research directions are discussed.

Key words: magnesium alloys, galvanic corrosion, boundary element method.

12.1 Introduction

Galvanic corrosion is a major concern in industrial design and material selection [1–4] for magnesium (Mg) because Mg is the most active structural metal and because of the increased Mg usage in the auto and aerospace industries [5,6]. The geometry determines the galvanic current. Geometric factors include the anode/cathode area ratio, insulation distance (s) between the anode and cathode, electrolyte film depth (d) and the shape of the anode and cathode [3,7]. Mechanical components can take various forms. Insulation spacers can be used between components made from different metals. The thickness of these spacers can be varied to alter the insulation distance between the anode and cathode. Additionally, the solution film depth can be different. Furthermore, there can be more than one galvanic couple such as many steel fasteners for one magnesium part, and there can be an interaction of the current caused by each galvanic couple. Waber [8] studied the influence of galvanic cell size theoretically, but had no experimental validation. Song *et al.* [4] developed a galvanic corrosion assembly (GCA) to study the influence of cathode material, distance between anode and cathode and anode/cathode area ratio. This study identified important effects such as 'alkalization', 'passivation', 'poisoning' and 'short-cut' as well as the effectiveness of an insulating spacer in reducing galvanic corrosion. This laid a foundation for studies on Mg galvanic corrosion.

Numerical methods are promising for studying galvanic corrosion [4,9–14], and in particular, for predicting the galvanic current density distribution. BEASY, a boundary element method (BEM) program, has become widely used to study galvanic corrosion [15]. Qualitative simulation of galvanic corrosion would facilitate efficient design of structural components incorporating galvanic corrosion. Total galvanic corrosion is considered to comprise two components: (1) galvanic corrosion and (2) self-corrosion. Galvanic corrosion is that part of the corrosion, directly caused by the coupling of the Mg to a steel fastener. The self-corrosion is defined as the extra corrosion. Both the galvanic corrosion and the self-corrosion may take the form of more or less general corrosion, or the form of localized corrosion.

This chapter reviews numerical modelling of Mg galvanic corrosion based on work by Jia *et al.* [16–20], in particular, the prediction of the galvanic current density distribution of a typical Mg alloy like AZ91 in contact with steel in a typical corrosive solution such as 5% NaCl. Figure 12.1 presents an idealized one-dimension (1D) galvanic couple. Section 12.2 describes the BEM model. Section 12.3 deals with idealized 1D galvanic corrosion and describes the BEM model calculations compared with experimental measurements using a modified GCA based on Song *et al.* [4] (Fig. 12.2), which captures the galvanic current densities over the limited testing surface area. Section 12.4 describes the interaction of two idealized galvanic couples, for example the situations where there are a number of fasteners, such as when there are a number of steel bolts for an Mg component. The galvanic interaction

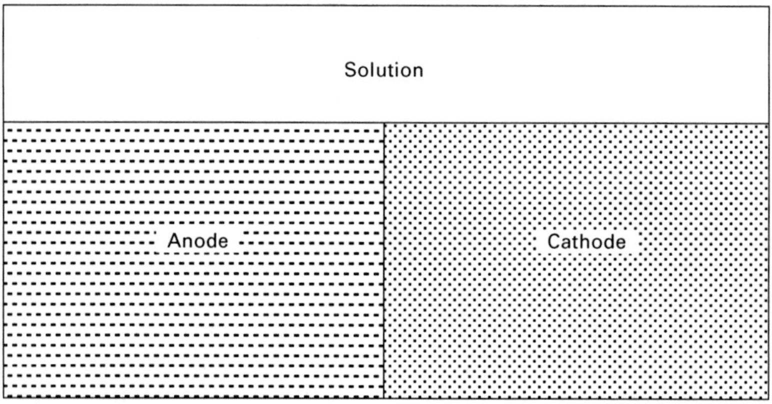

12.1 Schematic of idealized one dimension galvanic couple [16].

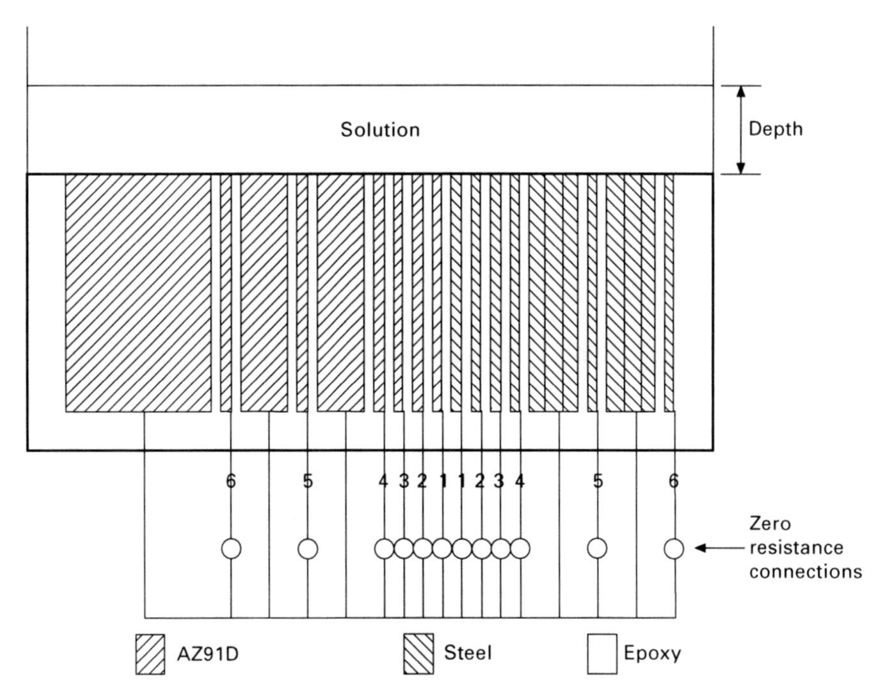

12.2 Section through multi-electrode Mg–steel galvanic corrosion assembly (GCA) [17].

assembly (GIA), shown schematically in section in Fig. 12.3 provides an experimental approach to measure the current density distribution. Section 12.5 describes modelling a practical component of an AZ91D sheet coupled to a steel fastener. The total corrosion rate was modelled and measured as a function of distance from the Mg–steel interface. Issues and future research directions are discussed in Sections 12.6 and 12.8.

12.2 Boundary element method (BEM) model

Adey and Niku [21] indicated that for a uniform, isotropic electrolyte domain Ω, as illustrated in Fig. 12.4, in steady state, the potential obeys the Laplace equation:

$$\nabla^2 \phi = 0 \qquad 12.1$$

The Laplace equation is solved using the boundary conditions:

$$\phi = \phi_0 \quad \text{on} \quad \Gamma_1 \qquad 12.2$$

$$I = I_0 \quad \text{on} \quad \Gamma_2 \qquad 12.3$$

$$I_a = -f_a(\phi) \quad \text{on} \quad \Gamma_{3a} \qquad 12.4$$

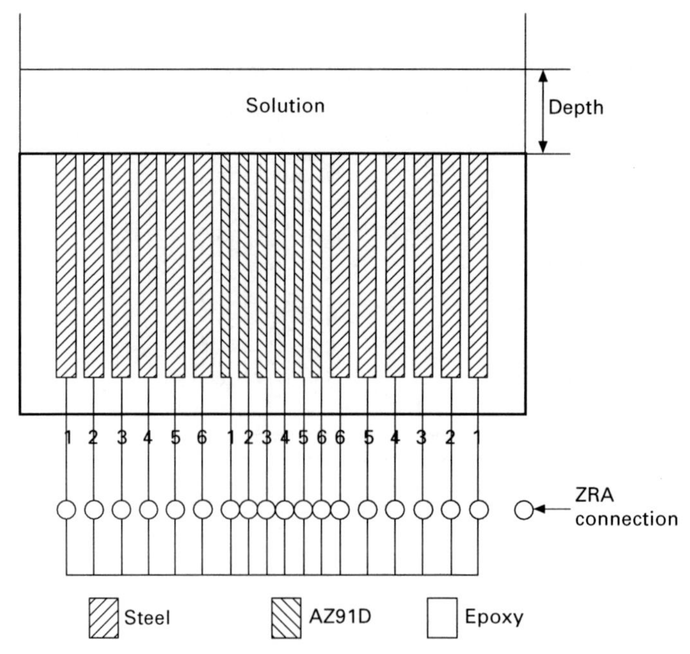

12.3 Section through galvanic interaction assembly (GIA) [17].

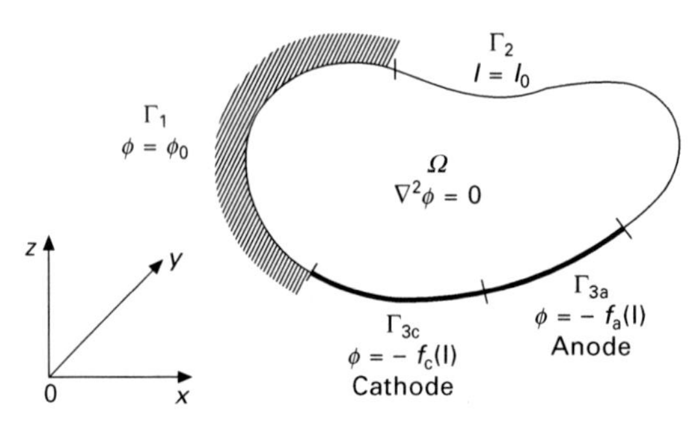

12.4 Basic equations and boundary conditions for the BEM model [19].

$$I_c = -f_c(\phi) \quad \text{on} \quad \Gamma_{3c} \qquad 12.5$$

where $\Gamma\,(\equiv \Gamma_1 + \Gamma_2 + \Gamma_{3a} + \Gamma_{3c})$ is the surface of the electrolyte Ω, I is the current density across the boundary, ϕ is the potential, ϕ_0 and I_0 are constants, and $f_a(\phi)$ and $f_c(\phi)$ are non-linear functions for the anode and cathode electrode kinetics. In the present case, there is no region of applied potential Γ_1, and $I = 0$ for all regions other than the AZ91D and steel electrodes in Figs 12.2 and 12.3. The boundary integral equations for all elements are assembled into a system of linear simultaneous equations, which is expressed in a matrix form as follows [21]:

$$H\phi = GI \tag{12.6}$$

where H and G are problem influence matrices, and ϕ and I are potential and current density vectors. Partitioning ϕ and I into the nodes, which form the anode and the cathode regions, and applying the boundary condition (equations (12.4) and (12.5)), gives [21]:

$$\begin{bmatrix} h_{aa} & h_{ac} \\ h_{ca} & h_{cc} \end{bmatrix} \begin{bmatrix} \phi_a \\ \phi_c \end{bmatrix} = \begin{bmatrix} g_{aa} & g_{ac} \\ g_{ca} & g_{cc} \end{bmatrix} \begin{bmatrix} f_a(\phi_a) \\ f_c(\phi_c) \end{bmatrix} \tag{12.7}$$

where $\begin{bmatrix} h_{aa} & h_{ac} \\ h_{ca} & h_{cc} \end{bmatrix}$ and $\begin{bmatrix} g_{aa} & g_{ac} \\ g_{ca} & g_{cc} \end{bmatrix}$ are elements of the problem influence matrices, and $\begin{bmatrix} \phi_a \\ \phi_c \end{bmatrix}$ and $\begin{bmatrix} f_a(\phi_a) \\ f_c(\phi_c) \end{bmatrix}$ describe the anode and cathode electrode kinetics. Equation (12.7) is solved by the Newton–Raphson iterative method.

Jia *et al.* [16–20] used BEASY to model the AZ91D–steel galvanic couples illustrated in Figs 12.2 and 12.3. A gradient of increasingly fine mesh size was used towards the AZ91D–steel interface where there was a sharp potential gradient. Otherwise, the mesh size of electrolyte elements was evenly divided to reduce the element number. Figure 12.5 is a schematic

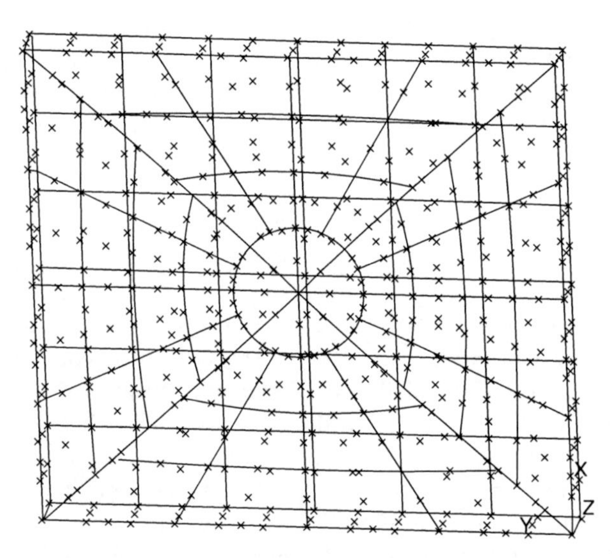

12.5 Schematic of the mesh for the BEM model for a steel fastener modelled as a steel inset (diameter = 30 mm) in a horizontal AZ91D sheet covered with an electrolyte of uniform depth [20].

of the BEASY mesh for a steel insert in an Mg sheet. The input parameters for the BEASY BEM model were the physical geometry, the electrolyte conductivity and the boundary conditions: $\begin{bmatrix} \phi_{a0} \\ \phi_{c0} \end{bmatrix}$ and $\begin{bmatrix} f_a'(\phi) \\ f_c'(\phi) \end{bmatrix}$, where ϕ_{a0} is the anode open circuit potential; ϕ_{c0} is the cathode open circuit potential; $f_a'(\phi)$ is the function for the anode polarization curve; similarly $f_c'(\phi)$ is the cathode polarization curve (Fig. 12.6). Functions $f_a'(\phi)$ and $f_c'(\phi)$ were defined in a piecewise linear manner using the values in Table 12.1, based on Fig. 12.6. The polarization curve was divided into small segments, with each segment approximated by a linear relationship between the potential and current:

$$I = f(\phi) = k(\phi - \phi_a) + I_a \qquad 12.8$$

where k, ϕ_a, I_a and were constants for each line segment.

12.3 One-dimensional (1D) galvanic corrosion

12.3.1 Experimental measurement

The GCA is shown schematically in section in Fig. 12.2. The top surface of the GCA models a galvanic couple of AZ91D–steel exposed to the 5% NaCl solution. The AZ91D is on the left-hand side and the steel is on the right-hand side. Electrical connections through the bottom of the GCA allow measurement of the galvanic current density distribution. AZ91D and mild steel were the materials. AZ91D is the most widely used Mg alloy. AZ91D contains ~9% Al and ~1% Zn. Mild steel is the most widely used metal in industry. A typical severe environment was provided by the 5% NaCl solution.

Metal plates (of AZ91D and steel) were moulded in epoxy as shown in Fig. 12.2, with the cross-sectional area of the top of each plate exposed to the solution. Moulding of the plates into the GCA was such that there was no electrical contact between each plate through the epoxy (which is a good insulator). Electrical contact was made to each plate via a lead through the bottom of the GCA through the epoxy. The GCA was equivalent to a galvanic couple of AZ91D coupled to steel when the top surface of the GCA was exposed to a solution and electrical contact was made between all the plates as shown in Fig. 12.2. Moreover, the current in each plate could be individually measured by the use of a zero resistance ammeter (ZRA) in the electrical circuit as shown in Fig. 12.2 (and the current density for each plate could be calculated by dividing the measured current by the area of the plate exposed to the solution). This allowed measurement of the current density distribution across the AZ91–steel galvanic couple.

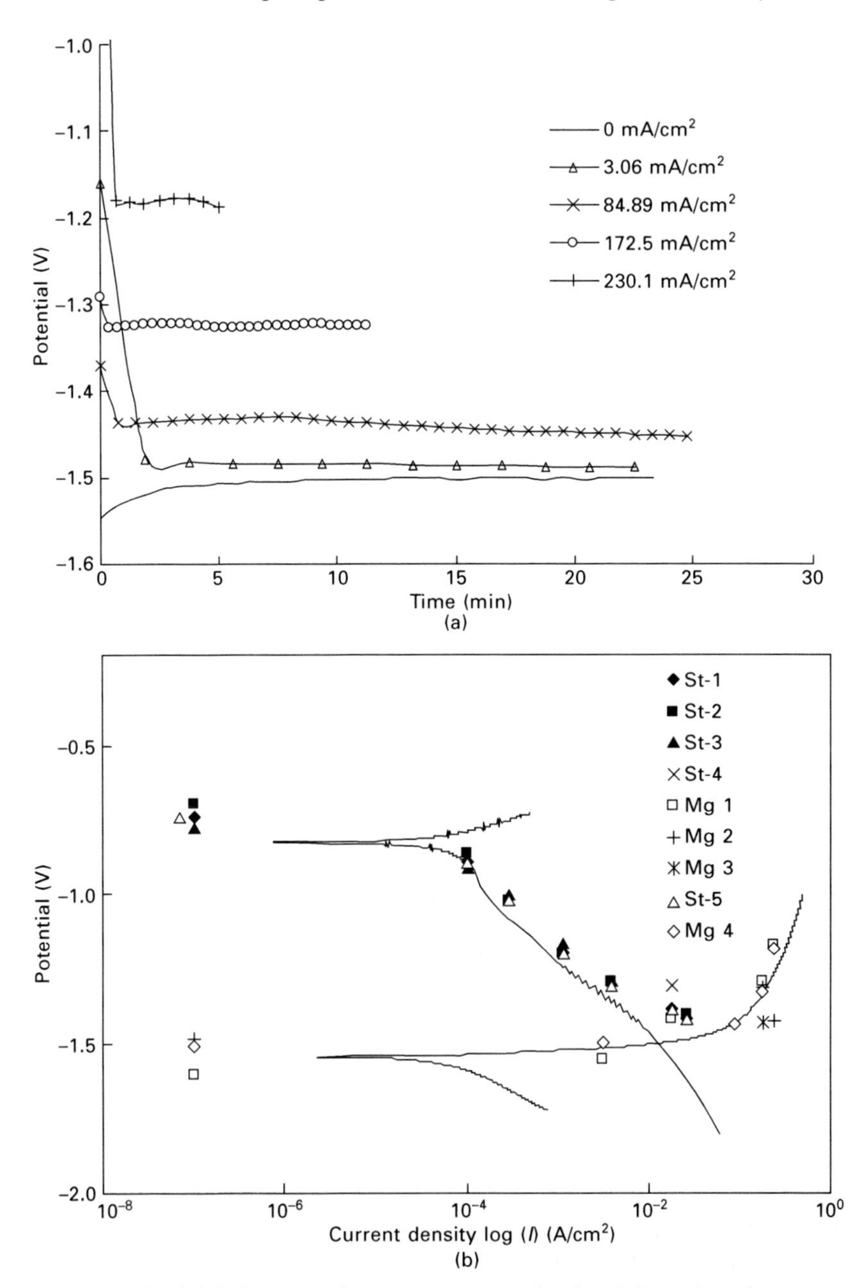

12.6 (a) Galvanostatic measurement of potential vs. time for AZ91D in 5% NaCl solution at room temperature. The potential of AZ91D rapidly reached steady state values [19]. (b) Galvanostatic polarization curve measurements in 5% NaCl, data points (from measurements such as those shown in (a)) compared with curves representing typical potentiodynamic polarization curves, at a scan rate of 2 mV/s [19].

Table 12.1 Boundary conditions in 5% NaCl solution used for BEM calculations, related to the data of Fig. 12.6

Current density (mA/cm^2)	Potential (V) BEM 1	Potential (V) BEM 2	Potential (V) BEM 3
Steel			
0.000	−0.771	−695	−0.741
0.100	−0.906	−0.863	−0.896
0.290	−1.003	−1.021	−1.018
1.100	−1.165	−1.202	−1.194
3.860	−1.298	−1.296	−1.305
17.700	−1.307	−1.413	−1.386
25.400	−1.418	−1.403	−1.417
AZ91D			
0.000	−1.480	−1.600	−1.506
−3.058	−1.430	−1.550	−1.492
−84.880	−1.431	−1.431	−1.431
−172.500	−1.310	−1.290	−1.325
−230.100	−1.420	−1.170	−1.183

A zero insulating distance between adjacent plates would be preferable, but an insulation distance of 0.5 mm was the smallest distance found to be practical. The dimension of the assembly into the plane of the figure was large compared with the thickness of each plate, so that the assembly is close to an ideal 1D galvanic couple of AZ91D in contact with steel.

A 12 channel ZRA allowed 12 simultaneous measurements, and the standard configuration was that shown in Fig. 12.2. Twelve plates of identical size were used for the galvanic current measurements: six plates of Mg and six plates of steel. Each of the metal plates had the following dimensions: 1.2 mm thickness × 35 mm width, with 42 mm^2 surface area exposed to the electrolyte. These plates were numbered 1 to 6 for both the AZ91D plates and the steel plates, with the numbers increasing as the plates were further from the AZ91D–steel interface. To ensure that plates 5 and 6 were able to provide current measurements at a sufficient distance from the AZ91D–steel interface, an extra plate of considerable thickness was added between plates 4 and 5, and also between plates 5 and 6. Moreover, to ensure there were no end effects for the AZ91D plate 6, there was an additional AZ91D plate moulded after AZ91D plate 6.

This GCA allowed investigation of various anode/cathode ratios and different insulation spacing between the AZ91D and steel by appropriate electrical connections of the plates. The GCA with the connections as shown in Fig. 12.2 allowed measurement of the galvanic current density distribution for an 'infinite' plate of AZ91D coupled to a finite plate of steel as an idealization of a large plate of AZ91D adjacent to a small plate of steel. The configuration was chosen to maximize the extent of the current density distribution that could be measured for the AZ91D side of the couple, that is, the current density in the

AZ91D plate 6 should be small compared with the current density in AZ91D plate 1. The plates 1 to 6 for AZ91D and steel were arranged in a symmetrical pattern with respect to the AZ91D–steel interface. Figure 12.7(a) illustrates the measurement of the galvanic current density for AZ91D electrodes in 5% NaCl; in all cases steady state values were reached in 40–55 minutes. The steady state values for the current density distribution were plotted (Fig. 12.7(b)) in terms of the actual distance (in cm) from AZ91D plate 6, so that the AZ91D–steel interface was at a distance $D = 2\,\text{cm}$.

For these connections, the area ratio of anode/cathode was designated as 1:1 as the current density measurement extended for 2 cm over AZ91D from AZ91D plate 1 to AZ91D plate 6, and likewise the current density measurement extended for 2 cm over the steel from steel plate 1 to steel plate 6. A larger area ratio of anode/cathode, designated as 2:1, was produced by removing the electrical connections to steel electrode 5, steel electrode 6, and the steel electrode between these two. The current density measurement extended for 2 cm over AZ91D from AZ91D plate 1 to AZ91D plate 6, and only for about 1 cm over the steel from steel plate 1 to steel plate 4. A smaller area ratio of anode/cathode, designated as 1:2, was effected by keeping the electrical connections to all the steel electrodes as in Fig. 12.2 and removing the electrical connections to AZ91D electrode 5, AZ91 electrode 6 and the AZ91D electrodes between these two and after AZ91D electrode 6. The current density measurement extended for about 1 cm over AZ91D from AZ91D plate 1 to AZ91D plate 4, and for 2 cm over the steel from steel plate 1 to steel plate 6.

The influence of solution depth was studied using the arrangement as shown in Fig. 12.2 (this has a solution depth $d = 10\,\text{mm}$), and with solution depths of $d = 1$ mm and $d = 3\,\text{mm}$.

With the connections shown in Fig. 12.2, the AZ91D–steel galvanic couple has an insulating distance $s = 0.5\,\text{mm}$. An insulating distance $s = 2.2\,\text{mm}$ can be produced by removing the electrical connection to either AZ91D plate 1 or steel plate 1 and by covering the plate with epoxy resin. $s = 3.9\,\text{mm}$ can be produced by removing the electrical connection to both AZ91D plate 1 and steel plate 1 and covering these plates with epoxy resin. Similarly, $s =$ 7.3 mm, 10.7 mm … can be produced by removing extra electrical connections and covering these plates with epoxy resin. This approach allows study of the influence of insulating distance between the anode and cathode. Inherent in this approach is that the area of anode and cathode decreases as the insulating distance is increased.

12.3.2 Effect of area ratio of anode/cathode

Figure 12.7 illustrates the galvanic current density distribution for the AZ91D–steel galvanic couple in 5% NaCl solution. (The galvanic current

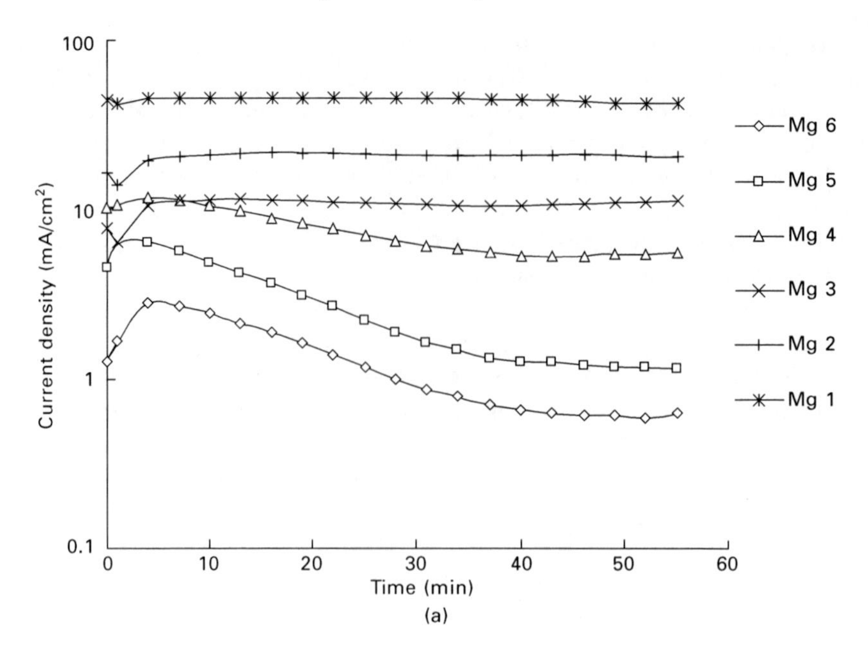

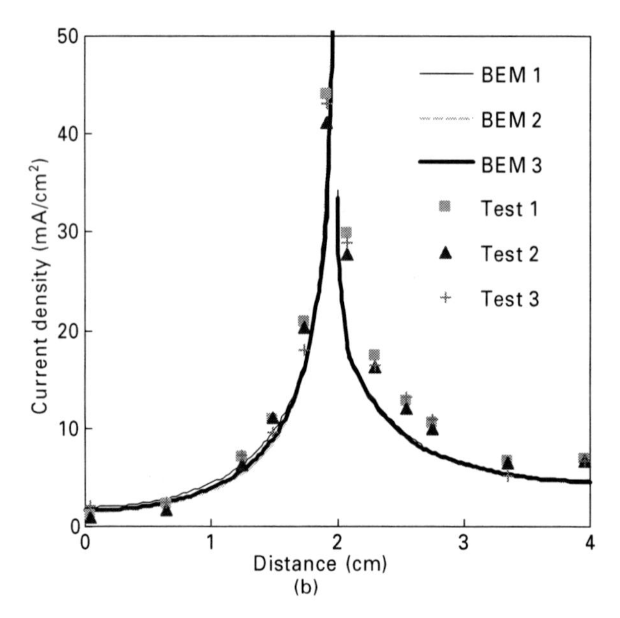

12.7 (a) Galvanic density at AZ91D electrodes (in the GCA, see Fig. 12.2, designated Mg 1 to Mg 6) in 5% NaCl solution [19] (b) Comparison of BEM model (curves) and experiment data (data points) for an area ratio of A/C of 1:1. d = 10 mm. AZ91D is on the left-hand side, steel on the right-hand side.

density was measured over a timespan of 55 min [16–19], sufficient to ensure steady state, and steady state current density values are used throughout this chapter.) The current density distributions from the BEM model are compared with the experimental measurement for an area ratio of anode/cathode of 1:1 in Fig. 12.7(b). The BEM model used the three boundary conditions as given in Table 12.1. These are the three curves designated BEM 1, BEM 2 and BEM 3. The results of three independent experimental measurements are plotted as the data points and labelled Test 1, Test 2 and Test 3. There was good agreement between the BEM model and the experimental measurements.

The galvanic current density of the AZ91D electrode increased (as the area ratio was increased from 2:1 to 1:2) from 33 to 44 mA/cm^2 for electrode Mg 1 (closest to the AZ91D–steel interface), from 14 to 22 mA/cm^2 for electrode Mg 2, from 11 to 17 mA/cm^2 for electrode Mg 3 and from 10 to 14 mA/cm^2 for electrode Mg 4. The galvanic current density of the AZ91D electrodes increased with decreasing anode/cathode area ratio as expected both experimentally and theoretically from prior study [4]. As the area ratio of anode/cathode increases, there a higher relative cathodic current density is available. Since the anode current equals the cathode current, the anode current density increases. The influence of anode/cathode area ratio on the galvanic current density distribution for AZ91D–steel in 5% NaCl can be predicted using the BEM model.

12.3.3 Influence of solution film depth, *d*

The galvanic current density distributions from the BEM model were compared with the experimental measurements for solution film depth values of 10, 3 and 1 mm. The BEM models were calculated using the three different boundary conditions as presented in Table 12.1. The galvanic current density of each electrode increased with increasing solution film depth as expected because the increase of the solution film depth resulted in a larger area for the current and thus reduced the resistance against current flow. There was good agreement both in trend and in current density value between the BEM model and experimental measurement. The scatter in the experimental measurement of the galvanic current density was comparable to or greater than the scatter in the BEM curves.

12.3.4 Effect of insulation distance, *s*

The prior sections have shown that the BEASY BEM program gave calculated values in good agreement with the experimental measurements. As a consequence, the BEASY BEM program was used to model the GCA and to calculate the galvanic current density distribution for the AZ91D–steel galvanic couple in 5% NaCl for insulation distance s = 0.5, 2.2, 3.9, 7.3, 10.7 and 25 mm. Figure 12.8 presents the calculated galvanic current density distribution for each case. The symbol n designates the general curve; n has the values 1, 2, 3, 4 5 and 6. This approach studied the influence of insulating distance between the anode and cathode and obtained the following: (1) the current density distributions, presented in Fig. 12.8, (2) the maximum galvanic current density for the AZ91D electrode immediately adjacent to the steel, (3) the total galvanic current $I_T(n)$ and (4) the average galvanic current density I_a. The total galvanic current was equal to the integration of the current density values over the area $A(n)$ of each current density distribution. The average galvanic current was equal to $I_T(n)/A(n)$.

The galvanic current density of the AZ91D electrodes decreased as the insulating distance increased (Fig. 12.8). Furthermore, the maximum galvanic current density for the AZ91D electrode adjacent to the steel decreased significantly with increasing insulating distance. Moreover, the total galvanic current for AZ91D electrodes also decreased as the insulating distance increased. Interpretation of the significance of the total current needs to take into account that two variables were changed simultaneously: the area was

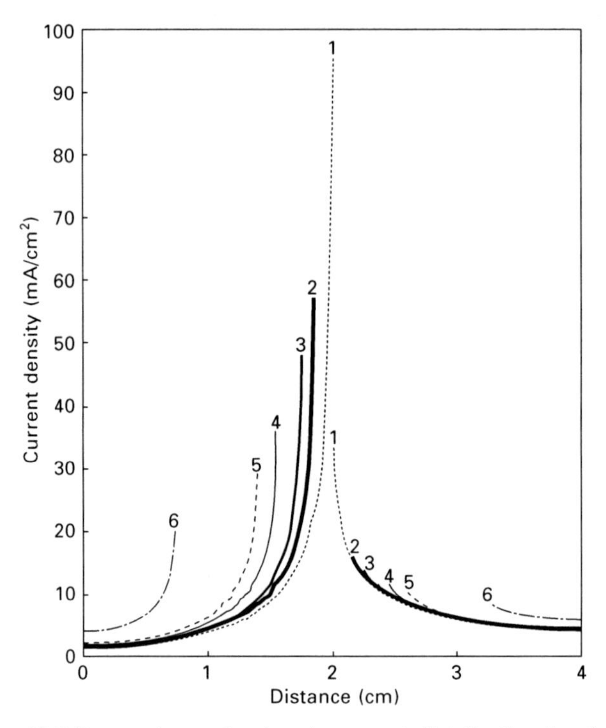

12.8 Comparison of galvanic current distribution for the following distances, d = 0.5 mm (curve 1), d = 2.2 mm (curve 2), d = 3.9 mm (curve 3), d = 7.3 mm (curve 4), d = 10.7 mm (curve 5) and d = 25 mm (curve 6) [16].

decreased and the insulating distance was increased. The average current density decreased with increasing insulating distance.

In another words, increasing insulating distance leads to less galvanic corrosion both locally and on average. The influence of the insulating distance on the galvanic current distribution of steel electrodes was the same as that for the AZ91D electrodes. This BEM simulation was within expectations. It is also consistent with the results and the theoretical predictions of Song *et al.* [4]. Increasing insulating distance increases the length of the circuit, thus increases the resistance of the circuit and decreases the galvanic current density according to Ohm's law.

12.3.5 Solution

In 5% NaCl solution, the galvanic current was well predicted by the BEM-based BEASY program for the AZ91D–steel couple. Five percent NaCl solution is an aggressive environment and intense hydrogen evolution stirred up the corrosion product. The galvanic corrosion of AZ91 in 5% NaCl solution was unlikely to be influenced by the formation of a surface film. The galvanostatic polarization curve of AZ91D in 5% NaCl solution rapidly approached steady state, was in good agreement with the potentiodynamic polarization curve and was likely to be representative of the galvanic corrosion behaviour of the AZ91D.

12.3.6 Steady state

Measurements of polarization curves were carried out in a potentiostatic manner. For each data point for each curve, a constant current was applied, and the potential was measured. Care was taken to ensure that steady state had been reached (Fig. 12.6(a)) so that the measured potentiostatic curves (Fig. 12.6(b)) represented steady state conditions [16–19]. Similarly, for the measurement of galvanic current density using the GCA, care was taken to ensure that the galvanic current density measurements represented steady state conditions (Fig. 12.7(a)) [16–19]. This meant that it was valid to compare the experimentally measured galvanic current density with the predictions of the BEM model because both were for steady state conditions, and were thus directly comparable. The fact that there was good agreement between the experimental measurements and the BEM calculations provides assurance that the approach is sound to use the GCA and the GIA to measure galvanic current density distributions.

12.3.7 Influence of experimental error

The influence of the experimental error in the measurement of each polarization curve as input to the BEM model meant that the BEM model gave three different curves, which had slightly different values. Similarly, there was experimental scatter in the experimental measurement of the current density for each metal plate of the GCA. Figures 12.7(b) and 12.9 indicated that the directly measured galvanic current density had an experimental scatter larger than the differences in the BEM curves caused by the scatter in the measurements of the polarization curves.

12.3.8 Experimental approach

Figure 12.7(b) indicates that there was good agreement between the BEM model and the experimental measurement of the current density. The experimental measurements corresponded to measurements at discrete positions from individual metal plates that were separated from each other by an insulation material 0.5 mm in thickness. In comparison, the BEM model was for a continuous electrode with no interruptions due to insulating spacers. The agreement between the BEM model and the experimental measurements

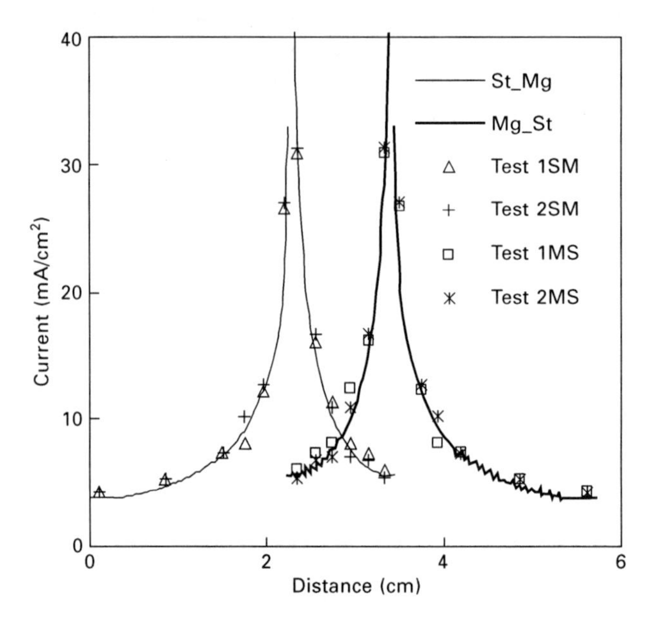

12.9 Comparison between BEM model (curves) and the experimental measurements (data points) of galvanic current without galvanic current interaction. Separate calculations were first made for the steel–Mg couple, i.e. connecting only the six steel and the six AZ91D plates on the left of the galvanic corrosion interaction specimen illustrated in Fig. 12.3 to give the curve St_Mg and the data (Test 1SM and Test 2SM). Subsequently calculations and measurements were made for Mg_St by connecting the six AZ91D plates to the six steel plates on the right. d = 10 mm [17].

indicated that the GCA provides a sound methodology for the measurements of galvanic current density distributions, and that the BEM model provided good predictions of the galvanic current density distribution.

12.4 Galvanic interaction

12.4.1 Experimental

The GIA [16,17] allowed investigation of the interaction of two (independent) galvanic couples. A section through the GIA is shown schematically in Fig. 12.3. The GIA consisted of a steel–AZ91D–steel arrangement (designated as St_Mg_St), with current measurement as for the GCA. The arrangement was symmetrical, so the extent of the steel was identical on both sides of the central AZ91D. Furthermore, there was flexibility in making electrical connections. The concept was that the AZ91D would experience current from two identical galvanic couples, namely there was (1) the left-hand side galvanic couple St_Mg and (2) the right-hand side galvanic couple Mg_St. Furthermore, the left-hand galvanic couple St_Mg was identical to the right-hand galvanic couple Mg_St. The GIA allowed measurement of the current density distribution for the full interaction situation, namely steel–AZ91D–steel, and moreover, allowed independent measurement of the two independent galvanic couples: (1) St_Mg and (2) Mg_St. This allowed investigation of the kind of interaction that would describe how the current density for the case of the two interacting identical galvanic couples was combined from the individual galvanic couples.

The GIA consisted of 12 mild steel plates and 6 AZ91D plates as shown in Fig. 12.3. The surface area of each plate exposed to the electrolyte was 56 mm^2 for each steel plate with 1.6 mm (thickness) × 35 mm (width) and 42 mm^2 for each AZ91D plate with 1.2 mm (thickness) × 35 mm (width). The connections for the measurement of the galvanic current for the GIA could be made connecting all the plates as shown Fig. 12.3. This arrangement is designated St_Mg_St, and allows the measurement of the interaction of the galvanic corrosion from the galvanic couple on the left (i.e. St_Mg) and from the galvanic couple on the right side (i.e. Mg_St). The current density was plotted against distance from the left-hand steel electrode 1, as shown in Fig. 12.9. This same convention was used throughout when presenting the current density for the GIA regardless of the details of the electrical connections. This convention was adopted to facilitate comparison between the plots.

There are 18 connections for the arrangement with all electrodes connected as illustrated in Fig. 12.3; however, the ZRA had only 12 channels, so that only 12 currents could be measured simultaneously. The measuring procedure adopted was to measure the currents from the 12 electrodes from the left, or the 12 electrodes from the right. A total of five tests were carried out.

In addition to the arrangement with all electrodes connected as illustrated in Fig. 12.3, the electrical connections could also be made to separately measure the galvanic currents from the left-hand side couple (St_Mg) or alternatively separately measure the galvanic currents associated with the right-hand side couple (Mg_St). For these measurements, there were no electrical connections to the six right-hand side electrodes (RHS steel electrodes 1 to 6) for the St_Mg couple; or alternatively there were no electrical connections for the left steel electrodes 1 to 6 for the Mg_St couple. In each case, all currents could be measured simultaneously as there were 12 electrodes and also 12 channels for the ZRA. The data are plotted as shown in Fig. 12.9.

12.4.2 Linear superposition

Figure 12.9 presents both (a) the galvanic current density distribution for the steel–AZ91D galvanic couple (curve St_Mg) without any galvanic interaction, and also (b) the AZ91D–steel galvanic couple (curve Mg_St) without any galvanic interaction. There was a good agreement between the BEM model and the experimental measurements for both cases (a) and (b).

Figure 12.10 presents the curve 'Addition' as the linear addition of the galvanic current density for the AZ91D electrodes calculated for steel–

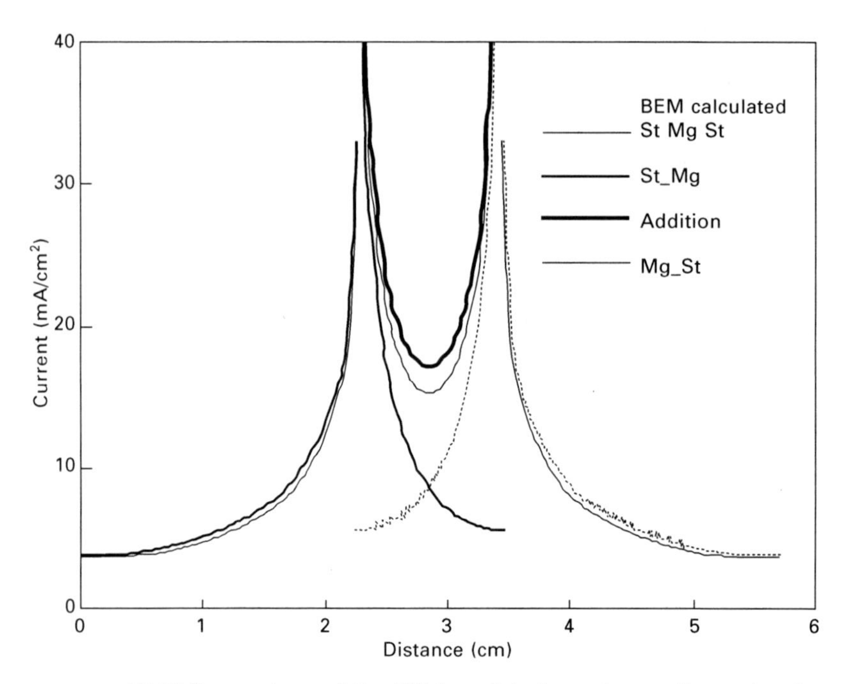

12.10 Comparison of the BEM model of non-interacting galvanic current (St_Mg and Mg_St) and linear addition of non-interacting galvanic current (curve 'Addition') [17].

AZ91D couple and AZ91D–steel couple, without galvanic interaction. The superimposed galvanic current density for the AZ91D electrodes was in good agreement with the BEM model for the steel–AZ91D–steel galvanic interaction assembly.

12.4.3 Theoretical background

The experimental result of linear superposition is a somewhat surprising outcome. The boundary conditions are non-linear as is clear from Fig. 12.6. Linear superposition normally operates in a linear system. In a non-linear system, a theoretical error of second order of deviation is introduced, although this may be a small error.

Furthermore, linear-superposition is not expected from a theoretical treatment of galvanic corrosion by Song [14]; the galvanic current of AZ91D in the galvanic system 'steel–AZ91D–steel' cannot be simply expressed as the sum of the galvanic current densities of AZ91D in two separate galvanic systems 'steel–AZ91D' and 'AZ91D–Steel'. The physical reason for non-linear behaviour may lie in the fact that the steel plates on both side of AZ91D can interact with each other. This 'remote' influence by an indirectly connected remote material is not considered in the 'steel–AZ91D' and 'AZ91D–steel' systems. Therefore, when the galvanic current densities of two simple couples are added together, the 'remote effect' will not be involved in the sum of these two current densities. A lack of the 'remote' effect in the sum of the galvanic current densities could be responsible for the inadequacy of the superposition principle in the 'a–b–c' system.

The apparent linear-superposition of the experimental data and BEM model predictions is attributed to the fact that the amount of interaction is small, so that it is not possible to measure or discern any non-linear effects.

12.5 Steel fastener

This section describes the analysis of a practical component modelled [20] as an AZ91D sheet coupled to a steel fastener. The total corrosion rate was modelled and measured as a function of distance from the Mg–steel interface. Figure 12.11 illustrates the physical appearance of the AZ91D plate with a 30 mm steel insert after 48 h immersion in 5% NaCl. The AZ91D sample exhibited unevenly distributed corrosion due to the galvanic couple with steel. Corrosion damage was deeper at the AZ91D–steel interface, and included localized corrosion further away from the interface. Corrosion damage to AZ91D was concentrated around the steel and was largely confined to within ~2 cm radial distance from the interface. The radial extent of the corrosion was similar in all cases. There was some localized corrosion distributed randomly on the remaining surface of the AZ91D. However, most of the remaining area

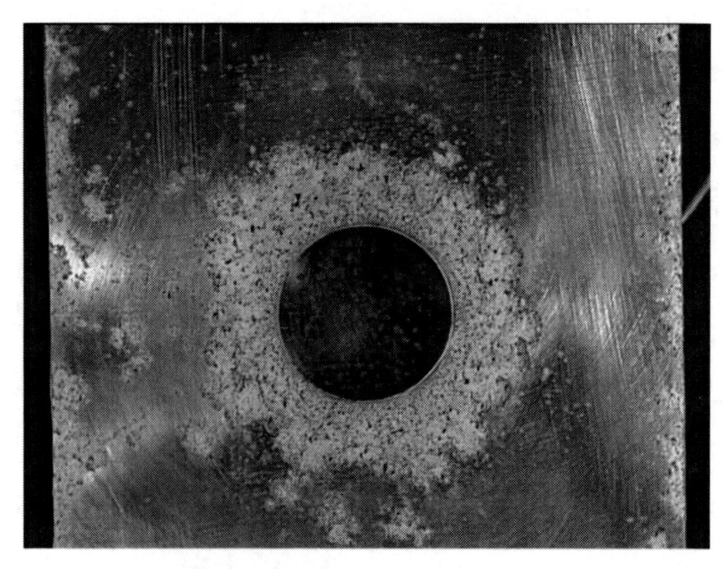

12.11 Appearance of the fastener assembly (30 mm diameter steel cathode) after immersion horizontally in the 5% NaCl solution [20].

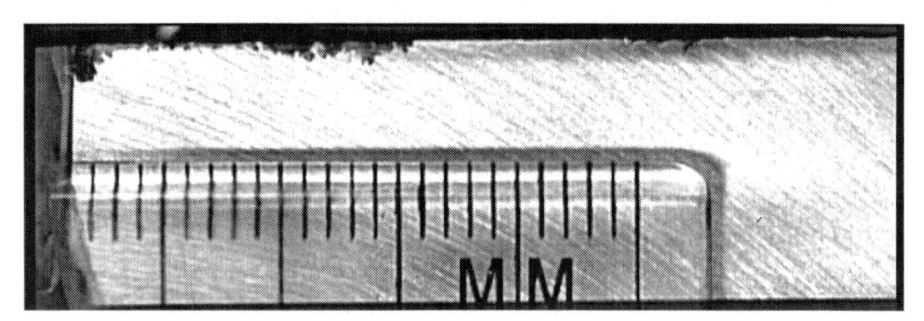

12.12 Image of corrosion on Mg surface coupled with 10 mm diameter steel insert [20].

was free from corrosion and retained its original appearance. This indicated that the effective distance of galvanic corrosion was limited to a distance of ~ 2 cm. Most of the remaining area seemed to be protected as a cathode. A similar finding was also reported by Hawke [22], who derived an effective insulating space of about 5 mm for Mg coupled with cast iron.

Comparison of the physical appearance of the corrosion, with the area of galvanic corrosion as simulated by the BEM model, showed that the computer model provided good predictions of the effective galvanic corrosion area for each of the three different sizes of steel cathodes in 5% NaCl solution.

Figure 12.12 illustrates the typical distribution of total corrosion on a random section through the Mg sheet. Figure 12.13 provides a comparison between the total corrosion rate on the corroded area for each small cross-

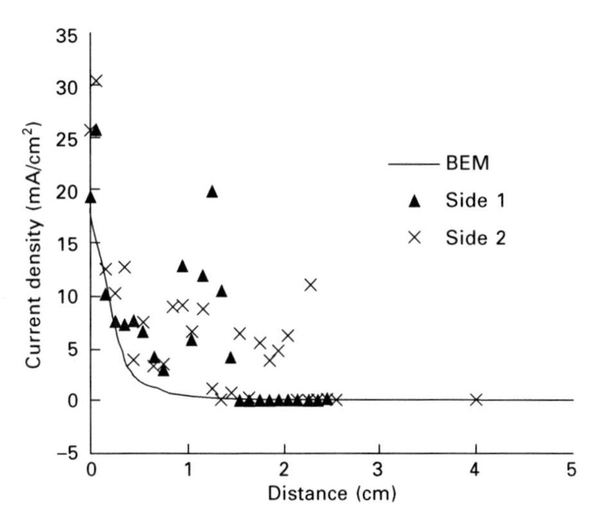

12.13 Comparison of corrosion current density between the BEM model and the experiments for Mg couple with 10 mm diameter steel insert [20].

section, and the galvanic current derived from the BEM model calculation for AZ91D coupled with a 10 mm diameter steel insert. The total corrosion rate for each small cross-section was calculated from the corresponding measured corroded area. This total corrosion rate includes galvanic and self-corrosion, both of which may be in the form of localized corrosion and general corrosion. Analysis of the cross-sectional damage indicated that the corrosion of AZ91D included unevenly distributed localized corrosion. The maximum corrosion depth occurred close to but not always immediately adjacent to the Mg–steel interface. Ault and Meany [23] reported a similar finding. The maximum depth occurred within about 1 mm away from the interface. The depth distribution of the galvanic localized corrosion was quite randomly scattered on the surface and there was no obvious relationship between the depth and the pit location within a 2 cm distance.

Figure 12.13 illustrates the comparison of the BEM model predicted galvanic corrosion and the experimental measurement of the total corrosion for AZ91D coupled with a 10 mm steel insert. Both the BEM model and the experimental measurements indicate maximum corrosion at the Mg–steel interface, and both indicated a decrease in corrosion rate with increasing distance from the interface.

The experimental measurements indicate that the measured corrosion rate was significantly higher than the galvanic corrosion rate, and this was interpreted as self-corrosion of Mg. This self-corrosion was estimated by subtracting the galvanic corrosion (as predicted by the BEM model) from the total measured corrosion. The value as estimated was ~10 mA/cm^2 at the

Mg–steel interface and similar values were estimated at a distance of 1 cm from the interface. This self-corrosion corresponds to a penetration rate of ~230 mm/yr.

12.6 Discussion

12.6.1 Self-corrosion

The postulation of self-corrosion in Section 12.5 is perhaps not desirable. It is useful to consider whether the postulation of self-corrosion indicates some fundamental flaw in the methodology of predicting galvanic corrosion as presented in Sections 12.2–12.5. Self-corrosion of Mg was postulated [20] to explain the fact that the measured corrosion rate (Fig. 12.13) was significantly higher than the BEM model prediction of the galvanic current [20]. However, the measured galvanic current (Figs 12.7(b) and 12.9 [16–20]) was in good agreement with the galvanic current predicted by the BEM model, based on the experimentally measured polarization curves. So perhaps there is something strange in the Mg corrosion mechanism, maybe this is another manifestation of the negative difference effect (NDE) [1–3]. Alternatively, there may be an issue with the measured polarization curves for Mg, which cannot correctly represent the self-corrosion of an Mg alloy.

12.6.2 Uni-positive Mg^+ ion

The corrosion rate predicted by electrochemistry is lower than the actual rate because of the uni-positive Mg^+ ion [1–3,24–28]. The corrosion of Mg converts metallic Mg to the stable ion Mg^{++} in two electrochemical steps, involving the uni-positive ion Mg^+ as an intermediate as given by Eqs (12.9) and (12.10). The electrochemical reactions are balanced by hydrogen evolution, Eq. (12.11).

$$Mg \rightarrow Mg^+ + e^- \qquad \text{anodic reaction} \qquad 12.9$$

$$kMg^+ \rightarrow kMg^{++} + ke^- \qquad \text{anodic reaction} \qquad 12.10$$

$$(1 + k)H_2O + (1 + k)e^- \rightarrow \tfrac{1}{2}(1 + k)H_2 + (1 + k)OH^- \qquad \text{cathodic reaction} \qquad 12.11$$

The uni-positive ion, Mg^+, is reactive, and can react chemically with water. Thus, a fraction, k, of the uni-positive Mg^+, reacts electrochemically via Eq (12.10) to Mg^{++}, and the complement reacts chemically via Eq. (12.12):

$$(1 - k)Mg^+ + (1 - k)H_2O + (1 - k)OH^- \rightarrow (1 - k)Mg(OH)_2 + \tfrac{1}{2}(1 - k)H_2 \qquad \text{chemical reaction} \qquad 12.12$$

The overall reaction is given by:

$$Mg + 2H_2O \rightarrow Mg(OH)_2 + H_2 \qquad \text{overall reaction} \qquad 12.13$$

The corrosion of Mg is thus only partly electrochemical, and electrochemical measurements predict [31] corrosion rates, P_i, lower than the real corrosion rate, P_H as determined for example by hydrogen evolution. The apparent electrochemical valence, $(1 + k)$, is determined by the quantity $(2P_i/P_H)$. For example Petty *et al.* [29] measured the apparent valence to be 1.5 in 150 g/L NaCl. If this was the only important effect, electrochemical measurements should always underestimate the actual corrosion rate by a constant fraction (which might depend on solution).

Figure 12.13 indicates that the corrosion rate is not simply a constant factor greater than the corrosion rate predicted by the BEM model, so the negative difference effect (NDE) by itself may not be sufficient to explain Fig. 12.13. There could be other factors.

12.6.3 Underestimation of galvanic current density

As the NDE is a phenomenon due to the unique electrochemical behaviour of Mg alloys, the error caused by the NDE effect as discussed above is not present in steel, nickel, or copper galvanic corrosion. Apart from the NDE-induced error, another source of error is that anodic polarization or the galvanic current does not always reflect the real galvanic corrosion rate of a metal. This is particularly true for any metal under weak anodic polarization. It is well known that:

$$i_p = i_g = i_a - i_c \qquad 12.14$$

where i_p and i_g are the measured anodic polarization and galvanic currents, respectively; i_a represents the real (theoretical) anodic dissolution or galvanic dissolution rate of the metal; i_c is the real cathodic current. In theory, there is always:

$$i_p = i_g < i_a \qquad 12.15$$

Only under a strong anodic polarization condition, $i_c \rightarrow 0$, is there:

$$i_p = i_g \approx i_a \qquad 12.16$$

Therefore, in a galvanic system, it is a normal phenomenon that the real galvanic corrosion is more severe than that estimated by a measured anodic polarization or galvanic current, particularly in the zone relatively far away from the cathode where anodic polarization has become weak and the cathodic current is not negligible.

In summary, NDE caused computer-modelling underestimation of the galvanic corrosion damage mainly occurs in the zone next to the cathode, and significant i_c is responsible for the corrosion damage more severe than computer prediction in areas far away from the cathode.

12.7 Conclusions

The GCA and GIA provide sound methodologies for the measurements of galvanic current density distributions, within their inherent limitations. The BEM-based BEASY program can reasonably predict the galvanic current density distribution for AZ91D–steel in 5% NaCl solution.

The directly measured galvanic current density had an experimental scatter larger than the differences in the BEM curves caused by the scatter in the measurements of the polarization curves. The galvanic current density of the AZ91D electrodes increased with decreasing area ratio of anode/cathode.

The galvanic current density of each electrode increased with increasing solution film depth as expected because the increase of the solution film depth resulted in a larger area for the current to pass and thus reduced the resistance against the current flow. The galvanic current density of the AZ91D electrodes decreased as the insulating distance increased. The maximum galvanic current density for the AZ91D electrode adjacent to the steel decreased significantly with increasing insulating distance. Furthermore, the average galvanic current for AZ91D electrodes decreased as the insulating distance increased. In other words, increasing insulating distance leads to less galvanic corrosion attack both locally and on average over the distribution.

The galvanic current density distribution on AZ91D electrodes caused by galvanic interaction can be experimentally estimated as the linear superposition of current caused by each individual galvanic couple. However, this is not expected to be always the case.

Both the BEM model and the experimental measurements of the galvanic corrosion of AZ91D sheet coupled to a steel insert showed a similar distribution of the current density distribution: a maximum at the interface and decreasing rapidly to zero within ~2 cm from the interface.

The total corrosion was interpreted as being due to galvanic corrosion plus self-corrosion for the galvanic corrosion of Mg sheet coupled to a steel insert. The self-corrosion was evaluated on the basis that the BEM model provides a good evaluation of the galvanic corrosion. On this basis, the self-corrosion rate was typically ~230 mm/yr for the area surrounding the interface and to a distance of about 2 cm from the interface. However, the necessity to postulate self-corrosion indicates shortcomings in the methodology.

12.8 Future trends

12.8.1 Methodology of computer simulation

The galvanic current density of the AZ91D electrodes decreased as the insulating distance increased. The increase of the insulating distance increases the length of the circuit and thus increases the resistance of the circuit. Therefore the galvanic current density of the adjacent anode and cathode decreases according to Ohm's law. It was suggested that these results could produce simple empirical formulations to provide rough estimates for service conditions. However, a much better way forward is the development of a user-friendly BEM package that could be used as a design tool to give quantitative evaluations of the galvanic current density distribution associated with a particular design. The research summarized herein [16–20] gives confidence that such a BEM package is possible. However, additional research is needed to address a number of points.

Thin solution films

It is conceivable that Mg corrosion under thin (~ μm) surface solution films is different from that in a bulk solution [1,2]. In particular, oxygen reduction may be an important cathodic reaction. Local alkalization in the thin solution film facilitates formation of a partially protective surface film. These considerations are of particular importance when considering realistic exposure of auto components to salt spray. Thin solution films and high solution surface area imply high availability of oxygen for the cathodic reaction. There is also the issue that the corrosion rate for Mg in intermittent salt spray is much lower than the corrosion rate in immersion tests [30] (Table 12.2).

Surface film composition

Use of a 5% NaCl solution gave BEM predictions in good agreement with experimental measurements. This solution represents an aggressive environment, and may be realistic for service conditions in North America or Europe where there is significant de-icing salt usage and there is consequently an issue from the resulting salt spray. An NaCl solution could also be considered as providing a simple analogue to marine exposure. Moreover, the NaCl solution provides a worst-case situation. If a design is formulated

Table 12.2 Corrosion rate (mm/yr) measured using 10 day exposure to ASTM B117 5% NaCl salt spray test for high-purity AZ91 in various metallurgical conditions (F is as-cast; T4 is solution treated: 16 h at 410 °C and water quench, 4 h at 215 °C) compared with the corrosion rate measured using 96 h exposure to 1 M NaCl solution for AZ91 in similar metallurgical conditions (F is as-cast; T4 is solution treated: 100 h at 410 °C and water quench; T6 is solution treated and aged: 100 h at 410 °C and water quench, 5 h at 200 °C). Corrosion rate for pure magnesium is given for comparison

Test for measurement of corrosion rate	F	T4	T6	Pure Mg
Salt spray corrosion rate (mm/yr)	0.64	4	0.15	–
Salt immersion corrosion rate (mm/yr)	16	24	6	1

to provide adequate performance in 5% NaCl, then it would be expected to perform better in a less aggressive environment. Nevertheless, it is important to have an adequate approach to less aggressive environments as Mg is widely used in environments much less aggressive than salt spray.

Surface topography

So far our research has largely addressed 1D galvanic corrosion. Ahead is the challenge of 2D influences.

12.8.2 Galvanic corrosion modelling theory

The development of galvanic corrosion modelling theory is another important future direction. Owing to the complexity in geometric shape of mixed-metal components, predicting the galvanic current and its distribution over a complicated component is difficult. Although computer simulation can estimate the galvanic corrosion damage of such a complicated system, computation is in nature a 'computer experimental process'. It can only demonstrate the influence of the variation of input parameters on the output results. However, to fully understand a system, clearly defined analytical relationships between experimental parameters and experimental phenomena are essential. Thus, computer modelling cannot replace the merit of an analytical solution.

Song [14] recently proposed that many practical galvanic corrosion systems can be reasonably simplified into a 1D model, of which an analytical description of the galvanic current is possible. For example, at least two types of galvanic systems can be treated in one dimension: (1) a fine tube containing electrolyte with cross section area far smaller than its length, and (2) a surface covered by a thin electrolytic film with film thickness much smaller than the length of electrolyte coverage. These two types of simplified 1D model represent a large number of practical galvanic corrosion systems, such as (i) a panel with an organic coating covered by an electrolyte, (ii) a panel exposed to the atmosphere with some condensed liquid on its surface, (iii) a panel with water splashed over its surface, (iv) a water or oil pipeline, (v) a hemmed edge of a vehicle body closure with moisture accumulated in the hem crevice, (vi) a fuel or brake oil pipeline in a vehicle, (vii) coolant and oil paths in an engine block, etc. This insight makes advancement of galvanic corrosion theory possible.

Song [14] indicated that any 1D system can be simplified into two groups of segments: (1) a 'dead end' system and (2) an 'open end' system. The potential and Faraday current density can be written as follows. For a piece of metal (c) having a length c, if a current or potential is applied at its right end, then $I_f = 0$ at $x = 0$ which is a 'dead end'. The potential and current equations for this system can be obtained. The same for a piece of metal

(a) having a length 'a', if a current or potential is applied at its right end, then $I_f = 0$ at $x = 0$ which is a 'dead end'. For a section of metal having length b, if both of its boundary conditions are known, e.g. at the left end, the potential of the electrolyte is Ψ_0^b and current is i_0^b; and at the right end, Ψ_b^b and i_b^b, respectively, then it is an 'open end' system. Its potential and current expressions can be mathematically deduced. The above 'dead end' and 'open end' system can comprise a complicated galvanic corrosion system. The potential and current equations deduced for these elementary systems lay a foundation for deducting potential and current equations of complicated galvanic systems.

For example, if there are two pieces of metal (a) and (c) joined together, each metal will be equivalent to a 'dead end' single piece metal as discussed above, and the current i_g^a flowing out of the electrolytic film over metal (a) should be equal to the current i_g^c into the electrolytic film over metal (c) across the point interface:

$$i_g^a = i_g^{ac} = i_g^c = i_p \qquad 12.17$$

or

$$i_g^a - i_g^c = 0 \qquad 12.18$$

For these two pieces of metal, the potential E^a of metal (a) should be equal to the potential E^c of metal (c) at their joint interface:

$$E^a|_{x=a} = E^c|_{x=0} \qquad 12.19$$

With these relationships, the potential and current of the system can be obtained.

Similarly, if there are three pieces of materials (a), (b) and (c) joined in series such that (a) is joined to (b) which is joined to (c), then materials (a) and (c) are equivalent to a system having a 'dead end', and material (b) that is inserted between metals (a) and (c) can be treated as a section with 'open ends'. Therefore, the potentials and flowing currents over different metals must be equal at their joints. These lead to their theoretical potential and current expressions.

The most important contribution of Song's theory [14] is the extension of the 'dead end' and 'open end' system to a complicated system. The methodology adopted to deduce the potentials and currents for the galvanic systems with two joints and three pieces of material is straightforward; simply let the flowing currents and potentials of the two materials in direct connection be equal at their joint point. This generates a sufficient number of equations to determine the unknown parameters of these galvanic systems. In fact, the methodology can be extended to work out the potentials and current densities for systems containing more than three pieces of material

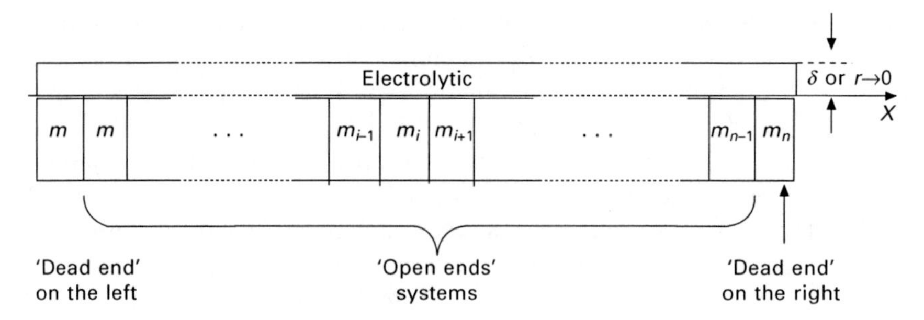

12.14 Schematic illustration of a multi-piece system (complicated system) [14].

joined together in series. For a multiple-material system, in addition to the left and right 'dead end' pieces m_0 and m_n, there are n–1 'open ends' pieces between m_0 and m_n (see Fig. 12.14). In this case, the polarization currents flowing into/out of these 'dead end' and 'open end' pieces can be obtained according to the above discussion:

$$i_b^0 = I_f^0\Big|_{x=0} = \frac{\psi_b^0 \tanh\left(\dfrac{m_0}{L^0}\right)}{\sqrt{\rho_s \rho_p^0}} \qquad 12.20$$

$$i_0^n = I_f^n\Big|_{x=\sum_{i=0}^{n-1} m_i} = -\frac{\psi_0^n}{\sqrt{\rho_s \rho_p^n}} \tanh\left(\frac{m_n}{L^a}\right) \qquad 12.21$$

$$i_0^i = I_f^i\Big|_{x=\sum_{ji=0}^{i-1} m_j} = \frac{\psi_0^i \cosh\left(\dfrac{m_i}{L^i}\right) - \psi_b^i}{\sqrt{\rho_s \rho_p^i}\, \sinh\left(\dfrac{m_i}{L^b}\right)} \qquad 12.22$$

$$i_b^i = I_f^b\Big|_{x=\sum_{ji=0}^{i} m_j} = \frac{\psi_0^i - \psi_b^i \cosh\left(\dfrac{m_i}{L^i}\right)}{\sqrt{\rho_s \rho_p^i}\, \sinh\left(\dfrac{m_i}{L^i}\right)} \qquad 12.23$$

where, i_b^0 is the flowing current at the right end of the first segment of the system, i_0^n is the flowing current at the left end of the last segment of the system, i_0^i and i_b^i are the flowing current at the left and right ends of segment i of the system. These currents should be equal between every two adjacent pieces:

$$i_b^0 = i_0^1$$

$$i_b^1 = i_0^2$$

$$\ldots$$

$$i_b^{i-1} = i_0^i$$

$$\ldots$$

$$i_b^{n-2} = i_0^{n-1}$$

$$i_b^{n-1} = i_n^n \qquad 12.24$$

Parameters m_i, L^i, ρ_s and ρ^i are constants that are known or can be measured, so in total $2n$ variables ψ_0^i and ψ_b^i are involved in the equation set (12.24). They are all simple linear equations. The number of equations is n. At the same time, we know:

$$\Psi_0^0 - \Psi_0^1 = E_{\text{corr}}^0 - E_{\text{corr}}^1$$

$$\Psi_b^1 - \Psi_0^2 = E_{\text{corr}}^1 - E_{\text{corr}}^2$$

$$\Psi_b^{i-1} - \Psi_0^i = E_{\text{corr}}^{i-1} - E_{\text{corr}}^i$$

$$\Psi_b^{n-2} - \Psi_0^{n-1} = E_{\text{corr}}^{n-2} - E_{\text{corr}}^{n-1} \qquad 12.25$$

Equation (12.25) provides an additional n linear equations. Therefore, all the potentials ψ_0^i and ψ_b^i can be evaluated and all have their analytical expressions. Correspondingly, the galvanic potentials and currents over each piece of the element can be obtained. In other words, the galvanic potential and current distributions over a multi-piece system is evaluated using analytical expressions. Although the expressions are more complicated than those for a three-piece system, it is a great advantage to analytically calculate a theoretical galvanic potential or current directly based on system parameters, rather than to simulate them through a numeric modelling process.

The significance of these theoretical expressions for galvanic currents and potentials is as follows:

- This multi-piece system can be easily extended to a more complicated non-uniform system. A non-uniform system can be divided into many small pieces and each of the small pieces can be regarded as a uniform system. Therefore, the complicated non-uniform system becomes a multi-piece system.
- This multi-piece system can also be extended to a micro-galvanic corrosion system. For example, an alloy sometimes consists of different

phases. The distribution of these phases can be regarded as segments of a multi-piece system. Hence, this multi-piece theory can be directly used to deal with a micro-galvanic corrosion of an alloy.

- Based on these theoretical equations for galvanic corrosion systems, computer simulation becomes much easier, as the theoretical calculation significantly reduces the computer trial-and-error iteration process and the physical meanings of the computer simulated results become clear. It is expected that the combination of the theory with computer modelling will be a future tendency.

12.9 Acknowledgements

The work was supported by the Australian Research Council, Centre of Excellence, Design of Light Alloys. The prior work was supported by the Cooperative Research Centre for Cast Metals Manufacturing (CAST). CAST was established and is funded in part by the Australian Government's Cooperative Research Centres Program.

12.10 References

1 G Song and A Atrens, *Advanced Engineering Materials*, **5** (2003) 837.
2 G Song and A Atrens, *Advanced Engineering Materials*, **1** (1999) 11.
3 G Song and A Atrens, *Advanced Engineering Materials*, **9** (2007) 177.
4 G Song, B Johannesson, S Hapugoda and D StJohn, *Corrosion Science*, **46** (2004) 955.
5 G Song, D StJohn, C Bettle and G Dunlop, *Journal of Metals* **57** (2005) 54.
6 G Song and D StJohn, *Materials and Corrosion*, **56** (2005) 15.
7 HP Hack, *Corrosion Tests and Standards Application and interpretation,* ed. R Baboian (1995) Fredericksburg, VA: American Society for Testing and Materials. 186.
8 JT Waber, *Corrosion*, **13** (1957) 95t.
9 DJ Astley, ASTM Spec. Tech. Publ. **978** Galvanic Corros. (1988) 53.
10 DJ Astley, *Chem. Ind. (London), (13, Suppl., Seawater Cool. Chem. Plant)*, (1977) 4.
11 JW Fu, ASTM Spec. Tech. Publ. **978** Galvanic Corros. (1988) 79.
12 RA Adey and SM Niku, *ASTM Spec. Tech. Publ. **978** Galvanic Corros.*, (1988) 96.
13 RG Kasper and EM Valeriote, *Galvanic Corrosion*, ed. H.P. Hack. 1988, Philadelphia: American Society for Testing and Material.
14 GL Song, Potential and current distributions of one-dimensional galvanic corrosion systems, *Corrosion Science* **52** (2010) 455–480.
15 HP Hack, *Corrosion Review.*, **15** (1997) 195.
16 JX Jia, GL Song and A Atrens, Influence of geometry on galvanic corrosion of AZ91D coupled to steel, *Corrosion Science* **48** (2006) 2133–2153.
17 JX Jia, G Song and A Atrens, Experimental measurement and computer simulation of galvanic corrosion of magnesium coupled to steel, *Advanced Engineering Materials* **9** (2007) 65–74.

18 JX Jia, G Song, A Atrens, D StJohn, J Baynham and G Chandler, Evaluation of the BEASY program using linear and piecewise linear approaches for the boundary conditions, *Materials and Corrosion* **55** (2004) 845–852.
19 JX Jia, G Song and A Atrens, Boundary element predictions of the influence of the electrolyte on the galvanic corrosion of AZ91D coupled to steel, *Materials and Corrosion* **56** (2005) 259–270.
20 JX Jia, A Atrens, G Song and T Muster, Simulation of galvanic corrosion of magnesium coupled to a steel fastener in NaCl solution, *Materials and Corrosion* **56** (2005) 468–474.
21 RA Adey and SM Niku, *ASTM Spec. Tech. Publ. STP 1154 Comput. Model. Corros.* (1992) 248.
22 DL Hawke, in 14th International Die Casting Cogress and Exposition, 1987, Toronto, Canada.
23 JP Ault Jr. and JJ Meany Jr., *12th Int. Corros. Congr.*, **5A**, (1993) 3519.
24 G Song, A Atrens, D StJohn, J Nairn and Y Li, *Corrosion Science*, **39** (1997) 855.
25 G Song, A Atrens, D StJohn, X Wu and J Nairn, *Corrosion Science*, **39** (1997) 1981.
26 A Atrens, W Dietzel, The negative difference effect and unipositive Mg^+, *Advanced Engineering Materials* **9** (2007) 292–297.
27 S Bender, J Goellner and A Atrens, *Advanced Engineering Materials* **10** (2008) 583.
28 M Liu, P Schmutz, S Zanna, A Seyeux, H Ardelean, G Song, A Atrens and P Marcus, Electrochemical reactivity, surface composition and corrosion mechanisms of the complex metallic alloy Al_3Mg_2, *Corrosion Science*, **52** (2010) 562–578.
29 RL Petty, AW Davidson and J Kleinberg, The anodic oxidation of magnesium metal: evidence for the existence of unipositive magnesium, *Journal of the American Chemical Society* **76** (1954) 363–366.
30 MC Zhao, P Schmutz, S Brunner, M Liu, G Song and A Atrens, An exploratory study of the corrosion of Mg alloys during interrupted salt spray testing, *Corrosion Science*, **51** (2009) 1277–1292.
31 Z Shi, A Artens, *Corrosion Science*, **53** (2011) 226–246.

13
Non-aqueous electrochemistry of magnesium (Mg)

D. AURBACH and N. POUR, Bar Ilan University, Israel

Abstract: Several important aspects of non-aqueous magnesium electrochemistry are reviewed. It is important to develop effective magnesium (Mg) deposition processes. There is also a strong interest in research and development (R&D) of rechargeable Mg batteries. This chapter describes non-aqueous solutions that may be relevant to Mg electrochemistry, including conventional polar aprotic solvents and ionic liquids. It also describes several basic aspects of active metal surface chemistry and their possible passivation processes. The electrochemistry of magnesium in conventional electrolyte solutions and in ionic liquids is discussed. The chapter then describes electrolyte solutions possessing wide electrochemical windows in which Mg electrodes behave reversibly. These systems are based on ether solvents and complexes of the (RMgCl or $R_2Mg)_x$–$(AlCl_{3-n}R_n)_y$ type in which R = alkyl or aryl groups. Finally, the chapter deals with another important aspect of non-aqueous Mg electrochemistry, i.e. the possibility of electrochemical intercalation of Mg ions into inorganic hosts. The development of electrolyte solutions with wide electrochemical windows in which Mg electrodes behave reversibly and hosts that can intercalate Mg ions are important milestones on the way to developing practical rechargeable Mg batteries.

Key words: magnesium, rechargeable batteries, intercalation, magnesium deposition, non-aqueous electrolyte solutions.

13.1 Introduction

There are two major interests in the non-aqueous electrochemistry of magnesium (Mg):

- electroplating of magnesium for purposes of surface finishing, cathodic protection against corrosion, etc.;
- the use of magnesium as an anode in high energy density, non-aqueous batteries.

Over the years much work has been done on these topics in addition to the publication of papers describing attempts to electroplate magnesium [1–5]. There have been reports on the study of the corrosion of magnesium in non-aqueous systems [6] and the behavior of the Mg^{2+} | Mg(Hg) couple [6–10].

There is no question that one of the most important challenges in modern electrochemistry is the development of novel, high energy density, rechargeable batteries. In the research and development (R&D) of such battery systems, environmental aspects are becoming increasingly important alongside the demand for high-performance and low-cost systems.

Batteries are electrochemical devices that store energy in the form of chemical bonds, and they convert this energy directly into electricity. In general, a battery consists of a cathode that is the positive pole of a cell, an anode that constitutes the negative pole, and an electrolyte matrix that is the medium which carries current internally, through which ions flow. Cathodes are usually made of materials which possess high oxidation states, and are thus oxidation agents. Conversely, anodes are usually made of materials in a low oxidation state, and are thus reduction agents. Therefore, a cell assembly is a metastable system far from equilibrium. The voltage of such a cell depends on the difference between the oxidation and reduction power of electrodes, and can be calculated from the thermodynamic properties of any selected couple by using Eq. 13.1:

$$\Delta E_0 = -\Delta G_0/nF \qquad 13.1$$

where ΔE_0 is the standard cell voltage, n is the number of electrons that pass during the electrochemical reaction (discharge or charge), F is the Faraday constant, which is 96500 coulombs per mole of electrons, and ΔG_0 is the standard free energy of the redox reaction of the entire cell.

A battery's performance is evaluated from many factors, the most important of which are its energy content, the maximum power it can deliver, its cycleability and its cost. No single battery system can be considered as a 'dream machine' that possesses all the desirable properties, which is why the market offers so many different types of battery. This is also the motivation behind the development of new types of battery systems. An examination of the periodic table clearly reveals that lithium metal would definitely provide the highest energy content for an anode. However, its selection as an anode has drawbacks: the metal is relatively expensive and its use is hazardous. Hence, although it might be the best anode material for high-end, small portable products, it is too expensive and not safe enough for large-scale applications such as large power backups (uninterruptible power supply, UPS) and battery systems for load leveling applications (e.g. storage of sustainable energy from large fields of solar panels or wind turbines).

In that respect, one of the great successes in electrochemical science in recent years was the development and commercialization of Li ion batteries [11,12]. In addition to R&D of Li and Li ion batteries over the past few years, the development and commercialization of other novel battery systems, such as nickel metal–hydride [13], zinc–air [13], and practical fuel cells [14] have taken place within the framework of a general effort to develop alternative means of propulsion to gasoline and internal combustion engines, namely, electric vehicles.

Following the successful implementation of lithium battery technology, the search for other promising battery materials focused on Mg metal as a significant and desirable candidate for the anode material in high energy density batteries.

Figure 13.1 presents the natural trend in the consideration of Mg metal as a very promising candidate for high energy density battery anodes that can be advantageous over Li–metal anodes, due to improved safety features.

In terms of battery application, the redox potential of the Mg/Mg^{2+} couple is ca. 1 V higher than that of the Li/Li^{+} couple. In addition, the charge capacity of magnesium, 2233 mA h/g, is lower than that of lithium, 3829 mA h/g. Thus, it is obvious that batteries based on magnesium would be inferior to Li batteries in terms of energy density. However, assuming that high-capacity cathode materials consisting of Mg, M, O or Mg, M, S atoms (M = transition metal) could be developed for these batteries, analogous to those developed for Li and Li ion batteries [15], one can predict that Mg batteries may have an energy density of >100 W h/kg. This is more than twice the energy density of the leading 'low-tech' rechargeable batteries,

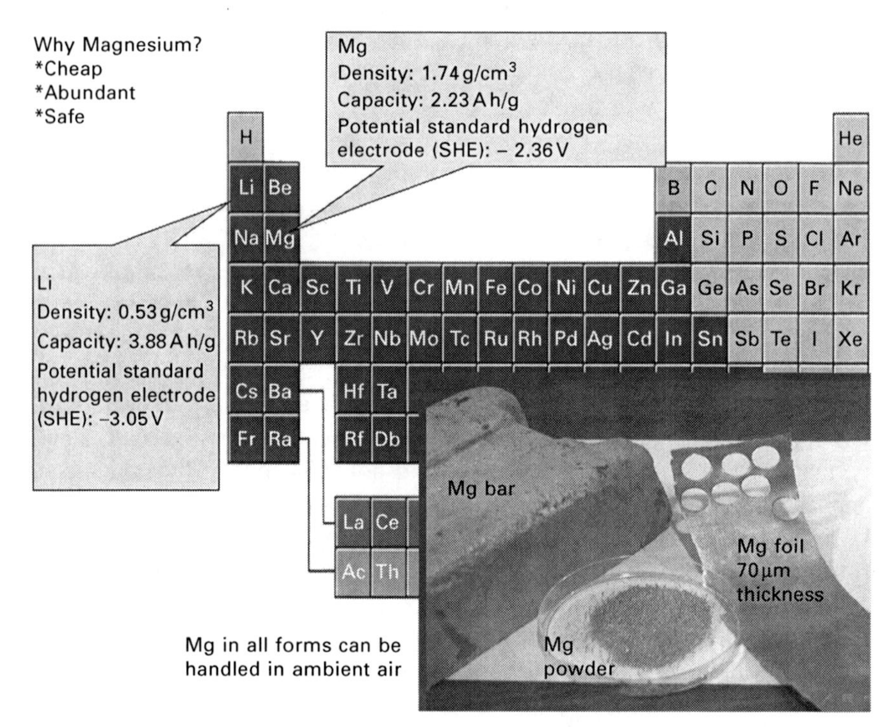

13.1 Properties of Mg metal that make it of interest as an anode material for batteries.

including lead–acid and Ni–Cd battery systems [13], the 'workhorses' of the industry. The enhanced safety, lower prices, and simple waste management of Mg batteries may compensate for their expected lower energy density compared with Li ion batteries. Furthermore, Mg compounds are abundant in the Earth's crust, and most of them are non-toxic.

The scientific literature of recent years contains a few reports of R&D efforts on Mg batteries. In addition to primary magnesium batteries and those developed for commercial and military applications [13], several research groups tried to develop primary aqueous and non-aqueous Mg batteries [16–19]. The key to operation with a metal-based anode is an appropriate electrolyte solution in which reversible electrochemical metal deposition can occur, i.e. as the battery supplies power, the anode dissolves electrochemically, yielding metallic ions in the solution. In a rechargeable battery this process must be reversible, enabling the re-deposition of metal from metallic ions in the solution onto the anode by electrolysis, using an external power source (a charger). Since magnesium is an active metal, aqueous solutions are obviously impractical for rechargeable systems; magnesium would react with water and corrode. Thus, as in the case of lithium, organic polar materials must be utilized as solvents for electrolytes. It is generally known that magnesium can be electrochemically and reversibly deposited and dissolved in ethereal Grignard solutions (RMgX, R = alkyl, aryl groups, and X = Cl, Br) [20]. Despite the very high reversibility of Mg electrodes in these solutions, they are not at all suitable as electrolyte solutions for battery applications owing to their extremely poor conductive and anodic stability [21]. Anodic stability is a term pertaining to the capability of a solution with a salt solvent to withstand the presence of an oxidizing agent, namely, the cathode.

Hence, when reviewing important aspects of non-aqueous magnesium electrochemistry, it is important to consider the reversibility of Mg deposition processes, to map possible corrosion process of Mg electrodes, and to determine the anodic stability of electrolyte solutions in which Mg electrodes behave reversibly. In the following sections, we will briefly review conventional non-aqueous electrolyte solutions and the passivation of active metals in non-aqueous solutions, after which we will describe systematically the behavior of Mg electrodes in various types of conventional and non-conventional non-aqueous electrolyte solutions. Finally, we will review in brief another important aspect of non-aqueous magnesium electrochemistry, which is the electrochemical intercalation of Mg ions into inorganic hosts.

13.2 A short review of non-aqueous electrolyte solutions

Figure 13.2 presents formulae of a variety of polar aprotic solvents that may be relevant to non-aqueous active metal electrochemistry. This figure also

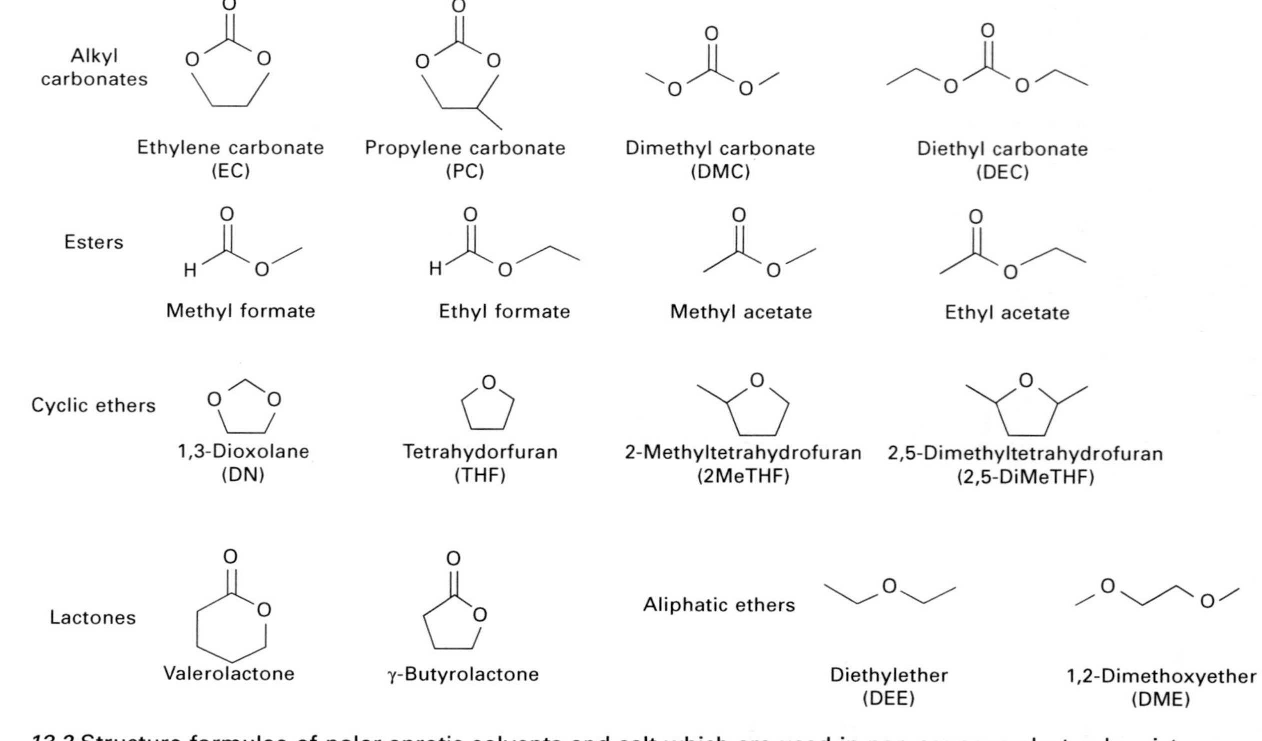

13.2 Structure formulae of polar-aprotic solvents and salt which are used in non-aqueous electrochemistry.

presents formulae of commonly used Li salts and commercially available Mg salts that can be dissolved in these solvents. It should be noted that polar aprotic solutions for electrochemistry were explored in depth. References [22–24] provide comprehensive reviews which include extensive physical data.

The first important topic for discussion is how salts are dissolved in aprotic solvents, and to what extent the ions are separated in solutions. The next issue is the ionic conductivity of certain combinations of electrolyte and solvents, and their temperature dependence. Intensive work has been devoted over the years to the discovery of representative solvent parameters, to which electrolyte solution properties such as solubilizing ability, ion separation and ion conductivity can be correlated [25]. Trivial solvent parameters that may be important in this respect are dipole moments and dielectric constants. However, over the years, parameters such as donor and acceptor numbers [26] were proven to provide better probes for solvent characterization. In general, polar solvents may allow appropriate ion separation of a wide variety of electrolytes. However, polar solvents are also more viscous and hence the high viscosity of a solvent may counter-balance good ion separation. Thereby, using highly polar solvents does not necessarily ensure high ionic conductivity. Consequently, it is possible to obtain high ionic conductivity by the use of mixtures that contain both highly polar viscous solvents and less polar, but less viscous, solvents. The impact of the salt concentration is also important. A high salt concentration means a high amount of ions for charge transport, but also pronounced inter-ion interactions that inhibit their fast mobility in solutions. Therefore, for most of the practical solutions, an optimum electrolyte concentration exists for maximizing ionic conductivity [22–24].

Another critical property of electrolyte solutions is their electrochemical window. This property can be measured with only inert electrodes. These may include carbonaceous materials: glassy carbon and conducting (by doping) diamonds and precious metals, mostly gold and platinum.

Figure 13.3 shows typical electrochemical windows of important families of electrolyte solutions. Several important aspects should be noted:

- Solvents are polar due to functional groups that can be electron-withdrawing or electron-pushing groups. Solvents with electron-withdrawing groups (e.g. carbonates, nitriles, sulfones) are presumed to have high anodic stability and low cathodic stability. In turn, solvents that demonstrate high cathodic stability (such as ethers) because they do not possess electron-withdrawing groups, should have limited anodic stability [27].
- The nature of the electrolyte cations may be critical for the apparent cathodic stability of polar aprotic solutions. The solubility of the reduction processes that limits the solution's cathodic stability depends on the cations. When the solutions' reduction products are insoluble, they precipitate

on the electrodes' surfaces and passivate them. Ions such as Li^+, Mg^{2+}, Ca^{2+} and Na^+ precipitate together with solvent reduction products such as RO^- (ethers), $RCOO^-$ (esters), RCO_2O^- (alkyl carbonates) to form passivating surface films. When the cations are tetraalkyl ammonium, passivating surface films are not formed by the reduction of the above solvents [28].

When considering the conventional polar aprotic families of solvents (mentioned in Figs 13.2 and 13.3), esters, alkyl carbonates, sulfones, amides and dimethylsulfoxide (DMSO), can be suspected as being reactive with active metals such as magnesium (see later discussion). Only ethers seem to be suitable for reversible Mg electrodes and electrochemical deposition of magnesium metal.

Another important class of polar aprotic systems that can be relevant for magnesium electrochemistry is that of ionic liquids (ILs). Figure 13.4 presents formulae of relevant ionic liquids. This figure also includes tabulated physical properties of commonly used ILs in electrochemistry. The main important features of ILs are the possibility to obtain highly ionically conducting solutions with wide electrochemical windows, low

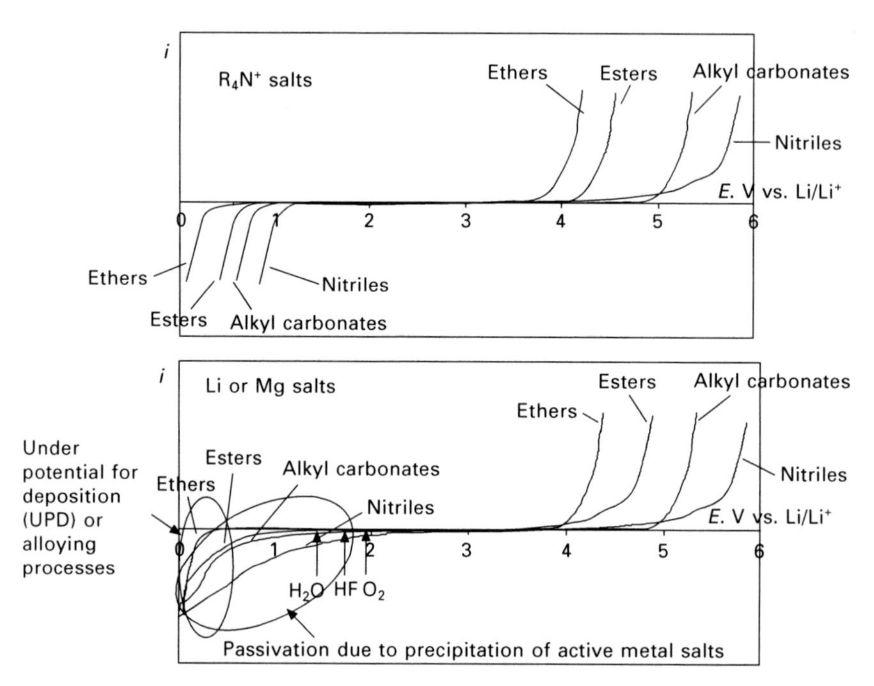

13.3 Schematic potentiodynamic behavior of several families of polar aprotic solutions, which demonstrate their electrochemical windows (obtained by voltammetric studies of solutions with inert electrodes such as platinum).

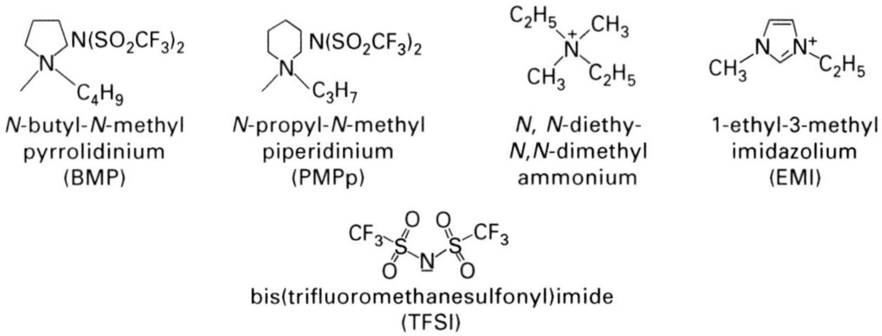

$N(SO_2CF_3)_2$ C_4H_9
N-butyl-*N*-methyl pyrrolidinium (BMP)

$N(SO_2CF_3)_2$ C_3H_7
N-propyl-*N*-methyl piperidinium (PMPp)

C_2H_5 CH_3 CH_3 C_2H_5
N, N-diethy-*N,N*-dimethyl ammonium

CH_3 C_2H_5
1-ethyl-3-methyl imidazolium (EMI)

CF_3 CF_3
bis(trifluoromethanesulfonyl)imide (TFSI)

Cation	Anion	Melting point (°C)	Density (g cm^3) (20 °C)	Viscosity (mPa s) (25 °C)	Conductivity σ (10^{-4} S cm^{-1}) (25 °C)
BMP	TFSI	−18	1.41	85	22
PMPp	TFSI	8.7	1.43	117	15.1
$(CH_3)_2(C_2H_5)_2N$	TFSI	−14	1.41	83	12
EMI	TFSI	−15	1.52	34	87
EMI	BF_4	13	1.28	37	140

13.4 Structural formulae and some physical properties of ionic liquids which are important for non-aqueous electrochemistry.

volatility and, hence, promising safety features of energy storage devices that use them as electrolytes [29]. Non-aromatic ILs based on quaternary ammonium cations, especially cyclic systems such as the derivatives of piperidinium and pyrrolidinium rings, demonstrate high cathodic stability, and may even be suitable for Li electrochemistry. Aromatic systems such as imidazolium-based ILs have a lower cathodic stability. However, since the redox potential of magnesium is higher than 1 V, compared with that of lithium, piperidinium, pyrrolidinium and imidazolium-based systems are apparently suitable for Mg electrochemistry. Hence, one can expect that in highly pure ILs, free from active contaminants such as H_2O, acidic species, O_2, CO_2, etc., Mg electrodes may behave reversibly.

In conclusion, it is expected that both ethereal solutions and a variety of IL-based electrolyte systems may be suitable for reversible Mg electrochemistry.

13.3 A short review of the passivation phenomena of active metals in non-aqueous electrolyte solutions

Active metals are always covered by native surface films formed by reactions between the metal and atmospheric components. The native surface films on active metals usually have a bilayer structure. The inner layer comprises mostly metal oxide, MO_x, and the outer layer comprises metal hydroxide $M(OH)_x^-$ and metal carbonate M_xCO_3. Active metals such as lithium also react with nitrogen to form lithium nitride (Li_3N), and thereby, the native films on lithium may contain lithium nitride (Li_3N) as well. When an active metal is introduced into polar aprotic solutions, a variety of reactions and interactions take place, as illustrated in Fig. 13.5. Some of the native surface species dissolve, and the solution components may percolate through the native film and react with the active metal. Trace water hydrates the metal oxide and hydroxide, and hence, can diffuse through the native surface films and react with the active metal. Metal oxides and hydroxides can nucleophilically attach organic esters and alkyl carbonates. Metal dissolution processes may break down the surface films and expose fresh active metal to the solution. The active metal thus exposed (unprotected by surface films) reacts with components of the electrolyte solutions.

Extensive studies of lithium surface chemistry have been carried out for more than three decades, and have enabled the mapping of a wide variety of surface reactions that occur in the most important families of solvents, salt anions and active atmospheric contaminants [30,31]. Based on these studies, it is possible to understand the surface reactions of other active metals in important polar aprotic electrolyte solutions. In each family of solvents there are dominant surface species that are formed on active metals: metal oxides in ethers, metal carboxylates in esters and metal alkyl carbonates in alkyl carbonates. However, the situation on active metal surfaces in solutions is

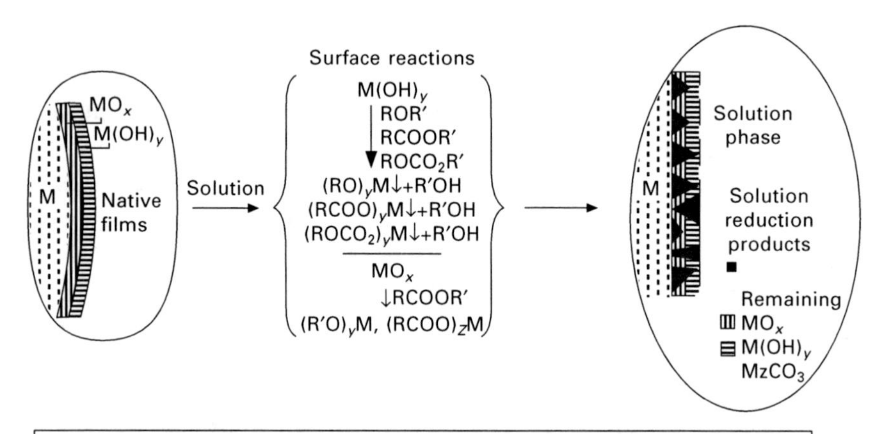

13.5 A schematic illustration of the surface chemistry of an active metal when introduced into non-aqueous (polar-aprotic) solutions. Active metals are usually initially covered by bi-layered surface films (due to reaction with atmospheric gases) which are converted to complicated surface films due to reactions with solution species.

very dynamic. Surface species that are initially formed, e.g. organic metal salts, can be further reduced to carbides and oxides. Trace water can be reduced stepwise, forming metal hydroxide, oxide and hydride. Trace water reacts with $ROCO_2M$ to form lithium carbonate, carbon dioxide and ROH. Carbon dioxide, when formed, reacts with $M(OH)_x$ or $M(OR)_x$ to form metal carbonates. Acids such as HF react with metal carbonate, $ROCO_2M$ and ROM to form lithium fluoride (LiF) and their parent protic precursor (carbonic acid (H_2CO_3), $ROCO_2H$ and ROH).

To these sets of primary and secondary reactions related to solvents, one has to add the contributions of salt anion reduction, which usually forms metal halides and M_xAX_y species (A is the main high oxidation-state element in the salt anion and X is a halide, such as chloride or fluoride). Most of the products of active metal surface reactions are ionic compounds that are insoluble in the mother solution, and therefore, precipitate as surface films. It should be added to this picture that possible polymeric species can be formed, especially in alkyl carbonate solvents, whose reduction forms polymerizable species such as ethylene or propylene. Hence, the surface films formed on active metal electrodes are very complicated. They have a multilayer structure perpendicular to the metal surface, and a lateral, mosaic-type composition and morphology (i.e. containing mixtures and islands of different compounds and grains). Such a structure may induce very non-uniform current distribution upon metal deposition or dissolution processes, which leads to dendrite formation, a breakdown of the surface films, etc. These situations are demonstrated in Fig. 13.6: active metal dissolution leads to the break-and-repair of the surface films, thus forming mosaic-type structures.

Turning specifically to magnesium, we explored in depth the surface chemistry of this metal [31,32]. As can be learnt from the value of the standard potential of the Mg^{2+}/Mg couple in the electrochemical series, –2.37 V, magnesium is a highly reactive metal. Therefore it is expected that any exposed metal surface will react promptly with most of the atmospheric gases. Oxygen and water are, naturally, the greatest concern. Therefore, even under the atmosphere of a high-purity glove box, the metal is expected to react with oxygen and humidity traces and form surface films. The expected reaction products are MgO and $Mg(OH)_2$, as well as some hydrated forms of these compounds. We used X-ray photoelectron spectroscopy (XPS) extensively for these studies. Indeed, all of the O(1s) spectra could be deconvoluted into two peaks, at ~529.6 and 530.5 eV (calibrated vs. C(1s = 285.0 eV)), the first one attributed to MgO and the other to $Mg(OH)_2$. Furthermore, in all in-depth profiles, it was demonstrated that the $Mg(OH)_2$ rich layer lies on top of the MgO one. Moreover, and no less important, there is no indication for the formation of $MgCO_3$.

One of the most important questions is whether magnesium metal reactions

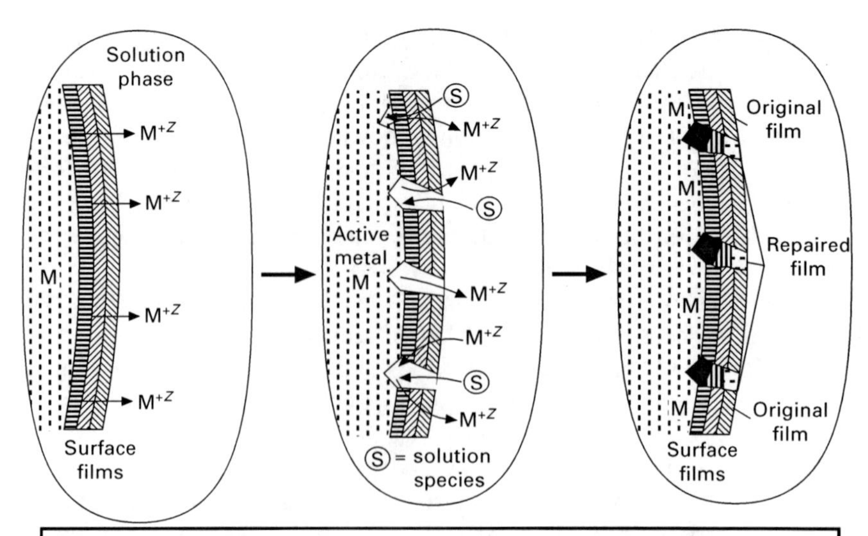

13.6 A schematic presentation of break and repair of surface films upon irreversible electrochemical dissolution processes of active metals such as magnesium.

with ethereal solvents, such as tetrahydrofuran (THF) and related solutions to form stable surface layers, in the same matter as lithium does. Such reaction products are expected to be some magnesium salts of the reduction products of the organic species. The main spectroscopic (by XPS) feature of such products should be the carbon C(1s) peak. Although all Mg samples exposed to ether solutions may show C(1s) XPS peaks, they contain only the minor features expected from stable, organic, salt-like compounds, if any. First, the C(1s) signal originates in all the samples from a carbonaceous material. Second, Ar^+ sputtering shows that this signal relates to weakly adsorbed molecules rather than to a strongly adherent species. Interestingly, even with Mg samples that were in contact with a reactive solvent such as propylene carbonate (PC), no evidence was found for either organic or inorganic carbonates, the expected reduction products of this solvent. Our XPS studies of Mg surfaces never provided unambiguous proof that there is no reaction whatsoever between magnesium metal and alkyl carbonate solutions, but they do indicate that if there is a surface film formation, it must be confined to a monolayer scale.

Similar conclusions can be drawn for other components, such as dibutylmagnesium and BuMgCl in ethers. The corresponding elements in these compounds were found at the outer surface and were, again, easily removed by sputtering. Further proof for the adsorption scheme can be found on the analysis of gold electrodes that were immersed in ether solutions with no polarization. Even without any electrochemical manipulations, and after rigorous washing, chlorine, aluminum, magnesium and carbon were clearly detected on the surface. Since gold is inert to these solutions, only tenacious adsorption processes could have led to the presence of these elements.

Probably the most important information regarding the electrochemistry of magnesium in complex ethereal solutions such as $Mg(AlCl_2R_2)_2$/THF, which are important for Mg batteries (see later), arises from the analysis of gold electrodes after electrodeposition of magnesium. In these spectra, we could observe that the quality of the magnesium deposits is identical to that of the pure Mg metal that was in contact with the same solutions. This fact indicates that even under the most sensitive conditions, i.e. a fine growth by slow electrochemical processes, the magnesium grows as pure metal with an interface that is probably free of surface films, in complex ether solutions.

On the basis of the above conclusions, we believe that upon immersion in ether solutions, Mg metal electrodes become covered with adsorbents composed of the various solution species, with no chemical reactions.

13.4 Magnesium (Mg) electrodes in conventional polar aprotic solvents and in Grignard solutions

In 2000, we disclosed a description of rechargeable Mg battery systems that include new solutions based on Mg–Al–Cl–R (R=alkyl, aryl groups) complexes in ether solvents. These solutions enable both the fully reversible behavior of Mg electrodes and wide electrochemical windows (>2.2–3.3 V) [33]. Below we review in brief extensive work that was carried out by others on non-aqueous Mg electrochemistry before the year 2000. The behavior of Mg electrodes in thionyl chloride has been investigated by Peled and Straze [34]. Genders and Pletcher have studied the basic electrochemical behavior of the $Mg^{2+}|Mg$ couple in THF and PC, using microelectrodes [35]. A comprehensive report on the feasibility of rechargeable non-aqueous magnesium batteries has been published by Gregory *et al.* [36]. Reviews of the anodic behavior of magnesium in non-aqueous electrolytic solutions [37], plating of magnesium from organic solovents [38], and the reversibility of Mg electrodes in Grignard solution [39] should also be acknowledged.

In general, the solubility of Mg salts such as $Mg(ClO_4)_2$, $Mg(SO_3CF_3)_2$, etc. in commonly used solvents (ethers, alkyl carbonates and esters) is lower than that of Li salts. From the extensive work of Lossius and Emmenegger [38], it appears that $MgCl_2$ and $Mg(CF_3SO_3)_2$ in dimethylformamide (DMF), dimethylacetamide (DMA), diethylacetamide, γ-butyrolactone (γ-BL), or binary mixtures of them, have acceptable solubility and conductivity. It was further discovered by these authors [38] that Mg dissolves with low overpotential and high dissolution efficiency in many systems with a $Mg(CF_3SO_3)_2$ solute. In solutions with DMF and DMA, a Mg^{2+} | Mg reduction process was identified at potential <–3.0 V (vs. ferrocene/ferricenic picrate, FC). However, in none of the systems mentioned above did the Mg^{2+} | Mg reduction lead to a measurable Mg deposition. It has been suggested that transient Mg^0 is too reactive for these solvents and undergoes a reaction with the solvent, instead of forming a metal layer on the electrode.

Mg electrodes were found to be highly reversible in solutions containing Grignard reagents. Most of the commonly used polar aprotic solvents, such as alkyl carbonates, esters and acetonitrile, are too electrophilic for Grignard reagents (which are strong nucleophiles), and thus react with them readily. The only solvents in which stable Grignard solutions can be prepared are ethers. The highly reversible behavior of Mg electrodes, which includes the deposition of metallic magnesium at high coulombic efficiency in THF containing Grignard salts, has been reported by a number of authors [1–4,35,36,39]. Typical Grignard reagents tested in this respect were RMgCl (R = methyl, ethyl or butyl) [36] and CH_3CH_2MgBr [35,39]. There are reports on the reversible behavior of Mg electrodes in high-temperature molten salts such as $MgCl_2$ (740 °C) [40], $MgCl_2$ + NaCl (700–800 °C) [41] and $MgCl_2$ + MgF_2 (700 °C) [42]. The attempts to prepare electrolyte solutions of Mg salts based on room temperature organic molten salts should also be acknowledged [43,44].

Figure 13.7 shows a typical voltammetric (black line) and electrochemical quartz crystal microbalance (EQCM) (grey line) response of a typical

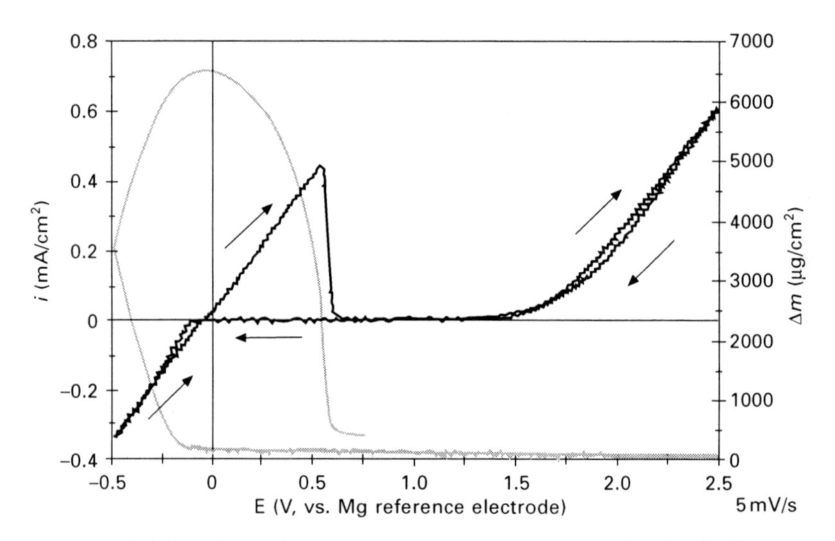

13.7 Typical voltammetric and mass accumulation/depletion responses (EQCM experiments) of Grignard reagent solutions such as THF/BuMgCl solution, with inert electrodes such as platinum. These reflect fully reversible Mg deposition/dissolution processes.

Grignard reagent solution: THF/C_4H_9–MgCl [45]. This voltammetric response demonstrates a reversible Mg deposition process at relatively low nucleating overvoltage and a very narrow electrochemical window, below 1.5 V. The EQCM responses of Mg deposition–dissolution processes in THF/RMgCl (R = CH_3, CH_3CH_2, C_4H_9) solutions indicate, in most cases, a cycling efficiency close to 100%. Mg deposition in these solutions forms micrometric-sized crystallites that are uniformly dispersed on the substrates' surfaces (either Mg or other metallic surfaces). Prolonged and repeated deposition-dissolution processes maintain a uniform morphology of the Mg deposits, with no dendritic formation.

Surface studies by Fourier transform infrared (FTIR) spectroscopy and XPS proved that there is no surface film formation on the electrodes (or Mg deposits) in ether–Grignard reagent solutions [32,46]. It is clear that Mg deposition takes place only on bare non-passivated surfaces. In general, the RMgX species in ethereal solutions undergoes the following equilibria [47]:

$$2RMgX \rightleftharpoons R_2Mg + MgX_2$$

$$2RMgX \rightleftharpoons RMg^+ + RMgX_2^-$$

Hence, RMg^+ is the moiety that interacts with the electrode's surface upon Mg deposition and $RMgX_2^-$ is the moiety whose oxidation process marks the anodic limit of the narrow electrochemical window of these solutions.

The behavior of Mg electrodes in many of the polar aprotic solvents described above resembles that of calcium electrodes [48,49]:

- Both metals in most of the commonly used polar aprotic electrolyte systems are passivated by surface films.
- These surface films do not conduct the bivalent metal ions, and thus, Ca or Mg dissolution may occur via the breakdown of the surface films, while Ca or Mg deposition in these systems is impossible.

However, a remarkable difference between the electrochemical behavior of the two metals lies in the fact that Mg deposition is possible in the above-mentioned Grignard salt solutions, while there are no parallel systems in which Ca deposition is possible. The electrochemical behavior of both these metals is quite different from that of Li electrodes. In the latter case, the surface films formed on Li in most of the commonly used electrolyte solutions are Li ion conducting. Hence, electrochemical Li deposition and dissolution is possible in most of the commonly used aprotic electrolyte solutions via the transport of Li ions through the surface films that always cover these electrodes in solutions.

Our studies on non-aqueous Mg electrochemistry in conventional electrolyte solutions and ether/RMgX solutions can be concluded as follows [31]. The electrochemical behavior of Mg electrodes is controlled by surface films that are formed spontaneously by the reaction of the active metal with atmospheric contaminants, solvents such as alkyl carbonates and acetonitrile, and salt anions such as ClO_4^- and BF_4^-. These surface films thus comprise insoluble Mg salts, including Mg halides, MgO, $Mg(OH)_2$, $(ROCO_2)_2$ Mg (in the case of alkyl carbonate solvents), and possibly species such as $Mg(N{=}C{=}CH_2)_2$ in acetonitrile (ACN) solutions. These surface species normally passivate Mg electrodes in solution and prevent a continuous spontaneous reaction between the active metal and the solution components. In this respect, the electrochemical behavior of Mg electrodes resembles that of Li electrodes in polar aprotic electrolyte solutions. In contrast to the case of Li electrodes, where the surface films that cover them allow Li ion migration through them, most, if not all surface films formed on Mg electrodes in polar aprotic solutions do not transport Mg^{2+} ions. Thus, Mg dissolution can only take place via the breakdown of the passivating surface films when the electrodes are polarized anodically. Cathodic polarization repairs the films and even thickens them, making it impossible to obtain Mg deposition in reactive solvents such as alkyl carbonates or acetonitrile and, probably, esters. The presence of acidic species in solutions also leads to the breakdown of the passivation, and thus enables Mg dissolution to occur at relatively low overpotentials.

A class of aprotic solvents that is not reactive with Mg is that of ethers such as THF. Thus, when using Mg salts whose anions are not reactive towards Mg metal, it is possible to obtain reversible Mg deposition and dissolution. Grignard reagents of the RMgX (R = alkyl, X = Cl, Br) type are good examples of such salts. In ether and RMgX solutions, stable passivating surface films are not formed on Mg electrodes. However, Mg deposition in these solutions does not occur via a simple two-electron transfer to Mg^{2+} ions, and involves the adsorption of species formed by the reversible dissociation processes of the Grignard salts.

13.5 Ionic liquids (ILs) for magnesium (Mg) electrochemistry

In the search for inert polar aprotic solutions suitable for reversible Mg electrochemistry having wide electrochemical windows, ILs appear to be promising candidates. Indeed, recent work has demonstrated the possibility of reversible Mg deposition–dissolution in ILs comprising derivatives of imidazolium salts [50,51]. In parallel to those studies, we also studied the possibility of using ILs as electrolyte systems for Mg electrodes [52]. Figure 13.8 shows the various IL systems studied and the relevant Mg salts. It should be noted that with none of these systems was it possible to obtain reversible Mg deposition and dissolution, even in systems that could dissolve Mg salts up to high concentrations.

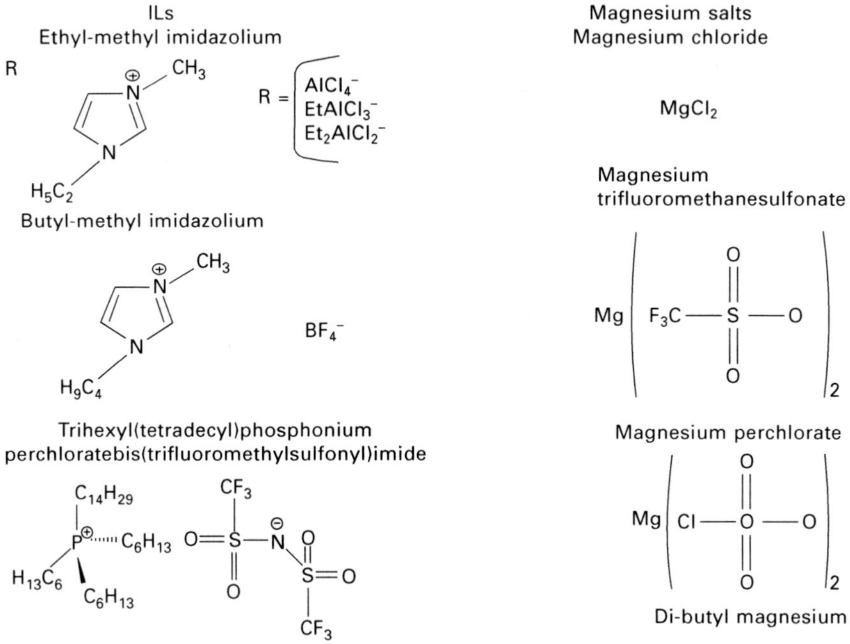

ILs
Ethyl-methyl imidazolium

R = $AlCl_4^-$, $EtAlCl_3^-$, $Et_2AlCl_2^-$

Butyl-methyl imidazolium

BF_4^-

Trihexyl(tetradecyl)phosphonium perchloratebis(trifluoromethylsulfonyl)imide

Magnesium salts
Magnesium chloride

$MgCl_2$

Magnesium trifluoromethanesulfonate

Magnesium perchlorate

Di-butyl magnesium

13.8 Formulae of the ionic liquids and relevant salts tested as potential electrolyte solutions for rechargeable Mg batteries.

It appears that Mg electrodes, even if initially active, reach passivation in all the IL-based systems presented in Fig. 13.8, no matter which Mg salts are used. Passivation occurs due to reactions of magnesium with several types of ILs (as is the case for imidazolium-based systems), or to reactions with unavoidably present trace water. Our previous studies [31] may indicate the reactivity of bare Mg metal with anions such as ClO_4^- and BF_4^-. The results of such reactivity are the formation of insoluble Mg halides that completely block the Mg electrodes.

In certain IL systems, such as ethyl-methyl imidazolium-$AlCl_4$, Mg metal visibly dissolves. We tried to repeat the experiments described by Nuli *et al.* [50,51] but were unable to obtain any reversible behavior of Mg electrodes or reversible Mg deposition–dissolution processes or noble metal electrodes in any of the IL systems described in Fig. 13.8. Hence, we have to conclude that most commonly used ILs, including those showing apparent high cathodic stability, are not suitable solvents for reversible Mg electrodes. Thereby, ILs cannot be considered as compatible/promising electrolyte solutions for non-aqueous magnesium electrochemistry.

13.6 On solutions with a wide electrochemical window (>2 V) in which magnesium (Mg) deposition is reversible

Over the years, some group have succeeded in developing electrolyte systems of wider electrochemical windows, based on Mg boranes [33] or magnesium silicon complexes in ethers [34]. For instance, solutions of $Mg(BPh_2Bu_2)_2$ in THF (Ph and Bu are phenyl and butyl groups) could provide an electrochemical window close to 1.8 V, whereas Mg electrodes may behave reversibly in them [33]. There were attempts to study and develop Mg insertion materials [33,35] in a similar manner to that of the intensive and successful R&D efforts related to positive electrodes for Li batteries.

The electrochemistry group at Bar-Ilan University was the first to develop rechargeable magnesium battery systems based on Mg or Mg alloy negative electrodes, $Mg_xMo_6S_8$ ($0<x<2$) Chevrel-phase (CP) positive electrodes, and electrolyte solutions comprising THF or glyme ethereal solvents and complexes of the $Mg(AlCl_{4-n-n}R_n)_2$ type (R = alkyl and aryl groups) [36,37]. It was also possible to compose solid state, rechargeable Mg batteries comprising the same electrodes and gel electrolytes, based on polyvinylidene difluoride (PVdF) or polyethylene oxide (PEO), $Mg(AlCl_2R_2)_2$ complexes, and tetraglyme $CH_3{-}(OC_2H_4)_4{-}OCH_3$ as a plasticizer [38]. The most important property of the electrolyte solutions is the lack of chemical reactivity between them and the Mg electrodes.

Figure 13.9 compares the electrochemical behavior of several types of Mg-based electrolytes in THF steady state voltammograms of Pt electrodes in THF/EtMgCl, THF/$Mg(BBu_2Ph_2)_2$, THF/$Mg(AlCl_2EtBu)_2$ (Et, Bu = $-C_2H_5$ and $-C_4H_9$) and THF/$Mg(AlCl_4)_2$. As can be seen, reversible Mg deposition is obtained in all four solutions. However, the use of organo-boron magnesium salts increases the electrochemical window by 0.6 V, compared with the use of Grignard reagents (1.8–1.9 vs. 1.2–1.3 V) and the use of Mg halo-aluminate complexes increases the electrochemical windows of these Mg complexes electrolyte solutions > 2.2 V.

Figure 13.10 demonstrates the effect of the synthesis of the $MgAl(Cl_{4-n}R_n)_2$ (formal stoichiometry) electrolytes on the electrochemical window of the solutions. The complex electrolytes are in fact products of the reaction between the R_2Mg Lewis base and an AlR_nCl_{3-n} Lewis acid. As seen in Fig. 13.10, the anodic stability of these solutions depends mostly on the ratio between the Lewis base and the Lewis acid in the complex electrolyte. The higher the amount of the acid in the reactant mixtures, the higher is the anodic stability. One should note that the anodic stability depends also on the R groups in the order: $-CH_3 > -C_2H_5 > -C_4H_9$. However, the effect of the acid/base ratio on the electrochemical window is the most pronounced. As will be discussed later, these differences in the anodic stability of the

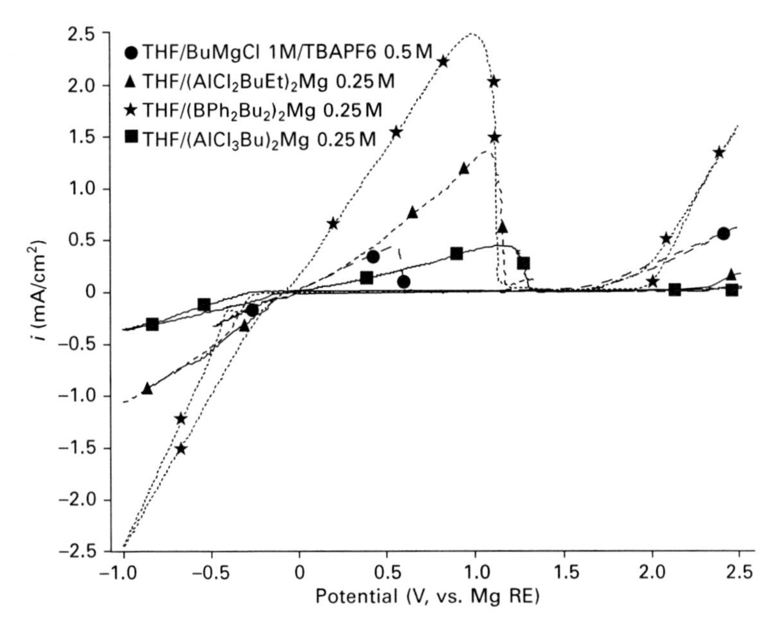

13.9 Comparison of typical steady state voltammograms of THF solutions with Pt electrodes, from which magnesium can be reversibly deposited. Note the reversible response of Mg deposition and dissolution and the difference in the anodic stability of the four solutions. The composition of the four solutions measured is indicated herein (near the V curves).

solutions are due to the different structures formed as a function of the Lewis acid/Lewis base ratios. It should also be noted that as the acid–base ratio is higher, or contains more chloride (e.g. $AlCl_2R_2$ vs. $AlCl_3R$), the overvoltage for Mg deposition is higher. The impact of the electrolyte composition on the morphology of Mg deposition is highly interesting.

The Lewis acid/Lewis base ratio determines the type of Mg crystallites that are deposited. At low and high acid/base ratios (e.g. 0.5, 3), Mg deposition is a submicronic-nanonic phenomenon at a wide distribution of crystal size, while at moderate A/B ratios (1, and 2), Mg is deposited in micrometer-size crystallites. Another important factor that determines the morphology of Mg deposition is the current density. The higher current density, the smaller is the crystallite size. Thus, while at moderate current densities (1–2 mA/cm^{-2}), Mg deposition is a micrometric phenomenon, while at higher current densities (e.g. 4 mA/cm^2), Mg is deposited in nanometer-size crystallites.

In any event, Mg deposition on bare metallic surfaces (oxide-film and passivation free) is not at all dendritic and is highly reversible. Figure 13.11 shows the typical Mg deposition morphology scanning electron microscopy, (SEM, images) in two types of Mg–Al–Cl–R complex solutions in THF. Mg

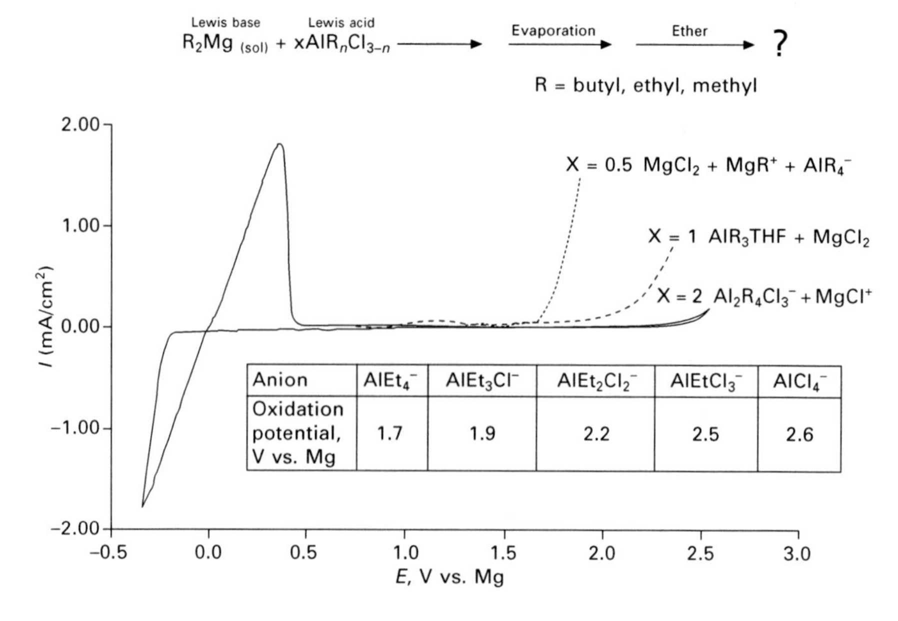

Anion	$AlEt_4^-$	$AlEt_3Cl^-$	$AlEt_2Cl_2^-$	$AlEtCl_3^-$	$AlCl_4^-$
Oxidation potential, V vs. Mg	1.7	1.9	2.2	2.5	2.6

13.10 The effect of the $MgR_2/AlCl_2R$ ratio in the complex ether (THF) solutions on the anodic stability of the solutions. The main species formed and the oxidation potentials of the various anions that can be formed in these solutions, are indicated herein.

is deposited uniformly in micrometer size crystallites. The electrochemical dissolution of magnesium deposits on bare metallic surfaces leaves no residual magnesium. This behavior is due to the fact that there are no chemical reactions between bare magnesium and these electrolyte solutions. Hence, Mg deposition (although it may be affected by some adsorption processes) takes place (only!) under passivation-free conditions. Since the properties of Mg deposition processes and the electrochemical properties of the solutions depend so strongly on the composition of the electrolyte, we review below our understanding of the structure of the THF/$Mg(AlCl_{4-n}R_n)_2$ solutions.

The electrolyte solutions discussed herein, namely, the magnesium-chloro-organo-aluminates, are prepared as acid–base reactions in which the organomagnesium reagent (the Lewis base) and the organohalo aluminum (the Lewis acid) compounds react to form THF soluble complex salts. The most valuable salt, yielding solutions with good electrochemical properties, is prepared by reacting 1 mol of dibutyl magnesium with 2 mol of ethyl aluminum dichloride, to yield the 1:2 product whose formal stoichiometry is $Mg(AlCl_2BuEt)_2$. In contrast to the low conductivity of the THF solutions of the starting materials, the solutions of the products display ionic conductivity in a milli-siemens/cm range. This substantiates the claim for a chemical reaction that produces ionic species.

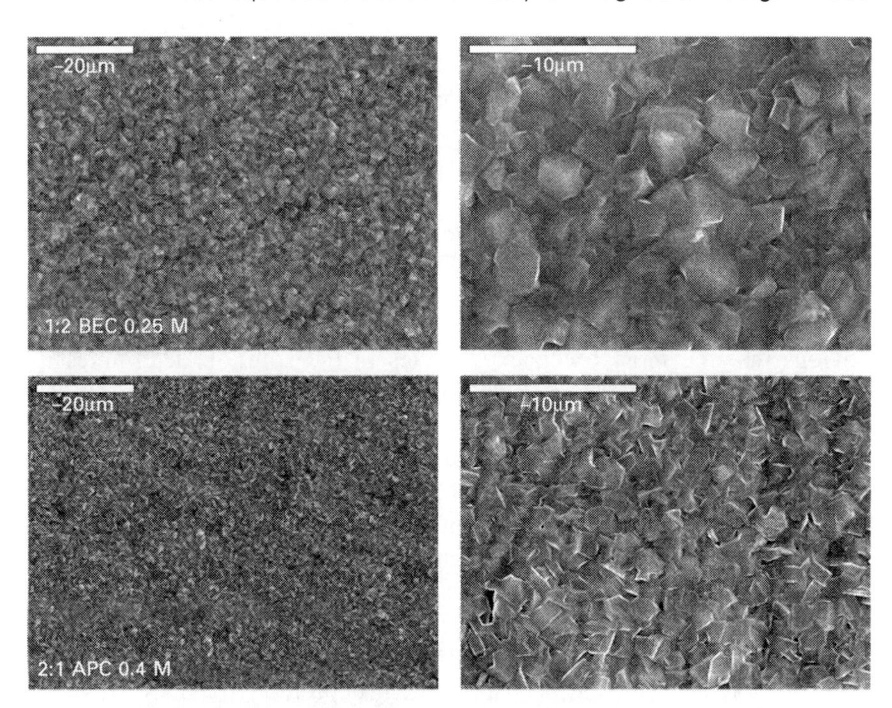

13.11 Scanning electron microscope (SEM) images of Mg deposition morphology from two different THF solutions: BEC (butyl-ethyl complex) = $MgBu_2$ + $2AlCl_2Et$/THF solutions, APC (all phenyl complex) = 2PhMgCl + $AlCl_3$/THF solutions. Particle size: BEC 2–2.5 μm; APC 1 μm.

For the analysis of these solutions, nuclear magnetic resonance (NMR) and Raman spectroscopies were found to be the most appropriate. The major constituents of the electrolyte solutions possess carbon, hydrogen, magnesium and aluminum, which have isotopes that are NMR active.

From the ^{13}C and the ^{1}H spectra it was clear that no bridging alkyl groups are present, and in the 1:2 and 1:1 complexes no organic ligands are bonded to magnesium core. From ^{27}Al NMR, it was clear that all the aluminum atoms in the complex are tetra-coordinated. From the ^{25}Mg NMR, it was inferred that magnesium is always hexa-coordinated.

It appears that the main reactions that form the complex electrolytes for the magnesium battery solutions are trans-metalation, i.e. the exchange of the ligands between the magnesium and aluminum, as presented in Fig. 13.12. The detailed studies are described in Gizbar *et al.* [53] and Vestfried *et al.* [54].

This reaction scheme is also consistent with the electrochemical windows presented in Fig. 13.10. The anodic limit of the electrochemical window, namely the potential at which the solution undergoes irreversible oxidation,

$2RAlCl_2 + R_2Mg + 6THF \leftrightarrow R_2AlCl_2^- + R_2AlCl{\cdot}THF + MgCl^{+}{\cdot}5THF$	13.2
$RAlCl_2 + R_2Mg + 5THF \leftrightarrow R_3Al{\cdot}THF + MgCl_2{\cdot}4THF$	13.3
$RAlCl_2 + 2R_2Mg + 9THF \leftrightarrow R_4Al^- + MgCl^{+}{\cdot}5THF + Et_2Mg{\cdot}4THF$	13.4
R = methyl, ethyl or butyl group	

13.12 The reactions between MgR_2 Lewis acid and $AlCl_2R$ Lewis acid in THF at different stoichiometries.

represents the susceptibility of the solution towards oxidation. Thus, it is obvious that the most oxidizable species in the solution determines its electrochemical window. The solution species specified in Eqs 13.2–13.4 in Fig. 13.12 contain organometallic aluminate anions and molecules, with an ever-increasing tendency to oxidize. $R_2AlCl_2^-$ is less prone to oxidization than the others, as the two electron-withdrawing chlorine ligands increase the actual oxidation state of the aluminum core. On the other hand, the product of reaction 13.4, R_4Al^-, not only contains four σ-bonded organic ligands, but is also negatively charged. Both these factors, increase the electron density around the aluminum core, weakening these bonds and increasing the number of the oxidizable bonds.

We explored in depth the mechanism of Mg deposition from these solutions [55]. The fact that the conductivity of the solutions in which magnesium electrodes are reversible is only a few millisiemens per centimeter, and the overvoltage for magnesium deposition may not be negligible, makes the application of a stationary microelectrode technique for the study of the electrochemical behavior of magnesium electrodes very advantageous.

From rigorous measurements of the potentiodynamics of magnesium deposition processes using microelectrodes and analyses of the Tafel slopes of the *i–E* curves [55] it was possible to suggest Fig. 13.13 as showing the basic mechanisms of the multistage process of magnesium deposition/dissolution. This mechanism includes the adsorption processes of $MgCl^+$. This suggestion is in line with previous *in situ* FTIR measurements in the same solutions, which revealed the presence of surface (sf) species containing Mg–Cl bonds on magnesium electrodes [46].

An understanding of the Mg deposition mechanisms as reflected by Fig. 13.13, explains the effect of the acid/base ratio of the electrolyte in solution on the morphology, which is very pronounced (e.g. in solutions comprising the electrolyte that is the product of MgR_2–$(AlCl_2R)_2$, uniform, micrometer-size deposits with pyramidal morphology are formed). Since magnesium deposition is a multistep process in which adsorption also plays a role, the nature of the adsorbed species influences the deposition process due to secondary current distribution consideration. Since different acid/base ratios mean different active moieties (Fig. 13.12), it is clear that different species are adsorbed in each case. The diversity in the morphology of Mg

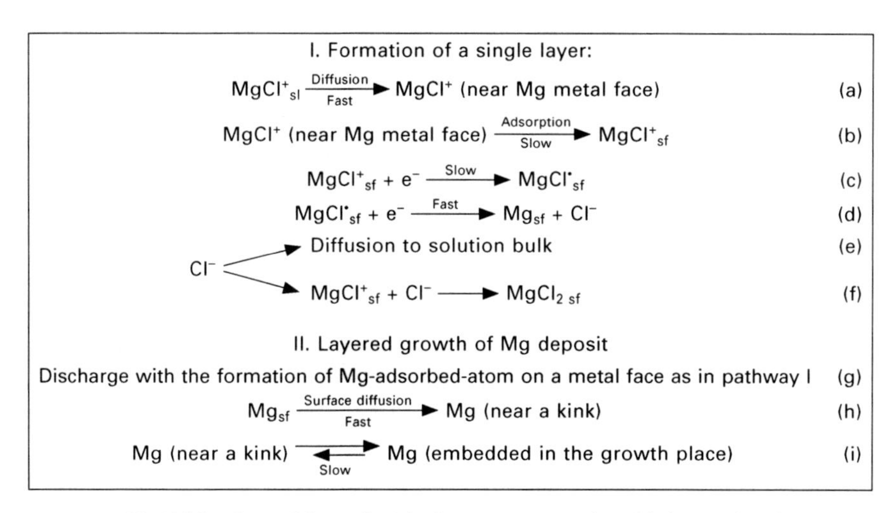

13.13 Mg deposition–dissolution processes (on Mg): mechanisms and reaction pathways.

deposition at different compositions of the electrolyte solutions results from the different adsorption processes in each system.

In a further development, we explored and demonstrated solutions that were prepared by reactive PhMgCl and $AlCl_3$ (Ph-phenyl group, $C_6H_5^-$) in an optimized stoichiometry of 2:1 (0.1–0.5 M $AlCl_3$) in THF [56]. These solutions have electrochemical windows > 3 V, conductivity twice as high as that of solutions comprising complexes with alkyl groups, and Mg deposition in them is fully reversible. We also demonstrated that polyethers from the 'glyme' family, such as tetraglyme ($CH_3(O{-}CH_2CH_2{-})OCH_3$ can also serve as single solvents or co-solvents with THF for reversible Mg electrodes using the above-described complexes as electrolyte [52]. It is important to note that Mg allows a few percent of Zn and Al (e.g. AZ31 alloys) to behave as reversible Mg electrodes in all the above-described solutions [57].

13.7 On magnesium (Mg) ions insertion into inorganic hosts

The main interest in Mg ion insertion into hosts (e.g. inorganic crystals, redox polymers, ceramic matrices), relates to efforts to develop rechargeable Mg batteries. In general, cathodes for Mg batteries could be based on Mg^{2+} ion insertion (intercalation) into the crystal structure of active materials, which are quite similar to those used in Li batteries. Extensive work on compounds, including TiS_2, SrS_2, RuO_2, Co_3O_4 and V_2O_5 as positive electrode materials for magnesium batteries has been carried out by Novak and co-workers. In recent years, VO_x compounds have received much attention regarding rechargeable magnesium batteries. In addition to Novak's work, there have been several reports by other groups on the study of vanadium oxides as a potential positive electrode material for magnesium batteries. Other inorganic hosts that were explored in recent years in connection with rechargeable magnesium batteries include MoO, Mg_xMnO_2, TiS_2 and MoS_2.

However, during efforts to develop rechargeable Mg batteries in recent years, it was realized that the selection of materials suitable for the Mg ion insertion presents a great challenge. In fact, in spite of the expected similarity between Li and Mg ion intercalation, almost all inorganic compounds, which prove themselves as suitable cathode materials for Li batteries, show very poor electrochemical performance regarding Mg ion insertion.

According to the literature, it is clear that the main problem of Mg-ion insertion into the usual hosts is its slow kinetics. A very good example is the insertion of Li and Mg ions into nanocrystalline V_2O_5, as illustrated in Fig. 13.14 [58]. The simplest way to estimate the kinetics of ionic transport in the intercalation compounds is the comparison of the intercalation and deintercalation potentials obtained upon the electrochemical response. In the case of fast kinetics, these potentials should be close to each other and to the equilibrium potential of the insertion reaction. The slower the process, the bigger is the difference between the insertion/deinsertion potentials (discharge/charge of the battery). The voltammograms of nanocrystalline V_2O_5 electrodes in Li and Mg ion solutions presented in Fig. 13.14 are typical and significant. Li ion insertion into V_2O_5 is fast and efficient, and the voltammograms of these processes reflect a very small hysteresis. In turn, Mg ions insertion to the same host is slow and the relevant CV, as seen in this figure, indeed shows a pronounced hysteresis. A relatively successful Mg intercalation was observed for nanocrystalline materials, thin films or nanotubes [59,60]. In such materials, the intercalation kinetics should be *a priori* much higher than with the micro-size materials. It is also clear that the reason for the slow kinetics is the divalent character of the inserted ions, resulting in strong interactions between the inserted divalent cations and the anions and the cations of the host, or high activation barriers for site changes in the case of inserted ions with high charge densities. As a result, Mg insertion was more successful in hydrates [61] because water or hydroxyl species can shield the strong coulombic interaction between the polyvalent guest species and the cation of the host.

We discovered that Chevrel phases, $M_XMo_6T_8$, T = S or Se, can insert reversibly two Mg atoms per cluster (i.e. M = Mg, $0<X<2$), and hence can serve as cathode materials for secondary Mg batteries [33]. Figure 13.15 shows the basic structure of $M_xMo_6S_8$ Chevrel phases which comprise octahedral clusters of six molybdenum ions confines in cubes of eight sulfur anions. Between each of two clusters there are 12 sites (two rings of six sites each) for ions insertion, two of which (inner and outer sites) can be occupied

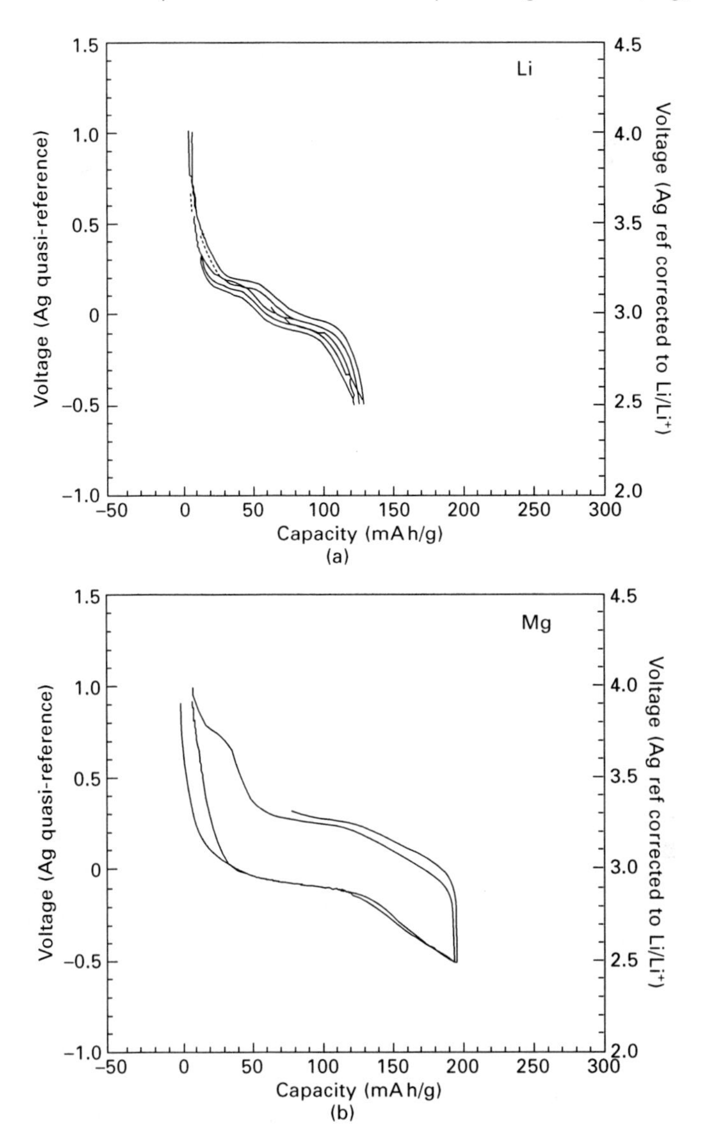

13.14 A comparison of the electrochemical Li and Mg ions insertion into porous V_2O_5 electrodes. The relevant solution compositions are indicated in the charts. Repeated galvanostatic experiments, voltage profiles are presented.

with Mg ions. This figure also shows a typical voltage profile of Mg ions insertion into this material, from the complex electrolyte solutions described in the previous section. As indicated by the voltage profile in Fig. 13.15 (two consecutive plateaux), Mg ions insertion into this material occurs via two first order phase transition processes. Note that full capacity intercalation

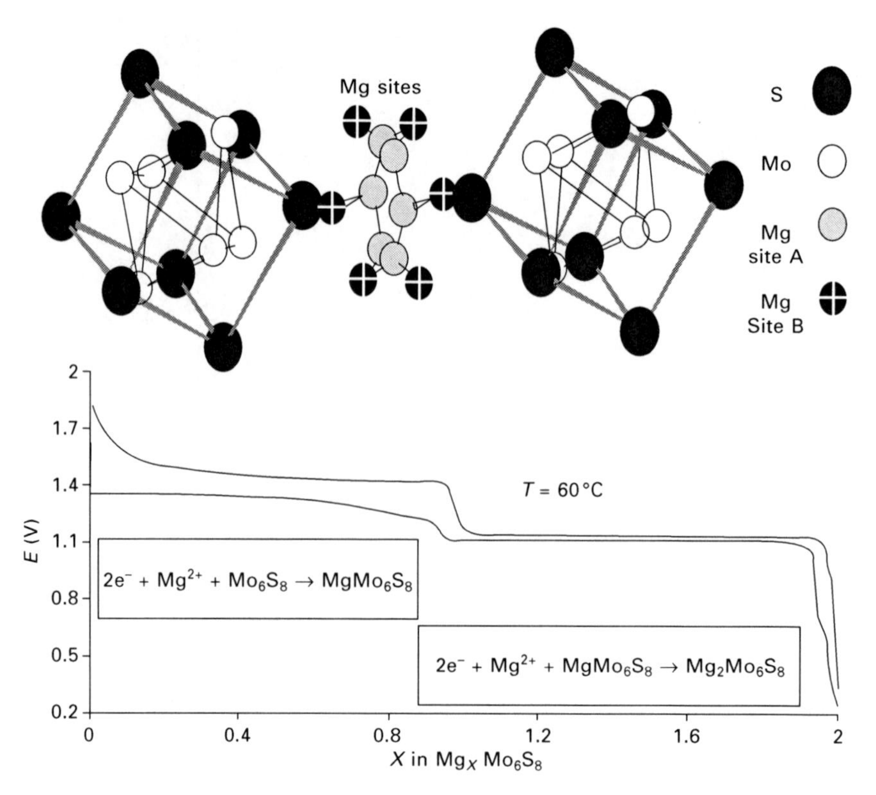

13.15 A schematic presentation of the structure of $Mg_xMo_6S_8$ Chevrel phases and the typical voltage profile of electrochemical Mg ions insertion/deinsertion into them in galvanostatic experiments with THF/$Mg(AlCl_2BuEt)_2$ solutions at 60 °C (fully reversible Mg ions intercalation in two consecutive phase transition processes is indicated).

into Mo_6S_8 required elevated temperatures while at room temperatures the capacity of the first insertion stage is somewhat limited and a partial charge (Mg ions) trapping is recorded. Thereby, the reversible capacity of Mg intercalation into this material at low temperatures is around 80% of the theoretical one (122 mA h/g, corresponding to insertion of two Mg ions per Mo_6S_8 unit).

The discovery of Chevrel phases as interesting hosts for reversible Mg ions intercalation resulted from many unsuccessful experiments of Mg insertion into well-known Li^+ ion hosts, as well as from the literature concerning the possibility of divalent ion intercalation in inorganic materials. This analysis revealed that Chevrel phases are unique materials that allow a relatively fast insertion of divalent cations [62] such as Zn^{2+}, Cd^{2+}, Ni^{2+}, Mn^{2+}, Co^{2+} and Fe^{2+}. In the intercalation processes into Chevrel phases, the six metal Mo ions can be regarded as a single ion that can accommodate up to four electrons (compared with one or two electrons for usual transition metal ions). Upon insertion of one Mg^{2+} ion per formula unit, the formal charge of an individual Mo ion in the cluster changes only by one-third of an electron. Moreover, in the Chevrel crystal structure, 12 vacant sites per formula unit are available for the inserting ions. The distances between them are very short (1.1–1.4 Å), and thus, only two of the sites can be occupied simultaneously by divalent ions. Therefore, the crystal structure of Chevrel phases is ideal for ion mobility because of the large number of the vacant sites, the short distances between them and the metallic cluster that ensures the local electroneutrality of the intercalation compound. Hence, the high activity of Chevrel phases in the process of Mg ions insertion/extraction can be attributed to the unusual crystal structure of these materials [63]. The electrochemical window of Mg ion insertion into Chevrel phases matches the electrochemical windows of the solutions in which the Mg electrodes behave reversibly (described in the previous section).

It was highly interesting to explore the effect of increasing the polarizability of the anionic framework of Chevrel phases on the nature of Mg^{2+} ion intercalation into these compounds. Consequently, Mg^{2+} ion insertion into Mo_6Se_8 has also been studied [64]. The replacement of sulfur by selenium as the anionic element in the Chevrel phase may reduce the intensity of attractive interactions between the intercalated Mg ions and the anionic framework of the host, thus reducing the diffusion barriers.

It should be noted that the Mo_6S_8 material is thermodynamically unstable and can be obtained only indirectly by chemical or electrochemical leaching of the more stable, metal-containing Chevrel phases, e.g. $Cu_2Mo_6S_8$. which can be synthesized from the Cu, Mo and S elements (or the metal sulfides) at high temperatures in quartz tubes. In contrast, Mo_6Se_8 can be synthesized directly from the elements.

Figure 13.16 shows a comparison between the cyclic voltammograms (CVs) of $Mg_xMo_6S_8$ and $Mg_xMo_6Se_8$ ($0<x<2$) measured with similar electrodes in similar solutions and experimental conditions [64]. The difference between the behavior of the two materials is spectacular. As expected, and as reflected by the CVs in Fig. 13.16, the kinetics of Mg ions into the selenide Chevrel phase is much faster than the sulfide (note also the smaller peak-potential differences for the selenide compound in Fig. 13.15). In addition, the capacity of Mg ion insertion into $Mg_xMo_6Se_8$ in repeated cycling is very close to the theoretical one (i.e. $0<x\leq2$), even at room temperature. Nevertheless, Mo_6S_8 is, of course, much more preferred as a cathode material for rechargeable Mg batteries than the Mo_6Se_8 due to the higher redox potentials and the theoretical capacity.

One possibility for increasing the performance of the Mo_6S_8 cathode material namely, to make the first Mg ions insertion stage faster (reduce the partial charge trapping in this stage) is to reduce the size of the particles

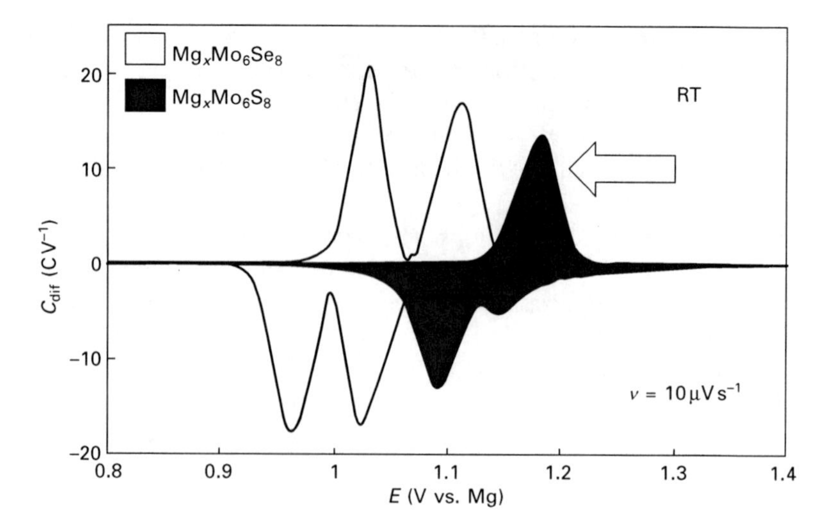

13.16 Typical steady state, slow scan rate (10 μV/s) CVs of Mg ion insertion into composite electrodes containing Mo_6S_8 and Mo_6Se_8, as indicated. The solution was THF/0.25 M $Mg(AlCl_2BuEt)_2$, 25 °C.

from micrometric to nanometric size. The fact that this compound does not react with ethers and with Mg–Al–Cl–R complexes ensures the lack of complications due to an increase in the surface area of the active mass. The easiest approach for that is milling. However, it should be noted that milling the precursor, $Cu_2Mo_6S_8$, in any atmosphere, or Mo_6S_8 in air, leads to irreversible mechanochemical reactions [65]. We demonstrated that it is possible to prepare nano-$MO_6S_{8-n}Se_n$, n = 1, 2, materials that serve as promising cathode materials for rechargeable Mg batteries [66]. It appears that the replacement of one or two sulfur atoms by Se increases the polarizability of the anionic framework of the Chevrel phases, and allows the faster kinetics of the problematic, first stage of Mg ion insertion into these materials. Another important recent discovery was the possibility of obtaining a very fast reaction of Mg ion insertion into $CuMo_6S_8$ [67]. The process includes the displacement of Cu^+ by Mg^{2+} and the reduction, which form nano-clusters of Cu^0 during the course of the process. This displacement reaction is fully reversible, and hence, the reversible process is:

$$2Mg^{2+} + 2e^- + CuMo_6S_8 \rightleftharpoons Mg_2Mo_6S_8 + Cu^0$$

In the search for new cathode materials for Mg batteries, we synthesized materials of the following compositions: $M'M_zS_xSe_y$, in which M, M′ = Cu, Ti, Fe, Ni or Cu and z, x and y assume various proportions. The compounds that we tested included CuS, $CuFeS_2$, $CuFe_2S_3$, Cu_9S_8, Cu_2S, NiS, c-Ti_2S_4, l-Ti_2S_4, TiSSe and NiS_xSe_y (solid solutions in various x/y ratios). In general, all of these compounds, except for P-T_2S_4, showed some reversible electrochemical activity in THF solutions containing complexes of the formal stoichiometry – $Mg(AlCl_2BuEt)_2$. In all cases, repeated magnesium insertion–deinsertion could be observed. However, the capacity gradually declines from cycle to cycle [52].

There is a continuous search by research groups throughout the world for new cathode materials for rechargeable Mg batteries. We can mention reports demonstrating the insertion of Mg ions into MoO_3 [68], $MSiO_x$ [69] and TiS_2 [70] hosts.

13.8 Future trends

Non-aqueous electrochemistry is a highly developed field. There are many thousands of reports on electrochemical reactions in non-aqueous systems. There are several families of relevant polar aprotic solvents, as well as an impressive arsenal of electrolytes that form ionically conductive media with polar aprotic solvents. These also include a wide variety of ionic liquids that are now available as electrochemical systems. Nevertheless, it seems that only ether-based solutions are really relevant for non-aqueous Mg electrochemistry. These solvents are sufficiently polar to form Mg ion-conducting solutions, and yet are not reactive with this active metal (while other solvent families such as esters, alkyl carbonates, nitriles and sulfones are reactive with magnesium). These ethers may include THF and glymes, e.g. $RO–(CH_2CH_2–O)R$. Even as thin films, ionic Mg compounds cannot conduct Mg ions, and thereby, whenever Mg electrodes become covered by surface films, they are electrochemically blocked. This blocking passivation situation for Mg electrodes also occurs in polar aprotic systems that contain trace water or oxygen that can react on the surface of the active metal (thus forming MgO or $Mg(OH)_2$). In ethers containing reducing agents such as RMgCl or MgR_2 (Grignard reagents), trace CO_2, O_2 or H_2O react with the Mg compounds in solutions, and hence cannot affect the surfaces of the Mg electrodes. Reacting MgR_2 or RMgCl, which can be considered as Lewis bases, with Lewis acids such as $AlCl_{3-n}R_n$ leads to trans-metallation that forms $MgCl^+$ (or $Mg_2Cl_3^+$) cations and $AlCl_{4-n}R_n^-$ anions. The ether molecules are strongly involved in such reactions and stabilize the above ions. The ether solutions of these complexes may allow fully (100%) reversible Mg deposition processes and wide electrochemical windows up to 3 V vs. Mg.

Another interesting topic dealt with herein is Mg ion insertion into inorganic hosts. While there are reports on Mg ion insertion into transition metal oxides (e.g. V_2O_5, MnO_2, Co_3O_4) and sulfides (e.g. TiS_2, MoS_2), these processes are very slow because of the strong interactions of the bivalent Mg ions with the anionic framework of these compounds, which leads to their very slow diffusion. Chevrel phases with the general stoichiometry of Mo_6T_8 (T = S, Se) are exceptions. They insert Mg ions reversibly at relatively fast rates, due to their unique structure in which clusters of molybdenum ions that may have several oxidation states are confined in cubes of S or Se anions. Chevrel phases can be considered as molecular sponges that allow the multi-directional diffusion of ions whose cationic clusters accommodate well the positive charge of inserted ions via changes in the oxidation state of the molybdenum ions. A rigorous study of these systems may allow the development of other families of hosts that can insert multivalent ions reversibly.

Further efforts in the field may include development of novel electrolyte solutions that enable reversible Mg deposition and have also a wide electrochemical window (i.e. high anodic stability).

There seems to be only two main classes of solvents, relevant to reversible Mg electrodes: ethers and ionic liquids. In our studies we concentrated mainly on THF as a suitable ether solvent for Mg electrochemistry. However, some preliminary work showed that ether oligomers from the 'glyme' family $[CH_3{-}OCH_2{-}CH_2{-})_n{-}R]$ are also suitable for Mg electrochemistry. Such solvents have a great advantage from a safety point of view due to their low volatility and high boiling point. Although we could not demonstrate reversible Mg deposition in IL-based systems, further work is needed in this direction. With the continuous improvement in the production and purification of commercially available ionic liquids from various families, there is a chance to find IL-based electrolyte solutions in which Mg is truly stable and can be deposited electrochemically reversibly. Many ionic liquids demonstrate very wide (may be beyond 5 V) electrochemical windows and hence may enable the use of high-voltage cathode materials. The development of new complex electrolytes for Mg electrochemistry is possible in the future; for instance, complexes of the $(MgR_2)_n{*}(AlRCl_2)_m$ type in which R is aryl groups substituted by electron withdrawing groups. Such complexes may have extended anodic stability.

A great challenge is development of cathodes for rechargeable Mg batteries. In general, there should be transition metal oxides and sulfide that may insert Mg ions reversibly in a similar manner as Chevrel phases do. For Mg electrochemistry based on ether or ionic liquid solvents, nano-materials of the MO_x or MS_x type (M = multi-valent transition metal) may be very suitable as inorganic host materials. Hence, there is plenty of scope for further work on the relevant material science. Another class of interesting materials in this respect is electronically conducting and redox polymers containing active sites such as S–S bonds. Such bonds can be reversibly reduced in the present of active metal ions. There are already interesting publications which deal with poly-RSSR materials as cathodes in rechargeable Li batteries. Similar materials may undergo reversible reduction in the presence of Mg ions as well. Highly interesting is the preparation of conducting polymers based on thiophen or para-phenylen as the backbone and side RSSR groups as the active redox centers. Such polymers may serve as universal cathodes for both Li and Mg batteries. Hence, the field of non-aqueous Mg electrochemistry is open to more innovative work on both solutions and solid state chemistries.

13.9 References

1. L.W. Gaddum, H.E. French, *J. Am. Chem. Soc.* **49** (1927) 129.
2. W. Evans, *J. Am. Chem. Soc.* **64** (1942) 2965.
3. J.H. Connor, W.E. Reid, G.B. Wood, *J. Electrochem. Soc.* **104** (1957) 38.
4. A. Brenner, J.J. Singh, *Trans Inst. Met. Finishing* **49** (1971) 71.
5. M.M. Baizer, H. Lund (Eds), *Organic Electrosynthesis*, Marcel Dekker, New York, 1983.
6. O.R. Brown, R. McIntyre, *Electrochim. Acta* **30** (1985) 627.
7. O.R. Brown, R. McIntyre, *Electrochim. Acta* **29** (1984) 995.
8. W.R. Fawcett, J.S. Jaworskii, *J. Chem. Soc. Faraday Trans.* **78** 1 (1971) 1982.
9. T. Psarras, R.E. Dessy, *J. Am. Chem. Soc.* **89** (1967) 5132.
10. J. Bittrich, R. Landsberg, W. Gaube, Wiss. Z. Hochsch, *Chem. Carl Schorlemmer, Lenna-Merseburg* **60** 2 (1959) 449.
11. N. Eda, A. Ohta. In *New Trends in Electrochemical Technology*, Vol. 1: *Energy Storage Systems for Electronics*, T. Osaka, M. Datta, Eds, Gordon and Breach Science Publishers, Amsterdam/Singapore, 2000, 9.
12. K. Kinoshita, M.M. Thackeray. In *Handbook of Battery Materials.*, J.O. Besenhard, Ed, Wiley-VCH Weinheim, New York, Toronto, 1999, 231–243, 293–322.
13. D. Linden. In *Handbook of Batteries*, D. Linden, Ed, McGraw Hill, New York, 1994, and chapters related to different battery system therein.
14. S. Gottesfeld, T.A. Zawodzinski. In A*dvances in Electrochemical Science and Engineering*, R.C. Alkire, H. Gerischer, D.M. Kolb, C.W. Tobias, Eds, Wiley-VCH, New York, 1997.
15. Y. Nishi. In *Li-ion Batteries*, M. Wakihara, O. Yamamoto, Eds, Wiley-VCH, New York, 1998, Ch. 18 and references therein.
16. A. Meitav, E. Peled, *J. Electrochem. Soc.* **128** (1981) 3536.
17. G.G. Kumar, N. Munichandraian, *Solid State Ionics* **128** (2000) 203.
18. G.G. Kumar, N. Munichandraian, *J. Power Sources* **102** (2001) 46.
19. G.G. Kumar, N. Munichandraian, *Electrochim. Acta* **47** (2002) 1013.
20. J.D. Genders, D. Fletcher, *J. Electroanal. Chem.* **199** (1986) 92.
21. C.J. Liebenow, *J. Appl. Chem.* **27** (1997) 221.
22. D. Aurbach, I. Weismann in *Nonaqueous Electrochemistry*, D. Aurbach, Ed, Marcel Dekker, Inc., New York 1999, Ch. 1.
23. D. Aurbach, A. Schechter, in: *Science and Technology of Advanced Lithium Batteries*, G. Pistoia and G. Nazri, Eds, Kluwer Academic Press, Plenum Publishers, New York, London, Boston 2004, Ch 18, pp. 530–568.
24. K. Xu, *Chem. Rev.*, **104** (2004) 4303.
25. C. Reichardt, '*Solvents and Solvent Effects in Organic Chemistry*', 2nd edtion, VCH, New York, 1988.
26. W. Linert, A. Camard, M. Armand, C. Michot, *Coordination Chem. Rev.*, **226** (2002) 137.
27. D. Aurbach, Y. Gofer in '*Nonaqueous Electrochemistry*', D. Aurbach, Ed, Marcel Dekker, Inc., New York, 1999, Ch. 6.

28. D. Aurbach, H. Gottlieb, *Electrochim. Acta* **34** (1989) 141.
29. M. Buzzeo, R. Evans, R. Compton, *Chem. Phys. Chem.* **5** (2004) 1106.
30. D. Aurbach, Y. Cohen, in *Lithium-ion Batteries: Solid-Electrolyte Interfacial Film*,' P. Balbuena, Ed, Imperial College Press, Oxford, 2004, Ch 2, pp. 70–139.
31. Z. Lu, A. Schechter, M. Moshkovich, D. Aurbach, *J. Electroanal. Chem.* **466** (1999) 203.
32. Y. Gofer, R. Turgeman, H. Cohen, D. Aurbach, *Langmuir* **19** (2003) 2344.
33. D. Aurbach, Z. Lu, A. Schechter, Y. Gofer, H. Gizbar, R. Turgeman, Y. Cohen, M. Moskovich, E. Levi, *Nature* **407** (2000) 724.
34. E. Peled, H. Straze, *J. Electrochem. Soc.* **124** (1997) 1030.
35. J.D. Genders, D. Pletcher, *J. Electroanal. Chem.* **199** (1986) 93.
36. T. Gregory, R. Hoffman, R. Winterton, *J. Electrochem. Soc.* **137** (1999) 775.
37. W. Vonau, F. Berthold, *J. Praktische Chem.-Chem.* **336** (1994) 140.
38. L.P. Lossius, F. Emmenegger, *Electrochim. Acta* **41** (1996) 445.
39. C. Liebenow, *J. Appl. Chem.* **27** (1997) 221.
40. A. Kisza, J. Kazmierczak, B. Børressen, H.M. Haarberg, R. Tunold, *J. Appl. Chem.* **25** (1995) 940.
41. A. Kisza, J. Kazmierczak, *J. Electrochem. Soc.* **144** (1997) 5.
42. B. Børressen, G.M. Harberg, R. Tunold, *Electrochim Acta* **42** (1997) 1613.
43. A. Brenner, *J. Electrochem. Soc.* **118** (1971) 99.
44. A. Brenner, J.L. Sligh, *Trans. Int. Met. Finish* **49** (1971) 71.
45. D. Aurbach, M. Moshkovich, A. Schechter, R. Turgeman, *Electrochem. and Solid State Letts.* **3** (2000) 31.
46. D. Aurbach, R. Turgeman, O. Chusid, Y. Gofer, *Electrochem. Com.* **3** (2001) 252.
47. W. Shlenk, W. Shlenk, Jr., *Chem. Ber.* **62** (1990) 920.
48. D. Aurbach, R. Skaletsky, Y. Gofer, *J. Electrochem. Soc.* **130** (1991) 3536.
49. A. Meitav, E. Peled, *J. Electrochem. Soc.* **118** (1971) 99.
50. Y. NuLi, J. Yang, J. Wang, J. Xu, P. Wang, *Electrochem. Solide State. Lett.* **8** (2005) C166.
51. Y. Nuli, J. Yang, R. Wu, *Electrochem. Commun.* **7** (2005) 1105.
52. D. Aurbach, O. Chusid, Y. Vestfried, Y. Gofer, N. Amir, *J. Power Sources* **174** (2007) 1234.
53. H. Gizbar, Y. Viestfrid, O. Chusid, Y. Gofer, H. E. Gottlieb, V. Marks, D. Aurbach, *Organometallics* **23** (2004) 3826–3831.
54. Y. Vestfried, O. Chusid, Y. Gofer, D. Aurbach, *Organometallics* **26** (2007) 3130.
55. Y. Viestfried, M. D. Levi, Y. Gofer, D. Aurbach, *J. Electroanal. Chem.* **57** (2005) 183.
56. O. Mizrahi, N. Amir, E. Pollak, O. Chusid, V. Marks, H. Gottlieb, E. Pollak, O. Chusid, V. Marks, L. Larush, E. Zinigrad, D. Aurbach, *J. Electrochem. Soc.* **155** (2008) A103.
57. O. Chusid, Y. Gofer, H. Gizbar, Y. Vestfried, E. Levi, D. Aurbach, *Adv. Mater.* **15** (2003) 627.
58. G. Amatucci, F. Badway, A. Sighal, B. Beaudoin, G. Skandan, T. Bowmer, I. Plitza, N. Pereira, T. Chapman, R. Jaworski, *J. Electrochem. Soc.* **148** (2001) A940.
59. G. Agarwal, G. B. Reddy, *J. Electrochem. Soc.* **157** (2007) A417.
60. L. Jiao, H. Yuan, Y. Wang, J. Cao, Y. Wang, *Electrochem Comm.* **7** (2005) 431.
61. P. Novak, R. Imhof, O. Haas, *Electrochim Acta* **45** (1999) 351.
62. E. Levi, G. Gershinsky, G. Ceder, D. Aurbach, *Chem. Mater.* **7** (2009) 1390–9.

63. D. Aurbach, E. Levi, O. Chusid, M. Levi, *J. Electroceramics* **22** (2009) 13.
64. M. D. Levi, E. Lancry, E. Levi, H. Gizbar, Y. Gofer, D. Aurbach, *Solid State Ionics* **176** (2005) 1695.
65. E. Levi, Y. Gofer, Y. Vestfried, E. Lancry, D. Aurbach, *Chem. Mater.* **14** (2002) 2767.
66. D. Aurbach, G.S. Suresh, E. Levi, A. Mitelman, O. Mizrahi, O. Chusid, M. Brunell, *Adv. Mat.* **19** (2007) 4260.
67. A. Mitelman, M.D. Levi, E. Lancry, E. Levi, D. Aurbach, *Chem. Commun.* **41** (2007) 4212.
68. T. S. Sian, G. B. Reddy, *Solid State Ionics* **167** (2004) 399.
69. Z. Feng, J. Yang, Jun, Y. NuLi, J. Wang, X. Wang, Z. Wang, *Electrochem. Comm.* **10** (2008) 1291.
70. Z.L. Tao, L.N. Xu, X. L. Gou, J. Chen, H.T. Yuan, *Chem. Comm.* **18** (2004) 2080.

Part IV

Corrosion protection

14
Electrodeposition of aluminum (Al) on magnesium (Mg) alloy in ionic liquid

W.-T. TSAI and I.-W. SUN, National Cheng Kung University, Taiwan

Abstract: To improve the corrosion resistance of magnesium (Mg) alloys, surface modification is applied in an attempt to produce a corrosion resistant barrier. In this chapter, a new method to deposit aluminum (Al) film on an Mg alloy surface is demonstrated. Electrodeposition of Al on AZ91D Mg alloy, using an acidic aluminum chloride–1-ethyl-3-methylimidazolium chloride ionic liquid ($AlCl_3$–EMIC), has been shown to be feasible. The existence of Al coating can cause a substantial increase in corrosion resistance, reducing the susceptibility of Mg alloy to aqueous corrosion. The results of electrochemical impedance spectroscopy show that Al coating leads to an increase in the polarization resistance of a bare Mg alloy by one order of magnitude in 3.5 wt% NaCl solution. Furthermore, the potentiodynamic polarization results show that Al-coated Mg alloy can be passivated, and a wider passive region with a lower passive current density can be obtained if the Al is electrodeposited at a lower applied current or a low cathodic overpotential. The passivity of the co-deposited Al/Zn film is slightly inferior to that of the pure Al coating.

Key words: magnesium alloy, ionic liquid, electrodeposition, aluminum, corrosion.

14.1 Introduction

Magnesium (Mg) and its alloys are increasingly implemented into a number of components where weight reduction is of great concern. However, the poor corrosion resistance of magnesium and its alloys has limited their applications in corrosive environments [1–5]. Extensive reviews on the forms of magnesium corrosion and the influences of composition and microstructure have been summarized by Song [5]. To mitigate the susceptibility to corrosion, surface treatment to form a protective barrier layer is always required. In general, the methods for surface treatment can be classified into two categories, dry and wet processes. A number of possible protective coating techniques for magnesium and its alloys have been reviewed by Gray and Luan [6]. Certainly, each method has its own advantages and disadvantages, regarding the material properties and the manufacturing processes, etc. Since the wet methods generally have a high yield, they have merit and potential for large-scale production.

Among the various wet processing techniques, anodization [7–10] and conversion coating [11–17] are most commonly applied. The coatings resulting from these treatments are mainly oxides or inorganic compounds. When electromagnetic shielding is required, metallic coating is recommended. Since electro- and electroless deposition are well established and have been applied to various substrates for surface modification, they are considered for metallic coating on Mg alloys. Electroless plating of Ni or its alloys such as Ni–P and Ni–B, employing aqueous electrolytes, has been attempted for Mg alloy metallization [18–24]. Electrodeposition of pure metal and/or alloys such as Cu, Ni, Zn, Cu–Ni, Zn–Sn and Zn–Ni on Mg alloys, using aqueous electrolytes, has also been investigated [20,25–28]. In order to improve the cohesion of the deposited layer, electroless Ni coating and/or Zn immersion process is always applied before electrodeposition. Nevertheless, the active nature of magnesium may always cause the formation of loose MgO or Mg(OH) on surface before electrodeposition when aqueous electrolyte is used. The liberation of hydrogen molecules during electrodeposition of metals from aqueous electrolytes may also produce pinholes in the deposits. To overcome these problems, non-aqueous solutions with organic solvents may be used. In a review article [29], Simka and her co-workers indicate that several metals and alloys can be electrodeposited from conventional organic solvents. However, to the authors' knowledge, metal coating on magnesium substrate from organic baths has yet to be established.

For combined corrosion resistance and electromagnetic shielding, Al coating is of interest because of the following reasons. First, being a light metal, an Al coating does not significantly increase the overall density of the material. Second, both anodization and electrolysis coloration techniques for Al are already well established, they can be directly applicable to the Al-coated Mg alloys. Third, the low electric resistivity and electromagnetic shielding property of the Al-coated Mg alloy can be maintained even after the anodization of the surface Al film. Finally, as one of the major alloying elements, Al coating is more chemically compatible with the substrate.

Various methods are available for Al coating such as sputter deposition, vapor deposition, thermal spraying, hot dipping and electrodeposition. Among these methods, electrodeposition offers some advantages including easy control of the coating thickness through the experimental parameters, lower operating temperature and relatively low cost. However, Al cannot be electrodeposited from aqueous solution because the reduction potential of Al metal ion is well below that of the hydrogen evolution. Therefore, aprotic electrolytes such as organic solvents or molten salts must be employed for Al deposition. Electrodeposition of Al has been successfully performed using three classes of organic solvents: ethers, aromatic hydrocarbons, and dimethylsulfone [30–35]. Diethyl ether containing $AlCl_3$ and $LiAlH_4$ was first employed in the NBS (National Bureau of Standards) bath [36]. Although this bath has been adopted in industrial application, it possessed several drawbacks including low Al anode current efficiency, limited lifetime, and hydrogen evolution [30]. Tetrahydrofuran (THF) was used to replace the diether to improve the anode current efficiency and longer bath lifetime [30]. On the other hand, owing to its high solubility compared with AlF_3 and $AlCl_3$, $AlBr_3$ has been used as the Al(III) source in plating baths using aromatic hydrocarbons such as benzene, toluene, xylene and their mixtures and derivatives [31–33]. The electroactive species responsible for the deposition of Al is $Al_2Br_7^-$. Alkali, quaternary ammonium or pyridinium halide salts were added to improve the conductivities and current efficiencies of the baths. Process using aromatic solvents containing alkylaluminum has also been reported to be successful [37]. Nevertheless, alkylaluminum compounds possessed the drawbacks of self-ignition in air. The utility of the aromatic hydrocarbons can be, however, limited by their toxicity. Another organic solvent that has been studied for electrodeposition of Al is dimethylsulfone (DMSO) [34]. DMSO bath shows good conductivity, thermal stability, and ability to dissolve metallic salts. The complex compound responsible for Al deposition from $AlCl_3/LiCl/DMSO_2$ bath is $Al[(CH_3)_2SO_2]_3^{3+}$. A recent study indicates that good Al coatings can be obtained in a temperature range of 110–140 °C [35].

Although Al can be successfully electrodeposited from organic solvents, the volatility, toxicities and flammability of the organic electrolytes create safety concerns. Such safety concerns can be circumvented by using inorganic molten salts such as mixtures of NaCl and $AlCl_3$ which are non-volatile and non-flammable. The inorganic molten salts, however, normally require a high operating temperature (> 130 °C) due to their high melting points. On the other hand, ambient temperature molten salts that are liquid below 100 °C can be obtained by replacing the inorganic salts with organic salts. To distinguish them from the high-temperature inorganic molten salts, the ambient temperature molten salts are classified as ionic liquids. The ionic liquids have proven to be versatile electrolytes for electrodeposition due to their advantageous properties such as wide electrochemical window, low melting temperature, high thermal stability, non-volatility, good ionic conductivity and non-flammability. In this chapter, the background of low-temperature electroplating of Al in aluminum chloride–1-ethyl-3-methylimidazolium chloride ionic liquids ($AlCl_3$–EMIC) is introduced. The deposition of Al and co-deposition of Al and Zn on an Mg alloy, namely AZ91D from the $AlCl_3$–EMIC baths are demonstrated, and the associated advantages in enhancing the corrosion resistance are described.

14.2 Basics for ionic liquid plating

Although numerous kinds of ionic liquids have been developed, as far as Al deposition is concerned, the most studied ionic liquids are the chloroaluminates which are obtained by reaction of $AlCl_3$ with a quaternary ammonium chloride salt (RCl) such as 1-ethyl-3-methylimidazolium chloride (EMIC) [38,39], 1-butyl-3-methylimidazolium chloride (BMIC) [40–42], trimethylphenylammonium chloride (TMPAC) [43,44] and benzyltrimethyl ammonium chloride (BTMAC) [45]. The fundamental studies reported by Osteryoung an co-workers [46,47], Lai and Skyllas-Kazacos [48] and Hussey and co-workers [49], have classified the chloroaluminates into Lewis basic, neutral and acidic depending on the relative mole fraction of the organic salt (X_R) and $AlCl_3$ (X_{Al}). In the Lewis basic ($X_{Al} < 0.5$) and the neutral ($X_{Al} = 0.5$) melts the major aluminum species is $AlCl_4^-$ in which Al(III) is fully coordinated by Cl^- ions, rendering the reduction potential of Al(III)/Al couple more negative than the cathodic electrochemical window of the melts and thus, metallic Al cannot be electrodeposited from the basic melts. On the other hand, in the Lewis acidic melts ($0.5 < X_{Al} < 0.67$), the major aluminum species is $Al_2Cl_7^-$ of which the number of Cl^- ions coordinated with Al(III) is less than that of $AlCl_4^-$, making the reduction of Al(III)/Al shift toward much positive potential so that the electrodeposition of Al metal is possible according to the equation:

$$4Al_2Cl_7^- + 3e^- \leftrightarrow Al + 7AlCl_4^- \qquad 14.1$$

The deposition process normally involves with either instantaneous or progressive three-dimensional nucleation with hemispherical diffusion-controlled growth depending on the substrate. The morphology of the Al deposit is affected, in addition to the deposition potential or current, by the type of cation forming the ionic liquid. For example, compact, shining and adherent Al films have been successfully electrodeposited from EMI–$AlCl_3$ [38] but the Al obtained from TMPAC–$AlCl_3$ is fairly coarse [43]. Reviews on the electrodeposition of Al and alloys from chloroaluminates are available [50,51].

Recently nanocrystalline Al has been electrodeposited from air- and water-stable ionic liquids based on the bis(trifluoromethylsulfonyl)amide (Tf_2N) anion with cations such as *N*-butyl-*N*-methyl pyrrolidinum (BMP–Tf_2N) [38,52–54] and trihexyl-tetradecyl phosphonium ($P_{14,6,6,6,6}$–Tf_2N) [52] saturated with $AlCl_3$. The cation of the ionic liquid influences profoundly the chemical and electrochemical behavior; biphasic behavior was observed for the EMI–Tf_2N and BMP–Tf_2N ionic liquids upon addition of $AlCl_3$ within certain concentration ranges at temperature < 80 °C, and Al films could be deposited from the upper phase. Shining, dense, and adherent nanocrystalline Al were deposited from $AlCl_3$-saturated BMP–Tf_2N, but coarse cubic-shaped Al particles were obtained from the $AlCl_3$-saturated EMI–Tf_2N ionic liquid. The $P_{14,6,6,6,6}$–Tf_2N ionic liquid does not show biphasic behavior upon addition of $AlCl_3$. However, Al could only be electrodeposited from this ionic liquid–$AlCl_3$ mixture containing more than 4M of $AlCl_3$ after being heated to 150 °C. More recently, a preliminary study shows that Al can also be electrodeposited from an *N*-butyl-*N*-methyl pyrrolidinum dicyanamide (BMP–DCA) ionic liquid containing $AlCl_3$ [55]. It is noteworthy that in all these processes, the highly water and air-sensitive $AlCl_3$ is used exclusively as the Al(III) source, and thus, the experiments need to be performed in an glove box filled with inert gas such as Ar and N_2. Some typical ionic liquids employed for electrodeposition of Al are summarized in Table 14.1.

Although the deposition of Al from ionic liquids has been intensively studied, corrosion behavior of the as-deposited Al films has not yet been fully explored. Recently, we have initiated a study on the deposition of Al and Al–Zn on Mg alloy, AZ91D, and demonstrated that the corrosion behavior of Al-coated AZ91D is greatly improved [56–59]. The results from these studies are described in the following section.

14.3 Electrochemical characteristics of $AlCl_3$–EMIC ionic liquids

Figure 14.1 shows the cyclic voltammogram of a tungsten electrode performed in room temperature $AlCl_3$–EMIC with a molar ratio of 60:40 ionic liquid. For brevity, the $AlCl_3$–EMIC mixture, consisting of a 60% molar fraction of $AlCl_3$, is denoted as 60 m/o ionic liquid. Similarly, the ionic liquids with $AlCl_3$–EMIC molar ratios of 50:50, 53:47 and 57:43 are denoted in this chapter as 50 m/o, 53 m/o and 57 m/o ionic liquid, respectively. As the potential is scanned from the open-circuit potential (approximately 0.9 V) towards the negative direction, the only cathodic reaction takes place around –0.2 V, corresponding to the reduction of $Al_2Cl_7^-$ precursor in the ionic liquid to form Al, according to reaction (14.1) specified above [49,60–63]. On reverse scanning, the sharp anodic peak corresponds to Al dissolution. Further scanning to the more positive potential, the oxidation of $AlCl_4^-$ according to the following reaction occurs at 2.6 V:

$$4AlCl_4^- \rightarrow 2Al_2Cl_7^- + Cl_2 + 2e^- \qquad 14.2$$

By decreasing the concentration of $AlCl_3$ to 53 m/o, the cathodic peak

Table 14.1 Examples of ionic liquids containing $Al_2Cl_7^-$ anion for electrodeposition of aluminum

Ionic liquid	References
EMI^+ $Al_2Cl_7^-$	38, 39, 46–49, 56–59
BMI^+ $Al_2Cl_7^-$	40–42
$TMPA^+$ $Al_2Cl_7^-$	43, 44
$BMP^+Tf_2N^-/AlCl_3$ (saturated)	38, 52–54
$P_{14,6,6,6,6}{}^+$ $Tf_2N^-/AlCl_3$ (saturated)	52

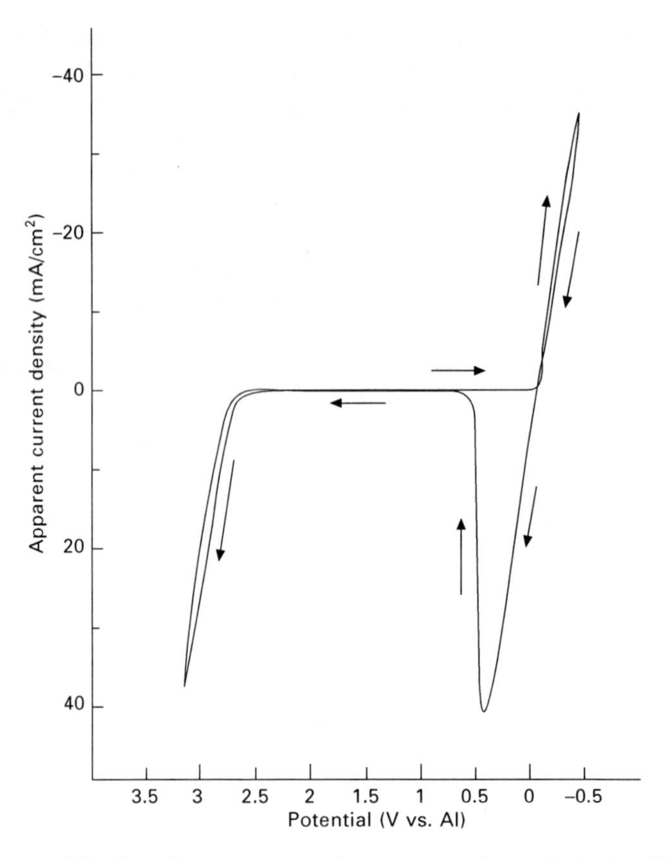

14.1 Cyclic voltammogram of a tungsten electrode in the 60 m/o $AlCl_3$–EMIC ionic liquid, room temperature, potential scan rate: 50 mV/s.

can still be observed but with a decreasing intensity. At 50 m/o $AlCl_3$, the absence of the cathodic peak indicates that Al cannot be deposited at such composition. It is also noted that the Al deposition potential shifts to a more positive value with increasing $AlCl_3$ molar fraction, indicating a thermodynamic favor of Al deposition.

Co-deposition of Al and Zn is also feasible. Figure 14.2 shows the cyclic voltammograms of a glassy carbon electrode obtained at room temperature in the 60 m/o ionic liquid with 1 wt% $ZnCl_2$ addition. Curve **a** in Fig. 14.2 displays the cyclic voltammogram obtained in 60 m/o $AlCl_3$–EMIC without $ZnCl_2$ addition, in the potential range of –0.5 to 2.5 V, consistent with that shown in Fig. 14.1.

Curve **b** in Fig. 14.2 depicts the cyclic voltammogram obtained at potentials ranging from 0 ~ +1.5 V. The cathodic scan shows that an abrupt increase in current density occurs at about +0.19 V, indicating the reduction

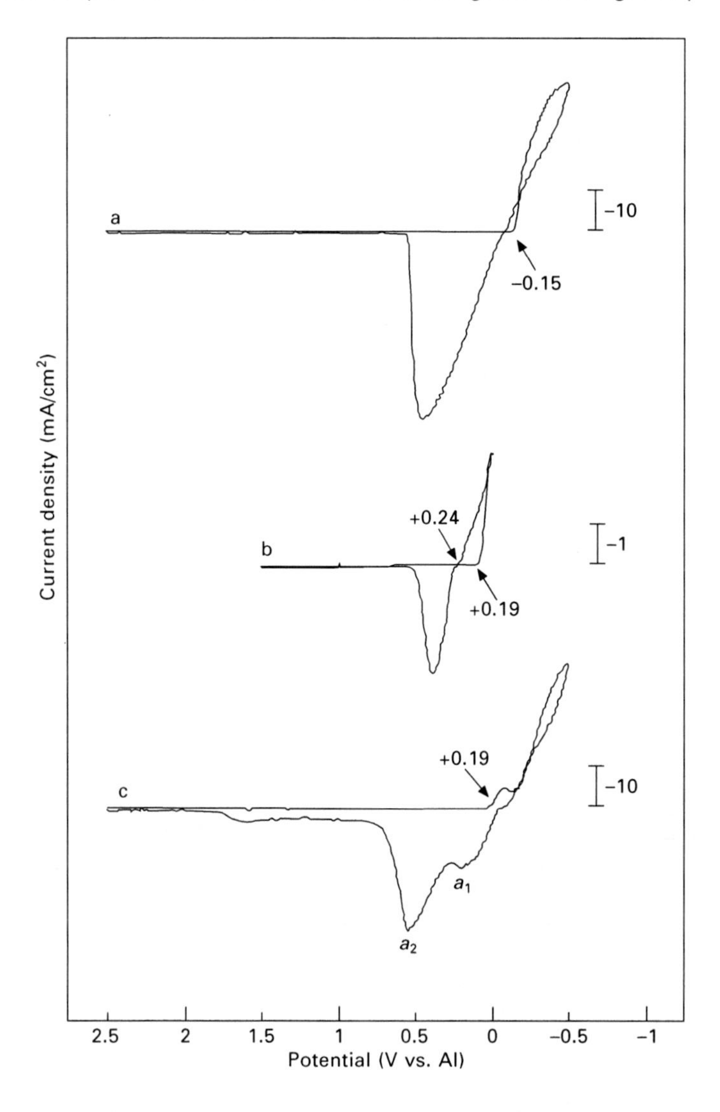

14.2 Cyclic voltammograms on a glassy carbon electrode at room temperature and at a scan rate of 50 mV/s: curve a: 60 m/o $AlCl_3$–EMIC (scan rage: –0.5 ~ +2.5 V); curve b: 1 wt% $ZnCl_2$ + 60 m/o $AlCl_3$–EMIC (scan range: 0 ~ +1.5 V), and curve c: 1 wt% $ZnCl_2$ + 60 m/o $AlCl_3$–EMIC (scan rage: –0.5 ~ +2.5 V).

of Zn. The intercept of the reverse scan at zero current density is found at +0.24 V, above which Zn cannot be reduced. The peak in the anodic scan is associated with the dissolution of Zn. The cyclic voltammogram for an enlarged potential scan range of –0.5 to +2.5 V is shown as curve **c** in Fig. 14.2. Two cathodic peaks are observed showing the sequential reduction of

Zn and Al. The initial potential for Zn reduction, as indicated in curve **c** is +0.19 V, identical to that revealed in curve **b**. The second peak at the more negative potential, where the current density begins to rise again, corresponds to Al reduction. The results shown in Fig. 14.2 suggest that co-deposition of Zn and Al can occur at potentials below –0.08 V. The two anodic peaks at a_1 and a_2 represent the sequential anodic reactions for Al and Zn respectively.

14.4 Material characteristics

Electrodeposition of Al, either under constant potential or constant current density condition, greatly depends on the acidity or chemical composition of the ionic liquid employed. In the $AlCl_3$–EMIC ionic liquid with 50 m/o of $AlCl_3$, electrodeposition of Al is hindered. As the concentration of $AlCl_3$ increases above 53 m/o, Al coating can take place. The surface morphologies of the uncoated and Al-coated AZ91D Mg alloy, observed under a scanning electron microscope (SEM), are demonstrated in Fig. 14.3. Figure 14.3(a) shows a micrograph of the bare AZ91D Mg alloy, demonstrating its dual-phase microstructure. Figure 14.3 (b)–(d) display the Al-coated sample deposited at –0.2 V (vs. Al) in the 53 m/o, 57 m/o and 60 m/o ionic liquids; respectively. It

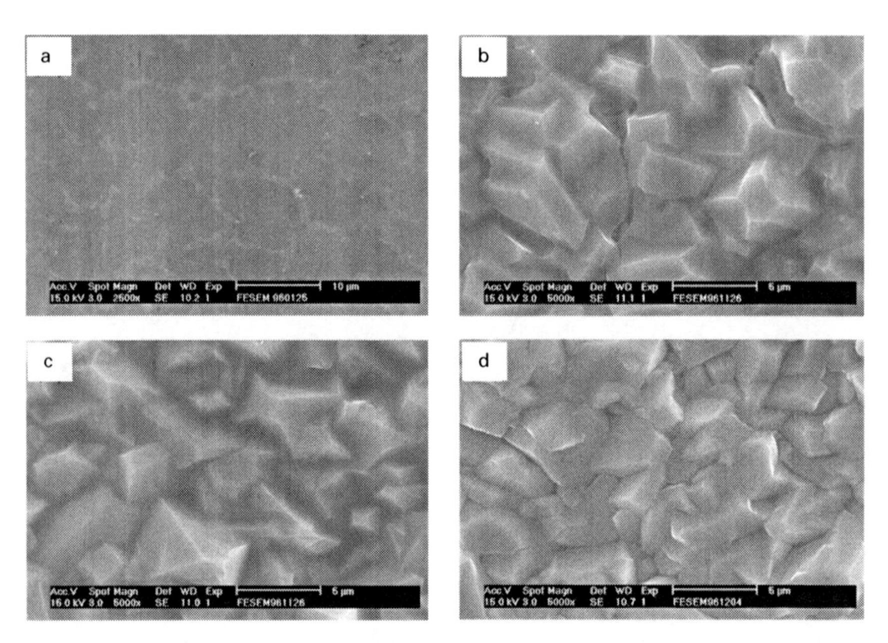

14.3 SEM top-view micrographs of (a) the bare AZ91D Mg alloy; (b), (c) and (d) the Al-coated Mg samples deposited at –0.2 V (vs. Al) in the 53 m/o, 57 m/o and 60 m/o ionic liquids, respectively.

is found that the entire surface of the Mg alloy, regardless of α or β-phase, is covered by a layer of Al with granular appearance. The properties of deposited film are greatly affected by the deposition condition. Figure 14.4 shows the effect of deposition potential on the surface morphology and cross-section micrograph of the Al-coated AZ91D Mg alloy. Figure 14.4(a) gives the SEM micrograph revealing a fine and uniform granular surface morphology of Al film formed at –0.2 V. At a more negative potential, namely –0.4 V, the Al film exhibits a coarse and non-uniform surface feature with crevices existing in the film (Fig. 14.4(b)). The corresponding cross-section SEM micrographs of the Al-coated AZ91D Mg alloy are displayed in Fig. 14.4(c) and (d). The energy dispersive spectroscopy (EDS) line scans showing in each figure confirm the formation of Al film on the Mg alloy surface. The thickness of Al film formed at –0.2 V is rather uniform as compared with that formed at –0.4 V. More defects, such as voids and crevices etc., are found in the film electrodeposited at a more negative potential. The quality of the Al film also depends on the deposition current density. As reported elsewhere [57], a porous and less compact Al film is observed if it is electrodeposited from an ionic liquid at a higher current density.

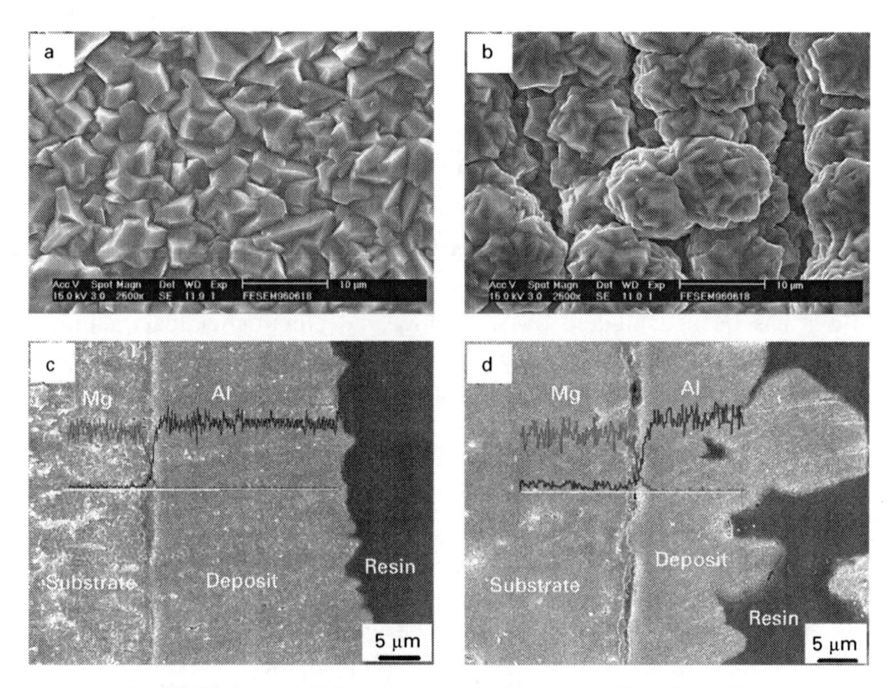

14.4 SEM surface micrographs of Al-coated AZ91D Mg alloy, electrodeposited at (a) –0.2 V (vs. Al) and (b) –0.4 V; and the cross-section images and energy dispersive spectroscopy (EDS) line scans of the Al film formed at (c) –0.2 V and (d) –0.4 V.

The deposition of Al from various $AlCl_3$–EMIC ionic liquids on the Mg alloy surface is confirmed by EDS and X-ray diffraction analysis (XRD). The EDS result for each of the above Al-coated AZ91D Mg alloy only reveals the spectrum for Al element. The absence of the alloying elements of the substrates indicates that the coated layer can be quite thick. Figure 14.5 compares the XRD patterns of the Mg substrate (curve **a**) with those electroplated in the ionic liquid containing 53 m/o (curve **b**), 57 m/o (curve **c**) and 60 m/o (curve **d**) $AlCl_3$. The characteristic peaks for metallic Al are clearly displayed in curves **c** and **d** in this figure. Similar results have been reported elsewhere [56].

Co-deposition of Al and Zn can be obtained by electrodeposition from $AlCl_3$–EMIC ionic liquids containing $ZnCl_2$. Figure 14.6(a) demonstrates the SEM micrograph showing the surface morphology of the AZ91D alloy after electrodeposition in 1 wt% $ZnCl_2$ + 60 m/o $AlCl_3$–EMIC ionic liquid at –0.2 V. The corresponding cross-section SEM micrograph shown in Fig. 14.6(b) indicates that the deposit is uniform in thickness. EDS results show that the distributions of Al, Zn and Mg across the whole thickness are displayed in Fig. 14.6(c). As shown in this figure, co-deposition of Al and Zn with uniform chemical composition can be obtained by electrodeposition in the above ionic liquid at an appropriate controlled potential. The surface roughness and the uniformity of chemical composition of the electrodeposited film are affected by the deposition potential as reported elsewhere [59].

14.5 Electrochemical and corrosion resistance of aluminum (Al) and aluminum/zinc (Al/Zn)-coated magnesium (Mg) alloys

The corrosion resistance of various Al- and Al/Zn-coated AZ91D Mg alloys has been evaluated by salt spray and electrochemical methods. In the following, the results from salt spray test, polarization curve and electrochemical impedance measurements manifesting the beneficial effects of Al and/or Al–Zn coating on AZ91D Mg alloy are demonstrated.

14.5.1 Salt spray test

The salt spray test was performed in accordance with the specification of the American Society of Testing and Materials B117. For the as-polished AZ91D Mg alloy, the digital micrograph showing the shiny surface appearance is depicted in Fig. 14.7(a). After exposing in the salt spray chamber for 1 h, fast corrosion occurs on the bare Mg alloy surface causing the loss of metallic luster, as can be seen in Fig. 14.7(b). The digital micrograph of the Al as-coated AZ91D Mg alloy is demonstrated in Fig. 14.7(c). After 8 h salt spray test, the Al-coated Mg sample maintains its integrity and silver-gray surface

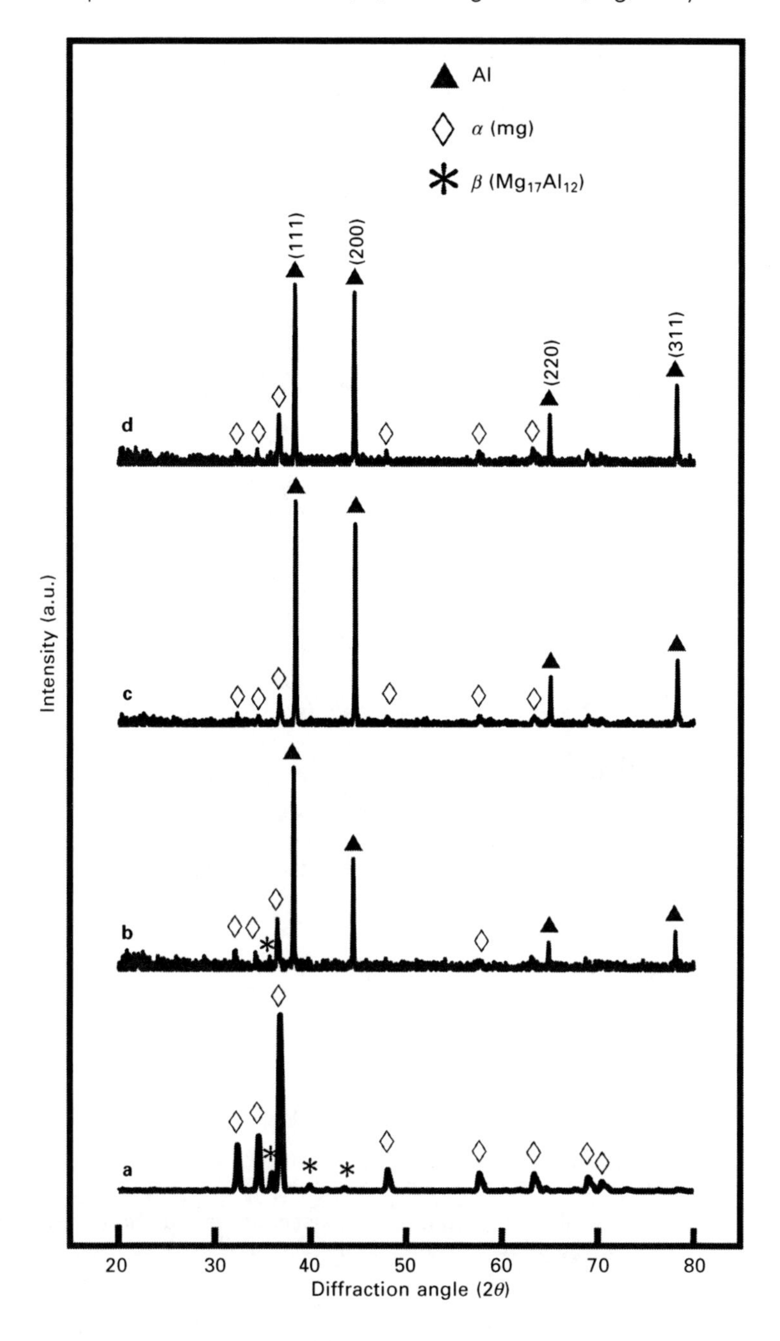

14.5 X-ray diffraction patterns of the various samples. Curve a presents the bare AZ91D Mg alloy. Curves b, c and d present the Al-coated Mg samples deposited in the ionic liquids containing 53, 57 and 60 m/o $AlCl_3$, respectively.

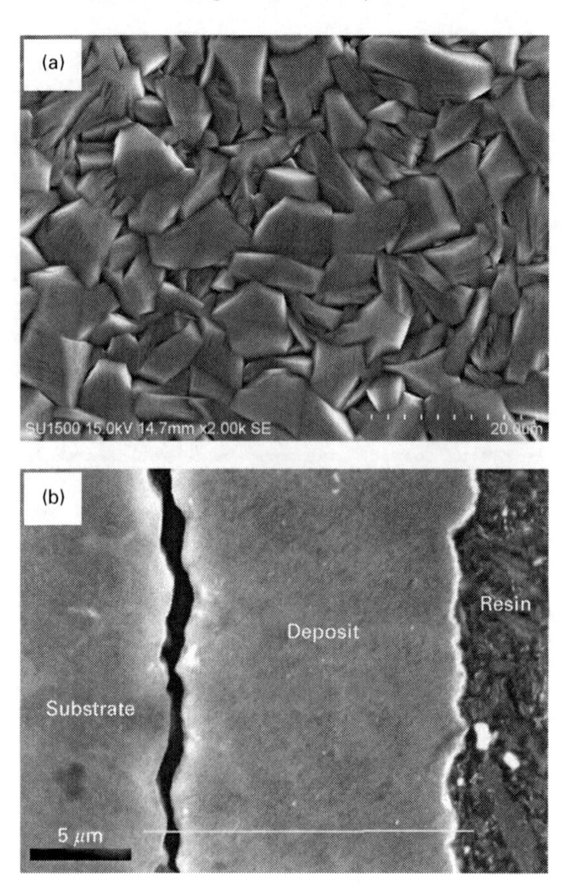

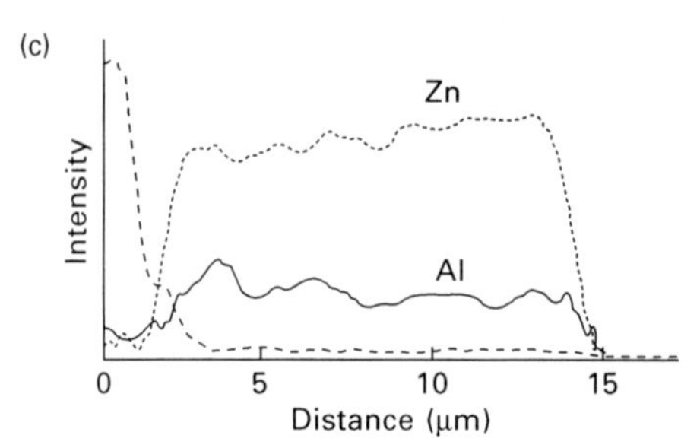

14.6 (a) SEM micrograph, (b) cross-section image and (c) EDS line scans for the deposit formed in 1 wt% $ZnCl_2$ + 60 m/o $AlCl_3$–EMIC ionic liquid at −0.2 V.

appearance as revealed in Fig. 14.7(d). The improved corrosion resistance by Al coating is clearly demonstrated. More detailed results can be found in a previous investigation [57].

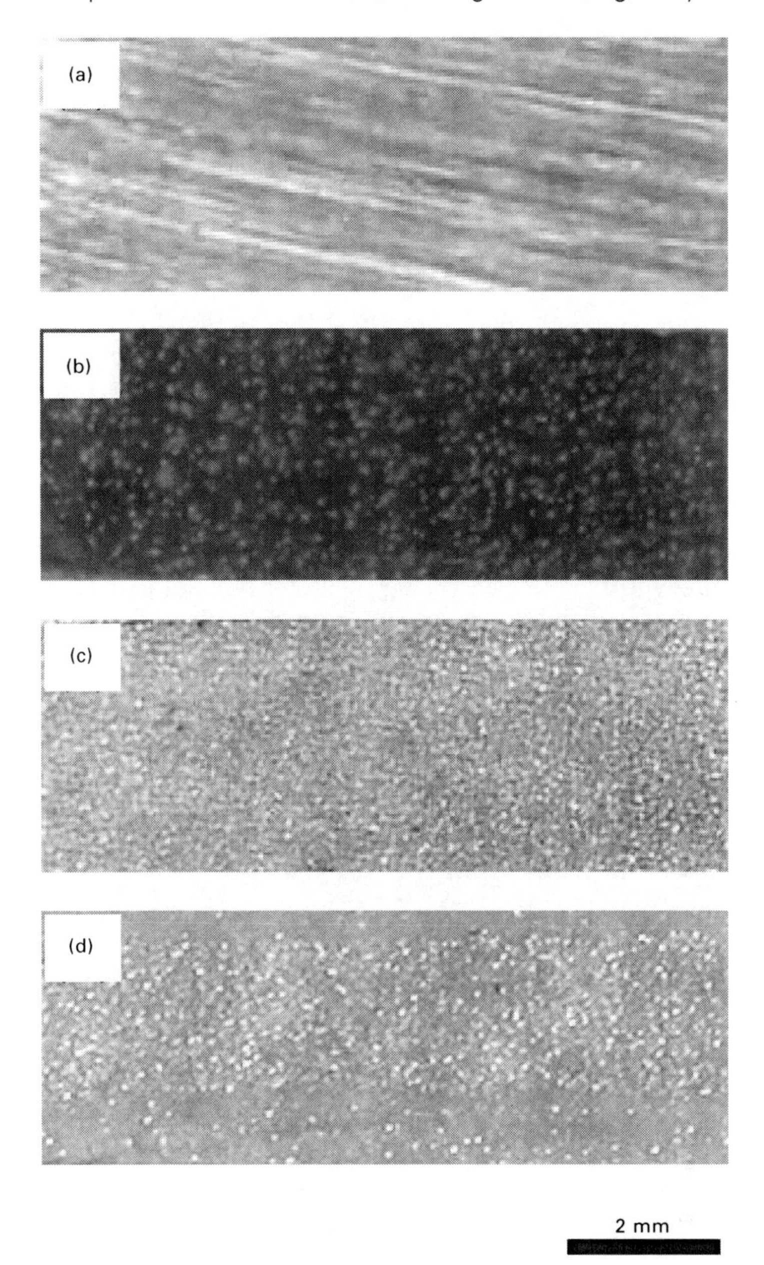

14.7 Digital micrographs of AZ91D Mg alloy (a) as-polished, (b) after salt spray for 1 h; and Al-coated AZ91D Mg alloy (c) as-coated, (d) after salt spray for 8 h.

14.5.2 Polarization behavior

The polarization behavior of the bare and Al-coated AZ91D Mg alloy has been compared in 3.5 wt% NaCl solution, as can be seen in Fig. 14.8. As illustrated in curve **a** of this figure, the anodic curve of the bare Mg alloy exhibits an active dissolution behavior. The passive region can hardly be

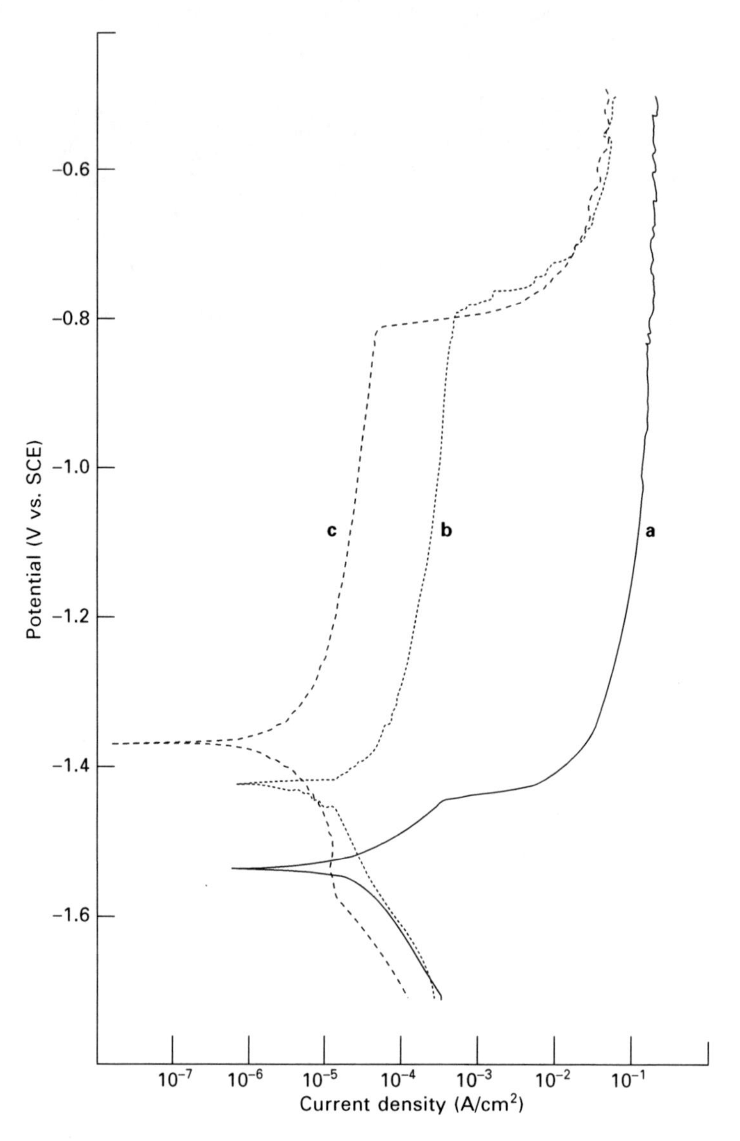

14.8 Potentiodynamic polarization curves of the bare Mg alloy (curve a) and the Al-coated Mg samples deposited in the 53 m/o (curve b) and 60 m/o (curve c) ionic liquids, respectively.

seen. The plateau region at very high current density corresponds to the limiting current density. For the Mg alloy with Al electrodeposited at –0.2 V either from the 53 m/o or the 60 m/o ionic liquid, a wide passive region with sufficient low passive current density is observed as demonstrated in curves **b** and **c** in Fig. 14.8. The passivity obtained for the Al-coated Mg alloy is attributed to the formation of aluminum oxide or hydroxide on the surface of Al film, which can consequently provide adequate corrosion protection of the Mg alloy substrate. As shown in Fig. 14.8, curve **b** has a high passive current density than that of curve **c**. The difference is attributed to the more protective film formed in 60 m/o ionic liquid as compared with that formed in 53 m/o ionic liquid. The passivity of Al-coated Mg alloy can vary if Al film formed in ionic liquid is electrodeposited at different conditions. It has been found that a wider passive region with a lower passive current density can be obtained if the Al is electrodeposited at a lower applied current [57] or a low cathodic overpotential [56]. Under such conditions, the deposition rate is slow, leading to the formation of a more protective film with low defect concentration as manifested in Fig. 14.4.

The potentiodynamic polarization curve of an Al/Zn-coated AZ91D Mg alloy in 3.5 wt% NaCl solution is demonstrated in Fig. 14.9, in comparison with those of the bare and the Al-coated Mg alloys. The Al/Zn coating was formed under the same condition as shown in Fig. 14.6. The results show that Al/Zn film can also be passivated in 3.5 wt% NaCl solution. However, the passive region becomes narrower and the passive current density is lower. The presence of Zn in the coating causes a slightly deterioration of the passivity as compared with that of a pure Al film.

14.5.3 Electrochemical impedance spectroscopy (EIS)

The Electrochemical impedance spectroscopy (EIS) results for the Mg alloy without and with surface Al coated from the 53 m/o and the 60 m/o ionic liquid, respectively, are depicted in Fig. 14.10. For bare Mg alloy, the polarization resistance was about $470\,\Omega\,cm^2$. A substantial increase in the polarization resistance, as evidenced by an enlarged diameter of the semicircle of the Nyquist plot, can be obtained for Mg alloy if it is electroplated with Al. For those with surface Al electrodeposited at –0.2 V from the 53 m/o and the 60 m/o ionic liquid, the polarization resistance in 3.5 wt% NaCl solution are 3000 and $5200\,\Omega\,cm^2$, respectively. The results were consistent with those revealed in the polarization curves demonstrated in Fig. 14.8. The improved polarization resistance of AZ91D Mg alloy with Al coating from ionic liquid is clearly demonstrated. However, the passivity or the polarization resistance of the Al-coated Mg alloy depends on the deposition conditions. The Al film formed in more acidic $AlCl_3$–EMIC and at a lower deposition rate renders a better passivation behavior.

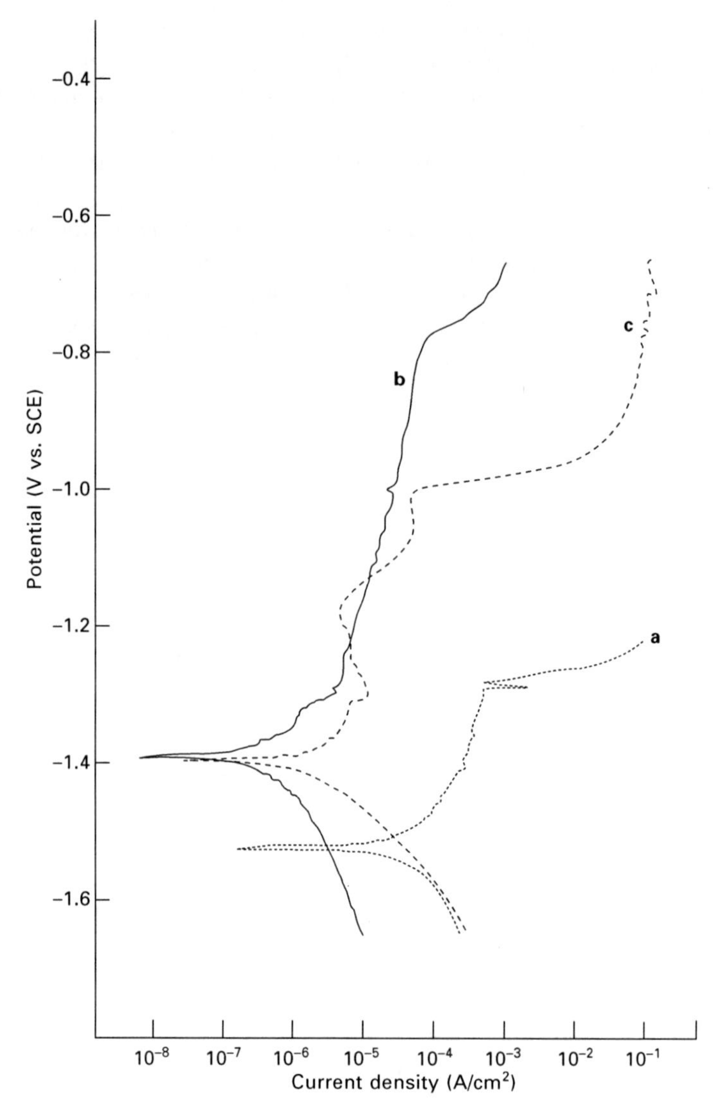

14.9 Comparison of the potentiodynamic polarization curves measured in 3.5 wt% NaCl solution: curve a, AZ91D alloy; curve b, Al deposit formed in $AlCl_3$–EMIC at –0.2 V (vs. Al); curve c, Al-Zn co-deposited at –0.2 V (vs. Al) in $AlCl_3$–EMIC + 1 wt% $ZnCl_2$.

14.6 Summary

The $AlCl_3$–EMIC ionic liquid with proper chemical composition can be used as a successful electrolyte for Al electrodeposition on Mg alloy. Above the critical chemical composition, the deposition efficiency increases with increasing acidity of the ionic liquid. With the presence of Zn(II) ion in

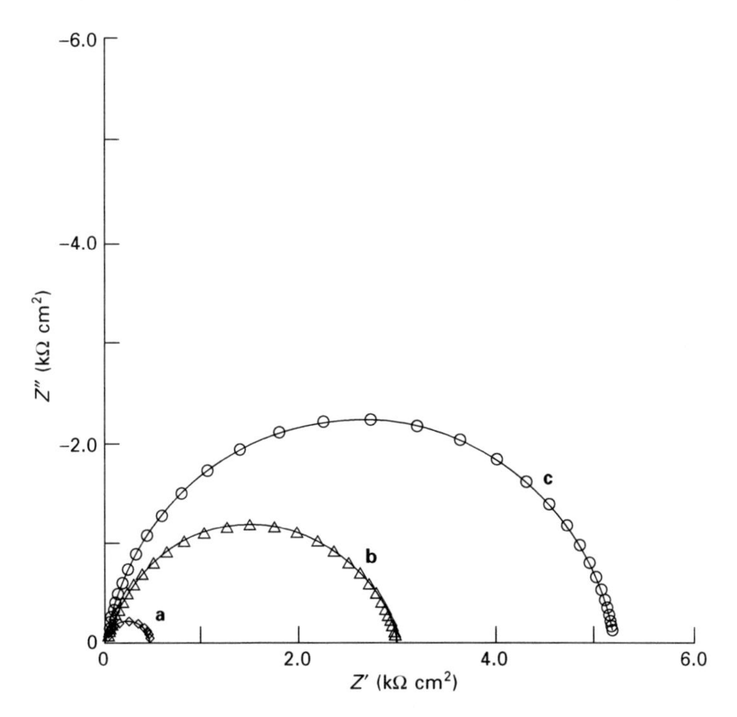

14.10 Nyquist plots of the bare Mg alloy (curve a) and the Al-coated Mg samples deposited in the 53 m/o (curve b) and 60 m/o (curve c) ionic liquids, respectively.

the $AlCl_3$–EMIC ionic liquid, co-deposition of Al and Zn on the Mg alloy can be achieved. The formation of a barrier Al and/or Al/Zn surface film can provide adequate corrosion resistance of the Mg alloy substrate. In 3.5 wt% NaCl solution, Al coating can cause an increase in the polarization resistance of AZ91D Mg alloy from 470 to 5200 $\Omega\,cm^2$, about one order of magnitude higher. The potentiodynamic polarization curves show that the passivity of the Al-coated Mg alloy depends on the deposition condition. A wider passive region with a lower passive current density can be obtained if the Al is electrodeposited at a lower applied current or a low cathodic overpotential. The passivity of the Al-coated Mg alloy is slightly degraded with the co-existence with Zn.

In conclusion, the electrodeposition of Al on AZ91D from the EMI–$AlCl_3$ ionic liquid is feasible. Since the deposition is performed at ambient temperature, the thermal stress in the substrate can be avoided. Without the presence of water, the oxidation of Mg and the formation of hydrogen pinhole during electrodeposition can be prevented. With these advantages, the formation of a more compact and adherent Al and/or Al-Zn coating on Mg alloy surface can be achieved. Recently, electroless plating of Al on glass substrate in EMI–$AlCl_3$ ionic liquid has been reported [64,65]. Furthermore, it has been demonstrated that corrosion protective Mg–Al intermetallic surface layer can be formed on AZ91D by immersing the AZ91D specimens in a NaCl–$AlCl_3$ molten salt at 300 °C for 7 h [66]. Taken together, the electrodeposition combining with electroless plating and conversion coating may provide directions for developing corrosion protective Al and alloy coatings on Mg alloys using ionic liquids.

14.7 Acknowledgement

The authors would like to thank Dr Jeng-Kuei Chang and Ms Szu-Jung Pan for assisting the preparation of this manuscript.

14.8 References

[1] G.L. Makar and J. Kruger, 'Corrosion studies of rapidly solidified magnesium alloys', *Journal of the Electrochemical Society*, **137**, (1990), 414–421.
[2] G. Song, A. Atrens, D. StJohn, X. Wu, and J. Nairn, 'The anodic dissolution of magnesium in chloride and sulphate solutions', *Corrosion Science*, **39**, (1997), 1981–2004.
[3] R. Ambat, N.N. Aung and W. Zhou, 'Evaluation of microstructural effects on corrosion behaviour of AZ91D magnesium alloy', *Corrosion Science*, **42**, (2000), 1433–1455.
[4] G. Song, A. Atrens and M. Dargusch, 'Influence of microstructure on the corrosion of diecast AZ91D', *Corrosion Science*, **41**, (1999), 249–273.
[5] G. Song, 'Recent progress in corrosion and protection of magnesium alloys', *Advanced Engineering Materials*, **7**, (2005), 563–586.
[6] J.E. Gray and B. Luan, 'Protective coatings on magnesium and its alloys – a critical review', *Journal of Alloys and Compounds*, **336**, (2002), 88–113.
[7] Y. Zhang, C. Yan, F. Wang, H. Lou and C. Cao, 'Study on the environmentally friendly anodizing of AZ91D magnesium alloy', *Surface and Coatings Technology*, **161**, (2002), 36–43.
[8] H.Y. Hsiao and W.T. Tsai, 'Characterization of anodic films formed on AZ91D magnesium alloy', S*urface Coatings and Technology*, **190/2–3**, (2005), 299–308.
[9] H.Y. Hsiao, H.C. Tseng and W.T. Tsai, 'Anodization of AZ91D magnesium alloy in silicate-containing electrolytes', *Surface and Coatings Technology*, **199/2–3** (2005), 127–134.
[10] H.Y. Hsiao and W.T. Tsai, 'Effect of heat treatment on anodization and electrochemical behavior of AZ91D magnesium alloy', *Journal of Materials Research*, **20**, (2005), 2763–2771.
[11] M.A. Gonzalez-Nunez, C.A. Nunez-Lopez, P. Skeldon, G.E. Thompson, H. Karimzadeh, P. Lyon and T.E. Wilks, 'A non-chromate conversion coating for magnesium alloys and magnesium-based metal matrix composites', *Corrosion Science*, **37**, (1995), 1763–1772.
[12] H. Huo, Y. Li and F. Wang, 'Corrosion of AZ91D magnesium alloy with a chemical

conversion coating and electroless nickel layer', *Corrosion Science*, **46**, (2004), 1467–1477.

[13] L.Y. Niu, Z.H. Jiang, G.Y. Li, C.D. Gu and J.S. Lian, 'A study and application of zinc phosphate coating on AZ91D magnesium alloy', *Surface and Coatings Technology*, **200**, (2006), 3021–3026.

[14] X. Cui, Y. Li, Q. Li, G. Jin, M. Ding and F. Wang, 'Influence of phytic acid concentration on performance of phytic acid conversion coatings on the AZ91D magnesium alloy', *Materials Chemistry and Physics*, **111**, (2008), 503–507.

[15] X. Cui, Q. Li, Y. Li, F. Wang, G. Jin and M. Ding, 'Microstructure and corrosion resistance of phytic acid conversion coatings for magnesium alloy', *Applied Surface Science*, **255**, (2008), 2098–2103.

[16] Z. Yong, J. Zhu, C. Qiu and Y. Liu, 'Molybdate/phosphate composite conversion coating on magnesium alloy surface for corrosion protection', *Applied Surface Science*, **255**, (2008), 1672–1680.

[17] X. Chen, G. Li, J. Lian and Q. Jiang, 'An organic chromium-free conversion coating on AZ91D magnesium alloy', *Applied Surface Science*, **255**, (2008), 2322–2328.

[18] C. Gu, J. Lian, G. Li, L. Niu and Z. Jiang, 'Electroless Ni–P plating on AZ91D magnesium alloy from a sulfate solution', *Journal of Alloys and Compounds*, **391**, (2005), 104–109.

[19] J. Chen, G. Yu, B. Hu, Z. Liu, L. Ye and Z. Wang, 'A zinc transition layer in electroless nickel plating', *Surface and Coatings Technology*, **201**, (2006), 686–690.

[20] C. Gu, J. Lian, J. He, Z. Jiang and Q. Jiang, 'High corrosion-resistance nanocrystalline Ni coating on AZ91D magnesium alloy', *Surface and Coatings Technology*, **200**, (2006), 5413–5418.

[21] H. Zhao, Z. Huang and J. Cui, 'A new method for electroless Ni–P plating on AZ31 magnesium alloy', *Surface and Coatings Technology*, **202**, (2007), 133–139.

[22] N. El Mahallawy, A. Bakkar, M. Shoeib, H. Palkowski and V. Neubert, 'Electroless Ni–P coating of different magnesium alloys', *Surface and Coatings Technology*, **202**, (2008), 5151–5157.

[23] W.X. Zhang, Z.H. Jiang, G.Y. Li, Q. Jiang and J.S. Lian, 'Electroless Ni-P/Ni-B duplex coatings for improving the hardness and the corrosion resistance of AZ91D magnesium alloy', *Applied Surface Science*, **254**, (2008), 4949–4955.

[24] H.H. Elsentriecy and K. Azumi, 'Electroless Ni–P deposition on AZ91D magnesium alloy prepared by molybdate chemical conversion coatings', *Journal of the Electrochemical Society*, **156**, (2009), D70–D77.

[25] Y.F. Jian, L.F. Liu, C.Q. Zhai, Y.P. Zhu and W.J. Ding, 'Corrosion behavior of pulse-plated Zn-Ni alloy coatings on AZ91D magnesium alloy in alkaline solutions', *Thin Solid Films*, **484**, (2005), 232–237.

[26] Y.F. Jian, C.Q. Zhai, L.F. Liu, Y.P. Zhu and W.J. Ding, 'Zn–Ni alloy coatings pulse-plated on magnesium alloy', *Surface and Coatings Technology*, **191**, (2005), 393–399.

[27] L. Zhu, W. Li and D. Shan, 'Effects of low temperature thermal treatment on zinc and/or tin plated coatings of AZ91D magnesium alloy', *Surface and Coatings Technology*, **201**, (2006), 2768–2775.

[28] C.A. Huang, T.H. Wang, T. Weirich and V. Neubert, 'Electrodeposition of a protective copper/nickel deposit on the magnesium alloy (AZ31)', *Corrosion Science*, **50**, (2008), 1385–1390.

[29] W. Simka, D. Puszczyk and G. Nawrat, 'Electrodeposition of metals from non-aqueous solutions', *Electrochimica Acta*, **54**, (2009), 5307–5319.

[30] Y. Zhao and T.J. VanderNoot, 'Electrodeposition of aluminium from nonaqueous organic electrolytic systems and room temperature molten salts', *Electrochimica Acta*, **42**, (1997), 3–13.

[31] E. Peled and E. Gileadi, 'The electrodeposition of aluminum from aromatic hydrocarbon', *Journal of the Electrochemical Society*, **123**, (1976), 15–19.

[32] W. Kautek and S. Birkle, 'Aluminum-electrocrystallization from metal–organic electrolytes', *Electrochimica Acta*, **34**, (1989), 1213–1218.

[33] S.P. Shavkunov and T.L. Strugova, 'Electrode processes upon electrodeposition of aluminum in aromatic solvents' *Elektrokhimiya*, **39**, (2003), 714–721.

[34] L. Legrand, A. Tranchant and R. Messina, 'Aluminium behaviour and stability in $AlCl_3/DMSO_2$ electrolyte', *Electrochimica Acta*, **41**, (1996), 2715–2720.

[35] T. Jiang, M.J. Chollier Brym, G. Dubé, A. Lasia and G.M. Brisard, 'Studies on the $AlCl_3$/dimethylsulfone ($DMSO_2$) electrolytes for the aluminum deposition processes', *Surface and Coatings Technology*, **201**, (2007), 6309–6317.

[36] D.E. Couch and A. Brenner, 'A hydride bath for the electrodeposition of aluminum', *Journal of the Electrochemical Society*, **99**, (1952), 234–244.

[37] K. Ziegler and H. Lehmkuhl, 'Elektrolytische abscheidung von aluminium', *Angewandte Chemie*, **67**, (1955), 424–425.

[38] Q.X. Liu, S. Zein El Abedin and F. Endres, 'Electroplating of mild steel by aluminium in a first generation ionic liquid: a green alternative to commercial Al-plating in organic solvents', *Surface and Coatings Technology*, **201**, (2006), 1352–1356.

[39] T. Jiang, M.J. Chollier Brym, G. Dubé, A. Lasia and G.M. Brisard, 'Electrodeposition of aluminium from ionic liquids: Part I – electrodeposition and surface morphology of aluminium from aluminium chloride ($AlCl_3$) – 1-ethyl-3-methylimidazolium chloride ([EMIm]Cl) ionic liquids', *Surface and Coatings Technology*, **201**, (2005), 1–9.

[40] C.L. Aravinda, B. Burger and W. Freyland, 'Nanoscale electrodeposition of Al on *n*-Si(111): H from an ionic liquid', *Chemical Physics Letters*, **434**, (2007), 271–275.

[41] S. Caporali, A. Fossati, A. Lavacchi, I. Perissi, A. Tolstogouzov and U. Bardi, 'Aluminium electroplated from ionic liquids as protective coating against steel corrosion', *Corrosion Science*, **50**, (2008), 534–539.

[42] G. Yue, X. Lu, Y. Zhu, X. Zhang and S. Zhang, 'Surface morphology, crystal structure and orientation of aluminium coatings electrodeposited on mild steel in ionic liquid', *Chemical Engineering Journal*, **147**, (2009), 79–86.

[43] T. Jiang, M.J. Chollier Brym, G. Dubé, A. Lasia and G.M. Brisard, 'Electrodeposition of aluminium from ionic liquids: Part II – studies on the electrodeposition of aluminum from aluminum chloride ($AlCl_3$) – trimethylphenylammonium chloride (TMPAC) ionic liquids', *Surface and Coatings Technology*, **201**, (2006), 10–18.

[44] Y. Zhao and T.J. VanderNoot, 'Electrodeposition of aluminium from room temperature $AlCl_3$–TMPAC molten salts', *Electrochimica Acta*, **42**, (1997), 1639–1643.

[45] A.P. Abbott, C.A. Eardley, N.R.S. Farley, G.A. Griffith and A. Pratt, 'Electrodeposition of aluminium and aluminium/platinum alloys from $AlCl_3$/benzyltrimethylammonium chloride room temperature ionic liquids', *Journal of Applied Electrochemistry*, **31**, (2001), 1345–1350.

[46] R.A. Osteryoung and J. Robinson, 'The electrochemical behavior of aluminum in the

low temperature molten salt system n-butylpyridinium chloride: aluminum chloride and mixtures of this molten salt with benzene', *Journal of the Electrochemical Society*, **127**, (1980) 122–128.

[47] R.A. Osteryoung and B.J. Welch, 'Electrochemical studies in low temperature molten salt systems containing aluminum chloride', *Journal of Electroanalytical Chemistry*, **118**, (1981), 455–466.

[48] P.K. Lai and M. Skyllas-Kazacos, 'Aluminium deposition and dissolution in aluminium chloride – n-butylpyridinium chloride melts', *Electrochimica Acta*, **32**, (1987), 1443–1449.

[49] C.L. Hussey, Q. Liao, W.R. Pitner, G. Stewart and G.R. Stafford, 'Electrodeposition of aluminum from the aluminum chloride – 1-methyl-3-ethylimidazolium chloride room temperature molten salt + benzene', *Journal of the Electrochemical Society*, **144**, (1997), 936–943.

[50] G.R. Stafford and C.L. Hussey, 'Electrodeposition of transition metal-aluminum alloys from chloroaluminatr molten salts', in *Advances in Electrochemical Science and Engineering*, Vol. 7, R.C. Alkire and D.M. Kolb (Ed.), Wiley-VCH Verlag GmbH, (2002).

[51] *Electrodeposition from Ionic Liquids*, F. Endres, A.P. Abbott and D.R. MacFarlane (Ed.), Wiley-VCH Verlag GmbH, (2008).

[52] S. Zein El Abedin, E.M. Moustafa, R. Hempelmann, H. Natter and F. Endres, 'Electrodeposition of nano- and microcrystalline aluminium in three different air and water stable ionic liquids', *ChemPhysChem*, **7**, (2006), 1535–1543.

[53] Q.X. Liu, S. Zein El Abedin and F. Endres, 'Electrodeposition of nanocrystalline aluminum: Breakdown of imidazolium cations modifies the crystal size', *Journal of the Electrochemical Society*, **155**, (2008), D357–D362.

[54] P. Eiden, Q. Liu, S. Zein El Abedin, F. Endres and I. Krossing, 'An experimental and theoretical study of the aluminium species present in mixtures of $AlCl_3$ with the ionic liquids [BMP]Tf_2N and [EMIm]Tf_2N', *Chemistry – A European Journal*, **15**, (2009), 3426–3434.

[55] M.J. Deng, P.Y. Chen, T.I. Leong, I.W. Sun, J.K. Chang and W.T. Tsai, 'Dicyanamide anion based ionic liquids for electrodeposition of metals', *Electrochemistry Communications*, **10**, (2008), 213–216.

[56] J.K. Chang, S.Y. Chen, W.T. Tsai, M.J. Deng and I.W. Sun, 'Electrodeposition of aluminum on magnesium alloy in aluminum chloride ($AlCl_3$) – 1–ethyl–3-methylimidazolium chloride (EMIC) ionic liquid and its corrosion behavior', *Electrochemistry Communications*, **9**, (2007), 1602–1606.

[57] J.K. Chang, S.Y. Chen, W.T. Tsai, M.J. Deng and I.W. Sun, 'Improved corrosion resistance of magnesium alloy with a surface aluminum coating electrodeposited in ionic liquid', *Journal of the Electrochemical Society*, **155**, (2008), C112–C116.

[58] J.K. Chang, I.W. Sun, S.J. Pan, M.H. Chuang, M.J. Deng and W.T. Tsai, 'Electrodeposition of Al coating on Mg alloy from Al chloride – 1-ethyl-3-methylimidazolium chloride ionic liquids with different Lewis acidity', *Transactions of the Institute of Metal Finishing*, **86**, (2008), 227–233.

[59] S.J. Pan, W.T. Tsai, J.K. Chang and I.W. Sun, 'Co-deposition of Al–Zn on AZ91D magnesium alloy in $AlCl_3$ – 1-ethyl-3-methylimidazolium chloride ionic liquid', *Electrochimica Acta*, **55**, (2010), 2158–2162.

[60] P.K. Lai and M. Skyllas-Kazacos, 'Electrodeposition of aluminium in aluminium chloride – 1-methyl-3-ethylimidazolium chloride', *Journal of Electroanalytical Chemistry*, **248**, (1988), 431–440.

[61] R.T. Carlin and R.A. Osteryoung, 'Aluminum anodization in a basic ambient temperature molten salt', *Journal of the Electrochemical Society*, **136**, (1989), 1409–1415.

[62] T.J. Melton, J. Joyce, J.T. Maloy, J.A. Boon and J.S. Wikes, 'Electrochemical studies of sodium chloride as a Lewis buffer for room temperature chloroaluminate molten salts', *Journal of the Electrochemical Society*, **137**, (1990), 3865–3869.

[63] R.L. Perry, K.M. Jones, W.D. Scott, Q. Liao and C.L. Hussey, 'Densities, viscosities, and conductivities of mixtures of selected organic cosolvents with the Lewis basic aluminum chloride + 1-methyl-3-ethylimidazolium chloride molten salt', *Journal of Chemical and Engineering Data*, **40**, (1995), 615–619.

[64] I. Shitanda, A. Sato, M. Itagaki, K. Watanbe and N. Koura, 'Electroless plating of aluminum using diisobutyl aluminum hydride as liquid reducing agent in room-temperature ionic liquid', *Electrochimica Acta*, **54**, (2009), 5889–5893.

[65] N. Koura, H. Nagase, A. Sato, S. Kumakura, K. Takeuchi, K. Ui, T. Tsuda and C. K. Loong, 'Electroless plating of aluminum from a room-temperature ionic liquid electrolyte', *Journal of the Electrochemical Society*, **155**, (2008), D155–D157.

[66] M. He, L. Liu, Y. Wu, Z. Tang and W. Hu, 'Improvement of the properties of AZ91D magnesium alloy by treatment with a molten $AlCl_3$–NaCl salt to form Mg–Al intermetallic surface layer', *Journal of Coating Technology Research*, **6**, (2009), 407–411.

15
Corrosion protection of magnesium (Mg) alloys using conversion and electrophoretic coatings

B. L. LUAN, D. YANG, X. Y. LIU, National Research Council of Canada, Canada and G.-L. SONG, General motors Corporation, USA

Abstract: Magnesium and its alloys have excellent physical and mechanical properties for a number of applications. In particular magnesium's high strength/weight ratio makes it an ideal metal for automotive and aerospace applications, where weight reduction is of significant concern. Unfortunately, magnesium and its alloys are highly susceptible to corrosion, which has limited their use in the automotive and aerospace industries, where exposure to harsh service conditions is unavoidable. The simplest way to mitigate corrosion is to coat the magnesium-based substrate to prevent its direct contact with the environment. This chapter discusses the state-of-the-art of two important coating processes for magnesium protection that are usually applied consecutively for vehicular bodies in the automotive industry: conversion coating and electrophoretic coating (E-coating).

Key words: conversion coating, E-coating, surface treatment, Mg alloys, corrosion.

15.1 Introduction

One of the main challenges in the use of magnesium (Mg) alloys, particularly for outdoor applications, is their poor corrosion resistance. Magnesium and its alloys are extremely susceptible to galvanic corrosion, which can cause severe localized damage in the metal resulting in decreased mechanical stability and an unattractive appearance. The mechanism of corrosion that occurs on the surface of magnesium and its alloys is shown in Fig. 15.1.

The electrochemical reactions that occur on the surface of magnesium and its alloys are described below.

Anodic reaction causes the dissolution of Mg metal:

$$Mg \rightarrow Mg^{2+} + 2e^- \qquad 15.1$$

and/or

$$Mg(s) + 2(OH)^- \rightarrow Mg(OH)_2(s) + 2e^- \qquad 15.2$$

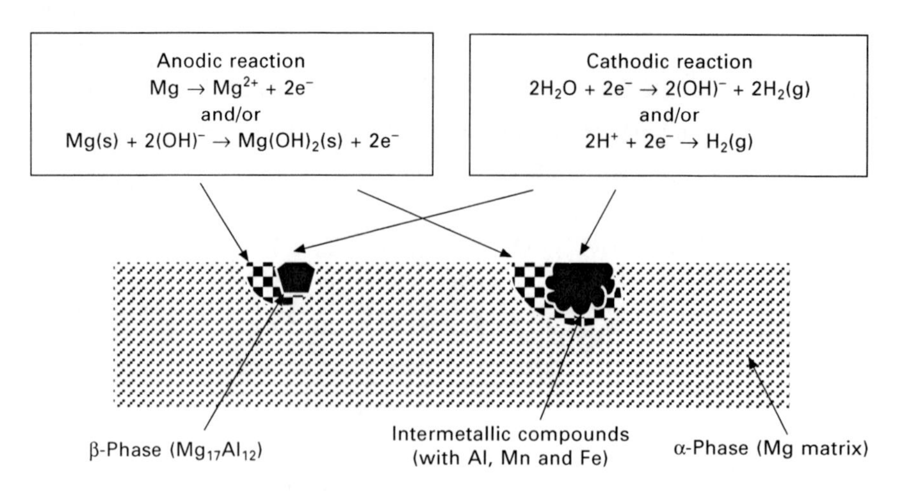

15.1 Corrosion processes occurring on an Mg alloy surface.

Cathodic reaction generates H_2 gas:

$$2H^{+} + 2e^{-} \rightarrow H_2(g) \qquad 15.3$$

and/or

$$2H_2O + 2e^{-} \rightarrow H_2(g) + 2(OH)^{-} \qquad 15.4$$

The overall corrosion reaction can be described as:

$$Mg(s) + 2H_2O \rightarrow Mg(OH)_2(s) + H_2(g) \qquad 15.5$$

Mg and its alloys spontaneously form a hydroxide layer in aqueous solutions or humid air. This layer is stable under alkaline conditions but unstable in acidic or neutral solutions, particularly in the presence of a very small concentration of aggressive ions, under which conditions there is dissolution of the $Mg(OH)_2$ film and corrosion develops at a high rate [1].

Surface coatings are used to protect the surface of magnesium and its alloys to prevent their direct contact with air and moisture so that the electrochemical corrosion reactions described above will not occur or occur at a very low rate. Most of the time, even blocking the cathodic reaction sites can significantly reduce the corrosion rate. Chemical and electrochemical methods such as chrome pickling and chrome-free treatments, anodic treatments and cathodic treatments are widely used to grow surface protective coatings for Mg alloys to protect their surfaces. Other coating processes such as sol–gel [2–5], silane treatments [6,7] and plasma vapor deposition [8] were also used to treat the surface of magnesium and its alloys to improve their corrosion resistance. In this chapter, special emphasis is made to review two important coating processes: conversion coating and electrophoretic coating.

15.2 Conversion coating for magnesium (Mg) and its alloys

Conversion coating solutions chemically react with the metal surface to create a physical surface that allows for better corrosion resistance and paint adhesion. Hexavalent chromium compounds used to be used for producing chemical conversion coatings to protect Mg alloy; however, they have been shown to be highly toxic carcinogens. New environmental regulations have resulted in considerable research activities in developing environmentally acceptable alternative chemical conversion coatings for magnesium and its alloys. Among all the chemical conversion processes patented for magnesium and its alloys, phosphate and phosphate–permanganate-related conversion processes are the most numerous, following by conversion coatings based on compounds of V, Zr, Mo, W, Ti and Co. Rare earth and organic polymer-based conversion coatings have also received significant attention. Based on the number of patents by country of origin, Japan, the USA and China are the top three most active countries conducting chemical conversion research and development (R&D) for magnesium and its alloys. This chapter intends to summarize these non-chromium chemical conversion processes that have been developed or currently are under development worldwide.

15.2.1 State-of-the-art of chemical conversion processes

A typical conversion process includes many steps such as degreasing, water rinsing, alkaline cleaning, acid picking, conversion treatment, post-treatment in organic chelating agents and heat drying as shown in Fig. 15.2. In this section, those steps will be described, reviewed and discussed.

Sample pre-treatment

As-cast Mg alloy parts first go through machining, deburring and grinding treatments, and are subsequently degreased. An organic solvent such as trichloroethylene, ethanol, methanol, isopropyl alcohol, and acetone or an alkali solution such as alkali hydroxide, carbonate, phosphate and silicate was typically used for degreasing Mg alloy parts to release oil, soap, abrasive and sand. Normally the degreasing treatment using an alkali solution is more preferred for Mg alloys.

Acid pickling and alkaline cleaning are the two important pre-treatment steps following degreasing. An acid etching process is to remove the segregation layer which consists of aluminum, zinc and manganese on the surface of an Mg alloy as well as residual grease agents. Nitric acid, phosphoric acid, sulfuric acid, hydrofluoric acid and hydrochloric acid are the common acids used. Alkali solution cleaning removes smut generated by the acid

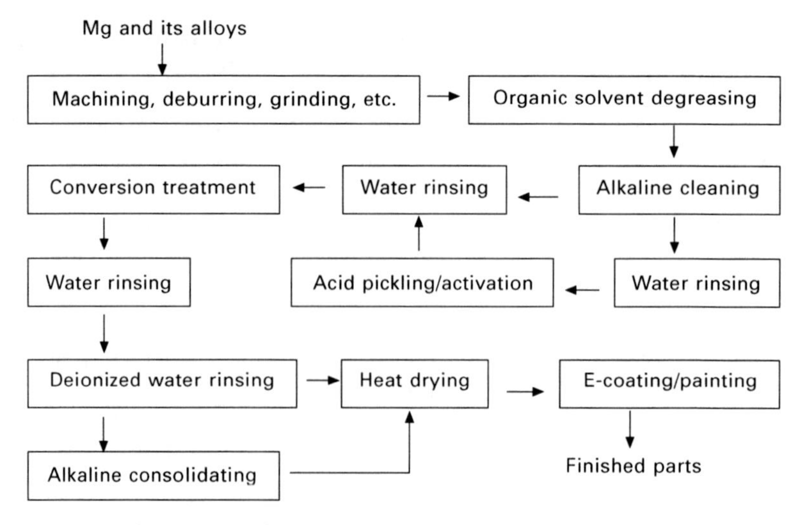

15.2 A typical conversion process.

etching process. Alkali and alkaline earth hydroxides, silicate, carbonate and phosphate are commonly used alkali solutions; however, hydroxides are preferred. Sometimes, chelating agents are added into the alkali solutions to improve the efficiency of removing surface contaminations of Mg alloys. The chelating agents are chosen from aminocarboxylic acid type, dithiocarbamic acid type, organic-phosphonic acid type and hydroxycarboxylic acid type chelating agents. Surface treatment in KOH was found to improve corrosion resistance of ceria conversion coatings on AZ91D [9]. Surface treatment prior to ceria treatments plays an important role in inhibiting the active surface sites, rejecting the chloride ions from the surface, and forming uniformly distributed oxide film. Acid pickling can also significantly improve the Cr-free conversion coatings. Acid pickling with HCl was found to improve the homogeneity, adherence and thickness of the cerium conversion coatings for Mg, AZ91 and AM50 as reported by Brunelli *et al.* [10].

Conversion process

Chemical conversion solutions could consist of either only a few simple main ingredients or many chemical compounds with complete recipe. Besides the main chemical compound such as H_2CrO_4 for the chromate coating, a typical conversion solution contains oxidants, promoters, corrosion inhibitors, wetting agents and pH buffer regulators. The important roles of those additives are described in the following:

- Oxidants (e.g. nitrate and perchlorate) help to speed up the cathodic reaction (equations 15.3 and 15.4). They consume a lot of H^+, raise the

OH^- concentration and speed up the dissolution of magnesium and the formation of conversion film.

- Promoters (e.g. zirconium and vanadic salts) precipitate firstly on the surface of Mg alloys to form seeding nuclei for the deposition of conversion layer.
- Corrosion inhibitors (e.g. pyridine, thiourea, and phytic acid) adsorbed on the surface of Mg alloys to slow down the dissolution of Mg^{2+}. They also form complexes with Mg^{2+} to stabilize the solution and control the conversion layer formation rate.
- Wetting agents (e.g. sodium dodecyl benzene sulfonate) reduce the surface tension of Mg alloys and make the precipitation of conversion layer easier. They also improve the adhesion of the conversion layer.
- pH buffer regulators control the pH of solutions which affects the speed of conversion layer formation as well as the quality of conversion layer. High pH speeds up the deposition of conversion layer but causes the poor adhesion and softness of the conversion layer.

Based on their main ingredient, chemical conversion processes currently used for magnesium and its alloys can be classified as the following:

- chromate conversion coatings;
- stannate conversion coatings;
- rare earth (e.g. cerium, lanthanum and praseodymium) oxide conversion coatings;
- phosphate, phosphate–permanganate and fluoride-related conversion coatings;
- conversion coatings based on compounds of V, Zr, Mo, W, Ti, Co.

A detailed literature review for each of the above conversion coatings is given in the following paragraphs. Other less studied conversion coatings such as organic-based [11] coatings will not be included in the review since this chapter focuses on inorganic conversion coating.

Chromate conversion coatings

Chromate conversion solution consists of chromic acid H_2CrO_4 or $H_2Cr_2O_7$, chromate salts and certain activator ions such as sulfates, chlorides, fluorides, phosphates and complex cyanides with pH around 1~2. As the solutions for chromate conversion treatment are acidic which cause the dissolution of Mg into the solutions as Mg^{2+}, there is a local rise in pH in the immediate vicinity of the metal–solution interface. Mg^{2+} ions combine with chromate ions to form a compound that is insoluble at the locally higher pH region. This compound precipitates on the metal surface as an adherent coating. Chromate conversion treatment is a very fast process (30–60 s) and can operate at room temperature [12]. Cr-conversion coatings have self-healing and inhibiting effect, which Cr-free coatings do not have with the exceptions of V and Mo oxide coatings [12]. Commercially available chromate conversion solutions such as DOW (developed by the DOW Chemical Co.) and JIS (H8651) have been widely applied to improve the corrosion resistance of Mg alloys [13–15]. Chromate conversion treatment is the most mature and reliable process that provides excellent corrosion protection for many metals and alloys including Mg and its alloys. New environmental regulations, however, have limited its applications, even though there are still efforts in developing low chromic content conversion solutions to treat Mg alloys to minimize the environmental impact.

Rare earth (cerium) conversion coating

Cerium conversion solution typically consists of cerium salt such as $CeCl_3$, $Ce(NO_3)_3$, $CePO_4$, $Ce(SO_4)_2$, $Ce_2(SO_4)_3$ or $Ce(NH_4)_2(NO_3)_6$, and sometimes with additives of other salts such as $La(NO_3)_3$ and $Pr(NO_3)_3$ [16], $ZrO(NO_3)_3$ and $Nb_xO_yF_z$ [17] and $Al(NO_3)_3$ and $Ca(OH)_2$ [18]. Wetting agents, buffer pH adjustors, corrosion inhibitors, activators and oxidants are always added into the conversion solutions too. When magnesium and its alloys are immersed in the cerium conversion solution and dissolved, the local pH at the metal–solution interface increases resulting from the reduction of proton or dissolved O_2 in the solution by the reaction:

$$O_2\ (aq.) + 2H_2O + 4e^- \rightarrow 4OH^-\ (aq.)$$

which enables precipitation of $Ce(OH)_3$ and $Ce(OH)_4$ coatings. Finally, $Ce(OH)_3$ and $Ce(OH)_4$ are precipitated as insoluble Ce_2O_3 and CeO_2. On AZ63 alloy [19], it was found that cerium-based conversion process improves the corrosion resistance in chloride media, however, the structure of the cerium-based conversion layer is not homogeneous and contains large agglomerates with dry-mud morphology. Magnesium–cerium mixed oxides were detected above the surface of cathodic intermetallic particles which contain Al, Mn and Zn since these particles favor hydrogen evolution or oxygen reduction to OH^-, inducing a local pH rise, which in turn enables the precipitation of hydrated cerium oxide over these intermetallics. The thickness of the cerium conversion coating rapidly grows during the first 30 s to about ~0.16 μm, then the thickness remains relatively constant. Too many repeated immersions reduce the corrosion resistance of the conversion layer probably because of a partial dissolution of the conversion layer. Mg alloy, WE43, when treated in cerium solution containing La and Pr, has led to a significant increase in the corrosion resistance when tested in a pH 8.5 buffer solution as reported by Rudd *et al.* [16]. Mixed Ce–Zr–Nb conversion coatings which consist of CeO_2, Ce_2O_3, ZrO_2, Nb_2O_5, MgO and MgF_2 exhibited greater corrosion potential than pure Ce conversion coating as reported by Ardelean *et al.* [17], while addition of $Al(NO_3)_3$ and $Ca(OH)_2$ to the cerium-based solution, as reported by Salman *et al.* [18], actually enhanced the surface properties (smooth and no-cracking) and corrosion resistance. The addition of trace amount of other metallic elements was also found to significantly increase the corrosion resistance of cerium conversion coatings.

Anions of cerium salts were also found to be an important factor that affects the formation and corrosion protection of cerium conversion layer [19]. Conversion treatment of AZ31 Mg alloy in solutions of different salts of trivalent cerium – cerium nitrate, cerium chloride, cerium sulphate and cerium phosphate – leads to formation of different layer thickness in the same immersion time: thicker film was formed in the $CeCl_3$ bath while thinner film in the $CePO_4$ bath. At short treatment time, conversion layer formed in $CeCl_3$ solutions is richer in Ce(IV) when compared to those formed in $CePO_4$ and $Ce_2(SO_4)_3$ solutions, but at longer treatment times (1 hour) the composition becomes nearly identical, with Ce(IV) oxides predominately existing over hydroxides in all the conversion films. The conversion layer formed in $CePO_4$ and $Ce_2(SO_4)_3$ revealed a higher content of magnesium in the surface layer. Addition of other anions, such as CO_3^{2-}, can also improve the corrosion resistance of cerium coatings. For example, addition of Na_2CO_3 into cerium conversion solution (i.e. 0.05 M $Ce(NO_3)_3$ + 25 mL • L^{-1} H_2O_2) formed Mg–Ce hydrotalcite film (i.e. $Mg_6Ce_2(CO_3)(OH)_{16}$•$4H_2O$) on Mg–Gd–Y–Zr magnesium alloy surface that showed better corrosion resistance than pure cerium conversion film. The better corrosion resistance in chloride solution is attributed to the high affinity of carbonate preventing attack by Cl^- [20].

Sometimes addition of a strong oxidant such as hydrogen peroxide H_2O_2 can accelerate the conversion process, resulting in rapid deposition of conversion coatings onto the surfaces of magnesium alloys (particularly copper-containing alloys) since H_2O_2 can be easily reduced on the magnesium alloy surface to create alkaline environment for $Ce(OH)_x$ or CeO_x to precipitate. Although an excess of hydrogen peroxide concentration can increase the amount of Ce(IV) species included in the conversion layer, too much H_2O_2 decreases the corrosion resistance of the conversion layer as reported by Dabalà *et al.* [19]

Stannate conversion coating

Stannate conversion solution typically consists of K_4SnO_4, NaOH, $C_2H_3NaO_2$ and $Na_4P_2O_7$ with alkaline pH such as pH=11.6 and processing temperature at around 80 °C [20,21]. Stannate coatings on Mg and its alloys are formed by initial dissolution of substrate Mg followed by deposition of coatings after concentrations of Mg^{2+} and stannate ions reach critical concentrations at the substrate/electrolyte interface. It was also found that the coatings are mainly composed of hydrated magnesium stannate particles, $MgSnO_3$•$3H_2O$ [22,23]. On AZ31 alloy [6], $MgSnO_3$ was initially formed on the surface of β-phase ($Mg_{17}Al_{12}$) and then extended to the α-phase.

It was found that stannate coatings do not block electrolyte penetration. The corrosion rate of stannate coated AZ31B was found to be higher than cerium coated samples. Chromate and galvanic black coated samples have the best corrosion resistance when compared with stannate and cerium coated samples in a comparative study by Shashikala *et al.* [21], which will be discussed later. Elsentriecy *et al.* [23] found that stannate coatings formed by potentiostatic technique (i.e. by potentiostatic polarization at $E = -1.1$ V vs. Ag/AgCl) were more uniform, dense and corrosion resistant than those formed by the simple immersion method. Figure 15.3 shows a comparison of samples treated by potentiostatic polarization method and by simple immersion in a stannate conversion bath containing 0.125 M NaOH at 353 K. The coating deposited by the simple immersion method was less uniform with pinholes present, while the coating formed by potentiostatic polarization method was relatively uniform and continuous. The difference in coating morphology was explained by the difference in the substrate dissolution

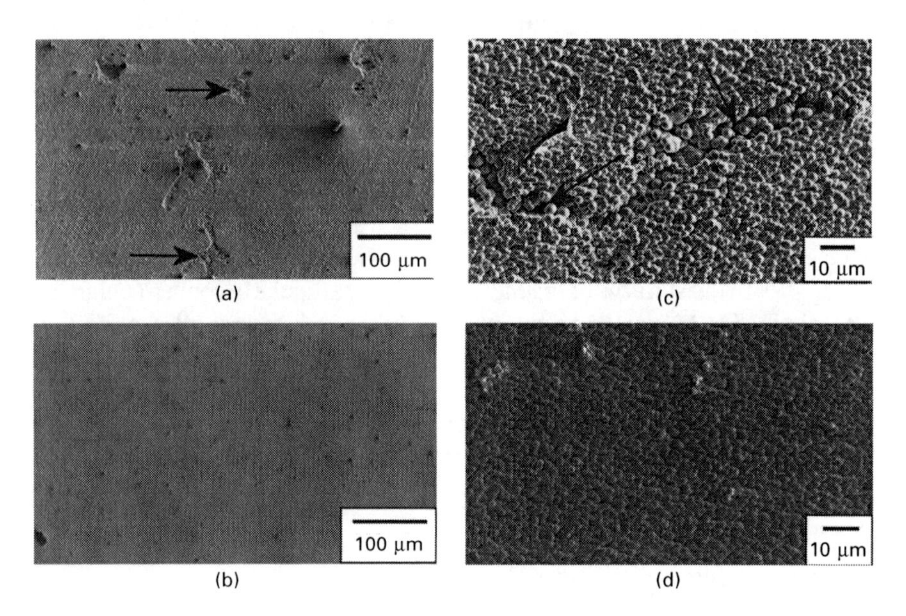

15.3 SEM images with different magnifications of AD91D alloy coated by simple immersion (a and c) and by potentiostatic polarization at $E = -1.1$ V (b and d) in a stannate bath containing 0.125 M NaOH at 353 K. Immersion time for the simple immersion method was 3.6 ks (from H.H. Elsentriecy *et al.*, *Electrochimica Acta*, **53** (2008) 4267–4275).

mechanism of the two methods during the coating formation. This example of holding the Mg alloy samples at the magnesium dissolution potential during conversion treatment is a simple modification of the conversion process yet can significantly improve the quality of conversion coatings. Such a modified conversion process could also be used for other types of conversion coatings such as cerium-based coatings.

Stannate conversion coatings were also combined with other surface coating processes to improve the corrosion protection of the Mg alloy. For example, electroless Ni–P coating was successfully deposited on the chemical stannate conversion coating for the AZ91D alloy by Huo *et al.* [24]; the corrosion resistance of the alloy was enhanced greatly and passivation occurred during anodic polarization during a potentiodynamic polarization test. The improvement in corrosion resistance was due to the presence of the conversion coating between the nickel electroless layer and the alloy substrate which reduced the electrochemical potential difference.

Phosphate and phosphate–permanganate related conversion coatings

Phosphate–permanganate conversion solution typically consists of $KMnO_4$, $MnHPO_4$, K_2HPO_4, and H_3PO_4 with a solution pH of 3~6 [25]. When magnesium or its alloys are placed in the phosphoric acid, Mg reaction with the acid takes place which locally depletes the H^+ ions, thus raising the pH and causing the dissolved Mg^{2+} to fall out of solution and precipitate on the surface as $Mg_2(PO_4)_3$. Permanganate ions (MnO_4^-) that also exist in the solution are reduced from Mn^{7+} to form manganese oxides on the surface of the Mg samples as follows:

$$MnO_4^- + 4H^+ + 3e^- \rightarrow MnO_2(s) + 2H_2O$$

or

$$2MnO_2 + 2H^+ + 2e^- \rightarrow Mn_2O_3 + H_2O$$

or

$$MnO_4^- + 2H_2O + 3e^- \rightarrow MnO_2(s) + 4OH^-$$

As the OH^- concentration increases at the anode locations which raises the pH value, some manganese products such as MnO_2, Mn_2O_3, Mn_3O_4 and MnOOH can be formed and co-exist within the conversion coating layer.

Corrosion resistance of Mg alloys after the phosphate–permanganate conversion treatment was found to be comparable with the chromate treatment. Chong and Shih [26] studied permanganate–phosphate conversion coating treatment for commercial pure magnesium, AZ61A, AZ80A, and AZ91D alloys. For the AZ series alloys, they found coated layer containing products of $Mg_2(PO_4)_3$, MgO, $Mg(OH)_2$, $MgAl_2O_4$, Al_2O_3, $Al(OH)_3$, MnO_2 or Mn_2O_3 and amorphous oxy-hydroxides, but on pure magnesium metal, $Mg_2(PO_4)_3$, MgO, $Mg(OH)_2$, MnO_2, amorphous oxy-hydroxides and $MgMn_2O_8$ were identified as the main products. They also found permanganate–phosphate conversion treatment for the AZ series alloys had an equivalent corrosion protection potential as the JIS H 8651 MX-1 chrome-based method. On AZ31, however, as reported by Zucchi *et al.* [22], permanganate–phosphate coatings have through cracks (more than stannate coatings). The co-existence of pure α-Mg and β-$Mg_{17}Al_{12}$ alloy phases of AZ series alloy has direct influence on the permanganate–phosphate deposition process [27]. The macroscopic galvanic effect between the two phases dominated the initial deposition. Network cracks typically existed on the permanganate–phosphate coating surface as shown in Fig. 15.4. These cracks may possibly be related to hydrogen release via the chemical reaction during the conversion treatment and/or the dehydration of the surface layer after treatment [26]. The cracks could also be caused by different deposition rates of α and β-phases [27].

Phosphate–permanganate conversion solution was also used to treat Mg–Li alloy to improve its corrosion resistance [28]. Recently, a new conversion coating using molybdate/phosphate (Mo/P) was developed for Mg alloys which showed better corrosion-resistant performance than pure molybdenum

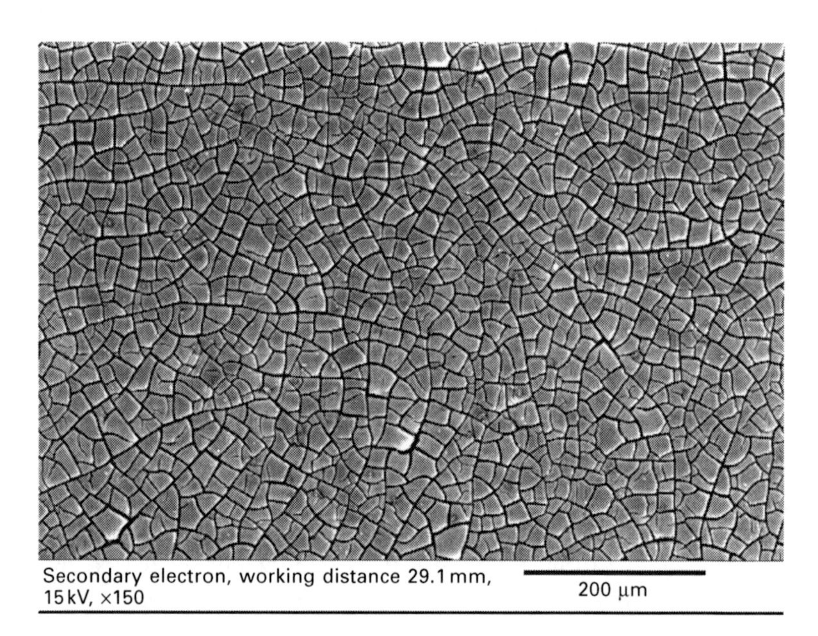

15.4 Conversion-coating morphology of AZ61A Mg alloys processed by immersing in a permanganate–phosphate conversion bath for 600 s at 323 K (from K.Z. Chong *et al.*, *Materials Chemistry and Physics*, **80** (2003) 191–200).

(Mo) conversion coating, with comparable corrosion protection for Mg alloy as the traditional chromate-based coating [29]. Other interesting works are related to phytic acid ($H_{12}P_6O_{24}C_6$), which was used to replace/combine with phosphate ions for the conversion treatment of Mg alloys and showed very promising results [30–32].

Conversion coatings based on compounds of V, Zr, Mo, W, Ti, Co

The vanadate conversion coating was obtained by immersion of Mg alloys in aqueous solution (pH 8) with different amounts of $NaVO_3$ at optimized temperature and immersion time. When the pH at the metal–solution interface increases due to surface-catalyzed proton reduction or dissolved oxygen reduction, rapid monomeric hydrolysis of V^{5+} is triggered:

$$VO_3^- + 2H_2O \rightarrow VO(OH)_3 + H^+,$$

and

$$VO(OH)_3 + 2H_2O \rightarrow VO(OH)_3(OH_2)_2$$

forming insoluble vanadium hydroxides or oxides. Yang *et al.* [33] showed that the conversion coating formed on the surface of AZ61 alloy by dipping the sample into the vanadium bath provided the best corrosion resistance when compared with phosphate (in solutions of H_3PO_4 5 ml L^{-1}, Na_2HPO_4 15 g L^{-1}, $CaNO_3$ 20 g L^{-1}) and cerium conversion (in solution $Ce(NH_4)_2(NO_3)_6$ 20 g L^{-1}) treated AZ61 samples, as indicated by the most positive corrosion potentials and lowest corrosion current densities in the polarization curves shown in Fig. 15.5.

Yang *et al.* [33] also found that the thickness of the vanadate conversion coating grows with increase of immersion time, however, excessive immersion induces the formation of cracks in the coating, leading to decrease of corrosion resistance of the coatings. The crack density can be reduced with increasing bath temperatures. They concluded that better corrosion resistance can be obtained by immersing the AZ61 samples in the conversion bath containing $NaVO_3$ with concentration of 30 g L^{-1} for 10 min at 80 °C. Other conversion treatments similar to vanadate conversion process include Zr-based, Mo-based [34], W-based, Ti-based, Co-based and Fe-based coatings as well as combinations of those coatings.

Sample post-treatment

The post-treatment process for coated Mg alloy parts normally includes a simple hot air drying in an oven before sending the parts for E-coating or painting. Temperatures, heating rate, drying time and heating atmosphere are important parameters to be controlled. In addition to heat drying

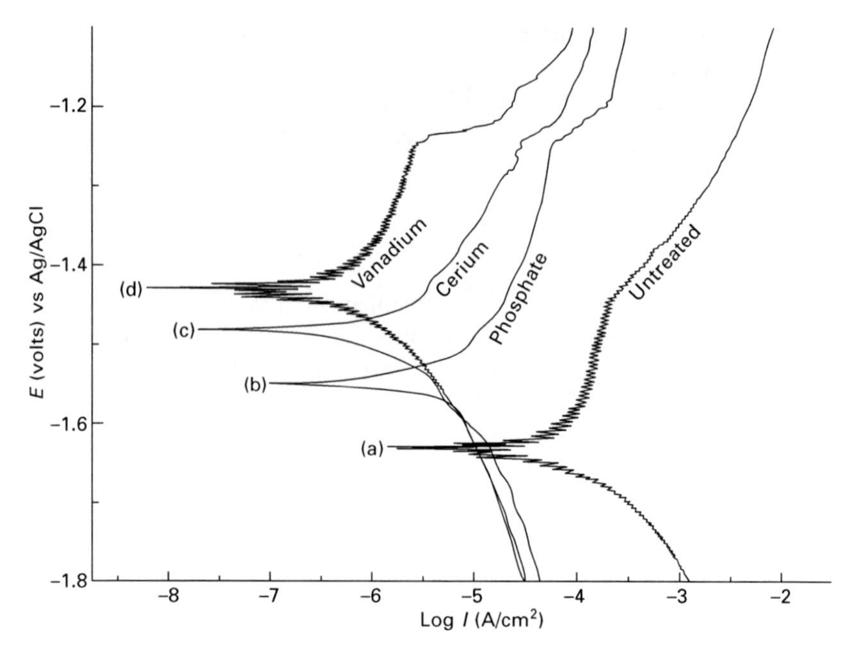

15.5 Polarization curves for Mg AZ61 treated with different conversion baths: (a) untreated, (b) phosphate, (c) cerium and (d) vanadium (from K.H. Yang *et al., Materials Chemistry and Physics,* **101** (2007) 480–485).

treatment, other post-treatments are used sometimes, for example, by dipping conversion coatings in a solution containing organic chelating agent such as aminocarboxylic acid and dithiocarbamic acid type chelating agents, phenolic compounds or alkaline such as NH_3–H_2O to stabilize and consolidate the conversion coatings.

Comparative studies

Comparative studies on chemical conversion coatings of stannate, cerium oxide, chromate and galvanic black anodizing on Mg alloy AZ31B were conducted by Shashikala *et al.* [21]. Scanning electron microscopy (SEM) analysis showed that the stannate coating has uniform spherical and cubic grains. Cerium coating consists of a thin and cracked coating with 'dry mud' morphology. A gel-like structure with a 'mud crack' pattern was observed in the case of chromate and black galvanic coating. Corrosion resistance of the various coatings was found in the following order: galvanic black anodizing > chromating > cerium oxide coating > stannate coating, as shown by the polarization curves in Fig. 15.6.

It should be noted that the above comparison is only for AZ31B under

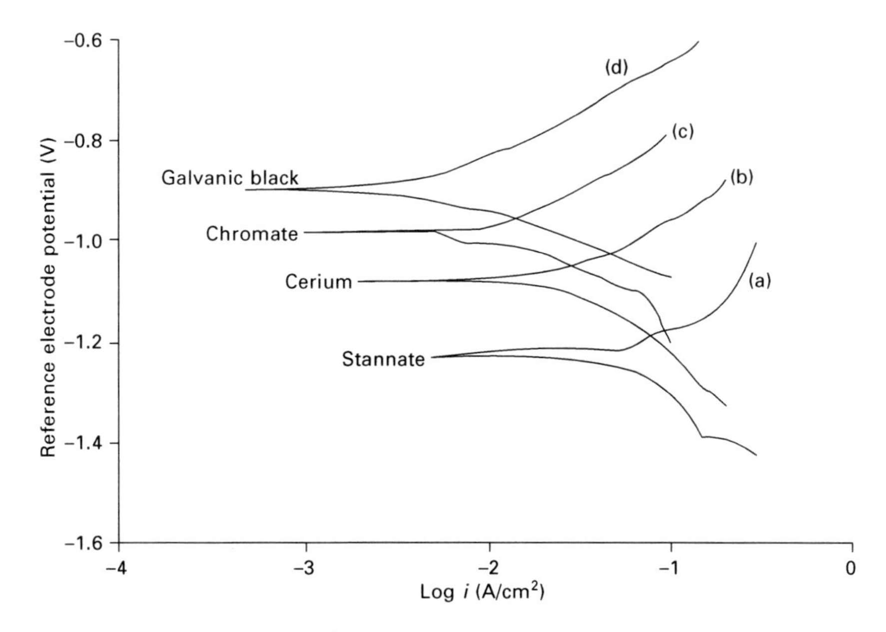

15.6 Polarization diagram of conversion coatings on Mg-AZ31B in 5% NaCl solution, (a) stannate (b) cerium (c) chromate (d) galvanic black anodizing (from A.R. Shashikala *et al.*, *International Journal of Electrochemical Science*, **3** (2008) 993–1004).

certain experimental conditions such as solution compositions, pH, bath temperatures, immersion time, etc. The conclusion above might not be applicable to other types of Mg alloys under different conversion treatment conditions.

15.2.2 Characterization of chemical conversion coatings

Chemical composition and microstructures

SEM, field emission SEM and X-ray diffraction (XRD) are the most common techniques used to characterize surface morphology, microstructure, and the phase composition of conversion coatings, while energy-dispersive X-ray analysis (EDS), X-ray photoelectron spectroscopy (XPS) and Auger electron spectroscopy (AES) are the most common techniques to characterize the chemical composition and oxidation state of conversion coatings. Other characterizations include transmission electron microscopy (TEM), atomic force microscopy (AFM), scratch test, and hardness and elastic modulus measurements.

Corrosion resistance

Corrosion resistance measurements for the conversion coatings are usually conducted using electrochemical polarization of the coated sample immersed in electrolyte such as 0.5 M Na_2SO_4 or 5% NaCl at room temperature with a potential scan range of –2.4 to –1.4 V/ saturated calomel electrode (SCE) and at scan rate of 0.2–1.0 mV/s. Impedance spectra of the conversion coated sample in electrolytes (e.g. 5% NaCl at pH 5.0 or 0.15 mol dm^{-3} H_3BO_3/0.05 mol dm^{-3} $Na_2B_4O_7$ buffer solution at pH 8.5) were measured with a sinusoidal signal of no more than 10 mV amplitude and a frequency range of typically 10 kHz to 100 mHz at the corrosion potential. Diameters of the capacitative semicircle in Nyquist diagrams obtained by impedance measurement represent the Faradic resistance of the coatings. Accelerated atmospheric corrosion test was typically done in humid (85%) atmosphere containing 0.5 ppm SO_2, while salt spray test uses 5 of NaCl solution spray onto the surface of coated sample at a temperature of about 40 °C. In some cases, thermal cycling in thermostatically controlled hot and cold chamber was performed if the coated Mg alloy was to be used in hot environments.

15.3 Electrocoat

Electrophoretic coating (E-coat) is a surface coating technology for conductive materials and is applied to magnesium alloys for corrosion protection. The process is also referred to as electrocoat, electrodeposition or electrophoretic painting. The process is based on electrophoretic motion of charged particles suspended in an electrolyte towards an electrode under an applied electrical field. According to the setup, electrocoating can be divided into anodic and cathodic processes. In anodic electrocoating, the workpiece to be coated acts as an anode, and an electrical field will drive negatively charged particles in the electrolyte towards the workpiece and deposit on the surface. In the cathodic electrocoating process, on the other hand, the workpiece to be coated acts as a cathode, and attracts positively charged particles under an electrical field.

The electrocoating process has many advantages compared with other coating processes. First, the electrocoating process produces uniform coatings with very low porosity, and hence provides excellent corrosion protection. Second, the process is capable of coating the surfaces of complicated shapes due to its unique coating formation mechanisms and high throwing power. Third, the process is cost effective for mass production due to its high degree of automation, high throughput and high paint transfer efficiency (95–98%). Furthermore, the process is environmentally friendly. The E-coat processes are predominantly water based, and feature very low level of volatile organic compounds (VOC), and free from hazardous air pollutants (HAP). Heavy metals such as lead have been completely eliminated from the bath chemistries without compromising the corrosion protection performance of the coatings. In addition, there is minimal waste discharge in the electrocoat plant. Some closed-loop operations have completely eliminated waste water discharge. These key advantages have made the E-coat process a key technology for the automotive industry, and a desirable surface protection approach for magnesium components.

The E-coat process involves four key steps:

- surface pre-treatment;
- electrodeposition of the paint in the bath;
- post-rinse; and
- baking.

15.3.1 Surface pre-treatment

Surface conditions, including surface chemistry, morphology and contamination, have very important effects on the deposition, bonding strength and the durability of the E-coatings. Proper surface preparation of the parts before coating is therefore critical for ensuring quality of the coatings. Various treatment methods for the substrates, such as surface cleaning, anodizing, chemical etching, conversion coating and plasma surface treatment, have been explored for modifying the surface conditions, with the aims of removing the undesirable contaminations from the surface, changing the surface chemistry and morphologies.

Surface cleaning

Surface contaminations such as grease, oil and dirt have a significant adverse effect on the E-coating formation and bonding strength. Silicone, in particular, is harmful to coatings even if only a trace amount (parts per million) exists on the metal surface. The benefits of surface cleaning are well known to the coating industry [35]. Alkaline solution treatment under controlled conditions is a well established and also the most commonly used industrial practice for the cleaning of metallic substrates before electrocoating, due to its low cost and suitability for mass production.

Wet glass beading was explored by Hsu as a surface preparation method for magnesium alloys [36]. The treatment can remove contaminants and rust products, as well as some casting defects from the surface of Mg alloys. The surface of Mg alloys prepared by such treatment can be used directly for various subsequent coatings, including electrocoating, electroplating or painting. An added benefit of the process is that Mg particles generated by glass beading will be oxidized in the liquid and hence the fire hazard of such particles is minimized.

Conversion coating and phosphating

Conversion coating is the most commonly used method of forming a base coating layer for electrocoating, and has been described in detail in the previous section and by Gray and Luan [37]. Chromate conversion coating is a conventional way of generating base coatings on magnesium and Mg alloys. Such processes, however, involve toxic Cr^{6+} ions and are being phased out due to environmental concerns. Many alternative chrome-free conversion coating methods are being developed to replace the chromate conversion coating processes. Some examples include phosphate coatings [38], conversion coating using an alkaline stannate bath to produce $MgSnO_3$ or $MgSnO_3$–H_2O coating [39,40], and permanganate coating forming various oxide and hydrides [41]. Silica-modified phosphates can replace conversion coatings containing chrome or lead in the paint solids for improved corrosion resistance without compromising the environmental concerns. Also, adding rare earth elements, such as cerium, to the phosphate layer can minimize the cracking of the phosphate layer and increase the stability of the phosphate, leading to improved corrosion resistance and coating durability [42]. It was also reported that adding molybdate to the phosphate coating both results in the formation of a conversion coating containing nanocrystalline zinc phosphate and $MgMoO_4$, which significantly refined the microstructure, and reduced crack formation in the coating [43].

Oxidization

By forming a magnesium oxide layer with a thickness of 5–20 μm using micro-arc oxidation before electrocoating, the corrosion resistance of the magnesium alloys can be effectively increased [44]. Oxide produced by anodizing also has a significant impact on the corrosion behavior of magnesium alloys. Compared with direct application a thin layer of organic coating on the bare magnesium substrate, the corrosion resistance of the magnesium alloys with a thin layer of oxide formed by anodizing and before organic coating has much improved corrosion resistance [45,46].

Plasma treatment of magnesium surface before electrocoating has been employed for improving the bonding strength of the coatings (Table 15.1) [47,48]. It is shown that delaminating time of the cathodic coating as revealed by the N-methylpyrrolidone (NMP) test increased from less than 1 hour to over 22 hours by appropriate plasma treatment of the magnesium surface [36], although such treatment will add to cost and may compromise productivity for commercial applications.

The oxide layer produced by micro-arc oxidation or anodizing has pores that can form interlocks with the paint deposited by subsequent electrocoating, which is believed to have excellent bonding with the magnesium substrate [49].

Table 15.1 NMP test results of bonding strength of coatings produced using different combinations of surface treatment on an AZ31B alloy [48]

Treatments	NMP time (h)
Cathodic E-coating without plasma treatment	0.07 + 0.018
Ar plasma surface cleaning + trimethylsilane plasma polymer coating + E-coating	0.25 + 0.002
(Ar + H_2) plasma surface cleaning + trimethylsilane plasma polymer coating + E-coating + Ar plasma post-treatment	39.16 + 0.200
Ar plasma surface cleaning + TMS plasma polymer coating + Ar plasma post-treatment + E-coating	0.19 + 0.197
Ar plasma surface cleaning + trimethylsilane plasma polymer coating + Ar plasma post-treatment + E-coating + Ar plasma post-treatment	27.52 + 0.495

Research work has shown that the initial bonding strength of the polymer to metal substrates depends critically on the nature of the oxide on the surface. The long-term durability of the coatings, however, depends strongly on the environmental stability of the underlying oxide. Therefore, by applying a proper stabilizing treatment of the surface oxide, the durability of the coating can be significantly improved [49].

15.3.2 Deposition

During a coating deposition process, a DC voltage is applied across the working electrode (parts) and the counter-electrode. The movement of the charged particles in the coating bath towards electrodes with opposite charges is accompanied by electrolysis of water. The water electrolysis will result in oxygen gas liberation on the anode and hydrogen gas evolution on the cathode. The evolution of these gases disturbs the hydrogen ion equilibrium in the water immediately surrounding the electrodes. This will lead to a corresponding pH change which will in turn de-stabilize the paint components of the solution and make them coagulate onto the appropriate electrode.

The key factors affecting the coating process include the chemistry of the coating bath, and the process parameters such as applied voltage, coating time and bath temperature.

Electrolytes

Owing to their low cost and environmental friendliness, aqueous electrolytes have been used predominantly in the electrocoating industry. Non-aqueous electrophoretic deposition processes have also been explored for use in the fabrication of electronic components and the production of ceramic coatings

[50,51]. Non-aqueous processes have the advantage of avoiding the electrolysis of water and the related gas evolution.

The aqueous-based electrolyte consists of an emulsion of polymer and deionized water in a stable condition. The coating bath normally contains 10–20% of paint solids and 80–90% of deionized water [52], although a solid content as high as 26% has been reported in recent publications [45,46]. The deionized water acts as the carrier of the paint solids which consist of resins, pigments and small amount of solvents. The resin is the backbone of the final paint film and provides protection of the substrates from various damages such as corrosion and chipping. The pigment mainly provides the color and gloss of the coating. The solvents serve the function of improving the smoothness and appearance of the coating film.

There are primarily two types of chemistries for the coating materials. One is epoxy-based and the other is acrylic-based. In general, epoxy-based coatings provide superior corrosion resistance and are used in applications where high corrosion resistance is required. This chemistry, however, does not have good resistance to ultraviolet light, and is commonly used in primer applications where a topcoat will be subsequently applied, particularly if the coated object needs to withstand sunlight.

Acrylic polymers are based on free radical initiated polymers containing monomers based on acrylic acid and methacrylic acid and their esters. Such polymers often also include styrene as a monomer. Acrylic-based coatings are more resistant to ultraviolet radiation and are widely used in applications where outstanding durability and color control are required. In recent years, hybrid coating systems are also being explored to satisfy more demanding performance needs for the coatings [52].

Process parameters

The most important process parameter in the electrocoating process is the applied voltage. Higher voltage provides higher throwing power for the process and increases the film growth rate, but too high a voltage can lead to the 'rupture' of the coatings, resulting in a very thick and porous film unacceptable from both a functional and a cosmetic perspective. Although DC voltages of 25–400 V have been cited in the literature for electrocoating, typical voltage used in the industry is in the range of 100–300 V. A higher applied voltage will normally lead to a thicker coating film [35]. Since the molar amount of hydrogen generated at the cathode is twice as much as that of oxygen generated at the anode for the same amount of water electrolysis, and gas has much higher electrical resistance than the electrolyte, the electrical resistance due to gas generation at the cathode is much higher than that at the anode. Therefore, the required voltage in the cathodic process is generally higher than in the anodic process.

The coating time is another important factor in the electrocoating process. A typical current–time curve under a constant applied voltage is shown in Fig. 15.7 [35]. The initial coating current is high for a minute or so at the beginning of the process and then drops sharply as coating grows. The coating time for the process is therefore normally short, varying from a few seconds to a few minutes. This fast process dynamics renders the process the suitability for high throughput. This is significantly different from the current characteristics in electroplating or anodizing where current is kept essentially constant.

The bath temperature also has a significant effect on the electrocoating process. First, the conductivity of the bath and the deposited coatings both increase with increase in the bath temperature. The increased bath conductivity will lead to decreased 'rupture' voltage; second, the viscosity of the coating film also changes with the bath temperature, thus affecting the ability of the coating film to release gas bubbles formed inside the coating.

An recent interesting development in electrocoating is the 'electroless' E-coating of Mg alloys in a conventional electrocoating bath [45,46,53]. In this novel process, pre-treated Mg alloy is immersed in an electrocoating

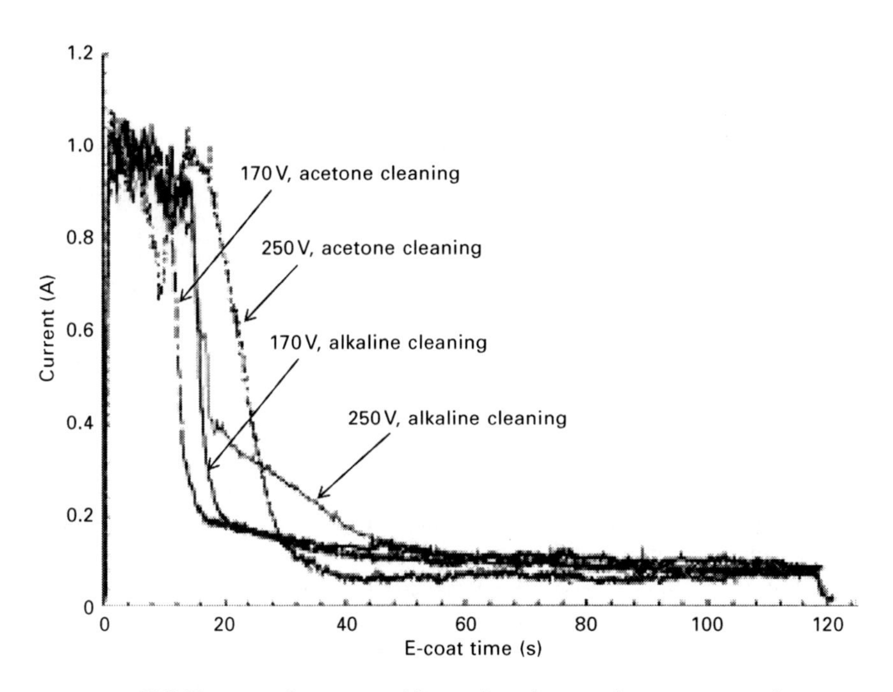

15.7 Change of current with coating time under constant voltages. The current was kept at <1 A at the beginning, and the voltage was increased to the set values after current dropped (from C. Reddy *et al.*, *Progress in Organic Coating*, **33** (1998), pp. 225–231).

bath for a few seconds without applying the electrical field and then pulled out and dry. A thin layer of stable paint film can be formed on the surface of a various Mg alloys, even after the surface is anodized. The coating film significantly improves the corrosion resistance of the magnesium alloys [45,46,53]. Further development in this process is expected to make the process more attractive for generating organic protective coatings on Mg alloys for various applications.

15.3.3 Post-treatment and baking

After deposition, the coated part is normally rinsed to clean the surface. The rinse process may utilize an ultrafilter to acquire water from the coating bath solution. The water, after rinsing, will be returned to the coating bath, allowing for high utilization efficiency of the coating materials, as well as reducing the amount of waste discharged into the environment.

After rinse, the coated parts are placed in an oven to cure the coated film. During the baking process, the polymer undergoes further crosslinking and becomes hard and resistant to chemical attacks. In addition, the baking process also allows the coating to flow out to fill the gas pores formed during the deposition process to make the coating film dense and continuous. The baking temperature is normally in the range of 82–177 °C (180–350 °F).

15.4 Concluding remarks

Conversion coating and e-coating are simple, low-cost and scalable coating schemes for the protection of Mg alloys. However, the information here has been based on a limited number of reviews and therefore does not contain all the information in the public domain. In addition, the current coating schemes are mostly complex, multi-layer systems that incorporate many different technologies and must be conducted very carefully in order to achieve optimum results.

Chemical conversion coatings that can effectively protect magnesium and its alloys from corrosion require excellent coating properties such as high density (low porosity), defect-free, good adherence, chemical inertness and self-healing characteristics. How to precisely control the physical, chemical and mechanical properties of the conversion coatings is the main challenge in developing chromium-free conversion coatings because many processing parameters can affect the properties of the coatings. It is very challenging to predict which processing parameter will play a key role under different circumstances. Trial and error is sometimes used in the determination of processing parameters. Important processing parameters that can affect the quality of conversion coatings besides the chemical compositions and microstructures of the Mg alloys include sample preparation methods, sample pre-treatment conditions such as degreasing, alkaline cleaning and acid pickling, conversion solution recipes, pH of the solutions, temperatures of the bath, immersion time and number of immersions, sample drying conditions and so on. Some processing parameters will play a more important role under certain circumstances while other factors could become more important under different circumstances. Generally speaking, an inert, dense, good adherence, defect-free (no cracking, pinholes, etc.) and self-healing conversion layer is desirable for good corrosion protection. The question is how to control chemical conversion processing parameters to achieve high-quality coatings with such desirable properties. To answer this question, it is critically important to understand the fundamentals of conversion coating processes in detail, which include chemical/electrochemical reaction mechanisms such as the rate of dissolution of magnesium and its alloy elements, the reduction of proton or water or dissolved oxygen at metal solution interface, the diffusion of reactants and products near the metal surface, the nuclei formation and growth on the surface of different phases, the influence of local pH and temperatures.

Another challenge is the pollution issue. Conversion treatment generates a number of wastes, including spent abrasives, solvents and/or aqueous cleaning baths, and conversion treatment baths, air emissions from abrasives and solvents, rinse waters, etc. New environmental regulations will force industry to continue to develop and improve the current chemical conversion processes.

Although the electrocoating process is well established and has been commercialized for several decades, there are ongoing demands for further improvements, in terms of better performance, lower cost, higher productivity and environmental friendliness. This is particularly true for magnesium alloys because they are relatively new for engineering applications compared with steels and aluminum alloys. The key performance enhancements required include improved coating adhesion to the substrate, better corrosion resistance and durability, better chipping resistance and color control. Some of these requirements need to be addressed by improving the coating chemistry through the modification of the bath compositions. Others can be dealt with by improving the coating process such as pre-treatment method and optimizing coating process parameters.

Cost reduction is an ongoing demand for every industry, including electrocoating. A few areas can be explored for further cost reduction. These include reducing baking temperature for the coated films because baking is a key energy consumption step in the entire process. The 'electroless' organic coating is another promising approach in saving energy and improving productivity.

15.5 References

1. G. Song, *Adv. Eng. Mater.* **7** (2005) 563.
2. X. K. Zhong *et al.*, *Corrosion Sci.* **50** (2008) 2304–2309.
3. A.R. Phani, F.J. Gammel, T. Hack, *Surf. Coat. Technol.* **201** (2006) 3299–3306.
4. A.N. Khramov, V.N. Balbyshev, L.S. Kasten, R.A. Mantz, *Thin Solid Films* **514** (2006) 174–181.
5. A.L.K. Tana, A.M. Soutar, I.F. Annergren, Y.N. Liu, *Surf. Coat. Technol.* **198** (2005) 478.
6. M.F. Montemor, M.G.S. Ferreira, *Electrochim. Acta* **52** (2007) 6976.
7. F. Zucchi, V. Grassi, A. Frignani, C. Monticelli, G. Trabanelli, *Surf. Coat. Technol.* **200** (2006) 4136.
8. H. Hoche, H. Scheerer, D. Probst, E. Broszeit, C. Berger, *Surf. Coat.Technol.* **174–175** (2003) 1018.
9. A.S. Hamdy, Symposium Green Engineering for Materials Processing, Materials Science & Technology Conference and Exhibition (MS&T'06), Cinergy Center, Cincinnati, OH (October 15–19, 2006), pp. 141–150.
10. K. Brunelli, M. Dabala, I. Calliari, M. Magrini, *Corrosion Sci.* **47**(4) (2005) 989–1000.
11. X. Yang , F. Pan, D.F. Zhang, *Appl. Surf. Sci.* **255**(5) (2008) 1782–1789.
12. A.R. Shashikala, R. Umarani, S.M. Mayanna, A.K. Sharma, *Int. J. Electrochem. Sci.*, **3** (2008) 993–1004.
13. F. Sato, Y. Asakawa, T. Nakayama, H. Satoh, *Jpn. Inst. Light Met.* **42** (12) (1992) 752–758.
14. S. Ono, T. Osaka, K. Asami, N. Masuko, *Corros. Rev.* **16** (1998) 175–190.
15. S. Ono, K. Asami, N. Masuko, *Mater. Trans. JIM* **42**(7) (2001) 1225–1231.
16. A. Rudd, C.B. Breslin, F. Mansfled, *Corrosion Sci.* **42**(2) (2000) 275–288.
17. H. Ardelean, I. Frateur and P. Marcus, *Corrosion Sci.* **50**(7) (2008) 1907–1918.
18. S. A. Salman, R. Ichino, M. Okido, *Chem. Let.* **36**(8) (2007) 1024–1025.
19. M. Dabalà, K. Brunellia, E. Napolitani, M. Magrini, *Surf. Coati. Technol.* **172**(2–3) (2003) 227–232.
20. J.-L. Yi, X.-M. Zhang, M.-A. Chen, R. Gu, *Scripta Materialia*, **59**(9) (2008) 955–958.
21. A.R. Shashikala, R. Umarani, S. M. Mayanna, A. K. Sharma, *Int. J. Electrochem. Sci.*, **3** (2008) 993–1004.
22. F. Zucchi, A. Frignani, V. Grassi, G. Trabanelli, C. Monticelli, *Corrosion Sci.* **49**(12) (2007) 4542–4552.
23. H. H. Elsentriecy, K. Azumi, H. Konno, *Electrochim. Acta* **53** (2008) 4267–4275.
24. H.-W. Huo, Y. Li, F.-H. Wang, *Corrosion Sci.* **46** (2004) 1467–1477.
25. M. Zhao, S.-S. WU, J.-R. Luo, Y. Fukuda, H. Nake, *Surf. Coat. Technol.* **200** (2006) 5407–5412.
26. K.Z. Chong, T.S. Shih, *Mater. Chem. Phys.* **80** (2003) 191–200.
27. W. Zhou, D. Shan, E.H. Han, W. Ke, *Corrosion Sci.* **50**(2) (2008) 329–337.
28. H. Zhang, G. C. Yao, S.L. Wang, Y.H. Liu, H.J. Luo, *Surf. Coat. Technol.* **202**(9) (2008) 1825–1830.
29. Z.-Y. Yong, J. Zhu, C. Qiu, Y.-L. Liu, *Appl. Surf. Sci.* **255** (2008) 1672–1680.
30. F. Pana, X. Yang and D. Zhang, *Appl. Surf. Sci.* **255** (2009) 8363–8371.
31. X. Cui, Y. Li, Q. Li, G. Jin, M. Ding, F. Wang, *Mater. Chem. Phys.* **111** (2008) 503–507.

32. H.-F. Cui, Q.-F. Li, Y. Li, F.-H. Wang, G. Jin, M.-H. Ding, *Appl. Surf. Sci.* **255** (2008) 2098–2103.
33. K.H. Yang, M.D. Ger, W.H. Hwu, Y. Sung, Y.C. Liu, *Mater. Chem. Phys.* 101 (2007) 480–485.
34. H.H. Elsentriecy, A. Kazuhisa, *J. Electrochem. Soc.* **156**(2) (2009) D70–D77.
35. C. Reddy, R. Gaston, C. Weikart, H. Ysuda, Influence of surface pretreatment and electrocoating parameters on the adhesion of cathodic electrocoat to the Al alloy surfaces, *Prog. Organic Coat.* **33** (1998) 225–231.
36. F. Hsu, Taiwan Patent 524857 2001TW-0120331, Surface treatment of magnesium alloys, August 20, 2001.
37. J. Gray, B. Luan, Protective coatings on magnesium and its alloys – a critical review, *J. Alloys Compounds* **336** (2002) 88–113.
38. L. Kouisni, M. Azzi, M. Zertoubi, F. Dalard, S. Maximovitch, Phosphate coatings on magnesium alloy AM60 part 1: study of the formation and the growth of zinc phosphate films, *Surf. Coat. Technol.* **185** (2004) 58–67.
39. M.A. Gonzalez-Nunez, C.A. Nunez-Lopez, P. Skeldon, G.E. Thompson, H. Karimzadeh, P. Lyon, T.E. Wilks, A non-chromate conversion coating for magnesium alloys and magnesium-based metal matrix composites, *Corrosion Sci.* **37** (1995) 1763–1772.
40. H. Huo, Y. Li, F. Wang, Corrosion of AZ91D magnesium alloy with a chemical conversion coating and electroless nickel layer, *Corrosion Sci.* **46** (2004) 1467–1477.
41. K. Chong, T. Shih, Conversion-coating treatment for magnesium alloys by a permanganate-phosphate solution, *Mater. Chem. Phys.* **80** (2003) 191–200.
42. J. Lian, G, Li, L. Niu, Zinc phosphate film and cathodic electro-coat deposition on magnesium alloy, *J. Jinagsu University (Natural Science Edition)* **28** (2007) 37–40.
43. G.Y. Li, J.S. Liang, L.Y. Niu, Z.H. Jiang, H. Dong, Effect of zinc-phosphate-molybdate conversion precoating on performance of cathode epoxy electrocoat on AZ91D alloy, *Surfc. Eng.* **23** (2007) 56–61.
44. B. Jiang, H. Shi, J. Li, W. Yang, Chinese patent CN 1908246A, Method for surface treatment of magnesium alloys using micro arc oxidation and electrocoating, February 7, 2007.
45. G. Song, 'Electroless' E-coating: an innovative surface treatment for magnesium alloys, *Electrochem. Solid State Lett.* **12** (2009) D77–D79.
46. G. Song, An irreversible dipping sealing technique for anodized ZE41 Mg alloy, *Surfc. Coat. Technol.* **203** (2009) 3618–3625.
47. J. Zhang, Y. Chan, Q. Yu, Plasma interface engineered coating systems for magnesium alloys. *Prog. Organic Coat.* **61** (2008) 28–37.
48. J. Zhang, R. Li, Q. Yu, Effect of plasma interface treatment and cathodic electrophoretic coating on Mg alloys, *Mater. Res.* **610–613** (2009) 984–990.
49. J.D. Venables, Review: Adhesion and durability of metal-polymer bonds, *J. Mater. Sci.* **19** (1984) 2431–2453.
50. J. Chen, H. Fan, X. Chen, P. Fang, C. Yang, S. Qiu, Fabrication of pyrochlore-free PMN-PT thick films by electrophoretic deposition, *J. Alloys Compounds* **471** (2009) 151–153.
51. L. Besra, M. Liu, A review on fundamentals and applications of electrophoretic deposition (EPD), *Prog. Mater. Sci.* **52** (2007) 1–61.
52. K.A. Follet, Electrocoat: The environomic solution, Proceedings of the WMA's 97th

Annual Conference and Exhibition; Sustainable Development: Gearing up for the Challenge. 2004.

53. G. Song, 'Electroless' deposition of a pre-film of electrophoresis coating and its corrosion resistance on a Mg alloy, *Electrochim. Acta* **55** (2010) 2258–2268.

16
Anodization and corrosion of magnesium (Mg) alloys

G.-L. SONG, General Motors Corporation, USA and
Z. SHI, The University of Queensland, Australia

Abstract: Anodization is one of the most important and effective surface pre-treatments for Mg alloys. This chapter systematically summarizes Mg alloy anodizing behavior, the compositions and microstructures of anodized films on Mg alloys and the anodization-influencing factors. Based on the anodizing voltage variation, gas evolution and sparking behavior in a typical anodizing process and the characteristic composition and microstructure of an anodized coating, a four-stage anodizing mechanism is postulated. Moreover, the corrosion performance of anodized Mg alloys is systematically reviewed and a corrosion model is proposed to explain the corrosion performance and electrochemical behavior. It is believed that some of the measured electrochemical features can be utilized to rapidly evaluate or compare the corrosion resistance of anodized Mg alloys.

Key words: Mg alloy, corrosion, anodizing, surface treatment, coating.

16.1 Overview of anodizing techniques

It is well known that magnesium (Mg) alloys are not corrosion resistant and require surface treatments or coatings in many applications. A variety of surface finishing processes are being used to protect magnesium alloys from corrosion, which include surface conversion treatments (e.g. chromating, phosphating), anodizing (e.g. DOW 17, HAE, Anomag, Keronite, Tagnite, Magoxid-Coat); galvanizing/plating (Zn, Cu, Ni, Cr). These processes can be used alone or in combination with application of organic coatings. In addition, there are also chemical vapor deposition (CVD), physical vapor deposition (PVD), flame or plasma spraying and laser/electron/ion beam surface treatments, etc., as options for corrosion protection of Mg alloys.

Among these techniques, anodizing appears to be one of the most effective corrosion protection techniques for Mg alloys. This is because a successful coating should meet the following requirements:

- be very stable in corrosive environments to protect the substrate;
- be resistant to scratching and flexible to follow the deformation of the substrate without cracking or delaminating;
- be a non-conductive layer to isolate the substrate and in case the layer

is damaged, no significant galvanic effect of the coating to accelerate the corrosion of the substrate;
- have a relatively rough morphology for top coating, if necessary;
- have high throwing power to treat components from all sides.

Anodizing is one of the most popular industrial processes (Blawert *et al.*, 2006), which applies an anodic current or voltage to a substrate metal to produce an oxide film. The nature of anodizing is an oxidation reaction in an aqueous environment driven by a current or voltage. Although many different terms, such as 'arc', 'sparking', 'plasma' and 'anodic deposition', are often used in various cases to describe this technique, they only refer to the phenomena exhibited in a certain stage of anodizing. The anodized coating has a porous morphology and ceramic-like composition, which is chemically stable. An anodized coating can be further sealed and/or top-coated for additional corrosion protection.

Anodizing is usually adopted as one of the important industrial surface treatment technologies for light metals, and it has been successfully used on aluminum (Al) alloys over many decades. It has also been further developed for use on Mg alloys. Although both Al and Mg alloys can be anodized, there are significant differences in their anodizing processes and anodized coatings. First, because of the high temperature caused by sparking during Mg alloy anodizing, there are a large number of species participating in the anodizing. The process on a Mg alloy is more complicated than that on an Al alloy. The latter normally does not have sparking. Second, the electrolytes used in the anodizing processes of an Al alloy and a Mg alloy are different. Third, the voltage used in anodizing a Mg alloy normally is much higher than in anodizing an Al alloy. Fourth, on an Al alloy, the formed anodized film is regular in structure and consists of a barrier layer adjacent to the metal surface as well as a layer containing uniform parallel pores normal to the surface, while the anodized film on a Mg alloy is irregularly porous in structure.

There are already several commercial anodized coatings that have been developed for magnesium alloys, such as Tagnite (Bartak *et al.*, 1993, 1995; Barton and Johnson, 1995), Anomag (Barton, 1998; Barton *et al.*, 2001), Magoxid (COMA, 2001; Schmeling *et al.*, 1990), Keronite (COMA, 2008; Shatrov, 1999), HAE (Abulsain *et al.*, 2004; Evangelides, 1955), Dow 17 (Abulsain *et al.*, 2004; COMA, 1956), MGZ (Hillis and Murray, 1987; Kotler *et al.*, 1976a; Leyendecker, 2001), etc. The first anodizing coating for Mg alloys is Dow 17 developed by the Dow Chemical Company. It uses an AC or DC current at a voltage below 100 V in a solution comprising sodium dichromate, ammonium acid fluoride and phosphoric acid at pH ~5 above 70 °F. The composition of the formed film is mainly MgF_2, $NaMgF_3$, $Mg_{x+y/2}O_x(OH)_y$ and smaller amounts of Cr_2O_3. However, the use of chromate in the electrolyte is a big concern limiting its applications. After that, the

HAE anodic coating technique was developed for Mg alloys. The anodizing electrolyte contains potassium permanganate, potassium fluoride, trisodium phosphate, potassium hydroxide and aluminum hydroxide with a high pH value (approximately 14) and anodizing is operated between 20 and 30 °C with an AC current at a voltage below 125 V.

Anomag is a technique commercialized by Magnesium Technology Ltd in New Zealand. The anodizing bath consists of ammonia, sodium ammonium phosphate, etc. No chromium or other heavy metals are used. Anodizing is obtained without formation of high-energy plasma discharges (spark discharge), so it is a non-sparking process. The Magoxid-Coat is formed in an electrolyte containing borate or sulfate, phosphate and fluoride or chloride at pH 5~11 (preferably 8–9) buffered by amines at a DC current preferably with a voltage up to 400 V. The coating contains MgO, $Mg(OH)_2$, MgF_2 and $MgAl_2O_4$. Tagnite is another chromate-free commercial coating system. The coating has better corrosion and abrasion resistance than any chromate-based coating. The electrolyte is an alkaline solution containing hydroxide, fluoride and silicate species operated below room temperature (4–15 °C). The anodizing current has a unique waveform optimized at voltages exceeding 300 V DC. The coating is white, consisting mostly of hard Mg oxide with minor deposition of hard fused silicates. Keronite process transforms Mg alloy surfaces into a hard, wear-resistant ceramic oxide coating at a temperature between 20 and 50 °C. This coating is a good corrosion and thermal barrier. Anodizing is performed with a bipolar (positive and negative) pulsed electrical current with a specific wave form in a proprietary, chrome and ammonia-free, low concentration alkaline (98% demineralized water) non-hazardous electrolyte. The coating is mainly composed of spinel ($MgAl_2O_4$) together with SiO_2 and SiP. Moreover, the high-voltage Cr-22 treatment is also commercially available, although the anodizing bath contains chromate, vanadate, phosphate and fluoride compounds (Hillis, 1995).

In addition to the commercially available anodizing processes, numerous new anodizing techniques are currently being developed (Blawert *et al.*, 2006). For example, a white glossy anodized oxide film formed in an aqueous solution was reported to contain silicate, carboxylate and alkali hydroxide (Kobayashi *et al.*, 1988). A film consisting of chemically stable and hard spinel compound of $MgO–Al_2O_3$ was found on Mg surfaces after anodizing (Kobayashi and Takahata, 1985). It was patented that magnesium substrate should be first pickled in an aqueous hydro-fluoric acid bath and then anodized in an aqueous bath composed of an alkali metal silicate and an alkali metal hydroxide to form an anodized film for magnesium (Kozak, 1980, 1986). Corrosion and wear-resistant protective layers on magnesium or Mg alloys can also be formed by anodic oxidation in an aqueous electrolyte bath containing borate or sulfate ions, phosphate ions fluoride ions, and alkali ions (Tanaka, 1993).

It should be noted that anodizing is a hot topic in the area of corrosion and protection of Mg alloys. It is expected that in near future, more corrosion-resistant, low cost, easily operated, environment-friendly and non-toxic anodizing techniques will become available for Mg alloys.

16.2 Characteristics of anodizing behavior

Although various anodizing techniques have been developed, which differ from one another for their different anodizing phenomena or coating performance, they do share some common characteristics.

16.2.1 Anodizing stages

Generally speaking, different stages can be observed in an anodizing process. For example, it was found that the DC anodizing of AM60 Mg alloy in 1.5 M KOH + 0.5 M KF + 0.25 M $Na_2HPO_4 \cdot 12H_2O$ with addition of various concentrations of $NaAlO_2$ has three anodizing stages: traditional anodizing followed by micro-arc anodizing and finally arcing (Verdier *et al.*, 2004). Khaselev and co-worker studied pure Mg (Khaselev and Yahalom, 1998a,b; Khaselev *et al.*, 1999), binary Mg–Al alloys (2 at.%, 5 at.% and 8 at.% of Al) and intermetallic $Mg_{17}Al_{12}$ in solution 3 M KOH + 0.6 M KF + 0.21 M Na_3PO_4, with and without addition of 0.4 M and 1.1 M of aluminate ($Al(OH)_3$) under constant voltage and current conditions, and found that the anodizing behavior showed clearly different stages. It is also reported that the anodizing voltage of diecast AZ91D in a 40 °C sodium aluminate + potassium fluoride electrolyte varied in the range of 240–600 V, the current density varied in the range of 0.5–5 A/dm^2, and micro-arcing was observed in the anodizing process. The process showed two stages. In the first stage, the cell voltage increased linearly at a very high rate of 80–300 V/min. With higher anodizing current densities and higher concentrations of the electrolyte components, the voltage increased more quickly. Approximately 3–20 min later, this process entered into the second stage. A steady state sparking was established on the surface and the cell voltage reached a relative stable value of 520–570 V. The formed ceramic coatings were composed of the spinel phase $MgAl_2O_4$. A few circular pores and micro-cracks formed on the ceramic coating surface; the number of the pores decreased, while the diameter of the pores increased as the anodizing treatment proceeded. Variation of treatment time in the range of 10–40 min caused no changes in the phase structure of the ceramic coatings.

An anodizing process may display different stages under potential control and current control modes. Generally speaking, a complete voltage-controlled anodic polarization curve of a Mg alloy under a potential control mode may be classified into four regions: (1) primary passivity; (2) breakdown of primary passivity and metal dissolution; (3) secondary passivity; and (4) breakdown of the secondary passivity. In practice, current-controlled anodizing is more popular. In this mode, an anodizing process may also be generally classified into four different stages: I-linear growing (the anodizing voltage increases quickly and linearly with time; there is no sparking, and gas evolution is not significant either); II-gas evolving (voltage keeps increasing, but the rate decreases with time; meanwhile, the gas evolution becomes much faster; sparking is insignificant); III-uniform sparking (anodizing voltage linearly increases with time again; sparking is clearly visible randomly over the specimen surfaces; the sparking activity together with gas evolution becomes more and more intense as the anodizing voltage increases with time); and IV-localized sparking (much more localized, intense vivid sparking arcs appear on the specimen surfaces; vigorous gas evolves from the sparking areas; the increasing of anodizing voltage with time slows down). In some cases, stages III and IV cannot be clearly separated and are combined together. Sometimes, stage I is too quick or short and may be overlooked. Therefore, two or three anodizing stages are more often reported in various studies. The four stages under current control mode do not necessarily correspond to the stages in potential control mode.

A schematic illustration of the anodizing stages is presented in Fig. 16.1. To obtain a steady micro-arcing and sparking process, the applied current density if under a current-controlled mode should be far higher than the active current peak in stage (2). Stage IV sometimes may correspond to stage (4). After stage (4) or IV, the surface film may collapse and cannot be anodized again; no more sparking can occur on a collapsed surface.

Anodizing voltage is a primary indicator of the formation of an anodized coating on Mg alloys under a current control mode. Initially, the voltage increases almost linearly with time until it reaches values of breakdown, at which a sharp deflection occurs and voltage oscillation and sparking start. The breakdown voltage increases with the Al content in the alloy and aluminate concentration in the bath solution. The highest breakdown voltage of 80 V was observed for intermetallic $Mg_{17}Al_{12}$ in a solution containing aluminate. This value is about 20 V higher than that observed for pure Mg (Khaselev and Yahalom, 1998a). In two-phase alloys, the anodizing is not uniform. Anodizing starts at α-phase while growth on β-phase requires voltages above 80 V. A uniform layer only forms above 120 V (Khaselev *et al.*, 1999).

Mg–Al alloys are passive at voltages up to 3 V (Takaya, 1988). Higher voltages lead to activation and dissolution of Mg–Al alloys accompanied by oxygen evolution. The dissolution is initiated in the form of separate pits. The number of pits gradually increases and the pits eventually join together to form large dissolution areas. The process continues with time until the Mg alloy surface is completely activated. When this happens, the surface of

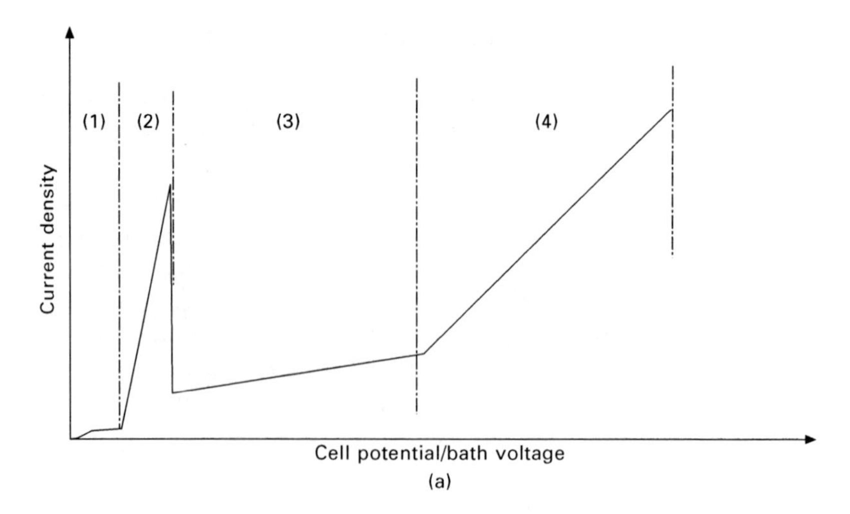

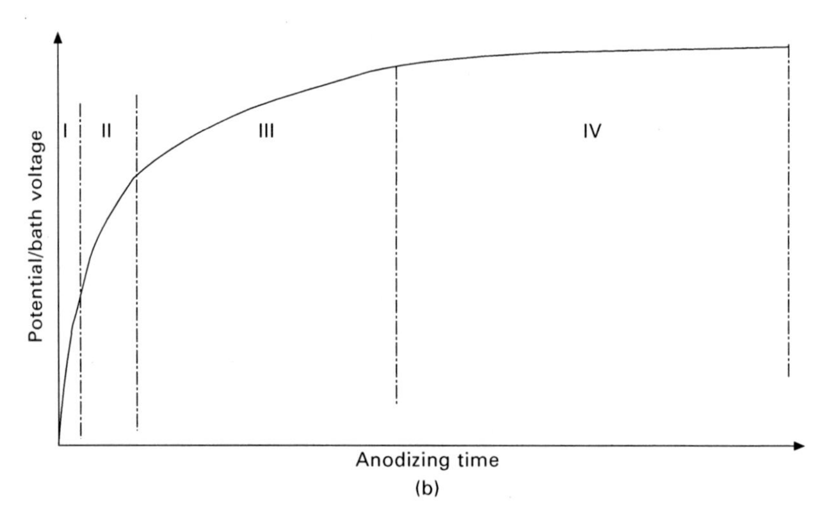

16.1 Schematic illustration of anodizing stages: (a) anodic polarization curve under a potential control and (b) dependence of cell potential (bath voltage) on anodizing time under a current control mode.

the alloy is fully black. Mg–Al alloys become passive again at an applied voltage of ~ 20 V. The secondary passivity is related to the formation of a MgO film. The presence of aluminate ion in the bath solution can promote the passivity. The aluminum content in the alloy is beneficial for the passivity of magnesium. Single phase Mg solid-solution alloy (Mg–0.8 at.% Cu and Mg–1.4 at.% Zn) was found to change appearance of its anodic film with applied voltage when anodized at up to 250 V at 10 mA/cm^{-2} in a 3 M ammonia

+ 0.05 M tri-ammonium orthophosphate electrolyte (pH 10.7) (Abulsain *et al.*, 2004). A relatively uniform film was developed at 26 V with dendritic-like and plate-like features; patches of porous material were formed at 160 V, which expanded as the voltage increased; and the final film formed at 240 V had typical sparking feature.

The anodizing stages vary from alloy to alloy, bath solution to bath solution and are influenced by applied potential and current density levels. It is well known that a film can be formed on a Mg alloy surface at a low voltage or current density. For example, for Mg in 1 M NaOH at voltages up to 3 V, when the current density remained low, a light grey protective film of $Mg(OH)_2$ is formed; a thick dark film of $Mg(OH)_2$ will be formed at higher voltages; above 20 V a protective coating will be produced (Huber, 1953). A barrier-type film or a semi-barrier film can be formed in an alkaline fluoride solution, which breaks down at around 5 V; above the breakdown potential, porous films were formed (Ono and Masuko, 2003; Ono *et al.*, 2003). Anodic films incorporating silicon can also be formed on AZ91D under a constant potential of 4 V in 3 M KOH solution with Na_2SiO_3. The films are uniform and thicker than the films formed in the bath solution without Na_2SiO_3. A few at.% of silicon can be detected in the films, although the main component of the films was $Mg(OH)_2$ (Fukuda and Matsumoto, 2004).

16.2.2 Sparking

Sparking can be observed in many anodizing processes for Mg alloys at high voltages or current densities. For example, in 3 M NaOH at 24 °C, anodizing voltage suddenly jumps to high levels when the applied anodic current density exceeds 20 mA/cm^2. After that, the potential oscillates in an irregular manner, and sparks appear and move around on the specimen surface (Carter *et al.*, 1999). Micro-arc, sparking or plasma oxidation of Mg alloys refers to the stage when applied voltage and current density are relatively high in an anodizing process. Normally, sparking process occurs when the applied voltage is over the dielectric breakdown potential of the anodized coating. The sparking voltage changes with bath composition and concentration. In most electrolytes, sparking does not occur on the surface of a magnesium alloy before the anodizing voltage exceeds 50 V.

Although an anodized film starts to form on Mg or a Mg alloy at a very low potential or current density, the film changes with potential or current density and breaks down at higher voltages. In other words, anodizing at high voltage or current density is a breaking down and repairing process of a film formed at a low current density or voltage, which accompanies a micro-arcing process. Initially the sparks are very small and are extinguished quickly. As the potential increases, the sparks become larger and begin to move over the specimen surface. During the whole sparking process, the specimen is actually being repeatedly activated and passivated. The passivation and activation correspond to the events that the film breaks down resulting in sparking and then it is rapidly repaired after sparking. As the rebuilt film at a sparked spot is thicker than the original film before sparking, the film keeps growing significantly during the sparking stage. The breakdown and repairing of the film can lead to a dramatic change in potential and current density. Thus, the current and potential fluctuations are always observed during sparking. If an anodized specimen remains at a constant potential for a long time, sparks may gradually diminish. If the potential was further increased slightly, the sparking will resume. A large positive potential step during an anodizing process can lead to suddenly intensified sparking. If sparking occurs continuously at a spot, not moving along, for example, at a sharp edge or a defective point on a specimen, there would be inevitably a large burned pit formed there. If this damaged area is too large, it may extinguish the sparking, resulting in failure of the anodizing process.

16.2.3 Gas evolution and anodizing efficiency

Apart from sparking, oxygen evolution is another important phenomenon in an anodizing process for Mg alloys. Oxygen evolution is closely related to sparking behavior. Before sparking, the oxygen evolution is insignificant. Vigorous oxygen evolution is normally observed when sparking occurs. For example, under a current-controlled mode, in the initial stage at a low voltage the evolved gas volume is very small; then the evolved gas volume increases gradually with voltage, and finally significantly at a high voltage.

The efficiency of film growth by anodizing is relatively low, only about 30% (Abulsain *et al.*, 2004). There is a close relationship between oxygen evolution and anodizing efficiency. It was found (Shi *et al.*, 2006b) that for AZ91D anodized in a silicate containing bath solution, apart from film formation, the anodizing current was also used for electrochemical oxygen evolution. Consequently, a dramatic decrease in anodizing efficiency down to a negative value was measured after the anodizing getting into an intensive sparking stage (see Fig. 16.2). This result implies that a higher current density may produce a thicker coating quickly, but it is not economical in practice.

16.3 Anodized coating/film

An anodized coating/film can be characterized by its thickness, composition and microstructure. These parameters are important to its corrosion protection performance.

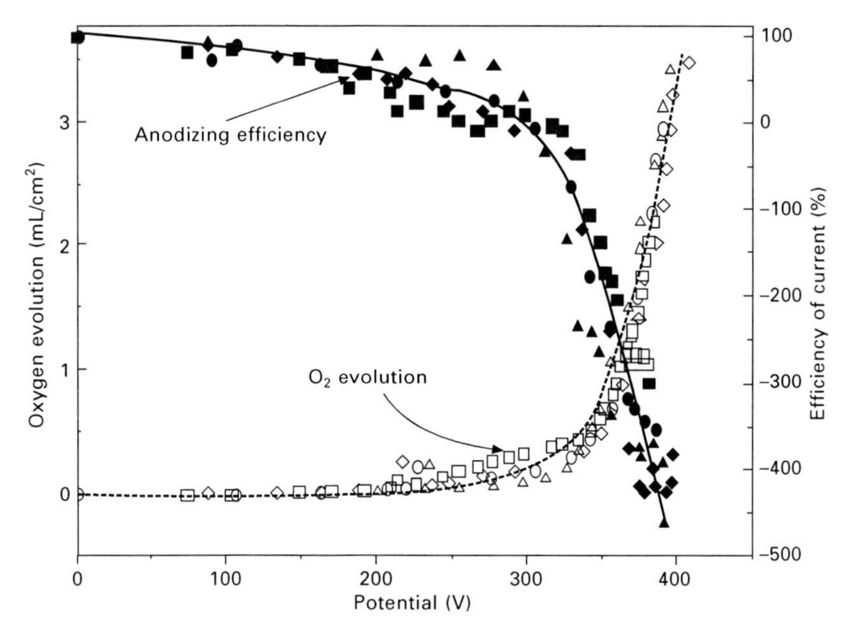

16.2 Dependence of oxygen evolution and current efficiency on cell voltage under constant electric charge (18 C/cm^2) (Shi *et al.*, 2006b).

16.3.1 Coating thickness

During anodizing, the substrate magnesium is consumed by oxidation and the oxide film is deposited on the surface. The surface film layer grows about 50% into the original magnesium and 50% above the original surface level (Kurze and Banerjee, 1996). The thickness of the coating is dependent on the anodizing voltage or current density. Figure 16.3 shows the dependence of the thickness of an anodized coating under a current control mode on the final anodizing voltage. It is a linear relationship, i.e. a higher cell voltage means a thicker anodized coating is formed. The increase in film thickness with final anodizing voltage can be easily understood if the resistance of the film is believed to be linearly dependent on the film thickness.

16.3.2 Coating composition

The elements in a formed anodized film are mainly magnesium, oxygen and aluminum (if the substrate or the bath solution contains aluminum), which can usually be further specified as MgO and some $MgAl_2O_4$. Detailed composition varies widely, depending on many factors. For example, a coating formed through a Dow17 process on pure magnesium is basically composed of MgF_2, $Mg_{x+y/2}O_x(OH)_y$ and $NaMgF_3$. The first two components are formed at the metal/film interface and the latter formed in the middle of

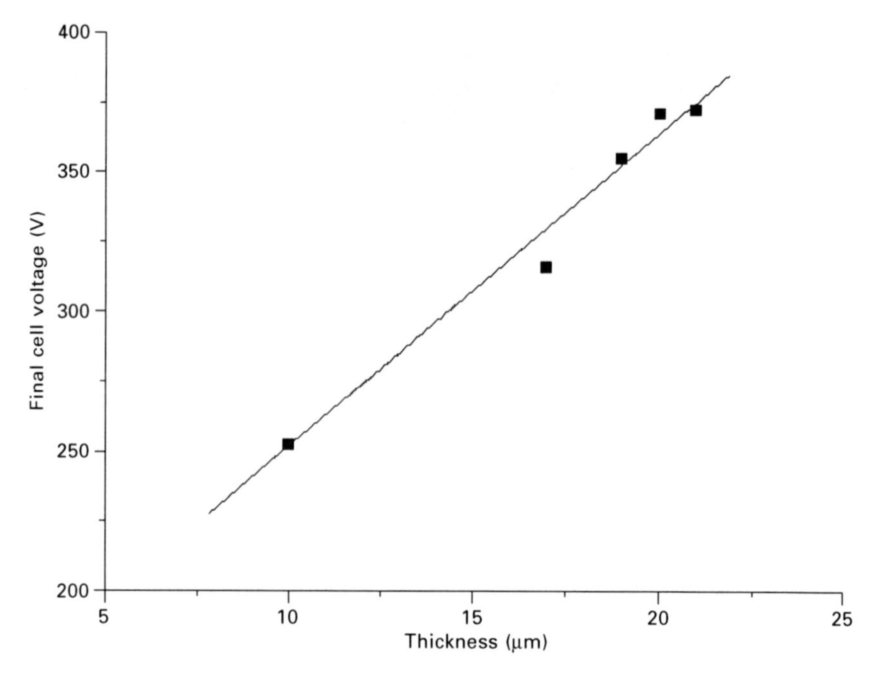

16.3 Dependence of the thickness of anodized coating on the final anodizing voltage (Shi *et al.*, 2006b).

the film by a chemical reaction in presence of Na^+ in the electrolyte (Ono, 1998). Recently, it was found that a significant amount of Si from an SiO_3^{2-}-containing anodizing bath can be incorporated into the anodized coating after anodizing at a high voltage (Shi *et al.*, 2006b). The Si/Mg ratio of the anodized coating increased as the anodizing proceeded. In the first stage, the atomic Si/Mg ratio was only 0.5, indicating that the deposition rate of $Mg(OH)_2$ was the same as that of $MgSiO_3$. In the second stage, the atomic ratio of Si/Mg was higher (around 1), suggesting more silicate deposited into the coating from the bath solution. In stage III, the ratio increased to 2, implying more silicates are incorporated in the coating. In stage IV, the amount of silicate deposited was much greater than $Mg(OH)_2$ with a ratio of Si/Mg = 5.7, indicating that the silicate is the main composition of the coating. Figure 16.4 summarizes the chemical compositions of films formed at different anodizing voltages.

For an Al-containing Mg alloy, Al-enrichment can normally be detected in the anodized coating (Shi *et al.*, 2006b). Some other alloying elements may also be enriched in an anodized film after anodizing. For example, alloying element enrichment was found on sputter-deposited Mg–0.4 at.% W and Mg–1.0 at.% W model alloys (Bonilla *et al.*, 2002a,b). The alloys were anodized at 10 mA/cm^2 up to 150 V in a 3 M ammonium hydroxide +

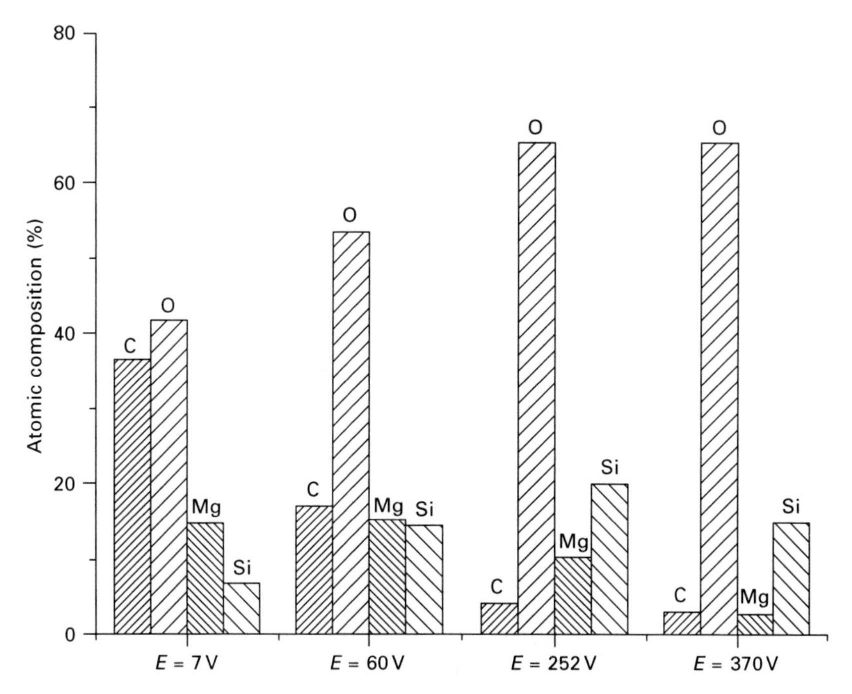

16.4 Chemical composition of anodized coatings formed at different anodizing voltages (based on Shi *et al.*, 2006b).

0.05 M ammonium phosphate electrolyte at 293 K. During anodizing to about 50 V, a relatively smooth but finely porous film developed. With increasing voltage, the film transformed gradually to a coarse porous morphology due to dielectric breakdown. Enrichments of tungsten to at least 1.7×10^{15} and 2.9×10^{15} W atoms/cm^2 for Mg–0.4 at.% W and Mg–1.0 at.% W alloys, respectively were found. The enrichment mechanism may be analogous to that of dilute Al alloys, in which enrichment of Al oxides occurs in its surface film due to the Gibbs free energies of the alloying elements oxides being greater than that for the formation of alumina. Enrichment can also be observed in solid-solution Mg–0.8 at.% Cu and Mg–1.4 at.% Zn alloys anodized up to 250 V at 10 mA/cm^{-2} in 3 M ammonia + 0.05 M tri-ammonium orthophosphate electrolyte (pH 10.7) at 293 K (Abulsain *et al.*, 2004). Rutherford backscattering spectroscopy (RBS) revealed enrichments to about 4.1×10^{15} Cu atoms cm^{-2} and 5.2×10^{15} Zn atoms cm^{-2}, which can be correlated with the higher standard Gibbs free energies for formation of copper and zinc oxides than that for formation of MgO. The enriched layers were about 1.5–4.0 nm thick as measured by medium energy ion scattering (MEIS). The anodized films, composed mainly of magnesium hydroxide, contained copper and zinc species throughout their thicknesses; the Cu:Mg and Zn:Mg atomic ratios were about 18% and 25% in those of the alloys, respectively. Phosphorus species were present in most of the film regions, with a P:Mg atomic ratio of about 0.16 (Abulsain *et al.*, 2004).

16.3.3 Coating microstructure

Generally speaking, in an anodizing process, a thin amorphous film can be formed during the first stage of anodizing. Prolonged anodizing causes local breakdown and crystallization of the initially formed anodic films. Anodized coatings formed at high voltages are porous. All the commercial anodized coatings that formed at high anodizing voltages, such as Dow17, HAE, Tagnite, Magoxide, Anomag and Keronite (most of them have got into the sparking stage) have been revealed to have a porous microstructure. A typical porous microstructure image of an anodized coating formed on AZ91D in a KOH and SiO_3^{2-}-containing solution is shown in Fig. 16.5.

In fact, anodized coatings are multi-layered. For example, using a Dow17 process (without sparking/electric breakdown), a cylindrical pore structure in the outer layer and barrier layer in the inner layer in contact with metal substrate can be formed on pure Mg (Ono, 1998). Figure 16.6 schematically illustrates the multi-layer structure of a Mg oxide coating.

It should be noted that the relatively compact inner barrier layer is normally very thin and can barely be revealed by scanning electron microscopy (SEM) in experiments. Compared with the barrier layer in the anodized film on an Al alloy, it is much less corrosion resistant and one cannot expect it to significantly retard the ingress of aggressive species.

The microstructure can vary with anodizing conditions. Mato *et al.* (2003) studied the anodizing behavior of sputter-deposited alloy, Mg–40% Ta, in ammonium pentaborate and sodium silicate electrolytes; in the pentaborate electrolyte, formation of an amorphous single layer composed of Ta_2O_5 and

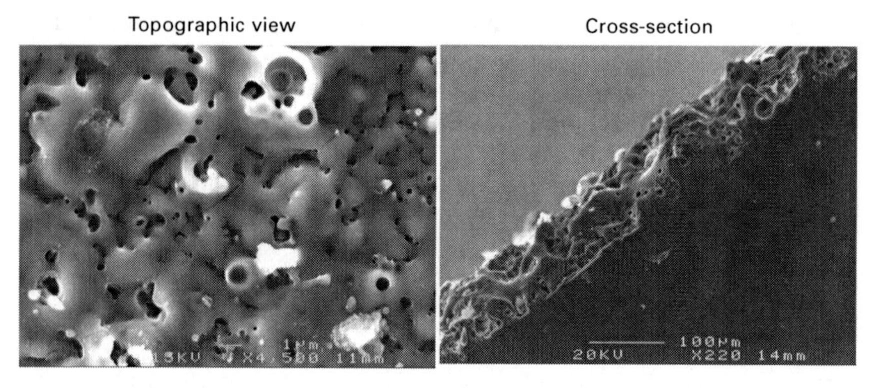

16.5 Typical microstructure of an anodized coating formed on AZ91D in a silicate-containing solution.

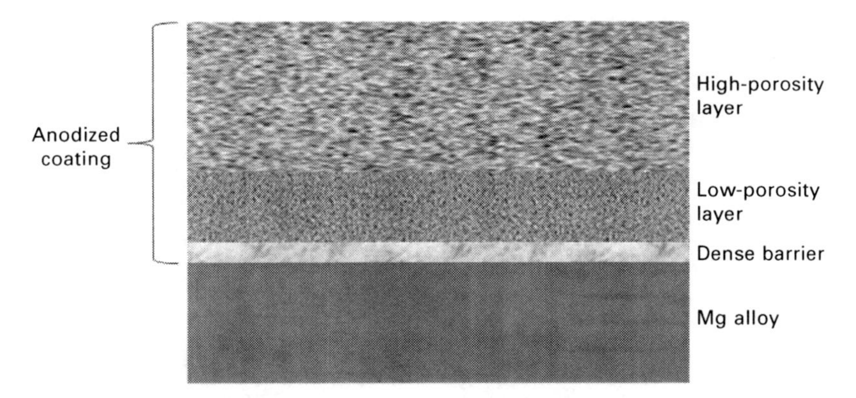

16.6 Schematic illustration of the multi-layer structure of Mg oxide coating (based on Blawert *et al.*, 2006).

MgO was observed, while in the silicate electrolyte, a two-layered structure was obtained. It was found that faster outward diffusing Mg ions resulted in a magnesium-rich top layer on the inner Ta_2O_5/MgO layer.

A compact anodized coating with less defects, higher thickness and stable composition in aggressive environments would be expected to provide favorable corrosion protection to the Mg alloy substrate.

16.4 Influencing factors

Theoretically, the main factors influencing the anodizing of magnesium alloys include the composition and microstructure of the substrate magnesium alloy, the composition of the anodizing bath solution, and the operational parameters, such as anodizing current density, voltage and temperature. For example, the dimension and distribution of porosity are a function of the electrolyte composition, anodizing current density and voltage, which further determine the corrosion resistance of an anodized magnesium alloy (Blawert *et al.*, 2006; Shi, 2005; Shi *et al.*, 2006a,b).

16.4.1 Anodizing current/charge and voltage

Anodizing current or charge and anodizing voltage can significantly affect the anodizing process. It has been found that the microstructure of an anodized coating is determined by the anodizing current charge used for the coating formation (Fig. 16.7(a)–(d)).

The pores in the coating formed in stage III are smaller in diameter (1–2 μm) than those in stage IV. With the same amount of total anodizing charge, the microstructure of a formed coating is relatively less significantly affected by anodizing current density (see Fig. 16.7(e)–(h)).

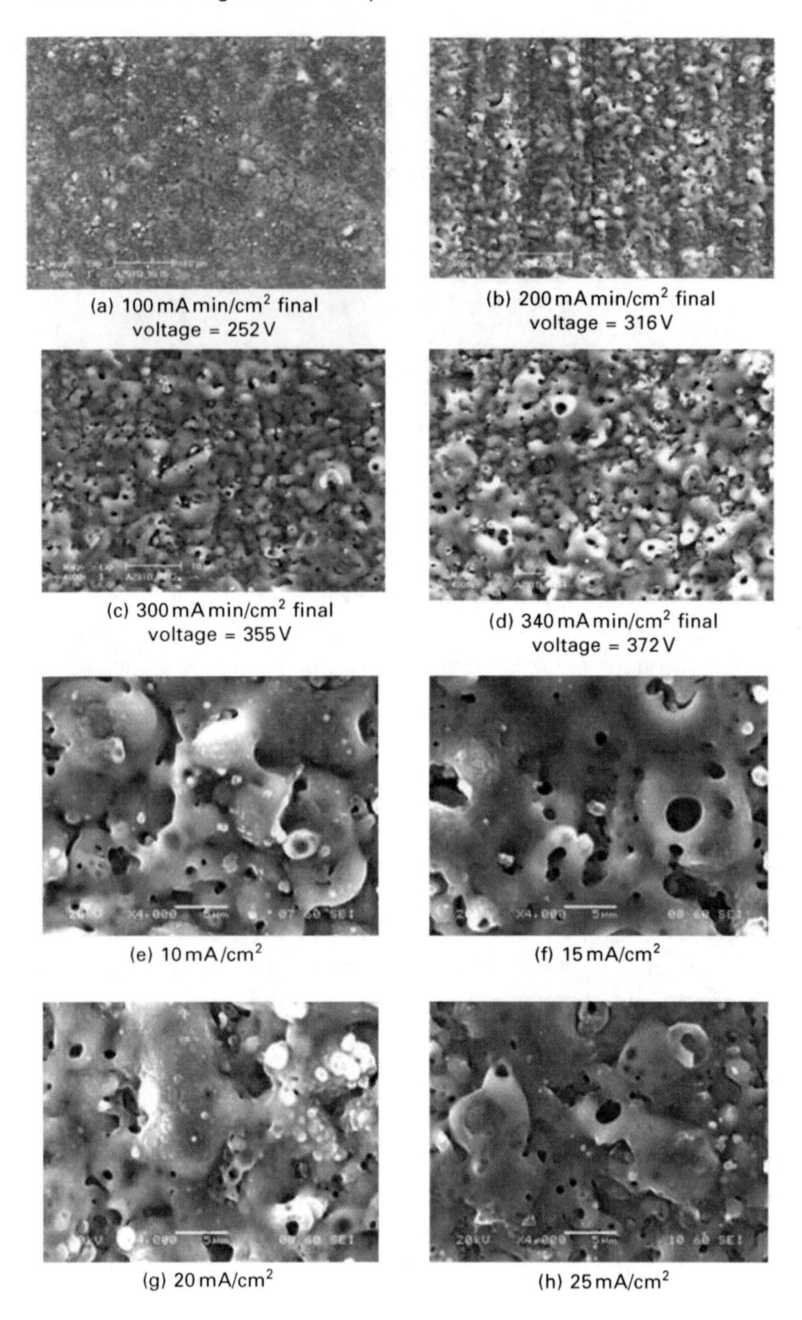

(a) 100 mA min/cm^2 final voltage = 252 V

(b) 200 mA min/cm^2 final voltage = 316 V

(c) 300 mA min/cm^2 final voltage = 355 V

(d) 340 mA min/cm^2 final voltage = 372 V

(e) 10 mA/cm^2

(f) 15 mA/cm^2

(g) 20 mA/cm^2

(h) 25 mA/cm^2

16.7 SEM images of anodized coatings formed in different stages (a), (b), (c) and (d) (with different anodizing current charges and final anodizing voltages) and under different current densities (e), (f), (g) and (h) with the same amount of total electric charge (18 C/cm^2) on AZ91D (Shi *et al.*, 2006b).

However, if the current charge is not controlled, then a high current density can lead to a thicker and more porous anodic film. It has also been found that (Chai *et al.*, 2008) the corrosion resistance of an anodic film can be related to the applied current density. An anodized coating formed at a higher current density should be thicker and have higher corrosion resistance.

16.4.2 Anodizing bath solution

The composition of anodizing bath solution is the core part of many patents in this area, such as DOW17 (Hillis, 1994), HAE (Hillis, 1994), MGZ (Kotler *et al.*, 1976a,b, 1977), Tagnite (Bartak *et al.*, 1995; Zozulin and Bartak, 1994), Magoxid (COMA, 2001), ANOMAG (Barton, 1998; Barton *et al.*, 2001) and Keronite (COMA, 2008), etc., because it can critically influence the coating composition, microstructure and performance. For example, although the Dow17, HAE and Tagnite anodized coatings are all porous, due to the different bath solutions used in these processes, their pore size and distribution characteristics are quite different. Figure 16.8 compares Anomag and Tagnite coatings on ZE44 Mg alloy, showing their different microstructures and thickness. Anomag bath normally contains phosphate and ammonia while Tagnite uses silicate and fluoride.

The effect of electrolyte on the characteristics of oxide coatings on Mg alloys has previously been studied (Barton, 1998; Fukuda and Matsumoto, 2004; Hsiao and Tsai, 2005; Khaselev *et al.*, 1999). In early studies, anodizing electrolytes mainly contained chromic acid or fluoride (Ono and Masuko, 2003; Zhang *et al.*, 2005). Recent research focused on chromate-free electrolytes, such as alkaline solution with additives such as phosphate, silicate, borate and organic substances (Cai *et al.*, 2006; Sharma *et al.*, 1997; Wu *et al.*, 2007). The effect of electrolyte on anodizing behavior aimed at improving coating quality is an important topic in investigations (Barton and Johnson, 1995; Hagans, 1984; Khaselev and Yahalom, 1998a; Khaselev *et al.*, 1999, 2001; Kotler *et al.*, 1976b, 1977; Takaya, 1987, 1988, 1989b; Takaya *et al.*, 1998; Zozulin and Bartak, 1994).

Although the influence of anodizing electrolyte on anodizing of Mg alloy is complicated, some common knowledge has been gained from existing studies. Generally speaking, some constituents from an anodizing electrolyte can be combined in the anodized film (Hillis, 2001; Ono *et al.*, 2003; Ono and Masuko, 2003; Takaya, 1989a). If an anodizing bath contains aluminates, then aluminum can normally get into the anodized coating (Khaselev and Yahalom, 1998b) as a significant component. The form of the aluminum in the coating is likely to be magnesium aluminate, combined with magnesium oxide and hydroxide. For example, when the electrolyte contains 1.1 M aluminate, the films on Mg are mainly composed of a single oxide phase – $MgAl_2O_4$ spinel (Khaselev *et al.*, 1999). Verdier further confirmed that

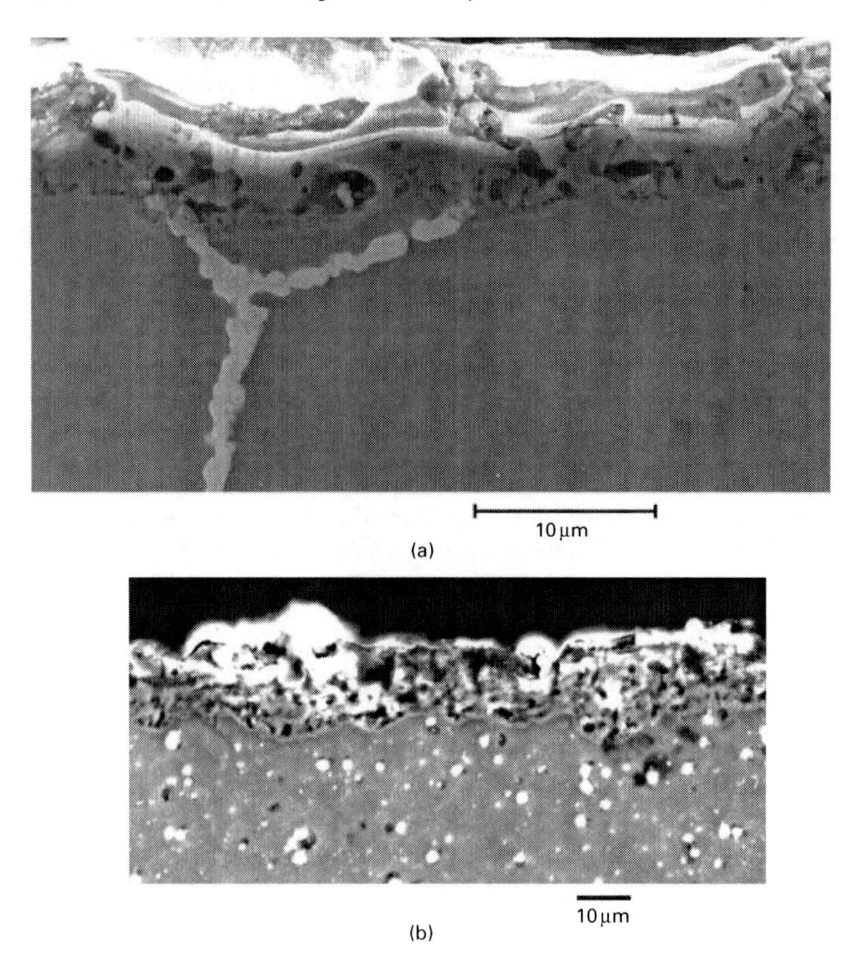

16.8 Cross-sectional microstructures of (a) Anomag and (b) Tagnite coatings.

the Al content in the film was a function of the aluminate concentration in the anodizing electrolyte (Verdier *et al.*, 2004). The addition of $Al(NO_3)_3$ into a 3 M KOH+0.21 M Na_3PO_4+0.6 M KF base electrolyte was found to contribute to a uniform sparking in anodizing AZ91 (Hsiao and Tsai, 2003). Without or with a low concentration of $Al(NO_3)_3$, a porous and non-uniform anodic film was formed. As the concentration of $Al(NO_3)_3$ was increased up to 0.15 M, uniform and compact anodic films were obtained. The addition of $Al(NO_3)_3$ into the base electrolyte resulted in the formation of Al_2O_3 and $Al(OH)_3$ in the anodic film. The maximum amount of Al_2O_3 was found in the anodic film when the alloy was anodized in an electrolyte containing 0.15 M $Al(NO_3)_3$. However, the film thickness decreased as the concentration of $Al(NO_3)_3$ was increased.

Addition of fluoride, aluminate, phosphate or tetraborate to the electrolyte solution results in a mixed composition of various magnesium compounds in the anodized layer (Barton and Johnson, 1995). The presence of sodium fluoride in a bath solution leads to improved surface texture and opacity, but does not affect the thickness significantly (Shi *et al.*, 2003). Citrate in the electrolyte is helpful in controlling sparking process, and preventing pit formation during sparking. Iodide is a damaging electrolyte and can lead to uncontrolled pitting. Barton (Carter *et al.*, 1999) reported that certain electrolytes including aluminate and tetraborate could influence the coating thickness and microstructure, while other electrolytes, such as fluoride and phosphate, contributed to the color, opacity and uniformity of the coating. Tetraborate contributes both to coating thickness and colour, and lowers sparking voltage (Shi *et al.*, 2003). Kotler *et al.* (1976a, 1977) investigated an anodizing process on magnesium using an alternating current in an aqueous solution comprising chromate, vanadate, phosphate and fluoride and found that the coating was better than the earlier magnesium anodized coatings DOW17 (Hillis, 1994) and HAE (Abulsain *et al.*, 2004) in terms of the corrosion resistance and wear resistance.

Obviously, in an anodizing bath, anions play a critical role in anodizing. The effects of SiO_3^{2-}, PO_4^{3-}, MoO_4^{2-} and AlO_2^- on anodizing behavior have been compared (Chai *et al.*, 2008). Among the selected oxysalts, sodium silicate could contribute to formation of an anodic film with the best anti-corrosion property, and the different concentrations of sodium silicate would cause the change of anodic film structure. Ma *et al.* (2004) showed that anodized coatings formed in Na_2HPO_4 solutions had better corrosion resistance than those produced from Na_2SiO_3 solutions, and the latter had slightly better erosion resistance than the former. They have different phase composition and corrosion failure processes (Cai *et al.*, 2006). Phosphate, fluoride and borate additives were reported to be beneficial to the improvement of corrosion resistance of anodized films on Mg alloy produced in silicate-based solutions (Duan *et al.*, 2007). Fluoride and phosphate were useful to enhance the corrosion resistance of inner layer of anodized films, while borate contributed to the growth of anodized film on Mg alloy.

Apart from inorganic chemicals in the anodizing bath, some organic additives can also influence the anodizing process and coating quality. For example, the addition of silica sol increased the thickness of the anodic film and improved the roughness of the film surface (Li *et al.*, 2006), and the anodic film formed in the electrolyte with addition of silicon–aluminum sol was more uniform and compact (Zhu and Liu, 2005). Titania sol can also lead to more uniform morphology with fewer structural imperfections and better corrosion resistance (Liang *et al.*, 2007).

Furthermore, anodizing of magnesium is even possible in organic electrolytes as demonstrated by Asoh and Ono (2003). In triethylamine/ethylene glycol solution, layers with transparent appearance and excellent corrosion resistance were obtained in the region between 10–40% of water content. However, the water content strongly affected the formation of the anodic film. Above 10% water content, a compact, transparent and enamel-like barrier-type film structure was observed. At higher water contents, the films became opaque and lost their good corrosion resistance. The properties were related to the incorporation of organic species (C, N, O) into the layer.

The solution temperature has a negative effect on the anti-corrosion property of an anodized film. This is probably because the large quantity of heat caused by sparking could not be released effectively (Chai *et al.*, 2008) when anodized at a higher temperature.

16.4.3 Influence of the substrate

Different Mg alloys have slightly different anodized films in the same bath solution. It has been reported (Hagans, 1984) that the coatings formed on Mg–Al alloys in a borate solution are enriched in aluminum compared with the base alloys and the increase in aluminum content in the alloy is beneficial. The enrichment of aluminum in the anodized coating becomes more evident with increasing aluminum content in the substrate alloy. It has also been found that the anodized coating formed on Mg–Al alloys in a bath containing aluminates is composed of $MgAl_2O_4$ spinel and that on intermetallic $Mg_{17}Al1_2$ (β-phase) comprised γ-Al_2O_3 and $MgAl_2O_4$ (Bonilla *et al.*, 2002a,b; Clapp *et al.*, 2001; Verdier *et al.*, 2004; Wright *et al.*, 1999).

The difference in substrate can affect the anodizing behavior, which can be directly reflected by an anodizing voltage curve. Ono and co-workers (Ono and Masuko, 2003; Ono *et al.*, 2003) found a difference in anodizing behavior under a potential control mode between AZ31, AZ91 and pure Mg in 3 M KOH, 0.6 M KF and 0.2 M Na_3PO_4 (pH 13) at 29.8 °C; the peak current at 5 V decreased with increasing Al content of the substrate; and the critical breakdown voltage was 60 V for 99.95% Mg, 99.6% Mg and AZ31B, but 70 V for AZ91D. Figure 16.9 shows that an Al-free alloy appears to have a lower anodizing voltage than an Al-containing alloy under a current control mode.

To some degree, even the microstructure of an anodized coating is dependent on the substrate, too. Figure 16.10 shows the different microstructures of the anodized coatings formed on α-matrix (Mg–1Al) and β-phase (Mg–41Al). It seems that the coating on the β-phase is more porous than the one on the α-phase.

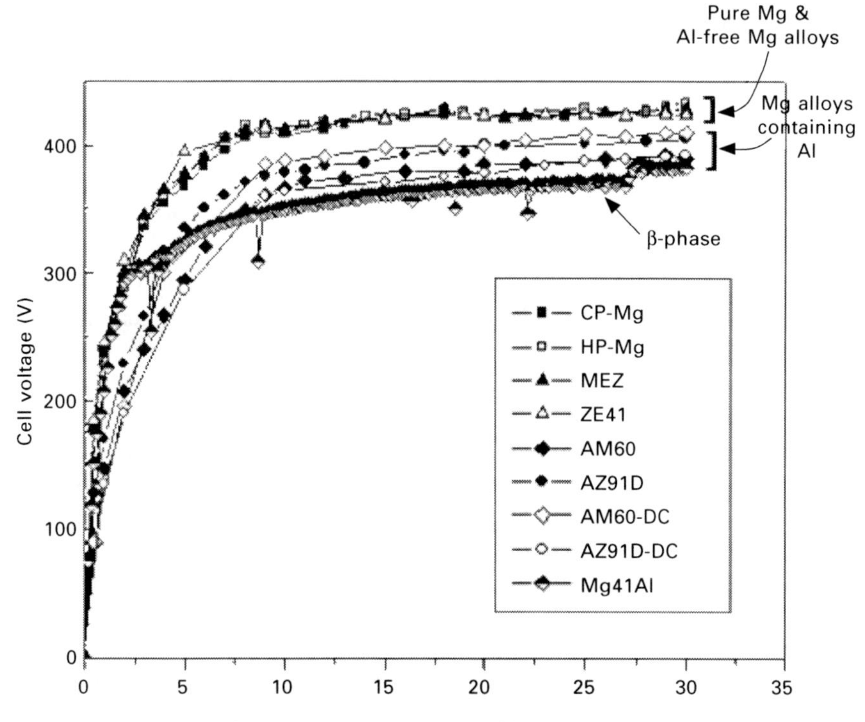

16.9 Anodizing voltage curves for different Mg alloys (Shi *et al.*, 2006c).

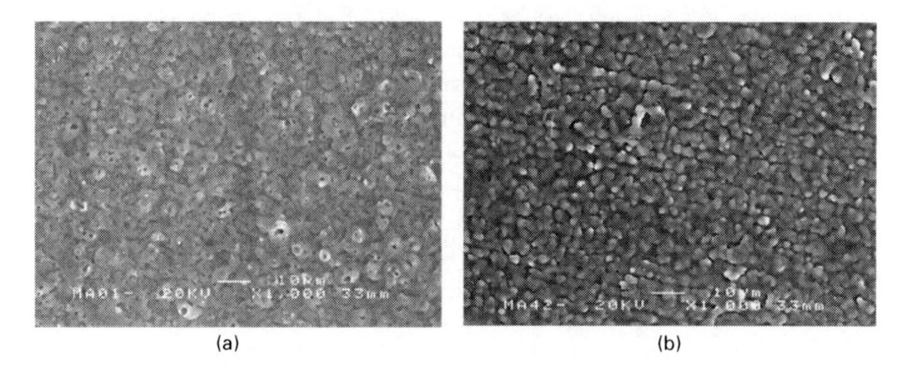

16.10 SEM images of the anodized coatings on (a) Mg–1Al (α-phase) and (b) Mg–41Al (β-phase) (Shi *et al.*, 2005).

16.5 Anodizing mechanism

16.5.1 Anodizing reactions

The formation of an anodized coating on a Mg alloy is a result of electrochemical and chemical reactions between the Mg alloy and an anodizing bath solution at an anodizing voltage (Barton and Johnson, 1995). The anodizing process of a Mg alloy includes complicated reactions between the alloy and the bath electrolyte.

Generally speaking, the electrochemical decomposition reaction of water occurs and oxygen from the anode and hydrogen from the cathode are generated before sparking. The intensity of the electrolysis reaction varies widely, depending on the electrolyte composition. Oxygen evolution from the anode is initially vigorous, then decreases with time as if the magnesium anode were somewhat passivated before sparking occurred (Barton and Johnson, 1995). When the voltage increases above a certain potential (also called initial sparking potential), sparking begins and a complex coating is generated with the arc travelling randomly over the surface. At the same time, oxygen evolution becomes more intense, because the evolution mechanism changes from electrochemical to chemical decomposition of water.

An anodizing process should include at least oxidation of the substrate Mg alloy, oxygen evolution and heat-decomposition of some electrolytes (e.g. aluminate, silicate and borate). The anodized coating on a Mg alloy is a mixture of the products from anodizing reactions, such as oxides and salt deposits of the parent metals and other high temperature decomposed oxide deposits from the solution (Avedesin and Baker, 1999). For example, during a Magoxid anodizing process (Kurze, 1998), at higher potentials (>100 V) the barrier layer is locally destroyed by injection of charged particles. The charge transfer and the diffusion of metal ions cause heat, and the release of energy is high enough to start the plasma chemical processes (gas discharges) in the metal surface/gas/electrolyte interfaces. Discharge channels are formed between the electrolyte/gas film (quasi cathode) interface and the metal surface (anode). This kind of arcing generates plasma similar to ionization of oxygen, melting and oxidizing the metal surface. Simultaneously, areas close to the discharge channels are thermally activated. Hence, on the entire surface a large number of partial anodes form, converting the whole surface into Mg oxide.

To further illustrate the anodizing mechanism and the reactions involved in Mg anodization, an anodizing process of a Mg alloy in a silicate containing bath is analyzed as an example (Shi, 2005; Shi *et al.*, 2003, 2005, 2006a,b,c).

First stage (stage I)

The anodizing voltage increased linearly with time. Dissolution of Mg, formation and deposition of $Mg(OH)_2$ and $MgSiO_3$ and oxygen evolution occur. The possible detailed reactions include:

$$4OH^- \rightarrow 2H_2O + O_2 + 4e^- \qquad 16.1$$

$$Mg \rightarrow Mg^{2+} + 2e^- \qquad 16.2$$

$$Mg^{2+} + 2OH^- \rightarrow Mg(OH)_2 \downarrow \quad 16.3$$

$$Mg^{2+} + SiO_3^{2-} \rightarrow MgSiO_3 \downarrow \quad 16.4$$

The electrochemically evolved oxygen is the main component of the gas from the anode in the alkaline silicate solution. The evolution of oxygen from the anode in anodizing has been reported in previous work (Khaselev and Yahalom, 1998a,b; Khaselev *et al.*, 1999). It is not surprising for oxygen to be electrochemically evolved at such a positive potential from an electrochemical point of view. In practice, the oxygen evolution was not significant in this stage because the overall voltage was mainly distributed in the bath solution and thus only a relatively small potential drop is distributed across the interface of electrode and solution where reaction (16.1) takes place.

Reactions (16.2) and (16.3) have been reported by Mizutani *et al.* (2003) to result in a film formed at a low anodizing voltage in alkaline solution consisting of $Mg(OH)_2$. Mg is first dissolved in solution and then deposited with hydroxyls on the electrode surface to form the primary coating. The most important evidence for reactions (16.3) and (16.4) is the chemical composition of the coating formed in this stage. This is determined by reaction (16.2). The X-ray photoelectron spectroscopy (XPS) results of the coating formed in this stage show that the coating contains $Mg(OH)_2$ and $MgSiO_3$. The postulation of reaction (16.3) is inspired by Khaselev's work (Khaselev *et al.*, 2001) in which $MgAl_2O_4$ was found to be the composition of the anodized coating in an aluminate containing solution. Most of the anodizing current in this stage is consumed by the reaction (16.2) because reaction (16.1) is insignificant. The formed coating in this stage is thin because the anodizing time is only a few seconds.

Second stage (stage II)

The anodizing voltage further increases to about 190 V. Within this potential range, evolution of some tiny gas bubbles was observed. There was no sparking. It is postulated that in addition to anodizing reactions (16.1), (16.2), (16.3) and (16.4), another reaction starts to take place:

$$Mg + 2OH^- \rightarrow MgO \downarrow + H_2O + 2e^- \quad 16.5$$

This is an electrochemical reaction resulting in coating formation directly. Mg is directly electrochemically oxidized into MgO.

Mizutani *et al.* (2003) reported that the anodized coating on Mg alloys formed at 60 V in an alkaline solution contained a mixture of magnesium hydroxide and magnesium oxide which has been confirmed by X-ray diffraction (XRD) analysis. Therefore, the involvement of reaction (16.5) in this stage is reasonable. The reasonability of reactions (16.3), (16.4) and (16.5) involved in this stage of anodizing can also be supported by experimental results. For reaction (16.4), magnesium silicate has been found by XPS analysis in the anodized coating. Moreover, the limited oxygen evolution observed in this stage indicates that reaction (16.1) is still slow in the second stage of the anodization. The high-resolution XPS results imply $Mg(OH)_2$, MgO and $MgSiO_3$ in this stage.

Third stage (stage III)

The anodizing voltage continues to increase to about 330 V. The most significant phenomena in this stage are uniform sparking and significant gas evolution from the anodized specimen surface. It is postulated that reactions (16.1), (16.2), (16.3), (16.4) and (16.5) continue in this stage and at the same time the following new reactions also occur:

$$Mg(OH)_2 \rightarrow MgO \downarrow + H_2O \tag{16.6}$$

$$2Mg + O_2 \rightarrow 2MgO \downarrow \tag{16.7}$$

In other words, the new reactions are mainly chemical precipitation processes. Reaction (16.6) is a dehydration process that can occur when temperature is higher than 350 °C (Weast, 1976–1977). It has been reported (Yahalom and Zahavi, 1971) that the temperature of the sparking arc during the anodization of Mg alloys was far above 1000 °C. Because of the sparking in this stage, some local areas where sparking occurs can easily exceed this dehydration temperature. However, not all the $Mg(OH)_2$ was changed into MgO. High-resolution XPS results show that there is still some $Mg(OH)_2$ in the coating. At the high temperature, direct oxidation of magnesium is also possible, particularly with the active oxygen freshly generated by reaction (16.1). Therefore, reaction (16.7) is proposed in this stage of anodizing, and this is proved by the increased amount of MgO in the coating in this stage.

Fourth stage (stage IV)

After the cell voltage is over 330 V, oxygen evolution becomes vigorous, and sparking much more intense and localized. It is proposed that the following two reactions are also involved in the fourth stage of anodizing in addition to the reactions listed above.

$$MgO + yMgSiO_3 \rightarrow (MgO)_x{\cdot}(SiO_2)_y \tag{16.8}$$

$$2H_2O \rightarrow 2H_2 \uparrow + O_2 \uparrow \tag{16.9}$$

Reaction (16.8) is supposed to be a melting and solidification process of MgO and $MgSiO_3$ in the coating in the sparking spots. In fact, at high temperatures during sparking, the anodized coating could be locally melted by sparking spots and mixed together with some components or decomposed products of the bath solution, then rapidly solidified to reform coating. The melting points of MgO and $MgSiO_3$ are 2852 and 1910 °C, respectively (Weast, 1976–1977). Under a very intense sparking condition, some sparking arcs may generate sufficient heat to melt coating and the components of the bath solution in some particularly confined local sparking areas. The sparking cannot last long. It stops shortly, and thus the melted coating is rapidly cooled by the surrounding solution. In this way, the coating deposited in the first three stages is locally melted by sparking to form a thick and coarse coating. The significant increase in the ratio of Si/Mg in the coating formed in stage IV can be interpreted as the significant deposition of the silicates from the bath solution directly into the coating through the melting and solidification process.

The dramatically increased gas evolution in this stage cannot be simply ascribed to the electrochemical oxygen evolution. Thermal decomposition of water, reaction (16.9), should be considered. The normal thermal decomposition temperature of water was reported to be over 2000 °C (Brown *et al.*, 2002). Since the temperature of the sparking arc during the anodization of magnesium alloys was reported to be far above 1000 °C (Yahalom and Zahavi, 1971), the possibility exists that in some particular areas where the sparking is extremely intense, the local temperature becomes high enough to decompose water. This postulation is further verified by the anodizing current efficiency calculated below.

Based on the above model, the changes in the anodizing voltage, sparking, oxygen evolution, thickness, composition and microstructure of the coating will be explained as follows.

16.5.2 Anodizing voltage and coating growth

In the initial stage, the coating resistance increases with the formation and thickening of the anodized coating. The coating formation reactions (16.3) and (16.4) determined by reaction (16.2) should be proportional to the anodizing current density. Hence, at a constant anodizing current density, the film thickness increases linearly with time, and the increasing rate should be proportional to the anodizing current density.

In the second stage, the coating resistance continues to increase with the coating thickening. Therefore, the anodizing voltage also increases quickly in this stage. The thickness of the anodized coating continues to increase with the deposition of $Mg(OH)_2$, $MgSiO_3$ and MgO from reactions (16.3), (16.4) and (16.5). As pores start to appear in the coating, the coating becomes less dense than in the first stage. Therefore, the increase of anodizing voltage starts to deviate from the linear increasing tendency and becomes relatively slower.

In the third stage, as the voltage is high enough to cause dielectric breakdown of the surface coating and sparking starts, more reactions occur than in the second stage. Due to the local high temperature, all the electrochemical reactions (16.1), (16.2) and (16.5) involved in this stage become relatively easy. Meanwhile, because of the sparking resulting in porosity of the formed coating, the coating resistance cannot increase significantly, even though the thickness of the coating keeps increasing. Therefore, an even slower increasing rate of the anodizing potential can be observed in this stage. Since sparking results in significantly increased deposition of Mg oxides, film thickening is evident in this stage. Dramatic potential oscillation always accompanies sparking behavior in this stage. For example, during sparking the potential oscillating randomly in the range of 60 to 90 V is often seen. This can be explained by the film breakdown and repair during sparking. Wright *et al.* (1999) anodized pure magnesium and some alloys of magnesium in alkaline solutions, using constant applied current, and found that the potential increased linearly with time to about 70 V as a thin barrier film grew on the metal surface. The electric field in the barrier film was found to be 9×10^8 V/m and there was no variation of field with the applied current. When the barrier film reached a critical thickness of around 80 nm, the observed potential dropped abruptly to a relatively low value of about 10 V. The potential then began to rise again linearly, until another sudden drop was observed. This pattern continued for a number of cycles, generating a sawtooth pattern of potential as a function of time. The abrupt drop in potential is due to the rupture of the barrier film, forming a porous secondary layer. This rupture is caused by the tensile stresses in the barrier film, which has a molar volume much smaller than the metal from which it is formed.

When anodizing develops into the fourth stage, due to the intense sparking, relatively coarser pores are formed in the coating, which significantly decreases the coating resistance. So the anodizing voltage does not increase with time. The localized intensive sparking can cause rapid deposition of the coating in the local area. This significantly roughens the film and also leads to a thicker coating.

16.5.3 Sparking and microstructure

Although the bulk electrolyte temperature might be only 50 °C, the local temperature in the plasma zone would probably be in excess of 1000 °C, which can lead to formation of 'glassy' or "ceramic" anodic coatings. The local high temperature can speed up reactions, such as deposition of the coating, gas evolution and oxidation of Mg alloy. It can also cause high-temperature decomposition of the components in the bath solution.

The high local temperature in an anodic film is a result of an extremely high current density passing through the pores or defects in the film. These elevated temperatures cause vaporization of the electrolyte and plasma formation in localized areas on the surface. The plasma or sparking in return produces even more heat and much higher temperatures locally. Nykyforchyn *et al.* (2003) investigated the plasma formation. Several hundred volts was applied to a metal–electrolyte system, resulting in electric discharges. Consequently, plasma develops in discharge channels and oxide deposition occurs on the metal surface. The density of the plasma electrons was determined to be $n_e \approx 10^{22}$ m^{-3} and the plasma electron temperature during the synthesis of oxides in the electrolyte plasma of spark discharges was $T_e \approx 10^4$K. The calculation of temperature distribution in the spark zone suggests that the temperature in centre of the spark zone can be higher than the melting point of $MgAl_2O_4$; that should be a reason for the porous morphology of anodic films formed under continuous sparking (Khaselev *et al.*, 2001).

Sparking is normally involved in an anodizing process at high voltages. Magnesium is observed to exhibit spark discharges that move randomly over the surface. The sparking could start from some 'weak' sites of the film formed in the second stage, such as thin areas or areas with defects or pores. At a sparking spot, the sparking has two consequences. First, the dielectric breakdown during sparking results in pores. The direct evidence for this is the porous morphology of the formed coatings in this stage. As the sparking becomes more intense with increasing cell voltage, the pore size and the porosity of the coating increase. Second, the high local temperature resulting from sparking can to some extent consolidate the coating and seal the pores caused by dielectric breakdown of the coating. The heat generated by sparking could melt the coating surrounding the sparking spots. After the sparking stops at those spots, the melted coating rapidly solidifies there. This leads to the overlapped pearl-like morphology surrounding the pores and to some extent sealing the porosity of the coating. Therefore, it is difficult for sparking to reoccur in a sparked spot. It always travels or occurs randomly over a Mg surface. In some cases, after intense sparking causes too large a pore or damaged area in an anodized coating, sparking and anodizing will stop, because the large pore or damage area will allow the passage of all the anodizing current through (all the anodizing current will be leaking through there) and thus a sufficiently high potential drop across the interface between the specimen and solution cannot be sustained for sparking. This is why the anodizing voltage sometimes suddenly drops to a very low value and sparking stops after a couple of blinding flashes on the specimen surface in practical anodizing.

Since the porosity of an anodized coating formed at high voltage results from sparking, the intensity of the sparking process can to a great extent determine the coating porosity. If the intensity of sparking is low, the anodized coating is formed slowly. Correspondingly, the sparking area is small and the breakdown and melting reaction are not severe. As a result, the coating microstructure becomes fine and compact because the pores formed during the sparking process were small. On the other hand, more intense sparking can lead to a thicker, coarser and more porous anodized coating. All of these expectations have been proved by SEM observation of the microstructure.

16.5.4 Oxygen evolution and anodizing efficiency

Oxygen evolution is found to occur after the cell voltage exceeds 3 V (Shi, 2005; Shi *et al.*, 2005, 2006a,b), and its rate increases with voltage. The trend of oxygen evolution dramatically accelerates after the anodizing develops into sparking stage, indicating that different reactions are responsible for the evolved gas before and after sparking. Before sparking, the evolved oxygen simply results from electrochemical decomposition of water (16.1). However, in the sparking stage, thermal decomposition of water (reaction (16.9)), is the major reaction responsible for the gas evolution. The gas evolution caused by the thermal decomposition of water is dependent on the heat or temperature. At a high temperature, the rate of the gas evolution can be very high.

The current efficiency can be defined by:

$$\eta = \frac{I_F}{I_T} \times 100\% \qquad 16.10$$

where I_T is the total applied current and I_F is that part of the total current resulting in film formation. If it is assumed that the by-product forming reaction is simply the production of oxygen by the reaction (16.1), then I_F may be estimated from:

$$I_F = I_T - I_{O_2} \qquad 16.11$$

where I_{O_2} is the rate of electrochemical evolution of oxygen (reaction (16.1)). Hence,

$$\eta = \frac{I_T - I_{O_2}}{I_T} \times 100\% \qquad 16.12$$

Theoretically, I_{O_2}, the electrochemical evolution of oxygen, cannot be greater than I_T. If the anodizing efficiency is calculated based on an assumption that the evolved gas is electrochemically evolved oxygen, a negative efficiency could be obtained. In other words, a negative η means a non-electrochemical evolution of gas involved in the anodizing process.

Figure 16.2 shows that the anodizing efficiency decreases with the increasing cell voltage, particularly in the fourth stage down to a negative value. It confirms that in the fourth stage (an intense sparking stage), a non-electrochemical gas evolution mechanism starts to operate. This non-electrochemical process is the thermal decomposition of water due to the intense sparking in this stage.

16.6 Corrosion of anodized magnesium (Mg) alloys

16.6.1 Corrosion performance

An anodized Mg alloy is much more corrosion resistant than an un-anodized one. A commercially anodized Mg alloy without any post-treatments or sealing can survive in standard salt spray (ASTM B117) for about 400–600 hours. Figure 16.11 presents the corrosion performance of a recently developed anodized coating (Song, 2004), which shows that the anodized AZ91D survives in 5 wt% NaCl environments for one month and no obvious corrosion damage can be visualized on the surfaces.

The corrosion of an anodized coating usually starts from tiny pitting corrosion and then progresses into localized or filiform corrosion. For example, in a 5% NaCl solution, specks and pits can be seen on the surface of an anodized AZ91D specimen after 16 hours of immersion; after 88 hours, the specks become larger and develop into localized corrosion and filiform corrosion (Shi *et al.*, 2006b). According to the corrosion morphologies of anodized AZ91D in 5 wt% NaCl, the corrosion resistance of the anodized coating formed at stage IV is better than that formed at stage III, which can mainly be ascribed to the larger thickness of the coating formed at stage IV.

The corrosion performance of an anodized magnesium alloy depends on the corrosion performance of the substrate alloy. For example, the corroded areas of the anodized commercial alloys are measured and plotted versus the corrosion rates of these alloys in Fig. 16.12. There is a good correlationship in corrosion damage degree between the anodized specimens and the corresponding un-anodized alloy.

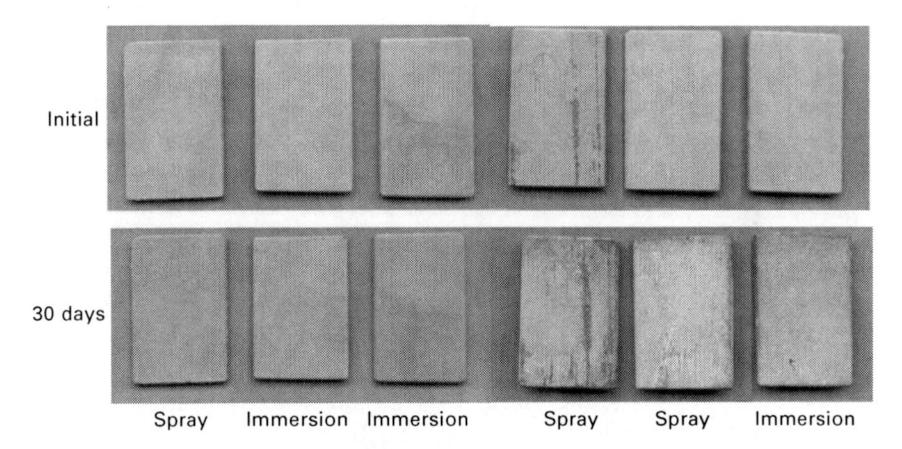

16.11 Corrosion damage of anodized AZ91D (Song, 2004) after 5 wt% NaCl immersion and salt spray.

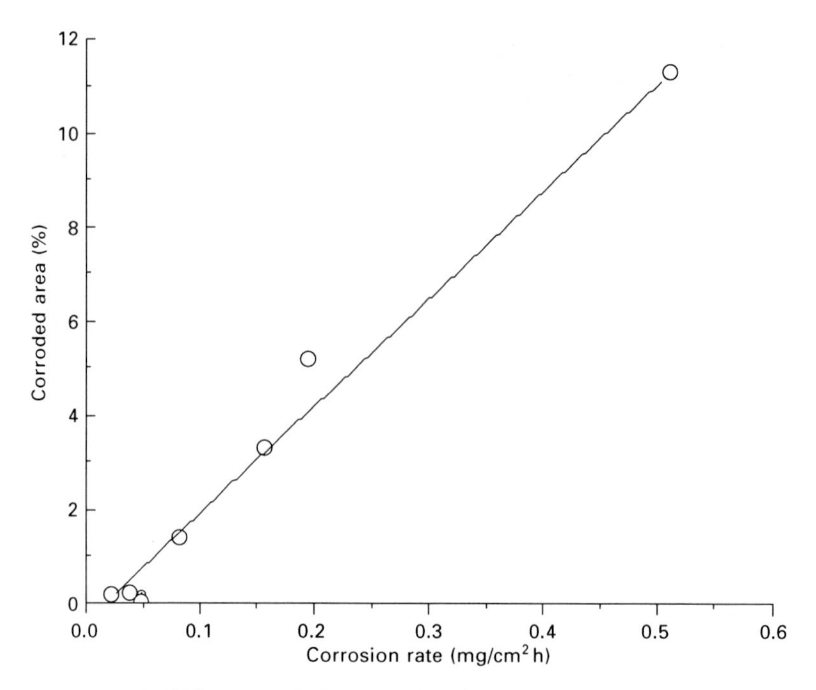

16.12 The corroded area ratio of the anodized specimens vs. the corrosion rate of the substrate alloy under 5 wt% NaCl immersion condition (Shi, 2005).

16.6.2 Protectiveness of an anodized coating

The penetration of corrosive species through the anodized coating is a critical step to the corrosion performance of an anodized Mg alloy if the coating is relatively compact. The corrosion resistance of a coated electrode should mainly be determined by the resistance of the coating to the corrosive species penetration. However, for a porous anodized coating on a magnesium alloy (Blawert *et al.*, 2006; Shi *et al.*, 2005, 2006a,b,c), the ingress of aggressive ions through the pores or defects in the coating to the substrate/coating interface would not be too difficult. In this case, both the resistance of the coating to the ingress of corrosive species and the resistance of the substrate/coating interface to corrosion would be responsible for the corrosion of the anodized alloy. The main role of an anodized coating is to retard the ingress of corrosive species and delay the corrosion of magnesium alloys, which is termed as 'retarding effect' in this study.

In an anodized coating, not all the pores are through from the top surface of the coating to the substrate/coating interface. The cross-sections of the anodized coatings (Figs 16.5 and 16.8) indicate that the possibility of through-pores is in effect not very high. Although the porosity of an anodized coating appears to be high, most of the pores are discontinuous and the number of the through pores is very small. There may be a dense inner layer in an anodized coating next to the substrate (Blawert *et al.*, 2006), but this layer could be too thin to be observed in this study, and it could be in nature similar to the surface film spontaneously formed on a Mg alloy, which might be more conveniently considered as a surface layer of the substrate, rather than a layer in the anodized coating. In the event that a corrosive solution finally reaches the substrate/coating interface through the small number of the through pores, the actual total area of the substrate exposed to the aggressive solution at the bottom of the through pores is very limited. Therefore, an anodized coating can effectively restrict the area of the substrate Mg alloy exposed to corrosive solutions, which is called the 'blocking effect'.

Apart from retarding the ingress of corrosive species and blocking the exposed substrate of the alloy to solution, the most important effect of an anodized coating is suppressing the substrate surface defects or active points that are susceptible to corrosion. It is well known that the corrosion of metals normally starts from some active or defective sites. For Mg alloys, the active sites can be grain boundaries or grain central areas (Song, 2005a) depending on their alloying elements. During anodizing, these active areas are likely anodized preferentially. If there is an anodized coating on a Mg alloy, the chance of the active sites to be right at the bottom of the through pores will be extremely low. In other words, even after corrosive solution has arrived at the substrate/coating interface underneath a through-pore, it is unlikely that the substrate at the bottom of the through pore happens to be an active or defective site. By this mechanism, the corrosion resistance of the substrate alloy at the bottom of the through pore should be much higher than the alloy without an anodized coating. Therefore, the presence of an anodized coating on a magnesium alloy will eliminate or suppress most active or defective sites. This is a kind of 'passivating effect'.

In summary, it is proposed that anodizing can improve the corrosion resistance of a magnesium alloy through three different mechanisms: (1) retarding effect, i.e. slowing down the ingress of corrosive ions, (2) blocking effect, reducing the possibility or the area of magnesium alloys directly exposed to corrosive media, and (3) passivating effect, i.e. suppressing the active sites for corrosion.

16.6.3 Corrosion model of an anodized Mg alloy

The ingress of corrosive ions into an anodized coating through its pores is not a slow process for a porous anodized coating. After corrosive ions reach a critical concentration threshold at the film/substrate interface and trigger the corrosion of the substrate Mg alloy, the corrosion of the anodized magnesium starts. Mg is dissolved into the solution in the pores, which gradually makes the solution saturated with $Mg(OH)_2$ due to the dissolved Mg^{2+}. Thus, the substrate magnesium alloy is actually mostly exposed to a $Mg(OH)_2$ saturated solution in the through pores, and the corrosion of an anodized Mg alloy in this sense can be simply regarded as the corrosion of a very small surface area of the alloy in a $Mg(OH)_2$ saturated corrosive solution.

After corrosion starts at the bottom of the through pores of an anodized coating, the corrosion will further penetrate and spread out from there, leading to collapse of the adjacent coating due to the expansion of the corrosion products. In this stage, the corrosion of an anodized coating covered substrate is in nature similar to the corrosion of the substrate alloy un-anodized.

Based on the above analyses, the typical microstructure and corrosion initiation of an anodized coating can be schematically illustrated as in Fig. 16.13.

16.6.4 Explanation of corrosion behavior

According to the corrosion mechanism, a porous anodized coating can only slow the ingress of aggressive species, reduce the areas of the substrate exposed to a corrosive solution and suppress the defective sites of the substrate for corrosion. All these are dependent on the porosity or more specifically the number of the through-pores in the anodized coating. It implies that a porous anodized coating itself may only slow down the corrosion and mitigate the corrosion damage, but not prevent the corrosion of the anodized Mg alloy. A decrease in porosity or the number of the through-pores of an anodized coating is the most effective way to improve its corrosion resistance. Sealing and top-coating with a primer or paint are effective ways of reducing the porosity or the number of the through-pores. It has also been found that the applied anodizing current density can significantly alter the porosity of an anodized coating and thus the corrosion resistance of an anodized Mg alloy (Shi *et al.*, 2006b).

The thickness also affects the penetration of corrosive solution due to its retarding effect and blocking effect. The ingress of corrosive solution into a thicker anodized coating is more difficult or slower than into a thinner coating. Moreover, the thickness of an anodized coating can affect the number of the through-pores; a pore is more likely blocked or becomes discontinuous somewhere in a thick coating than in a thin one. Therefore, a thick anodized coating is usually more corrosion resistant than a thin one. Therefore, when the Anomag coating was found to be thinner than Tagnite under a certain anodizing condition (see Fig. 16.8), it displayed relatively lower corrosion resistance under atmospheric exposure conditions (Song *et al.*, 2006).

Another implication of this corrosion model is the influence of the substrate on the corrosion resistance of an anodized Mg alloy. According to the corrosion model, after the corrosive solution reaches the film/Mg

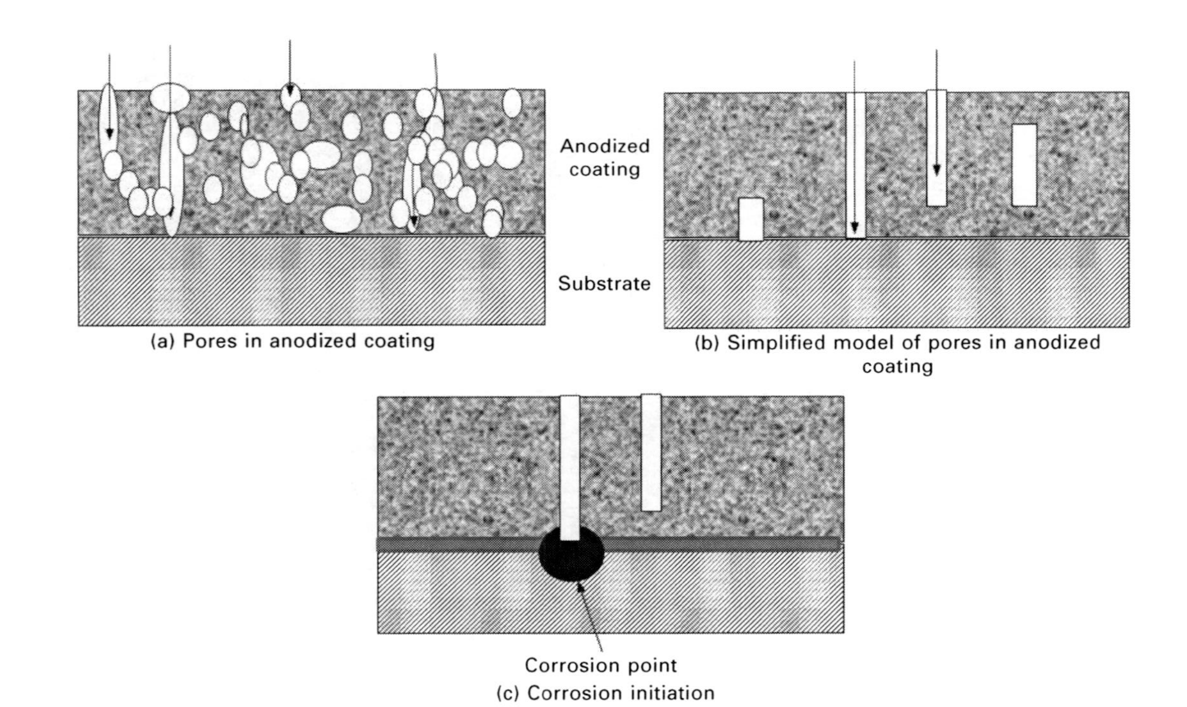

16.13 Schematic diagrams of (a) the microstructure of an anodized coating on the magnesium alloys (based on Shi *et al.*, 2005), (b) the simplified microstructure of the coating (based on Shi *et al.*, 2005) and (c) the corrosion of anodized magnesium alloy.

interface, the corrosion resistance of the substrate will directly influence the corrosion behavior of the anodized Mg alloy. As mentioned earlier, the porous anodized coating cannot very effectively retard and block the ingress of corrosive species. The corrosion performance of an anodized Mg alloy should be to a great extent determined by the passivating effect, which is closely related to the intrinsic corrosion resistance of the Mg alloy. Therefore, the corrosion of anodized Mg alloys can be significantly affected by the corrosion resistance of these alloys before anodizing (see Fig. 16.12). It has been found that the corrosion resistance of anodized AZ91 was better than that of anodized AZ31 (Kotler *et al.*, 1977), because AZ91 alloy in general is more corrosion resistant than AZ31.

Strictly speaking, the surface of a magnesium alloy is not uniform. The anodized coating in different areas of the surface could be different in property. Therefore, the corrosion performance of an anodized Mg alloy can have different corrosion resistance in different areas. For ZE41, the grain boundaries are less corrosion resistant than the grain central areas before anodizing (Fig. 16.14(a)). In other worlds, the defective sites of the alloy are mainly distributed along the grain boundaries. After anodizing, if there happens to be some through-pores in the anodized coating formed over the grain boundaries, then the anodized alloy will be preferentially corroded there. That is why the anodized ZE41 still preferentially corroded along its grain boundaries and has corrosion morphology in microstructure similar to the un-anodized ZE41 (Fig. 16.14(b)). Along some of the grain boundaries or the secondary phases, corrosion penetrates deeper into the substrate.

As the coatings on α and β-phases are porous (see Fig. 16.10), once the corrosive media penetrate the coating through the pores or the broken area and arrive at the substrate, the galvanic effect between the α and β-phases will

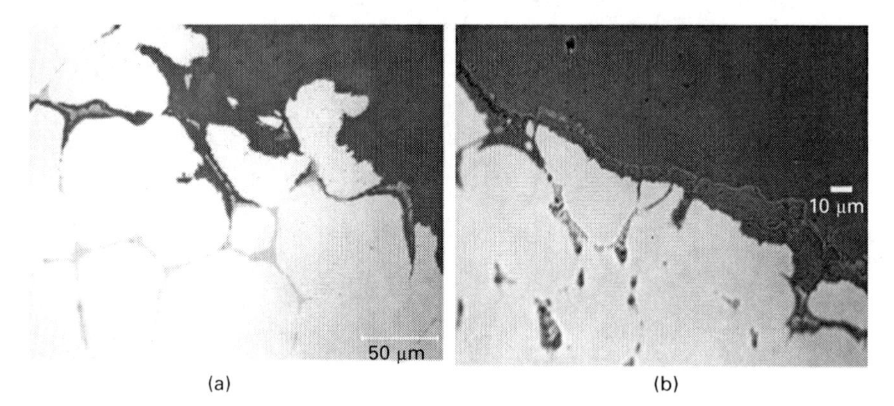

16.14 Optical photo of the cross-sections of (a) un-anodized ZE41 after 75 hours of immersion in 5 wt% NaCl solution and (b) Anomag anodized ZE41 after 20 hours in $Mg(OH)_2$ saturated 0.1 wt% NaCl solution.

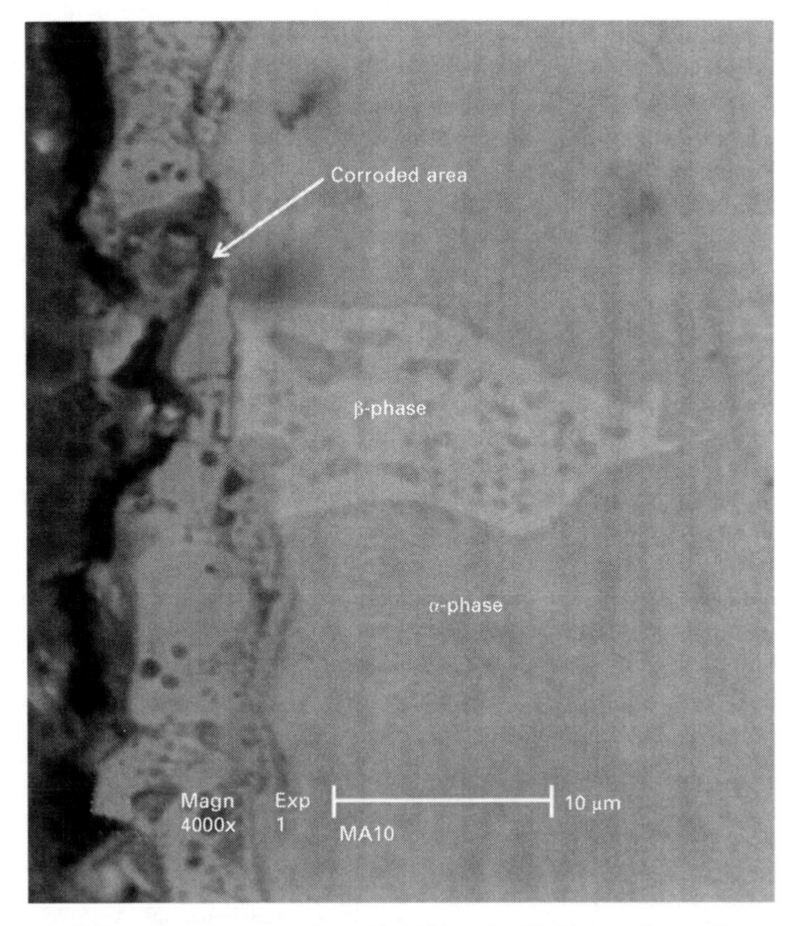

16.15 Corrosion damage of anodized Mg–10wt% Al two-phase alloy immersed in 5 wt% NaCl solution (Shi *et al.*, 2005).

govern the corrosion of the anodized Mg alloy, although the galvanic effect is to a certain extent mitigated by the coating. In this sense, the corrosion mechanisms of the anodized and non-anodized Mg–Al alloys are similar, governed by the galvanic current density between the α and β-phases, and the distribution of the β-phase in a substrate Mg alloy can affect the corrosion of the anodized alloy. The most likely or serious corrosion zone should be in the α matrix adjacent to β phase, which has been experimentally observed (see Fig. 16.15).

16.6.5 Electrochemical behavior

An un-anodized Mg alloy and its anodized counterpart usually display similar polarization curves in the same (Cl^- containing) corrosive solution, except that the latter has lower polarization current densities and a more positive breakdown potential (Shi *et al.*, 2005) than the former. An example is provided in Fig. 16.16.

According to the corrosion model proposed above, the anodic polarization of the anodized ZE41 should be mainly the dissolution of the substrate at the bottom of the through-pores in the anodized coating in response to an applied potential. The different passive current densities and pitting corrosion potentials of anodized ZE41 from unanodized ZE41 can be ascribed to the blocking effect and passivating effect of the anodized coating on ZE41. Because of the blocking effect, the anodic dissolution of an anodized Mg alloy can occur only in a very small area of the substrate at the bottom of the through-pores, so the passive current density is dramatically limited. At the same time, the passivating effect significantly reduces the number of active sites and thus leads to a lower passive current density. Therefore, the anodic or passive current density of an anodized Mg alloy is much lower than that of an un-anodized Mg alloy. Since passive sites normally have more positive pitting potentials than defective sites, the onset of pitting corrosion of an anodized alloy would be more difficult. In addition, at the pitting potential, the ohm potential (IR) drop caused by the anodic dissolution current or passive current flowing through the anodized coating is significant, which is added to the measured pitting potential. Therefore, the breakdown potential of an anodized Mg alloy appears to be more positive than that of an un-anodized Mg alloy.

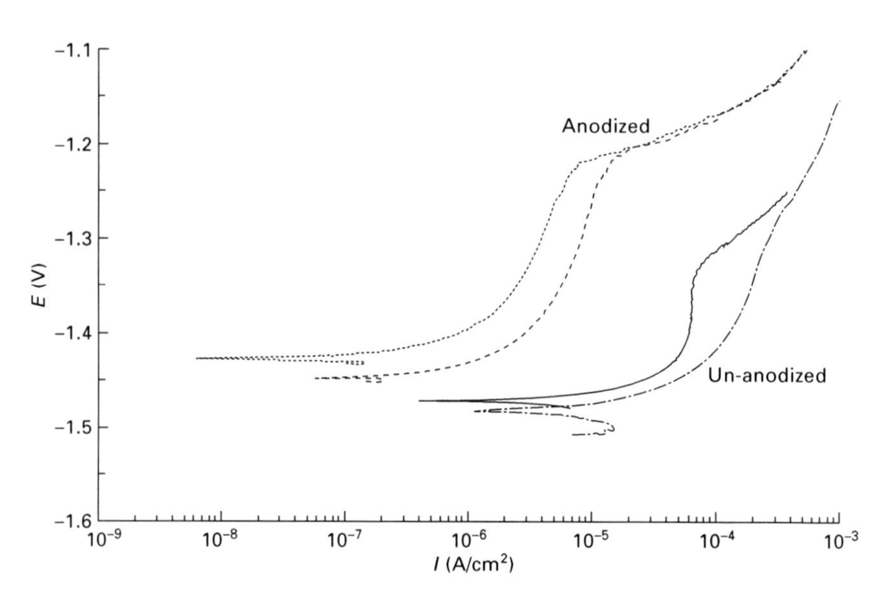

16.16 Polarization curves for un-anodized ZE41 and anodized ZE41 in $Mg(OH)_2$ saturated 0.1wt% NaCl solution (Song *et al.*, 2006).

The presence of an anodized coating on a Mg alloy can significantly alter the AC impedance behaviour of this alloy. For example, un-anodized ZE41 in $Mg(OH)_2$ saturated 0.1 wt% NaCl solution displays an AC impedance spectrum (EIS) containing two capacitive loops in the high- and intermediate-frequency ranges and an inductive loop in the low-frequency range (Fig. 16.17(a)). Mg and AZ91D have also similar AC impedance spectra in a chloride-containing solution (Song *et al.*, 1997, 2004). For an anodized ZE41 in the same corrosive solution basically, only one much larger capacitive loop in the high and intermediate frequency ranges and some inductive characteristic in the low-frequency range can be detected.

The similarity of un-anodized ZE41 to Mg (Song *et al.*, 1997) and AZ91D (Song *et al.*, 2004) in EIS suggests that the corrosion of ZE41 and AZ91D follows the same corrosion mechanism: the Mg alloy surface has a spontaneously formed surface film, but it has some broken areas; Mg is oxidized into Mg^+ and dissolved into solution from the surface film broken

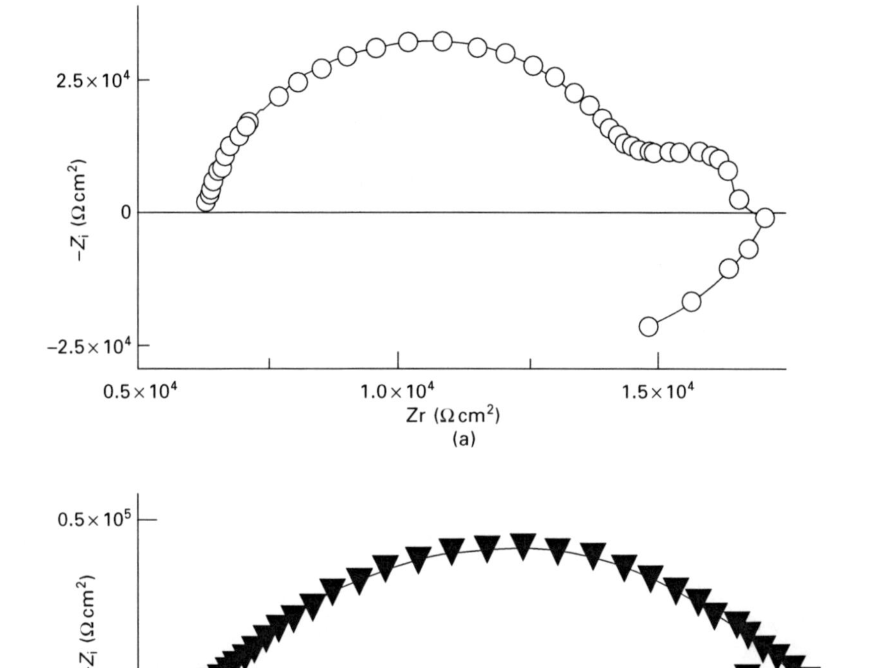

16.17 A typical AC impedance spectrum of (a) un-anodized and (b) anodized ZE41 immersed in $Mg(OH)_2$ saturated 0.1 wt% NaCl solution.

areas; Mg^+ further reacts with water to be oxidized into Mg^{2+}. Based on this understanding, the electrochemical interpretation of the EISs of Mg and its alloys was given by Song *et al.* (1997). The first capacitive loop in the high-frequency range can be ascribed to the charge transfer resistance R_t and the capacitance $C_{s/m}$ at the solution/Mg interface. The second capacitive loop in the mediate frequency range is attributed to the involvement of monovalent Mg^+ ions in the anodic dissolution at surface film broken areas. The inductive loop in the low frequency range is a result of the response of broken areas of the surface film to the applied potential or Faraday current density.

For an anodized Mg alloy with a porous coating formed on its surface, according to the coating model (Fig. 16.13) proposed earlier, the anodized coating cannot change the electrochemical reactions involved in the corrosion of the alloy, but it can significantly reduce the rates of the reactions. The porosity of the anodized coating allows the penetration of aggressive solution through the coating and the arrival of the solution at the substrate surface. After this film breaks down by the corrosive solution in the through-pores, Mg is first oxidized into Mg^+ from the substrate Mg alloy and further reacts with water to form Mg^{2+}. All these electrochemical reactions are the same as those on an unanodized Mg alloy. However, due to the blocking and passivating effects, the charge transfer step is significantly slowed down. Based on this, a corroding anodized Mg alloy can be depicted by an equivalent circuit as shown in Fig. 16.18. In the equivalent circuit, R_s is solution resistance between the reference electrode and Mg alloy specimen; $C_{s/m}$ is the capacitance between solution and magnesium substrate, which is equal to the coating capacitance for an anodized magnesium alloy; $R_{coating}$ is resistance of the anodized coating and its value is dependent on the coating thickness and porosity, particularly the number of through-pores; R_t is the charge transfer resistance at the bottom of the through-pores of the anodized coating; C_{Mg^+} and R_{Mg^+} are pseudo-capacitance and resistance caused by the involvement of Mg^+ in the corrosion (Song *et al.*, 1997); R_f and L_f are also

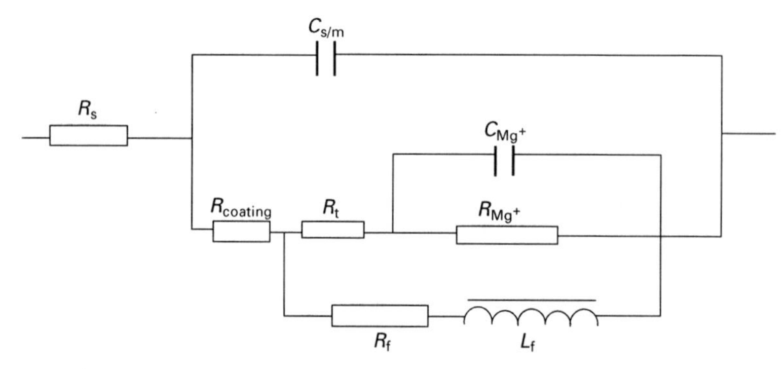

16.18 Equivalent circuit for corroding anodized Mg alloy.

pseudo-resistance and inductance resulting from breakdown of the surface film during corrosion (Song *et al.*, 1997).

Theoretically, a corroding anodized Mg alloy system with this equivalent circuit should have three time constants. On the Nyquist plane, there should be two capacitive loops in the high- and intermediate-frequency ranges and an inductive loop in the low-frequency range. In practice, the values of the components in the equivalent circuit may not be in suitable ranges. For example, when $R_{coating}$ is very large and $C_{s/m}$ is very small, the capacitive characteristic in the middle frequency range associated with the involvement of Mg^+ in corrosion can be overwhelmed by the coating related capacitive loop, and hence the displayed EIS on the Nyquist plane does not always have three clear loops, but simply one apparent capacitive loop in the high- and intermediate-frequency ranges and an inductive loop in the low-frequency range.

The reasonability of the equivalent circuit for a corroding anodized Mg alloy can be verified by regressing it into a simple one for an un-anodized Mg alloy. Simply let $R_{coating} = 0$ and slight modify the values of other equivalent components; the equivalent circuit will represent the electrochemical processes of a corroding Mg alloy. As there is no coating between the substrate and solution, $C_{s/m}$ becomes larger. For a film-free substrate with more defective sites, R_t should be considerably smaller. Certainly, the values of the other corresponding equivalent components C_{Mg^+}, R_{Mg^+}, R_f and L_f for the unanodized alloy will be different from those of its anodized counterpart. After these modifications, the equivalent circuit produces an EIS spectrum with two capacitive loops in the high- and intermediate-frequency ranges and an inductive loop in the low-frequency range.

16.6.6 Evaluation of corrosion resistance

The corrosion resistance of most commercial anodized coatings is normally measured and compared according to salt spray test (ASTM B-117). In fundamental studies, the immersion corrosion test is also used to assess the corrosion performance of anodized coatings. The corrosion resistance of anodized Mg alloys is compared according to their corrosion damage degrees (corrosion morphologies) after the tests. For example, Fig. 16.19 shows the corrosion damage of anodized Mg and its alloys after immersion and salt spray, based on which it can be concluded that the anodized AZ91D is much better than the anodized Mg.

However, both salt spray and immersion tests are time consuming and only provide the final corrosion damage information. A rapid method of evaluating corrosion performance of anodized coatings for Mg alloys is needed and electrochemical technique can serve this purpose.

The electrochemical behavior of anodized Mg alloys has been measured

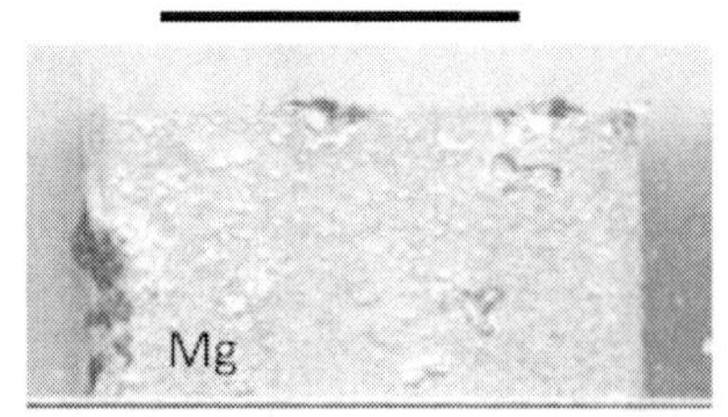

(a) Anodized Mg immersed in 5 wt% NaCl for 23 hours

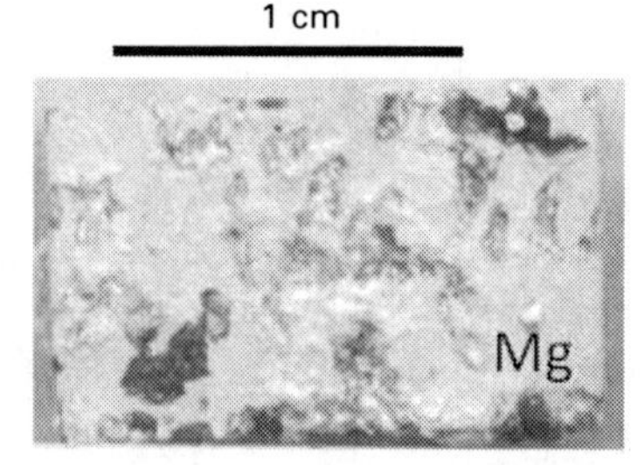

(d) Anodized Mg exposed to 5 wt% NaCl salt spray for 23 hours

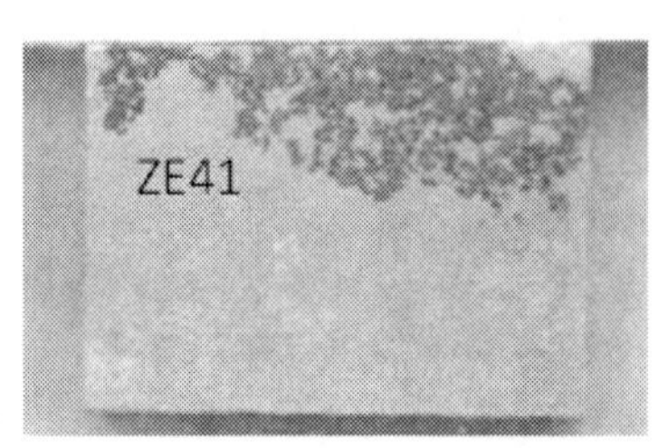

(b) Anodized ZE41 immersed in 5 wt% NaCl for 48 hours

(e) Anodized ZE41 exposed to 5 wt% NaCl salt spray for 87 hours

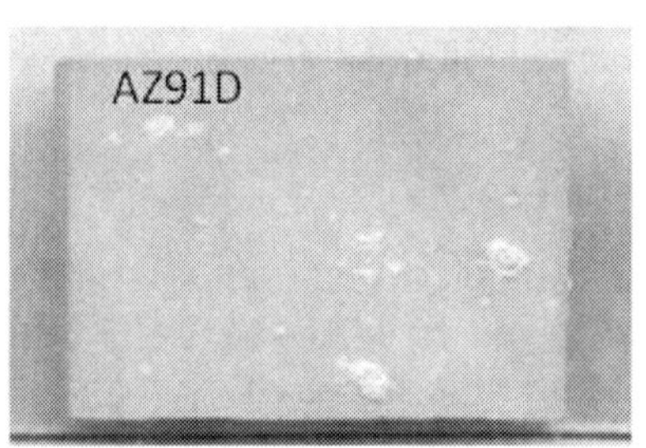

(c) Anodized AZ91D immersed in 5 wt% NaCl for 48 hours

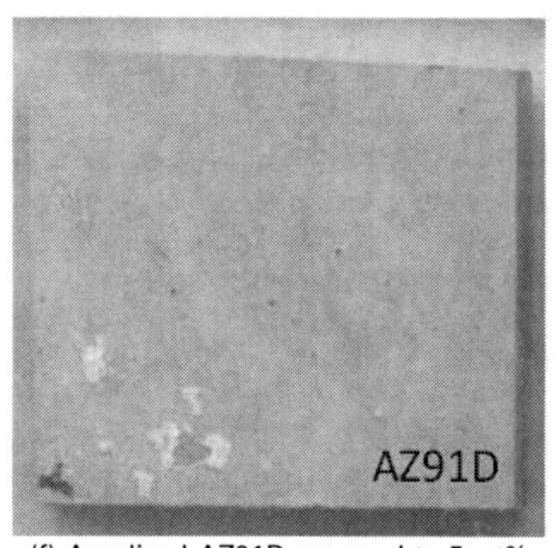

(f) Anodized AZ91D exposed to 5 wt% NaCl salt spray for 87 hours

16.19 Appearance of anodized Mg and alloys after salt immersion and salt spray tests (based on Shi *et al.*, 2005).

using cyclic voltammogram (CV) and following the EIS techniques. Anodized AZ91 and AM50 after exposure in salt spray for 21 days displayed different corrosion performance on their two side surfaces. The anodized coating on side B of the specimen was better than side A. There were a few corroded pits on side B, while side A was intact; side B was still under the protection of the anodized coating while side A suffered from pitting or filiform corrosion. Figure 16.20 shows the CV polarization curves of anodized AM50 measured in 5% NaCl solution after 21 days of salt spray testing.

The forward scanned curves show that the open circuit potential of the coating on side A was more negative than that of side B. For side A, the pitting potential is equal to the open circuit potential on the forward scanned curve. For side B, the pitting potential is more positive than its open circuit potential before the coating breaks down. After forward scanning, the coating on side B was also broken, similar to the damage to side A. Therefore, the backward scanned curve of the polarization curve of side B is almost the same as that of side A. The anodic polarization current on the reversed curves is larger than that on the forward scanned curves due to the increased corrosion area of the anodized coating. The zero current potential of side A and side B is at the same position on the reversed curves.

Figure 16.21 presents typical EIS spectra of an anodized coating before and after corrosion damage. The EIS of side A consists of a capacitance arc and an inductance arc, but only one large capacitance arc for side B. A single large capacitive arc means that the coating can be treated as a good barrier layer on the electrode, and hence indicates that the coating should be undamaged. An inductive arc at low frequencies should be an indication that localized or pitting corrosion is taking place. According to the equivalent

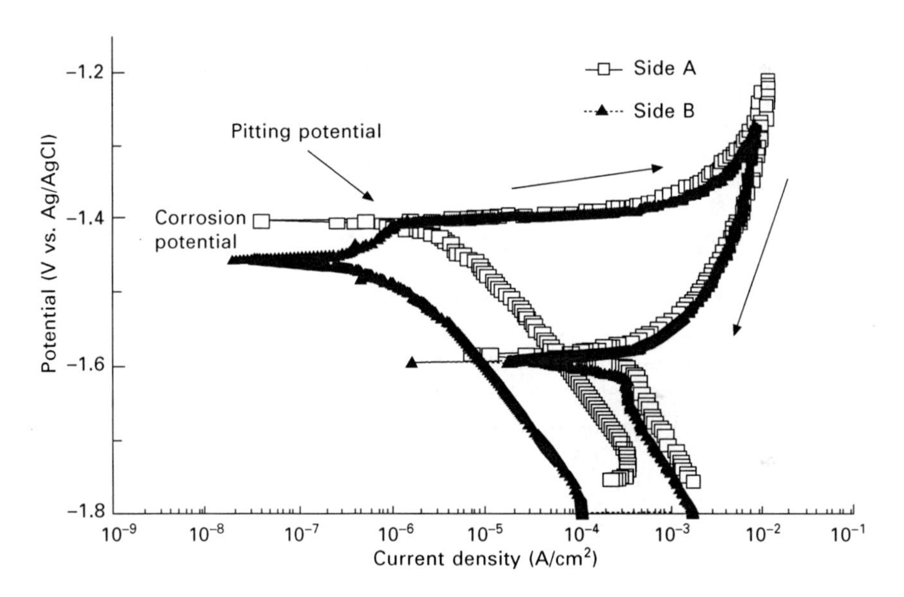

16.20 Polarization curves of anodized AM50 after 21 days salt spray testing. -□- side A, corrosion damaged; -▲- side B, no corrosion damage; OCP, open circuit potential; and PP, pitting potential (Song *et al.*, 2006).

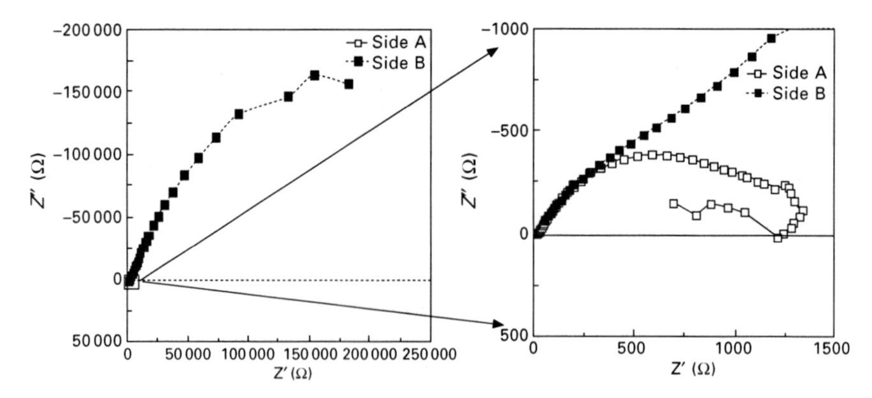

16.21 EIS of anodized AM50 in 5% NaCl solution after 21 days of salt spray testing (HF, high-frequency range; LF, low-frequency range; CP, capacitance loop; ID, inductance loop) (Song *et al.*, 2006).

circuit and the corrosion mechanism of an anodized Mg alloy proposed earlier, the inductive characteristic is closely associated with the breakdown of surface film and the initiation of corrosion. This interpretation is also consistent with the corrosion models of other film-covered metals (Cao, 1990; Song, 2005b). Therefore, the inductive characteristic of an anodized coating should signify possible breakdown of surface film or initiation of the corrosion of an anodized magnesium alloy. Therefore, the EISs in Fig. 16.21 imply that the anodized side B was in good condition and side A suffered localized corrosion. Furthermore, it can be seen that the diameter of the capacitive loop for the anodized coating on side B is several orders larger than that of side A. This further confirms that the anodized coating on side A was corrosion damaged while that on side B was not.

16.7 Application examples

Commercial anodized coatings without any sealing or post-treatments can offer a few hundred hours of protection for the substrate Mg alloy in a standard salt spray testing environment. Although an anodized coating is much better than a conversion coating or phosphated coating, it is rarely used alone for corrosion protection. More often, an anodized coating is used as a surface treatment to provide a base for top paint or coat. Sealing and post-treatments are strongly recommended for an anodized Mg alloy component for long service life when no organic coating is applied after anodizing.

16.7.1 Sealing

In principle, sealing is a process for species from a sealing solution to deposit on the anodized coating, particularly in the defective sites (such as the pores and cracks in the coating) to seal the active points in order to prevent preferential dissolution in these sites or aggressive solutions from penetrating the coating to attack the substrate. Existing sealing techniques normally include immersing an anodized Mg alloy specimen in a phosphate, silicate or borate-containing solution to allow these anions to react with Mg or the coating and form low-solubility salts in the defect pores or cracks, thereby blocking the corrosion paths (Blawert *et al.*, 2006). It is believed that sealing can significantly enhance the corrosion resistance of an anodized coating.

The above-mentioned sealing techniques are usually a reversible process. Salts can be deposited or formed in the defect pores or cracks when the coating is immersed in a sealing solution that contains a high concentration of those anions. The deposited salts may be dissolved or leached from the pores or cracks in service environments where there are no high concentrations of those particular anions to stabilize the deposited salts. Therefore, the improvement of the corrosion resistance of an anodized coating by this kind of sealing technique is not very significant and the durability of the sealing effect cannot last long.

Recently, an irreversible sealing process for anodized Mg alloys was reported (Song, 2009b), which was claimed to maintain a long-lasting corrosion resistant sealing effect. The deposition of sealant in the defect pores or cracks of an anodized coating is irreversible. The deposition mechanism has been systematically investigated and is believed to have chemical reactions similar to an E-coating process (Song, 2008, 2010; Song *et al.*, 2009). As long as the sealant is deposited in the defective sites, it becomes insoluble and very stable even if the environmental solution is significantly changed.

A comparison in corrosion morphology between sealed and unsealed anodized ZE41 specimens after corrosion test is presented in Fig. 16.22. The entire surface of both specimens (a) and (b) was originally anodized. The right side of specimen (a) and the left side of specimen (b) are as-anodized. The left side of specimen (a) and the right side of specimen (b) were sealed after anodizing, which appear to be darker than the as-anodized parts. The bottom parts of both specimens (a) and (b) have been immersed in 5 wt% NaCl solution for 2 days. It is clearly shown that corrosion only occurred in the as-anodized surface. Particularly on specimen (b) corrosion stopped exactly before the edge of the sealed surface. No corrosion damage can be detected in the sealed sections after immersion.

These corrosion results suggest that the sealing treatment can significantly improve the corrosion resistance of an anodized Mg alloy. In fact, this sealing technique can also be applied to a phosphated coating (Song, 2009a) to obtain improved corrosion performance.

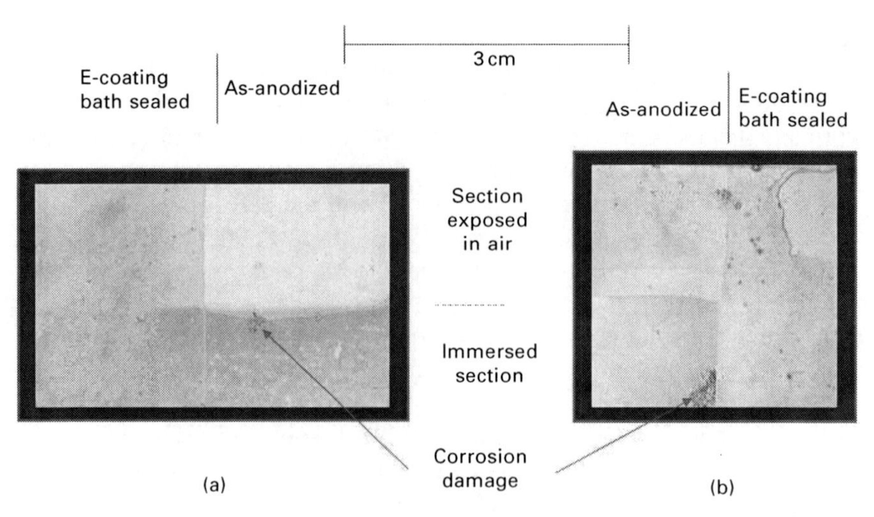

16.22 Corrosion morphologies of anodized ZE41 half sealed after half-immersion in 5 wt% NaCl solution for 2 days (black and white, no color) (Song, 2009b).

16.7.2 Organic top-coating

More commonly, an anodized coating is used together with an organic coating, such as an E-coating or a powder coating on its top. This is because its porous feature favors the formation of these organic coatings. In fact, these organic coatings can very well penetrate into the pores, defective points and cracks in an anodized coating and greatly seal the anodized coating, resulting in a slightly smoother surface and significantly higher corrosion resistance than the original anodized coating. Figure 16.23 shows the cross-section of an anodized coating + E-coating system as an example to illustrate the beneficial sealing and bonding effect of an E-coating on a commercial anodized coating.

A coating system consisting an anodized coating and an E-coating or a powder coating can usually survive in the standard salt spray test (ASTM B117) for a few thousand hours. In theory, this coating system should offer sufficient protection for Mg alloys. However, owing to the bad reproducibility and some unexpected defects in the coating system, premature failure can occur. Similar to organic coatings on other materials, the most common damage of such a coating system on a Mg alloy is still delamination. An example is given in Fig. 16.24.

16.7.3 Corrosion rate control

In some particular cases, we need to control corrosion rate rather than completely stop corrosion. For those applications, an anodized coating may

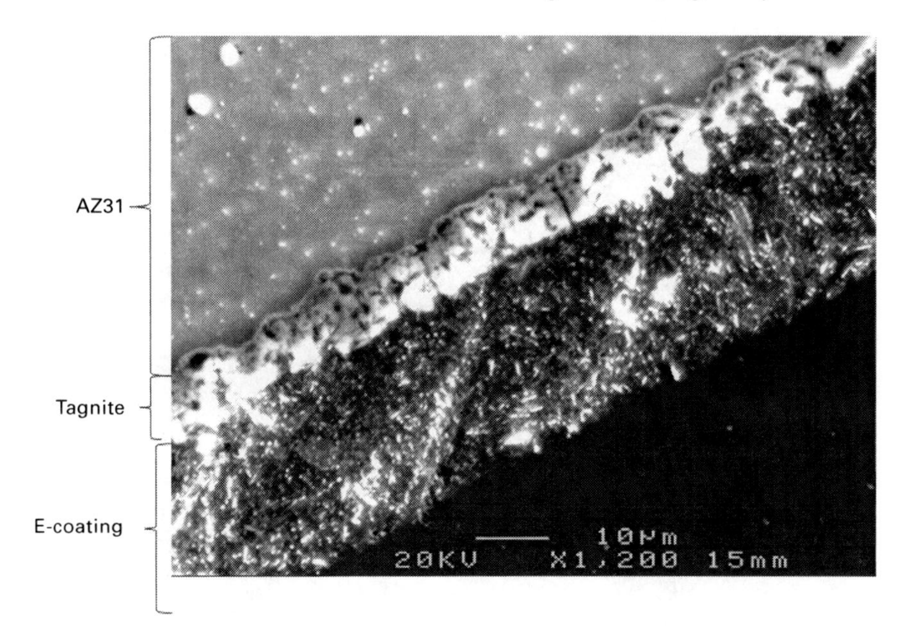

16.23 Cross-section of a commercial Tagnite anodized specimen with an E-coating on the top.

be used alone without sealing or organic coating as a top-paint. An example is the control of degradation rate of Mg alloy as a biodegradable material.

Magnesium is potentially a wonderful implant biomaterial because of its non-toxicity to the human body. Since rapid degradation is almost an intrinsic response of magnesium to a chloride containing solution (Song and Atrens, 2003), like the human body fluid or plasma, we can utilize its poor corrosion performance to make magnesium into a bio-degradable implant material (Song, 2007a,b, Song and Song, 2007; Song *et al.*, 2007). Theoretically, a bio-degradable material should have a controllable dissolution rate or a delayed degradation process (Hassel *et al.*, 2005). An implant made of such a material should fully function before the surgical region recovers or heals. After that, the implant should gradually dissolve, or be consumed or absorbed. Unfortunately, magnesium corrodes rapidly in simulated body fluids (Fonternier *et al.*, 1975; Kaesel *et al.*, 2003). The rapid biodegradation of magnesium can lead to a large amount of dissolved Mg^{2+}, a large volume of hydrogen gas and remarkable local alkalization of body fluid in the human body. Therefore, a straightforward strategy to solve these problems is to control the biodegradation rate of magnesium, so that the human body would have sufficient time to gradually consume, adsorb or release those degradation products (particularly hydrogen) of magnesium. The control of corrosion rate can be realized through a corrosion-resistant coating.

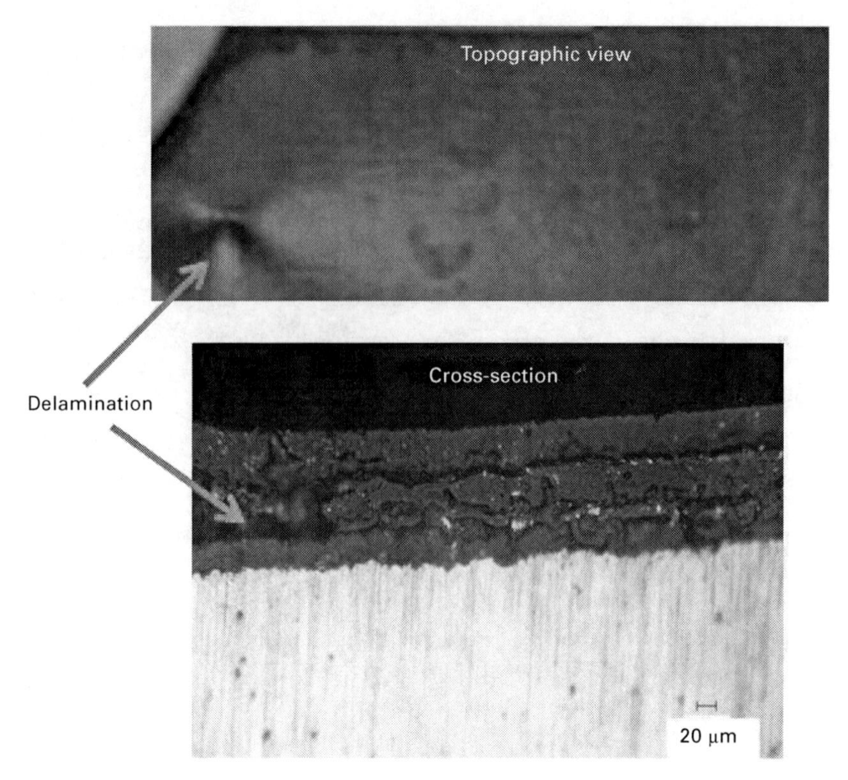

16.24 Delamination of E-coating on an anodized AZ31 alloy.

An example for such an application has been given by Song (2007b) who used an anodized coating containing less than 30% silicon oxides and hydroxides in addition to magnesium oxides and hydroxides which is hard, non-toxic, and in some sense similar to a ceramic layer. Although its microstructure is porous (Fig. 16.25(a)), it is corrosion resistant in the simulated body fluid (SBF) solution. From an undamaged anodized coupon (no scratch), no detectable hydrogen evolution was measured after one month testing (Fig. 16.25(b)). This means that the anodized coating can significantly delay the biodegradation process of Mg.

However, this anodized coating may be too protective, which may lead to insufficient degradation and a magnesium implant with such a coating may not biodegrade within a designated period. Fortunately, the slow biodegradation process can be accelerated by slightly polishing the anodized surface, making it thinner or less perfect. The slightly faster hydrogen evolution process shown in Fig. 16.25(b) verifies this idea.

Moreover, while implanting, the surface of an implant could be scratched. It was found that the degradation rate of an anodized coupon with a mechanically damaged corner was significantly increased (see Fig. 16.25(b)).

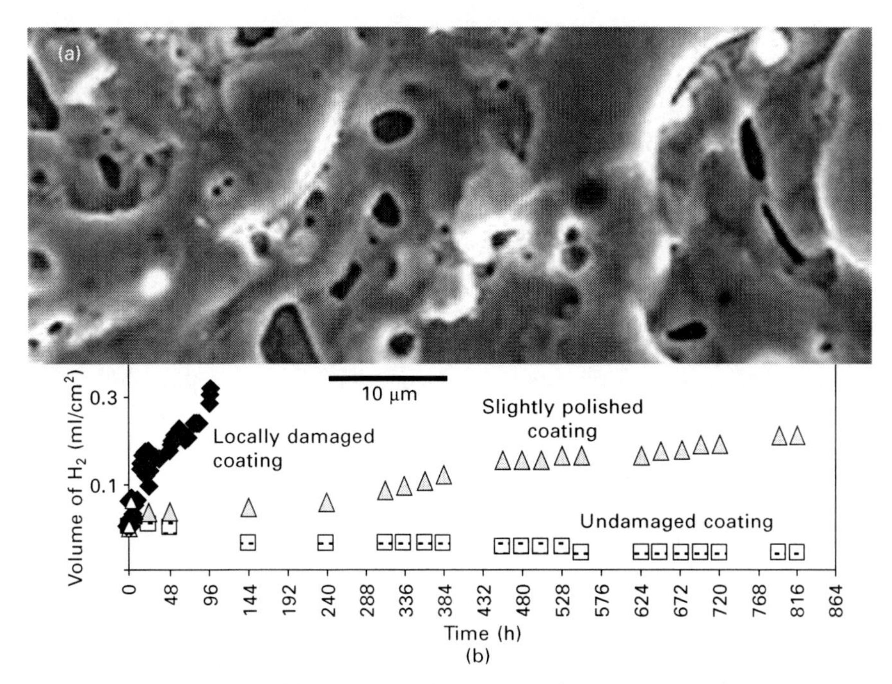

16.25 Anodized Mg: (a) microstructure of anodized coating and (b) hydrogen evolution from anodized coupons in the simulated body fluid (based on Song, 2007b).

Nevertheless, if compared with an unanodized coupon, its biodegradation rate is still evidently low.

16.8 References

Abulsain, M, Berkani, A, Bonilla, F A, Liu, Y, Arenas, M A, Skeldon, P, Thompson, G E, Bailey, P, Noakes, T C Q, Shimizu, K & Habazaki, H (2004), 'Anodic oxidation of Mg–Cu and Mg–Zn alloys'. *Electrochimica Acta*, **49**, 899–904.

Asoh, H & Ono, S (2003), 'Anodizinhg of Mg in amine-ethylene glycol electrolyte', *Materials Science Forum*, 419–422, 957–962.

Avedesin, M & Baker, H (eds.) (1999), *Magnesium and Magnesium Alloys*, Materials Park, OH: ASM International.

Bartak, D E, Lemieux, B E & Woolsey, E R (1993), *Two-step chemical/electrochemical process for coating magnesium alloys*. US Patent: 5240589.

Bartak, D E, Lemieux, B E & Woolsey, E R (1995), *Hard anodic coating for magnesium alloys*. US Patent: 5470664.

Barton, T F (1998), *Anodization of magnesium and magnesium based alloys*. US Patent: 5792335.

Barton, T F & Johnson, C B (1995), 'Effect of electrolyte on the anodized finish of magnesium alloy'. *Plating and Surface Finishing*, **82**, 138–141.

Barton, T F, Macculloch, J A & Ross, P N (2001), *Anodization of magnesium and magnesium based alloys*. US Patent: 6280598 (B1).

Blawert, C, Dietzel, W, Ghali, E & Song, G (2006), 'Anodizing treatments for magnesium alloys and their effect on corrosion resistance in various environments'. *Advanced Engineering Materials*, **8**, 511–533.

Bonilla, F, Berkani, A, Skeldon, P, Thompson, GE, Habazaki, H, Shimizu, K, John, C & Stevens, K (2002a), 'Enrichment of alloying elements in anodized magnesium alloys'. *Corrosion Science*, **44**, 1941–1948.

Bonilla, F A, Berkani, A, Liu, Y, Skeldon, P, Thompson, G E, Habazaki, H, Shimizu, K, John, C & Stevens, K (2002b), 'Formation of anodic films on magnesium alloys in an alkaline phosphate electrolyte'. *Journal of the Electrochemical Society*, **149**, B4–B13.

Brown, L, Besenbruch, G, Schultz, K, Showalter, S, Marshall, A, Pickard, P & Funk, J (2002), 'High efficiency generation of hydrogen fuels using thermochemical cycles and nuclear power'. *In:* AIChE 2002 Spring National Meeting, New Orleans, LA. 139b.

Cai, Q Z, Wang, L S, Wei, B K & Liu, Q X (2006), 'Electrochemical performance of micro-arc oxidation films formed on AZ91D magnesium alloy in silicate and phosphate electrolytes'. *Surface & Coating Technology*, **200**, 3727–3733.

Cao, C N (1990), 'On the impedance plane displays for irreversible electrode reactions based on the stability conditions of the steady-state – I. One state variable besides electrode potential'. *Electrochimica Acta*, **35**, 831.

Carter, E A, Barton, T F & Wright, G A (1999), *Journal of Surface Treatments*, 169–177.

Chai, L, Yu, X, Yang, Z, Wang, Y & Okido, M (2008), 'Anodizing of magnesium alloy AZ31 in alkaline solutions with silicate under continuous sparking'. *Corrosion Science*, **50**, 3274–3279.

Clapp, C, Kilmartin, P A & Wright, G A (2001). *In:* Corrosion & Prevention-2000. Australasian Corrosion Association, 5.

COMA (1956), *DOW17*. GB Patent: 762195.

COMA (2001), *A hard anodic coating for magnesium* [Online]. Luke Engineering & MFG Co. Available: http://www.lukeng.com/magoxid.htm [Accessed April 19 2001].

COMA (2008), *Keronite Plasma Electrolytic Oxidation (PEO) technology* [Online]. Keronite Advanced Surface Technology. Available: http://www.keronite.com/technology_centre.asp [Accessed 2008].

Duan, H, Yan, C & Wang, F (2007), 'Effect of electrolyte additives on performance of plasma electrolyte oxidation films formed on magnesium alloy AZ91D', *Electrochimica Acta*, **52**, 3785.

Evangelides, H A (1955), *Method of Electrochemically coating magnesium and electrolyte there-for*. US Patent: 2723952.

Fonternier, G, Freschard, R & Mourot, M (1975), 'Study of the corrosion *in vitro* and *in vivo* of magnesium anodes involved in an implantable bioelectric battery', *Medical and Biological Engineering*, **15**, 683.

Fukuda, H & Matsumoto, Y (2004), 'Effects of Na_2SiO_3 on anodization of Mg–Al–Zn alloy in 3 M KOH solution'. *Corrosion Science*, **46**, 2135–2142.

Hagans, P L (1984), Surface modification of magnesium for corrosion. *In:* 41st World Magnesium Conference, London, England. International Magnesium Association, 30–38.

Hassel, T, Bach, F W, Krause, C & Wilk, P (2005), Corrosion protection and repassivation after the deformation of magnesium alloys coated with a protective magnesium fluoride layer. *In:* Neelameggham, N R, Kaplan, H I & Powell, B R, eds. *Magnesium Technology 2005*, TMS. 485–490.

Hillis, J E (1994), 'Surface engineering of magnesium alloys'. *ASM Handbook, Surface Engineering*. ASM International. 819.

Hillis, J E (1995), Corrosion testing and standards: application and interpretation *In:* Baboian, R (ed.) *ASTM Manual Series: MNL 20, Magnesium, Chapter 45*. Philadelphia, ASTM. 438–446.

Hillis, J E (2001). *In:* Proc. of 40th Annual Conf. of Metallurgists of CIM. 3–26.

Hillis, J E & Murray, R W (1987), 'Finishing alternatives for high purity magnesium alloys'. *In:* SDCE 14th International. Society of Die Casting Engineers, Inc., #G–T87–003.

Hsiao, H Y & Tsai, W T (2003), *In:* Corrosion 2003. NACE Int., Paper No: 03212.

Hsiao, H Y & Tsai, W T (2005), 'Characterization of anodic films formed on AZ91D magnesium alloy'. *Surface and Coatings Technology*, **190**, 299–308.

Huber, K (1953), 'Anodic formation of coatings on magnesium, zinc, and cadmium'. *Journal of the Electrochemical Society*, **100**, 376–382.

Kaesel, V, Tai, P T, Bach, F W, Haferkamp, H, Witte, F & Windhagen, H (2003), Approach to control the corrosion of magnesium by alloying. *In:* Kainer, K U, ed. Magnesium, Proccedings of the 6th International Conference on magnesium Alloys and their applications. 534–539.

Khaselev, O & Yahalom, J (1998a), 'The anodic behavior of binary Mg–Al alloys in KOH–aluminate solutions'. *Corrosion Science*, **40**, 1149–1160.

Khaselev, O & Yahalom, J (1998b), 'Constant voltage anodizing of Mg–Al alloys in KOH–Al(OH)$_3$ solutions'. *Journal of the Electrochemical Society*, **145**, 190–193.

Khaselev, O, Weiss, D & Yahalom, J (1999), 'Anodizing of pure magnesium in KOH-aluminate solutions under sparking'. *Journal of the Electrochemical Society*, **146**, 1757–1761.

Khaselev, O, Weiss, D & Yahalom, J (2001), 'Structure and composition of anodic films formed on binary Mg–Al alloys in KOH–aluminate solutions under continuous spaking'. *Corrosion Science*, **43**, 1295–1307.

Kobayashi, W & Takahata, S (1985), *Aqueous anodizing solution and process for colouring article of magnesium or magnesium–base alloy*. US Patent: 4551211.

Kobayashi, W, Uehori, K & Furuta, M (1988), *Anodizing solution for anodic oxidation of magnesium or its alloys*. US Patent: 4744872.

Kotler, G L, Hawke, D L & Aqua, E N (1976a), 'MGZ coating process for magnesium and its alloys'. *Light Metal Age*, **34**, 20–21.

Kotler, G R, Hawke, D L, Aqua, E N & Kotler, G L (1976b), MGZ coating process for magnesium and its alloys. *In:* 33rd Annual Meeting- International Magnesium Association Proc., Montreal, Que, Canada, 45–48.

Kotler, G R, Hawke, D L & Aqua, E N. (1977), A new anodic coating process for magnesium and its alloys. *In:* The Society of Die Casting Engineers, ed. SDEC-77: 9th SDEC International Die Casting Exposition & Congress, MECCA Convention Center, Milwaukee, Wisconsin. G-T77–022.

Kozak, O (1980), *Anti-corrosive coating on magnesium and its alloys*. US Patent: 4184926.

Kozak, O (1986), *Method of coating articles of magnesium and an electrolytic bath therefor*. US Patent: 4620904.

Kurze, P (1998), 'Ceramic coatings on light metals by plasma-chemical treatment'. *Materialwiss. Werkstofftech.*, **29**, 85–89.

Kurze, P & Banerjee, D (1996), 'Eine neve anodische Beschichtung zur Verbesserung der korrosions-und VerschleiBbeständigkeit von Magneslumwerkstoffen', *Gießerei-Praxis*, **11/12**, 211–217.

Leyendecker, F (2001), Neue Entwicklungen Leichtmetall-Anwendungen. *In:* der Oberflächentechnik, Conf., Dt. Forschungsges. f. Oberflächenbehandlung, Dt. Ges. f. Galvano- u. Oberflächentech., DGO, Münster. Berichtsband DFO, 132–142.

Li, W, Zhu, L & Li, Y (2006), 'Electrochemical oxidation characteristic of AZ91D magnesium alloy under the action of silica sol'. *Surface and Coatings Technology*, **201**, 1085–1092.

Liang, J, Hu, L & Hao, J (2007), 'Preparation and characterization of oxide films containing crystalline TiO2 on magnesium alloy by plasma electrolytic oxidation'. *Electrochim. Acta*, **52**, 4836–4840.

Ma, Y, Nie, X, Northwood, D O & Hu, H (2004), 'Corrosion and erosion properties of silicate and phosphate coatings on magnesium'. *Thin Solid Films*, **469–470**, 472.

Mato, S, Alcala, G, Skeldon, P, Thompsen, G E, Masheder, D, Habazaki, H & Shimizu, K (2003), 'High resistivity magnesium-rich layers and current instability in anodizing a Mg/Ta alloy'. *Corrosion Science*, **45**, 1779–1792.

Mizutani, Y, Kim, S J, Ichino, R & Okido, M (2003), 'Anodizing of Mg alloys in alkaline solutions'. *Surface and Coatings Technology*, **169–170**, 143–146.

Nykyforchyn, H M, Dietzel, W, Klapkiv, M D & Blawert, C (2003), Corrosion properties of Conversion Plasma Coated Magnesium Alloys. *In:* Kainer, K U (ed.) *Magnesium, 6th International Conference Magnesium Alloys and their Applications*. Wolfsburg (D).

Ono, S (1998), 'Anodic film growth on magnesium die-cast AZ91D (12)'. *Metallurgical Science and Technology*, **16**, 91–104.

Ono, S & Masuko, N (2003), 'Anodic films grown on magnesium and magnesium alloys in fluoride solutions'. *Materials Science Forum*, **419–422**, 897–902.

Ono, S, Kijima, H & Masuko, N (2003), 'Microstructure and voltage-current characteristics of anodic films formed on magnesium in electrolytes containing fluoride'. *Materials Transactions, JIM*, **44**, 539–545.

Schmeling, E L, Roschenbleck, B & Weidemann, M H (1990), *Method of producing protective coatings that are resistant to corrosion and wear on magnesium and magnesium alloys*. US & EP Patent: EP: 333049 US: 4978432.

Sharma, A K, Rnai, R U & Giri, K (1997), 'Studies on anodization of magnesium alloy for thermal control applications'. *Metal Finishing*, **95**, 43–51.

Shatrov, A S (1999), *Method for Producing Hard Protection Coatings on Articles made of Aluminium Alloys*. World Intellectual Property Organization (WO) Patent: WO 9931303.

Shi, Z (2005), *The corrosion performance of anodized magnesium alloys*. PhD Thesis, University of Queensland.

Shi, Z, Song, G & Atrens, A (2003), Effects of zinc on the corrosion resistance of anodized coating on Mg–Zn alloys. *In:* Dahle, A, ed. Proceedings of the 1st International Light Metals Technology Conference, Brisbane, Australia. CAST, 393–396.

Shi, Z, Song, G & Atrens, A (2005), 'Influence of the δ phase on the corrossion performance of anodized coatings on magnesium–aluminium alloys'. *Corrosion Science*, **47**, 2760–2777.

Shi, Z, Song, G & Atrens, A (2006a), 'Influence of alloying on the corrosion performance of anodised single phase Mg alloys'. *Surface & Coatings Technology*, **201**, 492.

Shi, Z, Song, G & Atrens, A (2006b), 'Influence of anodizing current on the corrosion resistance of anodized AZ91D magnesium alloy'. *Corrosion Science*, **48**, 1939–1959.

Shi, Z, Song, G & Atrens, A (2006c), 'The corrosion performance of anodised magnesium alloys'. *Corrosion Science*, **48**, 3531–3546.

Song, G (2004), *Magnesium Anodisation* Australia Patent: 2004904949.

Song, G (2005a), 'Recent progress in corrosion and protection of magnesium alloy', *Advanced Engineering Materials*, **7**, 563–586.

Song, G (2005b), 'Transpassivation of Fe–Cr–Ni stainless steels'. *Corrosion Science*, **47**, 1953.

Song, G (2007a), 'Control of biodegradation of biocompatible magnesium alloys'. *Corrosion Science*, **49**, 1696–1701.

Song, G (2007b), 'Control of degradation of biocompatible magnesium in a pseudo-physiological environment by a ceramic like anodized coating'. *Advanced Materials Research*, **29–30**, 95–98.

Song, G (2008), *Self-deposited coatings on magnesium alloys*. US Patent: 61/047766.

Song, G (2009a), 'A prelimary quantative – XPS study of the surface films formed on pure magnesium and on magnesium-aluminium intermetallics by exposure to high purity water'. *Materials Science Forum*, **618–619**, 269–271.

Song, G (2009b), 'An irreversible dipping sealing technique for anodized ZE41 Mg alloy'. *Surface & Coatings Technology*, **203**, 3618–3625.

Song, G (2010), '"Electroless" deposition of pre-film of electrophoresis coating anits corrosion resistance on a Mg alloy'. *Electrochimica Acta*, **55**, 2258–2268.

Song, G & Atrens, A (2003), 'Understanding magnesium corrosion – a framework for improved alloy performance'. *Advanced Engineering Materials*, **5**, 837.

Song, G & Song, S (2007), 'A possible biodegradable magnesium implant material'. *Advanced Engineering Materials*, **9**, 298–302.

Song, G, Atrens, A, StJohn, D, Wu, X & Nairn, J (1997), 'The anodic dissolution of magnesium in chloride and sulphate solutions'. *Corrosion Science*, **39**, 1981.

Song, G, Bowles, A L & StJohn, D (2004), 'Corrosion resistance of aged die cast magnesium alloy AZ91D'. *Materials Science & Engineering – A*, **366**, 74–86.

Song, G, Shi, Z, Hinton, B, McAdam, G, Talevski, J & Gerrard, D. (2006), Electrochemical evaluation of the corrosion performance of anodized magnesium alloys. *In:* 14th Asian-Pacific Corrosion Control Conference, Shanghai, China. Keynote-11.

Song, G, Song, S & Li, Z (2007), Corrosion control of magnesium as an implant biomaterial in simulated body fluid. *In:* Ultralight2007 – 2nd International Symposium on Ultralight Materials and Structures, Beijing (Invited Lecture).

Song, G, Atrens, A, StJohn, D, Wu, X & Nairn, J (2009), '"Electroless" E-coating – an innovative surface treatment for magnesium alloys'. *Electrochemical and Solid-State Letters*, **12**, D77–D79.

Takaya, M (1987), 'Anodizing of magnesium alloy in potassium hydroxide–aluminum hydroxide solutions'. *Keikinzoku*, **37**, 581–586.

Takaya, M (1988), **マグネシウムの陽極酸化処理膜** *Journal of the Metal Finishing Society*, **35**, 290.

Takaya, M (1989a), 'Anodizinsg of magnesium alloys in KOH-AL$(OH)_3$ solutions', *Aluminium*, **65**, 1244–1248.

Takaya, M (1989b), **マグネシウム合金の陽極酸化処理膜と組成＊** *Gypsum & Lime*, **223**, 40.

Takaya, M, Inoue, T, Nakazato, D & Sugano, K (1998), *Surface treatment of magnesium-based metal material for forming anticorrosive and rustproofing coating*. Japan Patent: 10219496.

Tanaka, K (1993), *Anodization with the Colouring of magnesium and magnesium alloys*. European Patent: 0333048 BI.

Verdier, S, Boinet, M, Maximovitch, S & Dalard, F (2004), 'Formation structure and composition of anodic films on AM60 magnesium alloy obtained by DC plasma anodising'. *Corrosion Science*, **47**, 1429–1444.

Weast, R (ed.) (1976–1977), *Handbook of Chemistry and Physics*: CRC Press.

Wright, G A, Carter, E A & Barton, T F (1999), *Surface Treatment IV: Computer Methods and Experimental Measurements*: WIT Press, 169–177.

Wu, C S, Zhang, Z, Cao, F H, Zhang, L J, Zhang, J Q & Cao, C N (2007), 'Study on the anodizing of AZ31 magnesium alloys in alkaline borate solutions'. *Applied Surface Science*, **253**, 3893–3898.

Yahalom, J & Zahavi, J (1971), 'Experimental evaluation of some electrolytic breakdown hypotheses'. *Electrochimica Acta*, **16**, 603.

Zhang, Y, Yan, C, Wang, F & Li, W (2005), 'Electrochemical behavior of anodized Mg alloy AZ91D in chloride containing aqueous solution'. *Corrosion Science*, **47**, 2816–2831.

Zhu, L & Liu, H (2005), 'The effect of sol ingredients to oxidation film on magnesium alloys'. *Function Materials*, **36**, 923–926.

Zozulin, A J & Bartak, D E (1994), 'Anodized coatings for magnesium alloys'. *Metal Finishing*, **92**, 39–44.

17 Corrosion of magnesium (Mg) alloys: concluding remarks

G.-L. SONG, General Motors Corporation, USA

Apart from the specific conclusions of the 16 chapters in *Corrosion of Magnesium Alloys*, more general statements about the corrosion behavior of magnesium (Mg) alloys can also be made as follows.

Firstly, Mg has an electrochemistry unlike that of other conventional metals. Mg alloys thus display unique activity and passivity, as well as 'strange' electrochemical polarization and dissolution behaviors.

Secondly, the corrosion of an Mg alloy is the result of anodic dissolution of the matrix phase. The other phases in the alloy are relatively inert. However, their presence can significantly influence the dissolution of the matrix phase. Hence the chemical composition, the amount and the distribution of the other phases all play an important role in the corrosion of the alloy. Following this theory, the corrosion behavior of a Mg alloy is predictable after the electrochemistry of each phase constituent in the alloy is understood. For the same reason, the corrosion performance of an Mg alloy can be improved by modifying its chemical composition and microstructure through a metallurgical approach. Currently, the development of innovative alloys, including metallic glasses, is based on this idea.

Thirdly, the corrosion performance of an Mg alloy is also particularly dependent on the environment. It may vary dramatically in different service environments. Some species in solution are now known to cause severe corrosion damage to Mg alloys, but there are still many ions that have unknown interactions with these alloys. Generally, an aqueous solution is more aggressive than an atmospheric medium. A non-aqueous liquid has an electrochemical behavior totally different from that of an aqueous solution. Statistic or dynamic stress can also accelerate the corrosion process. It is noted that galvanic corrosion caused by other engineering metals bonded to magnesium is the biggest threat to Mg alloy parts in many practical applications. Fortunately, theoretical prediction or computer modeling of this well-studied galvanic corrosion damage is now becoming possible.

Fourthly, coating or surface treatment should be the most cost-effective way of retarding the corrosion of an Mg alloy at this stage. However, in reality coating or surface treating an Mg alloy is more difficult than for other conventional engineering metals. Most existing coating and surface treatment techniques that have succeeded in protecting other metals from corrosion attack cannot be directly applied to Mg alloys. New developments and modifications are required.

Based on findings such as these, perspective in the field of corrosion and corrosion protection of Mg alloys is presented in the following aspects:

- The fundamental understanding of the corrosion mechanisms of Mg and its alloys is very limited compared with that of other conventional engineering metals. However, such knowledge is essential if Mg applications are to expand. This knowledge is also a foundation for the development of cost-effective corrosion mitigation techniques for Mg alloys. In particular, as more innovative Mg alloys emerge for new applications, a comprehensive understanding of corrosion mechanisms will be one of the most important long-term research goals in the field of corrosion and protection of Mg alloys.
- The aim of a corrosion study is to solve a particular corrosion problem. It would be ideal if a material had a corrosion-resistant composition and microstructure and thus performed, for example, as stainless steel does in service. However, it is extremely challenging to change the nature of Mg from chemically active to passive through alloying with small amounts of some 'miracle' elements. Nevertheless, modification of the composition and microstructure to improve the corrosion performance of an Mg alloy to some degree will continue to be a significant research area. It will be a long journey to develop and commercialize a corrosion-resistant Mg alloy.
- Before a 'stainless' or corrosion-resistant Mg alloy is commercially available, opportunities to use Mg alloys in selective environments should not be overlooked. However, knowledge about how Mg alloys behave in particular environments is currently lacking. The application of Mg alloys in industry requires such information, particularly information on long-term corrosion. Therefore, collection of data on long-term exposure will continue to be one of the areas of interest to Mg scientists as well as industrial engineers. In the lab, the first step will be estimation or prediction of the corrosion behavior of Mg alloys under a few typical corrosion environments.
- Surface treatment and coating techniques are the most cost-effective approaches of improving the corrosion performance of Mg alloys and enabling their increased use. In fact, new surface treatment and coating techniques suitable for Mg alloys are urgently needed in many critical applications. This is currently a 'hot' area in the field of corrosion and protection of Mg alloys. It can be predicted that, before a 'stainless' or corrosion-resistant Mg alloy becomes a realistic possibility, the development of coatings and surface treatments for Mg alloys, as well

as the relevant fundamental research, will be a high-priority research area.

Overall, this book provides a picture of Mg alloys in the overall family of materials and offers an optimistic view that sustained research will steer this material to many applications that can benefit from its unique attributes as outlined in the Preface.

Index

CPSIA information can be obtained at www.ICGtesting.com
Printed in the USA
LVOW092125100712

289534LV00013B/46/P